Lecture Notes in Computer Science   12143

More information about this series at http://www.springer.com/series/7407

Valeria V. Krzhizhanovskaya ·
Gábor Závodszky · Michael H. Lees ·
Jack J. Dongarra · Peter M. A. Sloot ·
Sérgio Brissos · João Teixeira (Eds.)

Computational Science – ICCS 2020

20th International Conference
Amsterdam, The Netherlands, June 3–5, 2020
Proceedings, Part VII

Springer

Editors
Valeria V. Krzhizhanovskaya (iD)
University of Amsterdam
Amsterdam, The Netherlands

Michael H. Lees
University of Amsterdam
Amsterdam, The Netherlands

Peter M. A. Sloot (iD)
University of Amsterdam
Amsterdam, The Netherlands

ITMO University
Saint Petersburg, Russia

Nanyang Technological University
Singapore, Singapore

João Teixeira
Intellegibilis
Setúbal, Portugal

Gábor Závodszky (iD)
University of Amsterdam
Amsterdam, The Netherlands

Jack J. Dongarra (iD)
University of Tennessee
Knoxville, TN, USA

Sérgio Brissos
Intellegibilis
Setúbal, Portugal

ISSN 0302-9743 ISSN 1611-3349 (electronic)
Lecture Notes in Computer Science
ISBN 978-3-030-50435-9 ISBN 978-3-030-50436-6 (eBook)
https://doi.org/10.1007/978-3-030-50436-6

LNCS Sublibrary: SL1 – Theoretical Computer Science and General Issues

This Springer imprint is published by the registered company Springer Nature Switzerland AG
The registered company address is: Gewerbestrasse 11, 6330 Cham, Switzerland

Preface

Twenty Years of Computational Science

Welcome to the 20th Annual International Conference on Computational Science (ICCS – https://www.iccs-meeting.org/iccs2020/).

During the preparation for this 20th edition of ICCS we were considering all kinds of nice ways to celebrate two decennia of computational science. Afterall when we started this international conference series, we never expected it to be so successful and running for so long at so many different locations across the globe! So we worked on a mind-blowing line up of renowned keynotes, music by scientists, awards, a play written by and performed by computational scientists, press attendance, a lovely venue... you name it, we had it all in place. Then corona hit us.

After many long debates and considerations, we decided to cancel the physical event but still support our scientists and allow for publication of their accepted peer-reviewed work. We are proud to present the proceedings you are reading as a result of that.

ICCS 2020 is jointly organized by the University of Amsterdam, NTU Singapore, and the University of Tennessee.

The International Conference on Computational Science is an annual conference that brings together researchers and scientists from mathematics and computer science as basic computing disciplines, as well as researchers from various application areas who are pioneering computational methods in sciences such as physics, chemistry, life sciences, engineering, arts and humanitarian fields, to discuss problems and solutions in the area, to identify new issues, and to shape future directions for research.

Since its inception in 2001, ICCS has attracted increasingly higher quality and numbers of attendees and papers, and 2020 was no exception, with over 350 papers accepted for publication. The proceedings series have become a major intellectual resource for computational science researchers, defining and advancing the state of the art in this field.

The theme for ICCS 2020, "Twenty Years of Computational Science", highlights the role of Computational Science over the last 20 years, its numerous achievements, and its future challenges. This conference was a unique event focusing on recent developments in: scalable scientific algorithms, advanced software tools, computational grids, advanced numerical methods, and novel application areas. These innovative novel models, algorithms, and tools drive new science through efficient application in areas such as physical systems, computational and systems biology, environmental systems, finance, and others.

This year we had 719 submissions (230 submissions to the main track and 489 to the thematic tracks). In the main track, 101 full papers were accepted (44%). In the thematic tracks, 249 full papers were accepted (51%). A high acceptance rate in the thematic tracks is explained by the nature of these, where many experts in a particular field are personally invited by track organizers to participate in their sessions.

ICCS relies strongly on the vital contributions of our thematic track organizers to attract high-quality papers in many subject areas. We would like to thank all committee members from the main and thematic tracks for their contribution to ensure a high standard for the accepted papers. We would also like to thank Springer, Elsevier, the Informatics Institute of the University of Amsterdam, the Institute for Advanced Study of the University of Amsterdam, the SURFsara Supercomputing Centre, the Netherlands eScience Center, the VECMA Project, and Intellegibilis for their support. Finally, we very much appreciate all the Local Organizing Committee members for their hard work to prepare this conference.

We are proud to note that ICCS is an A-rank conference in the CORE classification.

We wish you good health in these troubled times and hope to see you next year for ICCS 2021.

June 2020   Valeria V. Krzhizhanovskaya
                                                     Gábor Závodszky
                                                     Michael Lees
                                                     Jack Dongarra
                                                     Peter M. A. Sloot
                                                     Sérgio Brissos
                                                     João Teixeira

Organization

Thematic Tracks and Organizers

Advances in High-Performance Computational Earth Sciences: Applications and Frameworks – IHPCES

Takashi Shimokawabe
Kohei Fujita
Dominik Bartuschat

Agent-Based Simulations, Adaptive Algorithms and Solvers – ABS-AAS

Maciej Paszynski
David Pardo
Victor Calo
Robert Schaefer
Quanling Deng

Applications of Computational Methods in Artificial Intelligence and Machine Learning – ACMAIML

Kourosh Modarresi
Raja Velu
Paul Hofmann

Biomedical and Bioinformatics Challenges for Computer Science – BBC

Mario Cannataro
Giuseppe Agapito
Mauro Castelli
Riccardo Dondi
Rodrigo Weber dos Santos
Italo Zoppis

Classifier Learning from Difficult Data – CLD²

Michał Woźniak
Bartosz Krawczyk
Paweł Ksieniewicz

Complex Social Systems through the Lens of Computational Science – CSOC

Debraj Roy
Michael Lees
Tatiana Filatova

Computational Health – CompHealth

Sergey Kovalchuk
Stefan Thurner
Georgiy Bobashev

Computational Methods for Emerging Problems in (dis-)Information Analysis – DisA

Michal Choras
Konstantinos Demestichas

Computational Optimization, Modelling and Simulation – COMS

Xin-She Yang
Slawomir Koziel
Leifur Leifsson

Computational Science in IoT and Smart Systems – IoTSS

Vaidy Sunderam
Dariusz Mrozek

Computer Graphics, Image Processing and Artificial Intelligence – CGIPAI

Andres Iglesias
Lihua You
Alexander Malyshev
Hassan Ugail

Data-Driven Computational Sciences – DDCS

Craig C. Douglas
Ana Cortes
Hiroshi Fujiwara
Robert Lodder
Abani Patra
Han Yu

Machine Learning and Data Assimilation for Dynamical Systems – MLDADS

Rossella Arcucci
Yi-Ke Guo

Meshfree Methods in Computational Sciences – MESHFREE

Vaclav Skala
Samsul Ariffin Abdul Karim
Marco Evangelos Biancolini
Robert Schaback

Rongjiang Pan
Edward J. Kansa

Multiscale Modelling and Simulation – MMS

Derek Groen
Stefano Casarin
Alfons Hoekstra
Bartosz Bosak
Diana Suleimenova

Quantum Computing Workshop – QCW

Katarzyna Rycerz
Marian Bubak

Simulations of Flow and Transport: Modeling, Algorithms and Computation – SOFTMAC

Shuyu Sun
Jingfa Li
James Liu

Smart Systems: Bringing Together Computer Vision, Sensor Networks and Machine Learning – SmartSys

Pedro J. S. Cardoso
João M. F. Rodrigues
Roberto Lam
Janio Monteiro

Software Engineering for Computational Science – SE4Science

Jeffrey Carver
Neil Chue Hong
Carlos Martinez-Ortiz

Solving Problems with Uncertainties – SPU

Vassil Alexandrov
Aneta Karaivanova

Teaching Computational Science – WTCS

Angela Shiflet
Alfredo Tirado-Ramos
Evguenia Alexandrova

Uncertainty Quantification for Computational Models – UNEQUIvOCAL

Wouter Edeling
Anna Nikishova
Peter Coveney

Program Committee and Reviewers

Ahmad Abdelfattah
Samsul Ariffin
   Abdul Karim
Evgenia Adamopoulou
Jaime Afonso Martins
Giuseppe Agapito
Ram Akella
Elisabete Alberdi Celaya
Luis Alexandre
Vassil Alexandrov
Evguenia Alexandrova
Hesham H. Ali
Julen Alvarez-Aramberri
Domingos Alves
Julio Amador Diaz Lopez
Stanislaw
   Ambroszkiewicz
Tomasz Andrysiak
Michael Antolovich
Hartwig Anzt
Hideo Aochi
Hamid Arabnejad
Rossella Arcucci
Khurshid Asghar
Marina Balakhontceva
Bartosz Balis
Krzysztof Banas
João Barroso
Dominik Bartuschat
Nuno Basurto
Pouria Behnoudfar
Joern Behrens
Adrian Bekasiewicz
Gebrai Bekdas
Stefano Beretta
Benjamin Berkels
Martino Bernard

Daniel Berrar
Sanjukta Bhowmick
Marco Evangelos
   Biancolini
Georgiy Bobashev
Bartosz Bosak
Marian Bubak
Jérémy Buisson
Robert Burduk
Michael Burkhart
Allah Bux
Aleksander Byrski
Cristiano Cabrita
Xing Cai
Barbara Calabrese
Jose Camata
Mario Cannataro
Alberto Cano
Pedro Jorge Sequeira
   Cardoso
Jeffrey Carver
Stefano Casarin
Manuel Castañón-Puga
Mauro Castelli
Eduardo Cesar
Nicholas Chancellor
Patrikakis Charalampos
Ehtzaz Chaudhry
Chuanfa Chen
Siew Ann Cheong
Andrey Chernykh
Lock-Yue Chew
Su Fong Chien
Marta Chinnici
Sung-Bae Cho
Michal Choras
Loo Chu Kiong

Neil Chue Hong
Svetlana Chuprina
Paola Cinnella
Noélia Correia
Adriano Cortes
Ana Cortes
Enrique
   Costa-Montenegro
David Coster
Helene Coullon
Peter Coveney
Attila Csikasz-Nagy
Loïc Cudennec
Javier Cuenca
Yifeng Cui
António Cunha
Ben Czaja
Pawel Czarnul
Flávio Martins
Bhaskar Dasgupta
Konstantinos Demestichas
Quanling Deng
Nilanjan Dey
Khaldoon Dhou
Jamie Diner
Jacek Dlugopolski
Simona Domesová
Riccardo Dondi
Craig C. Douglas
Linda Douw
Rafal Drezewski
Hans du Buf
Vitor Duarte
Richard Dwight
Wouter Edeling
Waleed Ejaz
Dina El-Reedy

Amgad Elsayed
Nahid Emad
Chriatian Engelmann
Gökhan Ertaylan
Alex Fedoseyev
Luis Manuel Fernández
Antonino Fiannaca
Christos
  Filelis-Papadopoulos
Rupert Ford
Piotr Frackiewicz
Martin Frank
Ruy Freitas Reis
Karl Frinkle
Haibin Fu
Kohei Fujita
Hiroshi Fujiwara
Takeshi Fukaya
Wlodzimierz Funika
Takashi Furumura
Ernst Fusch
Mohamed Gaber
David Gal
Marco Gallieri
Teresa Galvao
Akemi Galvez
Salvador García
Bartlomiej Gardas
Delia Garijo
Frédéric Gava
Piotr Gawron
Bernhard Geiger
Alex Gerbessiotis
Ivo Goncalves
Antonio Gonzalez Pardo
Jorge
  González-Domínguez
Yuriy Gorbachev
Pawel Gorecki
Michael Gowanlock
Manuel Grana
George Gravvanis
Derek Groen
Lutz Gross
Sophia
  Grundner-Culemann

Pedro Guerreiro
Tobias Guggemos
Xiaohu Guo
Piotr Gurgul
Filip Guzy
Pietro Hiram Guzzi
Zulfiqar Habib
Panagiotis Hadjidoukas
Masatoshi Hanai
John Hanley
Erik Hanson
Habibollah Haron
Carina Haupt
Claire Heaney
Alexander Heinecke
Jurjen Rienk Helmus
Álvaro Herrero
Bogumila Hnatkowska
Maximilian Höb
Erlend Hodneland
Olivier Hoenen
Paul Hofmann
Che-Lun Hung
Andres Iglesias
Takeshi Iwashita
Alireza Jahani
Momin Jamil
Vytautas Jancauskas
João Janeiro
Peter Janku
Fredrik Jansson
Jirí Jaroš
Caroline Jay
Shalu Jhanwar
Zhigang Jia
Chao Jin
Zhong Jin
David Johnson
Guido Juckeland
Maria Juliano
Edward J. Kansa
Aneta Karaivanova
Takahiro Katagiri
Timo Kehrer
Wayne Kelly
Christoph Kessler

Jakub Klikowski
Harald Koestler
Ivana Kolingerova
Georgy Kopanitsa
Gregor Kosec
Sotiris Kotsiantis
Ilias Kotsireas
Sergey Kovalchuk
Michal Koziarski
Slawomir Koziel
Rafal Kozik
Bartosz Krawczyk
Elisabeth Krueger
Valeria Krzhizhanovskaya
Pawel Ksieniewicz
Marek Kubalcík
Sebastian Kuckuk
Eileen Kuehn
Michael Kuhn
Michal Kulczewski
Krzysztof Kurowski
Massimo La Rosa
Yu-Kun Lai
Jalal Lakhlili
Roberto Lam
Anna-Lena Lamprecht
Rubin Landau
Johannes Langguth
Elisabeth Larsson
Michael Lees
Leifur Leifsson
Kenneth Leiter
Roy Lettieri
Andrew Lewis
Jingfa Li
Khang-Jie Liew
Hong Liu
Hui Liu
Yen-Chen Liu
Zhao Liu
Pengcheng Liu
James Liu
Marcelo Lobosco
Robert Lodder
Marcin Los
Stephane Louise

Frederic Loulergue
Paul Lu
Stefan Luding
Onnie Luk
Scott MacLachlan
Luca Magri
Imran Mahmood
Zuzana Majdisova
Alexander Malyshev
Muazzam Maqsood
Livia Marcellino
Tomas Margalef
Tiziana Margaria
Svetozar Margenov
Urszula
  Markowska-Kaczmar
Osni Marques
Carmen Marquez
Carlos Martinez-Ortiz
Paula Martins
Flávio Martins
Luke Mason
Pawel Matuszyk
Valerie Maxville
Wagner Meira Jr.
Roderick Melnik
Valentin Melnikov
Ivan Merelli
Choras Michal
Leandro Minku
Jaroslaw Miszczak
Janio Monteiro
Kourosh Modarresi
Fernando Monteiro
James Montgomery
Andrew Moore
Dariusz Mrozek
Peter Mueller
Khan Muhammad
Judit Muñoz
Philip Nadler
Hiromichi Nagao
Jethro Nagawkar
Kengo Nakajima
Ionel Michael Navon
Philipp Neumann

Mai Nguyen
Hoang Nguyen
Nancy Nichols
Anna Nikishova
Hitoshi Nishizawa
Brayton Noll
Algirdas Noreika
Enrique Onieva
Kenji Ono
Eneko Osaba
Aziz Ouaarab
Serban Ovidiu
Raymond Padmos
Wojciech Palacz
Ivan Palomares
Rongjiang Pan
Joao Papa
Nikela Papadopoulou
Marcin Paprzycki
David Pardo
Anna Paszynska
Maciej Paszynski
Abani Patra
Dana Petcu
Serge Petiton
Bernhard Pfahringer
Frank Phillipson
Juan C. Pichel
Anna
  Pietrenko-Dabrowska
Laércio L. Pilla
Armando Pinho
Tomasz Piontek
Yuri Pirola
Igor Podolak
Cristina Portales
Simon Portegies Zwart
Roland Potthast
Ela Pustulka-Hunt
Vladimir Puzyrev
Alexander Pyayt
Rick Quax
Cesar Quilodran Casas
Barbara Quintela
Ajaykumar Rajasekharan
Celia Ramos

Lukasz Rauch
Vishal Raul
Robin Richardson
Heike Riel
Sophie Robert
Luis M. Rocha
Joao Rodrigues
Daniel Rodriguez
Albert Romkes
Debraj Roy
Katarzyna Rycerz
Alberto Sanchez
Gabriele Santin
Alex Savio
Robert Schaback
Robert Schaefer
Rafal Scherer
Ulf D. Schiller
Bertil Schmidt
Martin Schreiber
Alexander Schug
Gabriela Schütz
Marinella Sciortino
Diego Sevilla
Angela Shiflet
Takashi Shimokawabe
Marcin Sieniek
Nazareen Sikkandar
  Basha
Anna Sikora
Janaína De Andrade Silva
Diana Sima
Robert Sinkovits
Haozhen Situ
Leszek Siwik
Vaclav Skala
Peter Sloot
Renata Slota
Grazyna Slusarczyk
Sucha Smanchat
Marek Smieja
Maciej Smolka
Bartlomiej Sniezynski
Isabel Sofia Brito
Katarzyna Stapor
Bogdan Staszewski

Jerzy Stefanowski
Dennis Stevenson
Tomasz Stopa
Achim Streit
Barbara Strug
Pawel Strumillo
Dante Suarez
Vishwas H. V. Subba Rao
Bongwon Suh
Diana Suleimenova
Ray Sun
Shuyu Sun
Vaidy Sunderam
Martin Swain
Alessandro Taberna
Ryszard Tadeusiewicz
Daisuke Takahashi
Zaid Tashman
Osamu Tatebe
Carlos Tavares Calafate
Kasim Tersic
Yonatan Afework
   Tesfahunegn
Jannis Teunissen
Stefan Thurner

Nestor Tiglao
Alfredo Tirado-Ramos
Arkadiusz Tomczyk
Mariusz Topolski
Paolo Trunfio
Ka-Wai Tsang
Hassan Ugail
Eirik Valseth
Pavel Varacha
Pierangelo Veltri
Raja Velu
Colin Venters
Gytis Vilutis
Peng Wang
Jianwu Wang
Shuangbu Wang
Rodrigo Weber
   dos Santos
Katarzyna
   Wegrzyn-Wolska
Mei Wen
Lars Wienbrandt
Mark Wijzenbroek
Peter Woehrmann
Szymon Wojciechowski

Maciej Woloszyn
Michal Wozniak
Maciej Wozniak
Yu Xia
Dunhui Xiao
Huilin Xing
Miguel Xochicale
Feng Xu
Wei Xue
Yoshifumi Yamamoto
Dongjia Yan
Xin-She Yang
Dongwei Ye
Wee Ping Yeo
Lihua You
Han Yu
Gábor Závodszky
Yao Zhang
H. Zhang
Jinghui Zhong
Sotirios Ziavras
Italo Zoppis
Chiara Zucco
Pawel Zyblewski
Karol Zyczkowski

Contents – Part VII

**Smart Systems: Bringing Together Computer Vision,
Sensor Networks and Machine Learning**

Software Engineering for Computational Science

Solving Problems with Uncertainties

Simulations of Flow and Transport: Modeling, Algorithms and Computation

Decoupled and Energy Stable Time-Marching Scheme for the Interfacial Flow with Soluble Surfactants

Guangpu Zhu[iD] and Aifen Li[(⊠)]

School of Petroleum Engineering, China University of Petroleum (East China),
Qingdao 266580, China
Aifenli64@gmail.com

Abstract. In this work, we develop an efficient energy stable scheme for the hydrodynamics coupled phase-field surfactant model with variable densities. The thermodynamically consistent model consists of two Cahn–Hilliard–type equations and incompressible Navier–Stokes equation. We use two scalar auxiliary variables to transform nonlinear parts in the free energy functional into quadratic forms, and then they can be treated efficiently and semi-implicitly. A splitting method based on pressure stabilization is used to solve the Navier–Stokes equation. By some subtle explicit-implicit treatments to nonlinear convection and stress terms, we construct a first-order energy stable scheme for the two-phase system with soluble surfactants. The developed scheme is efficient and easy-to-implement. At each time step, computations of phase-field variables, the velocity and pressure are decoupled. We rigorously prove that the proposed scheme is unconditionally energy stable. Numerical results confirm that our scheme is accurate and energy stable.

Keywords: Surfactant · Interfacial flow · Phase-field modeling · Navier–stokes

1 Introduction

Surfactants, interface active agents, are known to lower the interfacial tension and allow for the formation of emulsion [1, 2]. Commonly-used surfactants are amphiphilic compounds, meaning they contain both hydrophilic heads and hydrophobic tails [1, 3]. This special molecular composition enables surfactants to selectively absorb on fluid interfaces. Surfactants play a crucial role in everyday life and many industrial processes, such as the cleanser essence, the crude oil recovery and pharmaceutical materials, thus having an understanding of their behavior is a necessity. Numerical simulation is taking an increasingly significant position in investigating the interfacial phenomena, as it can provide easier access to some quantities such as surfactant concentration, pressure and velocity, which are difficult to measure experimentally. However, the computational modeling of interfacial dynamics with surfactants remains a challenging task.

© Springer Nature Switzerland AG 2020
V. V. Krzhizhanovskaya et al. (Eds.): ICCS 2020, LNCS 12143, pp. 3–17, 2020.
https://doi.org/10.1007/978-3-030-50436-6_1

The phase-field model is an effective modeling and simulation tool in investigating interfacial phenomena and it has been extensively used with much successes [4]. This method introduces a phase-field variable to distinguish two pure phases. The interface is treated as a thin layer, inside which the phase-field variable varies continuously [5, 6]. Unlike shape interface models, the phase-field model does not need to track the interface explicitly, and the interface can be implicitly and automatically captured by the evolution of phase-field variable. Therefore, the computations and analysis of the phase-field model are easier than other models [7, 8].

The phase-field model was first used to study the dynamics of phase separation with surfactants in [9]. Two phase-field variables were introduced in their work. Since then, a variety of phase-field surfactant models have been proposed and reviews of these models can refer to [10–12]. Here we only highlight two representative works. The authors in [13] introduced the logarithmic Floy-Huggins potential to restrict the range of surfactant concentration. A nonlinear coupling surface energy potential was used to account for the high surfactant concentration along the fluid interface. An enthalpic term was also adopted to stabilize the phase-field model and control the surfactant solubility in the bulk phases. Their model can describe realistic adsorption isotherms, e.g., Langmuir isotherm, in thermodynamic equilibrium. In [14], the authors analyzed the well-posedness of the phase-field surfactant model proposed in [13], and provided strong evidence that the model was mathematically ill-posed for a large set of physically relevant parameters. They made critical modifications to the model and substantially increased the domain of validity. In this study, we will use this modified model to describe a binary fluid-surfactant system.

Numerically, it is a challenging issue to discretize the strong couplings between two phase-field variables. The introduction of hydrodynamics will further increase the complexity for the development of numerical schemes. Several attempts have been made to solve the interfacial flows with surfactants [15–18], but none of them can provide the energy stability for numerical schemes in theory. Most recently, we constructed a first-order and a second-order schemes, which are linear and totally decoupled, for a phase-field surfactant model with fluid flow [19]. However, this study only considered the case of matched density and viscosity, which greatly reduces difficulties in algorithm developments. Thus, the main purpose of this study is to construct an efficient, easy-to-implement and energy stable scheme for the hydrodynamics coupled phase-field surfactant model with variable densities.

The rest of this paper is organized as follows. In Sect. 2, we describe a hydrodynamics coupled phase-field surfactant model with variable densities. In Sect. 3, we develop an efficient energy stable scheme carry out the energy stability for the proposed scheme. Several numerical experiments are investigated in Sect. 4 and the paper is finally concluded in Sect. 5.

2 Governing Equation

In this section, we consider a typical phase-field surfactant model in [14, 19] for a two-phase system with surfactants

$$E_f(\mathbf{u}, \phi, \psi) = \int \left(\frac{Cn^2}{4}|\nabla\phi|^2 + F(\phi) + PiG(\psi) - \frac{\psi(\phi^2-1)^2}{4} + \frac{\psi\phi^2}{4Ex} \right) d\Omega, \quad (2.1)$$

where $F(\phi)$ is the double well potential and $G(\psi)$ the logarithmic Flory–Huggins potential,

$$F(\phi) = \frac{(\phi^2-1)^2}{4}, \quad G(\psi) = \psi\ln\psi + (1-\psi)\ln(1-\psi).$$

Two phase-field variables are used in the free energy functional. The first phase-field variable ϕ uses two constants (–1 and 1) to distinguish two phases, and it varies continuously across the interface between –1 and 1. The other phase-field variable ψ is used to represent the surfactant concentration. The parameter Cn determines the interfacial thickness and Pi is a temperature-dependent parameter. More details of the free energy functional can refer to [6] and [19].

Although both the double well potential and the Flory–Huggins potential are bounded from below, the latter is not always positive in the whole domain. Thus, we add a zero term $PiB - PiB$ to the free energy functional, and rewrite (2.1) into

$$E_f(\phi, \psi) = \int \left(\frac{Cn^2}{4}|\nabla\phi|^2 + F(\phi) + Pi(G(\psi) + B) + \frac{\psi\phi^2}{4Ex} - \frac{\psi(\phi^2-1)^2}{4} \right) d\Omega$$
$$- PiB|\Omega|,$$
$$(2.2)$$

where the positive constant B ensures $G(\psi) + B > 0$, and $B = 1$ is adopted in this study. Note that the free energy is not changed due to the introduction of the zero term $PiB - PiB$. We now use the scalar auxiliary variable (SAV) approach [12, 20] to transform the free functional into a new form. Through the simple substitution of scalar variables, the nonlinear parts of the free energy are transformed into quadratic forms of new scalar variables. More precisely, we define two scalar variables

$$U = \sqrt{E_u(\phi)}, \quad V = \sqrt{E_v(\psi)}, \tag{2.3}$$

Where

$$E_u(\phi) = \int F(\phi)d\Omega, \quad E_v(\psi) = \int (G(\psi) + B)d\Omega.$$

Then the free energy can be transformed into

$$E_f(\phi, \psi, U, V) = \int \left(\frac{We}{2}\rho|\mathbf{u}|^2 + \frac{Cn^2}{4}|\nabla\phi|^2 - \frac{\psi(\phi^2-1)^2}{4} + \frac{\psi\phi^2}{4Ex} \right) d\Omega + U^2 + PiV^2 - PiB|\Omega|,$$
$$(2.4)$$

Through the functional derivatives of E_f with respect to phase-field variables ϕ and ψ, we can obtain chemical potentials w_ϕ and w_ψ

$$w_\phi = -\frac{\text{Cn}^2}{2}\Delta\phi + \frac{U}{\sqrt{E_u(\phi)}}F'(\phi) + \frac{\psi\phi}{2\text{Ex}} - \psi\phi W, \quad U_t = \frac{1}{2\sqrt{E_u(\phi)}}\int F'(\phi)\phi_t d\Omega,$$

(2.5)

$$w_\psi = \frac{\text{Pi}V}{\sqrt{E_v(\psi)}}G'(\psi) + \frac{\phi^2}{4\text{Ex}} - \frac{W^2}{4}, \quad V_t = \frac{1}{2\sqrt{E_v(\psi)}}\int G'(\psi)\psi_t d\Omega. \quad (2.6)$$

Note that $\phi^2 - 1$ are denoted as Win (2.5) and (2.6).

. Evolutions of phase-field variables ϕ and ψ can be described by the conserved Cahn–Hilliard–type equations [14, 21],

$$\phi_t + \nabla \cdot (\mathbf{u}\phi) = \frac{1}{Pe_\phi}\Delta w_\phi, \quad (2.7)$$

$$\psi_t + \nabla \cdot (\mathbf{u}\psi) = \frac{1}{Pe_\psi}\nabla \cdot M_\psi \nabla w_\psi, \quad (2.8)$$

where Pe_ϕ and Pe_ψ are Péclet numbers. A degenerate mobility $M_\psi = \psi(1 - \psi)$, which vanishes at the extreme points $\psi = 0$ and $\psi = 1$, is adopted to combine with the logarithmic chemical potential w_ψ. Eqs. (2.6)–(2.9) are coupled to the Navier–Stokes equation in the form [4, 14]

$$\rho\mathbf{u}_t + \rho\mathbf{u} \cdot \nabla\mathbf{u} + \mathbf{J} \cdot \nabla\mathbf{u} - \frac{1}{\text{Re}}\nabla \cdot \eta D(\mathbf{u}) + \nabla p + \frac{1}{\text{ReCaCn}}(\phi\nabla w_\phi + \psi\nabla w_\psi) = 0, \quad (2.9)$$

$$\nabla \cdot \mathbf{u} = 0, \quad (2.10)$$

where $D(\mathbf{u}) = \nabla\mathbf{u} + \nabla^T\mathbf{u}$, and $\mathbf{J} = (\lambda_\rho - 1)\nabla w_\phi/2Pe_\phi$. $\mathbf{u}$ is the velocity field, p is the pressure, Re is the Reynolds number and Ca is the Capillary number. We usually assume the density ρ and viscosity η has the following linear relations,

$$\rho = \frac{1 - \lambda_\rho}{2}\phi + \frac{1 + \lambda_\rho}{2}, \quad \eta = \frac{1 - \lambda_\eta}{2}\phi + \frac{1 + \lambda_\eta}{2}.$$

where λ_ρ and λ_η are density and viscosity ratios, respectively.

In particular, if we consider the body force, e.g., the gravitational force, the dimensionless momentum equation read

$$\rho\mathbf{u}_t + \rho\mathbf{u} \cdot \nabla\mathbf{u} + \mathbf{J} \cdot \nabla\mathbf{u} - \frac{1}{\text{Re}}\nabla \cdot \eta D(\mathbf{u}) + \nabla p + \frac{1}{\text{BoCn}}(\phi\nabla w_\phi + \psi\nabla w_\psi) - \rho\mathbf{g} = 0,$$

(2.11)

where Bo = ReCa is the Bond number, and g is the unit vector denoting the direction of body force.

Periodic boundary conditions or the following boundary conditions

$$\partial_n \phi^{n+1} = \nabla w_\phi^{n+1} \cdot \mathbf{n} = \nabla w_\psi^{n+1} \cdot \mathbf{n} = \mathbf{u} = \partial_n p^{n+1} = 0, \text{ on } \Gamma,$$

can be used to close the above governing system. Here Γ denotes boundaries of the domain.

The total energy E_{tot} of the hydrodynamic system (2.5)–(2.10) is the sum of kinetic energy E_k and free energy E_f

$$E_{tot}(\mathbf{u}, \ \phi, \ \psi, U, V) = \int \left(\frac{We}{2} \rho |\mathbf{u}|^2 + \frac{Cn^2}{4} |\nabla \phi|^2 + \frac{\psi \phi^2}{4Ex} - \frac{\psi(\phi^2 - 1)^2}{4} \right) d\Omega$$
$$+ U^2 + PiV^2 - PiB|\Omega|,$$

where $We = ReCaCn$, and we can easily derive the following energy dissipation law.

$$\frac{d}{dt} E_{tot} = -\frac{1}{Pe_\phi} \int |\nabla w_\phi|^2 d\Omega$$
$$- \frac{1}{Pe_\psi} \int |\sqrt{M_\psi} \nabla w_\psi|^2 d\Omega - \frac{CaCn}{2} \int |\sqrt{\eta} D(\mathbf{u})|^2 d\Omega \leq 0.$$

Next, we will develop an efficient time-marching scheme for the above governing system and carry out the energy estimate. To simplify the presentation, in the next section, we will take (2.9) as an example to construct the desired scheme.

3 Numerical Scheme

3.1 Energy Stable First-Order Scheme

We now present a first-order time-marching scheme to solve the governing system in Sect. 2. To deal with the case of nonmatching density, a cut-off function [4] is defined as

$$\tilde{\phi}^{n+1} = \begin{cases} \phi^{n+1} & |\phi^{n+1}| \leq 1, \\ sign(\phi^{n+1}) & |\phi^{n+1}| > 1. \end{cases}$$

Given ψ^n, ϕ^n, $\mathbf{u}^n$ and p^n, the scheme (3.1) calculates ψ^{n+1}, ϕ^{n+1}, $\mathbf{u}^{n+1}$ and p^{n+1} for $n \geq 0$ in three steps.

In step 1, we update ψ^{n+1} and ϕ^{n+1} by solving

$$\frac{\psi^{n+1} - \psi^n}{\delta t} + \nabla \cdot (\mathbf{u}_*^n \psi^n) - \frac{1}{Pe_\psi} \nabla \cdot M_\psi^n \nabla w_\psi^{n+1} = 0, \tag{3.1a}$$

$$w_\psi^{n+1} = \frac{\text{Pi}V^{n+1}}{\sqrt{E_v(\psi^n)}}G'(\psi^n) + \frac{(\phi^n)^2}{4\text{Ex}} - \frac{(W^n)^2}{4}, \tag{3.1b}$$

$$\frac{V^{n+1} - V^n}{\delta t} = \frac{1}{2\sqrt{E_v(\psi^n)}}\int G'(\psi^n)\frac{\psi^{n+1} - \psi^n}{\delta t}d\Omega, \tag{3.1c}$$

$$\frac{\phi^{n+1} - \phi^n}{\delta t} + \nabla \cdot (\mathbf{u}_*^n \phi^n) - \frac{1}{\text{Pe}_\phi}\Delta w_\phi^{n+1} = 0, \tag{3.1d}$$

$$w_\phi^{n+1} = -\frac{\text{Cn}^2}{2}\Delta\phi^{n+1} + \frac{U^{n+1}}{\sqrt{E_u(\phi^n)}}F'(\phi^n) + \frac{\psi^{n+1}\phi^{n+1}}{2\text{Ex}} - \frac{1}{2}\psi^{n+1}W^n(\phi^{n+1} + \phi^n),$$

$$\tag{3.1e}$$

$$\frac{U^{n+1} - U^n}{\delta t} = \frac{1}{2\sqrt{E_u(\phi^n)}}\int F'(\phi^n)\frac{\phi^{n+1} - \phi^n}{\delta t}d\Omega, \tag{3.1f}$$

with periodic boundary conditions or the following boundary conditions

$$\partial_n w_\psi^{n+1} = \partial_n \phi^{n+1} = \partial_n w_\phi^{n+1} = 0, \text{ on } \Gamma,$$

where $\mathbf{u}_*^n$ is the intermediate velocity

$$\mathbf{u}_*^n = \mathbf{u}^n - \frac{\delta t \psi^n}{\text{We}\rho^n}\nabla w_\psi^{n+1} - \frac{\delta t \phi^n}{\text{We}\rho^n}\nabla w_\phi^{n+1}. \tag{3.1g}$$

In step 2, we update $\mathbf{u}^{n+1}$ by solving [22]

$$\begin{cases} \rho^n\dfrac{\mathbf{u}^{n+1} - \mathbf{u}^n}{\delta t} + (\rho^n\mathbf{u}_*^n) \cdot \nabla\mathbf{u}^{n+1} + \mathbf{J}^{n+1} \cdot \nabla\mathbf{u}^{n+1} - \dfrac{1}{\text{Re}}\nabla \cdot \eta^n D(\mathbf{u}^{n+1}) + \nabla(2p^n - p^{n-1}) \\ \qquad + \dfrac{1}{\text{We}}\left(\phi^n\nabla w_\phi^{n+1} + \psi^n\nabla w_\psi^{n+1}\right) + \dfrac{1+\lambda_\rho}{4}(\nabla \cdot \mathbf{u}_*^n)\mathbf{u}^{n+1} = 0, \\ \mathbf{u}^{n+1} = 0, \text{ on } \Gamma, \end{cases}$$

$$\tag{3.1h}$$

where

$$\mathbf{J}^{n+1} = \frac{\lambda_\rho - 1}{2\mathrm{Pe}_\phi} \nabla w_\phi^{n+1}, \quad \rho^{n+1} = \frac{1 - \lambda_\rho}{2} \tilde{\phi}^{n+1} + \frac{1 + \lambda_\rho}{2},$$

$$\eta^{n+1} = \frac{1 - \lambda_\eta}{2} \tilde{\phi}^{n+1} + \frac{1 + \lambda_\eta}{2}. \tag{3.1i}$$

In step 3, we update p^{n+1} by solving the pressure Poisson equation with a constant coefficient [4, 23]

$$\begin{cases} -\Delta(p^{n+1} - p^n) = -\dfrac{\chi}{\delta t} \nabla \cdot \mathbf{u}^{n+1}, \\ \nabla p^{n+1} \cdot \mathbf{n} = 0, \text{ on } \Gamma, \end{cases} \tag{3.1j}$$

where $\chi = \frac{1}{2}\min(1_1, \lambda_\rho)$.

Remark 3.1. (1) Computations of (ψ^{n+1}, ϕ^{n+1}), $\mathbf{u}^{n+1}$ and p^{n+1} are decoupled, which indicate that the scheme (3.1) is efficient and easy-to-implement. At each time step, u^{n+1} and p^{n+1} can be obtained by solving only two elliptic equations; Moreover, V^{n+1} and U^{n+1} do not involve any extra computational cost, since they can be calculated explicitly once we obtain ψ^{n+1} and ϕ^{n+1}. (2) In the explicit convective velocity $\mathbf{u}_*^n$, we introduce a first-order stabilization term [24], which plays a dominant role in decoupling the computation of (ψ^{n+1}, ϕ^{n+1}) from u^{n+1} and constructing the unconditionally energy stable scheme.

Theorem 3.1. The scheme (3.1) is unconditionally energy stable, and satisfies the following discrete energy dissipation law:

$$E_{tot}^{n+1} - E_{tot}^n \le$$
$$-\frac{\delta t}{Pe_\psi} \left\| \sqrt{M_\psi^n} \nabla w_\psi^{n+1} \right\|^2 - \frac{\delta t}{Pe_\phi} \left\| \nabla w_\phi^{n+1} \right\|^2 - \frac{\delta t\mathrm{CaCn}}{2} \left\| \sqrt{\eta^n} D(\mathbf{u}^{n+1}) \right\|^2 \le 0, \tag{3.2}$$

where

$$E_{tot}^n = \frac{\mathrm{We}}{2}\left(\rho^n, |\mathbf{u}^n|^2\right) + \frac{\delta t^2 \mathrm{We}}{2\chi} \|\nabla p^n\|^2 + \frac{\mathrm{Cn}^2}{4} \|\nabla \phi^n\|^2 + (U^n)^2 + \mathrm{Pi}(V^n)^2$$
$$+ \frac{1}{4\mathrm{Ex}}\left(\psi^n, |\phi^n|^2\right) - \frac{1}{4}\left(\psi^n, |W^n|^2\right) - \mathrm{PiB}|\Omega|,$$

here $\|\cdot\|$ denotes the L^2-norm in Ω. Now we will rigorously prove the discrete energy dissipation law in (3.2). We first introduce an intermediate kinetic energy [25] as

$$E_{k,*}^n = \frac{\mathrm{We}}{2}\left(\rho^n \mathbf{u}_*^n, \mathbf{u}_*^n\right). \tag{3.3}$$

The difference between E_k^{n+1} and $E_{k,*}^n$ is estimated as

$$E_k^{n+1} - E_{k,*}^n = \frac{\mathrm{W\,e}}{2}\left(\rho^{n+1},\ |\mathbf{u}^{n+1}|^2\right) - \frac{\mathrm{W\,e}}{2}\left(\rho^n,\ |\mathbf{u}_*^n|^2\right)$$

$$= \frac{\mathrm{W\,e}}{2}\left(\rho^n,\ |\mathbf{u}^{n+1}|^2 - |\mathbf{u}_*^n|^2\right) + \frac{\mathrm{W\,e}}{2}\left(\rho^{n+1} - \rho^n,\ |\mathbf{u}^{n+1}|^2\right)$$

$$= \mathrm{W\,e}(\rho^n(\mathbf{u}^{n+1} - \mathbf{u}_*^n),\ \mathbf{u}^{n+1}) - \frac{\mathrm{W\,e}}{2}\left(\rho^n,\ |\mathbf{u}^{n+1} - \mathbf{u}_*^n|^2\right) + \frac{\mathrm{W\,e}}{2}\left(\rho^{n+1} - \rho^n,\ |\mathbf{u}^{n+1}|^2\right).$$

$$(3.4)$$

Substituting (3.1i) into (3.1a), we obtain the following identity

$$\rho^{n+1} - \rho^n = -\frac{\delta t(1 - \lambda_\rho)}{2}\nabla\cdot(\phi^n \mathbf{u}_*^n) - \delta t\nabla\cdot\mathbf{J}^{n+1}. \qquad (3.5)$$

We can also easily derive from (3.1g) that

$$\mathrm{W}\,e\rho^n\left(\mathbf{u}^{n+1} - \mathbf{u}^n\right) + \delta t\left(\psi^n\nabla w_\psi^{n+1} + \phi\nabla w_\phi^{n+1}\right) = \mathrm{W}\,e\rho^n\left(\mathbf{u}^{n+1} - \mathbf{u}_*^n\right). \qquad (3.6)$$

Using the identity (3.6), we have

$$\mathrm{W}\,e\rho^n\left(\mathbf{u}^{n+1} - \mathbf{u}_*^n\right) = \delta t\mathrm{CaCn}\nabla\cdot\eta^n D(\mathbf{u}^{n+1}) - \delta t\mathrm{W}\,e\nabla(2p^n - p^{n-1}) - \delta t\mathrm{W}\,e(\rho^n\mathbf{u}_*^n)\cdot\nabla\mathbf{u}^{n+1}$$

$$-\delta t\mathrm{W}\,e\mathbf{J}^{n+1}\cdot\nabla\mathbf{u}^{n+1} - \frac{\delta t\mathrm{W}\,e(1 + \lambda_\rho)}{4}(\nabla\cdot\mathbf{u}_*^n)\mathbf{u}^{n+1}.$$

$$(3.7)$$

By taking the L^2 inner product of (3.7) with u^{n+1}, and using (3.4) and the following identities

$$-\delta t\mathrm{W}\,e\left((\rho^n\mathbf{u}_*^n)\cdot\nabla\mathbf{u}^{n+1},\ \mathbf{u}^{n+1}\right) = \frac{\delta t\mathrm{W}\,e}{2}\left(\nabla\cdot(\rho^n\mathbf{u}_*^n),\ |\mathbf{u}^{n+1}|^2\right)$$

$$= \frac{\delta t\mathrm{W}\,e(1 - \lambda_\rho)}{4}\left(\nabla\cdot(\phi^n\mathbf{u}_*^n),\ |\mathbf{u}^{n+1}|^2\right) + \frac{\delta t\mathrm{W}\,e(1 + \lambda_\rho)}{4}\left(\nabla\cdot\mathbf{u}_*^n,\ |\mathbf{u}^{n+1}|^2\right),$$

$$\left((\mathbf{J}^{n+1}\cdot\nabla)\mathbf{u}^{n+1} + \frac{1}{2}(\nabla\cdot\mathbf{J}^{n+1})\mathbf{u}^{n+1},\ \mathbf{u}^{n+1}\right) = 0.$$

we can derive that

$$
\begin{aligned}
E_k^{n+1} - E_{k,*}^n &= -\frac{\delta t\,Ca\,Cn}{2}\left\|\sqrt{\eta^n}D(\mathbf{u}^{n+1})\right\|^2 - \delta t\,We\left((\rho^n\mathbf{u}_*^n)\cdot\nabla\mathbf{u}^{n+1},\,\mathbf{u}^{n+1}\right) \\
&\quad - \delta t\,We\left(\mathbf{J}^{n+1}\cdot\nabla\mathbf{u}^{n+1},\,\mathbf{u}^{n+1}\right) - \frac{\delta t\,We(1+\lambda_\rho)}{4}\left(\nabla\cdot\mathbf{u}_*^n,\,\left|\mathbf{u}^{n+1}\right|^2\right) \\
&\quad - \delta t\,We\left(p^{n+1}-2p^n+p^{n-1},\,\nabla\cdot\mathbf{u}^{n+1}\right) + \delta t\,We\left(p^{n+1},\,\nabla\cdot\mathbf{u}^{n+1}\right) \\
&\quad - \frac{We}{2}\left(\rho^n,\,\left|\mathbf{u}^{n+1}-\mathbf{u}_*^n\right|^2\right) - \frac{\delta t\,We}{2}\left(\frac{1-\lambda_\rho}{2}\nabla\cdot(\phi^n\mathbf{u}_*^n)+\nabla\cdot\mathbf{J}^{n+1},\,\left|\mathbf{u}^{n+1}\right|^2\right) \\
&= -\frac{\delta t\,Ca\,Cn}{2}\left\|\sqrt{\eta^n}D(\mathbf{u}^{n+1})\right\|^2 - \delta t\,We\left(p^{n+1}-2p^n+p^{n-1},\,\nabla\cdot\mathbf{u}^{n+1}\right) \\
&\quad + \delta t\,We\left(p^{n+1},\,\nabla\cdot\mathbf{u}^{n+1}\right) - \frac{We}{2}\left(\rho^n,\,\left|\mathbf{u}^{n+1}-\mathbf{u}_*^n\right|^2\right).
\end{aligned}
\tag{3.8}
$$

Using the Eq. (3.1g), we obtain

$$
\begin{aligned}
E_{k,*}^n - E_k^n &= \frac{We}{2}\left(\rho^n,\,\left|\mathbf{u}_*^n\right|^2-\left|\mathbf{u}^n\right|^2\right) = \left(We\,\rho^n(\mathbf{u}_*^n-\mathbf{u}^n),\,\mathbf{u}^n\right) - \frac{We}{2}\left(\rho^n,\,\left|\mathbf{u}_*^n-\mathbf{u}^n\right|^2\right) \\
&= -\delta t\left(\psi^n\nabla\cdot w_\psi^{n+1}+\phi^n\nabla w_\phi^{n+1},\,\mathbf{u}^n\right) - \frac{We}{2}\left(\rho^n,\,\left|\mathbf{u}_*^n-\mathbf{u}^n\right|^2\right) \\
&= \delta t\left(\nabla\cdot(\psi\mathbf{u}_*^n),\,w_\psi^{n+1}\right) + \delta t\left(\nabla\cdot(\phi\mathbf{u}_*^n),\,w_\phi^{n+1}\right) - \frac{We}{2}\left(\rho^n,\,\left|\mathbf{u}_*^n-\mathbf{u}^n\right|^2\right).
\end{aligned}
\tag{3.9}
$$

Summing up Eqs. (3.8) and (3.9), we get

$$
\begin{aligned}
E_k^{n+1} - E_k^n &= -\frac{\delta t\,Ca\,Cn}{2}\left\|\sqrt{\eta^n}D(\mathbf{u}^{n+1})\right\|^2 - \delta t\,We\left(p^{n+1}-2p^n+p^{n-1},\,\nabla\cdot\mathbf{u}^{n+1}\right) \\
&\quad + \delta t\,We\left(p^{n+1},\,\nabla\cdot\mathbf{u}^{n+1}\right) + \delta t\left(\nabla\cdot(\psi\mathbf{u}_*^n),\,w_\psi^{n+1}\right) + \delta t\left(\nabla\cdot(\phi\mathbf{u}_*^n),\,w_\phi^{n+1}\right) \\
&\quad - \frac{We}{2}\left(\rho^n,\,\left|\mathbf{u}^{n+1}-\mathbf{u}_*^n\right|^2\right) - \frac{We}{2}\left(\rho^n,\,\left|\mathbf{u}_*^n-\mathbf{u}^n\right|^2\right).
\end{aligned}
\tag{3.10}
$$

By taking the L^2 inner product of (3.1j) with $\delta t^2 We(p^{n+1}-2p^n+p^n-1)/\chi$ and with $\delta t^2 We\,p^{n+1}/\chi$ separately, we obtain

$$
\begin{aligned}
-\frac{\delta t^2 We}{2\chi}&\left(\left\|\nabla(p^{n+1}-p^n)\right\|^2-\left\|\nabla(p^n-p^{n-1})\right\|^2+\left\|\nabla(p^{n+1}-2p^n+p^{n-1})\right\|^2\right) \\
&= \delta t\,We\left(p^{n+1}-2p^n+p^{n-1},\,\nabla\cdot\mathbf{u}^{n+1}\right),
\end{aligned}
\tag{3.11}
$$

and

$$\frac{\delta t^2 \mathrm{W}\,\mathrm{e}}{2\chi}\left(\|\nabla p^{n+1}\|^2 - \|\nabla p^n\|^2 + \|\nabla(p^{n+1}-p^n)\|^2\right) = -\delta t \mathrm{W}\,\mathrm{e}(p^{n+1},\,\nabla\cdot\mathbf{u}^{n+1}). \quad (3.12)$$

Combining (3.11) and (3.12), yields

$$-\delta t \mathrm{W}\,\mathrm{e}(p^{n+1} - 2p^n + p^{n-1},\,\nabla\cdot\mathbf{u}^{n+1}) + \delta t \mathrm{W}\,\mathrm{e}(p^{n+1},\,\nabla\cdot\mathbf{u}^{n+1})$$
$$= -\frac{\delta t^2 \mathrm{W}\,\mathrm{e}}{2\chi}\left(\|\nabla p^{n+1}\|^2 - \|\nabla p^n\|^2\right) - \frac{\delta t^2 \mathrm{W}\,\mathrm{e}}{2\chi}\|\nabla(p^n - p^{n-1})\|^2 \qquad (3.13)$$
$$+ \frac{\delta t^2 \mathrm{W}\,\mathrm{e}}{2\chi}\|\nabla(p^{n+1} - 2p^n + p^{n-1})\|^2.$$

We take the difference of (3.1j) at step t^{n+1} and t^n, pair the resulting equation with $\delta t^2 \mathrm{W}\mathrm{e}(p^{n+1} - 2p^n + p^n - 1)/(2\chi)$ then take integration by parts for both sides to derive

$$\frac{\delta t^2 \mathrm{W}\,\mathrm{e}}{2\chi}\|\nabla(p^{n+1} - 2p^n + p^{n-1})\|^2 \le \frac{\chi}{2}\mathrm{W}\,\mathrm{e}\|\mathbf{u}^{n+1} - \mathbf{u}^n\|^2 \le \frac{\mathrm{W}\,\mathrm{e}}{4}(\rho^n,\,|\mathbf{u}^{n+1} - \mathbf{u}^n|)^2. \quad (3.14)$$

Summing up Eqs. (3.10), (3.13) and (3.14), and using the triangle inequality

$$\frac{\mathrm{W}\,\mathrm{e}}{2}\left(\rho^n,\,|\mathbf{u}^{n+1} - \mathbf{u}_*^n|^2\right) + \frac{\mathrm{W}\,\mathrm{e}}{2}\left(\rho^n,\,|\mathbf{u}_*^n - \mathbf{u}^n|^2\right) \ge \frac{\mathrm{W}\,\mathrm{e}}{4}\left(\rho^n,\,|\mathbf{u}^{n+1} - \mathbf{u}^n|\right)^2.$$
$$(3.15)$$

we can derive that

$$E_k^{n+1} - E_k^n \le -\frac{\delta t \mathrm{CaCn}}{2}\|\sqrt{\eta^n}D(\mathbf{u}^{n+1})\|^2 - \frac{\delta t^2 \mathrm{W}\,\mathrm{e}}{2\chi}\left(\|\nabla p^{n+1}\|^2 - \|\nabla p^n\|^2\right)$$
$$+ \delta t\left(\nabla\cdot(\psi\mathbf{u}_*^n),\,w_\psi^{n+1}\right) + \delta t\left(\nabla\cdot(\phi\mathbf{u}_*^n),\,w_\phi^{n+1}\right). \qquad (3.16)$$

By taking the inner product of (3.1a) with $\delta t w_\psi^{n+1}$, we can easily derive that

$$\left(\psi^{n+1} - \psi^n,\,w_\psi^{n+1}\right) + \delta t\left(\nabla\cdot(\mathbf{u}_*^n\psi^n),\,w_\psi^{n+1}\right) = -\frac{\delta t}{Pe_\psi}\|\sqrt{M_\psi^n}\nabla w_\psi^{n+1}\|^2. \quad (3.17)$$

By taking the inner product of (3.1b) with $-(\psi^{n+1} - \psi^n)$, we can derive that

$$-\left(\psi^{n+1} - \psi^n,\,w_\psi^{n+1}\right) = -\mathrm{Pi}(V^{n+1}a^n,\,\psi^{n+1} - \psi^n) - \frac{1}{4\mathrm{Ex}}\left(|\phi^n|^2,\,\psi^{n+1} - \psi^n\right)$$
$$+ \frac{1}{4}\left(|W^n|^2,\,\psi^{n+1} - \psi^n\right). \qquad (3.18)$$

where $a^n = G'(\psi^n)/\sqrt{E_v(\psi^n)}$. Taking the inner product of (3.1c) with $2\delta t \mathrm{Pi} V^{n+1}$ to obtain

$$\mathrm{Pi}\left[(V^{n+1})^2 - (V^n)^2 + (V^{n+1} - V^n)^2\right] = \mathrm{Pi}(V^{n+1}a^n, \psi^{n+1} - \psi^n). \qquad (3.19)$$

Summing up Eqs. (3.17)–(3.19), we get

$$\mathrm{Pi}\left[(V^{n+1})^2 - (V^n)^2 + (V^{n+1} - V^n)^2\right] = -\delta t\left(\nabla(\mathbf{u}_*^n \psi^n), w_\psi^{n+1}\right) - \frac{\delta t}{Pe_\psi}\left\|\sqrt{M_\psi^n}\nabla w_\psi^{n+1}\right\|^2$$
$$- \frac{1}{4\mathrm{Ex}}\left(|\phi^n|^2, \psi^{n+1} - \psi^n\right) + \frac{1}{4}\left(|W^n|^2, \psi^{n+1} - \psi^n\right).$$
$$(3.20)$$

By taking the inner product of (3.1d) with $\delta t w_\phi^{n+1}$, we have

$$\left(\phi^{n+1} - \phi^n, w_\phi^{n+1}\right) + \delta t\left(\nabla \cdot (\mathbf{u}_*^n \phi^n), w_\phi^{n+1}\right) = -\frac{\delta t}{Pe_\phi}\left\|\nabla w_\phi^{n+1}\right\|^2. \qquad (3.21)$$

By taking the inner product of (3.1e) with $-(\phi^{n+1} - \phi^n)$, we can derive that

$$-\left(\phi^{n+1} - \phi^n, w_\phi^{n+1}\right) = -\frac{Cn^2}{2}(\nabla\phi^{n+1}, \nabla\phi^{n+1} - \nabla\phi^n) - (U^{n+1}b^n, \phi^{n+1} - \phi^n)$$
$$- \frac{1}{2\mathrm{Ex}}(\psi^{n+1}\phi^{n+1}, \phi^{n+1} - \phi^n) + \frac{1}{2}(\psi^{n+1}W^n(\phi^{n+1} + \phi^n), \phi^{n+1} - \phi^n)$$
$$= -\frac{Cn^2}{4}\left(\|\nabla\phi^{n+1}\|^2 - \|\nabla\phi^n\|^2 + \|\nabla\phi^{n+1} - \nabla\phi^n\|^2\right) - (U^{n+1}b^n, \phi^{n+1} - \phi^n)$$
$$- \frac{1}{4\mathrm{Ex}}\left[\left(\psi^{n+1}, |\phi^{n+1}|^2\right) - \left(\psi^{n+1}, |\phi^n|^2\right) + \left(\psi^{n+1}, |\phi^{n+1} - \phi^n|^2\right)\right]$$
$$+ \frac{1}{4}\left[\left(\psi^{n+1}, |W^{n+1}|^2\right) - \left(\psi^{n+1}, |W^n|^2\right) - \left(\psi^{n+1}, |W^{n+1} - W^n|^2\right)\right].$$
$$(3.22)$$

where $b^n = F'(\phi^n)/\sqrt{E_u(\phi^n)}$. Taking the inner product of (3.1f) with $2\delta t U^{n+1}$ to obtain

$$\left[(U^{n+1})^2 - (U^n)^2 + (U^{n+1} - U^n)^2\right] = (U^{n+1}b^n, \phi^{n+1} - \phi^n). \qquad (3.23)$$

Summing up Eqs. (3.20)–(3.23), and dropping off some positive terms, we have

$$\frac{Cn^2}{4}\left(\left\|\nabla\phi^{n+1}\right\|^2 - \left\|\nabla\phi^n\right\|^2\right) + \left[\left(U^{n+1}\right)^2 - \left(U^n\right)^2\right] + Pi\left[\left(V^{n+1}\right)^2 - \left(V^n\right)^2\right]$$

$$+ \frac{1}{4Ex}\left[\left(\psi^{n+1}, \left|\phi^{n+1}\right|^2\right) - \left(\psi^n, \left|\phi^n\right|^2\right)\right] - \frac{1}{4}\left[\left(\psi^{n+1}, \left|W^{n+1}\right|^2\right) - \left(\psi^n, \left|W^n\right|^2\right)\right]$$

$$\leq -\frac{\delta t}{Pe_\psi}\left\|\sqrt{M_\psi^n}\nabla w_\psi^{n+1}\right\|^2 - \frac{\delta t}{Pe_\psi}\left\|\nabla w_\psi^{n+1}\right\|^2 - \delta t\left(\nabla\left(\mathbf{u}_*^n\psi^n\right), w_\psi^{n+1}\right) - \delta t\left(\nabla\left(\mathbf{u}_*^n\phi^n\right), w_\phi^{n+1}\right).$$

$$(3.24)$$

Finally, combining (3.16) and (3.24), we arrive at the desired result.

4 Numerical Results

To implement the scheme (3.1), we use a finite difference method on staggered grids to discretize space. We pay special attention to the discretization of the convection terms in the Cahn-Hilliard and Navier-Stokes equations. A composite high resolution scheme, known as the MINMOD scheme, is used to reduce the undershoot and overshoot around the interface. The computations of ψ^{n+1}, ϕ^{n+1}, $\mathbf{u}^{n+1}$ and p^{n+1} can be totally decoupled if we replace ψ^{n+1} in (3.1e) with ψ^n. The simplified scheme is extremely efficient and easy-to-implement. However, this simplification will definitely destroy the unconditional energy stability of our scheme. The implementation of such a simplified scheme requires small time step-sizes to obtain the desired accuracy and energy stability. The above scheme is adopted in [26] and numerical results demonstrate the energy stability of the proposed scheme. Here we will not present these results due to the limit of article length.

We simulate the droplet deformation under the horizontal body force and a shear flow in a computational domain $\Omega = [0, 3] \times [0, 1]$. Periodic boundary conditions are applied on the left and right sides. A circular droplet with the radius of $r = 0.3$ is initially placed at $(1, 0.5)$. Other simulation parameters are listed as follows:

$$Pe_\phi = 10, Pe_\psi = 100, \ Re = 10, Bo = 1, \ Cn = 0.01, \ Ex = 1, \ Pi = 0.1227, \lambda_\rho$$
$$= \lambda_\nu = 10.$$

Figure 1 shows the time evolution plots of droplet deformation and surfactant concentration. The droplet continuously deforms and moves forward under the action of the shear flow and the body force. We can divide the whole process into two stages based on the droplet deformation and surfactant migration. At the first stage, the body force has limited effect on the droplet deformation compared with the shear flow. Surfactants gradually migrate toward droplet tips, as shown in Fig. 1(b), resulting in the non-uniformity of interfacial tension along the interface. As we mentioned before, the surfactant concentration gradient induces the Marangoni stress, which will resist the further migration of surfactants. However, the Marangoni stress is not large enough to balance the effect of shear flow, and surfactants continue to move toward tips. In Fig. 1 (c), surfactants are swept into the bulk phases when concentration reaches the maximum at the droplet tips. At the second stage, the body force plays an important role in the droplet deformation and surfactant migration. In Fig. 1(d), surfactants on the tip A are slowly swept towards the ABC segment under the effect of the body force. Surfactants along the ADC segment continuously move to the tips under the combined action of the shear flow and the body force.

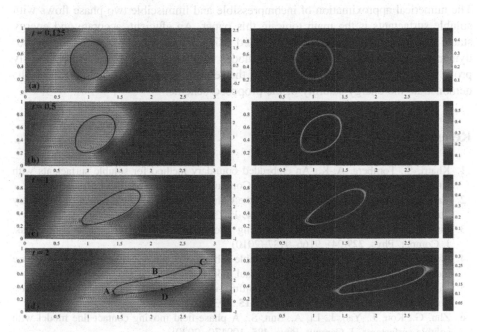

Fig. 1. Evolutions of pressure field (background color), quiver plot of velocity (u, v), phase-field variables ϕ and ψ. For each subfigure, the right is the profile of ψ. ($\psi_b = 1.5 \times 10^{-2}$).

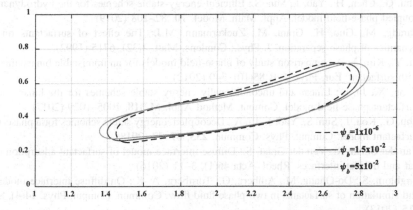

Fig. 2. Profiles of phase-field variable ϕ at $t = 1$ (left) and $t = 2$ (right). (black dash line: $\psi_b = 1 \times 10^{-6}$; blue solid line: $\psi_b = 1.5 \times 10^{-2}$; red solid line: $\psi_b = 5 \times 10^{-2}$) (Color figure online)

Figure 2 demonstrates the profiles of phase-field variable ϕ at three different ψ_b values. A more prolate profile of ϕ is observed for a higher surfactant bulk concentration, which confirms the effect of surfactants in reducing the interfacial tension.

5 Conclusion

The numerical approximation of incompressible and immiscible two-phase flows with soluble surfactants is the main topic in this paper. An efficient, accurate and energy stable time-marching scheme is constructed using the SAV approach for the hydrodynamics coupled phase-field surfactant model with variable densities. We rigorously prove the unconditional energy stability of the semi-implicit scheme. Numerical results demonstrate the energy stability of the proposed scheme.

References

1. Khatri, S., Tornberg, A.-K.: An embedded boundary method for soluble surfactants with interface tracking for two-phase flows. J. Comput. Phys. **256**, 768–790 (2014)
2. Yang, X.: Numerical approximations for the Cahn-Hilliard phase field model of the binary fluid-surfactant system. J. Sci. Comput. **74**, 1533–1553 (2018)
3. Liu, H., Zhang, Y.: Phase-field modeling droplet dynamics with soluble surfactants. J. Comput. Phys. **229**(24), 9166–9187 (2010)
4. Shen, J., Yang, X.: Decoupled, energy stable schemes for phase-field models of two-phase incompressible flows. SIAM J. Numer. Anal. **53**(1), 279–296 (2015)
5. Yue, P., Feng, J.J., Liu, C., Shen, J.: A diffuse-interface method for simulating two-phase flows of complex fluids. J. Fluid Mech. **515**, 293–317 (2014)
6. Zhu, G., Kou, J., Yao, J., Li, A., Sun, S.: A phase-field moving contact line model with soluble surfactants. J. Comput. Phys. **405**, 109170 (2020)
7. Xu, X., Di, Y., Yu, H.: Sharp-interface limits of a phase-field model with a generalized Navier slip boundary condition for moving contact lines. J. Fluid Mech. **849**, 805–833 (2018)
8. Zhu, G., Chen, H., Yao, J., Sun, S.: Efficient energy-stable schemes for the hydrodynamics coupled phase-field model. Appl. Math. Model. **70**, 82–108 (2019)
9. Laradji, M., Guo, H., Grant, M., Zuckermann, M.J.: The effect of surfactants on the dynamics of phase separation. J. Phys.: Condens. Matt. **4**(32), 6715 (1992)
10. Li, Y., Kim, J.: A comparison study of phase-field models for an immiscible binary mixture with surfactant. Eur. Phys. J. B **85**(10), 340 (2012)
11. Yang, X., Ju, L.: Linear and unconditionally energy stable schemes for the binary fluid–surfactant phase field model. Comput. Method Appl. M **318**, 1005–1029 (2017)
12. Zhu, G., Kou, J., Sun, S., Yao, J., Li, A.: Decoupled, energy stable schemes for a phase-field surfactant model. Comput. Phys. Commun. **233**, 67–77 (2018)
13. Van der Sman, R., Van der Graaf, S.: Diffuse interface model of surfactant adsorption onto flat and droplet interfaces. Rheol. Acta **46**(1), 3–11 (2016)
14. Engblom, S., Do-Quang, M., Amberg, G., Tornberg, A.-K.: On diffuse interface modeling and simulation of surfactants in two-phase fluid flow. Commun. Comput. Phys. **14**(4), 879–915 (2013)

15. Booty, M., Siegel, M.: A hybrid numerical method for interfacial fluid flow with soluble surfactant. J. Comput. Phys. **229**(10), 3864–3883 (2010)
16. Pätzold, G., Dawson, K.: Numerical simulation of phase separation in the presence of surfactants and hydrodynamics. Phys. Rev. E **52**(6), 6908 (1995)
17. Teigen, K.E., Song, P., Lowengrub, J., Voigt, A.: A diffuse-interface method for two-phase flows with soluble surfactants. J. Comput. Phys. **230**(2), 375–393 (2011)
18. Yun, A., Li, Y., Kim, J.: A new phase-field model for a water–oil-surfactant system. Appl. Math. Comput. **229**, 422–432 (2014)
19. Zhu, G., Kou, J., Sun, S., Yao, J., Li, A.: Numerical approximation of a phase-field surfactant model with fluid flow. J. Sci. Comput. **80**(1), 223–247 (2019)
20. Shen, J., Xu, J., Yang, J.: The scalar auxiliary variable (SAV) approach for gradient flows. J. Comput. Phys. **353**, 407–416 (2018)
21. Zhu, G., Kou, J., Yao, B., Wu, Y.-S., Yao, J., Sun, S.: Thermodynamically consistent modelling of two-phase flows with moving contact line and soluble surfactants. J. Fluid Mech. **879**, 327–359 (2019)
22. Feng, X., Kou, J., Sun, S.: A novel energy stable numerical scheme for Navier-Stokes-Cahn-Hilliard two-phase flow model with variable densities and viscosities. In: Shi, Y., et al. (eds.) ICCS 2018. LNCS, vol. 10862, pp. 113–128. Springer, Cham (2018). https://doi.org/10.1007/978-3-319-93713-7_9
23. Gao, M., Wang, X.-P.: An efficient scheme for a phase field model for the moving contact line problem with variable density and viscosity. J. Comput. Phys. **272**, 704–718 (2014)
24. Shen, J., Yang, X.: Decoupled energy stable schemes for phase-field models of two-phase complex fluids. SIAM J. Sci. Comput. **36**(1), B122–B145 (2014)
25. Kou, J., Sun, S.: Thermodynamically consistent modeling and simulation of multi-component two-phase flow with partial miscibility. Comput. Method Appl. M **331**, 623–649 (2018)
26. Zhu, G., Li, A.: Interfacial dynamics with soluble surfactants: a phase-field two-phase flow model with variable densities. Adv. Geo-Energy Res. **4**(1), 86–98 (2020)

A Numerical Algorithm to Solve the Two-Phase Flow in Porous Media Including Foam Displacement

Filipe Fernandes de Paula[1,2]([✉]) [iD], Thiago Quinelato[1] [iD], Iury Igreja[1,2] [iD], and Grigori Chapiro[1,2] [iD]

[1] Laboratório de Matemática Aplicada – LAMAP, Juiz de Fora, MG, Brazil
[2] Graduate Program on Computational Modeling, Federal University of Juiz de Fora, Juiz de Fora, MG, Brazil
filipe.paula@engenharia.ufjf.br,
{thiago.quinelato,iuryigreja,grigori}@ice.ufjf.br

Abstract. This work is dedicated to simulating the Enhanced Oil Recovery (EOR) process of foam injection in a fully saturated reservoir. The presence of foam in the gas-water mixture acts in controlling the mobility of the gas phase, contributing to reduce the effects of fingering and gravity override. A fractional flow formulation based on global pressure is used, resulting in a system of Partial Differential Equations (PDEs) that describe two coupled problems of distinct kinds: elliptic and hyperbolic. The numerical methodology is based on splitting the system of equations into two sub-systems that group equations of the same kind and on applying a hybrid finite element method to solve the elliptic problem and a high-order finite volume method to solve the hyperbolic equations. Numerical results show good efficiency of the algorithm, as well as the remarkable ability of the foam to increase reservoir sweep efficiency by reducing gravity override and fingering effects.

Keywords: EOR · Hybrid mixed methods · Finite volume methods · Foam injection · Mobility reduction

1 Introduction

The enhanced oil recovery by injection of gas is a technique that is used since the 1930's [14]. The sweep efficiency of gas, however, can be affected by gravity (by

This research was carried out in association with the R&D project registered as ANP 20715-9, "Modelagem matemática e computacional de injeção de espuma usada em recuperação avançada de petróleo" (Universidade Federal de Juiz de Fora (UFJF)/Shell Brasil/ANP) - Mathematical and computational modeling of foam injection as an enhanced oil recovery technique applied to Brazil pre-salt reservoirs, sponsored by Shell Brasil under the ANP R&D levy as "Compromisso de Investimentos com Pesquisa e Desenvolvimento". This project is carried out in partnership with Petrobras. G. Chapiro was supported in part by CNPq grant 303245/2019-0.

V. V. Krzhizhanovskaya et al. (Eds.): ICCS 2020, LNCS 12143, pp. 18–31, 2020.
https://doi.org/10.1007/978-3-030-50436-6_2

the *gravity override* phenomenon, that occurs when the injected gas accumulates in the upper layers of the reservoir) and by the development of preferential paths (*viscous fingering*), due to gas lower density and viscosity. These obstacles can be surpassed by the creation of foam, that can be defined as the agglomeration of gas bubbles separated by thin liquid films (lamellae), since foam apparent viscosity is much higher than the viscosity of gas [11,12,23]. The usage of foam in oil recovery is mainly motivated by the reduction of the gas phase mobility [15].

Population balance models can be used to simulate foam creation, destruction and flow through porous media. In this approach, it is common to define the foam texture (n_f), a quantity that represents the number of bubbles (or lamellae) per unit volume. Mechanisms of foam creation and coalescence play an important role in the model [15,17]. The major hypothesis adopted for the lamella-coalescence mechanism is that bubbles collapse near the limiting water saturation (S_w^*) or, equivalently, the limiting capillary pressure. In this context, a foam model based on the well-known steady-state behavior of foam in porous media was proposed by Ashoori et al. in [2]. It considers a large, nearly constant, reduction in gas mobility at high water saturation and an abrupt weakening or collapse of foam at a limiting water saturation. Foam texture in local equilibrium (n_D^{LE}), where bubble generation and destruction reach a local equilibrium state, depends only on the water saturation (S_w):

$$n_D^{\mathrm{LE}}(S_w) = \begin{cases} 0, & S_w \leq S_w^* \\ \tanh\left[A(S_w - S_w^*)\right], & S_w > S_w^* \end{cases}, \tag{1}$$

with constant A. The dynamic foam net generation is given by a first-order approach, introduced in [24] and later related to the local-equilibrium bubble texture in [2], with a time constant $1/K_c$, as follows:

$$r_g - r_c = K_c n_{\max} \left(n_D^{\mathrm{LE}}(S_w) - n_D \right), \tag{2}$$

where r_g and r_c are the foam generation and coalescence rates, respectively, $n_{\max}$ is the maximum foam texture and $n_D = n_f/n_{\max}$ is the dimensionless foam texture. This model represents a simplification of foam behavior in porous media without significantly sacrificing the physical phenomena [24]. Other models that associate bubble generation and destruction to the limiting capillary pressure or the gradient of gas pressure can be found in [15,17].

In this model, the reduction in the mobility of gas by foam is viewed as a reduction of the gas relative permeability k_{rg}:

$$k_{rg}(S_w, n_D) = \frac{k_{rg}^0(S_w)}{18500 n_D + 1}, \tag{3}$$

where k_{rg}^0 is the gas relative permeability when no foam is formed. Another view on the mobility reduction is based on the apparent viscosity of foam [13,15,17,24]. Comprehensive reviews on the mechanisms of bubble creation, destruction, and also on the reduction of gas mobility can be found in [9,21].

The following system of equations describes an incompressible two-phase flow in a porous medium:

$$\frac{\partial}{\partial t}(\phi S_{\text{g}}\, n_{\text{D}}) + \nabla \cdot (\boldsymbol{u}_{\text{g}}\, n_{\text{D}}) = \frac{\phi S_{\text{g}}(r_{\text{g}} - r_{\text{c}})}{n_{\text{max}}}, \tag{4}$$

$$\frac{\partial}{\partial t}(\phi S_\alpha) + \nabla \cdot \boldsymbol{u}_\alpha = 0, \quad \alpha \in \{\text{g,w}\}, \tag{5}$$

$$\boldsymbol{u}_\alpha = -\mathbb{K}\lambda_\alpha\left(\nabla p_\alpha - \rho_\alpha \boldsymbol{g}\right), \quad \alpha \in \{\text{g,w}\}, \tag{6}$$

where (4) is a population balance equation for foam texture [6] and (5)–(6) account for fractional flow and hydrodynamics. We use ϕ to denote the porosity of the medium, and ρ_α, S_α, $\boldsymbol{u}_\alpha$, and p_α to denote density, saturation, superficial velocity, and pressure, respectively, of phase α. Also, $\mathbb{K} = \mathbb{K}(\boldsymbol{x})$ is the intrinsic permeability tensor, and $\boldsymbol{g}$ is the gravity vector. From the fractional flow theory, $\lambda_{\text{w}} = k_{\text{rw}}/\mu_{\text{w}}$ and $\lambda_{\text{g}} = k_{\text{rg}}/\mu_{\text{g}}$ denote the mobility of water and gas phases, respectively, where the viscosity of water and gas are given by μ_{w} and μ_{g}. It is assumed that the porous medium is fully saturated, i.e., $S_{\text{w}} + S_{\text{g}} = 1$.

The numerical approach for solving this system of PDEs should be capable of handling several complexities due to discontinuity, non-linearity, stiffness, natural instabilities, among others. The numerical methods should also preserve important properties, such as local conservation of mass, shock capture, non-oscillatory solutions, accurate approximations, and reduced numerical diffusion effects.

To the extent of our knowledge, this problem is usually solved using explicit-in-time finite difference schemes [1,16,21,25]. Also, the use of commercial software is prevalent in the literature [7,20, and references within]. The most common approach in commercial software is to represent the effect of foam by a factor that reduces the mobility of the gas phase; therefore, bubble creation and destruction are not represented explicitly [9,21].

An effective numerical scheme to solve this kind of model and to address its inherent complexities is based on rewriting the problem in terms of global pressure, as in [4]. In this scheme, one has two distinct coupled problems: an elliptic problem and a degenerate hyperbolic problem. The next step is to decouple the system of PDEs into two subsystems of equations, each one of a different nature. In doing so, each subsystem can be solved by specialized methods, such as finite element and finite volume methods, according to their mathematical properties and the relation between precision and computational efficiency required in the resolution of each step. In this sense, for the spatial discretization of the hyperbolic problems we can employ the finite volume method, for instance; for time discretization, a common choice is a finite difference method, while a hybrid finite element method can be applied to the elliptic equations.

In this context, we develop a staggered algorithm to decouple the hydrodynamics from the hyperbolic system, resulting in a scheme that uses a locally conservative hybrid mixed finite element method to approximate the velocity and pressure fields and a high-order finite volume scheme to solve the hyperbolic equations. The two problems are solved in different time scales. Thus, the proposed staggered algorithm is employed to simulate two-phase (water and gas) flow in a heterogeneous porous medium. We compare pure gas-water injection with the gas-water-foam flow. The results show a reduction of gravity override and viscous fingering effects when the foam is present.

This work is organized as follows: in Sect. 2, we define a fractional flow formulation for Eqs. (4)–(6) using the concept of global pressure; in Sect. 3, we present an algorithm to solve the problem using a hybrid mixed finite element method for the elliptic problem and a high-order finite volume method methodology to solve the hyperbolic equations; numerical results are shown in Sect. 4. Finally, in Sect. 5, we present some concluding remarks.

2   Model Problem

To build a fractional flow model for the water-gas-foam flow in porous media we follow the global pressure approach from [4]. The global pressure is defined as

$$p = p_g - \int_{1-S_{gr}}^{S_w} f_w \frac{dP_c}{d\eta} d\eta \quad \text{and} \quad f_w = \frac{\lambda_w}{\lambda} = \frac{\lambda_w}{\lambda_w + \lambda_g}, \tag{7}$$

where $P_c(S_w) = p_g - p_w$ is the capillary pressure. From (6) and (7) the total velocity u is written as [4]:

$$u = u_g + u_w = -\mathbb{K}\lambda \left(\nabla p - G(S_w)\right), \quad \text{with} \quad G(S_w) = \frac{\lambda_g \rho_g + \lambda_w \rho_w}{\lambda} g, \tag{8}$$

where $g = -9.81\hat{j} \, m/s^2$ and $\hat{j}$ is the unitary vector in vertical direction.

It follows directly from the previous definitions that

$$u_w = f_w u + \mathbb{K}\lambda_g f_w \nabla P_c - b,$$
$$u_g = f_g u - \mathbb{K}\lambda_g f_w \nabla P_c + b,$$

where $b = \mathbb{K}\lambda_g f_w \left(\rho_g - \rho_w\right) g$.

Let $\Omega \subset \mathbb{R}^d$, $d = 2$ or 3, have Lipschitz boundary $\Gamma = \partial\Omega$. Using the hypothesis of rigid porous medium, the total fluid velocity u, the global pressure p, the water saturation S_w, and the foam texture n_D satisfy, in $\Omega \times (0, T]$, the following system of equations:

$$u = -\mathbb{K}\lambda \left(\nabla p - G(S_w)\right), \tag{9}$$

$$\nabla \cdot u = 0, \tag{10}$$

$$\phi \frac{\partial S}{\partial t} + \sum_{i=1}^{d} \frac{\partial f_i}{\partial x_i} - \nabla \cdot (\mathbb{C}\nabla S) = \Phi, \tag{11}$$

where i denotes a spatial direction, and

$$S = \begin{bmatrix} S_w \\ S_g n_D \end{bmatrix}, \quad f_i = \begin{bmatrix} f_w u_i - b_i \\ n_D f_g u_i + n_D b_i \end{bmatrix}, \quad \Phi = \begin{bmatrix} 0 \\ \phi S_g(r_g - r_c) \\ n_{\max} \end{bmatrix},$$

$$\mathbb{C}_{ijkl} = \mathbb{B}_{ik}\mathbb{D}_{jl}, \quad \mathbb{B} = \begin{bmatrix} -1 & 0 \\ n_D & 0 \end{bmatrix}, \quad \mathbb{D}_{jl} = \mathbb{K}_{jl}\lambda_g f_w \frac{dP_c}{dS_w},$$

and boundary and initial conditions

$$
\begin{aligned}
\boldsymbol{u} \cdot \boldsymbol{n} &= \bar{u} && \text{on } \Gamma_N \times (0, T], & p &= \bar{p} && \text{on } \Gamma_D \times (0, T], &&(12) \\
\boldsymbol{S} &= \bar{\boldsymbol{S}} && \text{on } \Gamma_N^- \times (0, T], & \boldsymbol{S}(\boldsymbol{x}, 0) &= \boldsymbol{S}^0 && \text{in } \Omega,
\end{aligned}
$$

where $\Gamma = \Gamma_N \cup \Gamma_D$, $\Gamma_N \cap \Gamma_D = \emptyset$, with Γ_N denoting the boundary region with Neumann condition (specified injection velocity), Γ_D defining the boundary region with Dirichlet condition on the potential, $\Gamma_N^- = \{\boldsymbol{x} \in \Gamma_N; \bar{u}(\boldsymbol{x}) < 0\}$ and $\boldsymbol{n}$ is the unit outer normal vector to Γ. For simplicity, we assume $\Gamma_D \neq \emptyset$ and $\Gamma_N^- \neq \emptyset$.

3   Numerical Method

In this section, we introduce the sequential algorithm that combines two kinds of numerical methods to solve (9)–(11).

The hydrodynamics (9)–(10) is approximated using a naturally stable mixed finite element method introduced in [22]. This method is locally conservative, relying on the strong imposition of the continuity of normal fluxes and on a discontinuous pressure field. The combination of a hybrid formulation with a static condensation technique reduces the number of degrees of freedom in the global problem.

The transport system (11) is solved using the KNP method, a conservative, high-order, central-upwind finite volume scheme introduced in [18] that shows reduced numerical diffusion effects. The KNP scheme is an extension of the KT method [19] that generalizes the numerical flux using more precise information about the local propagation velocities. At the same time, the KNP scheme has an upwind nature, since it respects the directions of wave propagation by measuring the one-sided local velocities. KNP is a semi-discrete method based on the REA (Reconstruct Evolve Average) algorithm of Godunov [8]. Furthermore, the KNP scheme allows for using small steps in time without requiring an excessive refinement of the spatial mesh, since the numerical diffusion does not depend on the time step. After discretization in space, the resulting system of ODEs is integrated in time using a BDF (Backward Differentiation Formula), an implicit, multi-step method that is especially indicated to solve stiff equations [5].

3.1   The Sequential Algorithm

The system of Eqs. (9)–(11) is strongly coupled. It is possible, however, to apply a staggered algorithm to solve an approximate problem composed of two sub-systems: an elliptic one, with time step Δt_u, and a hyperbolic one, with time step Δt_s. Each of them is solved separately using adaptive time steps, as described in Algorithm 1, i.e., one can use smaller time steps to bound the error in the approximations under a certain tolerance. In addition, it is often recommended that $\Delta t_u > \Delta t_s$, as the time scale of the hydrodynamics is usually much slower than of the transport. In each iteration, approximations for velocity $\boldsymbol{u}^{n+1}$ and pressure p^{n+1} fields at $t = t^{n+1}$ are computed from

$$\boldsymbol{u}^{n+1} = -\mathbb{K}\lambda\left(\nabla p^{n+1} - \boldsymbol{G}\left(S_{\mathrm{w}}^n\right)\right) \qquad\qquad \text{in } \Omega, \qquad (13a)$$

$$\nabla\cdot\boldsymbol{u}^{n+1} = 0 \qquad\qquad \text{in } \Omega, \qquad (13b)$$

supplemented by the boundary conditions (12).

Then an iterative algorithm is used to find approximations for water satura-
tion (S_{w}^{n+1}) and foam texture (n_{D}^{n+1}) by solving the following system of PDEs
in $\Omega \times (t^n, t^{n+1}]$ (for simplicity, we omit the superscript $n+1$):

$$\phi\frac{\partial \boldsymbol{S}^{n+1}}{\partial t} + \sum_{i=1}^{d}\frac{\partial \boldsymbol{f}_i^{n+1}}{\partial x_i} - \nabla\cdot\left(\mathbb{C}\nabla \boldsymbol{S}^{n+1}\right) = \boldsymbol{\Phi}^{n+1}, \qquad (14)$$

with boundary and initial conditions

$$\boldsymbol{S} = \bar{\boldsymbol{S}} \text{ on } \Gamma_N^- \times (t^n, t^{n+1}], \qquad\qquad \boldsymbol{S}(\boldsymbol{x}, t^n) = \boldsymbol{S}^n \text{ in } \Omega. \qquad (15)$$

Algorithm 1: Sequential algorithm to solve (13)–(15).

Set initial conditions S_{w}^0 and n_{D}^0;
$n \leftarrow 0$; $t \leftarrow 0$; $t_s \leftarrow 0$;
do
    Compute velocity $(\boldsymbol{u}^{n+1})$ and pressure (p^{n+1}) fields using (13);
    $t = t + \Delta t_{\mathrm{u}}$;
    $k \leftarrow 0$;
    $\boldsymbol{S}^{n+1,0} \leftarrow \boldsymbol{S}^n$;
    do
        Compute $(\boldsymbol{S}^{n+1,k+1})$ using (14) and (15) with $\boldsymbol{u}^{n+1}$;
        $t_s = t_s + \Delta t_s$;
        $k = k + 1$;
    while $t_s < t$;
    $\boldsymbol{S}^{n+1} \leftarrow \boldsymbol{S}^{n+1,k}$;
    $n = n + 1$;
while $t < T$;

In the following sections we comment on the methods employed to solve each
problem.

3.2 Hybrid Mixed Finite Element Method for Darcy Flow

When a mixed finite element formulation is used to approximate the Darcy
system (13), it is necessary to simultaneously fulfill the compatibility condition between spaces and to impose the continuity of the normal vector across

interelement edges. In addition, the resulting linear system is indefinite, which can restrict the numerical solvers that could be applied. By using a hybrid formulation, the continuity requirement is imposed via Lagrange multipliers, defined on the interelement edges. Furthermore, if the local problems are solvable, it is possible to eliminate all degrees of freedom related to local problems using a static condensation technique, resulting in a considerable reduction of the computational cost, since the global system involves only the degrees of freedom of the Lagrange multiplier. Also, in this case, the global problem is positive-definite. Once the approximation for the Lagrange multipliers is found, the original degrees of freedom (associated with velocity and pressure) can be computed in local, independent problems.

We first introduce some notations and definitions, for simplicity restricting ourselves to $\Omega \subset \mathbb{R}^2$. The three-dimensional case follows directly. Let $L^2(\Omega)$ denote the Hilbert space of square-integrable functions in Ω, with the usual inner product $(\cdot, \cdot)_\Omega$, and let $H(\text{div}, \Omega)$ be the space of vector functions having each component and divergence in $L^2(\Omega)$.

Assuming Ω is a polygon, we define a partition $\mathcal{T}_h$ of Ω composed of quadrilaterals and use K to denote an arbitrary element of the partition. The set of edges of K is denoted by ∂K, the set of edges in $\mathcal{T}_h$ is denoted by $\mathcal{E}_h$, and $\mathcal{E}_h^\partial$ denotes the set of all boundary edges, i.e., those with all points in Γ. Finally, the set of interior edges is denoted by $\mathcal{E}_h^0 = \mathcal{E}_h \setminus \mathcal{E}_h^\partial$. For every element $K \in \mathcal{T}_h$, there exists $c > 0$ such that $h \leq c h_e$, where h_e is the diameter of the edge $e \in \partial K$ and h, the mesh parameter, is the element diameter. For each edge of an element K we associate a unit outward normal vector $\mathbf{n}_K$.

The (discontinuous) $\mathcal{RT}$ spaces of index k [22] are here denoted by $\mathcal{U}_h^k \times \mathcal{P}_h^k$. We define the following sets of functions on the mesh skeleton:

$$\mathcal{M}_h^k = \left\{ \mu_h \in L^2(\mathcal{E}_h); \ \mu_h|_e \in p_k(e), \ \forall e \in \mathcal{E}_h, \ \mu_h|_e = \bar{p}, \forall e \in \mathcal{E}_h^\partial \cap \Gamma_D \right\}, \quad (16)$$

$$\bar{\mathcal{M}}_h^k = \left\{ \mu_h \in L^2(\mathcal{E}_h); \ \mu_h|_e \in p_k(e), \ \forall e \in \mathcal{E}_h, \ \mu_h|_e = 0, \forall e \in \mathcal{E}_h^\partial \cap \Gamma_D \right\}, \quad (17)$$

where $p_k(e)$ denotes the set of polynomial functions of degree up to k on e.

From these definitions, we can write the following hybrid mixed formulation for the hydrodynamics problem (13):

Given S_w^n and n_D^n, find the pair $[\mathbf{u}_h, p_h] \in \mathcal{U}_h^k \times \mathcal{P}_h^k$ and the Lagrange multiplier $\lambda_h \in \mathcal{M}_h^k$ such that, for all $[\mathbf{v}_h, q_h, \mu_h] \in \mathcal{U}_h^k \times \mathcal{P}_h^k \times \bar{\mathcal{M}}_h^k$,

$$\sum_{K \in \mathcal{T}_h} \left[(\mathbb{A}\mathbf{u}_h, \mathbf{v}_h)_K - (p_h, \nabla \cdot \mathbf{v}_h)_K + \int_{\partial K} \lambda_h \mathbf{v}_h \cdot \mathbf{n}_K \text{d}s \right] = (\mathbf{G}(S_w^n), \mathbf{v}_h)_\Omega \quad (18)$$

$$\sum_{K \in \mathcal{T}_h} -(q_h, \nabla \cdot \mathbf{u}_h)_K = -(f, q_h)_\Omega \quad (19)$$

$$\sum_{K \in \mathcal{T}_h} \int_{\partial K} \mu_h \mathbf{u}_h \cdot \mathbf{n}_K \text{d}s = \int_{\Gamma_N} \bar{u} \, \mu_h \text{d}s, \quad (20)$$

where $\mathbb{A} = \mathbb{A}(S_w^n, n_D^n) = (\mathbb{K}\lambda(S_w^n, n_D^n))^{-1}$.

To solve the hybrid formulation (18)–(19) we apply the static condensation technique that consists in a set of algebraic operations, done at the element level, to eliminate all degrees of freedom corresponding to the variables u_h and p_h, leading to a global system with the degrees of freedom associated with the multipliers only.

We can observe that static condensation causes a major reduction in the size of the global problem, which is now rewritten in terms of the multiplier only. Also, the new system of equations is positive-definite, allowing for using simpler and more robust solvers. In the end, a hybrid formulation associated with static condensation leads to a great reduction of the computational cost required to solve the global problem. In this work, the deal.II library [3] is used to solve this hydrodynamics problem.

3.3   High Order Central-Upwind Scheme for the Transport Problem

The numerical methodology used to approximate the water saturation and bubble texture Eqs. (14) and (15) is a high-order non-oscillatory central-upwind finite volume method proposed in [18] and here referred to as KNP. Like many other finite volume methods, the KNP scheme is based on a grid of control volumes (or cells).

The upwind nature of KNP is because it respects the directions of wave propagation by measuring the one-sided local speeds, given by

$$a_{l\pm 1/2,i}^{\max/\min} = \max/\min\left\{ \Lambda_{S_{l\pm 1/2,i}^-}^{\max/\min}, \Lambda_{S_{l\pm 1/2,i}^+}^{\max/\min}, 0\right\},$$

on direction i and a cell of index l, where $l+1/2$ is the right (resp. top) face and $j-1/2$ is the left (resp. bottom) face of a cell, $S_{l\pm 1/2,i}^-$ is the local reconstruction of S at the left (resp. bottom) side of a face, and $S_{l\pm 1/2,i}^+$ is the local reconstruction of S at the right (resp. top) side of a face; $\Lambda_X^{\max}$ and $\Lambda_X^{\min}$ are the maximum and minimum eigenvalues, respectively, of the Jacobian $\partial f_i/\partial S$ at $S - X$. The result of spatial discretization using KNP is the system of ODEs in conservative form:

$$\phi\frac{\mathrm{d}S_l}{\mathrm{d}t} = \sum_{i=1}^{d}\left(\frac{H_{l-1/2,i} - H_{l+1/2,i}}{h_i} + P_{l,i}\right) + \Phi_l, \tag{21}$$

where $\Phi_l = \Phi(S_l)$, h_i is the cell size in the i-th direction, with the convective numerical fluxes given by

$$H_{l\pm 1/2,i} = \frac{a_{l\pm 1/2,i}^{\max}f_i\left(S_{l\pm 1/2,i}^-\right) - a_{l\pm 1/2,i}^{\min}f_i\left(S_{l\pm 1/2,i}^+\right)}{a_{l\pm 1/2,i}^{\max} - a_{l\pm 1/2,i}^{\min}}$$

$$+ \frac{a_{l\pm 1/2,i}^{\max}a_{l\pm 1/2,i}^{\min}}{a_{l\pm 1/2,i}^{\max} - a_{l\pm 1/2,i}^{\min}}\left(S_{l\pm 1/2,i}^+ - S_{l\pm 1/2,i}^-\right), \tag{22}$$

and diffusive numerical fluxes given by

$$\mathbf{P}_{l,i}(t) = \frac{\tilde{\mathbb{D}}_{l+1/2,i}\boldsymbol{S}_{l+1,i} - \left(\tilde{\mathbb{D}}_{l+1/2,i} + \tilde{\mathbb{D}}_{l-1/2,i}\right)\boldsymbol{S}_l + \tilde{\mathbb{D}}_{l-1/2,i}\boldsymbol{S}_{l-1,i}}{h_i^2}, \quad (23)$$

where $\tilde{\mathbb{D}}_{l\pm1/2,i}$ is defined as the harmonic mean of $(\mathbb{DB})_l$ and $(\mathbb{DB})_{l\pm1,i}$. The scheme (21)–(23), combined with minmod reconstruction of the type

$$\tilde{\boldsymbol{S}}_l(\boldsymbol{x}) = \boldsymbol{S}_l^n + \sum_{i=1}^{d} \boldsymbol{d}_{l,i}^n(x_i - x_{l,i}),$$

$$\boldsymbol{d}_{l,i}^n = \text{minmod}\left(\theta\frac{\boldsymbol{S}_{l,i}^n - \boldsymbol{S}_{l-1,i}^n}{h_i}, \frac{\boldsymbol{S}_{l+1,i}^n - \boldsymbol{S}_{l-1,i}^n}{2h_i}, \theta\frac{\boldsymbol{S}_{l+1,i}^n - \boldsymbol{S}_{l,i}^n}{h_i}\right)$$

is a TVD scheme if $1 \leq \theta \leq 2$ [18], where $\tilde{\boldsymbol{S}}_l$ is a piecewise linear approximation to the solution at time t^n, i.e., $\tilde{\boldsymbol{S}}_l(\boldsymbol{x}) \approx \boldsymbol{S}_l^n(\boldsymbol{x})$. Then we can use the fact that $x_{l\pm1/2,i} = x_{l,i} \pm h_i/2$ to find $\boldsymbol{S}_{l\pm1/2,i}^\pm$.

Various numerical methods can be used to solve the system of ODEs (21). In this work, a variable order, adaptive step Backward Differentiation Formula (BDF) was chosen. This stable, implicit scheme allows for taking larger time steps than an explicit method would require, which reduces computational cost. In our numerical simulations we used the implementation of the BDF scheme from the CVode package, available in the SUNDIALS library [10].

4   Numerical Results

Applying the numerical methods described in Sect. 3, we now present results of numerical experiments that aim to assess the influence of foam and gravity effects in two-phase flow. Two scenarios are simulated: flow without and with foam. In the first scenario, we consider that only a mixture of water and gas is flowing through the porous medium, setting $n_D = 0$ in the hyperbolic problem and solving only for S_w in (11). The hydrodynamics and the mobility of the gas phase remain unchanged ($k_{rg} = k_{rg}^0$). In the second scenario, we assume that surfactant is readily available in the water phase, allowing for foam creation and changes in the mobility of gas phase. This scenario is simulated using the full problem (9)–(11). In both scenarios, the capillary pressure and relative permeabilities are

$$P_c = 330\frac{(1 - S_w - S_{gr})^{0.01}}{S_w - S_{wc}}, \quad k_{rw} = \left(\frac{S_w - S_{wc}}{1 - S_{wc} - S_{gr}}\right)^4, \quad k_{rg}^0 = \left(\frac{1 - S_w - S_{gr}}{1 - S_{wc} - S_{gr}}\right)^2.$$

The permeability is assumed isotropic $\mathbb{K} = \kappa(\boldsymbol{x})\mathbb{I}$, where $\kappa(\boldsymbol{x})$ is the permeability field of layers 1 (case A, Fig. 1(a)) and 36 (case B, Fig. 1(b)) of the SPE10 project[1], rotated to the xy plane. The right boundary is chosen to be the Dirichlet

---

[1] https://www.spe.org/web/csp/datasets/set02.htm.

type (Γ_D) with $\bar{p} = 0$, while left, top and bottom boundaries are set to Neumann condition (Γ_N) with $\bar{u} < 0$ for the left boundary and $\bar{u} = 0$ for the top and bottom boundaries. Coefficients and numerical parameters used in the simulations are shown in Table 1.

Table 1. Simulation parameters.

Parameter	Value	Parameter	Value
Water viscosity (μ_{w}) [Pa s]	1.0×10^{-3}	Porosity	0.25
Gas viscosity (μ_{g}) [Pa s]	2.0×10^{-5}	A	400
Water residual saturation (S_{wc})	0.2	K_{c} [$1/s$]	1.0×10^{-6}
Gas residual saturation (S_{gr})	0.0	Dimensions [m]	3.67×1.0
Critical water saturation (S_{w}^{*})	0.37	Final time [s]	1.0×10^{4}
Max foam texture (n_{max}) [m^{-3}]	8.0×10^{13}	$\Delta t_{\mathrm{u}}[s]$	20.0
Injection velocity ($\bar{u}$) [m s^{-1}]	3.0×10^{-5}	Number of cells	220×60
Initial water saturation (S_{w}^{0})	1.0	Minmod parameter (θ)	1.0
Injected water saturation ($\bar{S}_{\mathrm{w}}$)	0.372	Absolute tolerance	1.0×10^{-6}
Initial foam texture (n_{D}^{0})	0.0	Relative tolerance	1.0×10^{-4}
Injected foam texture ($\bar{n}_{\mathrm{D}}$)	0.0	$\mathcal{RT}$ index (k)	0

The water saturation profiles for case A at $t = 2\,000$ s and $t = 10\,000$ s are shown in Figs. 2 and 3, respectively. The gravity effects are much more pronounced in the no-foam simulation. Also, as expected, the water phase displacement occurs more slowly in the foam presence due to the gas mobility reduction caused by foam. Note that, without foam, the gas breakthrough has already taken place at $t = 10\,000$ s (Fig. 3), which does not occur when foam is present. Moreover, for the foam model adopted [2], viscous fingering and gravity override are reduced with foam as time advances. As a result, a better sweep efficiency of the medium is observed when foam is present, as can be seen in Fig. 6(a). In our experiments, foam injection increased total water recovery by approximately 100%.

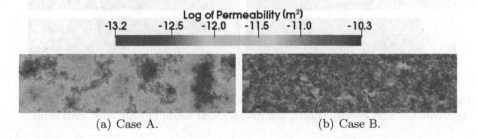

(a) Case A.   (b) Case B.

Fig. 1. Permeability map of layer 1 (a) and layer 36 (b) of the SPE10.

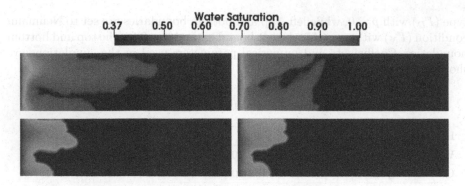

Fig. 2. Case A: water saturation at $t = 2\,000$ s. Left column: without gravity effects; right column: with gravity effects; top row: without foam; bottom row: with foam.

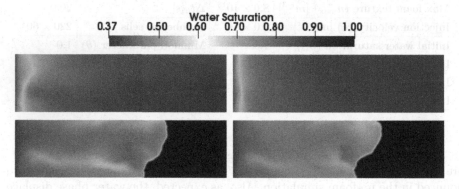

Fig. 3. Case A: water saturation at $t = 10\,000$ s. Left column: without gravity effects; right column: with gravity effects; top row: without foam; bottom row: with foam.

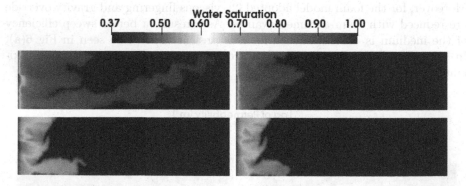

Fig. 4. Case B: water saturation at $t = 2\,000$ s. Left column: without gravity effects; right column: with gravity effects; top row: without foam; bottom row: with foam.

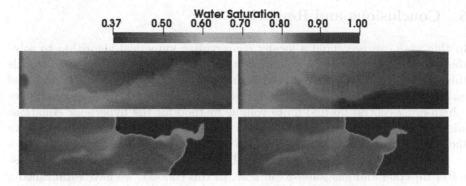

Fig. 5. Case B: water saturation at $t = 10\,000$ s. Left column: without gravity effects; right column: with gravity effects; top row: without foam; bottom row: with foam.

In case B, the permeability field has a more evident preferential channel in the lower region (see Fig. 1(b)). The results for this channelized porous formation (Figs. 4 and 5) reinforce the foam's ability to reduce the effects of gravity override and viscous fingering, according to the model used, even though this case presents a more pronounced preferential path. The water cumulative production curves for case B (Fig. 6(b)) show that gravity effects on the production are much more pronounced when no foam is present; this is due to the influence of gravity on diverting the flow from the high permeability channel, resulting in higher sweeping efficiency. Moreover, foam injection in this case increases total water production by approximately 50%, when gravity is considered, and by about 100% when gravity is neglected.

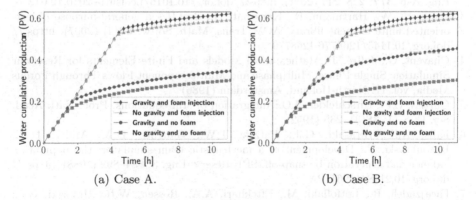

(a) Case A.   (b) Case B.

Fig. 6. Water cumulative production for cases A and B.

5   Conclusions and Remarks

In this work, we presented a locally conservative numerical algorithm to solve the gas-water flow including foam injection. The system of PDEs that models this phenomenon was derived considering a fractional flow formulation based on the global pressure. The numerical staggered approach proposed combines a high-order central upwind finite volume method for the hyperbolic equations adopting BDF time integrations with a hybrid finite element method to solve the Darcy's problem employing Raviart-Thomas spaces.

The proposed methodology was applied to simulate regimes with pure gas-water injection and gas-water-foam flow. In this context, we have established a comparison between these two regimes considering two layers of SPE10 project with different heterogeneous permeability fields. The results, based on the model proposed in [2], point to the foam's ability to reduce the gravity override and viscous fingering even in cases of porous media with rather pronounced preferential channels.

Acknowledgements. The authors are thankful to Professor Pacelli P. L. Zitha for fruitful preliminary discussions.

References

1. Afsharpoor, A.: Mechanistic foam modeling and simulations: gas injection during surfactant-alternating-gas processes using foam-catastrophe theory. Ph.D. thesis, Louisiana State University (2009)
2. Ashoori, E., Marchesin, D., Rossen, W.R.: Roles of transient and local equilibrium foam behavior in porous media: traveling wave. Colloids Surf. A Phys. Chemical Eng. Asp. **377**, 228–242 (2011). https://doi.org/10.1016/j.colsurfa.2010.12.042
3. Bangerth, W., Hartmann, R., Kanschat, G.: deal.II - a general-purpose object-oriented finite element library. ACM Trans. Math. Softw. **33**(4) (2007). https://doi.org/10.1145/1268776.1268779
4. Chavent, G., Jaffré, J.: Mathematical Models and Finite Elements for Reservoir Simulation: Single Phase, Multiphase and Multicomponent Flows Through Porous Media, vol. 17. North-Holland, Amsterdam (1986)
5. Curtiss, C.F., Hirschfelder, J.O.: Integration of stiff equations. Proc. Natl. Acad. Sci. U. S. A. **38**(3), 235 (1952)
6. Falls, A.H., Hirasaki, G.J., Patzek, T.W., Gauglitz, D.A., Miller, D.D., Ratoulowski, T.: Development of a mechanistic foam simulator: the population balance and generation by snap-off. SPE Reserv. Eng. **3**, 884–892 (1988). https://doi.org/10.2118/14961-PA
7. Farajzadeh, R., Lotfollahi, M., Eftekhari, A.A., Rossen, W.R., Hirasaki, G.J.: Effect of permeability on implicit-texture foam model parameters and the limiting capillary pressure. Energy Fuels **29**(5), 3011–3018 (2015). https://doi.org/10.1021/acs.energyfuels.5b00248
8. Godunov, S.K.: A difference method for numerical calculation of discontinuous solutions of the equations of hydrodynamics. Mat. Sb. **89**(3), 271–306 (1959)

9. Hematpur, H., Mahmood, S.M., Nasr, N.H., Elraies, K.A.: Foam flow in porous media: concepts, models and challenges. J. Nat. Gas Sci. Eng. **53**, 163–180 (2018). https://doi.org/10.1016/j.jngse.2018.02.017
10. Hindmarsh, A.C., et al.: SUNDIALS: suite of nonlinear and differential/algebraic equation solvers. ACM Trans. Math. Softw. **31**(3), 363–396 (2005)
11. Hirasaki, G.J.: A review of the steam foam process mechanisms (1989). SPE 19518
12. Hirasaki, G.J.: The steam-foam process. J. Pet. Technol. **41**(5), 449–456 (1989). https://doi.org/10.2118/19505-PA
13. Hirasaki, G.J., Lawson, J.B.: Mechanisms of foam flow in porous media: apparent viscosity in smooth capillaries. SPE J. **25**(02), 176–190 (1985)
14. Jones, S.A., Getrouw, N., Vincent-Bonnieu, S.: Foam flow in a model porous medium: I. The effect of foam coarsening. Soft Matter **14**, 3490–3496 (2018). https://doi.org/10.1039/C7SM01903C
15. Kam, S.I.: Improved mechanistic foam simulation with foam catastrophe theory. Colloids Surf. A Physicochem. Eng. Asp. **318**(1), 62–77 (2008). https://doi.org/10.1016/j.colsurfa.2007.12.017
16. Kam, S.I., Nguyen, Q.P., Li, Q., Rossen, W.R.: Dynamic simulations with an improved model for foam generation. In: SPE Annual Technical Conference and Exhibition. Society of Petroleum Engineers (2004)
17. Kovscek, A.R., Patzek, T.W., Radke, C.J.: A mechanistic population balance model for transient and steady-state foam flow in boise sandstone. Chem. Eng. Sci. **50**(23), 3783–3799 (1995). https://doi.org/10.1016/0009-2509(95)00199-F
18. Kurganov, A., Noelle, S., Petrova, G.: Semidiscrete central-upwind schemes for hyperbolic conservation laws and Hamilton-Jacobi equations. SIAM J. Sci. Comput. **23**(3), 707–740 (2001). https://doi.org/10.1137/S1064827500373413
19. Kurganov, A., Tadmor, E.: New high-resolution central schemes for nonlinear conservation laws and convection-diffusion equations. J. Comput. Phys. **160**(1), 241–282 (2000). https://doi.org/10.1006/jcph.2000.6459
20. Ma, K., Farajzadeh, R., Lopez-Salinas, J.L., Miller, C.A., Biswal, S.L., Hirasaki, G.J.: Non-uniqueness, numerical artifacts, and parameter sensitivity in simulating steady-state and transient foam flow through porous media. Transp. Porous Media **102**(3), 325–348 (2014). https://doi.org/10.1007/s11242-014-0276-9
21. Ma, K., Ren, G., Mateen, K., Morel, D., Cordelier, P.: Modeling techniques for foam flow in porous media. SPE J. **20**(3), 453–470 (2015)
22. Raviart, P.A., Thomas, J.M.: A mixed finite element method for 2-nd order elliptic problems. In: Galligani, I., Magenes, E. (eds.) Mathematical Aspects of Finite Element Methods. LNM, vol. 606, pp. 292–315. Springer, Heidelberg (1977). https://doi.org/10.1007/BFb0064470
23. Smith, D.H. (ed.): Surfactant-Based Mobility Control. No. 373 in ACS Symp. Ser., Am. Chem. Soc., Washington, D.C. (1988)
24. Zitha, P.L.J.: A new stochastic bubble population model for foam in porous media. In: SPE/DOE Symposium on Improved Oil Recovery. Society of Petroleum Engineers (2006)
25. Zitha, P.L.J., Du, D.X., Uijttenhout, M., Nguyen, Q.P.: Numerical analysis of a new stochastic bubble population foam model. In: SPE/DOE Symposium on Improved Oil Recovery. Society of Petroleum Engineers (2006)

A Three-Dimensional, One-Field, Fictitious Domain Method for Fluid-Structure Interactions

Yongxing Wang$^{(\boxtimes)}$ (iD), Peter K. Jimack (iD), and Mark A. Walkley (iD)

School of Computing, University of Leeds, Leeds LS2 9JT, UK
`scsywan@leeds.ac.uk`

Abstract. In this article we consider the three-dimensional numerical simulation of Fluid-Structure Interaction (FSI) problems involving large solid deformations. The one-field Fictitious Domain Method (FDM) is introduced in the framework of an operator splitting scheme. Three-dimensional numerical examples are presented in order to validate the proposed approach: demonstrating energy stability and mesh convergence; and extending two dimensional benchmarks from the FSI literature. New three dimensional benchmarks are also proposed.

Keywords: Fluid-Structure Interaction · Finite element · Fictitious domain · Immersed Finite Element · One-field · Monolithic scheme · Eulerian formulation

1 Introduction

Numerical simulation of Fluid-Structure Interaction (FSI) problems is a computational challenge due to its strong nonlinearity, especially in the case of large solid deformations. This challenge is exacerbated in three dimensions due to the need for efficient numerical algorithms to handle the large number of degrees of freedom that are inevitably required. In this paper we generalize our recent one-field Fictitious Domain Method (FDM) [22,23] from two to three dimensions, enhance the efficiency and robustness of the proposed time-stepping scheme, and demonstrate the resulting algorithm's capabilities on a number of challenging test problems. We also provide potential benchmark problems to allow results to be compared against those from other schemes in the future.

Lagrangian and Arbitrary Lagrangian-Eulerian (ALE) methods are widely adopted when considering a relatively small solid deformation [6,11,15]. Discrete remeshing can be used for large deformations [10,18], however this can be very costly in the case of three dimensions and can present challenges for mass conservation. The cut finite element method (cutFEM) [4,5] may also be applied to solve FSI problems [14,19], although it is not trivial to deal with the discontinuous integral across the elements cut by the moving fluid-solid interface, especially in three dimensional cases. The Fictitious Domain Method (FDM) [1,3,8,12,13] uses two meshes to represent the fluid and solid separately, which

© Springer Nature Switzerland AG 2020
V. V. Krzhizhanovskaya et al. (Eds.): ICCS 2020, LNCS 12143, pp. 32–45, 2020.
https://doi.org/10.1007/978-3-030-50436-6_3

can easily handle large deformation of the solid. However the FDM approach solves a very large equation system: both the velocity in the whole domain (fluid and solid) and the displacement in the solid domain, coupled via a distributed Lagrange multiplier (DLM) which is also an unknown variable. The Immersed Finite Element Method (IFEM) [2,17,24,27] also uses two meshes but only solves for velocity in the whole domain, while the solid information is assembled on the right-hand side of the fluid equation as a prescribed force term. This IFEM approach achieves FSI behaviour through this forcing term, and is therefore relatively efficient in three dimensional simulations compared to the DLM approach. Performance of the IFEM method depends strongly on the fluid and solid properties and usually works well when the solid behaves similarly to the fluid (such as a relatively soft solid) [20]. It has been successfully used, for example, in the area of biomechanics [16,27].

The one-field FDM approach [22,23] similarly only solves for one velocity field in the whole domain. However, this proposed method assembles the solid equations and implicitly includes them in the equation system. The one-field FDM approach has the same generality and robustness as the FDM/DLM: both of them solve the fluid equations and solid equations as one system. However the former needs to solve only for one velocity field while the latter solves for fluid velocity, solid displacement and Lagrange multiplier. The proposed one-field FDM may also be regarded as a special linearisation of the implicit IFEM, which however is more robust compared with explicit IFEM and more efficient compared with the implicit IFEM [24], allowing a wide range of solid parameters to be considered and naturally dealing with the case of different densities between fluid and solid [22,23]. In short, the one-field FDM combines the FDM/DLM advantage of robustness and the classical/explicit IFEM advantage of efficiency. The scheme has been validated through comparison with idealised two-dimensional test cases and against experimental data and simulation results drawn from the literature [21].

In this article, the one-field FDM is extended, implemented and validated in three dimensions for the first time through the use of a newly applied operator splitting scheme. The paper is organized as follows. The control equations and a general finite element weak formulation are introduced in Sect. 2.1 and 2.2 respectively, followed by time discretization in Sect. 2.3. The operator splitting scheme is introduced in Sect. 2.4, followed by the linearisation (implementation detail) in Sect. 2.5 and the numerical algorithm for the final linear equation system in Sect. 2.6. Several three-dimensional numerical tests are given in Sect. 3, and conclusions are presented in Sect. 4.

2   One-Field Fictitious Domain Method

In this section, we review the one-field fictitious domain method [22] and develop it further based upon a three-step operator splitting scheme and the case novel block-matrix preconditioners. The system is described in a manner that is independent of the spatial dimensions, thus ensuring its capability in three dimensions, which is the primary purpose of this paper.

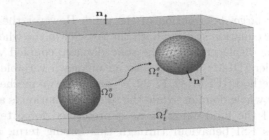

Fig. 1. Schematic diagram of FSI, $\overline{\Omega} = \overline{\Omega}_t^f \cup \overline{\Omega}_t^s$.

2.1   Control Partial Differentiation Equations

In the following context, $\Omega_t^f \subset \mathbb{R}^d$ and $\Omega_t^s \subset \mathbb{R}^d$ ($d = 3$ in this article) denote the moving fluid and solid domain respectively, with the moving interface $\Gamma_t = \overline{\Omega}_t^f \cap \overline{\Omega}_t^s$ as shown schematically in Fig. 1. $\Omega = \overline{\Omega}_t^f \cup \overline{\Omega}_t^s$ is a fixed domain with outer boundary $\Gamma = \Gamma_D \cup \Gamma_N$, with Γ_D and Γ_N being Dirichlet and Neumann boundaries respectively. We denote by $\mathbf{X}$ the reference coordinates of the solid, by $\mathbf{x} = \mathbf{x}(\cdot, t)$ the current coordinates of the solid, and by $\mathbf{x}_0$ the initial coordinates of the solid. Notice that we choose $\mathbf{X}$ to be the stress-free configuration, which may be different to the initial configuration $\mathbf{x}_0$.

Let $\rho, \mu, \mathbf{u}, \boldsymbol{\sigma}$ and $\mathbf{g}$ denote the density, viscosity, velocity, stress tensor and acceleration due to gravity respectively. We assume both an incompressible fluid in Ω_t^f and incompressible solid in Ω_t^s. The conservation of momentum and conservation of mass therefore take the same form as follows. Momentum equation:

$$\rho \frac{d\mathbf{u}}{dt} = \nabla \cdot \boldsymbol{\sigma} + \rho \mathbf{g}, \tag{1}$$

and continuity equation:

$$\nabla \cdot \mathbf{u} = 0. \tag{2}$$

An incompressible Newtonian constitutive equation in Ω_t^f can be expressed as:

$$\boldsymbol{\sigma} = \boldsymbol{\sigma}^f = \boldsymbol{\tau}^f - p^f \mathbf{I}, \tag{3}$$

with $\boldsymbol{\tau}^f = \mu^f \mathbf{D}\mathbf{u}^f$ being the deviatoric part of stress $\boldsymbol{\sigma}^f$, and $\mathbf{D}\mathbf{u} = \nabla \mathbf{u} + \nabla^T \mathbf{u}$. An incompressible neo-Hookean solid with viscosity μ^s is assumed in Ω_t^s [3], and the constitutive equation may be expressed as:

$$\boldsymbol{\sigma} = \boldsymbol{\sigma}^s = \boldsymbol{\tau}^s - p^s \mathbf{I}, \tag{4}$$

with $\boldsymbol{\tau}^s = c_1 \left(\mathbf{F}\mathbf{F}^T - \mathbf{I}\right) + \mu^s \mathbf{D}\mathbf{u}^s$ being the deviatoric part of stress $\boldsymbol{\sigma}^s$, and $\mathbf{F} = \frac{\partial \mathbf{x}}{\partial \mathbf{X}} = \frac{\partial \mathbf{x}}{\partial \mathbf{x}_0} \frac{\partial \mathbf{x}_0}{\partial \mathbf{X}} = \nabla_0 \mathbf{x} \nabla_{\mathbf{X}} \mathbf{x}_0$ being the deformation tensor of the solid, and c_1 is a solid material parameter. Finally the system is completed with continuity of velocity $\mathbf{u}^f = \mathbf{u}^s$ and normal stress $\boldsymbol{\sigma}^f \mathbf{n}^s = \boldsymbol{\sigma}^s \mathbf{n}^s$ on interface Γ_t, and standard Dirichlet/Neumann boundary (on Γ_D/Γ_N) and initial conditions.

2.2   Finite Element Weak Form

In the following context, let $L^2(\omega)$ be the square integrable functions in domain ω, and $H^1(\omega) = \left\{ u : u, \frac{\partial u}{\partial x_i} \in L^2(\omega) \quad \text{for} \quad i = 1, \cdots, d \right\}$. We also denote by $H_0^1(\omega)$ the subspace of $H^1(\omega)$ whose functions have zero values on the Dirichlet boundary of ω, and denote by $L_0^2(\omega)$ the subspace of $L^2(\omega)$ whose functions have zero mean value. Let $p = \begin{cases} p^f & in \quad \Omega_t^f \\ p^s & in \quad \Omega_t^s \end{cases}$. Given $\mathbf{v} \in H_0^1(\Omega)^d$, we perform the following symbolic operations:

$$\int_\Omega \text{Eq. (1)} (\boldsymbol{\sigma}) \cdot \mathbf{v} \equiv \int_{\Omega_t^f} \text{Eq. (1)} (\boldsymbol{\sigma}^f) \cdot \mathbf{v} + \int_{\Omega_t^s} \text{Eq. (1)} (\boldsymbol{\sigma}^s) \cdot \mathbf{v}$$

$$\equiv \int_\Omega \text{Eq. (1)} (\boldsymbol{\sigma}^f) \cdot \mathbf{v} + \int_{\Omega_t^s} \left(\text{Eq. (1)} (\boldsymbol{\sigma}^s) - \text{Eq. (1)} (\boldsymbol{\sigma}^f) \right) \cdot \mathbf{v}.$$

Integrating the stress terms by parts, the above operations, using constitutive equations (3) and (4), give:

$$\rho^f \int_\Omega \frac{d\mathbf{u}}{dt} \cdot \mathbf{v} + \int_\Omega \boldsymbol{\tau}^f : \nabla \mathbf{v} - \int_\Omega p\nabla \cdot \mathbf{v} + (\rho^s - \rho^f) \int_{\Omega_t^s} \frac{d\mathbf{u}}{dt} \cdot \mathbf{v}$$
$$+ \int_{\Omega_t^s} (\boldsymbol{\tau}^s - \boldsymbol{\tau}^f) : \nabla \mathbf{v} = \int_\Omega \rho^f \mathbf{g} \cdot \mathbf{v} + \int_{\Omega_t^s} (\rho^s - \rho^f) \mathbf{g} \cdot \mathbf{v} + \int_{\Gamma_N} \bar{\mathbf{h}} \cdot \mathbf{v}, \tag{5}$$

where $\bar{h}$ denotes the prescribed normal stress on Γ_N. Note that the integrals on the interface Γ_t are cancelled out due to the continuity of normal stress: $\boldsymbol{\sigma}^f \mathbf{n}^s = \boldsymbol{\sigma}^s \mathbf{n}^s$, because they are internal forces for the whole FSI system. Combining with the following symbolic operations for $q \in L^2(\Omega)$,

$$- \int_{\Omega_t^f} \text{Eq. (2)} q - \int_{\Omega_t^s} \text{Eq. (2)} q \equiv - \int_\Omega \text{Eq. (2)} q,$$

leads to the weak form of the FSI system as follows.

Given $\mathbf{u}_0$ and Ω_0^s, find $\mathbf{u}(t) \in H^1(\Omega)^d$, $p(t) \in L^2(\Omega)$ and Ω_t^s, such that for $\forall \mathbf{v} \in H_0^1(\Omega)^d$, $\forall q \in L^2(\Omega)$, the following equation holds:

$$\rho^f \int_\Omega \frac{\partial \mathbf{u}}{\partial t} \cdot \mathbf{v} + \rho^f \int_\Omega (\mathbf{u} \cdot \nabla) \mathbf{u} \cdot \mathbf{v} + \frac{\mu^f}{2} \int_\Omega D\mathbf{u} : D\mathbf{v} - \int_\Omega p\nabla \cdot \mathbf{v}$$
$$- \int_\Omega q\nabla \cdot \mathbf{u} + \rho^\delta \int_{\Omega_t^s} \frac{\partial \mathbf{u}}{\partial t} \cdot \mathbf{v} + \frac{\mu^\delta}{2} \int_{\Omega_t^s} D\mathbf{u} : D\mathbf{v} \tag{6}$$
$$+ c_1 \int_{\Omega_t^s} (\mathbf{F}\mathbf{F}^T - \mathbf{I}) : \nabla \mathbf{v} = \rho^f \int_\Omega \mathbf{g} \cdot \mathbf{v} + \rho^\delta \int_{\Omega_t^s} \mathbf{g} \cdot \mathbf{v} + \int_{\Gamma_N} \bar{\mathbf{h}} \cdot \mathbf{v},$$

where $\rho^\delta = \rho^s - \rho^f$ and $\mu^\delta = \mu^s - \mu^f$, and the integral over Ω_t^s, $\frac{\partial(\cdot)}{\partial t}$ is the time derivative with respect to a frame moving with the solid velocity $\mathbf{u}^s = \mathbf{u}|_{\Omega_t^s}$.

2.3   Discretisation in Time

Using the backward Euler method to discretise in time, Eq. (6) may be approximated as follows.

Given $\mathbf{u}_n$, p_n and Ω_n^s, find $\mathbf{u}_{n+1} \in H^1(\Omega)^d$, $p_{n+1} \in L^2(\Omega)$ and Ω_{n+1}^s, such that for $\forall \mathbf{v} \in H_0^1(\Omega)^d$, $\forall q \in L^2(\Omega)$, the following equation holds:

$$
\rho^f \int_\Omega \frac{\mathbf{u}_{n+1} - \mathbf{u}_n}{\Delta t} \cdot \mathbf{v} + \rho^f \int_\Omega (\mathbf{u}_{n+1} \cdot \nabla) \mathbf{u}_{n+1} \cdot \mathbf{v}
$$
$$
+ \frac{\mu^f}{2} \int_\Omega D\mathbf{u}_{n+1} : D\mathbf{v} - \int_\Omega p_{n+1} \nabla \cdot \mathbf{v} - \int_\Omega q \nabla \cdot \mathbf{u}_{n+1}
$$
$$
+ \rho^\delta \int_{\Omega_{n+1}^s} \frac{\mathbf{u}_{n+1} - \mathbf{u}_n}{\Delta t} \cdot \mathbf{v} + \frac{\mu^\delta}{2} \int_{\Omega_{n+1}^s} D\mathbf{u}_{n+1} : D\mathbf{v}
$$
$$
+ c_1 \int_{\Omega_{n+1}^s} \left(\mathbf{F}\mathbf{F}^T - \mathbf{I}\right) : \nabla \mathbf{v} = \rho^f \int_\Omega \mathbf{g} \cdot \mathbf{v} + \rho^\delta \int_{\Omega_{n+1}^s} \mathbf{g} \cdot \mathbf{v} + \int_{\Gamma_N} \bar{\mathbf{h}} \cdot \mathbf{v},
$$

$$(7)$$

and Ω_{n+1}^s is updated from Ω_n^s by the following formula:

$$
\Omega_{n+1}^s = \{\mathbf{x} : \mathbf{x} = \mathbf{x}_n + \Delta t \mathbf{u}_{n+1}, \mathbf{x}_n \in \Omega_n^s\}. \tag{8}
$$

2.4   An Operator Splitting Scheme

The formulation of (7) is implicit. However we shall solve it semi-implicitly via the following operator spitting scheme which is based upon [9].

(1) Convection step:

$$
\int_\Omega \frac{\mathbf{u}_{n+1/3} - \mathbf{u}_n}{\Delta t} \cdot \mathbf{v} + \int_\Omega \left(\mathbf{u}_{n+1/3} \cdot \nabla\right) \mathbf{u}_{n+1/3} \cdot \mathbf{v} = 0. \tag{9}
$$

(2) Diffusion step:

$$
\rho^f \int_\Omega \frac{\mathbf{u}_{n+2/3} - \mathbf{u}_{n+1/3}}{\Delta t} \cdot \mathbf{v} + \frac{\mu^f}{2} \int_\Omega D\mathbf{u}_{n+2/3} : D\mathbf{v}
$$
$$
+ \rho^\delta \int_{\Omega_n^s} \frac{\mathbf{u}_{n+2/3} - \mathbf{u}_n}{\Delta t} \cdot \mathbf{v} + \frac{\mu^\delta}{2} \int_{\Omega_n^s} D_n \mathbf{u}_{n+2/3} : D_n \mathbf{v}
$$
$$
+ c_1 \int_{\Omega_n^s} \left(\mathbf{F}_{n+2/3} \mathbf{F}_{n+2/3}^T - \mathbf{I}\right) : \nabla_n \mathbf{v}
$$
$$
= \rho^f \int_\Omega \mathbf{g} \cdot \mathbf{v} + \rho^\delta \int_{\Omega_n^s} \mathbf{g} \cdot \mathbf{v} + \int_{\Gamma_N} \bar{\mathbf{h}} \cdot \mathbf{v}.
$$

$$(10)$$

(3) Pressure step:

$$
\rho^f \int_\Omega \frac{\mathbf{u}_{n+1} - \mathbf{u}_{n+2/3}}{\Delta t} \cdot \mathbf{v} - \int_\Omega p_{n+1} \nabla \cdot \mathbf{v} - \int_\Omega q \nabla \cdot \mathbf{u}_{n+1} = 0. \tag{11}
$$

In the above, $\nabla_n(\cdot)$ represents the divergence in the current coordinates at $t = t_n$ and $D_n = \nabla_n + \nabla_n^T$. Note that the variables $\mathbf{u}_{n+1/3}$ and $\mathbf{u}_{n+2/3}$ are just intermediate values, not specifically the velocity at time $t = t_n + \frac{\Delta t}{3}$ or $t = t_n + \frac{2\Delta t}{3}$. The notation $\mathbf{F}_{n+1/3}$ or $\mathbf{F}_{n+2/3}$ is interpreted as follows:

$$\mathbf{F}_t = \frac{\partial \mathbf{x}_t}{\partial \mathbf{X}} = \nabla_{\mathbf{X}}\left(\mathbf{x}_n + \mathbf{u}_t \Delta t\right), \tag{12}$$

with $t = n + 1/3$ or $n + 2/3$.

Using this splitting scheme, standard approaches can be taken to solve the pure convection equation (9) (see [9]), and iterative methods with an efficient preconditioner can be applied to solve the "degenerate" Stokes Equations (11) (see Sect. 2.6 of [7,21]). The main challenge is in how to approximate the term $\mathbf{F}_{n+2/3}\mathbf{F}_{n+2/3}^T - \mathbf{I}$ in Eq. (10), which is nonlinearly related to the solid displacement and hence to the solid velocity. In the following subsection we focus on expressing and linearising $\mathbf{F}_{n+2/3}\mathbf{F}_{n+2/3}^T - \mathbf{I}$ in terms of velocity $\mathbf{u}_n$.

2.5   Linearisation of the Diffusion Step

The specific choice of linearisation is the core of this proposed one-field FDM approach, and is what makes it distinctive from all other schemes. Let $\mathbf{F}_{n+2/3}\mathbf{F}_{n+2/3}^T - \mathbf{I}$ be denoted by $\mathbf{F}_t\mathbf{F}_t^T - \mathbf{I} = \mathbf{s}_t$ with $t = n + 2/3$, then $\mathbf{s}_t$ can be computed as follows:

$$\mathbf{s}_t = \mathbf{F}_t\mathbf{F}_t^T - \mathbf{I} = \left(\nabla_{\mathbf{X}}\mathbf{x}_t\nabla_{\mathbf{X}}^T\mathbf{x}_t - \mathbf{I}\right). \tag{13}$$

Using the chain rule, this last equation can also be expressed as:

$$\mathbf{s}_t = \nabla_n\mathbf{x}_t\nabla_{\mathbf{X}}\mathbf{x}_n\nabla_{\mathbf{X}}^T\mathbf{x}_n\nabla_n^T\mathbf{x}_t - \mathbf{I} + \nabla_n\mathbf{x}_t\nabla_n^T\mathbf{x}_t - \nabla_n\mathbf{x}_t\nabla_n^T\mathbf{x}_t \tag{14}$$

or

$$\mathbf{s}_t = \nabla_n\mathbf{x}_t\nabla_n^T\mathbf{x}_t - \mathbf{I} + \nabla_n\mathbf{x}_t\left(\nabla_{\mathbf{X}}\mathbf{x}_n\nabla_{\mathbf{X}}^T\mathbf{x}_n - \mathbf{I}\right)\nabla_n^T\mathbf{x}_t. \tag{15}$$

Then $\mathbf{s}_t$ can be expressed based on the previous coordinate $\mathbf{x}_n$ as follows:

$$\mathbf{s}_t = \nabla_n\mathbf{x}_t\nabla_n^T\mathbf{x}_t - \mathbf{I} + \nabla_n\mathbf{x}_t\mathbf{s}_n\nabla_n^T\mathbf{x}_t. \tag{16}$$

Using $\mathbf{x}_t = \mathbf{x}_n + \Delta t\mathbf{u}_t$ (see (12)), this can finally be expressed as:

$$\begin{aligned}\mathbf{s}_t = \Delta t\left(\nabla_n\mathbf{u}_t + \nabla_n^T\mathbf{u}_t + \Delta t\nabla_n\mathbf{u}_t\nabla_n^T\mathbf{u}_t\right) + \mathbf{s}_n \\ + \Delta t^2\nabla_n\mathbf{u}_t\mathbf{s}_n\nabla_n^T\mathbf{u}_t + \Delta t\nabla_n\mathbf{u}_t\mathbf{s}_n + \Delta t\mathbf{s}_n\nabla_n^T\mathbf{u}_t.\end{aligned} \tag{17}$$

There are two nonlinear terms in this equation, which may be linearised as

$$\nabla_n\mathbf{u}_t\nabla_n^T\mathbf{u}_t = \nabla_n\mathbf{u}_t\nabla_n^T\mathbf{u}_n + \nabla_n\mathbf{u}_n\nabla_n^T\mathbf{u}_t - \nabla_n\mathbf{u}_n\nabla_n^T\mathbf{u}_n, \tag{18}$$

and

$$\nabla_n\mathbf{u}_t\mathbf{s}_n\nabla_n^T\mathbf{u}_t = \nabla_n\mathbf{u}_t\mathbf{s}_n\nabla_n^T\mathbf{u}_n + \nabla_n\mathbf{u}_n\mathbf{s}_n\nabla_n^T\mathbf{u}_t - \nabla_n\mathbf{u}_n\mathbf{s}_n\nabla_n^T\mathbf{u}_n. \tag{19}$$

Substituting $\mathbf{s}_{n+2/3} = \mathbf{F}_{n+2/3}\mathbf{F}_{n+2/3}^T - \mathbf{I}$, using expressions (17), (18) and (19), into the diffusion step (10), and neglecting terms of $O\left(\Delta t^2\right)$, after some algebra we produce the one-field FDM formulation:

$$
\rho^f \int_\Omega \frac{\mathbf{u}_{n+2/3} - \mathbf{u}_{n+1/3}}{\Delta t} \cdot \mathbf{v} + \frac{\mu^f}{2} \int_\Omega D\mathbf{u}_{n+2/3} : D\mathbf{v}
$$
$$
+ \rho^\delta \int_{\Omega_n^s} \frac{\mathbf{u}_{n+2/3} - \mathbf{u}_n}{\Delta t} \cdot \mathbf{v} + \frac{\mu^\delta + \Delta t c_1}{2} \int_{\Omega_n^s} D_n\mathbf{u}_{n+2/3} : D_n\mathbf{v}
$$
$$
+ \Delta t c_1 \int_{\Omega_n^s} \left(\nabla_n\mathbf{u}_{n+2/3}\mathbf{s}_n + \mathbf{s}_n\nabla_n^T\mathbf{u}_{n+2/3}\right) : \nabla_n\mathbf{v}
$$
$$
= \rho^f \int_\Omega \mathbf{g} \cdot \mathbf{v} + \rho^\delta \int_{\Omega_n^s} \mathbf{g} \cdot \mathbf{v} + \int_{\Gamma_N} \bar{\mathbf{h}} \cdot \mathbf{v} - c_1 \int_{\Omega_n^s} \mathbf{s}_n : \nabla_n\mathbf{v}.
$$

(20)

2.6 Iterative Linear Algebra Solver

In this section, we shall discuss the numerical algorithms in order to solve the final linear equations from the diffusion step (20) and pressure step (11). For convection step, we use the Taylor-Galerkin method in this paper [9]. Let us write Eq. (20) in an operator matrix form as follows:

$$
\mathbf{A}_n\mathbf{u}_{n+2/3} = \mathbf{b}_n,
$$

(21)

where

$$
\mathbf{A}_n = \mathbf{M}/\Delta t + \mathbf{K} + \mathbf{P}_n^T \left(\mathbf{M}_n^s/\Delta t + \mathbf{K}_n^s\right) \mathbf{P}_n,
$$

and

$$
\mathbf{b}_n = \mathbf{f} + \mathbf{P}_n^T\mathbf{f}_n^s + \mathbf{M}\mathbf{u}_{n+1/3}/\Delta t + \mathbf{P}_n^T\mathbf{M}_n^s\mathbf{P}_n\mathbf{u}_n/\Delta t.
$$

The above matrix operators are defined as:

$$
\int_\Omega \mathbf{M}\mathbf{u} \cdot \mathbf{v} = \rho^f \int_\Omega \mathbf{u} \cdot \mathbf{v}, \quad \int_\Omega \mathbf{K}\mathbf{u} \cdot \mathbf{v} = \frac{\mu^f}{2} \int_\Omega D\mathbf{u} : D\mathbf{v},
$$

$$
\int_{\Omega_n^s} \mathbf{K}_n^s\mathbf{u} \cdot \mathbf{v} = \frac{\mu^\delta + \Delta t c_1}{2} \int_{\Omega_n^s} D\mathbf{u} : D\mathbf{v} + \Delta t c_1 \int_{\Omega_n^s} \left(\nabla_n\mathbf{u}\mathbf{s}_n + \mathbf{s}_n\nabla_n^T\mathbf{u}\right) : \nabla_n\mathbf{v},
$$

$$
\int_\Omega \mathbf{f} \cdot \mathbf{v} = \rho^f \int_\Omega \mathbf{g} \cdot \mathbf{v} + \int_{\Gamma_N} \bar{\mathbf{h}} \cdot \mathbf{v}, \quad \int_{\Omega_n^s} \mathbf{f}_n^s \cdot \mathbf{v} = \rho^\delta \int_{\Omega_n^s} \mathbf{g} \cdot \mathbf{v} - c_1 \int_{\Omega_n^s} \mathbf{s}_n : \nabla_n\mathbf{v},
$$

where $\mathbf{u}, \mathbf{v} \in H^1(\omega)^d$ with ω being Ω or Ω_n^s. Finally $\mathbf{P}_n$ is a restriction from $H^1(\Omega)^d$ to $H^1(\Omega_n^s)^d$ ($\mathbf{P}_n^T$ is the corresponding injection from $H^1(\Omega_n^s)^d$ to $H^1(\Omega)^d$): $\mathbf{P}_n\mathbf{u} = \mathbf{u}^s = \mathbf{u}|_{\Omega_n^s}$. We use the finite element interpolation to approximate $\mathbf{P}_n$ after discretisation in space.

A preconditioned Conjugate Gradient method can efficiently solve Eq. (21). We use the incomplete Cholesky decomposition of matrix $\mathbf{M}/\Delta t + \mathbf{K}$ as a preconditioner in order to solve Eq. (21). Very good convergence performance can be observed from our numerical tests (although the precise performance of the linear algebraic solver is not the topic of this article).

Similarly, the "degenerate Stokes" problem (pressure step (11)) can also be expressed in an operator matrix form:

$$\begin{bmatrix} \mathbf{M}/\Delta t & \mathbf{B} \\ \mathbf{B}^T & 0 \end{bmatrix} \begin{pmatrix} \mathbf{u}_{n+1} \\ \mathbf{p}_{n+1} \end{pmatrix} = \begin{pmatrix} \mathbf{M}\mathbf{u}_{n+2/3}/\Delta t \\ 0 \end{pmatrix}, \tag{22}$$

where, $\forall \mathbf{v}$ in $H^1(\Omega)^d$ and $q \in L^2(\Omega)$, $\int_\Omega (\mathbf{B}\mathbf{v})\, q = - \int_\Omega q \nabla \cdot \mathbf{v}$. We use the MinRes algorithm [7] to solve the system with the following preconditioner:

$$\begin{bmatrix} \mathbf{M} & \\ & \mathbf{\Delta}_p \end{bmatrix}, \quad \text{where } \int_\Omega (\mathbf{\Delta}_p p)\, q = \int_\Omega \nabla p \cdot \nabla q, \quad \forall p, q \in H^1(\Omega). \tag{23}$$

We justify this since we can derive a Schur complement in the form of $\mathbf{S} = \mathbf{B}^T \mathbf{M}^{-1} \mathbf{B}$. The operators that are discretised in this form imply that $\mathbf{S}$ will be spectrally equivalent to a discrete Laplacian. Hence we expect this preconditioner will be effective for this system, similarly to analysis for Stokes equation [7].

3   Numerical Experiments

In the following numerical tests, the convection and diffusion steps are discretised with quadratic finite elements (tri-quadratic hexahedra and quadratic tetrahedra), and the pressure step is discretised with the Taylor-Hood element. For stability it is sufficient that $\mu^\delta \geq 0$ [21,22], however for simplicity, and to be consistent with [2,17,27] for example, we assume $\mu^\delta = 0$ in these tests.

3.1   Oscillating Ball

In this section, we consider a 3D oscillating ball which is an extension of the 2D disc in [23,28]. We use this example to test stability of the proposed approach by investigating the evolution of total energy:

$$E_{total}(t_n) = \frac{\rho^f}{2} \int_\Omega |\mathbf{u}_n|^2 + \frac{\rho^\delta}{2} \int_{\Omega_n^s} |\mathbf{u}_n|^2$$
$$+ \frac{\Delta t \mu^f}{2} \sum_{k=1}^n \int_\Omega \mathbf{Du}_k : \mathbf{Du}_k + \frac{c_1}{2} \int_{\Omega_\mathbf{X}^s} \left(tr_{\mathbf{F}_n \mathbf{F}_n^T} - d \right). \tag{24}$$

The four different energy contributions/terms in the above equation have the following respective meanings: Kinetic energy of fluid plus fictitious fluid, kinetic energy of solid minus fictitious fluid, viscous dissipation (over n time steps of size Δt) and the potential energy of the solid.

The ball is initially located at the centre of $\Omega = [0,1] \times [0,1] \times [0,0.6]$ with a radius of 0.2. Using the property of symmetry this computation is carried out on 1/8 of domain Ω: $[0,0.5] \times [0,0.5] \times [0,0.3]$. The initial velocities of x and y components are the same as that used in [22,28], which are prescribed by the stream function

$$\Phi = \Phi_0 \sin(ax)\sin(by), \tag{25}$$

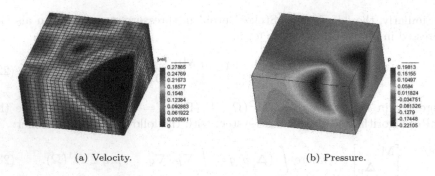

(a) Velocity.   (b) Pressure.

Fig. 2. Distribution of velocity norm and pressure on the fluid mesh at $t = 0.21$ (the ball is maximally stretched), $\Delta t = 5.0 \times 10^{-3}$.

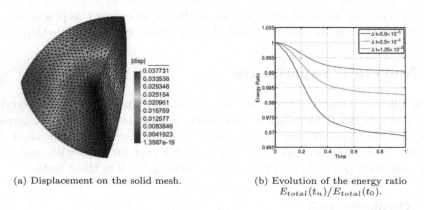

(a) Displacement on the solid mesh.   (b) Evolution of the energy ratio
                                        $E_{total}(t_n)/E_{total}(t_0)$.

Fig. 3. Solid deformation corresponding to Fig. 2 and Evolution of the energy ratio.

with $\Phi_0 = 5.0 \times 10^{-2}$ and $a = b = 2\pi$. The z component of velocity is initially set to be 0. In this test, $\rho^f = 1$, $\mu^f = 0.01$, $\rho^s = 1.5$ and $c_1 = 1$. In order to visualise the mesh and deformation of the solid, a snapshot of fluid velocity and pressure are presented in Fig. 2, and the corresponding deformed solid is displayed in Fig. 3(a). It can be seen from Fig. 3(b) that the total energy is nonincreasing, which is an indication of stability. In addition, the total energy converges to the initial system energy as we reduce the size of the time step, which shows the desired energy conservation property of the proposed scheme.

3.2   Oscillating Cylinder

In this test we consider a cylindrical pillar oscillating in a cuboid channel as shown in Fig. 4, which is a 3D extension of the 2D leaflet in [13,22,25]. We use this example to test the mesh convergence of the proposed scheme. The size of the cuboid is: length $L = 3$, height $H = 1$ and width $W = 0.5$. The cylinder is located at the center of the cuboid's base, with radius of $r = 0.05$ and height $h = 0.8$. We use a symmetry boundary condition on the top, front and back

Table 1. Material properties for the oscillating cylinder and oscillating tri-leaflets.

Fluid	Solid
$\rho^f = 100$ kg/m^3	$\rho^s = 100$ kg/m^3
$\mu^f = 10$ N $\cdot$ s/m^2	$c_1 = 10^7$ N/m^2

surfaces of the cuboid. All the velocity components are fixed to be zero at the bottom of the cuboid, and the inlet and outlet flow are defined by:

$$u_x = 15y\,(2 - y/H)\,sin\,(2\pi t)\,, \quad u_y = u_z = 0. \tag{26}$$

We use the same material properties as used in [13, 22, 25] for the 2D leaflet (see Table 1), which is a natural extension of the corresponding 2D problem with similar boundary conditions. We use a tri-quadratic hexahedras fluid mesh of size $10 \times 20 \times 60$ (*width* $\times$ *height* $\times$ *length*) for a coarse mesh, $16 \times 32 \times 96$ for a medium mesh and $20 \times 40 \times 120$ for a fine mesh. We use a linear tetrahedral solid mesh of 10304 elements with 2675 vertices for a coarse mesh, 19040 elements with 4786 vertices for a medium mesh and 38080 elements with 8883 vertices for a fine mesh. A stable small time step $\Delta t = 1.0 \times 10^{-4}$ is adopted for all the cases. In order to visualise the results of this simulation, snapshots of the velocity norm and stream lines in the background domain and the solid deformations are presented in Fig. 5 and 6 respectively. The displacement of initial point $(1.55, 0.8, 0.5)$ (the top of the cylinder) for three different meshes is plotted in Fig. 7 as a function of time, from which mesh convergence with regard to the displacement is observed (the medium and fine mesh results are almost indistinguishable in these plots).

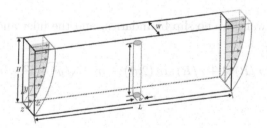

Fig. 4. Sketch of the oscillating cylinder in a cuboid.

3.3 Oscillating Tri-Leaflets

In this section we consider a 3D circular tube with flexible, opening tri-leaflets. A similar case has been studied in [26]. The computational domain is shown in Fig. 8 with $L = 2$ and $R = 0.5$ in this test. Note that there is a small gap (with the angle $\alpha = 0.4°$ as shown in Fig. 8(b)) between the three parts of the tri-leaflets in order to avoid contact, which is not currently included in our

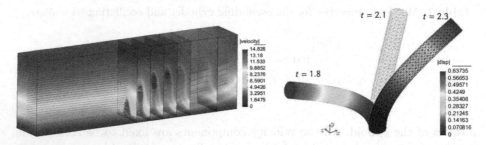

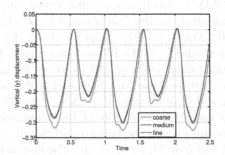

Fig. 5. Velocity norm and stream lines using a medium mesh at $t = 2.3$. The shadow shows the deformation of the solid corresponding to the case of $t = 2.3$ in Fig. 6.

Fig. 6. Solid deformation at three different stages.

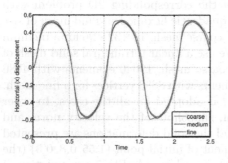

(a) Horizontal displacement against time.

(b) Vertical displacement against time.

Fig. 7. Displacement at point $(1.55, 0.8, 0.25)$ versus time.

model. The tube walls are no-slip boundaries, and the inlet and outlet flow are prescribed by:

$$u_x = 15r\left(1 - r/R\right)\left(1 + r/R\right)sin\left(2\pi t\right), \quad r = \sqrt{y^2 + z^2}, \quad u_y = u_z = 0,$$

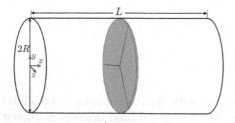

(a) Sketch of the oscillating tri-leaflets in a tube.

(b) Geometry of the tri-leaflets with $\alpha = 0.4°$.

Fig. 8. Computational domain for the test problem of tri-leaflets.

which is an extension of formula (26). The material properties for the fluid and solid are presented in Table 1. Two different meshes are used to test this problem: a coarse mesh of 12750 tri-quadratic hexahedra with 106151 nodes and a fine mesh of 90000 tri-quadratic hexahedra with 735861 nodes in order to discretise the cube; a coarse mesh of 7390 linear tetrahedra with 2657 vertices and a fine mesh of 27460 linear tetrahedra with 8917 vertices in order to discretise the tri-leaflets. The density of the background mesh can be observed in Fig. 9, which also presents a snapshot of the velocity norm and stream lines. The solid mesh can be observed in Fig. 11, which also shows the deformation of the tri-leaflets with horizontal velocity (x component) at different stages in order to visualise the pattern of the oscillation. The displacement at one of the tri-leaflet tips is plotted as a function of time in Fig. 10, from which it can be seen that the coarse mesh leads to small oscillation, however it is not present in the fine mesh simulation. The maximal fluid velocity is $\mathbf{u}_x = 18.2$ at the centre of the channel when the tri-leaflets are completely open, and the maximal solid velocity at the leaflet tip is $\mathbf{u}_x = 18.2$ when the tri-leaflets are completely close.

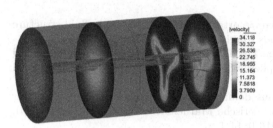

Fig. 9. Snapshot of velocity norm and stream lines in background domain and x-component velocity on the solid mesh at $t = 0.2$.

Fig. 10. The x-component displacement at the tip of each of the tri-leaflets.

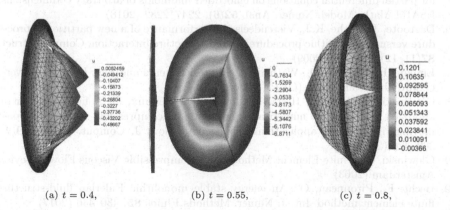

(a) $t = 0.4$,   (b) $t = 0.55$,   (c) $t = 0.8$,

Fig. 11. The velocity of x component at different times for the tri-leaflets using the coarse mesh.

4   Conclusions

In this article the one-field Fictitious Domain Method (FDM) [22] is extended in three ways: through an efficient operator splitting scheme, the implementation of block-matrix preconditioning and into three space dimensions. One numerical example is presented in order to validate the energy conservation, a second is used to test mesh convergence, and the last numerical example is extended from a two-dimensional benchmark for comparison and, we believe, can act as a 3D benchmark for future comparison. It can be seen from these tests that the one-field FDM approach may be adopted to simulate a variety of FSI problems with large solid deformation in three space dimensions. We know from our 2D tests that, for soft solids, execution times are comparable with an IFEM implementation: and considerably faster as the solid becomes more stiff (capable of using a larger time step). Consequently we propose the one-field FDM as a general approach that combines that robustness of FDM/DLM and the computational efficiency of IFEM.

References

1. Baaijens, F.P.: A fictitious domain/mortar element method for fluid-structure interaction. Int. J. Numer. Methods Fluids **35**(7), 743–761 (2001)
2. Boffi, D., Cavallini, N., Gastaldi, L.: The finite element immersed boundary method with distributed Lagrange multiplier. SIAM J. Numer. Anal. **53**(6), 2584–2604 (2015)
3. Boffi, D., Gastaldi, L.: A fictitious domain approach with Lagrange multiplier for fluid-structure interactions. Numerische Mathematik **135**(3), 711–732 (2016). https://doi.org/10.1007/s00211-016-0814-1
4. Burman, E., Hansbo, P.: Fictitious domain methods using cut elements: III. A stabilized Nitsche method for Stokes problem. ESAIM Math. Model. Numer. Anal. **48**(3), 859–874 (2014)
5. Bürman, E., Hansbo, P., Larson, M.G., Massing, A.: Cut finite element methods for partial differential equations on embedded manifolds of arbitrary codimensions. ESAIM Math. Model. Numer. Anal. **52**(6), 2247–2282 (2018)
6. Degroote, J., Bathe, K.J., Vierendeels, J.: Performance of a new partitioned procedure versus a monolithic procedure in fluid–structure interaction. Comput. Struct. **87**(11–12), 793–801 (2009)
7. Elman, H., Silvester, D., Wathen, A.: Finite Elements and Fast Iterative Solvers. Oxford University Press (OUP), Oxford (2014)
8. Glowinski, R., Pan, T., Hesla, T., Joseph, D., Périaux, J.: A fictitious domain approach to the direct numerical simulation of incompressible viscous flow past moving rigid bodies: application to particulate flow. J. Comput. Phys. **169**(2), 363–426 (2001)
9. Glowinski, R.: Finite Element Methods for Incompressible Viscous Flow. Elsevier, Amsterdam (2003)
10. Hecht, F., Pironneau, O.: An energy stable monolithic Eulerian fluid-structure finite element method. Int. J. Numer. Methods Fluids **85**, 430–446 (2017)
11. Heil, M., Hazel, A.L., Boyle, J.: Solvers for large-displacement fluid-structure interaction problems: segregated versus monolithic approaches. Comput. Mech. **43**(1), 91–101 (2008)

12. Hesch, C., Gil, A., Carreño, A.A., Bonet, J., Betsch, P.: A mortar approach for fluid–structure interaction problems: immersed strategies for deformable and rigid bodies. Comput. Methods Appl. Mech. Eng. **278**, 853–882 (2014)
13. Kadapa, C., Dettmer, W., Perić, D.: A fictitious domain/distributed Lagrange multiplier based fluid–structure interaction scheme with hierarchical B-Spline grids. Comput. Methods Appl. Mech. Eng. **301**, 1–27 (2016)
14. Kadapa, C., Dettmer, W., Perić, D.: A stabilised immersed framework on hierarchical b-spline grids for fluid-flexible structure interaction with solid-solid contact. Comput. Methods Appl. Mech. Eng. **335**, 472–489 (2018)
15. Küttler, U., Wall, W.A.: Fixed-point fluid-structure interaction solvers with dynamic relaxation. Comput. Mech. **43**(1), 61–72 (2008)
16. Liu, W.K., et al.: Immersed finite element method and its applications to biological systems. Comput. Methods Appl. Mech. Eng. **195**(13–16), 1722–1749 (2006)
17. Peskin, C.S.: The immersed boundary method. Acta Numerica **11**, 479–517 (2002)
18. Pironneau, O.: Numerical study of a monolithic fluid–structure formulation. In: Frediani, A., Mohammadi, B., Pironneau, O., Cipolla, V. (eds.) Variational Analysis and Aerospace Engineering. SOIA, vol. 116, pp. 401–420. Springer, Cham (2016). https://doi.org/10.1007/978-3-319-45680-5_15
19. Schott, B., Ager, C., Wall, W.A.: Monolithic cut finite element based approaches for fluid-structure interaction. Int. J. Numer. Methods Eng. **119**(8), 757–796 (2019)
20. Wang, X., Zhang, L.T.: Interpolation functions in the immersed boundary and finite element methods. Comput. Mech. **45**(4), 321–334 (2009)
21. Wang, Y.: A one-field fictitious domain method for fluid-structure interactions. Ph.D. thesis, University of Leeds (2018)
22. Wang, Y., Jimack, P.K., Walkley, M.A.: A one-field monolithic fictitious domain method for fluid–structure interactions. Comput. Methods Appl. Mech. Eng. **317**, 1146–1168 (2017)
23. Wang, Y., Jimack, P.K., Walkley, M.A.: Energy analysis for the one-field fictitious domain method for fluid-structure interactions. Appl. Numer. Math. **140**, 165–182 (2019)
24. Wang, Y., Jimack, P.K., Walkley, M.A.: A theoretical and numerical investigation of a family of immersed finite element methods. J. Fluids Struct. **91**, 102754 (2019)
25. Yu, Z.: A DLM/FD method for fluid/flexible-body interactions. J. Comput. Phys. **207**(1), 1–27 (2005)
26. Yu, Z., Shao, X.: A three-dimensional fictitious domain method for the simulation of fluid-structure interactions. J. Hydrodyn. Ser. B **22**(5), 178–183 (2010). https://doi.org/10.1016/S1001-6058(09)60190-6
27. Zhang, L., Gerstenberger, A., Wang, X., Liu, W.K.: Immersed finite element method. Comput. Methods Appl. Mech. Eng. **193**(21), 2051–2067 (2004)
28. Zhao, H., Freund, J.B., Moser, R.D.: A fixed-mesh method for incompressible flow–structure systems with finite solid deformations. J. Comput. Phys. **227**(6), 3114–3140 (2008)

Multi Axes Sliding Mesh Approach
for Compressible Viscous Flows

Masashi Yamakawa[1]([✉]), Satoshi Chikaguchi[1], Shinichi Asao[2],
and Shotaro Hamato[1]

[1] Kyoto Institute of Technology, Matsugasaki Sakyo-ku, Kyoto 606-8585, Japan
yamakawa@kit.ac.jp
[2] College of Industrial Technology, Nishikoya,
Amagasaki, Hyogo 661-0047, Japan

Abstract. To compute flows around a body with a rotating or movable part like
a tiltrotor aircraft, the multi axes sliding mesh approach has been proposed. This
approach is based on the unstructured moving grid finite volume method, which
has adopted the space-time unified domain for control volume. Thus, it can
accurately express such a moving mesh. However, due to the difficulty of mesh
control in viscous flows and the need to maintain the stability of computation, it is
restricted to only inviscid flows. In this paper, the multi axes sliding mesh
approach was extended to viscous flows to understand detailed flow phenomena
around a complicated moving body. The strategies to solve several issues not
present in inviscid flow computations are described. To show the validity of the
approach in viscous flows, it was applied to the flow field of a sphere in uniform
flow. Multiple domains that slide individually were placed around the sphere, and
it was confirmed that the sliding mesh did not affect the flow field. The usability
of the approach is expected to be applied to practical viscous flow computations.

Keywords: Computational fluid dynamics · Unstructured moving mesh ·
Sliding mesh approach · Viscous flows

1 Introduction

Numerical simulations of flows around a body with movable parts like a rotorcraft or
sports athlete has a high utility value for various fields. However, handling a moving
mesh is challenging in a body-fitted coordinate system. When the movable scope of its
parts is small, the moving mesh method using a tension spring [1] can be used. On the
other hand, for large motions, the mesh method is restricted. It is almost impossible to
express a rotary motion such as the rotor part of a helicopter by using the moving mesh
method with spring. To resolve this issue, the sliding mesh approach [2] was proposed.
In this approach, the motion of a body is expressed by sliding the boundary of adjacent
divided computational domains. This is different from the overset grid method in which
one domain is put on another domain. An information exchange of physical values
between domains is then conducted by interpolation, which might not satisfy physical
conservation laws. On the other hand, by using the sliding mesh approach for the
information exchange, the physical value can be conserved. One of the simplest

© Springer Nature Switzerland AG 2020
V. V. Krzhizhanovskaya et al. (Eds.): ICCS 2020, LNCS 12143, pp. 46–59, 2020.
https://doi.org/10.1007/978-3-030-50436-6_4

applications of the sliding mesh approach is the divided cylindrical computational domain for axial direction. Its rotating cylinder has been applied to, for example, the simulation of a flow around a multistage turbine cascade. Also, one domain can be also embedded in another domain. In this case, the embedded sub domain should be cylindrical or spherical. Furthermore, there should not be a gap between two domains during the rotation of the embedded domain.

Although the sliding mesh approach is very useful, it is difficult to express complicated motion. For example, the rotor part of a helicopter is expressed with comparative ease, but to express the rotor blade of a tiltrotor like the Osprey V-22 is impossible. This is because the rotor blade rotates, and moreover, an engine nacelle having a rotor blade also rotates on different axis to change the flight mode. In this case, the flows around a tiltrotor during rotor-blade mode and fixed-wing mode are computed [3] individually. In a simulation focused on changing flight modes, its computations [4] were conducted for fixed degrees of the engine nacelle at 0, 30, 60, and 90° as calculating a moving engine nacelle was quite difficult. For this issue, we proposed the multi axes sliding mesh approach [5], in which the moving engine nacelle is expressed in the middle size computational domain. The small size domain including the rotating blade is then embedded in the middle size domain with both domains embedded in the large size main domain. Furthermore, we succeeded in rotating the small and middle domains individually. However, the approach is conducted under inviscid flows to prioritize reproducibility of complicated motion. Therefore, the turbulent flow transition phenomenon in the wake of rotor could not be calculated.

The objective of this paper is to apply the multi axes sliding mesh approach to viscous flows. The formulation of the approach and its validity when applying a flow around a sphere will be shown.

2 Numerical Approach

2.1 Governing Equation

For the governing equation, the following three-dimensional (3D) Navier–Stokes equation for compressible flows written in conservation law form is adopted.

$$\frac{\partial \mathbf{q}}{\partial t} + \frac{\partial \mathbf{E}}{\partial x} + \frac{\partial \mathbf{F}}{\partial y} + \frac{\partial \mathbf{G}}{\partial z} = \frac{1}{Re}\left(\frac{\partial \mathbf{E_v}}{\partial x} + \frac{\partial \mathbf{F_v}}{\partial y} + \frac{\partial \mathbf{G_v}}{\partial z}\right) \tag{1}$$

Where

$$\mathbf{q} = \begin{bmatrix} \rho \\ \rho u \\ \rho v \\ \rho w \\ e \end{bmatrix}, \mathbf{E} = \begin{bmatrix} \rho u \\ \rho u^2 + p \\ \rho uv \\ \rho uw \\ u(e+p) \end{bmatrix}, \mathbf{F} = \begin{bmatrix} \rho v \\ \rho uv \\ \rho v^2 + p \\ \rho vw \\ v(e+p) \end{bmatrix}, \mathbf{G} = \begin{bmatrix} \rho w \\ \rho uw \\ \rho vw \\ \rho w^2 + p \\ w(e+p) \end{bmatrix}, \mathbf{E_v} = \begin{bmatrix} 0 \\ \tau_{xx} \\ \tau_{xy} \\ \tau_{xz} \\ f_{ES} \end{bmatrix}, \mathbf{F_v} = \begin{bmatrix} 0 \\ \tau_{yx} \\ \tau_{yy} \\ \tau_{yz} \\ f_{FS} \end{bmatrix}, \mathbf{G_v} = \begin{bmatrix} 0 \\ \tau_{zx} \\ \tau_{zy} \\ \tau_{zz} \\ f_{GS} \end{bmatrix} \tag{2}$$

The unknown variables ρ, u, v, w, and e show the gas density, velocity components in the x, y, *and* z directions, and total energy per unit volume, respectively. The working fluid is assumed to be a perfect gas, and the pressure p is defined by

$$p = (\gamma - 1)\left\{e - \frac{1}{2}\rho(u^2 + v^2 + w^2)\right\} \tag{3}$$

f_{E5}, f_{F5}, and f_{G5} are shown in Eq. (4). Here, μ and μ_t are the coefficients of molecular viscosity and eddy viscosity, respectively. Pr, Pr_t, and Re are the Prandtl number, turbulent Prandtl number, and Reynolds number, respectively. The ratio of specific heats γ is typically taken as being 1.4. In this study, $Pr = 0.72$ and $Pr_t = 0.9$ are obtained.

$$f_{E5} = u\tau_{xx} + v\tau_{xy} + w\tau_{xz} + \frac{1}{\gamma - 1}\left(\frac{\mu}{Pr} + \frac{\mu_t}{Pr_t}\right)\frac{\partial T^2}{\partial x}$$

$$f_{F5} = u\tau_{yx} + v\tau_{yy} + w\tau_{yz} + \frac{1}{\gamma - 1}\left(\frac{\mu}{Pr} + \frac{\mu_t}{Pr_t}\right)\frac{\partial T^2}{\partial y} \tag{4}$$

$$f_{G5} = u\tau_{zx} + v\tau_{zy} + w\tau_{zz} + \frac{1}{\gamma - 1}\left(\frac{\mu}{Pr} + \frac{\mu_t}{Pr_t}\right)\frac{\partial T^2}{\partial z}$$

2.2 Numerical Schemes

The sliding mesh approach is a type of moving mesh approach. In this study, the unstructured moving grid finite volume method [6] is adopted. The method assures a geometric conservation law [7] as well as a physical conservation law. A control volume in the space-time unified domain (x, y, z, t), which is four-dimensional (4D) for 3D flows, is then used. This approach has been mainly applied to Euler equations for inviscid compressible flows. In this paper, the approach is discretized for compressible viscous F_v flows. For the discretization, Eq. (1), which is written in divergence form, is integrated as

$$\int_{\Omega} \tilde{\nabla}\tilde{F}_v d\Omega = 0, \tag{5}$$

where

$$\tilde{F}_v = \left(E - \frac{1}{Re}E_v, F - \frac{1}{Re}F_v, G - \frac{1}{Re}G_v, q\right). \tag{6}$$

Since the approach is based on a cell-centered finite volume method, the flow variables are defined at the center of the cell in the (x, y, z) space. Thus, the control volume becomes a 4D polyhedron in the (x, y, z, t)-domain. For the control volume, Eq. (4) is rewritten using the Gauss theorem as:

$$\int_{\Omega} \nabla \widetilde{\mathbf{F}}_v d\Omega = \int_V \widetilde{\mathbf{F}}_v \cdot \widetilde{\mathbf{n}} \, dV = \sum_{l=1}^{Ns+2} (q n_t + \mathbf{\Phi})_l = 0, \tag{7}$$

where

$$\mathbf{\Phi} = \mathbf{H} - \mathbf{H}_v,$$

$$\mathbf{H} = \mathbf{E} n_x + \mathbf{F} n_y + \mathbf{G} n_z, \tag{8}$$

$$\mathbf{H}_v = \frac{1}{\mathrm{Re}} \left(\mathbf{E}_v n_x + \mathbf{F}_v n_y + \mathbf{G}_v n_z \right),$$

Here, Ns indicates the number of boundary surfaces of the element. l is the volume of trajectory generated by the moving boundary surface of the element from $t = n$ to $t = n + 1$. Then, Eq. (7) is rewritten as Eq. (9), and by solving Eq. (9), new $\mathbf{q}$ is obtained.

$$\mathbf{q}^{n+1}(n_t)_{N_s+2} + \mathbf{q}^n (n_t)_{N_s+1} + \sum_{l=1}^{N_s} \left[\mathbf{q}^{n+1/2} n_t + \mathbf{\Phi}^{n+1/2} \right]_l = 0$$

$$\mathbf{q}^{n+1/2} = \frac{1}{2} \left(\mathbf{q}^{n+1} + \mathbf{q}^n \right) \tag{9}$$

$$\mathbf{\Phi}^{n+1/2} = \frac{1}{2} \left(\mathbf{\Phi}^{n+1} + \mathbf{\Phi}^n \right)$$

The inviscid flux vectors are evaluated using the Roe flux difference splitting scheme [8] with the MUSCL scheme as well as the Venkatakrishnan limiter [9]. The vectors are discretized by central difference. To solve the implicit algorithm, the LU-SGS implicit scheme is adopted.

2.3 Evaluation on a Boundary

On a boundary, the first derivative of a physical value cannot be evaluated using central difference. For example, discretization of the first derivative for primitive variable u is described. Figure 1 shows a discretization outline of the first derivative.

The first derivative for primitive variable u is obtained by solving the follow equations.

$$\mathbf{A}\dot{\mathbf{u}} = \mathbf{b}, \tag{10}$$

where

$$\mathbf{A} = \begin{pmatrix} x_c - x_w & y_c - y_w & z_c - z_w \\ 0.5(x_{v2} - x_{v3}) - x_{v1} & 0.5(y_{v2} - y_{v3}) - y_{v1} & 0.5(z_{v2} - z_{v3}) - z_{v1} \\ 0.5(x_{v3} - x_{v1}) - x_{v2} & 0.5(y_{v3} - y_{v1}) - y_{v2} & 0.5(z_{v3} - z_{v1}) - z_{v2} \end{pmatrix}, \tag{11}$$

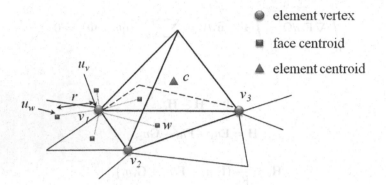

Fig. 1. Discretization outline of the first derivative

$$\dot{\mathbf{u}} = \begin{pmatrix} u_x \\ u_y \\ u_z \end{pmatrix}, \quad \mathbf{b} = \begin{pmatrix} u_c - u_w \\ 0.5(u_{v2} - u_{v3}) - u_{v1} \\ 0.5(u_{v3} - u_{v1}) - u_{v2} \end{pmatrix}.$$

Here, v indicates vertex, c indicates the center of an element, and w indicates the center of a boundary surface for an element. Also, u_w is evaluated as following equation,

$$u_w = \frac{1}{2}\left(u_c + u_{ghost}\right). \tag{12}$$

The vertex of the primitive variable u_{vi} is calculated using the following weighted average method, where u_{wj} is the physical value at the cell center of the triangle constructed by vertex $\psi v_{i\psi}$ and r_{ij} is the distance between the vertex and center point of each cell around it.

$$u_{vi} = \frac{\sum\limits_{j \in i}^{N} u_{wj}\frac{1}{r_{ij}}}{\sum\limits_{j \in i}^{N}\frac{1}{r_{ij}}}, \tag{13}$$

$$r_{ij} = \sqrt{\left(x_{wj} - x_{vi}\right)^2 + \left(y_{wj} - y_{vi}\right)^2 + \left(z_{wj} - z_{vi}\right)^2}. \tag{14}$$

3 Sliding Mesh Approach

3.1 Multi Axes Sliding Mesh Approach

In the sliding mesh approach, a sliding boundary surface exists. Here, the embedded sub computational domain is rotated in the main domain. In a 3D system, the embedded sub domain should have an almost spherical or cylindrical configuration. Although the computational cost using the approach is not expensive, the movable range of vertices is limited. In other words, the motions of an object are restricted. Thus, to improve flexibility, the axes of the rotating sub domain are added in the approach. However, to avoid an interaction between sub domains that have individual axes, one sub domain is embedded in the other sub domain, as shown in Fig. 2. In this figure, computational domain 3 is embedded in computational domain 2, which is embedded in computational domain 1. The whole domain can be moved using the moving computational domain (MCD) method [10]. The advantage of this method is that it does not require a spring method to move the object, so it is less likely to create extremely skewed elements. Basically, the multi axes sliding mesh approach has the potential to express any object motion combined with the MCD method without destroying computational mesh.

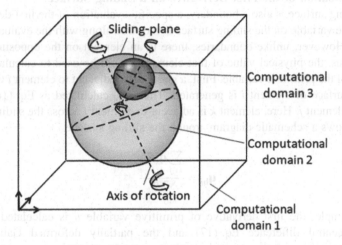

Fig. 2. Multi axes sliding mesh approach

The physical values on the sliding plane interpolate with each other through the plane. Interpolation values are determined depending on the area where domain elements overlap. Specifically, the value is calculated in accordance with the area of the overlapping part S_{ij} between the elements of the sliding plane, as shown in Fig. 3. The value of the part is defined with Eq. (15).

$$q_{\mathbf{b}i} = \frac{\sum\limits_{j \in i} q_j S_{ij}}{\sum\limits_{j \in i} S_{ij}} \tag{15}$$

Where $\sum_{j \in i}$ shows the sum of cell j adjacent to cell i. Then, q_j is the physical value of cell j.

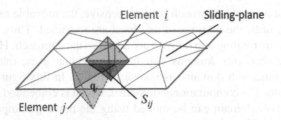

Fig. 3. Overlapping part in slide element

3.2 Multi Axes Sliding Mesh for Viscous Flow Computation

3.2.1 Evaluation of the First Derivative on a Sliding Surface

As the sliding surface is also a boundary, a specific evaluation of the first derivative of the primitive variable on the sliding surface is required along with the evaluation of the boundary. However, unlike boundaries, there is an element on the opposite side of a surface. Thus, the physical value of that element should be used to calculate the first derivative of the primitive variable. First, a ghost cell j adjacent to element i through the boundary surface of element i is generated. Then, $\mathbf{q}_{\mathrm{bj}}$ calculated as Eq. (16) is interpolated in element j. Here, element k is adjacent to element i across the sliding surface. Figure 4 shows a schematic diagram around the sliding surface.

$$\mathbf{q}_{\mathrm{b}j} = \frac{\sum\limits_{k \in j} \mathbf{q}_k S_{jk}}{\sum\limits_{k \in j} S_{jk}}. \tag{16}$$

For example, the first derivative of primitive variable u is calculated using the following central difference Eq. (17) and the partially deformed Gauss-Green's theorem (18).

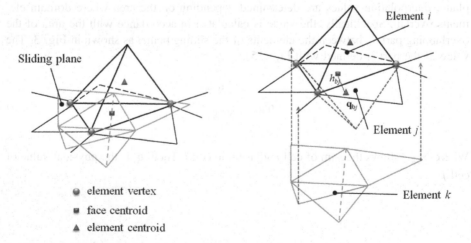

- ● element vertex
- ■ face centroid
- ▲ element centroid

Fig. 4. Evaluation of the first derivative on a sliding surface

$$\left(\frac{\partial u}{\partial x}\right)_{ij} = \frac{1}{2}\left\{\left(\frac{\partial u}{\partial x}\right)_i + \left(\frac{\partial u}{\partial x}\right)_j\right\}, \tag{17}$$

$$\left(\frac{\partial u}{\partial x}\right)_i = \frac{1}{V_{\Omega i}}\sum_{j \in i}^{N_s}\left(\frac{h_{fi}u_i + h_{fj}u_j\lambda(j) + h_{bj}u_{bj}(1 - \lambda(j))}{h_{fi} + h_{fj}\lambda(j) + h_{bj}(1 - \lambda(j))}\right)\mathbf{n}_{ij}. \tag{18}$$

Where h_{bj} is calculated using h_k, which is the distance between the center point of element k and the center of the adjacent surface of elements i and j, as shown in Eq. (19),

$$h_{bj} = \frac{\displaystyle\sum_{k \in j} h_k S_{jk}}{\displaystyle\sum_{k \in j} S_{jk}}. \tag{19}$$

In Eq. (18), if adjacent element j is a ghost cell, $\lambda(j) = 0$, else, $\lambda(j) = 1$.

3.2.2 Evaluation of the First Derivative on an Element Having both a Sliding Surface and Boundary

In this subsection, an evaluation of the first derivative on an element that has both a sliding surface and boundary is described. First, the primitive variable for a vertex located on both the sliding surface and boundary is calculated. The first derivative of the primitive variable is then calculated using Eqs. (10) to (14). For example, the calculation procedure of the primitive variable u_{vi} is shown in Eqs. (20) to (22). Its schematic figure of this case is shown in Fig. 5.

$$u_{vi} = \frac{\displaystyle\sum_{j \in i}^{N}\left(u_{wj}\frac{1}{r_{ij}}\lambda_v(j) + u_{bj}\frac{1}{r_{bij}}(1 - \lambda_v(j))\right)}{\displaystyle\sum_{j \in i}^{N}\left(\frac{1}{r_{ij}}\lambda_v(j) + \frac{1}{r_{bij}}(1 - \lambda_v(j))\right)}, \tag{20}$$

$$r_{bij} = \sqrt{\left(x_{bj} - x_{vi}\right)^2 + \left(y_{bj} - y_{vi}\right)^2 + \left(z_{bj} - z_{vi}\right)^2}, \tag{21}$$

$$x_{bj} = \frac{\displaystyle\sum_{k \in j} x_{ck} S_{jk}}{\displaystyle\sum_{k \in j} S_{jk}}, \quad y_{bj} = \frac{\displaystyle\sum_{k \in j} y_{ck} S_{jk}}{\displaystyle\sum_{k \in j} S_{jk}}, \quad z_{bj} = \frac{\displaystyle\sum_{k \in j} z_{ck} S_{jk}}{\displaystyle\sum_{k \in j} S_{jk}}. \tag{22}$$

Where u_{bj} is the primitive variable in the center of the ghost cell of element j that has vertex i. The variable is then calculated from Eq. (15). (x_{ck}, y_{ck}, z_{ck}) is the coordinates in the center of element k located adjacent to element j across the sliding surface. Finally, element j has vertex i. If element j is a ghost cell, $\lambda_v(j) = 0$, else, $\lambda_v(j) = 1$.

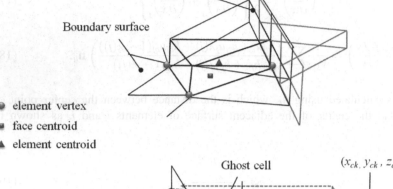

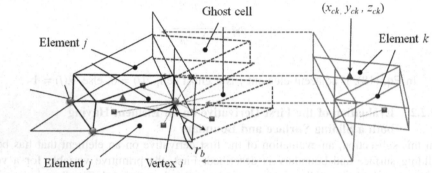

Fig. 5. Evaluation of the first derivative on a sliding surface and boundary

3.2.3 Prism Element on Sliding Surface

When viscous flows are computed using an unstructured mesh, it is necessary to use quite thin prism elements in the boundary layer. However, if the shape of the body boundary is curved, part of an element might overlap the sliding element and static element as shown in Fig. 6. If there is no overlap between the elements, the physical value cannot be interpolated. Such a problem occurs when the difference between both volumes is not small. Thus, the volume difference should be as small as possible.

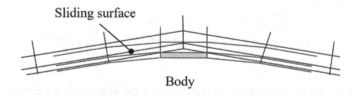

Fig. 6. Sliding mesh near a body surface

4 Verification of the Multi Axes Sliding Mesh Approach

4.1 Application to a Flow Around a Sphere

The multi axes sliding mesh approach is applied to a viscous flow around a sphere. Figure 7 shows a schematic figure of the flow. The sphere is placed in a uniform flow with two sliding cylinders, which have rotation axes in different directions. Each sliding cylindrical mesh rotates around the static sphere, so the sliding mesh must not affect the flow. To confirm the validity of the approach, it is compared with the flow around a sphere in a single mesh. In Fig. 8, case 1 shows a schema of multi axes sliding cylinders around a sphere and case 2 shows its comparison.

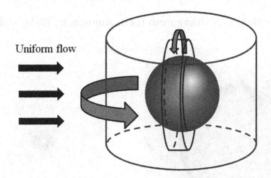

Fig. 7. Multi axes sliding mesh around a sphere

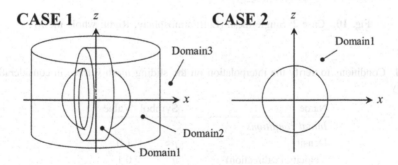

Fig. 8. Schema of comparative computation

4.2 Initial Mesh and Computational Conditions

Figure 9 shows the initial mesh for case 1. The total number of meshes is 4,219,268. Figure 10 shows a single mesh for comparison (case 2). The number of meshes is 4,578,854. Their elements were created by using MEGG3D [11]. The diameter of the computational domain (domain 3 in case 1, whole domain in case 2) is 40 times that of the sphere.

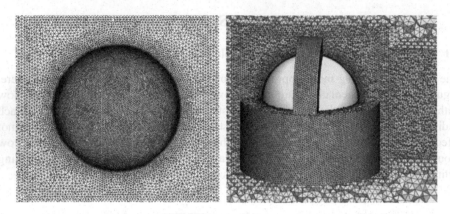

Fig. 9. Case 1: Multi axes sliding mesh (Left: atmosphere, Right: sliding cylinders)

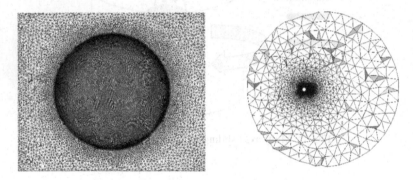

Fig. 10. Case 2: Single mesh (Left: atmosphere, Right: whole mesh)

Table 1 Conditions to verify the interpolation on the sliding mesh surface in consideration of viscosity

Name	Symbol	Value
Initial conditions		
Density	ρ	1.0
Velocity (x-direction)	u	0.1
Velocity (y-direction)	v	0.0
Velocity (z-direction)	w	0.0
Pressure	p	$1.0/\gamma$
Other conditions		
Time step size	Δt	0.001
Reynolds number	Re	10,000
Rotational speed of domains	ω_1, ω_2	0.05, 0.03
Radius (domain 1, domain 2)	r_1, r_2	0.7, 0.75
Height (domain 1, domain 2)	h_1, h_2	0.25, 1.5

Computational conditions are shown in Table 1. The rotations of domains 1 and 2 in case 1 are dominated by Eqs. (23) and (24), respectively. Therefore, while both domain 1 and its axis rotates, only domain 2 rotates and its axis remains fixed.

$$\begin{pmatrix} x' \\ y' \\ z' \end{pmatrix} = \begin{pmatrix} \cos \omega t & -\sin \omega t & 0 \\ \sin \omega t & \cos \omega t & 0 \\ 0 & 0 & 1 \end{pmatrix} \begin{pmatrix} 1 & 0 & 0 \\ 0 & \cos \omega t & -\sin \omega t \\ 0 & \sin \omega t & \cos \omega t \end{pmatrix} \begin{pmatrix} x \\ y \\ z \end{pmatrix}, \quad (23)$$

$$\begin{pmatrix} x' \\ y' \end{pmatrix} = \begin{pmatrix} \cos \omega t & -\sin \omega t \\ \sin \omega t & \cos \omega t \end{pmatrix} \begin{pmatrix} x \\ y \end{pmatrix}. \quad (24)$$

4.3   Computational Result

Figure 11 shows the conditions of the sliding mesh as the result. The sliding motion was confirmed to have no skewed and crushed elements. Under the sliding mesh environment, a flow around a sphere was computed. Figure 12 shows the velocity contours. Around the sphere, the flow in case 1 corresponded reasonably well with that in case 2. Thus, highly accurate interpolation was seen on the sliding surface, confirming that the sliding mesh did not affect the flow.

The pressure drag coefficient of the sphere surface in case 1 was compared with that of case 2, other calculation results [12], and experimental results [13] as shown in Table 2. The discrepancy between case 1 and other calculation results is around 1.0%. Furthermore, the deviation from the experimental results is less than 3.0%, which also shows the validity of the sliding approach. The discrepancy between case 2 and the other calculation and experimental results is larger than case 1 despite no moving and sliding mesh around the sphere. This is possibly due to the cylindrical sliding domain potentially generating a regular mesh.

Figure 13 shows the averaged pressure drag coefficient of case 1 and case 2 on a sphere surface. As the flow is unsteady, the time-averaged drag coefficient is used. Both match in front of the sphere, but there is a slight difference in wake. In general, a complicated flow containing vortices occurs behind a sphere. Thus, the mesh behind the sphere should be generated delicately. However, interpolation between the first layer of the static mesh and sliding mesh might affect such a sensitive flow.

Fig. 11. Conditions of the sliding mesh

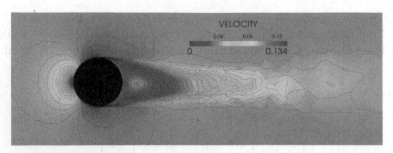

Case 1

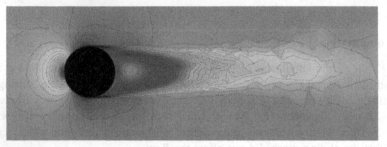

Case 2

Fig. 12. Velocity contours around sphere

Table 2. Drag coefficient of sphere

	Case 1	Case 2	Calculated value	Experimental value
Drag coefficient	0.389	0.379	0.393	0.40

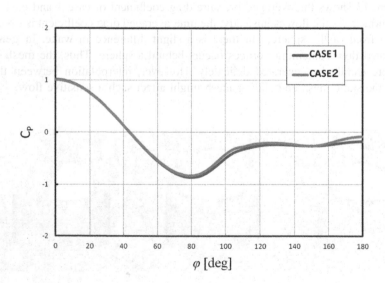

Fig. 13. Averaged pressure drag coefficient on sphere surface

5 Conclusion

In this paper, the multi axes sliding mesh approach for compressible viscous flows was formulated. In particular, the interpolation process between prism elements on sliding surfaces was described. As a result of the computation of the flow around a sphere, the sliding motion of multiple cylinders without skewed and crushed elements were confirmed. The results also showed that there the sliding mesh had no affect on flow. A comparison of other experimental and computational results showed the validity of the multi axes sliding mesh approach. This approach could potentially be applied to complicated motions like a bicycle rider is computing.

Acknowledgments. This publication was subsidized by JKA through its promotion funds from KEIRIN RACE.

References

1. Murayama, M., Nakahashi, K., Matsushima, K.: Unstructured dynamic mesh for large movement and deformation. AIAA Paper 2002-0122 (2002, to appear)
2. Bakker, A., et al.: Sliding mesh simulation of laminar flow in stirred reactors. Chem. Eng. Res. Des. **75**(1), 42–44 (1997)
3. Chaderjian, N.M., Field, M.: Advances in rotor performance and turbulent wake simulation using DES and adaptive mesh refinement. In: 7th International Conference on Computational Fluid Dynamics (2012)
4. Ying, Z., Liang, Y., Shuo, Y.: Numerical study on flow fields and aerodynamics of tilt rotor aircraft in conversion mode based on embedded grid and actuator model. Chin. J. Aeronaut. **28**(1), 93–102 (2015)
5. Takii, A., Yamakawa, M., Asao, S., Tajiri, K.: Six degrees of freedom numerical simulation of tilt-rotor plane. In: Rodrigues, J.M.F., Cardoso, P.J.S., Monteiro, J., Lam, R., Krzhizhanovskaya, V.V., Lees, M.H., Dongarra, J.J., Sloot, P.M.A. (eds.) ICCS 2019. LNCS, vol. 11536, pp. 506–519. Springer, Cham (2019). https://doi.org/10.1007/978-3-030-22734-0_37
6. Yamakawa, M., et al.: Unstructured moving-grid finite-volume method for unsteady shocked flows. J. Comput. Fluids Eng. **10**(1), 24–30 (2005)
7. Obayashi, S.: Freestream capturing for moving coordinates in three dimensions. AIAA J. **30**, 1125–1128 (1992)
8. Roe, P.L.: Approximate Riemann solvers parameter vectors and difference schemes. J. Comput. Phys. **43**, 357–372 (1981)
9. Venkatakrishnan, V.: On the accuracy of limiters and convergence to steady state solutions. AIAA Paper, 93-0880 (1993)
10. Asao, S., et al.: Simulations of a falling sphere with concentration in an infinite long pipe using a new moving mesh system. Appl. Thermal Eng. **72**, 29–33 (2014)
11. Ito, Y.: Challenges in unstructured mesh generation for practical and efficient computational fluid dynamics simulations. Comput. Fluids **85**, 47–52 (2013)
12. Constantinescu, G.S., Squires, K.D.: LES and DES investigations of turbulent flow over a sphere at Re = 10,000. Flow Turbul. Combust. **70**, 267–298 (2003)
13. Achenbach, E.: Experiments on the flow past spheres at very high Reynolds numbers. J. Fluid Mech. **54**, 565–575 (1972)

Monolithic Arbitrary Lagrangian–Eulerian Finite Element Method for a Multi-domain Blood Flow–Aortic Wall Interaction Problem

Pengtao Sun[1](✉), Chen-Song Zhang[2], Rihui Lan[1], and Lin Li[3]

[1] Department of Mathematical Sciences, University of Nevada Las Vegas,
4505 Maryland Parkway, Las Vegas, NV 89154, USA
pengtao.sun@unlv.edu
[2] LSEC and NCMIS, Academy of Mathematics and System Science, Beijing, China
[3] Department of Mathematics, Peking University, Beijing, China
https://sun.faculty.unlv.edu/

Abstract. In this paper, an arbitrary Lagrangian–Eulerian (ALE) finite element method in the monolithic approach is developed for a multi-domain blood flow–aortic wall interaction problem with multiple moving interfaces. An advanced fully discrete ALE-mixed finite element approximation is defined to solve the present fluid–structure interaction (FSI) problem in the cardiovascular environment, in which two fields of structures are involved with two fields of fluid flow, inducing three moving interfaces in between for the interactions. Numerical experiments are carried out for a realistic cardiovascular problem with the implantation of vascular stent graft to demonstrate the strength of our developed ALE-mixed finite element method.

Keywords: Cardiovascular fluid–structure interaction (FSI) · Multi-domain · Vascular aneurysm · Stent graft · Arbitrary Lagrangian–Eulerian (ALE) · Mixed finite element

1 Introduction

In this paper, we consider the problem of interactions between the incompressible free viscous fluid flow and the deformable elastic structure of multiple fields. This problem is of great importance in a wide range of applications, such as in the cardiovascular area of physiology, where a significant example of this type of problem can be described as the blood fluid–vessel interaction problem. During a cardiovascular cycle, any abnormal mechanical change on the elastic property of the aortic vessel could induce some severe cardiovascular diseases (CVDs), such

P. Sun and R. Lan were partially supported by NSF Grant DMS-1418806. C.-S. Zhang was supported by the National Key Research and Development Program of China (Grant No. 2016YFB0201304), NSFC 11971472, and the Key Research Program of Frontier Sciences of CAS.

V. V. Krzhizhanovskaya et al. (Eds.): ICCS 2020, LNCS 12143, pp. 60–74, 2020.
https://doi.org/10.1007/978-3-030-50436-6_5

as the vascular stenosis, the vascular rupture, the aneurysm, the aortic dissection, and etc. Hence, a comprehensive modeling study and an accurate numerical simulation on the interactions between the blood fluid flow and the elastic aortic wall become extremely important and urgent to help medical professionals understand how the CVDs occur, grow, and fatally affect the patients, as well as to help cardiovascular physicians to find out the best medical treatment.

In the past several decades, the vascular stent graft has become one of the most popular fixation devices to carry out surgery for endovascular aortic repair (EVAR) in order to rescue CVD patients. For instance, without the need for an open surgery, the implanted vascular stent graft may cure the vascular stenosis by expanding the constrictive aortic lumen back to normal and then supporting therein as the backbone of the aorta; can treat the aortic dissection (see the left of Fig. 1) by substituting for the dissected aortic wall that is induced by an intima tear; and can remedy the abdominal aortic aneurysm (AAA) (see the middle of Fig. 1) by replacing the bulgy part of the aortic wall that is caused by a lack of elasticity in the aorta. In all cases, because of the elastic expansion of the vascular stent and the residual elasticity energy remained in the diseased aortic wall, most of the blood fluid that is trapped in the void space between the stent graft and the pathologically changed aortic wall are supposed to be squeezed back to the blood vessel lumen through the stent graft. But, there might be still a small amount of blood fluid remaining in the corner of aorta and will be eventually solidified as the thrombi. Thus, all extra blood fluids that are trapped in the shrinking void spaces are supposed to be expelled. Then the diseased blood vessel would be gradually cured to a great extent, demonstrated as either bouncing back for the aortic aneurysm case, or no longer deteriorated and then eventually normalized in an improving cardiovascular environment for the aortic dissection case (see Fig. 1, right).

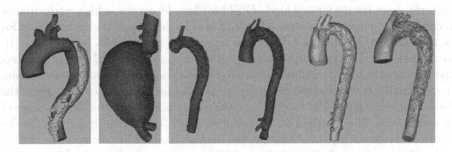

Fig. 1. The CT-scanned data of CVD patients. **Left:** The aortic dissection, where the green part illustrates the false aortic lumen. **Middle:** The abdominal aortic aneurysm. **Right:** Different aortic dissection cases treated by the implanted vascular stent graft. (Color figure online)

However, the EVAR surgery has been long lacking a quantitative study about how and under what circumstance the CVDs are given rise to and then grow, as well as how they are changed after the implantation of a vascular stent graft, i.e., an accurate and efficient mathematical modeling and numerical simulation of the hemodynamic FSI are still largely missing in the investigations before and after the EVAR surgery in the modern medical field of CVDs. Moreover, it has been increasingly recognized that a personalized vascular stent graft is crucially needed for CVD patients, which cannot be done either without a quantitative numerical study of the hemodynamic FSI on the feasibility of the vascular stent graft in the cardiovascular environment.

In this paper, we study the effect of using an elastic material model to describe the deformable aortic wall and stent graft, and their interactions with the blood fluid flow, where, due to the loss of elasticity of the aortic wall, which is however still within the range of linear elasticity, the blood vessel lumen is inflated to form an abdominal aortic aneurysm, and the vascular stent graft is implanted into the aneurysm to separate the blood fluid into two parts. Thus four domains, i.e., two fluid domains and two structure domains, are separated by three interfaces. To model the blood fluid, we consider the Navier–Stokes equations, under the assumptions of incompressibility and Newtonian rheology, which is conventionally defined in Eulerian description. The elastic structure equation, which is conventionally defined in Lagrangian description, can be generally defined by various elastic materials such as the classical Hookean-type material model. In addition, in such a set of governing equations of FSI problem, the time and space dependence of the primary unknowns and of the moving interfaces between the fluid and the elastic structure play a significant role for the dynamic interactions in between, where, the well defined interface conditions among four domains are crucial as well.

Regarding the numerical methodology to be studied in this paper, we develop an advanced numerical method based upon the arbitrary Lagrangian–Eulerian (ALE) mapping within a monolithic approach to solve the present FSI problem. In the first place, we prefer the monolithic approach [10], in view of its unconditional stability and the immunity of any systematic error in the implementation of interface conditions for any kind of FSI problem. Moreover, a high-performance preconditioning linear algebraic solver can also be developed and parallelized for the monolithic system without doing an alternating iteration by subdomains [1]. In contrast, the partitioned approach [2], which decouples the FSI system and iteratively solves the fluid and the structure equations via an iteration-by-subdomain, is conditionally stable and conditionally convergent under a particular range of the physical parameters of FSI model, e.g., when fluid and structural densities are of the same order. This is called the added-mass effect [4] specifically induced by the partitioned approach. Unfortunately, the hemodynamic FSI problems are within the particular range of the added-mass effect, making the partitioned approach very difficult to use. Hence again, the monolithic approach is the primarily reliable method to be studied in this paper.

On the top of the monolithic approach, we develop an ALE method for the present FSI problem. As a type of body-fitted mesh method, ALE technique [10] has become the most accurate and also the most popular approach for solving FSI problems and other general interface problems [6,8], where the mesh on the interface is accommodated to be shared by both the fluid and the structure, and thus to automatically satisfy the interface conditions across the interface.

The structure of the paper is the following: in Sect. 2 we define a type of dynamic FSI model. The ALE mapping and the weak form in ALE description are defined in Sect. 3. We further develop the monolithic ALE finite element method for the present FSI model in semi- and fully discrete schemes in Sect. 4. Numerical experiments for a realistic cardiovascular problem with the implantation of vascular stent graft are carried out in Sect. 5.

2   A Model of Fluid-Structure Interaction (FSI) Problem

Let Ω be an open bounded domain in $\mathbb{R}^d$ ($d = 2, 3$) with a convex polygonal boundary $\partial\Omega$, $\mathcal{I} = (0, T]$ ($T > 0$). Four subdomains, $\Omega_t^k := \Omega_i(t) \subset \Omega$ ($k = f_1, f_2, s_1, s_2$) ($0 \leq t \leq T$), satisfying $\overline{\Omega_t^{f_1}} \cup \overline{\Omega_t^{f_2}} \cup \overline{\Omega_t^{s_1}} \cup \overline{\Omega_t^{s_2}} = \overline{\Omega}$, $\Omega_t^{f_i} \cap \Omega_t^{s_j} = \emptyset$ ($i, j = 1, 2$), respectively represent two fluid domains: $\Omega_t^{f_1}$ is embraced by the implanted stent graft while $\Omega_t^{f_2}$ occupies the interior of the aortic aneurysm, both of which are filled with an incompressible and viscous blood fluid, and, two elastic structure domains: $\Omega_t^{s_1}$ is the aortic wall while $\Omega_t^{s_2}$ is formed by the implanted vascular stent graft. As illustrated in Fig. 2, four subdomains are separated by three interfaces: $\Gamma_t^{12} := \partial\Omega_t^{f_1} \cap \partial\Omega_t^{s_2}$, $\Gamma_t^{21} := \partial\Omega_t^{f_2} \cap \partial\Omega_t^{s_1}$, $\Gamma_t^{22} := \partial\Omega_t^{f_2} \cap \partial\Omega_t^{s_2}$, which may move/deform along with $t \in \mathcal{I}$, resulting that Ω_t^k ($k = f_1, f_2, s_1, s_2$) also change with $t \in \mathcal{I}$ and are termed as the current (Eulerian) domains with respect to x_k, in contrast to their initial (reference/Lagrangian) domains, $\hat{\Omega}^k := \Omega_0^k$ ($k = f_1, f_2, s_1, s_2$) with respect to $\hat{x}_k$, where, a *flow map* is defined from $\hat{\Omega}^k$ to Ω_t^k ($k = f_1, f_2, s_1, s_2$), as: $\hat{x}_k \mapsto x_k(\hat{x}_k, t)$ such that $x_k(\hat{x}_k, t) = \hat{x}_k + \hat{u}_k(\hat{x}_k, t), \forall t \in \mathcal{I}$, where $\hat{u}_k$ is the displacement field in the Lagrangian frame. In what follows, we set $\hat{\psi} = \hat{\psi}(\hat{x}_k, t)$ which equals $\psi(x_k(\hat{x}_k, t), t)$, and $\hat{\nabla} = \nabla_{\hat{x}_k}$ ($k = s_1, s_2$). Let $\Omega_t^f = \Omega_t^{f_1} \cup \Omega_t^{f_2}$, $\hat{\Omega}^s = \hat{\Omega}^{s_1} \cup \hat{\Omega}^{s_2}$, $\Gamma_t = \Gamma_t^{12} \cup \Gamma_t^{21} \cup \Gamma_t^{22}$.

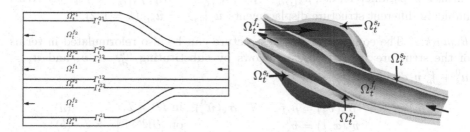

Fig. 2. Two schematic domains of multi-domain FSI for the CVD case of aneurysm, where, $\Omega_t^{f_1}$: the blood vessel lumen, $\Omega_t^{f_2}$: the blood fluid squeezed inside the aneurysm, $\Omega_t^{s_1}$: the aortic wall, $\Omega_t^{s_2}$: the vascular graft, and three interfaces $\Gamma_t^{12} := \partial\Omega_t^{f_1} \cap \partial\Omega_t^{s_2}$, $\Gamma_t^{21} := \partial\Omega_t^{f_2} \cap \partial\Omega_t^{s_1}$, $\Gamma_t^{22} := \partial\Omega_t^{f_2} \cap \partial\Omega_t^{s_2}$.

In this paper, we are interested in studying a pressure-driven flow through the deformable channel with a two-way coupling between the incompressible fluid and the elastic material, where the fluid motion is defined in Eulerian description and the structure motion is defined in Lagrangian description as follows

Fluid motion:

$$\begin{cases}
\rho_f \frac{Dv_f}{Dt} - \nabla \cdot \sigma_f = \rho_f f_f, & \text{in } \Omega_t^f \times \mathcal{I}, & (a) \\
\nabla \cdot v_f = 0, & \text{in } \Omega_t^f \times \mathcal{I}, & (b) \\
v_f = v_f^b, & \text{on } [\partial\Omega_t^f \backslash \Gamma_t]_D \times \mathcal{I}, & (c) \\
\sigma_f n_f = 0, & \text{on } [\partial\Omega_t^f \backslash \Gamma_t]_N \times \mathcal{I}, & (d) \\
v_f = v_f^0, & \text{in } \hat{\Omega}^f, & (e)
\end{cases}$$

Elastic structure motion:

$$\begin{cases}
\hat{\rho}_s \frac{\partial^2 \hat{u}_s}{\partial t^2} - \hat{\nabla} \cdot \hat{\sigma}_s = \hat{\rho}_s \hat{f}_s, & \text{in } \hat{\Omega}^s \times \mathcal{I}, & (f) \\
\hat{u}_s = \hat{u}_s^b, & \text{on } [\partial\hat{\Omega}^s \backslash \hat{\Gamma}]_D \times \mathcal{I}, & (g) \\
\hat{\sigma}_s \hat{n}_s = 0, & \text{on } [\partial\hat{\Omega}^s \backslash \hat{\Gamma}]_N \times \mathcal{I}, & (h) \\
\hat{u}_s = \hat{u}_s^0, \ \frac{\partial \hat{u}_s}{\partial t} = \hat{v}_s^0, & \text{in } \hat{\Omega}^s, & (i)
\end{cases}$$

Interface conditions:

$$\begin{cases}
v_f = v_s, & \text{on } \Gamma_t \times \mathcal{I}, & (j) \\
\sigma_f n_f = \sigma_s n_f, & \text{on } \Gamma_t \times \mathcal{I}, & (k)
\end{cases}$$

$$(1)$$

where, v_f is the velocity of the free fluid defined in Ω_t^f, $\hat{u}_s$ is the displacement of elastic structure that leads to $\frac{\partial \hat{u}_s}{\partial t} = \hat{v}_s(\hat{x}_s, t) = v_s(x_s(\hat{x}_s, t), t)$, ρ_f and $\hat{\rho}_s$ are the constant densities of the incompressible free fluid and the elastic structure, respectively. $\frac{D \cdot}{Dt}$ denotes the classical concept of material derivative as $\frac{D\psi}{Dt} = \frac{\partial\psi}{\partial t} + (\psi \cdot \nabla)\psi$. The stress tensor of each phase is defined as

$$\begin{aligned}
\sigma_f := \sigma_f(v_f) = 2\mu_f D(v_f) - p_f I, \qquad & D(v_f) := (\nabla v_f + (\nabla v_f)^T)/2, \\
\hat{\sigma}_s := \hat{\sigma}_s(\hat{u}_s) = 2\mu_s \varepsilon(\hat{u}_s) + \lambda_s \hat{\nabla} \cdot \hat{u}_s I, \quad & \varepsilon(\hat{u}_s) := (\hat{\nabla}\hat{u}_s + (\hat{\nabla}\hat{u}_s)^T)/2,
\end{aligned} \qquad (2)$$

where, p_f denotes the fluid pressure, μ_f, μ_s and λ_s are constant physical parameters representing the fluid viscosity, the shear modulus, and the Lamé constant of elastic structures, respectively. Here different elastic structures may have different elastic parameters, i.e., $\mu_s|_{\hat{\Omega}^{s_k}} = \mu_{s_k}$, $\lambda_s|_{\hat{\Omega}^{s_k}} = \lambda_{s_k}$, $\hat{\rho}_s|_{\hat{\Omega}^{s_k}} = \hat{\rho}_{s_k}$ $(k = 1, 2)$, inducing different structure displacements $\hat{u}_s|_{\hat{\Omega}^{s_k}} = \hat{u}_{s_k}$ $(k = 1, 2)$.

Remark 1. The equation of elastic structure can be also reformulated in terms of the structure velocity, v_s, as follows by substituting $\frac{\partial\hat{u}_s}{\partial t} = \hat{v}_s$ and $\hat{u}_s = \hat{u}_s^0 + \int_0^t \hat{v}_s d\tau$ into (1(f)) [10]

$$\begin{cases}
\hat{\rho}_s \frac{\partial\hat{v}_s}{\partial t} - \hat{\nabla} \cdot \hat{\sigma}_s(\hat{v}_s) = \hat{\rho}_s \hat{f}_s + \hat{\nabla} \cdot \hat{\sigma}_s(\hat{u}_s^0), & \text{in } \hat{\Omega}^s \times \mathcal{I}, & (a) \\
\hat{v}_s(\hat{x}, t) = \hat{v}_s^b, & \text{on } [\partial\hat{\Omega}^s \backslash \hat{\Gamma}]_D \times \mathcal{I}, & (b) \\
\hat{\sigma}_s \hat{n}_s = 0, & \text{on } [\partial\hat{\Omega}^s \backslash \hat{\Gamma}]_N \times \mathcal{I}, & (c) \\
\hat{v}_s(\hat{x}, 0) = \hat{u}_s^0, & \text{in } \hat{\Omega}^s, & (d)
\end{cases} \qquad (3)$$

where $\hat{\sigma}_s(\hat{v}_s) = 2\mu_s \varepsilon \left(\int_0^t \hat{v}_s d\tau \right) + \lambda_s \hat{\nabla} \cdot \left(\int_0^t \hat{v}_s d\tau \right) I$.

3   ALE Mapping and the Weak Form in ALE Description

The core of ALE method is the introduction of ALE mapping, which is essentially a type of time-dependent bijective affine mapping, $X_t \in (W^{1,\infty}(\hat{\Omega}^f))^d$, defined from the initial (Lagrangian) domain $\hat{\Omega}^f \subset \mathbb{R}^d$ to the current (Eulerian) domain $\Omega_t^f \subset \mathbb{R}^d$, as follows [8,9]

$$X_t : \hat{\Omega}^f \to \Omega_t^f,$$
$$\hat{x}_f \mapsto x_f(\hat{x}_f, t), \tag{4}$$

and $X_t^{-1} \in (W^{1,\infty}(\Omega_t^f))^d$. The key reason why this mapping works well for FSI problems and other standard interface problems is because the following Proposition 1 holds for any $v = v(x_f(\hat{x}_f, t), t)$ and its ALE time derivative, $\frac{dv}{dt}\big|_{\hat{x}_f} = \frac{\partial v}{\partial t}(x_f, t) + w_f(x_f, t) \cdot \nabla v(x_f, t)$, where w_f is the domain velocity of Ω_t and $w_f = \frac{\partial X_t}{\partial t} \circ (X_t)^{-1}(x_f)$.

Proposition 1 [9]. $v \in H^1(\Omega_t)$ and $\frac{dv}{dt}\big|_{\hat{x}_f} \in H^1(\Omega_t)$ if and only if $\hat{v} = \hat{v}(\hat{x}_f, t) = v \circ X_t \in H^1(\hat{\Omega})$, and vice versa.

Only the current (Eulerian) domain that involves a moving boundary/interface needs the ALE mapping to produce a moving mesh therein which can accommodate the motion of the boundary/interface without breaking the mesh connectivity along the time. In practice, one way to define the affine-type ALE mapping, $X_t : \hat{\Omega}^f \to \Omega_t^f$, is the harmonic extension technique. For instance, to compute the ALE mapping for the fluid mesh, X_t, in the present FSI problem, we solve the following Laplace equation

$$\begin{cases} \Delta_{\hat{x}_f} X_t = 0, & \text{in } \hat{\Omega}^f, & (a) \\ X_t = \hat{u}_s, & \text{on } \hat{\Gamma}, & (b) \\ X_t = 0, & \text{on } \partial\hat{\Omega}^f \backslash \hat{\Gamma}. & (c) \end{cases} \tag{5}$$

Once the ALE mapping X_t is computed, then the fluid mesh and its moving velocity, w_f, can be updated, accordingly, as: $x_f = \hat{x}_f + X_t$, $\hat{w}_f = \frac{\partial X_t}{\partial t}$ and $w_f = \hat{w}_f \circ (X_t)^{-1}$. Due to Proposition 1, such defined ALE mapping, X_t, can guarantee that $v_f \in (H^1(\Omega_t^f))^d$ and $\frac{dv_f}{dt}\big|_{\hat{x}_f} \in (H^1(\Omega_t^f))^d$ if and only if $\hat{v}_f = v_f \circ X_t \in (H^1(\hat{\Omega}^f))^d$, and vice versa, which is a sufficient and necessary condition for definitions of the following functional spaces

$$V_t^f := \{\psi_f : \Omega_t^f \to \mathbb{R}^d, \psi_f = \hat{\psi}_f \circ (X_t)^{-1}, \hat{\psi}_f \in (H^1(\hat{\Omega}^f))^d\},$$
$$V_{t,0}^f := \{\psi_f \in V_t^f : \psi_f = 0 \text{ on } [\partial\Omega_t^f \backslash \Gamma_t]_D\},$$
$$V_{t,b}^f := \{\psi_f \in V_t^f : \psi_f = v_f^b \text{ on } [\partial\Omega_t^f \backslash \Gamma_t]_D\},$$
$$\hat{V}^s := \{\hat{\psi}_s \in (H^1(\hat{\Omega}^s))^d : \hat{\psi}_s = \psi_f \circ X_t \text{ on } \hat{\Gamma}, \psi_f \in V_t^f \cap L^2(\Gamma_t)\},$$
$$\hat{V}_0^s := \{\hat{\psi}_s \in \hat{V}^s : \hat{\psi}_s = 0 \text{ on } [\partial\hat{\Omega}^s \backslash \hat{\Gamma}]_D\},$$
$$\hat{V}_b^s := \{\hat{\psi}_s \in \hat{V}^s : \hat{\psi}_s = \hat{v}_s^b \text{ on } [\partial\hat{\Omega}^s \backslash \hat{\Gamma}]_D\},$$
$$Q_t^f := \{q_f : \Omega_t^f \to \mathbb{R}, q_f = \hat{q}_f \circ (X_t)^{-1}, \hat{q}_f \in L_0^2(\hat{\Omega}^f)\},$$
$$\hat{W}_b^f := \{\hat{\xi} \in (H^1(\hat{\Omega}^f))^d : \hat{\xi} = 0 \text{ on } \partial\hat{\Omega}^f \backslash \hat{\Gamma}; \hat{\xi} = \hat{\psi}_s \text{ on } \hat{\Gamma}, \hat{\psi}_s \in \hat{V}^s \cap L^2(\hat{\Gamma})\},$$
$$\hat{W}_0^f := \{\hat{\xi} \in (H^1(\hat{\Omega}^f))^d : \hat{\xi} = 0 \text{ on } \partial\hat{\Omega}^f\}, \tag{6}$$

which make $v_f \in V_{t,b}^f \subset (H^1(\Omega_t^f))^d$, $p_f \in Q_t^f \subset L_0^2(\Omega_t^f) = \{\hat{q}_f \in L^2(\hat{\Omega}^f) : \int_{\hat{\Omega}^f} \hat{q}_f d\hat{x}_f = 0\}$, and $\hat{v}_s \in \hat{V}_b^s \subset (H^1(\hat{\Omega}^s))^d$.

In view of the definition of ALE time derivative $\frac{dv_f}{dt}\big|_{\hat{x}}$, the interface condition (1(k)) and the time differentiation of the harmonic ALE Eq. (5), the weak form of FSI problem (1)–(3) can be defined as follows.

Weak Form. *Find* $(v_f, \hat{v}_s, p_f, \hat{w}_f) \in (H^2 \cap L^\infty)(0, T; V_{t,b}^f \times \hat{V}_b^s \times Q_t^f \times \hat{W}_b^f)$ *such that*

$$\rho_f(\frac{dv_f}{dt}\big|_{\hat{x}_f}, \psi_f)_{\Omega_t^f} + \rho_f\left((v_f - \hat{w}_f \circ X_t^{-1}) \cdot \nabla v_f, \psi_f\right)_{\Omega_t^f}$$
$$+ 2\mu_f\left(D(v_f), D(\psi_f)\right)_{\Omega_t^f} - (p_f, \nabla \cdot \psi_f)_{\Omega_t^f} + (\nabla \cdot v_f, q_f)_{\Omega_t^f}$$
$$+ \hat{\rho}_s(\frac{\partial \hat{v}_s}{\partial t}, \hat{\psi}_s)_{\hat{\Omega}_s} + 2\mu_s\left(\varepsilon\left(\int_0^t \hat{v}_s d\tau\right), \varepsilon(\hat{\psi}_s)\right)_{\hat{\Omega}_s} \qquad (7)$$
$$+ \lambda_s\left(\hat{\nabla} \cdot \left(\int_0^t \hat{v}_s d\tau\right), \hat{\nabla} \cdot \hat{\psi}_s\right)_{\hat{\Omega}_s} + (\hat{\nabla}\hat{w}_f, \hat{\nabla}\hat{\xi}_f)_{\hat{\Omega}_f} = \rho_f(f_f, \psi_f)_{\Omega_t^f}$$
$$+ \hat{\rho}_s(\hat{f}_s, \hat{\psi}_s)_{\hat{\Omega}_s} - 2\mu_s\left(\varepsilon(\hat{u}_s^0), \varepsilon(\hat{\psi}_s)\right)_{\hat{\Omega}_s} - \lambda_s\left(\hat{\nabla} \cdot \hat{u}_s^0, \hat{\nabla} \cdot \hat{\psi}_s\right)_{\hat{\Omega}_s},$$
$$\forall(\psi_f, \hat{\psi}_s, q_f, \hat{\xi}_f) \in V_{t,0}^f \times \hat{V}_0^s \times Q_t^f \times \hat{W}_0^f.$$

Introduce the following trilinear function

$$\beta(u, v, w) = \frac{1}{2}\left((u \cdot \nabla v, w)_{\Omega_t^f} - (u \cdot \nabla w, v)_{\Omega_t^f}\right). \qquad (8)$$

Applying the Green's theorem, the incompressibility (1(b)), the interface conditions (1(j)) and (5(b)), we obtain

$$((v_f - w_f) \cdot \nabla v_f, \psi_f)_{\Omega_t^f} = \beta(v_f, v_f, \psi_f)_{\Omega_t^f} - \beta(w_f, v_f, \psi_f)_{\Omega_t^f} + \frac{1}{2}(v_f \nabla \cdot w_f, \psi_f)_{\Omega_t^f}.$$

Then (7) can be reformulated as

$$\rho_f\left[\left(\frac{dv_f}{dt}\big|_{\hat{x}_f}, \psi_f\right)_{\Omega_t^f} + \beta(v_f, v_f, \psi_f)_{\Omega_t^f} - \beta(\hat{w}_f \circ X_t^{-1}, v_f, \psi_f)_{\Omega_t^f}\right.$$
$$\left. + \frac{1}{2}(v_f \nabla \cdot (\hat{w}_f \circ X_t^{-1}), \psi_f)_{\Omega_t^f}\right] + 2\mu_f\left(D(v_f), D(\psi_f)\right)_{\Omega_t^f}$$
$$- (p_f, \nabla \cdot \psi_f)_{\Omega_t^f} + (\nabla \cdot v_f, q_f)_{\Omega_t^f} + (\hat{\nabla}\hat{w}_f, \hat{\nabla}\hat{\xi}_f)_{\hat{\Omega}_f}$$
$$+ \hat{\rho}_s(\frac{\partial \hat{v}_s}{\partial t}, \hat{\psi}_s)_{\hat{\Omega}_s} + 2\mu_s\left(\varepsilon\left(\int_0^t \hat{v}_s d\tau\right), \varepsilon(\hat{\psi}_s)\right)_{\hat{\Omega}_s} \qquad (9)$$
$$+ \lambda_s\left(\hat{\nabla} \cdot \left(\int_0^t \hat{v}_s d\tau\right), \hat{\nabla} \cdot \hat{\psi}_s\right)_{\hat{\Omega}_s} = \rho_f(f_f, \psi_f)_{\Omega_t^f} + \hat{\rho}_s(\hat{f}_s, \hat{\psi}_s)_{\hat{\Omega}_s}$$
$$- 2\mu_s\left(\varepsilon(\hat{u}_s^0), \varepsilon(\hat{\psi}_s)\right)_{\hat{\Omega}_s} - \lambda_s\left(\hat{\nabla} \cdot \hat{u}_s^0, \hat{\nabla} \cdot \hat{\psi}_s\right)_{\hat{\Omega}_s},$$
$$\forall(\psi_f, \hat{\psi}_s, q_f, \hat{\xi}_f) \in V_{t,0}^f \times \hat{V}_0^s \times Q_t^f \times \hat{W}_0^f.$$

4   The Monolithic ALE Finite Element Approximation

We first introduce finite element spaces to discretize the functional spaces defined in (6). Conventionally, the finite element approximation to the linear elasticity equation is defined in a Lagrange-type piecewise polynomial (finite element) space, $\hat{V}_h^s \subset \hat{V}_b^s$, to accommodate its numerical solution $\hat{v}_{s,h}$. In order to satisfy the interface condition (1(j)), we need not only the meshes in Ω_t^f and Ω_t^s are conforming through Γ_t, but also the finite element space of fluid (Navier-Stokes) equations, $V_{t,h}^f \times Q_{t,h}^f \subset V_t^f \times Q_t^f$, matches the finite element space of the structure equation, $\hat{V}^s$, through $\hat{\Gamma}$, i.e., a Lagrange-type finite element space shall be adopted to accommodate $(v_{f,h}, p_{f,h})$ as well. So in this paper, we use the stable Stokes-pair, i.e., $(H^1)^d$-type mixed element, to discretize the Navier-Stokes equations, e.g., the Taylor-Hood (P^2P^1) element, where the quadratic piecewise polynomials, P^2, form the finite element $V_{t,h}^f \subset V_t^f$ to hold $v_{f,h}$, and the linear piecewise polynomials, P^1, form $Q_{t,h}^f \subset Q_t^f$ to hold $p_{f,h}$. To match $V_{t,h}^f$, we choose the same quadratic piecewise polynomial set P^2 to form $\hat{V}_h^s \subset \hat{V}_b^s$ that holds $\hat{v}_{s,h}$ as the finite element solution to the linear elasticity equation. In addition, $V_{t,h,0}^f \subset V_{t,0}^f$, $\hat{V}_{h,0}^s \subset \hat{V}_0^s$, $\hat{W}_h^f \subset \hat{W}_b^f$, $\hat{W}_{h,0}^f \subset \hat{W}_0^f$.

For any $t \in [0, T]$, we consider a discretization of the mapping X_t by means of piecewise linear Lagrangian finite elements, denoted by $X_{h,t} : \hat{\Omega}^f \to \Omega_t^f$. We assume that $X_{h,t}$ is smooth and invertible, $X_{h,t} \in (W^{1,\infty}(\hat{\Omega}^f))^d$ and $X_{h,t}^{-1} \in (W^{1,\infty}(\Omega_t^f))^d$. Then, we have the discrete mesh velocity $w_{f,h} = \hat{w}_{f,h} \circ X_{h,t}^{-1} = \frac{\partial X_{h,t}}{\partial t} \circ (X_{h,t})^{-1}(x_f)$, and the discrete ALE time derivative $\frac{dv_f}{dt}\big|_{\hat{x}_f}^h (x_f, t) = \frac{\partial v_f}{\partial t}(x_f, t) + w_{f,h}(x_f, t) \cdot \nabla v_f(x_f, t)$. Clearly, $\hat{w}_{f,h} = w_{f,h} \circ X_{h,t} = \hat{v}_{s,h}$ on $\hat{\Gamma}$. So, differentiating (5) with time, we can similarly define a discrete ALE mapping for $\hat{w}_{f,h}$ as follows

$$\begin{cases} \Delta_{\hat{x}_f} \hat{w}_{f,h} = 0, & \text{in } \hat{\Omega}^f, & (a) \\ \hat{w}_{f,h} = \hat{v}_s, & \text{on } \hat{\Gamma}, & (b) \\ \hat{w}_{f,h} = 0, & \text{on } \partial\hat{\Omega}^f \backslash \hat{\Gamma}. & (c) \end{cases} \qquad (10)$$

4.1   The Semi-discrete Scheme of ALE Finite Element Method

Choosing P^2P^1 mixed element, P^2 element and P^1 element to construct the finite element spaces $V_{t,h}^f \times Q_{t,h}^f$, $\hat{V}_h^s$ and $\hat{W}_h^f$, respectively, we define the following semi-discrete ALE finite element approximation to FSI problem (1)–(3) based on the weak form (9).

ALE-FEM 1. *Find* $(v_{f,h}, \hat{v}_{s,h}, p_{f,h}, \hat{w}_{f,h}) \in (H^2 \cap L^\infty)(0, T; V_{t,h}^f \times \hat{V}_h^s \times Q_{t,h}^f \times \hat{W}_h^f)$ *such that*

$$
\rho_f \left[\left(\frac{\partial v_{f,h}}{\partial t} \Big|_{\hat{x}_f}^h, \psi_{f,h} \right)_{\Omega_t^f} + \beta(v_{f,h}, v_{f,h}, \psi_{f,h})_{\Omega_t^f} - \beta(\hat{w}_{f,h} \circ X_{h,t}^{-1}, v_{f,h}, \psi_{f,h})_{\Omega_t^f} \right.
$$

$$
\left. + \frac{1}{2} \left(v_{f,h} \nabla \cdot (\hat{w}_{f,h} \circ X_{h,t}^{-1}), \psi_{f,h} \right)_{\Omega_t^f} \right] + 2\mu_f \left(D(v_{f,h}), D(\psi_{f,h}) \right)_{\Omega_t^f}
$$

$$
- (p_{f,h}, \nabla \cdot \psi_{f,h})_{\Omega_t^f} + (\nabla \cdot v_{f,h}, q_{f,h})_{\Omega_t^f} + (\hat{\nabla} \hat{w}_{f,h}, \hat{\nabla} \hat{\xi}_{f,h})_{\hat{\Omega}_f}
$$

$$
+ \hat{\rho}_s \left(\frac{\partial \hat{v}_{s,h}}{\partial t}, \hat{\psi}_{s,h} \right)_{\hat{\Omega}_s} + 2\mu_s \left(\varepsilon \left(\int_0^t \hat{v}_{s,h} d\tau \right), \varepsilon(\hat{\psi}_{s,h}) \right)_{\hat{\Omega}_s}
$$

$$
+ \lambda_s \left(\hat{\nabla} \cdot \left(\int_0^t \hat{v}_{s,h} d\tau \right), \hat{\nabla} \cdot \hat{\psi}_{s,h} \right)_{\hat{\Omega}_s} = \rho_f(f_f, \psi_{f,h})_{\Omega_t^f} + \hat{\rho}_s(\hat{f}_s, \hat{\psi}_{s,h})_{\hat{\Omega}^s}
$$

$$
- 2\mu_s \left(\varepsilon(\hat{u}_s^0), \varepsilon(\hat{\psi}_{s,h}) \right)_{\hat{\Omega}^s} - \lambda_s \left(\hat{\nabla} \cdot \hat{u}_s^0, \hat{\nabla} \cdot \hat{\psi}_{s,h} \right)_{\hat{\Omega}^s},
$$

$$
\forall (\psi_{f,h}, \hat{\psi}_{s,h}, q_{f,h}, \hat{\xi}_{f,h}) \in V_{t,h,0}^f \times \hat{V}_{h,0}^s \times Q_{t,h}^f \times \hat{W}_{h,0}^f. \tag{11}
$$

4.2 The Fully Discrete Scheme of ALE Finite Element Method

Now we develop the fully discrete monolithic ALE finite element approximation for the FSI model (1). Introduce a uniform partition $0 = t_0 < t_1 < \cdots < t_N = T$ with the time-step size $\Delta t = T/N$. Set $t_n = n\Delta t$, $\varphi^n = \varphi(x^n, t_n)$ for $n = 0, 1, \cdots, N$. Define the following backward Euler time differences based on the discrete ALE mapping $X_{h,t}$:

$$
\begin{array}{ll}
\left(\frac{du}{dt} \Big|_{\hat{x}_f} \right)^{n+1} \approx d_t^{X_t} u^{n+1} = \frac{1}{\Delta t} \left(u^{n+1} - u^n \circ X_{n+1,n} \right), & (a) \\[2mm]
\left(\frac{\partial u}{\partial t} \right)^{n+1} \approx d_t u^{n+1} = \frac{1}{\Delta t} \left(u^{n+1} - u^n \right), & (b)
\end{array} \tag{12}
$$

where, $X_{m,n} = X_{h,t_n} \circ (X_{h,t_m})^{-1}$. By the Taylor's expansion in a subtle way, we can obtain the first-order convergence with respect to Δt for (12(a)) [5].

Based on the semi-discrete scheme (11), now we can define a fully discrete ALE finite element method (FEM) for (1) as follows.

ALE-FEM 2. *Find* $(v_{f,h}^{n+1}, \hat{v}_{s,h}^{n+1}, p_{f,h}^{n+1}, \hat{w}_{f,h}^{n+1}) \in V_{n+1,h}^f \times \hat{V}_h^s \times Q_{n+1,h}^f \times \hat{W}_h^f$
such that for $n = 0, 1, \cdots, N - 1,$

$$
\rho_f \left[\left(d_t^{X_t} v_{f,h}^{n+1}, \psi_{f,h} \right)_{\Omega_{n+1}^f} + \beta(v_{f,h}^{n+1}, v_{f,h}^{n+1}, \psi_{f,h})_{\Omega_{n+1}^f} \right.
$$

$$
\left. - \beta(\hat{w}_{f,h}^{n+1} \circ X_{h,t_{n+1}}^{-1}, v_{f,h}^{n+1}, \psi_{f,h})_{\Omega_{n+1}^f} + \frac{1}{2} \left(v_{f,h}^{n+1} \nabla \cdot (\hat{w}_{f,h}^{n+1} \circ X_{h,t_{n+1}}^{-1}), \psi_{f,h} \right)_{\Omega_{n+1}^f} \right]
$$

$$
+ 2\mu_f \left(D(v_{f,h}^{n+1}), D(\psi_{f,h}) \right)_{\Omega_{n+1}^f} - (p_{f,h}^{n+1}, \nabla \cdot \psi_{f,h})_{\Omega_{n+1}^f} + (\nabla \cdot v_{f,h}^{n+1}, q_{f,h})_{\Omega_{n+1}^f}
$$

$$
+ \hat{\rho}_s (d_t \hat{v}_{s,h}^{n+1}, \hat{\psi}_{s,h})_{\hat{\Omega}_s} + \mu_s \Delta t \left(\varepsilon \left(\hat{v}_{s,h}^{n+1} \right), \varepsilon(\hat{\psi}_{s,h}) \right)_{\hat{\Omega}_s} \quad (13)
$$

$$
+ \frac{\lambda_s \Delta t}{2} \left(\hat{\nabla} \cdot \hat{v}_{s,h}^{n+1}, \hat{\nabla} \cdot \hat{\psi}_{s,h} \right)_{\hat{\Omega}_s} + (\hat{\nabla} \hat{w}_{f,h}^{n+1}, \hat{\nabla} \hat{\xi}_{f,h})_{\hat{\Omega}_f} = \rho_f (f_f^{n+1}, \psi_{f,h})_{\Omega_t^f}
$$

$$
+ \hat{\rho}_s (\hat{f}_s^{n+1}, \hat{\psi}_{s,h})_{\hat{\Omega}_s} - \mu_s \Delta t \left(\varepsilon(\hat{v}_{s,h}^n), \varepsilon(\hat{\psi}_{s,h}) \right)_{\hat{\Omega}_s} - \frac{\lambda_s \Delta t}{2} \left(\hat{\nabla} \cdot \hat{v}_{s,h}^n, \hat{\nabla} \cdot \hat{\psi}_{s,h} \right)_{\hat{\Omega}_s}
$$

$$
- 2\mu_s \left(\varepsilon(\hat{u}_s^n), \varepsilon(\hat{\psi}_{s,h}) \right)_{\hat{\Omega}_s} - \lambda_s \left(\hat{\nabla} \cdot \hat{u}_s^n, \hat{\nabla} \cdot \hat{\psi}_{s,h} \right)_{\hat{\Omega}_s},
$$

$$
\forall (\psi_{f,h}, \hat{\psi}_{s,h}, q_{f,h}, \hat{\xi}_{f,h}) \in V_{n+1,h,0}^f \times \hat{V}_{h,0}^s \times Q_{n+1,h}^f \times \hat{W}_{h,0}^f.
$$

4.3 Nonlinear Iteration Algorithm

Assume that the mesh at the time step t_n is denoted by $\mathbb{T}_h^n = \mathbb{T}_{f,h}^n \cup \hat{\mathbb{T}}_{s,h}$, where $\hat{\mathbb{T}}_{s,h}$ is the Lagrangian structural mesh in $\hat{\Omega}_s$ which is always fixed, $\mathbb{T}_{f,h}$ is the ALE fluid mesh in Ω_t^f which needs to be updated all the time through the discrete ALE mapping, i.e., $\mathbb{T}_{f,h}^n = \hat{\mathbb{T}}_{f,h} + X_{h,t_n}$, where, X_{h,t_n} is only subject to the structure displacement $\hat{u}_{s,h}^n = \frac{\Delta t}{2}(\hat{v}_{s,h}^n + \hat{v}_{s,h}^{n-1}) + \hat{u}_{s,h}^{n-1}$ on $\hat{\Gamma}$ for the sake of a conforming mesh across Γ_t, or, $X_{h,t_n} = \frac{\Delta t}{2}(\hat{w}_{f,h}^n + \hat{w}_{f,h}^{n-1}) + X_{h,t_{n-1}}$. Due to a small deformation displacement of the structure, such specific ALE mapping always guarantees a shape-regular fluid mesh in Ω_t^f.

Suppose all the necessary solution variables at the time step t_n are given as:

$$
\chi^n := \left(v_{f,h}^n, \hat{v}_{s,h}^n, p_{f,h}^n, \hat{w}_{f,h}^n, X_{h,t_n} \right).
$$

In the following, we define an implicit iterative scheme for the FSI simulation at the current $(n + 1)$-th time step.

Algorithm 1. *Nonlinear iteration at the (n+1)-th time step for FSI.*

1. *Let* $j = 1$, $\chi^{n+1,0} = \chi^n$, *and* $\mathbb{T}_{f,h}^{n+1,0} = \mathbb{T}_{f,h}^n$ *which partitions* $\Omega_{n+1,0}^f = \Omega_n^f$.
2. *Solve the following linearized mixed finite element equation for* $\chi^{n+1,j}$ *with* $\chi^{n+1,j-1}$ *given at the* $j - 1$ *iteration:*
   Find $(v_{f,h}^{n+1,j}, \hat{v}_{s,h}^{n+1,j}, p_{f,h}^{n+1,j}, \hat{w}_{f,h}^{n+1,j}) \in V_{n+1,j,h}^f \times \hat{V}_h^s \times Q_{n+1,j,h}^f \times \hat{W}_h^f$ *such that*

$$
\rho_f \left[\left(d_t^{X_t} \boldsymbol{v}_{f,h}^{n+1,j}, \boldsymbol{\psi}_{f,h} \right)_{\Omega_{n+1,j-1}^f} + \beta(\boldsymbol{v}_{f,h}^{n+1,j-1}, \boldsymbol{v}_{f,h}^{n+1,j}, \boldsymbol{\psi}_{f,h})_{\Omega_{n+1,j-1}^f} \right.
$$

$$
- \beta(\hat{\boldsymbol{w}}_{f,h}^{n+1,j-1} \circ \boldsymbol{X}_{h,t_{n+1,j-1}}^{-1}, \boldsymbol{v}_{f,h}^{n+1,j}, \boldsymbol{\psi}_{f,h})_{\Omega_{n+1,j-1}^f}
$$

$$
\left. + \frac{1}{2} \left(\nabla \cdot (\hat{\boldsymbol{w}}_{f,h}^{n+1,j-1} \circ \boldsymbol{X}_{h,t_{n+1,j-1}}^{-1}) \boldsymbol{v}_{f,h}^{n+1,j}, \boldsymbol{\psi}_{f,h} \right)_{\Omega_{n+1,j-1}^f} \right]
$$

$$
+ 2\mu_f \left(\boldsymbol{D}(\boldsymbol{v}_{f,h}^{n+1,j}), \boldsymbol{D}(\boldsymbol{\psi}_{f,h}) \right)_{\Omega_{n+1,j-1}^f} - (p_{f,h}^{n+1,j}, \nabla \cdot \boldsymbol{\psi}_{f,h})_{\Omega_{n+1,j-1}^f}
$$

$$
+ (\nabla \cdot \boldsymbol{v}_{f,h}^{n+1,j}, q_{f,h})_{\Omega_{n+1,j-1}^f} + \hat{\rho}_s (d_t \hat{\boldsymbol{v}}_{s,h}^{n+1,j}, \hat{\boldsymbol{\psi}}_{s,h})_{\hat{\Omega}_s} \qquad (14)
$$

$$
+ \mu_s \Delta t \left(\varepsilon \left(\hat{\boldsymbol{v}}_{s,h}^{n+1,j} \right), \varepsilon(\hat{\boldsymbol{\psi}}_{s,h}) \right)_{\hat{\Omega}_s} + \frac{\lambda_s \Delta t}{2} \left(\hat{\nabla} \cdot \hat{\boldsymbol{v}}_{s,h}^{n+1,j}, \hat{\nabla} \cdot \hat{\boldsymbol{\psi}}_{s,h} \right)_{\hat{\Omega}_s}
$$

$$
+ (\hat{\nabla} \hat{\boldsymbol{w}}_{f,h}^{n+1,j}, \hat{\nabla} \hat{\boldsymbol{\xi}}_{f,h})_{\hat{\Omega}_f} = \rho_f (\boldsymbol{f}_f^{n+1,j}, \boldsymbol{\psi}_{f,h})_{\Omega_{n+1,j-1}^f} + \hat{\rho}_s (\hat{\boldsymbol{f}}_s^{n+1,j}, \hat{\boldsymbol{\psi}}_{s,h})_{\hat{\Omega}^s}
$$

$$
- \mu_s \Delta t \left(\varepsilon(\hat{\boldsymbol{v}}_{s,h}^n), \varepsilon(\hat{\boldsymbol{\psi}}_{s,h}) \right)_{\hat{\Omega}^s} - \frac{\lambda_s \Delta t}{2} \left(\hat{\nabla} \cdot \hat{\boldsymbol{v}}_{s,h}^n, \hat{\nabla} \cdot \hat{\boldsymbol{\psi}}_{s,h} \right)_{\hat{\Omega}^s}
$$

$$
- 2\mu_s \left(\varepsilon(\hat{\boldsymbol{u}}_s^n), \varepsilon(\hat{\boldsymbol{\psi}}_{s,h}) \right)_{\hat{\Omega}^s} - \lambda_s \left(\hat{\nabla} \cdot \hat{\boldsymbol{u}}_s^n, \hat{\nabla} \cdot \hat{\boldsymbol{\psi}}_{s,h} \right)_{\hat{\Omega}^s},
$$

$$
\forall (\boldsymbol{\psi}_{f,h}, \hat{\boldsymbol{\psi}}_{s,h}, q_{f,h}, \hat{\boldsymbol{\xi}}_{f,h}) \in \boldsymbol{V}_{n+1,j,h,0}^f \times \hat{\boldsymbol{V}}_{h,0}^s \times Q_{n+1,j,h}^f \times \hat{\boldsymbol{W}}_{h,0}^f.
$$

3. Let $\boldsymbol{X}_{h,t_{n+1,j}} = \frac{\Delta t}{2}(\hat{\boldsymbol{w}}_{f,h}^{n+1,j} + \hat{\boldsymbol{w}}_{f,h}^n) + \boldsymbol{X}_{h,t_n}$, and update the ALE fluid mesh $\mathbb{T}_{f,h}^{n+1,j} = \hat{\mathbb{T}}_{f,h} + \boldsymbol{X}_{h,t_{n+1,j}}$ which partitions $\Omega_{n+1,j}^f$.

4. For a given tolerance ε, determine whether or not the following stopping criteria hold for the relatively iterative errors:

$$
\frac{\|\boldsymbol{v}_{f,h}^{n+1,j} - \boldsymbol{v}_{f,h}^{n+1,j-1}\|_{(H^1(\Omega_t^f))^d}}{\|\boldsymbol{v}_{f,h}^{n+1,j-1}\|_{(H^1(\Omega_t^f))^d}} + \frac{\|\hat{\boldsymbol{v}}_{s,h}^{n+1,j} - \hat{\boldsymbol{v}}_{s,h}^{n+1,j-1}\|_{(H^1(\hat{\Omega}^s))^d}}{\|\hat{\boldsymbol{v}}_{s,h}^{n+1,j-1}\|_{(H^1(\hat{\Omega}^s))^d}}
$$

$$
+ \frac{\|p_{f,h}^{n+1,j} - p_{f,h}^{n+1,j-1}\|_{L^2(\Omega_t^f)}}{\|p_{f,h}^{n+1,j-1}\|_{L^2(\Omega_t^f)}} + \frac{\|\hat{\boldsymbol{w}}_{f,h}^{n+1,j} - \hat{\boldsymbol{w}}_{f,h}^{n+1,j-1}\|_{(H^1(\hat{\Omega}^f))^d}}{\|\hat{\boldsymbol{w}}_{f,h}^{n+1,j-1}\|_{(H^1(\hat{\Omega}^f))^d}} \leq \varepsilon. \qquad (15)
$$

If yes, go to Step 5; otherwise, let $j \leftarrow j+1$ and go back to Step 2 to continue the nonlinear iteration.

5. Let $\chi^{n+1} = \chi^{n+1,j}$, and update the ALE fluid mesh $\mathbb{T}_{f,h}^{n+1} = \mathbb{T}_{f,h}^{n+1,j}$ which partitions Ω_{n+1}^f.

5   Numerical Experiments

A hypothetical CVD patient with a growing aneurysm is used as an illustration example. The patient's CT imaging data (See Fig. 3) are collected to construct a geometrical model (See the right of Fig. 2). The cardiovascular parameters of this patient are estimated based on the patient's routine test results such as blood work, blood pressure, diabetes. Usually, if an aneurysm becomes about 2 in. in

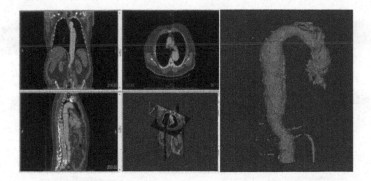

Fig. 3. The computational domain (Right) is generated from a patient's CT data (Left).

Fig. 4. Left: the overall computational mesh. **Right**: the cross section in flow direction from the left to the right, where a stent graft separates the blood fluid to two fields.

diameter and continues to grow, or begins to cause symptoms, the patient may need surgery to repair the artery before the aneurysm bursts. An endovascular aortic repair (EVAR) surgery involves replacing the weakened section of the vessel with an artificial tube, called a vascular stent graft. A computational mesh for an aneurysm CVD patient implanted with a stent graft is shown in Fig. 4, where the vascular stent graft, that is represented by an artificial blood vessel, expands inside the artery and eventually sticks to the inlet end of the aortic wall. Hence, a large blood fluid cavity is formed between the aortic aneurysm and the stent graft. Note that the blood fluid is thus separated by the stent graft into two parts: one inside the stent graft lumen driven by the incoming blood flow, while the other part is squeezed into the aneurismal cavity, waiting for being expelled due to the residual elasticity energy that remains in the diseased aorta.

To numerically model this multi-domain cardiovascular FSI problem in an accurate and effective fashion, an important technical issue needs to be clarified as follows. As shown on the right of Fig. 4, we extend the outlet end of the aortic wall further in order to connect two blood fluid fields together through the gap between the aortic wall and the stent graft. Thus, two blood fluid fields which respectively flow through the stent graft lumen and the aneurysm cavity, are eventually contained in one single connected fluid body and share the same inlet and outlet. This is crucial for the success of our FSI simulation, since the blood fluid that is contained inside the aneurysm cavity no longer owns an inlet thus no more incoming flow after the stent graft expands and squeezes onto the

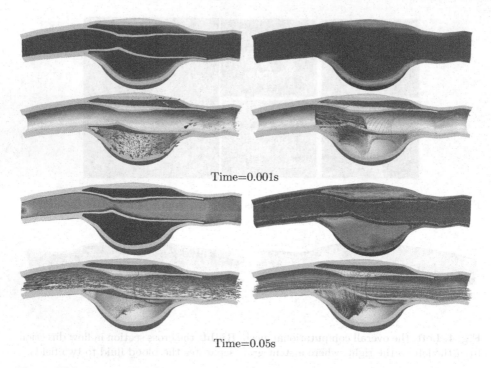

Time=0.001s

Time=0.05s

Fig. 5. Velocity magnitude, vector and streamline fields for the 4-field FSI between blood fluids, aortic wall and stent graft in the early stage.

inlet end of the aortic wall, therefore the incompressible fluid equations that is adopted to model the blood fluid flow will be violated in this fluid part if it is not connected with the main blood flow stream through the stent graft lumen.

Another important technical issue is that, to make ALE method successfully work for this multi-domain FSI simulation all the time, we need to avoid two structure parts, the aortic wall and the stent graft, to come into contact with each other. Otherwise, there is no space left for the blood fluid to flow through the contacting places of these two structure parts, which will induce a failure of ALE method since the ALE fluid mesh in these contacting area will lose many fluid mesh cells due to the physical contact of two structure parts, recalling that the preservation of mesh connectivity all the time is a unique feature and a warrant for success of ALE method. To deal with such a possible situation during the multi-domain FSI simulation, we set up a threshold for the distance in the normal direction between the aortic wall and the stent graft, namely, we do not let them physically get in touch with each other but just treat they "contact" at a pseudo zero distance in the normal direction once the threshold is reached. In addition, they also do not make any relative shift in the tangential direction since they are merged together on the inlet end of the blood vessel, as shown in Fig. 4. By that way, we do not have to handle the complicated contacting problem occurring between two structures at this time. Simultaneously, the ALE fluid mesh is still feasible within the threshold gap, thus our developed ALE mixed finite element method still works well for the multi-domain cardiovascular FSI problem. Such idea to avoid the contact between multiple structures is also similarly adopted in [3] where a thin layer of fluid is introduced around structures, such that there is no real "contact" between both phases. In the realistic numerical simulation, the forces are still transferred via the remaining small layer of fluid. Our approach is stable under refinement of the temporal and spatial discretization. We will carry out a more comprehensive study in the future for this approach by comparing with experiments and numerical benchmarks, in depth.

In this work, for the first time we successfully apply our monolithic ALE FEM [7, 10] and FSI codes to this multi-domain (artery, stent graft, blood fluid in the vessel lumen and the aneurysmal cavity) FSI problem with three moving interfaces in a much subtle fashion. Given a time-dependent incoming velocity profile on the inlet to simulate the pulsatile blood flow, some numerical results are gained for the velocity field of this multi-domain FSI problem in its early stage in forms of magnitude contour, vector and streamline, as shown in Fig. 5.

We can clearly see that the larger velocity magnitudes occur in the stent graft lumen rather than in the aneurysm cavity, which is because the incoming blood fluid continues to flow through the inlet to the blood vessel lumen while the contained blood fluid inside the aneurysm cavity is less active due to the lack of residual elasticity inside the diseased aortic wall, but after a realistically long term run, the aortic wall of the aneurysm, driven by the residual elasticity, shall be able to be shrunk back to some extent, and then squeeze the remaining blood out of the aneurysm cavity through the small gap, as illustrated in Fig. 5 after 0.05 s in the computation.

References

1. Barker, A., Cai, X.: Scalable parallel methods for monolithic coupling in fluid-structure interaction with application to blood flow modeling. J. Comput. Phys. **229**(3), 642–659 (2010)
2. Causin, P., Gerbeau, J., Nobile, F.: Added-mass effect in the design of partitioned algorithms for fluid-structure problems. Comput. Methods Appl. Mech. Eng. **194**(42), 4506–4527 (2005)
3. Frei, S., Richter, T., Wick, T.: Eulerian techniques for fluid-structure interactions: Part II – applications. In: Abdulle, A., Deparis, S., Kressner, D., Nobile, F., Picasso, M. (eds.) Numerical Mathematics and Advanced Applications - ENUMATH 2013. LNCSE, vol. 103, pp. 755–762. Springer, Cham (2015). https://doi.org/10.1007/978-3-319-10705-9_75
4. Idelsohn, S., Del, P., Rossi, R., Oñate, E.: Fluid-structure interaction problems with strong added-mass effect. Int. J. Numer. Methods Eng. **80**, 1261–1294 (2009)
5. Lan, R., Sun, P.: A monolithic arbitrary Lagrangian-Eulerian finite element analysis for a Stokes/parabolic moving interface problem. J. Sci. Comput. (2020, in press)
6. Lan, R., Sun, P., Mu, M.: Mixed finite element analysis for an elliptic/mixed-elliptic coupling interface problem with jump coefficients. Procedia Comput. Sci. **108**, 1913–1922 (2017)

7. Leng, W., Zhang, C., Sun, P., Gao, B., Xu, J.: Numerical simulation of an immersed rotating structure in fluid for hemodynamic applications. J. Comput. Sci. **30**, 79–89 (2019)
8. Martín, J.S., Smaranda, L., Takahashi, T.: Convergence of a finite element/ALE method for the Stokes equations in a domain depending on time. J. Comput. Appl. Math. **230**, 521–545 (2009)
9. Nobile, F., Formaggia, L.: A stability analysis for the arbitrary Lagrangian Eulerian formulation with finite elements. East West J. Numer. Math. **7**, 105–132 (2010)
10. Yang, K., Sun, P., Wang, L., Xu, J., Zhang, L.: Modeling and simulation for fluid-rotating structure interaction. Comput. Methods Appl. Mech. Eng. **311**, 788–814 (2016)

Morphing Numerical Simulation of Incompressible Flows Using Seamless Immersed Boundary Method

Kyohei Tajiri[✉][ID], Mitsuru Tanaka[ID], Masashi Yamakawa[ID], and Hidetoshi Nishida

Department of Mechanophysics, Kyoto Institute of Technology, Matsugasaki, Sakyo-ku, Kyoto, Japan
tajiri@kit.ac.jp

Abstract. In this paper, we proposed the morphing simulation method on the Cartesian grid in order to realize flow simulations for shape optimization with lower cost and versatility. In conventional morphing simulations, a simulation is performed while deforming a model shape and the computational grid using the boundary fitting grid. However, it is necessary to deform the computational grid each time, and it is difficult to apply to a model with complicated shape. The present method does not require grid regeneration or deformation. In order to apply the present method to models with various shapes on the Cartesian grid, the seamless immersed boundary method (SIBM) is used. Normally, when the SIBM is applied to a deformed object, the velocity condition on the boundary is imposed by the moving velocity of the boundary. In the present method, the velocity condition is imposed by zero velocity even if the object is deformed because the purpose of the present morphing simulation is to obtain simulation results for a stationary object. In order to verify the present method, two-dimensional simulations for the flow around an object were performed. In order to obtain drag coefficients of multiple models, the object was deformed in turn from the initial model to each model in the present morphing simulation. By using the present method, the drag coefficients for some models could be obtained by one simulation. It is concluded that the flow simulation for shape optimization can be performed very easily by using the present morphing simulation method.

Keywords: Computational fluid dynamics · Morphing simulation method · Immersed boundary method · Incompressible flow · Shape optimization

1 Introduction

There are many products around us that are closely related to the flow phenomenon. Improvements in the performance of these products are always

Supported by organization x.

expected. On the other hand, reducing the time and cost required to develop these products is also an important issue. Shape optimization through flow simulations at the stage of design is one of these efforts. By determining the optimum shape from many candidate product shapes (candidate models) at the early stage of product development, the effort of the redesign is reduced. As a result, development costs are reduced. Conventionally, flow simulations have been performed for each of these many candidate models. However, in recent years, the cost required for flow simulations has increased because the number of candidate models has increased in order to develop higher performance products. In order to reduce the number of these simulations, simulations are performed while deforming the model shape and the computational grid in shape optimization using flow simulations [1]. In this method, the number of flow simulations for shape optimization can be reduced, and the optimum shape can be determined in the flow simulation because results for many models can be obtained in one simulation. However, it is necessary to deform the computational grid each time, and it is difficult to apply to a model with complicated shape. In addition, the simulation on the boundary fitted grid can be expected to have high computational accuracy, however, the computational efficiency is inferior to the simulation on the Cartesian grid. In this paper, in order to realize flow simulations for shape optimization with lower cost and versatility, a method is proposed to perform simulation while deforming a model on the Cartesian grid that does not require grid regeneration or deformation. We call this method the morphing simulation method.

In order to apply the present method to models with various shapes on the Cartesian grid, the seamless immersed boundary method (SIBM) [2], which is an improved method of the immersed boundary method (IBM) [3] is used. In the IBM, additional force terms are added to the momentum equations to satisfy the velocity conditions on the virtual boundary points where the computational grid and the boundary of the object intersect. In order to apply the IBM to an object with arbitrary shape, it is only necessary to know the position of the virtual boundary on the grid. Therefore, the IBM can be easily applied to an object with a complicated shape. As for the estimation of the additional forcing term, there are mainly two methods, that is, the feedback [4,5] and direct [6] forcing term estimations. Generally, the direct forcing term estimation is adopted because of the simplicity of the algorithm. However, the conventional IBM with the direct forcing term estimation generates the unphysical pressure oscillations near the virtual boundary because of the pressure jump between inside and outside of the virtual boundary. The SIBM was proposed in order to remove these unphysical pressure oscillations. In the past study, the SIBM was applied not only to stationary objects but also to moving or scaling objects [7,8]. Therefore, it is possible to use the SIBM in the morphing simulation method proposed in this paper. Normally, when the SIBM is applied to a moving or scaling object, the velocity condition in the estimation of the additional forcing term is determined by the moving velocity of the object. In the present method, the additional forcing term is estimated under the condition that the velocity is zero even if the object is deformed because the purpose of the present morphing simulation is to obtain simulation results for a stationary object.

In this paper, the morphing simulation by the present method is performed for some models and compared with the conventional static SIBM where simulation is performed for each model and the effectiveness of the present method is discussed.

2 Morphing Numerical Simulation Using Seamless Immersed Boundary Method

2.1 Governing Equations

The governing equations are the continuity equation and the incompressible Navier-Stokes equations. Moreover, the forcing term is added to the Navier-Stokes equation for the SIBM. The non-dimensional continuity equation and incompressible Navier-Stokes equations are written as,

$$\frac{\partial u_i}{\partial x_i} = 0, \tag{1}$$

$$\frac{\partial u_i}{\partial t} = F_i - \frac{\partial p}{\partial x_i} + G_i, \tag{2}$$

$$F_i = -u_j \frac{\partial u_i}{\partial x_j} + \frac{1}{Re} \frac{\partial^2 u_i}{\partial x_j \partial x_j}, \tag{3}$$

where, Re denotes the Reynolds number defined by $Re = L_0 U_0 / \nu_0$. U_0, L_0 and ν_0 are the reference velocity, the reference length and the kinematic viscosity, respectively. $u_i = (u, v)$ and p are the velocity components and the pressure. G_i in Eq. 2 denotes the additional forcing term for the SIBM. F_i denotes the convective and diffusion terms.

2.2 Numerical Method

The incompressible Navier-Stokes equations (Eq. 2) are solved by the second order finite difference method on the collocated grid arrangement. The convective terms are discretized by the fully conservative finite difference method [9] and is written, for example, as,

$$
\begin{aligned}
v \frac{\partial u}{\partial y}\bigg|_{I,J} &= \frac{1}{2} \left(v \frac{\partial u}{\partial y}\bigg|_{I,J+\frac{1}{2}} + v \frac{\partial u}{\partial y}\bigg|_{I,J-\frac{1}{2}} \right) \\
&= \frac{1}{2} \left(v_{I,J+\frac{1}{2}} \frac{u_{I,J+1} - u_{I,J}}{\Delta} + v_{I,J-\frac{1}{2}} \frac{u_{I,J} - u_{I,J-1}}{\Delta} \right),
\end{aligned} \tag{4}
$$

where, v is the y component of velocity I, J are the grid index and Δ is grid spacing. The velocity at the midpoint (for example, $J + \frac{1}{2}$) of the grid is calculated by linear interpolation. The diffusive and pressure terms are discretized by the conventional second order centered finite difference method. For the time integration, the fractional step approach [12] based on the forward Euler method is applied. For the incompressible Navier-Stokes equations in the SIBM, the fractional step approach can be written by

$$u_i^* = u_i^n + \Delta t F_i^n, \tag{5}$$

$$u_i^{n+1} = u_i^* + \Delta t \left(-\frac{\partial p^n}{\partial x_i} + G_i^n \right), \tag{6}$$

where u_i^* denotes the fractional step velocity and Δt is the time increment. The resulting pressure equation is solved by the successive over-relaxation (SOR) method.

2.3   Seamless Immersed Boundary Method

In order to adopt the SIBM, the additional forcing term in the momentum equations, G_i, should be estimated. In the SIBM, the additional forcing term is estimated by the direct forcing term estimation [2]. The direct forcing term estimation is shown in Fig. 1. We explain in two-dimensions but the extension to three-dimensions is straightforward. For the forward Euler time integration, the forcing term can be determined by

$$G_i^n = -F_i^n + \frac{\partial p^n}{\partial x_i} + \frac{\bar{U}_i^{n+1} - u_i^n}{\Delta t}, \tag{7}$$

where $\bar{U}_i^{n+1}$ denotes the velocity linearly interpolated from the velocity on the near grid point and the velocity (u_{vb}) determined by the velocity condition on the virtual boundary. Namely, the forcing term is specified as the velocity components at next time step satisfy the relation, $u_i^{n+1} = \bar{U}_i^{n+1}$. In the IBM, the grid points added forcing term are restricted near the virtual boundary only (show Fig. 1(a)). In this approach, the non-negligible velocity appears inside the virtual boundary. Also, the pressure distributions near the virtual boundary show the unphysical oscillations because of the pressure jump. In the SIBM, the forcing term is added not only on the grid points near the virtual boundary but also in the region inside the virtual boundary shown in Fig. 1(b) in order to remove the unphysical oscillations near the virtual boundary. In the region inside the virtual boundary, the forcing term is determined by satisfying the relation, $\bar{U}_i^{n+1} = \bar{U}_b$, where $\bar{U}_b$ is the velocity which satisfies the velocity condition at the grid point. When applying the SIBM to a stationary object, the velocity condition on and inside the virtual boundary is zero velocity. As mentioned above, an algorithm of the SIBM is very simple and can easily be extended to three dimensions. Therefore, it is applied to flow around moving or scaling objects [7,8]. When applying the SIBM to a moving or scaling object, the velocity condition on and inside the virtual boundary are obtained by the moving velocity of the object at that point. Moreover, there are also examples of application to turbulence flow [10, 11].

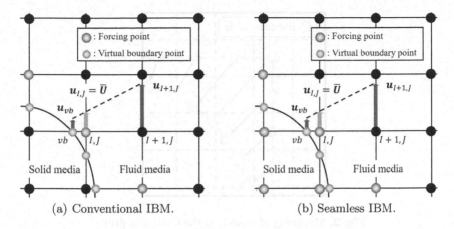

(a) Conventional IBM.   (b) Seamless IBM.

Fig. 1. Grid points added forcing terms.

2.4 Morphing Numerical Simulation

The morphing of a model on the Cartesian grid is shown in Fig. 2. In the present method, the object is deformed in turn from the first model to the model that requires simulation results. In the SIBM, only the position of the virtual boundary of the object on the fixed grid is updated even if the object deforms. In the SIBM, the virtual boundary of the object with arbitrary shape is represented by boundary nodes in the two-dimensional simulation as shown in Fig. 2. The boundary between these boundary nodes is approximated by straight lines. By determining the intersection between the boundary and the grid that is the virtual boundary point, SIBM can be applied to an object having an arbitrary shape. In the three-dimensional simulation, the virtual boundary of the object with arbitrary shape is represented by triangular polygons and boundary nodes [7]. In the present morphing simulation, the object is deformed from one model to another model by moving these boundary nodes every time step. Once the position of the boundary nodes at each time step is determined, it is easy to apply the SIBM to the model. In the present method, the boundary nodes for the model before deformation is linearly moved to the position of the boundary nodes for the next model. Therefore, the algorithm in the present method is extremely easy. Normally, in the SIBM for the moving or deforming object, the additional forcing term is determined by the moving velocity of the object or boundary. In the present method, the additional forcing term is estimated under the condition that the velocity is zero even if the object is deformed. It is because the purpose of the present morphing simulation is to obtain simulation results for a stationary each model.

3 Application to Two-Dimensional Model

In this paper, in order to verify the present method, two-dimensional simulations for the flow around an object are performed. In order to obtain drag coefficients

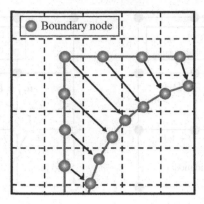

Fig. 2. Morphing of model on the Cartesian grid.

of multiple models, the object is deformed in turn from the initial model to each model in the present morphing simulation. In another case, the morphing has downtime in each model. In this paper, two-dimensional flows around square, circular and elliptic cylinder whose drag coefficients can be compared with the reference results are considered. The computational domain is shown in Fig. 3. In this simulation, a model is set as shown in Fig. 4 and the model is deformed in order from 1 to 5. Each process indicates morphing processes. For example, the model is deformed from 1 to 2 in the process 1, and the model is deformed from 3 to 2 in the process 3. In the present morphing simulation, firstly, the static SIBM simulation is performed for the model 1 and then the processes 1 to 4 are performed in the morphing simulation. This deformation may be larger than the deformation in general shape optimization. In each model, the length of the side of the square cylinder, the diameter of the circular cylinder, and the length of the major axis of the elliptic cylinder are the reference length $L = 1$. The length of the minor axis of the elliptic cylinder is 0.5. As a result, the processes 1 and 2 are scaling down the model and the processes 3 and 4 are scaling up the model. In addition, the processes 1 and 4 are two-dimensional deformations and the processes 2 and 3 are one-dimensional deformations. As for the computational conditions, the impulsive start determined by the uniform flow ($u = 1$, $v = 0$) is adopted. On the inflow boundary, the velocity is fixed by the uniform flow and the pressure is imposed by the Neumann condition obtained by the normal momentum equation. On the outflow and side boundaries (right, top and bottom boundaries), the velocity is extrapolated from the inner points and the pressure is obtained by the Sommerfeld radiation condition [13]. On the virtual boundary and inside the boundary, the velocity condition is the velocity is zero. The Reynolds number is set as $Re = 40$. The flow around each model is steady flow under this Reynolds number. In order to reduce the number of grid points, the hierarchical Cartesian grid with level 4 is introduced. The grid resolution near the model is $\Delta = 1/80$. The number of boundary nodes in each model is 400 and the distance between the nodes in the case of the square cylinder (models 1, 5) which is the largest model is smaller than the grid spacing. In addition, the boundary node also exists at each vertex of the square cylinder as shown in Fig. 2.

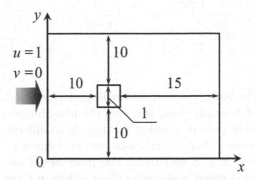

Fig. 3. Computational domain.

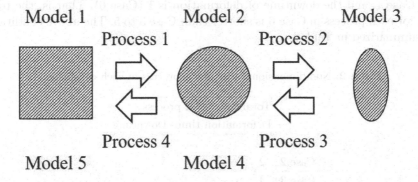

Fig. 4. Configuration of model morphing.

Firstly, in order to obtain the reference results for each model, the conventional static simulations by SIBM without morphing are performed under the above conditions. In Table 1, the drag coefficient is shown with the reference results [14–16]. In this paper, the drag coefficient is estimated by

$$C_D = \frac{-2\int_O (G_x - u_j \frac{\partial u_i}{\partial x_j} - \frac{\partial u_i}{\partial t})ds}{\rho_0 U_0^2 L}, \tag{8}$$

where O denotes the region to which the forcing term is added in the SIBM. ρ_0 and U_0 denote the reference density and velocity of the flow. The drag coefficient by the conventional static SIBM is in good agreement with the reference result in each model. Therefore, these drag coefficients are used as reference results for verifying the present morphing simulation method.

Table 1. Drag coefficient of each model in static SIBM.

	Square cylinder	Circular cylinder	Elliptic cylinder
Static SIBM (Presents)	1.728	1.568	1.631
Sen et al. [14]	1.787	–	–
Dennis et al. [15]	–	1.522	–
Sen et al. [16]	–	–	1.567

In this paper, the morphing simulation is performed under some deformation speeds. The deformation speed is set by the non-dimensional time for each process in Fig. 4. In the present simulation, there is no difference in deformation time between processes. Then, simulations are performed in the case of non-dimensional time is 1, 2, 4, 8 and 16 for the processes (Case 1 to 5). In other words, the deformation speed is slower in Case 5 than in Case 1. In addition, in order to investigate the possibility that the deformation time can be set shorter, a simulation is performed in which downtime of deformation is set after the deformation to each model. In this simulation, the deformation time is the same 1 as Case 1, and the downtime of deformation is 1 (Case 6). That is, the total time for each process in Case 6 is shorter than Case 3 to 5. The above conditions are summarized in Table 2.

Table 2. Non-dimensional time for a process in each condition.

	Total time for a process	
	Deformation time	Downtime
Case 1	1	–
Case 2	2	–
Case 3	4	–
Case 4	8	–
Case 5	16	–
Case 6	1	1

Figures 5, 6, 7 and 8 show the pressure contours of each model. Note that the pressure contours of the models 2 and 4 by the static SIBM is same. In all cases, the pressure contours obtained by the present method are similar to those obtained by the static SIBM. In particular, those in Cases 5 and 6 are in good agreement with those in the static SIBM.

Figure 9 shows time histories of the comparison of the drag coefficients at each deformation speed. The horizontal axis shows the non-dimensional time converted into the model number. For example, the model number is 2 when the horizontal axis is 2 and the model is being deformed from 2 to 3 (process 2)

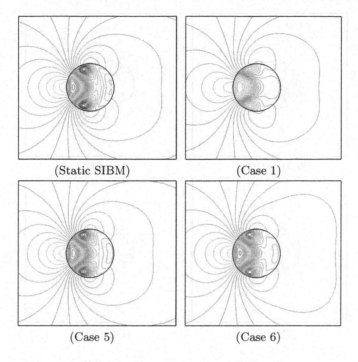

Fig. 5. Pressure contours of model 2.

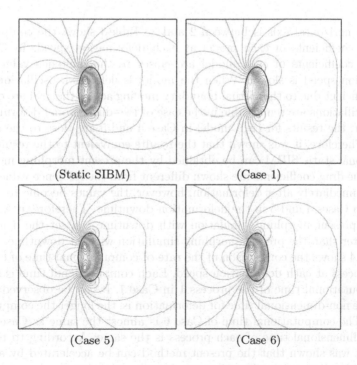

Fig. 6. Pressure contours of model 3.

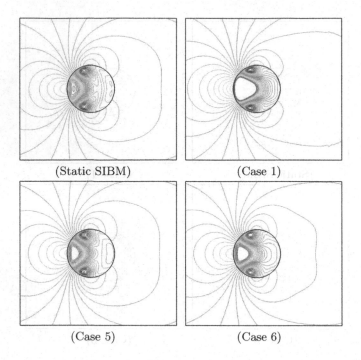

(Static SIBM)   (Case 1)

(Case 5)   (Case 6)

Fig. 7. Pressure contours of model 4.

when the horizontal axis is between 2 and 3. Table 3 shows the comparison of the drag coefficients of each model at each deformation speed. In Case 1–5, the drag coefficients of each model are closer to the reference values as the deformation speed is slower. When the model is deformed, oscillations of the drag coefficient due to the virtual boundary moving across the grid are observed. These oscillations are remarkable in the case of two-dimensional deformation. In particular, the results for each model in Case 4 and 5 are close to the reference results. Therefore, it was shown that the results equivalent to the results by the conventional static SIBM can be obtained by the present morphing method. In Case 6, the drag coefficients are shown different from the reference value just like Case 1 immediately after deformation, however, the values become to the same level as in Cases 4 and 5 in the deformation downtime. Therefore, it was shown that the present morphing simulation with downtime can set the deformation speed faster than the present morphing simulation without downtime.

Table 4 shows the comparison of the rate of computational time of each morphing process at each deformation speed. Each computational time is based on the computational time of the process 1 in Case 1. It can be observed that the longer the non-dimensional time for deformation is, the longer the computational time is. The computational time of Case 6 is almost the same as Case 2 where the non-dimensional time of each process is the same. According to the above results, it was shown that the present method can be accelerated by adding in the morphing process the downtime.

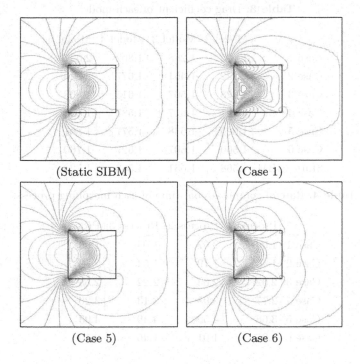

Fig. 8. Pressure contours of model 5.

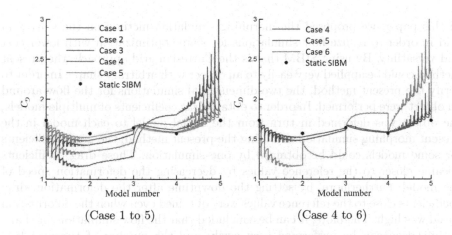

Fig. 9. Comparison of drag coefficients at each deformation speed.

Table 3. Drag coefficient of each model.

	Model 2	Model 3	Model 4	Model 5
Case 1	1.175	1.288	1.802	2.327
Case 2	1.342	1.461	1.677	2.048
Case 3	1.428	1.567	1.615	1.908
Case 4	1.487	1.617	1.587	1.840
Case 5	1.525	1.638	1.577	1.801
Case 6	1.500	1.607	1.527	1.760
Static SIBM	1.568	1.631	1.568	1.728

Table 4. Rate of computational time of each morphing process.

	Process 1	Process 2	Process 3	Process 4
Case 1	1.00	0.67	1.04	1.88
Case 2	1.46	1.19	1.54	2.19
Case 3	2.02	2.03	2.22	2.55
Case 4	2.77	3.05	3.13	3.20
Case 5	3.68	4.50	4.49	4.03
Case 6	1.25	1.01	1.46	2.10

4   Conclusions

In this paper, we proposed the morphing simulation method on the Cartesian grid in order to realize flow simulations for shape optimization with lower cost and versatility. By using SIBM that is the Cartesian grid approach, the present method could be applied very easily to an object with arbitrary shape. In order to verify the present method, the two-dimensional simulations for the flow around an object were performed. In order to obtain drag coefficients of multiple models, the object was deformed in turn from the initial model to each model in the present morphing simulation. By using the present method, the drag coefficients for some models could be obtained by one simulation. These drag coefficients became closer to the reference values by decreasing the deformation speed of the model. Furthermore, by setting the downtime after the deformation, drag coefficients close to the reference values were obtained even when the deformation speed was high. Therefore, it can be concluded that the flow simulation for shape optimization can be performed very easily and the number of times of flow simulation for many models can be significantly reduced by using the present morphing simulation method.

References

1. Hino, T.: Shape optimization of practical ship hull forms using Navier-Stokes analysis. In: Proceedings of the 7th International Conference on Numerical Ship Hydro (1999)
2. Nishida, H., Sasao, K.: Incompressible flow simulations using virtual boundary method with new direct forcing terms estimation. In: Deconinck, H., Dick, E. (eds.) Computational Fluid Dynamics 2006, pp. 371–376. Springer, Heidelberg (2009). https://doi.org/10.1007/978-3-540-92779-2_57
3. Peskin, C.S., McQueen, D.M.: A three-dimensional computational method for blood flow in the heart I. Immersed elastic fibers in a viscous incompressible fluid. J. Comput. Phys. **81**(2), 372–405 (1989)
4. Goldstein, D., Handler, R., Sirovich, L.: Modeling a no-slip flow boundary with an external force field. J. Comput. Phys. **105**(2), 354–366 (1993)
5. Saiki, E.M., Biringen, S.: Numerical simulation of a cylinder in uniform flow: application of a virtual boundary method. J. Comput. Phys. **123**(2), 450–465 (1996)
6. Fadlun, E.A., Verzicco, R., Orlandi, P., Mohd-Yosof, J.: Combined immersed-boundary finite-difference methods for three-dimensional complex simulations. J. Comput. Phys. **161**(1), 35–60 (2000)
7. Nishida, H., Tajiri, K.: Numerical simulation of incompressible flows around a fish model at low reynolds number using seamless virtual boundary method. J. Fluid Sci. Technol. **4**(3), 500–511 (2009)
8. Nishida, H., Tajiri, K., Tanaka, M.: Seamless immersed boundary method for flow around a scaled object. In: Proceedings 12th Asian Computational Fluid Dynamics Conference, pp. 1–9 (2018)
9. Morinishi, Y., Lund, T.S., Vasilyev, O.V., Moin, P.: Fully conservative higher order finite difference schemes for incompressible flow. J. Comput. Phys. **143**(1), 90–124 (1998)
10. Tajiri, K., Nishida, H., Tanaka, M.: Large eddy simulation of turbulent flow using seamless immersed boundary method. In: Proceedings of the 8th International Conference on Computational Fluid Dynamics, ICCFD8-197, pp. 1–13 (2014)
11. Tajiri, K., Nishida, H., Tanaka, M.: Property of seamless immersed boundary method for large eddy simulation of incompressible turbulent flows. J. Fluid Sci. Technol. **9**(2), 1–8 (2014)
12. Rhie, C.M., Chow, W.L.: Numerical study of the turbulent flow past an airfoil with trailing edge separation. AIAA J. **21**(11), 1525–1532 (1983)
13. Kawakami, K., Nishida, H., Satofuka, N.: An open boundary condition for the numerical analysis of unsteady incompressible flow using the vorticity-streamfunction formulation. Trans. Jpn. Soc. Mech. Eng. Ser. B **60**(574), 1891–1896 (1994). (Japanese)
14. Sen, S., Mittal, S., Biswas, G.: Flow past a square cylinder at low Reynolds numbers. Int. J. Numer. Methods Fluids **67**(9), 1160–1174 (2011)
15. Dennis, S.C.R., Chang, G.Z.: Numerical solutions for steady flow past a circular cylinder at Reynolds num-bers up to 100. J. Fluid Mech. **42**(3), 471–489 (1970)
16. Sen, S., Mittal, S., Biswas, G.: Steady separated flow past elliptic cylinders using a stabilized finite-element method. Comput. Model. Eng. Sci. **86**(1), 1–26 (2012)

deal.II Implementation of a Two-Field Finite Element Solver for Poroelasticity

Zhuoran Wang and Jiangguo Liu[✉]

Department of Mathematics, Colorado State University, Fort Collins, CO 80523, USA
{wangz,liu}@math.colostate.edu

Abstract. This paper presents a finite element solver for poroelasticity in the 2-field approach and its implementation on the deal.II platform. Numerical experiments on benchmarks are presented to demonstrate the accuracy and efficiency of this new solver.

Keywords: Darcy flow · deal.II · Finite element methods · Hexahedral meshes · Poroelasticity · Quadrilateral meshes · Weak Galerkin

1 Introduction

Poroelasticity is an important problem in science and engineering. The Biot's model for linear poroelasticity has been well accepted and commonly used. It couples solid displacement $\mathbf{u}$ and fluid pressure p through the following partial differential equations (PDEs)

$$\begin{cases} -\nabla \cdot (2\mu\varepsilon(\mathbf{u}) + \lambda(\nabla \cdot \mathbf{u})\mathbf{I}) + \alpha\nabla p = \mathbf{f}, \\ \partial_t (c_0 p + \alpha\nabla \cdot \mathbf{u}) + \nabla \cdot (-\mathbf{K}\nabla p) = s, \end{cases} \tag{1}$$

where $\varepsilon(\mathbf{u}) = \frac{1}{2}\left(\nabla\mathbf{u} + (\nabla\mathbf{u})^T\right)$ is the strain tensor with $\lambda > 0, \mu > 0$ being the Lamé constants, $\mathbf{f}$ a given body force, $\mathbf{K}$ a conductivity tensor, s a known fluid source, α (usually close to 1) the Biot-Williams constant, and $c_0 \geq 0$ the constrained storage capacity. Appropriate boundary and initial conditions are posed to close the system.

Finite element methods (FEMs) are common tools for solving the Biot's model. Depending on the unknown quantities to be solved, poroelasticity solvers are usually grouped into 3 types:

- *2-field*: Solid displacement and fluid pressure are to be solved;
- *3-field*: Solid displacement, fluid pressure and velocity are to be solved;
- *4-field*: Solid stress & displacement, fluid pressure & velocity are to be solved.

Liu and Wang were partially supported by US National Science Foundation grant DMS-1819252. We thank Dr. Wolfgang Bangerth for the computing resources.

V. V. Krzhizhanovskaya et al. (Eds.): ICCS 2020, LNCS 12143, pp. 88–101, 2020.
https://doi.org/10.1007/978-3-030-50436-6_7

A major issue in numerical solvers for poroelasticity is the poroelasticity locking, which usually appears as nonphysical pressure oscillations. This happens when the porous media are low-permeable or low-compressible [12, 28, 36].

Early on, the continuous Galerkin (CG) FEMs were applied respectively to solve for displacement and pressure. But it was soon recognized that such solvers were subject to poroelasticity locking and the 2-field approach was nearly abandoned. The mixed finite element methods can be used to solve for pressure and velocity simultaneously and meanwhile coupled with a FEM for linear elasticity that is free of Poisson-locking. Therefore, the 3-field approach has been the main stream [5, 25–27, 33, 34]. The 4-field approach is certainly worth of investigation, but it just involves too many unknowns (degrees of freedom) [35].

The weak Galerkin (WG) finite element methods [31] have emerged as a new class of numerical methods with nice features that can be applied to a wide variety of problems including Darcy flow and linear elasticity [14, 18, 24, 30]. Certainly, WG solvers can be developed for linear poroelasticity [17], they are free of poroelasticity locking but may involve a lot of degrees of freedom.

Recently, our efforts have been devoted to reviving the 2-field approach for development of efficient and robust finite element solvers for poroelasticity [13]. This may involve incorporation of WG FEMs with WG FEMs or classical FEMs. In this paper, we continue such efforts to develop a poroelasticity solver that couples the WG finite elements for Darcy flow and the classical Lagrangian elements with reduced integration for linear elasticity. Moreover, we provide an accessible efficient implementation of this new solver on deal.II, a popular finite element package [3].

2 Discretization of Linear Elasticity by Lagrangian Elements with Reduced Integration

This section discusses discretization of linear elasticity by Lagrangian Q_1^d finite elements ($d = 2, 3$) with reduced integration that is needed for our new FE solver for poroelasticity. For convenience of presentation, we consider the linear elasticity in its usual form

$$\begin{cases} -\nabla \cdot \sigma = \mathbf{f}(\mathbf{x}), & \mathbf{x} \in \Omega, \\ \mathbf{u}|_{\Gamma^D} = \mathbf{u}_D, & (\sigma\mathbf{n})|_{\Gamma^N} = \mathbf{t}_N, \end{cases} \tag{2}$$

where Ω is a 2d- or 3d-bounded domain occupied by a homogeneous and isotropic elastic body, $\mathbf{f}$ a body force, $\mathbf{u}_D, \mathbf{t}_N$ respectively Dirichlet and Neumann data, $\mathbf{n}$ the outward unit normal vector on the domain boundary that has a non-overlapping decomposition $\partial\Omega = \Gamma^D \cup \Gamma^N$. As mentioned in Sect. 1, $\mathbf{u}$ is the solid displacement, $\varepsilon(\mathbf{u}) = \frac{1}{2}\left(\nabla\mathbf{u} + (\nabla\mathbf{u})^T\right)$ the strain tensor, and $\sigma = 2\mu\,\varepsilon(\mathbf{u}) + \lambda(\nabla{\cdot}\mathbf{u})\mathbf{I}$ the Cauchy stress tensor with $\mathbf{I}$ being the identity matrix. The Lamé constants λ, μ are given by

$$\lambda = \frac{E\nu}{(1+\nu)(1-2\nu)}, \qquad \mu = \frac{E}{2(1+\nu)},$$

where E is the elasticity modulus and $\nu \in (0, \frac{1}{2})$ is Poisson's ratio.

One major issue in finite element solvers for linear elasticity is that as the elastic material becomes nearly incompressible or $\nu \to \frac{1}{2}$, mathematically as $\lambda \to \infty$, a FE solver may fail to produce correct results. This often appears as loss of convergence rates in displacement errors or spurious behaviors in numerical stress and dilation (divergence of displacement). This is the so-called "Poisson locking" [6]. It is well known that the classical linear (bilinear, trilinear) Lagrangian finite elements are subject to Poisson locking.

Many remedies for Poisson locking have been developed. Reduced integration is probably the easiest technique aiming at a quick fix for the classical Lagrangian elements, although the theory was less elegant [7,9,22].

In this paper, we adopt the remedy in [9] and extend it to 3-dim. In other words, we consider vector-valued Lagrangian bilinear and trilinear finite elements with reduced integration CG.Q_1^d (R.I.) (here $d = 2,3$) for solving linear elasticity and provide deal.II implementation of these solvers. Specifically, the 1-point Gaussian quadrature is employed for handling the dilation term.

Let E be a convex quadrilateral with vertices $P_i(x_i, y_i)(i = 1,2,3,4)$ that are oriented counterclockwise. A bilinear mapping F from $(\hat{x}, \hat{y})$ in the reference element $\hat{E} = [0,1]^2$ to $(x,y) \in E$ is established. Its Jacobian determinant is denoted as $J(\hat{x}, \hat{y})$. On $\hat{E}$, we have 4 scalar-valued bilinear basis functions

$$\begin{aligned} \hat{\phi}_4(\hat{x}, \hat{y}) &= (1 - \hat{x})\hat{y}, & \hat{\phi}_3(\hat{x}, \hat{y}) &= \hat{x}\hat{y}, \\ \hat{\phi}_1(\hat{x}, \hat{y}) &= (1 - \hat{x})(1 - \hat{y}), & \hat{\phi}_2(\hat{x}, \hat{y}) &= \hat{x}(1 - \hat{y}). \end{aligned} \tag{3}$$

They are mapped to the quadrilateral E as rational functions of x, y:

$$\phi_i(x, y) = \hat{\phi}_i(\hat{x}, \hat{y}), \quad i = 1,2,3,4. \tag{4}$$

On E, we have 8 node-based vector-valued local basis functions:

$$\begin{bmatrix} \phi_1 \\ 0 \end{bmatrix}, \begin{bmatrix} 0 \\ \phi_1 \end{bmatrix}, \begin{bmatrix} \phi_2 \\ 0 \end{bmatrix}, \begin{bmatrix} 0 \\ \phi_2 \end{bmatrix}, \begin{bmatrix} \phi_3 \\ 0 \end{bmatrix}, \begin{bmatrix} 0 \\ \phi_3 \end{bmatrix}, \begin{bmatrix} \phi_4 \\ 0 \end{bmatrix}, \begin{bmatrix} 0 \\ \phi_4 \end{bmatrix}. \tag{5}$$

They span CG.$Q_1^2(E)$. The notation is a bit confusing, since the shape functions are now rationals instead of polynomials. For any $\mathbf{v} \in$ CG.$Q_1^2(E)$, we consider

$$\overline{\nabla \cdot \mathbf{v}} = \frac{1}{|E|} \int_E \mathbf{v}(x, y)dxdy = \frac{1}{|E|} \int_{\hat{E}} \mathbf{v}(x, y)J(\hat{x}, \hat{y})d\hat{x}d\hat{y}, \tag{6}$$

where $|E|$ is the volume of E.

Let $\mathbf{V}_h$ be the space of vector-valued shape functions constructed from the CG.Q_1^2 elements on a quasi-uniform quadrilateral mesh $\mathcal{E}_h$. Let $\mathbf{V}_h^0$ be the subspace of $\mathbf{V}_h$ consisting of shape functions that vanish on Γ^D. A finite element scheme for linear elasticity in the strain-div formulation seeks $\mathbf{u}_h \in \mathbf{V}_h$ so that

$$\mathcal{A}_h^{SD}(\mathbf{u}_h, \mathbf{v}) = \mathcal{F}_h(\mathbf{v}), \quad \forall \mathbf{v} \in \mathbf{V}_h^0, \tag{7}$$

where

$$\mathcal{A}_h^{SD}(\mathbf{u}_h, \mathbf{v}) = \sum_{E \in \mathcal{E}_h} 2\mu \left(\varepsilon(\mathbf{u}_h), \varepsilon(\mathbf{v}) \right)_E + \lambda (\overline{\nabla \cdot \mathbf{u}_h}, \overline{\nabla \cdot \mathbf{v}})_E, \tag{8}$$

$$\mathcal{F}_h(\mathbf{v}) = \sum_{E \in \mathcal{E}_h} (\mathbf{f}, \mathbf{v})_E + \sum_{\gamma \in \Gamma_h^N} \langle t_N, \mathbf{v} \rangle. \tag{9}$$

3 WG Finite Element Discretization for Darcy Flow

This section briefly discusses the weak Galerkin finite element discretization for Darcy flow that is needed for our new 2-field solver for linear poroelasticity.

Among the existing finite element solvers for Darcy flow [4,8,10,11,15,18,19], [20,21,23,29,31,32], the newly developed weak Galerkin solvers have some nice features that are attractive for large-scale computing tasks. In particular, the $WG(Q_k, Q_k; RT_{[k]})$ methods (with integer $k \geq 0$) approximate the primal unknown pressure by using polynomial shape function of degree at most k separately defined in element interiors and on edges/faces. Their discrete weak gradients are reconstructed in the unmapped Raviart-Thomas spaces $RT_{[k]}$ and used to approximate the classical gradient in the variational form. The WG Darcy solvers based on these novel notions

(i)   are locally mass-conservative;
(ii) provide continuous normal fluxes;
(iii) result in SPD linear systems that are easy to be solved.

In [32], we discussed deal.II implementation of such WG Darcy solvers for $0 \leq k \leq 5$. The numerical tests on SPE10 Model 2 have demonstrated the aforementioned nice features and practical usefulness of the novel WG methodology.

In this section, we briefly review the basic concepts of weak Galerkin by recapping $WG(Q_0, Q_0; RT_{[0]})$ for Darcy flow on quadrilateral meshes. For ease of presentation, we consider the Darcy flow problem in its usual form

$$\begin{cases} \nabla \cdot (-\mathbf{K} \nabla p) \equiv \nabla \cdot \mathbf{u} = s, \\ p|_{\Gamma^D} = p_D, & \text{on } \Gamma^D \\ \mathbf{u} \cdot \mathbf{n} = u_N, & \text{on } \Gamma^N, \end{cases} \tag{10}$$

where Ω is a polygonal domain, p the primal unknown pressure, $\mathbf{u}$ the Darcy velocity, $\mathbf{K}$ conductivity tensor (medium permeability divided fluid dynamic viscosity) that is uniformly SPD over the domain, s a known source, p_D a Dirichlet boundary condition, u_N a Neumann boundary condition, $\mathbf{n}$ the outward unit normal vector on $\partial\Omega$, which has a nonoverlapping decomposition $\Gamma^D \cup \Gamma^N$.

First we define the lowest-order unmapped Raviart-Thomas space as

$$RT_{[0]}(E) = \text{Span}(\mathbf{w}_1, \mathbf{w}_2, \mathbf{w}_3, \mathbf{w}_4), \tag{11}$$

where

$$\mathbf{w}_1 = \begin{bmatrix} 1 \\ 0 \end{bmatrix}, \quad \mathbf{w}_2 = \begin{bmatrix} 0 \\ 1 \end{bmatrix}, \quad \mathbf{w}_3 = \begin{bmatrix} X \\ 0 \end{bmatrix}, \quad \mathbf{w}_4 = \begin{bmatrix} 0 \\ Y \end{bmatrix}, \tag{12}$$

and $X = x - x_c$, $Y = y - y_c$ are the normalized coordinates using the element center (x_c, y_c).

For a given quadrilateral element E, we consider 5 discrete weak functions $\phi_i (0 \leq i \leq 4)$ as follows:

- ϕ_0 for element interior: It takes value 1 in the interior E° but 0 on the boundary E^∂;
- $\phi_i (1 \le i \le 4)$ for the four sides respectively: Each takes value 1 on the i-th edge but 0 on all other three edges and in the interior.

The discrete weak gradient $\nabla_w \phi$ is established in $RT_{[0]}(E)$ via integration by parts [31]:

$$\int_E (\nabla_w \phi) \cdot \mathbf{w} = \int_{E^\partial} \phi^\partial (\mathbf{w} \cdot \mathbf{n}) - \int_{E^\circ} \phi^\circ (\nabla \cdot \mathbf{w}), \quad \forall \mathbf{w} \in RT_{[0]}(E). \tag{13}$$

For implementation, this involves solving a size-4 SPD linear system.

However, when E becomes a rectangle $[x_1, x_2] \times [y_1, y_2]$ with $\Delta x = x_2 - x_1$, $\Delta y = y_2 - y_1$, one can obtain these discrete weak gradients explicitly:

$$\begin{cases} \nabla_w \phi_0 = \ 0\mathbf{w}_1 + \ 0\mathbf{w}_2 + \frac{-12}{(\Delta x)^2}\mathbf{w}_3 + \frac{-12}{(\Delta y)^2}\mathbf{w}_4, \\ \nabla_w \phi_1 = \frac{-1}{\Delta x}\mathbf{w}_1 + \ 0\mathbf{w}_2 + \frac{6}{(\Delta x)^2}\mathbf{w}_3 + \ 0\mathbf{w}_4, \\ \nabla_w \phi_2 = \frac{1}{\Delta x}\mathbf{w}_1 + \ 0\mathbf{w}_2 + \frac{6}{(\Delta x)^2}\mathbf{w}_3 + \ 0\mathbf{w}_4, \\ \nabla_w \phi_3 = \ 0\mathbf{w}_1 + \frac{-1}{\Delta y}\mathbf{w}_2 + \ 0\mathbf{w}_3 + \frac{6}{(\Delta y)^2}\mathbf{w}_4, \\ \nabla_w \phi_4 = \ 0\mathbf{w}_1 + \frac{1}{\Delta y}\mathbf{w}_2 + \ 0\mathbf{w}_3 + \frac{6}{(\Delta y)^2}\mathbf{w}_4. \end{cases} \tag{14}$$

These discrete weak gradients are used to approximate the classical gradient in the variational form for the Darcy flow problem.

Let $\mathcal{E}_h$ be a quasi-uniform convex quadrilateral mesh for the given polygonal domain Ω. Let Γ_h^D be the set of all edges on the Dirichlet boundary Γ^D and Γ_h^N be the set of all edges on the Neumann boundary Γ^N. Let S_h be the space of discrete shape functions on $\mathcal{E}_h$ that are degree 0 polynomials in element interiors and also degree 0 polynomials on edges. Let S_h^0 be the subspace of functions in S_h that vanish on Γ_h^D. For (10), we seek $p_h = \{p_h^\circ, p_h^\partial\} \in S_h$ such that $p_h^\partial|_{\Gamma_h^D} = Q_h^\partial(p_D)$ (the L^2-projection of Dirichlet boundary data into the space of piecewise constants on Γ_h^D) and

$$\mathcal{A}_h(p_h, q) = \mathcal{F}(q), \quad \forall q = \{q^\circ, q^\partial\} \in S_h^0, \tag{15}$$

where

$$\mathcal{A}_h(p_h, q) = \sum_{E \in \mathcal{E}_h} \int_E \frac{\mathbf{K}}{\mu} \nabla_w p_h \cdot \nabla_w q, \tag{16}$$

$$\mathcal{F}(q) = \sum_{E \in \mathcal{E}_h} \int_{E^\circ} s q^\circ - \sum_{\gamma \in \Gamma_h^N} \int_\gamma u_N q^\partial. \tag{17}$$

Clearly, (15) is a large-size sparse SPD system.

After the numerical pressure p_h is solved from (15), an elementwise numerical velocity is obtained by a local L_2-projection back into the subspace $RT_{[0]}$:

$$\mathbf{u}_h = \mathbf{Q}_h(-\mathbf{K}\nabla_w p_h). \tag{18}$$

The projection can be skipped if $\mathbf{K}$ is an elementwise constant scalar matrix. Furthermore, the bulk normal flux on any edge is defined as

$$\int_{e \in E^{\partial}} \mathbf{u}_h \cdot \mathbf{n}_e. \tag{19}$$

It has been proved [21] that such a WG solver is locally conservative and guarantees normal flux continuity.

4 Coupling $WG(Q_0, Q_0; RT_{[0]})$ and $CG.Q_1^2$ (R.I.) for Poroelasticity

In this section, the continuous Galerkin Q_1^d ($d = 2, 3$) elements with reduced integration and the weak Galerkin $WG(Q_0, Q_0; RT_{[0]})$ elements are combined with the implicit Euler temporal discretization to solve linear poroelasticity problems.

Assume a given domain Ω is already partitioned into a quasi-uniform quadrilateral mesh $\mathcal{E}_h$. For a given time period $[0, T]$, let

$$0 = t^{(0)} < t^{(1)} < \ldots < t^{(n-1)} < t^{(n)} < \ldots < t^{(N)} = T$$

be a temporal partition. We denote $\Delta t_n = t^{(n)} - t^{(n-1)}$ for $n = 1, 2, \ldots, N$.

Let $\mathbf{V}_h$ and $\mathbf{V}_h^0$ be the spaces of vector-valued shape functions based on the first-order CG elements. Let $\mathbf{u}_h^{(n)}, \mathbf{u}_h^{(n-1)} \in \mathbf{V}_h$ be the approximations to solid displacement at time moments $t^{(n)}$ and $t^{(n-1)}$, respectively.

Let S_h and S_h^0 be the spaces of scalar-valued discrete weak functions constructed in Sect. 3 based on the $WG(Q_0, Q_0; RT_{[0]})$ elements. Similarly, let $p_h^{(n)}, p_h^{(n-1)} \in S_h$ be the approximations to fluid pressure at time moments $t^{(n)}$ and $t^{(n-1)}$, respectively. Note that the discrete weak trial function has two parts:

$$p_h^{(n)} = \{p_h^{(n),\circ}, p_h^{(n),\partial}\}, \tag{20}$$

where $p_h^{(n),\circ}$ lives in element interiors and $p_h^{(n),\partial}$ lives on the mesh skeleton.

Applying the implicit Euler discretization, we establish the following time-marching scheme, for any $\mathbf{v} \in \mathbf{V}_h^0$ and any $q \in S_h^0$,

$$\begin{cases} 2\mu\left(\varepsilon(\mathbf{u}_h^{(n)}), \varepsilon(\mathbf{v})\right) + \lambda(\overline{\nabla \cdot \mathbf{u}_h^{(n)}}, \overline{\nabla \cdot \mathbf{v}}) - \alpha(p_h^{(n),\circ}, \overline{\nabla \cdot \mathbf{v}}) = (\mathbf{f}^{(n)}, \mathbf{v}), \\ c_0\left(p_h^{(n),\circ}, q^\circ\right) + \Delta t_n \left(\mathbf{K}\nabla p_h^{(n)}, \nabla q\right) + \alpha(\overline{\nabla \cdot \mathbf{u}_h^{(n)}}, q^\circ) \\ \quad = c_0\left(p_h^{(n-1),\circ}, q^\circ\right) + \Delta t_n \left(s^{(n)}, q^\circ\right) + \alpha(\overline{\nabla \cdot \mathbf{u}_h^{(n-1)}}, q^\circ), \end{cases} \tag{21}$$

for $n = 1, 2, \ldots, N$, where $\overline{\nabla \cdot \mathbf{v}}$ is the elementwise average that represents the reduced integration technique. The above two equations are further augmented with appropriate boundary and initial conditions. This results in a large monolithic system at each time step.

Theses errors are calculated to assess the accuracy of our poroelasticity solver:

– $L_2([0,T]; L_2(\Omega))$-norm for interior pressure errors

$$\|p - p_h^\circ\|_{L_2(L_2)}^2 = \sum_{n=1}^N \Delta t_n \|p^{(n)} - p_h^{((n),\circ)}\|_{L_2(\Omega)}^2, \tag{22}$$

– $L_2([0,T]; L_2(\Omega))$-norm for displacement errors

$$\|\mathbf{u} - \mathbf{u}_h\|_{L_2(L_2)}^2 = \sum_{n=1}^N \Delta t_n \|\mathbf{u}^{(n)} - \mathbf{u}_h^{(n)}\|_{L_2(\Omega)}^2, \tag{23}$$

– $L_2([0,T]; H^1(\Omega))$-norm for displacement errors

$$\|\mathbf{u} - \mathbf{u}_h\|_{L_2(H^1)}^2 = \sum_{n=1}^N \Delta t_n \|\nabla\mathbf{u}^{(n)} - \nabla\mathbf{u}_h^{(n)}\|_{L_2(\Omega)}^2, \tag{24}$$

– $L_2([0,T]; L_2(\Omega))$-norm for stress errors

$$\|\sigma - \sigma_h\|_{L_2(L_2)}^2 = \sum_{n=1}^N \Delta t_n \|\sigma^{(n)} - \sigma_h^{(n)}\|_{L_2(\Omega)}^2. \tag{25}$$

5   Code Excerpts with Comments

This section provides some code excerpts with comments. More details can be found in our code modules for deal.II (subject to minor changes). We want to point that the elasticity discretization can also be replaced by the so-called EQ_1 or BR_1 elements [3,16], which are now available in deal.II Version 9.1.

5.1   Code Excerpts for WG($Q_0, Q_0; RT_{[0]}$)

There was a discussion on this in [32]. Here we recap the most important concepts very briefly. Note that FE_RaviartThomas is a Raviart-Thomas space for vector-valued functions, FESystem defines WG finite element spaces in the interiors and on edges/faces. Shown below is the code for the lowest-order WG finite elements.

```
88   FE_RaviartThomas<dim> fe_rt;
89   DoFHandler<dim> dof_handler_rt;
90   FESystem<dim> fe;
91   DoFHandler<dim> dof_handler;
```

```
227   fe_rt (0);
228   dof_handler_rt (triangulation);
229   fe (FE_DGQ<dim>(0), 1, FE_FaceQ<dim>(0), 1);
230   dof_handler (triangulation);
```

5.2   Code Excerpts for CG.Q_1^2 with Reduced Integration

This part shows how we use CG.Q_1^2 with reduced integration to discretize linear elasticity. FE_Q defines the finite element space for displacement vectors. Each component of the vector is in the FE_Q space.

```
88   FE_Q<dim>(1),dim;
```

Here, the reduced integration technique with one-point Gaussian quadrature is used to calculate the dilation (divergence of displacement).

```
88   QGauss<dim>  reduced_integration_quadrature_formula(1);
```

5.3   Code Excerpts for Coupled Discretizations for Poroelasticity

We couple CG.Q_1^2(R.I.) and WG($Q_0, Q_0; RT_{[0]}$) to solve linear poroelasticity. FESystem defines the finite element spaces for displacement, interior pressure, and face pressure. Shown below is the coupled finite elements.

```
88   FE_RaviartThomas<dim>   fe_rt;
89   DoFHandler<dim>         dof_handler_rt;
90   FESystem<dim>           fe;
91   DoFHandler<dim>         dof_handler;
```

```
88   fe_rt (0),
89   dof_handler_rt (triangulation),
90
91   fe (FE_Q<dim>(1),dim,
92     FE_DGQ<dim>(0), 1,
93     FE_FaceQ<dim>(0), 1),
94   dof_handler (triangulation),
```

We use block structures to store matrices and variables. The following piece defines the degrees of freedom associated with displacement, interior pressure, and face pressure.

```
88   std::vector<types::global_dof_index> dofs_per_block (3);
89   DoFTools::count_dofs_per_block
90   (dof_handler, dofs_per_block, block_component);
91   const unsigned int n_u = dofs_per_block[0],
92                 n_p_interior = dofs_per_block[1],
93                 n_p_face =  dofs_per_block[2],
94                 n_p = dofs_per_block[1]+ dofs_per_block[2];
```

The implementation for the WG Darcy solver discussed in [32] is naturally re-used and incorporated. The following piece calculates the coupling terms with reduced integration in the local matrix. However, we only use the reduced integration for divergence of vector-valued shape functions.

```
88   for (unsigned int q_index = 0;
89         q_index < n_q_points_reduced_integration; ++q_index){
90    for (unsigned int i = 0; i < dofs_per_cell; ++i){
91     const double div_i_reduced_integration =
92     fe_values_reduced_integration
93     [displacements_reduced_integration].divergence(i, q_index);
94      for (unsigned int j = 0; j < dofs_per_cell; ++j){
95       const double div_j_reduced_integration =
96       fe_values_reduced_integration
97       [displacements_reduced_integration].divergence(j, q_index);
98
99       local_matrix(i, j) +=
100      - alpha * fe_values_reduced_integration
101       [pressure_interior_reduced_integration].value(j, q_index)
102      * div_i_reduced_integration
103      + alpha* (div_j_reduced_integration
104       * fe_values_reduced_integration
105      [pressure_interior_reduced_integration].value (i,q_index)))
106       * fe_values_reduced_integration.JxW(q_index);
107  }}}
```

Finally, this piece hands the coupling term in the local right-hand side.

```
88   for (unsigned int q=0; q<n_q_points_reduced_integration; ++q){
89     for (unsigned int i=0; i<dofs_per_cell; ++i){
90       const double phi_i_q =
91         fe_values_reduced_integration
92           [pressure_interior_reduced_integration].value(i,q);
93       local_rhs(i) +=
94         (alpha*div_old_displacement_reduced_integration[q]
95         * phi_i_q)
96         * fe_values_reduced_integration.JxW(q);
97  }}
```

6   Numerical Experiments

This section presents numerical examples to demonstrate the accuracy and robustness of this new finite element solver for poroelasticity.

Example 1 (A 2-dim smooth example for convergence rates). Here our domain is $\Omega = (0,1)^2$. Analytical solutions for solid displacement and fluid pressure are given as

$$
\mathbf{u} = \sin\left(\frac{\pi}{2}t\right)
\begin{bmatrix}
\dfrac{\pi}{2}\sin^2(\pi x)\sin(2\pi y) + \dfrac{1}{\lambda}\sin(\pi x)\sin(\pi y) \\[2mm]
-\dfrac{\pi}{2}\sin(2\pi x)\sin^2(\pi y) + \dfrac{1}{\lambda}\sin(\pi x)\sin(\pi y)
\end{bmatrix},
\tag{26}
$$

$$
p = \frac{\pi}{\lambda}\sin\left(\frac{\pi}{2}t\right)\sin(\pi(x+y)).
\tag{27}
$$

It is interesting to see that

$$
\nabla \cdot \mathbf{u} = p,
\tag{28}
$$

and hence $\nabla \cdot \mathbf{u} \to 0$ as $\lambda \to \infty$. Dirichlet boundary conditions for both displacement and pressure are specified on the whole boundary using the exact solutions. For the parameters, we have $\mathbf{K} = \kappa\mathbf{I}$ with $\kappa = 1$, $\mu = 1$, $\alpha = 1$, and $c_0 = 0$. To examine the solver's locking-free property, we shall consider $\lambda = 1$ and $\lambda = 10^6$, respectively. The time period is $[0,T] = [0,1]$.

For numerical simulations, we consider uniform rectangular meshes. Shown in Tables 1 and 2 are the numerical results obtained with this new solver. Clearly, the convergence rates do not deteriorate as λ increases from 1 to 10^6. In other words, our new 2-field solver is locking-free.

Table 1. Ex.1 with $\lambda = 1$: Numerical results of $CG.Q_1^2(\text{R.I.}) + WG(Q_0, Q_0; RT_{[0]})$ solver on rectangular meshes

$1/h$	$1/\Delta t$	$\|p - p_h^\circ\|_{L_2(L_2)}$	$\|\mathbf{u} - \mathbf{u}_h\|_{L_2(L_2)}$	$\|\mathbf{u} - \mathbf{u}_h\|_{L_2(H^1)}$	$\|\sigma - \sigma_h\|_{L_2(L_2)}$
4	16	5.07478E−1	1.78798E−1	2.35598E−0	4.44080E−0
8	64	2.52365E−1	4.54880E−2	1.15497E−0	2.29855E−0
16	256	1.25983E−1	1.14071E−2	5.74435E−1	1.15784E−0
32	1024	6.29657E−2	2.85375E−3	2.86836E−1	5.79949E−1
Conv. rate		1.00	1.98	1.01	0.97

Example 2 (A 3-dim example with a sandwiched low permeability layer). The domain is the unit cube $\Omega = (0,1)^3$. The permeability is $\mathbf{K} = \kappa\mathbf{I}$. Specifically, the middle region $0.25 \leq z \leq 0.75$ has a low permeability $\kappa = 10^{-8}$, whereas $\kappa = 1$ in other parts, see Fig. 1(a). There is no body force for solid or source for fluid. Other parameters are $\lambda = 1$, $\mu = 1$, $\alpha = 1$, $c_0 = 0$.

The boundary conditions are as follows.

(i) For the solid, a downward traction (Neumann) condition $\mathbf{t}_N = (0,0,-1)^T$ is posed on the top face, whereas all five other faces are clamped, i.e., $\mathbf{u} = \mathbf{0}$;

Table 2. Ex.1 with $\lambda = 10^6$: Numerical results of CG.Q_1^2(R.I.) + WG($Q_0, Q_0; RT_{[0]}$) solver on rectangular meshes

$1/h$	$1/\Delta t$	$\|p - p_h^\circ\|_{L_2(L_2)}$	$\|\mathbf{u} - \mathbf{u}_h\|_{L_2(L_2)}$	$\|\mathbf{u} - \mathbf{u}_h\|_{L_2(H^1)}$	$\|\sigma - \sigma_h\|_{L_2(L_2)}$
4	16	5.07481E−7	1.76096E−1	2.30126E−0	1.36770E+6
8	64	2.52367E−7	4.48677E−2	1.12759E−0	7.66388E+5
16	256	1.25984E−7	1.12553E−2	5.60529E−1	3.92554E+5
32	1024	6.29658E−8	2.81600E−3	2.79849E−1	1.97411E+5
Conv. rate		1.00	1.98	1.01	0.93

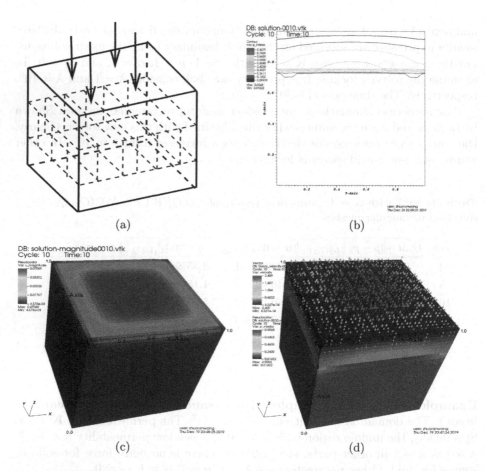

(a)

(b)

(c)

(d)

Fig. 1. Ex.2: Numerical results from the new FE solver at final time $T = 0.01$ with $h = 1/32$ and $\Delta t = 10^{-3}$. (a) Problem illustration; (b) Numerical pressure contours for $x = 0.5$; (c) Numerical displacement magnitude elementwise average; (d) Numerical pressure cell averages and velocity at element centers.

(ii) For the fluid, the top face ($z = 1$) has a Dirichlet condition $p = 0$; whereas all five other faces have a no-flow condition, in other words, zero Neumann boundary condition.

A similar 2-dim problem has been tested in [12,13,17]. But we shall observe richer features in this 3-dim problem.

For numerical simulations, we use uniform rectangular meshes and a uniform temporal partition. Specifically, $h = \frac{1}{32}$ and $\Delta t = 10^{-3}$ so that $\Delta t \approx h^2$. The final time is $T = 0.01$, which means 10 time steps for simulation. Shown in Fig. 1 are the profiles of numerical pressure and velocity along with solid displacement magnitude. There is no pressure oscillation, even though there is a layer with a very low permeability. A pressure steep front is observed near $z = 0.75$. The low permeability layer provides some kind of insulation. There is basically no solid deformation or fluid pressure change below this layer.

7 Concluding Remarks

A new finite element solver for poroelasticity is presented and proven numerically to be locking-free. This new solver is in the 2-field approach, i.e., only solid displacement and fluid pressure are treated as unknowns. Specifically, the new solver discretizes displacement using the classical Lagrangian Q-type elements with reduced integration, whereas the pressure is approximated by piecewise constants respectively defined inside elements and on inter-element boundaries. Discrete weak gradients of such piecewise constant shape functions are established in the unmapped lowest-order Raviart-Thomas spaces on quadrilaterals and hexahedra, which are required to be asymptotically parallelogram or parallelopiped. This new solver has been implemented in the dimension-independent paradigm on the `deal.II` platform. Our code modules are openly accessible.

The new solver in this paper is different than the one presented in [13]. Now the elasticity part is discretized using the classical Lagrangian Q-type elements with reduced integration. This results in even less degrees of freedom.

There are several directions one can go from here.

(i) Code optimization, especially, preconditioning and parallelization, shall make this new solver more efficient;

(ii) A rigorous analysis on this new solver is to be established for the locking-free property and convergence rates;

(iii) A similar solver can be developed for simplicial (triangular and tetrahedral) meshes; Implementation on `FEniCS` or `FreeFEM++` platforms are surely attractive for scientific computing tasks;

(iv) To remove the restriction *asymptotically parallelogram or parallelopiped*, we could utilize the newly developed Arbogast-Correa and Abogast-Tao elements [1,2] for more general convex quadrilaterals and cuboidal hexahedra. Again `deal.II` implementation will be attractive.

These are under our investigation and will be reported in our future work.

References

1. Arbogast, T., Correa, M.: Two families of mixed finite elements on quadrilaterals of minimal dimension. SIAM J. Numer. Anal. **54**, 3332–3356 (2016)
2. Arbogast, T., Tao, Z.: Construction of H(div)-conforming mixed finite elements on cuboidal hexahedra. Numer. Math. **142**, 1–32 (2019)
3. Arndt, D., et al.: The deal.II library, version 9.1. J. Numer. Math. (2019). https:// doi.org/10.1515/jnma-2019-0064. https://dealii.org/deal91-preprint.pdf
4. Bastian, P., Riviere, B.: Superconvergence and $H(div)$ projection for discontinuous Galerkin methods. Int. J. Numer. Meth. Fluids **42**, 1043–1057 (2003)
5. Berger, L., Bordas, R., Kay, D., Tavener, S.: Stabilized lowest-order finite element approximation for linear three-field poroelasticity. SIAM J. Sci. Comput. **37**, A2222–A2245 (2015)
6. Brenner, S., Scott, L.: The Mathematical Theory of Finite Element Methods, Texts in Applied Mathematics, vol. 15, 3rd edn. Springer, New York (2008)
7. Brenner, S., Sung, L.Y.: Linear finite element methods for planar linear elasticity. Math. Comput. **59**, 321–338 (1992)
8. Bush, L., Ginting, V.: On the application of the continuous Galerkin finite element method for conservation problems. SIAM J. Sci. Comput. **35**, A2953–A2975 (2013)
9. Cheng, X., Huang, H., Zou, J.: Quadrilateral finite elements for planar linear elasticity problem with large Lamé constant. J. Comput. Math. **16**, 357–366 (1998)
10. Cockburn, B., Gopalakrishnan, J., Wang, H.: Locally conservative fluxes for the continuous Galerkin method. SIAM J. Numer. Anal. **45**, 1742–1770 (2007)
11. Brezzi, F., Fortin, M.: Mixed and Hybrid Finite Element Methods. Springer, New York (1991). https://doi.org/10.1007/978-1-4612-3172-1
12. Haga, J., Osnes, H., Langtangen, H.: On the causes of pressure oscillations in low permeable and low compressible porous media. Int. J. Numer. Anal. Meth. Geomech. **36**, 1507–1522 (2012)
13. Harper, G., Liu, J., Tavener, S., Wang, Z.: A two-field finite element solver for poroelasticity on quadrilateral meshes. In: Shi, Y., Fu, H., Tian, Y., Krzhizhanovskaya, V.V., Lees, M.H., Dongarra, J., Sloot, P.M.A. (eds.) ICCS 2018. LNCS, vol. 10862, pp. 76–88. Springer, Cham (2018). https://doi.org/10.1007/978-3-319-93713-7_6
14. Harper, G., Liu, J., Tavener, S., Zheng, B.: Lowest-order weak Galerkin finite element methods for linear elasticity on rectangular and brick meshes. J. Sci. Comput. **78**, 1917–1941 (2019)
15. Harper, G., Liu, J., Zheng, B.: The THex algorithm and a simple Darcy solver on hexahedral meshes. Proc. Comput. Sci. **108C**, 1903–1912 (2017)
16. Harper, G., Wang, R., Liu, J., Tavener, S., Zhang, R.: A locking-free solver for linear elasticity on quadrilateral and hexahedral meshes based on enrichment of Lagrangian elements. Technical report, Colorado State University (2020)
17. Hu, X., Mu, L., Ye, X.: Weak Galerkin method for the Biot's consolidation model. Comput. Math. Appl. **75**, 2017–2030 (2018)
18. Lin, G., Liu, J., Mu, L., Ye, X.: Weak Galerkin finite element methdos for Darcy flow: anistropy and heterogeneity. J. Comput. Phys. **276**, 422–437 (2014)
19. Liu, J., Sadre-Marandi, F., Wang, Z.: Darcylite: a matlab toolbox for Darcy flow computation. Proc. Comput. Sci. **80**, 1301–1312 (2016)
20. Liu, J., Tavener, S., Wang, Z.: Lowest-order weak Galerkin finite element method for Darcy flow on convex polygonal meshes. SIAM J. Sci. Comput. **40**, B1229–B1252 (2018)

21. Liu, J., Tavener, S., Wang, Z.: The lowest-order weak Galerkin finite element method for the Darcy equation on quadrilateral and hybrid meshes. J. Comput. Phys. **359**, 312–330 (2018)
22. Malkus, D.S., Hughes, T.: Mixed finite element methods - reduced and selective integration techniques: a unification of concepts. Comput. Meth. Appl. Mech. Eng. **15**, 63–81 (1978)
23. Mu, L., Wang, J., Ye, X.: A weak Galerkin finite element method with polynomial reduction. J. Comput. Appl. Math. **285**, 45–58 (2015)
24. Mu, L., Wang, J., Ye, X.: Weak Galerkin finite element methods on polytopal meshes. Int. J. Numer. Anal. Model. **12**, 31–53 (2015)
25. Phillips, P., Wheeler, M.: A coupling of mixed with continuous Galerkin finite element methods for poroelasticity I: the continuous in time case. Comput. Geosci. **11**, 131–144 (2007)
26. Phillips, P., Wheeler, M.: A coupling of mixed with continuous Galerkin finite element methods for poroelasticity II: the-discrete-in-time case. Comput. Geosci. **11**, 145–158 (2007)
27. Phillips, P., Wheeler, M.: A coupling of mixed with discontinuous Galerkin finite element methods for poroelasticity. Comput. Geosci. **12**, 417–435 (2008)
28. Phillips, P.J., Wheeler, M.F.: Overcoming the problem of locking in linear elasticity and poroelasticity: an heuristic approach. Comput. Geosci. **13**, 5–12 (2009)
29. Sun, S., Liu, J.: A locally conservative finite element method based on piecewise constant enrichment of the continuous Galerkin method. SIAM J. Sci. Comput. **31**, 2528–2548 (2009)
30. Wang, C., Wang, J., Wang, R., Zhang, R.: A locking-free weak Galerkin finite element method for elasticity problems in the primal formulation. J. Comput. Appl. Math. **307**, 346–366 (2016)
31. Wang, J., Ye, X.: A weak Galerkin finite element method for second order elliptic problems. J. Comput. Appl. Math. **241**, 103–115 (2013)
32. Wang, Z., Harper, G., O'Leary, P., Liu, J., Tavener, S.: deal.II implementation of a weak galerkin finite element solver for Darcy flow. In: Rodrigues, J.M.F., et al. (eds.) ICCS 2019. LNCS, vol. 11539, pp. 495–509. Springer, Cham (2019). https://doi.org/10.1007/978-3-030-22747-0_37
33. Wheeler, M., Xue, G., Yotov, I.: Coupling multipoint flux mixed finite element methods with continuous Galerkin methods for poroelasticity. Comput. Geosci. **18**, 57–75 (2014)
34. Yi, S.Y.: A coupling of nonconforming and mixed finite element methods for Biot's consolidation model. Numer. Meth. PDEs **29**, 1749–1777 (2013)
35. Yi, S.Y.: Convergence analysis of a new mixed finite element method for Biot's consolidation model. Numer. Meth. PDEs **30**, 1189–1210 (2014)
36. Yi, S.Y.: A study of two modes of locking in poroelasticity. SIAM J. Numer. Anal. **55**, 1915–1936 (2017)

Numerical Investigation of Solute Transport in Fractured Porous Media Using the Discrete Fracture Model

Mohamed F. El-Amin[1,2(✉)], Jisheng Kou[3], and Shuyu Sun[4]

[1] Energy Research Laboratory, College of Engineering, Effat University,
Jeddah 21478, Kingdom of Saudi Arabia
momousa@effatuniversity.edu.sa
[2] Mathematics Department, Faculty of Science, Aswan University,
Aswan 81528, Egypt
[3] School of Mathematics and Statistics, Hubei Engineering University,
Xiaogan 432000, Hubei, China
[4] King Abdullah University of Science and Technology (KAUST),
Thuwal 23955–6900, Kingdom of Saudi Arabia

Abstract. In this paper, we investigate flow with solute transport in fractured porous media. The system of the governing equations consists of the continuity equation, Darcy's law, and concentration equation. A discrete-fracture model (DFM) has been developed to describe the problem under consideration. The multiscale time-splitting method was used to handle different sizes of time-step for different physics, such as pressure and concentration. Some numerical examples are presented to show the efficiency of the multi-scale time-splitting approach.

Keywords: Multiscale time-splitting · Discrete fracture model · Mass transfer · Fracture porous media · Reservoir simulation

1 Introduction

The flow and solute transport in fractured porous media is very important in many applications such as contaminants migration in fractured aquifer systems. The fractured porous medium consists of two domains, namely, matrix blocks and fractures. The fractures are more permeable than the matrix blocks, but they contain very little fluid than the matrix. Different scales are therefore invoked in fractured porous media such as discrete fracture models (DFMs) and dual continuum models, can describe flow and transport in fractured porous media. The model of solute transport in fractured and porous media has been solved analytical by Park and Lee [1] and Roubinet et al. [2]. In order to model solute transport in fractured porous media, Bodin et al. [3] and Graf and Therrien [4] have used the discrete network model; and Refs. [5–7] have used equivalent continuum models, while Refs. [8,9] have used the continuum model. In order

© Springer Nature Switzerland AG 2020
V. V. Krzhizhanovskaya et al. (Eds.): ICCS 2020, LNCS 12143, pp. 102–115, 2020.
https://doi.org/10.1007/978-3-030-50436-6_8

to represent the fractures explicitly in the fractured porous media the discrete-fracture model (DFM) has been used. This procedures can remove the contrast of the length-scale resulting of the direct representation of the fracture aperture as in the dual-porosity model.

Time discretization, on the other hand, has a significant impact on the efficiency of numerical solutions. Therefore, the use of traditional single-scale time schemes is limited by the rapid variation in pressure and concentration in matrix or fracture, where applicable. The multi-scale time-splitting technique is therefore considered to be one of the major improvements in the treatment of the gap between pressure and concentration. In a number of publications, such as [10–13], the multi-scale time splitting method was considered. El-Amin et al. [14] have developed a discrete-fracture-model with multi-scale time-splitting two-phase flow including nanoparticles transport in fractured porous media. In this work, we develop a discrete-fracture-model with multi-scale time-splitting of solute transport in fractured porous media. The modeling and mathematical formulation is considered in the second section. In Sect. 3, the time-stepping technique with spatial discretization have been presented. The fourth section is devoted to numerical test, and then, the conclusions are given in the last section.

2   Modeling and Formulation

2.1   Governing Equations

This paper considers the problem of mass transfer and flow in fractured porous media. The system of equations consists of continuity, momentum, and concentration.

Momentum Conservation (Darcy's Law):

$$\mathbf{u} = -\mathbf{K}\nabla\Phi \tag{1}$$

where $\mathbf{K}$ is the permeability tensor $\mathbf{K} = \frac{k}{\mu}\mathbf{I}$, $\mathbf{I}$ is the identity matrix and k/μ is a positive real number. $\mathbf{u}, \Phi, \mu, k$ are, respectively, the velocity, the pressure, the viscosity and the permeability.

Mass Conservation:

$$\nabla \cdot \mathbf{u} = q, \tag{2}$$

where q is the external mass flow rate.

Mass Transfer: The solute transport equation in porous media may be given as,

$$\phi\frac{\partial C}{\partial t} + \nabla \cdot (\mathbf{u}C - \phi D\nabla C) = Q_c, \tag{3}$$

where C is the solute concentrations, ϕ is the porosity, D is the diffusion coefficient, and Q_c is the rate of change of volume belonging to a source/sink term.

2.2   Initial and Boundary Conditions

Consider the computational domain Ω with the boundary $\partial\Omega$ which is subjected to Dirichlet Γ_D and Neumann Γ_N boundaries, where $\partial\Omega = \Gamma_D \cup \Gamma_N$ and $\Gamma_D \cap \Gamma_N = \emptyset$. At the beginning of the injection process, we have,

$$C = 0 \quad \text{in} \quad \Omega \quad \text{at} \quad t = 0, \tag{4}$$

The boundary conditions are given as,

$$P = P^D \quad \text{on} \quad \Gamma_D, \tag{5}$$

$$\mathbf{u} \cdot \mathbf{n} = q^N, \quad C = C_0, \quad \text{on} \quad \Gamma_N. \tag{6}$$

where $\mathbf{n}$ is the outward unit normal vector to $\partial\Omega$, P^D is the pressure on Γ_D and q^N the imposed inflow rate on Γ_N, respectively.

2.3   Discrete Fracture Model

In the discrete-fracture-model (DFM), the fracture gridcells are simplified to represent as the interfaces of the matrix gridcell and fractures are surrounded by matrix blocks. Thus, the dimension of fracture reduced by one than the dimension of matrix, i.e., if matrix is of n-dimension, then, fracture is of $(n-1)$-dimension. The domain is decomposed into the matrix domain, Ω_m and fracture domain, Ω_f. The pressure equation in the matrix domain is given by,

$$-\nabla \cdot \mathbf{K}_m \nabla \Phi_m = q_m, \tag{7}$$

Assuming that the pressure along the fracture width are constants, and by integration, the pressure equation in the fracture becomes,

$$-\nabla \cdot \mathbf{K}_f \nabla \Phi_f = q_f + Q_f, \tag{8}$$

The matrix-fracture interface condition is given by,

$$\Phi_m = \Phi_f, \quad \text{on} \quad \partial\Omega_m \cap \Omega_f. \tag{9}$$

where the subscript m represents the matrix domain, while the subscript f represents the fracture domain. Q_f is the mass transfer across the matrix-fracture interfaces.

The solute transport equation in the matrix domain may be expressed as,

$$\phi_m \frac{\partial C_m}{\partial t} + \nabla \cdot (\mathbf{u}_m C_m - \phi_m D \nabla C_m) = Q_{c,m}, \tag{10}$$

where C_m is the concentration in the matrix domain. $Q_{c,m}$ is the rate of change of volume belonging to a source/sink term in the matrix domain. On the other hand, the solute transport equation in fractures is represented by,

$$\phi_f \frac{\partial C_f}{\partial t} + \nabla \cdot (\mathbf{u}_f C_f - \phi_f D \nabla C_f) = Q_{c,f} + Q_{c,f}, \tag{11}$$

where C_f is the mass concentration in the fracture domain. $Q_{c,f}$ is the rate of change of particle belonging to a source/sink term in the fracture domain. $Q_{c,f}$ represents the rate of change of volume across the matrix-fracture interfaces. The interface condition of the solute concentration is,

$$C_m = C_f, \quad \text{on} \quad \partial\Omega_m \cap \Omega_f. \tag{12}$$

3   Multiscale Time-Splitting and Spatial Discretization

In the multiscale time-splitting method, we employ a different time step-size for each time derivative term as they have different physics. For example, the time-step size for the pressure can be larger than it of the solute concentration. Also, the fractures pressure may be has a larger time-step size than one for the matrix. So, we may use a small time step-size for the pressure in fractures, and so on. We use the CCFD method for the spatial discretization. The CCFD method is locally conservative and equivalent to the quadratic mixed finite element method.

3.1   Multiscale Time-Splitting Approach for Pressure

Now, let us introduce the time discretization for the pressure in the matrix domain. The total time interval $[0, T]$ is divided into $N_{p,m}$ steps, i.e., $0 = t^0 < t^1 < \cdots < t^{N_{p,m}} = T$ and the time step length is $\Delta t^i = t^{i+1} - t^i$. Therefore, we divide each subinterval $(t^i, t^{i+1}]$ into $N_{p,f}$ sub-subintervals as $(t^i, t^{i+1}] = \bigcup_{j=0}^{N_{p,f}-1}(t^{i,j}, t^{i,j+1}]$, where $t^{i,0} = t^i$ and $t^{i,N_{p,f}} = t^{i+1}$ and $\Delta t^{i,j} = t^{i,j+1} - t^{i,j}$. In the following, b refers to the boundary of the matrix gridcells K such that its area is $|K|$, and $|b|$ is its length. $\mathbf{n}_b$ is a unit normal vector pointing from K to K' on each interface $b \in \partial K \cap \partial K'$. The flux across the boundary b of the gridcell K is denoted by ξ. $d_{K,b}$ is the distance from the central points of the cell K and the cell boundary b. $d_{K,K'}$ is distance between the central points of the cells K and K'. When b is located on the entire domain boundary, the pressure is provided by Dirichlet boundary conditions, $b \in \Gamma^D$. Otherwise, the Neumann conditions $b \in \partial\Omega^N$ is used to calculate fluxes.

The pressure equation in the matrix domain and fractures is written, respectively as,

$$- \nabla \cdot \mathbf{K}_m \nabla \Phi_m^{i+1} = q_m^{i+1}, \tag{13}$$

and

$$- \nabla \cdot \mathbf{K}_f \nabla \Phi_f^{i+1} = q_f^{i+1} + Q_f^{i+1}. \tag{14}$$

Now, applying the CCFD scheme on (13), one obtains,

$$\sum_{b \in \partial K} \xi_{a,m,b}^{i+1} = \mathbf{q}_{m,K}^{i+1}|\mathbf{K}|, \tag{15}$$

If $b \in \partial K \cap \partial K'$ and $b \nsubseteq \Omega_f$, the fluxes in (15) are given by,

$$\xi_{a,m,b}^{i+1} = -|\mathbf{b}|\chi_{t,b}^i \frac{\Phi_{m,K'}^{i+1} - \Phi_{m,K}^{i+1}}{d_{K,K'}}, \tag{16}$$

where $\chi_{t,b}^i$ is given by

$$\chi_{t,b}^i = \frac{d_{K,K'}\mathbf{K}_{m,K}\mathbf{K}_{m,K'}}{d_{K,b}\mathbf{K}_{m,K'} + d_{K',b}\mathbf{K}_{m,K}}, \tag{17}$$

On the other hand, if $b \in \Omega_f \cap \partial K \cap \partial K'$ and b is a gridcell of the fracture system, we have,

$$\xi_{a,m,b}^{i+1} \equiv \xi_{a,m,b,K}^{i+1} = -|b|\chi_{t,mf,b}^i \frac{\Phi_{f,b}^{i+1} - \Phi_{m,K}^{i+1}}{d_{K,b} + \frac{\epsilon}{2}}, \tag{18}$$

where $\chi_{t,mf,b}^i$ is given by,

$$\chi_{t,mf,b}^i = \frac{(d_{K,b} + \frac{\epsilon}{2})\mathbf{K}_{f,K}\mathbf{K}_{m,K}}{\frac{\epsilon}{2}\mathbf{K}_{m,K} + d_{K,b}\mathbf{K}_{f,K}}, \tag{19}$$

Similarly, let b be a gridcell of the fracture network. Equation (14) may be discretized by the CCFD method to get,

$$\sum_{\gamma \in \partial b} \xi_{a,f,\gamma}^{i+1} = q_{f,b}^{i+1}|b| + Q_{f,b}^{i+1}|b|, \tag{20}$$

where γ is the face of the gridcell b in the fracture network. ξ is the flux across the boundary γ of the fracture gridcell b. The matrix-fracture transfer is treated as a source term in the fracture system.

$$Q_{f,b}^{i+1} = -(Q_{f,b,K}^{i+1} + Q_{f,b,K'}^{i+1})/\epsilon, \tag{21}$$

$$Q_{f,b,K}^{i+1} = \xi_{a,m,b,K}^{i+1}, \tag{22}$$

$$Q_{f,b,K'}^{i+1} = \xi_{a,m,b,K'}^{i+1}, \tag{23}$$

where $\xi_{a,m,b}^{i+1}$ is defined in (18).

In the case of multiple fractures that connected by the interface γ. Assume that Λ_γ is the set of the fracture grid cells joint by γ. The mass conservation equation discretization is,

$$\sum_{b \in \Lambda_\gamma} \xi_{a,f,\gamma,e}^{i+1} = 0, \tag{24}$$

where $\xi_{a,f,\gamma,b} = \xi_{a,f,\gamma}|_{\gamma \in b}$ and $\xi_{c,f,\gamma,b} = \xi_{c,f,\gamma}|_{\gamma \in b}$.

The discretization of the total mass conservation of the matrix domain and the fractures network is represented as,

$$\begin{bmatrix} \mathbf{A}_{a,m,m}^i & \mathbf{A}_{a,m,f}^i \\ \mathbf{A}_{a,f,m}^i & \mathbf{A}_{a,f,f}^i \end{bmatrix} \begin{bmatrix} \Phi_m^{i+1} \\ \Phi_f^{i+1} \end{bmatrix} = \begin{bmatrix} \mathbf{Q}_{ac,m}^{i+1} \\ \mathbf{Q}_{ac,f}^{i+1} \end{bmatrix}. \tag{25}$$

The pressure equation in the fractures at each time subtime-step is given by,

$$-\nabla \cdot \mathbf{K}_f \nabla \Phi_f^{i,j+1} = q_f^{i,j+1} + Q_f^{i,j+1}. \tag{26}$$

It is obtained by using the CCFD scheme to (26) that

$$\sum_{\gamma \in \partial e} \xi_{a,f,\gamma}^{i,j+1} = q_{f,e}^{i,j+1} |e| + Q_{f,e}^{i,j+1} |e|, \tag{27}$$

The matrix-fracture transfer is given by,

$$Q_{f,b}^{i,j+1} = -(Q_{f,b,K}^{i,j+1} + Q_{f,b,K'}^{i,j+1})/\epsilon, \tag{28}$$

$$Q_{f,b,K}^{i,j+1} = \xi_{a,m,e,K}^{i,j+1}, \tag{29}$$

$$Q_{f,e,K'}^{i,j+1} = \xi_{a,m,b,K'}^{i,j+1}, \tag{30}$$

where

$$\xi_{a,m,b,K}^{i,j+1} \equiv \xi_{a,m,b}^{i,j+1} = -|e|\chi_{t,mf,b}^{i,j} \frac{\Phi_{f,b}^{i,j+1} - \Phi_{m,K}^{i+1}}{d_{K,b} + \frac{\epsilon}{2}}, \tag{31}$$

along with,

$$\chi_{t,mf,b}^{i,j} = \frac{(d_{K,b} + \frac{\epsilon}{2})\mathbf{K}_{f,K}\mathbf{K}_{m,K}}{\frac{\epsilon}{2}\mathbf{K}_{m,K} + d_{K,b}\mathbf{K}_{f,K}}, \tag{32}$$

For the case of multiple fractures, the pressure equation may be given as,

$$\mathbf{A}_f^{i,j} \Phi_f^{i,j+1} = \mathbf{Q}_f^{i,j}, \tag{33}$$

where

$$\mathbf{A}_f^{i,j} = \mathbf{A}_{a,f,f}^{i,j}, \tag{34}$$

and

$$\mathbf{Q}_f^{i,j} = \mathbf{Q}_{ac,f}^{i,j+1} - \mathbf{A}_{a,f,m}^{i,j} \Phi_m^{i+1}. \tag{35}$$

At the time step (t^i, t^{i+1}), (25) is solved implicitly to get Φ_m^{i+1}. Then, at the time substep $(t^{i,j}, t^{i,j+1})$, we compute $\Phi_f^{i,j+1}$ using (33). After that, we calculate fluxes as explained below. For the boundary b of the matrix gridcell K, $\mathbf{n}_b$ is the unit normal vector pointing towards outside K. If $e \in \partial K \cap \partial K'$ and $b \not\subseteq \Omega_f$,

$$\xi_{a,m,b}^{i,j+1} = -|b|\chi_{t,b}^{i,j} \frac{\Phi_{m,K'}^{i+1} - \Phi_{m,K}^{i+1}}{d_{K,K'}}, \tag{36}$$

where

$$\chi_{t,b}^{i,j} = \frac{d_{K,K'}\mathbf{K}_{m,K}\mathbf{K}_{m,K'}}{d_{K,b}\mathbf{K}_{m,K'} + d_{K',b}\mathbf{K}_{m,K}}. \tag{37}$$

If $b \in \Omega_f \cap \partial K \cap \partial K'$ and b is a fracture gridcell.

3.2   Multiscale Time-Stepping of the Concentration Equation

On the other hand, as the solute concentration vary more rapidly than the pressures. We also use a smaller time-step size for the concentration in matrix domain and the smallest time-step size for the concentration in fractures. The backward Euler time discretization is used for the equation of concentration. Therefore, the system of governing equations is solved based on the multiscale time-splitting technique. Now, let us divide the time-step $(t^{i,j}, t^{i,j+1}]$ of the fractures pressure into $N_{c,m}$ sub-steps such that $(t^{i,j}, t^{i,j+1}] = \bigcup_{k=0}^{N_{s,m}-1}(t^{i,j,k}, t^{i,j,k+1}]$, $t^{i,j,0} = t^{i,j}$ and $t^{i,j,N_{s,m}} = t^{i,j+1}$. This time discretization is employed for the concentration in the matrix domain. Moreover, we use a smaller time-step size for the fracture concentration. Thus, we partition the time-step, $(t^{i,j,k}, t^{i,j,k+1}]$ into $N_{c,f}$ time sub-steps as $(t^{i,j,k}, t^{i,j,k+1}] = \bigcup_{l=0}^{N_{c,f}-1}(t^{i,j,k,l}, t^{i,j,k,l+1}]$, where $t^{i,j,k,0} = t^{i,j,k}$ and $t^{i,j,k,N_{c,f}} = t^{i,j,k+1}$. The concentration is computed implicitly as follow,

$$\phi_m \frac{C_m^{i,j,k+1} - C_m^{i,j,k}}{\Delta t^{i,j,k}} + \nabla \cdot \left(\mathbf{u}_m^{i+1} C_m^{i,j,k+1} - \phi_m D\nabla C_m^{i,j,k+1} \right) = Q_{c,m}^{i,j,k+1} \quad (38)$$

In a similar manner, we consider variation of the concentration in the fractures are faster than those in the matrix domain. So, the concentration in the fractures is expressed as follow,

$$\phi_f \frac{C_f^{i,j,k,l+1} - C_f^{i,j,k,l}}{\Delta t^{i,j,k,l}} + \nabla \cdot \left\{ \mathbf{u}_f^{i+1} C_f^{i,j,k,l+1} - \phi_f D\nabla C_f^{i,j,k,l+1} \right\} = Q_{c,f}^{i,j,k,l+1} + Q_{c,m,f}^{i,j,k,l+1} \quad (39)$$

We use the upwind CCFD method to discretize the concentration Eq. (38),

$$|K|\phi_{m,K} \frac{C_{m,K}^{i,j,k+1} - C_{m,K}^{i,j,k}}{\Delta t^{i,j,k}} + \sum_{b \in \partial K} \hat{C}_{m,K}^{i,j,k+1} \mathbf{F}_{a,m,b}^{i,j+1} + \sum_{b \in \partial K} \mathbf{F}_{D,m,b}^{i,j,k+1} = Q_{c,m,K}^{i,j,k+1}|K|.$$
$$\quad (40)$$

where

$$\mathbf{F}_{m,b}^{i,j+1} = \mathbf{u}_{m,b}^{i,j+1}|b|$$

Let b be the interface between the matrix gridcells K and K'; that is, $b = \partial K \cap \partial K'$. If $b \not\subseteq \Omega_f$, the term $\hat{C}_{m,K}^{i,j,k+1}$ in (40) is given by,

$$\hat{C}_{m,K}^{i,j,k+1} = \begin{cases} C_{m,K}^{i,j,k}, & \mathbf{F}_{m,b}^{i,j+1} > 0, \\ C_{m,K'}^{i,j,k}, & \mathbf{F}_{m,b}^{i,j+1} < 0. \end{cases} \quad (41)$$

Now for the diffusion term; if $b \in \partial K \cap \partial K'$ and $b \not\subseteq \Omega_f$, the fluxes in (40) are given by,

$$\mathbf{F}_{D,m,b}^{i,j,k+1} = -|b|\chi_{t,b}^{i,j,k+1} \frac{C_{m,K'}^{i,j,k+1} - C_{m,K}^{i,j,k+1}}{d_{K,K'}}, \quad (42)$$

where $\chi_{t,b}^{i,j,k}$ is given by the harmonic mean as,

$$\chi_{t,b}^{i,j,k+1} = \frac{d_{K,K'} D_{m,K}^{i,j,k} D_{m,K'}^{i,j,k} \phi_{m,K}^{i,j,k} \phi_{m,K'}^{i,j,k}}{d_{K,b} D_{m,K'}^{i,j,k} \phi_{m,K'}^{i,j,k} + d_{K',b} D_{m,K}^{i,j,k} \phi_{m,K}^{i,j,k}}, \quad (43)$$

On the other hand, if $b \in \Omega_f \cap \partial K \cap \partial K'$ and b is a gridcell of the fracture system, we have,

$$\mathbf{F}_{D,m,b}^{i,j,k,l+1} \equiv \mathbf{F}_{D,m,b,K}^{i,j,k+1} = -|b|\chi_{t,\mathrm{mf},b}^{i,j,k}\frac{C_{f,b}^{i,j,k+1} - C_{m,K}^{i,j,k+1}}{d_{K,b} + \frac{\epsilon}{2}}, \tag{44}$$

where $\chi_{t,\mathrm{mf},b}^{i,j,k,l+1}$ is defined as,

$$\chi_{t,\mathrm{mf},b}^{i,j,k,l+1} = \frac{(d_{K,b} + \frac{\epsilon}{2})D_{m,K}^{i,j,k}D_{f,K}^{i,j,k}\phi_{m,K}^{i,j,k}\phi_{f,K}^{i,j,k}}{\frac{\epsilon}{2}D_{m,K}^{i,j,k}\phi_{m,K}^{i,j,k} + d_{K,b}D_{f,K}^{i,j,k}\phi_{f,K}^{i,j,k}}, \tag{45}$$

Table 1. Physical and computational parameters

Parameter	Example (1)	Example (2)	Example (3)
Domain dimensions (m)	$10 \times 10 \times 1$	$10 \times 10 \times 1$	$20 \times 15 \times 1$
Fracture aperture (m)	0.01	0.01	0.01
ϕ_m	0.2	0.2	0.15
ϕ_f	1	1	1
K_m (md)	1	1	50
K_f (md)	10^5	10^6	10^6
μ (cP)	1	1	1
Injection rate (PVI)	0.1	0.15	0.1
Total gridcells	2500	2500	3300
$N_{p,f}$	5	5	5
$N_{c,m}$	2	2	2
$N_{c,f}$	8	8	8
c_0	0.1	0.1	0.1

4   Numerical Tests

In order to examine the proposed scheme, three examples of fractured media with different dimensions, namely, $20\,\mathrm{m} \times 15\,\mathrm{m} \times 1\,\mathrm{m}$, $10\,\mathrm{m} \times 10\,\mathrm{m} \times 1\,\mathrm{m}$, and $10\,\mathrm{m} \times 10\,\mathrm{m} \times 1\,\mathrm{m}$ are presented with different multiple interconnected fractures as shown in Fig. 1. The matrix permeability is taken as 1 md in Examples (1) and (2), while in Example (3) is taken as 50 md. On the other hand, the fractures permeability is taken as 10^6 md for Examples (2) and (3), while in Example (1) is taken as 10^5 md. The total number of gridcells is 2500 in Examples (1) and (2), while in Example (3) is 3300. The solute is injected into a water aquifer with a rate of 0.1 PV with initial concentration of 0.1. The remaining of the physical and computational parameters are given for the three examples in Table 1. The outer (pressure) time-step size is taken as $\Delta t = 1.9$, while we chose, $N_{p,f} = 5$, $N_{c,m} = 2$ and $N_{c,f} = 8$ for the three examples.

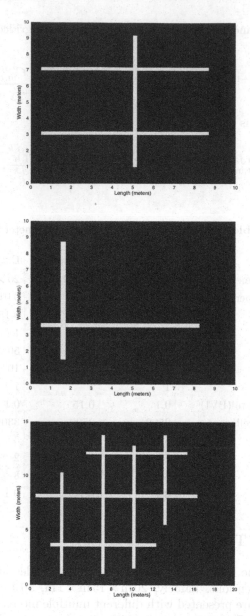

Fig. 1. Distribution of fractures: Example (1), Example (2), Example (3).

The distribution of the contours and profiles of solute concentration in the fractured medium of Example (1) at different dimensionless times 25, 35, 45 and 85 are shown in Fig. 2. This figure indicates that the solute-water mixture moves rapidly in the horizontal fractures due to their high permeability compare to the matrix permeability. The concentration profiles (shown in the right section

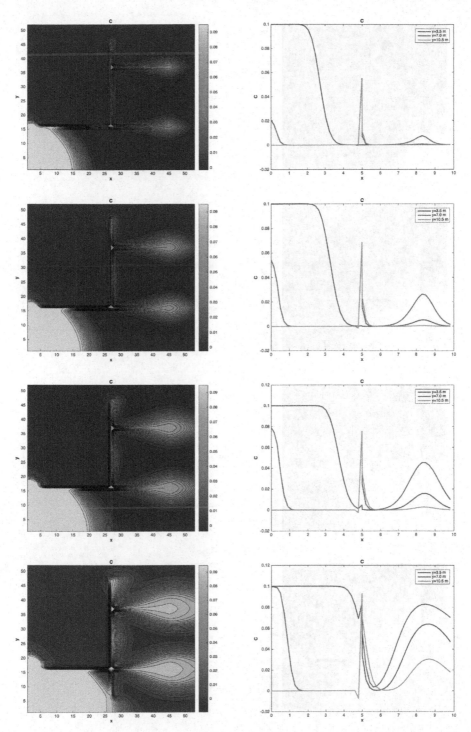

Fig. 2. Distribution of the contours and profiles of solute concentration at different dimensionless times 25, 35, 45 and 85.

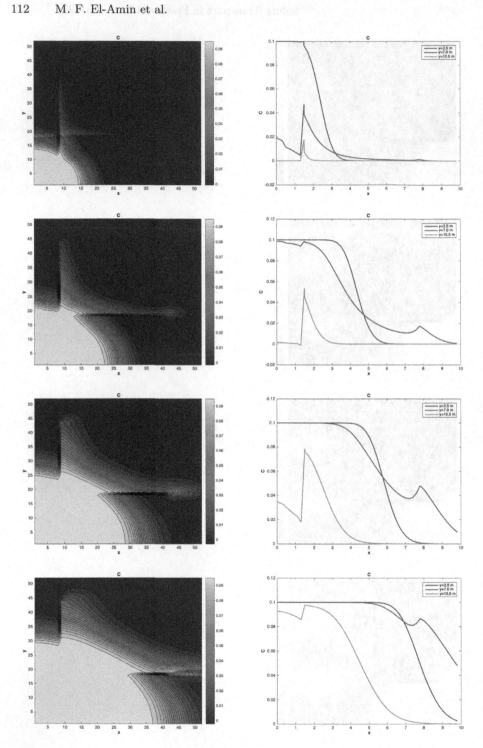

Fig. 3. Distribution of the contours and profiles of solute concentration at different dimensionless times 15, 35, 55 and 85.

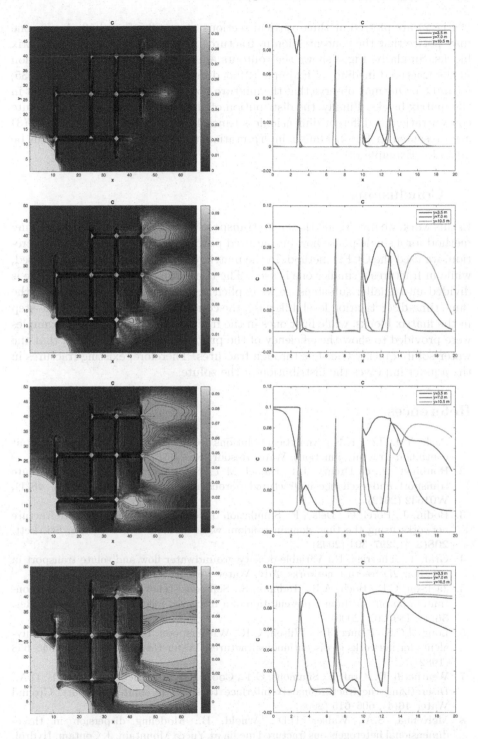

Fig. 4. Distribution of the contours and profiles of solute concentration at different dimensionless times 15, 25, 35 and 60.

of Fig. 2) are plotted at three vertical sections at $y = 3.5, 7$ and 10. Again, one may notice that the concentration in fractures is much higher than it in matrix blocks. Similarly, Fig. 3 shows the contours and profiles of solute concentration in the fractured medium of Example (2) at different dimensionless times 15, 35, 55 and 85. One may observe that the mixture flow in fractures is higher than it in the matrix blocks. Finally, the distribution of the contours and profiles of solute concentration at different dimensionless times 15, 25, 35 and 60 of Example (3) are presented in Fig. 4. Similar interpretation to Examples (1) and (2) can be given for Example (3).

5   Conclusions

In this work, we investigate the solute transport using a multi-scale time-stepping method for a single-phase flow in fractured porous media. For the spatial discretion, we used the CCFD method. In the matrix domain a large time-step is used, while in fractures a smaller one is used. The time-step of the fracture pressure is divided into smaller sub-steps, so we explicitly update the concentration at the same time-discretization level. Likewise, the concentration has a bigger time-step in the matrix blocks, while finer ones in the fractures. Three numerical examples were provided to show the efficiency of the proposed scheme. We found that the water-solute quickly transfers through fractures. Therefore, existing fractures in the aquifer improves the distribution of the solute.

References

1. Park, Y.J., Lee, K.K.: Analytical solutions for solute transfer characteristics at continuous fracture junctions. Water Resour. Res. **35**(5), 1531–1537 (1999)
2. Roubinet, D., de Dreuzy, J.R., Daniel, M.T.: Semi-analytical solutions for solute transport and exchange in fractured porous media. Water Resour. Res. **48**(1), W01542 (2012)
3. Bodin, J., Porel, G., Delay, F.: Simulation of solute transport in discrete fracture networks using the time domain random walk method. Earth Planet. Sci. Lett. **208**(3–4), 297–304 (2003)
4. Graf, T., Therrien, R.: Variable-density groundwater flow and solute transport in irregular 2D fracture networks. Adv. Water Resour. **30**(3), 455–468 (2007)
5. Jackson, C.P., Hoch, A.R., Todman, S.: Self-consistency of a heterogeneous continuum porous medium representation of a fractured medium. Water Resour. Res. **36**(1), 189–202 (2000)
6. Long, J.C.S., Remer, J.S., Wilson, C.R., Witherspoon, P.A.: Porous media equivalents for networks of discontinuous fractures. Water Resour. Res. **18**(3), 645–658 (1982)
7. Weatherill, D., Graf, T., Simmons, C.T., Cook, P.G., Therrien, R., Reynolds, D.A.: Discretizing the fracture-matrix interface to simulate solute transport. Ground Water **46**(4), 606–615 (2008)
8. McKenna, S.A., Walker, D.D., Arnold, B.: Modeling dispersion in three-dimensional heterogeneous fractured media at Yucca Mountain. J. Contam. Hydrol. **62–63**, 577–594 (2003)

9. Zhou, Z.F., Guo, G.X.: Numerical modeling for positive and inverse problem of 3-D seepage in double fractured media. J. Hydrodyn. **17**(2), 186–193 (2005)
10. Kou, J., Sun, S., Yu, B.: Multiscale time-splitting strategy for multiscale multiphysics processes of two-phase flow in fractured media. J. App. Math. **2011**, Article ID 861905, 24 pages (2011)
11. El-Amin, M.F., Khalid, U., Beroual, A.: Magnetic field effect on a ferromagnetic fluid flow and heat transfer in a porous cavity. Energies **11**(11), 3235 (2018)
12. El-Amin, M.F., Kou, J., Sun, S., Salama, A.: Adaptive time-splitting scheme for two-phase flow in heterogeneous porous media. Adv. Geo-Energy Res. **1**(3), 182–189 (2017)
13. El-Amin, M.F., Kou, J., Sun, S.: Adaptive time-splitting scheme for nanoparticles transport with two-phase flow in heterogeneous porous media. In: Shi, Y., et al. (eds.) ICCS 2018. LNCS, vol. 10862, pp. 366–378. Springer, Cham (2018). https://doi.org/10.1007/978-3-319-93713-7_30
14. El-Amin, M.F., Kou, J., Sun, S.: Discrete-fracture-model of multi-scale time-splitting two-phase flow including nanoparticles transport in fractured porous media. J. Comp. App. Math. **333**, 327–249 (2018)

Adaptive Multiscale Model Reduction for Nonlinear Parabolic Equations Using GMsFEM

Yiran Wang, Eric Chung$^{(\boxtimes)}$, and Shubin Fu

Department of Mathematics, The Chinese University of Hong Kong,
Shatin, Hong Kong SAR
tschung@math.cuhk.edu.hk

Abstract. In this paper, we propose a coupled Discrete Empirical Interpolation Method (DEIM) and Generalized Multiscale Finite element method (GMsFEM) to solve nonlinear parabolic equations with application to the Allen-Cahn equation. The Allen-Cahn equation is a model for nonlinear reaction-diffusion process. It is often used to model interface motion in time, e.g. phase separation in alloys. The GMsFEM allows solving multiscale problems at a reduced computational cost by constructing a reduced-order representation of the solution on a coarse grid. In [14], it was shown that the GMsFEM provides a flexible tool to solve multiscale problems by constructing appropriate snapshot, offline and online spaces. In this paper, we solve a time dependent problem, where online enrichment is used. The main contribution is comparing different online enrichment methods. More specifically, we compare uniform online enrichment and adaptive methods. We also compare two kinds of adaptive methods. Furthermore, we use DEIM, a dimension reduction method to reduce the complexity when we evaluate the nonlinear terms. Our results show that DEIM can approximate the nonlinear term without significantly increasing the error. Finally, we apply our proposed method to the Allen Cahn equation.

Keywords: Online adaptive model reduction · Discrete Empirical Interpolation Method · Flows in heterogeneous media · Exponential Time Differencing

1 Introduction

In this paper, we consider the Generalized Multiscale Finite element method (GMsFEM) for solving nonlinear parabolic equations. The main objectives of the paper are the following: (1) to demonstrate the main concepts of GMsFEM and brief review of the techniques; (2) to compare various online enrichment techniques; (3) to discuss the use of the Discrete Empirical Interpolation Method (DEIM) and present its performance in reducing complexity. GMsFEM is a flexible general framework that generalizes the Multiscale Finite Element Method

© Springer Nature Switzerland AG 2020
V. V. Krzhizhanovskaya et al. (Eds.): ICCS 2020, LNCS 12143, pp. 116–132, 2020.
https://doi.org/10.1007/978-3-030-50436-6_9

(MsFEM) by systematically enriching the coarse spaces. The main idea of this enrichment is to add extra basis functions that are needed to reduce the error substantially. Once the offline space is derived, it stays fixed and unchanged in the online stage. In [3,4], it is shown that a good approximation from the reduced model can be expected only if the offline information is a good representation of the problem. For time dependent problems, online enrichment is necessary. We compare two kinds of online enrichment methods: uniform and adaptive enrichment, where the latter focuses on where to add online basis. We will discuss it in numerical results with more details. When a general nonlinearity is present, the cost to evaluate the projected nonlinear function still depends on the dimension of the original system, resulting in simulation times that can hardly improve over the original system. One approach to reduce computational cost is the POD-Galerkin method [5–8], which is applied to many applications, for example, in [9–13]. DEIM focuses on approximating each nonlinear function so that a certain coefficient matrix can be precomputed and, as a result, the complexity in evaluating the nonlinear term becomes proportional to the small number of selected spatial indices. In this paper, we will compare various approximations of the DEIM projection. We will illustrate these concepts by applying our proposed method to the Allen Cahn equation. The remainder of the paper is organized as follows. In Sect. 2, we present the problem setting and main ingredients of GMsFEM. In Sect. 3, we consider the methods to solve the Allen-Cahn equation.

2 Multiscale Model Reduction Using the GMsFEM

In this section, we will give the construction of our GMsFEM for nonlinear parabolic equations. First, we present some basic notations and the coarse grid formulation in Sect. 2.1. Then, we present the construction of the multiscale snapshot functions and basis functions in Sect. 2.2. The online enrichment process is introduced in Sect. 2.3.

2.1 Preliminaries

Consider the following parabolic equation in the domain $\Omega \subset \mathbb{R}^d$

$$
\begin{aligned}
\frac{\partial u}{\partial t} - \mathrm{div}(\kappa \nabla u) &= f && \text{in } \Omega \times [0, T], \\
u(x, 0) &= g(x) && \text{in } \Omega, \\
u(x, t) &= 0 && \text{on } \partial\Omega \times [0, T].
\end{aligned}
\tag{1}
$$

Here, we denote the exact solution of (1) by u, $\kappa(x)$ is a high-contrast and heterogeneous permeability field, $f = f(x, u)$ is the nonlinear source function depending on the u variable, $g(x)$ is a given function and $T > 0$ is the final time. We denote the solution and the source term at $t = t_n$ by $u(\cdot, t_n)$ and $f(u(\cdot, t_n))$ respectively. The variational formulation for the problem (1) is: find $u(\cdot, t) \in H_0^1(\Omega)$ such that

$$\left\langle \frac{\partial u}{\partial t}, v \right\rangle + \mathcal{A}(u, v) = \langle f, v \rangle \quad \text{in } \Omega \times [0, T], \quad \forall v \in H_0^1(\Omega),$$

$$u(x, 0) = g(x) \quad \text{in } \Omega, \tag{2}$$

$$u(x, t) = 0 \quad \text{on } \partial\Omega \times [0, T].$$

where $\mathcal{A}(u, v) = \int_\Omega \kappa \nabla u \cdot \nabla v \, dx$.

In order to discretize (2) in time, we need to apply some time differencing methods. For simplicity, we first apply the implicit Euler scheme with time step $\Delta t > 0$ and in Sect. 3, we will consider the exponential time differencing method (ETD). We obtain the following discretization for each time $t_n = n\Delta t, n = 1, 2, \cdots, N$ $(T = N\Delta t)$,

$$\frac{u(\cdot, t_n) - u(\cdot, t_{n-1})}{\Delta t} = \text{div}(\kappa \nabla u(\cdot, t_n)) + f(u(\cdot, t_n)).$$

Let T^h be a partition of the domain Ω into fine finite elements. Here $h > 0$ is the fine grid mesh size. The coarse partition, T^H of the domain Ω, is formed such that each element in T^H is a connected union of fine-grid blocks. More precisely, $\forall K_j \in T^H$, $K_j = \bigcup_{F \in I_j} F$ for some $I_j \subset T^h$. The quantity $H > 0$ is the coarse mesh size. We will consider the rectangular coarse elements and the methodology can be used with general coarse elements. An illustration of the mesh notations is shown in the Fig. 1. We denote the interior nodes of T^H by $x_i, i = 1, \cdots, N_{\text{in}}$, where N_{in} is the number of interior nodes. The coarse elements of T^H are denoted by $K_j, j = 1, 2, \cdots, N_e$, where N_e is the number of coarse elements. We define the coarse neighborhood of the nodes x_i by $D_i := \cup \{K_j \in T_H : x_i \in \overline{K_j}\}$.

2.2   The GMsFEM and the Multiscale Basis Functions

In this paper, we will apply the GMsFEM to solve nonlinear parabolic equations. The method is motivated by the finite element framework. First, a variational formulation is defined. Then we construct some multiscale basis functions. Once the fine grid is given, we can compute the fine-grid solution. Let $\gamma_1, \cdots, \gamma_n$ be the standard finite element basis, and define $V_f = \text{span}\{\gamma_1, \cdots, \gamma_n\}$ to be the fine space. We obtained the fine solution denoted by u_f^n at $t = t_n$ by solving

$$\frac{1}{\Delta t} \langle u_f^n, v \rangle + \mathcal{A}(u_f^n, v) = \left\langle \frac{1}{\Delta t} u_f^{n-1} + f(u_f^n), v \right\rangle, \quad \forall v \in V_f,$$

$$u_f^0 = g_h, \tag{3}$$

where g_h is the V_f based approximation of g. The construction of multiscale basis functions follows two general steps. First, we construct snapshot basis functions in order to build a set of possible modes of the solutions. In the second step, we construct multiscale basis functions with a suitable spectral problem defined in the snapshot space. We take the first few dominated eigenfunctions as basis functions. Using the multiscale basis functions, we obtain a reduced model.

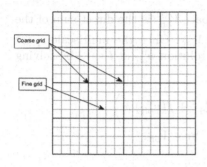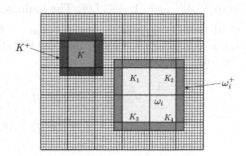

Fig. 1. Left: an illustration of fine and coarse grids. Right: an illustration of a coarse neighborhood, coarse element, and oversampled domain

More specifically, once the coarse and fine grids are given, one may construct the multiscale basis functions to approximate the solution of (2). To obtain the multiscale basis functions, we first define the snapshot space. For each coarse neighborhood D_i, define $J_h(D_i)$ as the set of the fine nodes of T^h lying on ∂D_i and denote the its cardinality by $L_i \in \mathbb{N}^+$. For each fine-grid node $x_j \in J_h(D_i)$, we define a fine-grid function δ_j^h on $J_h(D_i)$ as $\delta_j^h(x_k) = \delta_{j,k}$. Here $\delta_{j,k} = 1$ if $j = k$ and $\delta_{j,k} = 0$ if $j \neq k$. For each $j = 1, \cdots, L_i$, we define the snapshot basis functions $\psi_j^{(i)}$ ($j = 1, \cdots, L_i$) as the solution of the following system

$$-\mathrm{div}\left(\kappa \nabla \psi_j^{(i)}\right) = 0 \quad \text{in } D_i$$
$$\psi_j^{(i)} = \delta_j^h \quad \text{on } \partial D_i. \tag{4}$$

The local snapshot space $V_{snap}^{(i)}$ corresponding to the coarse neighborhood D_i is defined as follows $V_{snap}^{(i)} := \mathrm{span}\{\psi_j^{(i)} : j = 1, \cdots, L_i\}$ and the snapshot space reads $V_{snap} := \bigoplus_{i=1}^{N_{in}} V_{snap}^{(i)}$. In the second step, a dimension reduction is performed on V_{snap}. For each $i = 1, \cdots, N_{in}$, we solve the following spectral problem:

$$\int_{D_i} \kappa \nabla \phi_j^{(i)} \cdot \nabla v = \lambda_j^{(i)} \int_{D_i} \hat{\kappa} \phi_j^{(i)} v \quad \forall v \in V_{snap}^{(i)}, \quad j = 1, \ldots, L_i \tag{5}$$

where $\hat{\kappa} := \kappa \sum_{i=1}^{N_{in}} H^2 |\nabla \chi_i|^2$ and $\{\chi_i\}_{i=1}^{N_{in}}$ is a set of partition of unity that solves the following system:

$$-\nabla \cdot (\kappa \nabla \chi_i) = 0 \quad \text{in } K \subset D_i$$
$$\chi_i = p_i \quad \text{on each } \partial K \text{ with } K \subset D_i$$
$$\chi_i = 0 \quad \text{on } \partial D_i$$

where p_i is some polynomial functions and we can choose linear functions for simplicity. Assume that the eigenvalues obtained from (5) are arranged in ascending order and we may use the first $1 < l_i \leq L_i$ (with $l_i \in \mathbb{N}^+$) eigenfunctions (related to the smallest l_i eigenvalues) to form the local multiscale space $V_{\mathrm{off}}^{(i)} :=$

snap$\{\chi_i \phi_j^{(i)} : j = 1, \cdots, L_i\}$. The mulitiscale space $V_{\text{off}}^{(i)}$ is the direct sum of the local mulitiscale spaces, namely $V_{\text{off}} := \bigoplus_{i=1}^{N_{\text{in}}} V_{\text{off}}^{(i)}$. Once the multiscale space V_{off} is constructed, we can find the GMsFEM solution u_{off}^n at $t = t_n$ by solving the following equation

$$\frac{1}{\Delta t} \langle u_{\text{off}}^n, v \rangle + \mathcal{A}(u_{\text{off}}^n, v) = \left\langle \frac{1}{\Delta t} u_{\text{off}}^{n-1} + f(u_{\text{off}}^n), v \right\rangle, \tag{6}$$

$$\langle u_{\text{off}}^0, v \rangle = \langle g, v \rangle, \quad \forall v \in V_{\text{off}}.$$

2.3 Online Enrichment

We will present the constructions of online basis functions [1] in this section.

Online Adaptive Algorithm. In this subsection, we will introduce the method of online enrichment. After obtaining the multiscale space V_{off}, one may add some online basis functions based on local residuals. Let $u_{\text{off}}^n \in V_{\text{off}}$ be the solution obtained in (6) at time $t = t_n$. Given a coarse neighborhood D_i, we define $V_i := H_0^1(D_i) \cap V_{\text{snap}}$ equipped with the norm $\|v\|_{V_i}^2 := \int_{D_i} \kappa |\nabla v|^2$. We also define the local residual operator $R_i^n : V_i \to \mathbb{R}$ by

$$\mathcal{R}_i^n (v; u_{\text{off}}^n) := \int_{D_i} \left(\frac{1}{\Delta t} u_{\text{off}}^{n-1} + f(u_{\text{off}}^n) \right) v - \int_{D_i} \left(\kappa \nabla u_{\text{off}}^n \cdot \nabla v + \frac{1}{\Delta t} u_{\text{off}}^n v \right), \quad \forall v \in V_i. \tag{7}$$

The operator norm R_i^n, denoted by $\|R_i^n\|_{V_i^*}$, gives a measure of the quantity of residual. The online basis functions are computed during the time-marching process for a given fixed time $t = t_n$, contrary to the offline basis functions that are pre-computed.

Suppose one needs to add one new online basis ϕ into the space V_i. The analysis in [1] suggests that the required online basis $\phi \in V_i$ is the solution to the following equation

$$\mathcal{A}(\phi, v) = \mathcal{R}_i^n (v; u_{\text{off}}^{n,\tau}) \quad \forall v \in V_i. \tag{8}$$

We refer to $\tau \in \mathbb{N}$ as the level of the enrichment and denote the solution of (6) by $u_{\text{off}}^{n,\tau}$. Remark that $V_{\text{off}}^{n,0} := V_{\text{off}}$ for time level $n \in \mathbb{N}$. Let $\mathcal{I} \subset \{1, 2, \ldots, N_{in}\}$ be the index set over some non-lapping coarse neighborhoods. For each $i \in \mathcal{I}$, we obtain a online basis $\phi_i \in V_i$ by solving (8) and define $V_{\text{off}}^{n,\tau+1} = V_{\text{off}}^{n,\tau} \oplus$ span $\{\phi_i : i \in \mathcal{I}\}$. After that, solve (6) in $V_{\text{off}}^{n,\tau+1}$.

Two Online Adaptive Methods. In this section, we compare two ways to obtain online basis functions which are denoted by online adaptive method 1 and online adaptive method 2 respectively. Online adaptive method 1 is adding online basis using online adaptive method from offline space in each time step, which means basis functions obtained in last time step are not used in current time step. Online adaptive method 2 is keeping online basis functions in each time step. Using this accumulation strategy, we can skip online enrichment after a certain time period when the residual defined in (7) is under given tolerance. We also presents the results of these two methods in Fig. 3 and Fig. 4 respectively.

Numerical Results. In this section, we present some numerical examples to demonstrate the efficiency of our proposed method. The computational domain is $\Omega = (0,1)^2 \subset \mathbb{R}^2$ and $T = 1$. The medium κ_1 and κ_2 are shown in Fig. 2, where the contrasts are 10^4 and 10^5 for κ_1 and κ_2 respectively. Without special descriptions, we use κ_1.

For each function to be approximated, we define the following quantities e_a^n and e_2^n at $t = t_n$ to measure energy error and L^2 error respectively.

$$e_a^n = \frac{\|u_f^n - u_{\text{off}}^n\|_{V(\Omega)}}{\|u_f^n\|_{V(\Omega)}} \qquad e_2^n = \frac{\|u_f^n - u_{\text{off}}^n\|_{L^2(\Omega)}}{\|u_f^n\|_{L^2(\mathcal{D})}}$$

where u_f^n is the fine-grid solution (reference solution) and u_{off}^n is the approximation obtained by the GMsFEM method. We define the energy norm and L^2 norm of u by

$$\|u\|_{V(\Omega)}^2 = \int_\Omega \|\nabla u\|^2 \qquad \|u\|_{L^2}^2 = \int_\Omega \|u\|^2.$$

Example 2.1. In this example, we compare the error using adaptive online method 1 and uniform enrichment under different numbers of initial basis functions. We set the mesh size to be $H = 1/16$ and $h = 1/256$. The time step is $\Delta t = 10^{-3}$ and the final time is $T = 1$. The initial condition is $u(x,y,t)|_{t=0} = 4(0.5 - x)(0.5 - y)$. We set the permeability to be κ_1. We set the source term $f = \frac{1}{\epsilon^2}(u^3 - u)$, where $\epsilon = 0.01$. We present the numerical results for the GMsFEM at time $t = 0.1$ in Table 1, 2, and 3. For comparison, we present the results where online enrichment is not applied in Table 4. We observe that the adaptive online enrichment converges faster. Furthermore, as we compare Table 4 and Table 1, we note that the online enrichment does not improve the error if we only have one offline basis function per neighborhood. Because the first eigenvalue is small, the error decreases in the online iteration is small. In particular, for each iteration, the error decrease slightly. As we increase the number of initial offline basis, the convergence is very fast and one online iteration is sufficient to reduce the error significantly.

Example 2.2. We compare online Method 1 and 2 under different tolerance. We keep H, h and the initial condition the same as in Example 2.1. We choose intial number of basis to be 450, which means we choose two initial basis per neighborhood. We keep the source term as $f = \frac{1}{\epsilon^2}(u^3 - u)$. When $\epsilon = 0.01$, we choose the time step Δt to be 10^{-4}. We plot the error and DOF from online Method 1 in Fig. 3 and compare with results from online Method 2 in Fig. 4. From Fig. 3 and 4, we can see the error and DOF reached stability at $t = 0.01$. In Fig. 4, we can see the DOF keeps increasing before turning steady. The error remains at a relatively low level without adding online basis after some time. As a cost, online method 2 suffers bigger errors than method 1 with same tolerance. We also apply our online adaptive method 2 under permeability κ_2 in Fig. 5. The errors are relatively low for two kinds of permeability.

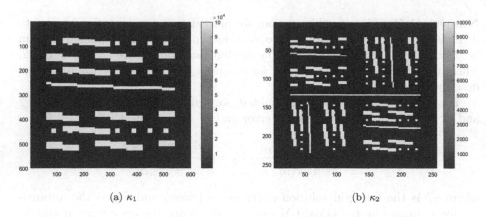

(a) κ_1 (b) κ_2

Fig. 2. Permeability field

Table 1. The errors for online enrichment when number of initial basis = 1. Left: Adaptive enrichment. Right: Uniform enrichment.

DOF	e_a	e_2	DOF	e_a	e_2
225	14.47%	19.55%	225	14.48%	19.56%
460	2.23%	1.14%	450	8.39%	6.54%
550	1.20%	0.6%	675	2.45%	1.1%

Table 2. The errors for online enrichment when number of initial basis = 2. Left: Adaptive enrichment. Right: Uniform enrichment

DOF	e_a	e_2	DOF	e_a	e_2
450	4.66%	2.64%	450	4.65%	2.64%
681	1.65%	0.52%	675	1.10%	0.669%

Table 3. The errors for online enrichment when number of initial basis = 3. Left: Adaptive enrichment. Right: Uniform enrichment

DOF	e_a	e_2	DOF	e_a	e_2
675	2.89%	1.07%	675	2.89%	1.07%
903	0.944%	0.511%	900	1.13%	0.894%

Table 4. The errors for different ϵ in source term without online enrichment. Up: Energy error. Down: L^2 error

Source function	$t = 0.1$	$t = 0.2$
$\epsilon = 0.1$	5.97%	5.94%
$\epsilon = 0.01$	15.1%	15.3%
Source function	$t = 0.1$	$t = 0.2$
$\epsilon = 0.1$	4.57%	4.57%
$\epsilon = 0.01$	11.9%	12.0%

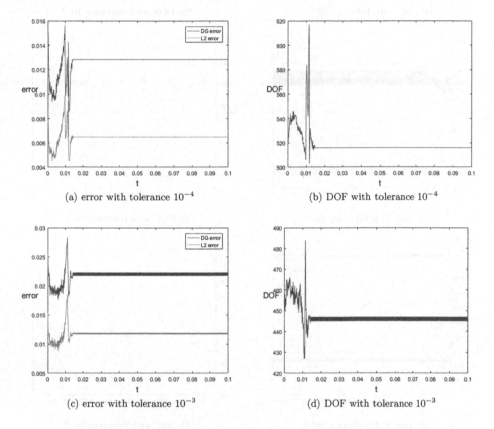

(a) error with tolerance 10^{-4}

(b) DOF with tolerance 10^{-4}

(c) error with tolerance 10^{-3}

(d) DOF with tolerance 10^{-3}

Fig. 3. Error and DOF obtained by online method 1 in Example 2.2

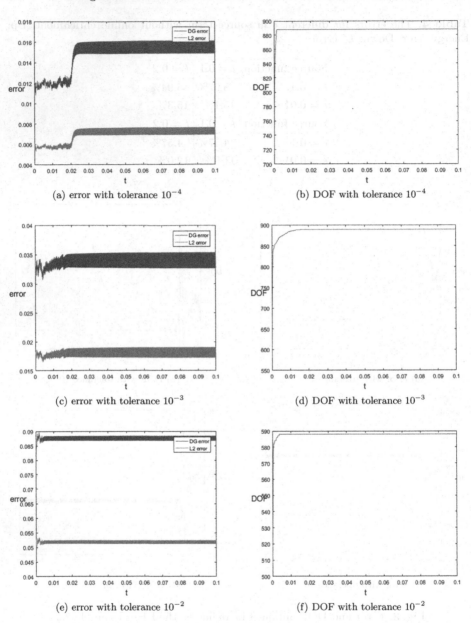

Fig. 4. Error and DOF obtained by online method 2 in Example 2.2

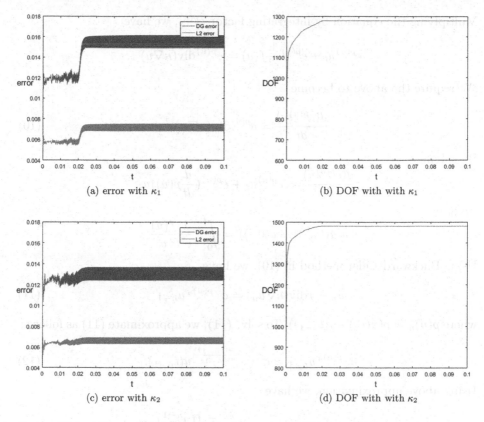

(a) error with κ_1

(b) DOF with with κ_1

(c) error with κ_2

(d) DOF with with κ_2

Fig. 5. Error and DOF obtained by online method 2 in Example 2.2

3 Application to the Allen-Cahn Equation

In this section, we apply our proposed method to the Allen Cahn equation. We use the Exponential Time Differencing (ETD) for time dsicretization. To deal with the nonlinear term, DEIM is applied. We will present the two methods in the following subsections.

3.1 Derivation of Exponential Time Differencing

Let τ be the time step. Using ETD, u_{off}^n is the solution to (9)

$$\langle u_{\text{off}}^n, v \rangle + \tau \mathcal{A}(u_{\text{off}}^n, v) = \langle \exp(-\frac{\tau}{\epsilon^2} \frac{f(u_{\text{off}}^{n-1})}{u_{\text{off}}^{n-1}}) u_{\text{off}}^{n-1}, v \rangle$$

$$\langle u_{\text{off}}^0, v \rangle = \langle g, v \rangle \quad \forall v \in V_{\text{off}} \tag{9}$$

Next, we will derive this equation. We have

$$u_t - \text{div}(\kappa \nabla u) + \frac{1}{\epsilon^2} f(u) = 0$$

Multiplying the equation by integrating factor $e^{p(u)}$, we have

$$e^{p(u)}u_t + e^{p(u)}\frac{1}{\epsilon^2}f(u) = e^{p(u)}\text{div}(\kappa\nabla u)$$

We require the above to become

$$\frac{d(e^{p(u)}u)}{dt} = e^{p(u)}\text{div}(\kappa\nabla u) \tag{10}$$

By solving

$$\frac{d(e^{p(u)}u)}{dt} = e^{p(u)}u_t + e^{p(u)}(\frac{d}{dt}p(u))u,$$

we have

$$p(u(t_n, \cdot)) - u(0, \cdot)) = \int_0^{t_n}\frac{1}{\epsilon^2}\frac{f(u)}{u}.$$

Using Backward Euler method in (10), we have

$$u_n - \tau\text{div}(\kappa\nabla u_n) = e^{-p(u)_n}u_{n-1} \tag{11}$$

where $p(u)_n = p(u(t_n) - u(t_{n-1}))$. To solve (11), we approximate (11) as follows:

$$e^{-p(u)_n}u_{n-1} \approx e^{-\frac{\tau}{\epsilon^2}\frac{f(u(t_{n-1}))}{u(t_{n-1})}}u(t_{n-1}). \tag{12}$$

Using above approximation, we have

$$u_{\text{off}}^n - \tau\text{div}(\kappa\nabla u_{\text{off}}^n) = \exp(-\frac{\tau}{\epsilon^2}\frac{f(u_{\text{off}}^{n-1})}{u_{\text{off}}^{n-1}})u_{\text{off}}^{n-1}. \tag{13}$$

3.2 DEIM Method

When we evaluate the nonlinear term, the complexity is $O(\alpha(n)+c\cdot n)$, where α is some function and c is a constant. To reduce the complexity, we approximate local and global nonlinear functions with the Discrete Empirical Interpolation Method (DEIM) [2]. DEIM is based on approximating a nonlinear function by means of an interpolatory projection of a few selected snapshots of the function. The idea is to represent a function over the domain while using empirical snapshots and information at some locations (or components). The key to complexity reduction is to replace the orthogonal projection of POD with the interpolation projection of DEIM in the same POD basis.

We briefly review the DEIM. Let $f(\tau)$ be the nonlinear function. We are desired to find an approximation of $f(\tau)$ at a reduced cost. To obtain a reduced order approximation of $f(\tau)$, we first define a reduced dimentional space for it. We would like to find m basis vectors (where m is much smaller than n), $\phi_1, \cdots, \phi_m$, such that we can write

$$f(\tau) = \Phi d(\tau),$$

where $\Phi = (\phi_1, \cdots, \phi_m)$. We employ POD to obtain Φ and use DEIM (refer Table 5) to compute $d(\tau)$ as follows. In particular, we solve $d(\tau)$ by using m rows of Φ. This can be formalized using the matrix P

$$P = [e_{\wp_1}, \ldots, e_{\wp_m}] \in \mathbb{R}^{n \times m},$$

where $e_{\wp_i} = [0, \cdots, 1, 0, \cdots, 0] \in \mathbb{R}^n$ is the $\wp_i^{th}$ column of the identity matrix $I_n \in \mathbb{R}^{n \times n}$ for $i = 1, \cdots, m$. Using $P^T f(\tau) = P^T \Phi d(\tau)$, we can get the approximation for $f(\tau)$ as follows:

$$f(\tau) \approx \tilde{f}(\tau) = \Phi d(\tau) = \Phi \left(P^T \Phi \right)^{-1} P^T f(\tau)$$

Table 5. DEIM algorithm

DEIM	Algorithm
Input	$\Phi = (\phi_1, \cdots, \phi_m)$ obtained by applying POD on a sequence of n_s functions evaluations
Output	The interpolation indices $\vec{\lambda} = (\lambda_1, \cdots, \lambda_m)^T$
	1. Set $[\rho, \lambda_1] = \max\{|\phi_1|\}$
	2. Set $\Phi = [\phi_1]$, $P = [e_{\lambda_1}]$, and $\vec{\lambda} = (\lambda_1)$
	3. for $i = 2, \cdots, m$, do
	Solve $(P^T \Phi)w = P^T \phi_i$ for some i
	Compute $r = \phi_i - \Phi w$
	Compute $[\rho, \lambda_i] = \max\{|r|\}$
	Set $\Phi = [\Phi, \phi_i]$, $P = [P, e_{\lambda_i}]$, and $\vec{\lambda} = \begin{pmatrix} \vec{\lambda} \\ \lambda_i \end{pmatrix}$
	end for

3.3   Numerical Results

Example 3.3. In this example, we apply the DEIM under the same setting as in Example 2.2 and we did not use the online enrichment procedure. We compare the results in Fig. 8. To test the DEIM, we first consider the solution using DEIM where the snapshot are obtained by the same equation. First, we set $\epsilon = 0.01$. We first solve the same equation and obtain the snapshot Φ. Secondly, we use DEIM to solve the equation again. The two results are presents in Fig. 6. The first picture are the errors we get when DEIM are not used while used in second one. The errors of these two cases differs a little since the snapshot obtained in the same equation. Then we consider the cases where the snapshots are obtained:

1. Different right hand side functions.
2. Different initial conditions.
3. Different permeability field.
4. Different time steps.

Different Right Hand Side. Since the solution for different ϵ can have some similarities, we can use the solution from one to solve the other. In particular, since it will be more time-consuming to solve the case when ϵ is smaller. We can use the $f(u)$ for $\epsilon = 0.09$ to compute the solution for $\epsilon = 0.1$ since solutions for these two cases can only vary a little. I show the results in Fig. 7.

Different Initial Conditions. In this section, we consider using the snapshot from different initial conditions, we record the results in Fig. 9. We first choose the initial condition to be compared Fig. 9 and Fig. 6, we can see that different initial conditions can have less impact on the final solution since the solution is close to the one where the snapshot is obtained in the same equation.

Different Permeability Field. In this section, we consider using the snapshot from different permeability, we record the results in Fig. 10. For reference, the first two figures plots the fine solution and multiscale solution without using DEIM. And we construct snapshot from another permeability κ_1 and we apply it to compute the solution in κ_2. The last figure shows the of using DEIM is relatively small.

Different Time Steps. In this section, we construct the snapshot by using nonlinear function obtained in previous time step for example when $t < 0.05$. Then we apply it to DEIM to solve the equation in $0.05 < t < 0.06$. We use these way to solve the equation with permeability κ_1 and κ_2 respectively. We plot the results in Fig. 11 and 12. From these figures, we can see that DEIM have different effects applied to different permeability. With κ_1, the error increases significantly when DEIM are applied. But with κ_2, the error decreased to a lower level when we use DEIM.

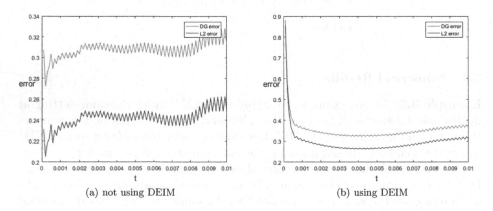

(a) not using DEIM   (b) using DEIM

Fig. 6. Error for same ϵ

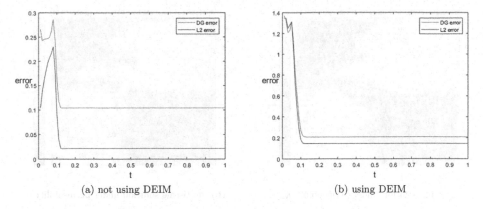

(a) not using DEIM   (b) using DEIM

Fig. 7. Error for different ϵ

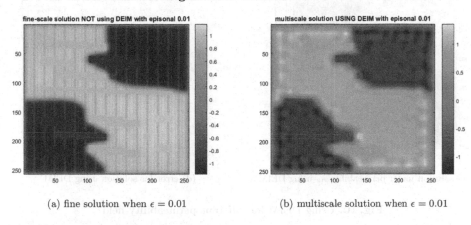

(a) fine solution when $\epsilon = 0.01$ (b) multiscale solution when $\epsilon = 0.01$

Fig. 8. Comparing fine and multiscale solutions.

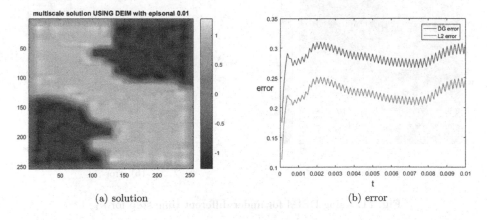

(a) solution   (b) error

Fig. 9. Using DEIM for different initial conditions

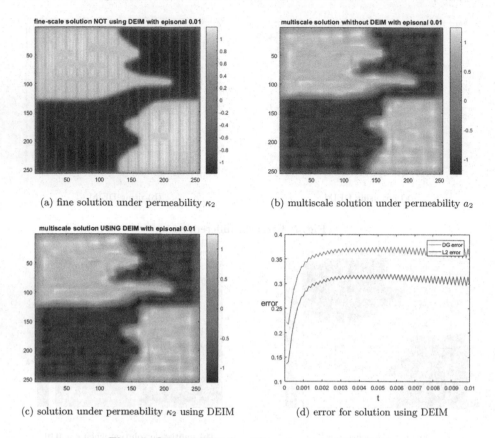

(a) fine solution under permeability κ_2

(b) multiscale solution under permeability a_2

(c) solution under permeability κ_2 using DEIM

(d) error for solution using DEIM

Fig. 10. Using DEIM for different permeability field

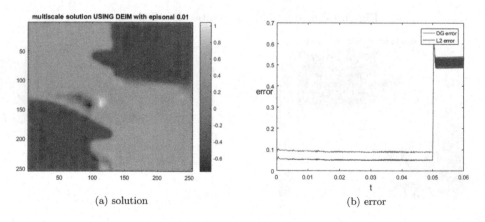

(a) solution

(b) error

Fig. 11. Using DEIM for under different time step for κ_1

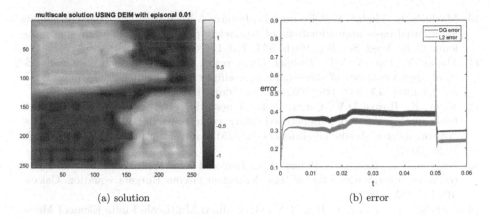

(a) solution

(b) error

Fig. 12. Using DEIM for under different time step for κ_2

Acknowledgement. The research of Eric Chung is partially supported by the Hong Kong RGC General Research Fund (Project numbers 14302018 and 14304719) and CUHK Faculty of Science Direct Grant 2018-19.

References

1. Chung, E.T., Efendiev, Y., Leung, W.T.: Residual-driven online generalized multiscale finite element methods. J. Comput. Phys. **302**, 176–190 (2015)
2. Chaturantabut, S., Sorensen, D.C.: Nonlinear model reduction via discrete empirical interpolation. SIAM J. Sci. Comput. **32**(5), 2737–2764 (2010)
3. Chung, E., Efendiev, Y., Hou, T.Y.: Adaptive multiscale model reduction with generalized multiscale finite element methods. J. Comput. Phys. **320**, 69–95 (2016)
4. Efendiev, Y., Gildin, E., Yang, Y.: Online adaptive local-global model reduction for flows in heterogeneous porous media. Computation **4**, 22 (2016)
5. Hinze, M., Volkwein, S.: Proper orthogonal decomposition surrogate models for nonlinear dynamical systems: error estimates and suboptimal control. In: Benner, P., Sorensen, D.C., Mehrmann, V. (eds.) Dimension Reduction of Large-Scale Systems. LNCS, vol. 45, pp. 261–306. Springer, Berlin (2005). https://doi.org/10.1007/3-540-27909-1_10
6. Kunisch, K., Volkwein, S.: Control of the Burgers equation by a reduced-order approach using proper orthogonal decomposition. J. Optim. Theory Appl. **102**, 345–371 (1999). https://doi.org/10.1023/A:1021732508059
7. Graham, W.R., Peraire, J., Tang, K.Y.: Optimal control of vortex shedding using loworder models. Part I-Open-loop model development. Int. J. Numer. Methods Eng. **44**, 945–972 (1999)
8. Ravindran, S.S.: A reduced-order approach for optimal control of fluids using proper orthogonal decomposition. Int. J. Numer. Methods Fluids **34**, 425–448 (2000)
9. Prud'homme, C., et al.: Reliable real-time solution of parametrized partial differential equations: reduced-basis output bound methods. J. Fluids Eng. **124**, 70–80 (2002)

10. Machiels, L., Maday, Y., Oliveira, I.B., Patera, A.T., Rovas, D.V.: Output bounds for reduced-basis approximations of symmetric positive definite eigenvalue problems. C. R. Acad. Sci. Ser. Math. **331**, 153–158 (2000)

11. Maday, Y., Patera, A.T., Turinici, G.: A priori convergence theory for reduced-basis approximations of single-parameter elliptic partial differential equations. J. Sci. Comput. **17**, 437–446 (2002). https://doi.org/10.1023/A:1015145924517

12. Veroy, K., Rovas, D.V., Patera, A.T.: A posterior error estimation for reduced-basis approximation of parametrized elliptic coercive partial differential equations: "convex inverse" bound conditioners. ESAIM Control Optim. Calc. Var. **8**, 1007–1028 (2002)

13. Nguyen, N.C., Rozza, G., Patera, A.T.: Reduced basis approximation dand a posteriori error estimationd for the time-dependent viscous Burgers' equation. Calcolo **46**, 157–185 (2009)

14. Efendiev, Y., Galvis, J., Hou, T.Y.: Generalized Multiscale Finite Element Methods. arXiv:1301.2866. http://arxiv.org/submit/631572

Parallel Shared-Memory Isogeometric Residual Minimization (iGRM) for Three-Dimensional Advection-Diffusion Problems

Marcin Łoś[1], Judit Munoz-Matute[2], Krzysztof Podsiadło[1], Maciej Paszyński[1(✉)], and Keshav Pingali[3]

[1] AGH University of Science and Technology, Kraków, Poland
maciej.paszynski@agh.edu.pl
[2] The University of the Basque Country, Bilbao, Spain
[3] The University of Texas at Austin, Austin, USA

Abstract. In this paper, we present a residual minimization method for three-dimensional isogeometric analysis simulations of advection-diffusion equations. First, we apply the implicit time integration scheme for the three-dimensional advection-diffusion equation. Namely, we utilize the Douglas-Gunn time integration scheme. Second, in every time step, we apply the residual minimization method for stabilization of the numerical solution. Third, we use isogeometric analysis with B-spline basis functions for the numerical discretization. We perform alternating directions splitting of the resulting system of linear equations, so the computational cost of the sequential LU factorization is linear $\mathcal{O}(N)$. We test our method on the three-dimensional simulation of the advection-diffusion problem. We parallelize the solver for shared-memory machine using the GALOIS framework.

Keywords: Isogeometric analysis · Implicit dynamics · Advection-diffusion problems · Linear computational cost · Direct solvers · GALOIS framework

1 Introduction

The alternating direction implicit method (ADI) is a popular method for performing finite difference simulations on regular grids. The first papers concerning the ADI method were published in 1960 [1,3,5,19]. This method is still popular for fast solutions of different classes of problems with finite difference method [8,9]. In its basic version, the method introduces intermediate time steps, and the differential operator splits into the x, y (and z in 3D) components. As a result of this operation, on the left-hand side, we only deal with derivatives in one direction, while the rest of the operator is on the right-hand side. The resulting system of linear equations has a multi-diagonal form, so the factorization of this system is possible with a linear $\mathcal{O}(N)$ computational cost. It is a common misunderstanding that the direction splitting solvers are limited to simple geometries.

© Springer Nature Switzerland AG 2020
V. V. Krzhizhanovskaya et al. (Eds.): ICCS 2020, LNCS 12143, pp. 133–148, 2020.
https://doi.org/10.1007/978-3-030-50436-6_10

They can be also applied to discretizations in extremely complicated geometries, as described in [10].

In this paper, we generalize this method for three-dimensional simulations of the time-dependent advection-diffusion problem with the residual minimization method. We use the basic version of the direction splitting algorithm, working on a regular computational cube, since this approach is straightforward and it is enough to proof our claims that the residual minimization stabilizes the advection-diffusion simulations. In particular, we apply the residual minimization method with isogeometric finite element method simulations over a three-dimensional cube shape computational grids with tensor product B-spline basis functions. The resulting system of linear equations can be factorized in a linear $\mathcal{O}(N)$ computational cost when executed in sequential mode.

We use the finite element method discretizations with B-spline basis functions. This setup, as opposed to the traditional finite difference discretization, allows us to apply the residual minimization method to stabilize our simulations.

The isogeometric analysis (IGA) [4] is a modern method for performing finite element method (FEM) simulations with B-splines and NURBS. In enables higher order and continuity B-spline based approximations of the modeled phenomena. The direction splitting method has been rediscovered to solve the isogeometric L^2 projection problem over regular grids with tensor product B-spline basis functions [6,7]. The direction splitting, in this case, is performed with respect to space, and the splitting is possible by exploiting the Kronecker product structure of the Gram matrix with tensor product structure of the B-spline basis functions. The L^2 projections with IGA-FEM were applied for performing fast and smooth simulations of explicit dynamics [11–16,20]. This is because the explicit dynamics with isogeometric discretization is equivalent to the solution of a sequence of isogeometric L^2 projections.

In this paper, we focus on the advection-diffusion equation used for simulation of the propagation of a pollutant from a chimney. We introduce implicit time integration scheme, that allows for the alternating direction splitting of the advection-diffusion equation. We discover that the numerical simulations are unstable, and deliver some unexpected oscillations and reflections. Next, we utilize the residual minimization method in a way that it preserves the Kronecker product structure of the matrix and enables stabilized linear computational cost solutions.

The actual mathematical theory concerning the stability of the numerical method for weak formulations is based on the famous "Babuśka-Brezzi condition" (BBC) developed in years 1971–1974 at the same time by Ivo Babuśka, and Franco Brezzi [25–27]. The condition states that a weak problem is stable when

$$\sup_{v \in V} \frac{|b(u,v)|}{\|v\|_V} \geq \gamma \|u\|_U, \forall u \in U. \tag{1}$$

However, the inf-sup condition in the above form concerns the abstract formulation where we consider all the test functions from $v \in V$ and look for solution at $u \in U$ (e.g. $U = V$). The above condition is satisfied also if we restrict to the space of trial functions $u_h \in U_h \subset U$

$$\sup_{v \in V} \frac{|b(u_h, v)|}{\|v\|_V} \geq \gamma \|u_h\|_{U_h}. \tag{2}$$

However, if we use test functions from the finite dimensional test space $V_h = \text{span}\{v_h\} \subset V$

$$\sup_{v_h \in V_h} \frac{|b(u_h, v_h)|}{\|v_h\|_{V_h}} \geq \gamma_h \|u_h\|_{U_h}, \tag{3}$$

we do not have a guarantee that the supremum (3) will be equal to the original supremum (1), since we have restricted V to V_h. The optimality of the method depends on the quality of the polynomial test functions defining the space $V_h = \text{span}\{v_h\}$ and how far are they from the supremum defined in (1). There are many method for stabilization of different PDEs [28–31]. In 2010, the Discontinuous Petrov Galerkin (DPG) method was proposed, with the modern summary of the method described in [32].

The DPG method utilizes the residual minimization with broken test spaces. In other words, it first generates a system of linear equations

$$\begin{bmatrix} G & -B \\ B^T & 0 \end{bmatrix} \begin{bmatrix} r \\ u \end{bmatrix} = \begin{bmatrix} l \\ 0 \end{bmatrix}. \tag{4}$$

This system of linear equations has the inner product block G over the test space, the two blocks with the actual weak form B and B^T, and the zero block 0. The test space is larger than the trial space, and the inner product and the weak form blocks are rather sparse matrices. Therefore, the dimension of the system of linear equations is at least two times larger than the original system of equations arising from standard Galerkin method. In the DPG method, the test space is broken in order to obtain a block-diagonal matrix G and the Schur complements can be locally computed over each finite element. The price to pay is the presence of the additional fluxes on the element interfaces, resulting from breaking the test spaces, so the system over each finite element looks like

$$\begin{bmatrix} G & -B_1 & -B_2 \\ B_1^T & 0 & 0 \\ B_2^T & 0 & 0 \end{bmatrix} \begin{bmatrix} r \\ u \\ t \end{bmatrix} = \begin{bmatrix} l \\ 0 \\ 0 \end{bmatrix}. \tag{5}$$

We do not know any other reason of breaking the test spaces in the DPG method other then reduction of the computational cost of the solver.

In this paper, we want to avoid dealing with fluxes and broken spaces since it is technically very complicated. Thus, we stay with the unbroken global system (4) and then we have to face one of the two possible methods. The first one would be to apply adaptive finite element method, but then the cost of factorization in 3D would be up to four times slower than in the standard finite element method and broken DPG (without the static condensation). This is because depending on the structure of the refined mesh, we will have a computational cost of the multi-frontal solver varying between $\mathcal{O}(N)$ to $\mathcal{O}(N^2)$ [33], and our N is two times bigger than in the original weak problem, and $2^2 = 4$. This could be an option that we will discuss in a future paper.

Another method that we exploit in this paper is to keep a tensor product structure of the computational patch of elements with tensor product B-spline basis functions, decompose the system matrix into a Kronecker product structure, and utilize a linear computational cost alternating directions solver. Even for the system (4) resulting from the residual minimization we successfully perform direction splitting to obtain a Kronecker product structure of the matrix to maintain the linear computational cost of the alternating directions method.

In order the stabilize the time-dependent advection-diffusion simulations, we perform the following steps. First, we apply the time integration scheme. We use the Douglas-Gunn second order time integration scheme [2]. Second, we stabilize a system from every time step by employing the residual minimization method [34–36]. Finally, we perform numerical discretization with isogeometric analysis [4], using tensor product B-spline basis functions over a three-dimensional cube shape patch of elements.

The novelties of this paper with regard to our previous work are the following. In [11], we described parallel object-oriented JAVA based implementation of the explicit dynamics version of the alternating directions solver, without any residual minimization stabilization, and for two-dimensional problems only. In [12], we described sequential Fortran based implementation of the explicit dynamics solver, with applications of the elastic wave propagation, without implicit time integration schemes and any residual minimization stabilization. In [16], we described the parallel distributed memory implementation of the explicit dynamics solver, again without implicit time integration scheme and residual minimization method. In [14], we described the parallel shared-memory implementation of the explicit dynamics solver, with the same restrictions as before. In [13,17] we applied the explicit dynamics solver for two and three-dimensional tumor growth simulations. In all of these papers, we did not used implicit time integration schemes, and we did not perform operator splitting on top of the residual minimization method. In [20], we investigate different time integration schemes for two-dimensional residual minimization method for advection-diffusion problems. We do not go for three-dimensional computations, and we do not apply parallel computations there.

In this paper, we apply the residual minimization with direction splitting for the first time in three-dimensions. We also investigate the parallel scalability of our solver, using the GALOIS framework for parallelization. For more details on the GALOIS framework itself, we refer to [21–24].

The structure of this paper is the following. We start in Sect. 2 with the derivation of the isogeometric alternating direction implicit method for the advection-diffusion problem. The following Sect. 3 derives the residual minimization method formulation of the advection-diffusion problem in three-dimensions. Next, in Sect. 4, we present the linear computational cost numerical results. We summarize the paper with conclusions in Sect. 5.

2 Model Problem of Three-Dimensional Advection-Diffusion

Let $\Omega = \Omega_x \times \Omega_y \times \Omega_z \subset \mathbb{R}^3$ an open bounded domain and $I = (0, T] \subset \mathbb{R}$, we consider the three-dimensional *linear advection-diffusion equation*

$$\begin{cases} u_t - \nabla \cdot (\alpha \nabla u) + \beta \cdot \nabla u = f & \text{in } \Omega \times I, \\ u = 0 & \text{on } \Gamma \times I, \\ u(0) = u_0 & \text{in } \Omega, \end{cases} \tag{6}$$

where Ω_x, Ω_y and Ω_z are intervals in $\mathbb{R}$. Here, $u_t := \partial u / \partial t$, $\Gamma = \partial \Omega$ denotes the boundary of the spatial domain Ω, $f : \Omega \times I \longrightarrow \mathbb{R}$ is a given source and $u_0 : \Omega \longrightarrow \mathbb{R}$ is a given initial condition. We consider constant diffusivity α and constant velocity field $\beta = [\beta_x \ \beta_y \ \beta_z]$.

We split the advection-diffusion operator $\mathcal{L}u = -\nabla \cdot (\alpha \nabla u) + \beta \cdot \nabla u$ as $\mathcal{L}u = \mathcal{L}_1 u + \mathcal{L}_2 u + \mathcal{L}_3 u$ where

$$\mathcal{L}_1 u := -\alpha \frac{\partial u}{\partial x^2} + \beta_x \frac{\partial u}{\partial x}, \quad \mathcal{L}_2 u := -\alpha \frac{\partial u}{\partial y^2} + \beta_y \frac{\partial u}{\partial y}, \quad \mathcal{L}_3 u := -\alpha \frac{\partial u}{\partial z^2} + \beta_z \frac{\partial u}{\partial z}.$$

Based on this operator splitting, we consider different Alternating Direction Implicit (ADI) schemes to discretize problem (6).

First, we perform a uniform partition of the time interval $\bar{I} = [0, T]$ as

$$0 = t_0 < t_1 < \ldots < t_{N-1} < t_N = T,$$

and denote $\tau := t_{n+1} - t_n$, $\forall n = 0, \ldots, N - 1$.

In the Douglas-Gunn scheme, we integrate the solution from time step t_n to t_{n+1} in three substeps as follows:

$$\begin{cases} (1 + \frac{\tau}{2}\mathcal{L}_1)u^{n+1/3} = \tau f^{n+1/2} + (1 - \frac{\tau}{2}\mathcal{L}_1 - \tau\mathcal{L}_2 - \tau\mathcal{L}_3)u^n, \\ (1 + \frac{\tau}{2}\mathcal{L}_2)u^{n+2/3} = u^{n+1/3} + \frac{\tau}{2}\mathcal{L}_2 u^n, \\ (1 + \frac{\tau}{2}\mathcal{L}_3)u^{n+1} = u^{n+2/3} + \frac{\tau}{2}\mathcal{L}_3 u^n. \end{cases} \tag{7}$$

The variational formulation of scheme (7) is

$$\begin{cases} (u^{n+1/3}, v) + \frac{\tau}{2}\left(\alpha \frac{\partial u^{n+1/3}}{\partial x}, \frac{\partial v}{\partial x}\right) + \frac{\tau}{2}\left(\beta_x \frac{\partial u^{n+1/3}}{\partial x}, v\right) = (u^n, v) - \frac{\tau}{2}\left(\alpha \frac{\partial u^n}{\partial x}, \frac{\partial v}{\partial x}\right) \\ \quad - \frac{\tau}{2}\left(\beta_x \frac{\partial u^n}{\partial x}, v\right) - \tau\left(\alpha \frac{\partial u^n}{\partial y}, \frac{\partial v}{\partial y}\right) - \tau\left(\beta_y \frac{\partial u^n}{\partial y}, v\right) - \tau\left(\alpha \frac{\partial u^n}{\partial z}, \frac{\partial v}{\partial z}\right) \\ \quad - \tau\left(\beta_z \frac{\partial u^n}{\partial z}, v\right) + \tau(f^{n+1/2}, v), (u^{n+2/3}, v) + \frac{\tau}{2}\left(\alpha \frac{\partial u^{n+2/3}}{\partial y}, \frac{\partial v}{\partial y}\right) + \frac{\tau}{2}\left(\beta_y \frac{\partial u^{n+2/3}}{\partial y}, v\right) \\ = (u^{n+1/3}, v) + \frac{\tau}{2}\left(\alpha \frac{\partial u^n}{\partial y}, \frac{\partial v}{\partial y}\right) + \frac{\tau}{2}\left(\beta_y \frac{\partial u^n}{\partial y}, v\right), (u^{n+1}, v) + \frac{\tau}{2}\left(\alpha \frac{\partial u^{n+1}}{\partial z}, \frac{\partial v}{\partial z}\right) \\ + \frac{\tau}{2}\left(\beta_z \frac{\partial u^{n+1}}{\partial z}, v\right) = (u^{n+2/3}, v) + \frac{\tau}{2}\left(\alpha \frac{\partial u^n}{\partial z}, \frac{\partial v}{\partial z}\right) + \frac{\tau}{2}\left(\beta_z \frac{\partial u^n}{\partial z}, v\right), \end{cases} \tag{8}$$

where $(\cdot, \cdot)$ denotes the inner product of $L^2(\Omega)$. Finally, expressing problem (8) in matrix form we have

$$
\begin{cases}
\left[M^x + \dfrac{\tau}{2}(K^x + G^x) \right] \otimes M^y \otimes M^z u^{n+1/3} \\
\quad = \left[M^x - \dfrac{\tau}{2}(K^x + G^x) \right] \otimes M^y \otimes M^z u^n \\
\qquad - \tau M^x \otimes (K^y + G^y) \otimes M^z u^n - \tau M^x \otimes M^y \otimes (K^z + G^z)u^n + \tau F^{n+1/2} \\
M^x \otimes \left[M^y + \dfrac{\tau}{2}(K^y + G^y) \right] \otimes M^z u^{n+2/3} \\
\quad = M^x \otimes M^y \otimes M^z u^{n+1/3} + M^x \otimes \dfrac{\tau}{2}(K^y + G^y) \otimes M^z u^n, \\
M^x \otimes M^y \otimes \left[M^z + \dfrac{\tau}{2}(K^z + G^z) \right] u^{n+1} \\
\quad = M^x \otimes M^y \otimes M^z u^{n+2/3} + M^x \otimes M^y \otimes \dfrac{\tau}{2}(K^z + G^z)u^n,
\end{cases}
\tag{9}
$$

where $M^{x,y,z}$, $K^{x,y,z}$ and $G^{x,y,z}$ are the 1D mass, stiffness and advection matrices, respectively.

3 Isogeometric Residual Minimization Method

In our method, in every time step we solve the problem with identical left-hand-side: Find $u \in U$ such that

$$
b(u, v) = l(v) \quad \forall v \in V,
\tag{10}
$$

$$
b(u, v) = (u, v) + \tau/2 \left(\left(\beta_{\underline{i}} \frac{\partial u}{\partial x_{\underline{i}}}, v \right) + \alpha_{\underline{i}} \left(\frac{\partial u}{\partial x_{\underline{i}}}, \frac{\partial v}{\partial x_{\underline{i}}} \right) \right),
\tag{11}
$$

Here $i \in \{1, 2, 3\}$, so we have denoted here $(x_1, x_2, x_3) = (x, y, z)$, and $\underline{i}$ means that we are not using the Einstein summation here. The right-hand-side depends on the sub-step and the time integration scheme used. In the Douglas-Gunn time integration scheme, in the first, second and third sub-step the right-hand side is defined as:

$$
\begin{cases}
l(w, v) = (w, v) - \dfrac{\tau}{2} \left(\alpha \dfrac{\partial w}{\partial x}, \dfrac{\partial v}{\partial x} \right) - \dfrac{\tau}{2} \left(\beta_x \dfrac{\partial w}{\partial x}, v \right) - \tau \left(\alpha \dfrac{\partial w}{\partial y}, \dfrac{\partial v}{\partial y} \right) - \tau \left(\beta_y \dfrac{\partial w}{\partial y}, v \right) \\
\qquad - \tau \left(\alpha \dfrac{\partial w}{\partial z}, \dfrac{\partial v}{\partial z} \right) - \tau \left(\beta_z \dfrac{\partial w}{\partial z}, v \right) + \tau(f^{n+1/2}, v), \\
l(w, v) = (w, v) + \dfrac{\tau}{2} \left(\alpha \dfrac{\partial w}{\partial y}, \dfrac{\partial v}{\partial y} \right) + \dfrac{\tau}{2} \left(\beta_y \dfrac{\partial w}{\partial y}, v \right), \\
l(w, v) = (w, v) + \dfrac{\tau}{2} \left(\alpha \dfrac{\partial w}{\partial z}, \dfrac{\partial v}{\partial z} \right) + \dfrac{\tau}{2} \left(\beta_z \dfrac{\partial w}{\partial z}, v \right).
\end{cases}
\tag{12}
$$

In our advection-diffusion problem we seek the solution in space

$$
U = V = \left\{ v : \int_\Omega \left(v^2 + \left(\frac{\partial v}{\partial x_{\underline{i}}} \right)^2 \right) < \infty \right\}.
\tag{13}
$$

where $i = 1, 2, 3$ denotes the spatial directions. The inner product in V is defined as

$$(u, v)_V = (u, v)_{L_2} + \left(\frac{\partial u}{\partial x_{\underline{i}}}, \frac{\partial v}{\partial x_{\underline{i}}} \right)_{L_2}, \tag{14}$$

where $i = 1, 2, 3$ depending on the sub-step index in the alternating directions method, and we do not use here the Einstein convention. For a weak problem, we define the operator

$$B : U \to V', \tag{15}$$

such that

$$\langle Bu, v \rangle_{V' \times V} = b(u, v), \tag{16}$$

so we can reformulate the problem as

$$Bu - l = 0. \tag{17}$$

We wish to minimize the residual

$$u_h = \operatorname{argmin}_{w_h \in U_h} \frac{1}{2} \| Bw_h - l \|_{V'}^2. \tag{18}$$

We introduce the Riesz operator being the isometric isomorphism

$$R_V : V \ni v \to (v, .) \in V'. \tag{19}$$

We can project the problem back to V

$$u_h = \operatorname{argmin}_{w_h \in U_h} \frac{1}{2} \| R_V^{-1}(Bw_h - l) \|_V^2. \tag{20}$$

The minimum is attained at u_h when the Gâteaux derivative is equal to 0 in all directions:

$$\langle R_V^{-1}(Bu_h - l), R_V^{-1}(B w_h) \rangle_V = 0, \quad \forall w_h \in U_h. \tag{21}$$

We define the error representation function $r = R_V^{-1}(Bu_h - l)$ and our problem is reduced to

$$\langle r, R_V^{-1}(B w_h) \rangle = 0, \quad \forall w_h \in U_h, \tag{22}$$

which is equivalent to

$$\langle Bw_h, r \rangle = 0, \quad \forall w_h \in U_h. \tag{23}$$

From the definition of the residual we have

$$(r, v)_V = \langle Bu_h - l, v \rangle, \quad \forall v \in V. \tag{24}$$

Our problem reduces to the following semi-infinite problem: Find $(r, u_h)_{V \times U_h}$ such as

$$(r, v)_V - \langle Bu_h, v \rangle = \langle l, v \rangle, \quad \forall v \in V,$$
$$\langle Bw_h, r \rangle = 0 \quad \forall w_h \in U_h. \tag{25}$$

We discretize the test space $V_m \in V$ to get the discrete problem: Find $(r_m, u_h)_{V_m \times U_h}$ such as

$$
\begin{aligned}
(r_m, v_m)_{V_m} - \langle B u_h, v_m \rangle &= \langle l, v \rangle \quad \forall v \in V_m \\
\langle B w_h, r_m \rangle &= 0 \quad \forall w_h \in U_h.
\end{aligned}
\tag{26}
$$

Note that the residual minimization method is a Petrov-Galerkin method (test and trial spaces are different). We stabilize the problem by increasing the dimension of the test space. Notice that the residual minimization system here is of the following form

$$
\begin{bmatrix} G & -B \\ B^T & 0 \end{bmatrix} \begin{bmatrix} r \\ u \end{bmatrix} = \begin{bmatrix} l \\ 0 \end{bmatrix},
\tag{27}
$$

where the right-top and left-bottom matrices B and B^T can be split according to (9), and the inner product (14) part G can be split in the following way:

$$
G = \begin{cases} [\tilde{M}^x + \tilde{K}^x] \otimes \tilde{M}^y \otimes \tilde{M}^z, \\ \tilde{M}^x \otimes [\tilde{M}^y + \tilde{K}^y] \otimes \tilde{M}^z, \\ \tilde{M}^x \otimes \tilde{M}^y \otimes [\tilde{M}^z + \tilde{K}^z], \end{cases}
\tag{28}
$$

where we consider three different splittings for three sub-steps, and $\tilde{M}^{x,y,z}$, and $\tilde{K}^{x,y,z}$ are the 1D mass and stiffness matrices over the test space in direction x, y, or z, respectively.

Now, in the first sub-step, we approximate the solution with tensor product of one dimensional B-splines basis functions of order p, $u_h = \sum_{i,j,k} u_{i,j,k} B^x_{i;p}(x) B^y_{j;p}(y) B^z_{k;p}(z)$. We test with tensor product of one dimensional B-splines basis functions, where we enrich the order in the direction of the x axis from p to $o \geq p$, and we enrich the test space only in the direction of the alternating splitting $v_m \leftarrow B^x_{i;o}(x) B^y_{j;p}(y) B^z_{k;p}(z)$. We approximate the residual with tensor product of one dimensional B-splines basis functions of order p, $r_m = \sum_{s,t,q} r_{s,t,q} B^x_{s;t}(x) B^y_{t;p}(y) B^z_{t;p}(z)$, and we test with tensor product of 1D B-spline basis functions of order o and p, in the corresponding directions $w_h \leftarrow B^x_{k;o}(x) B^y_{l;p}(y) B^z_{m;p}(z)$.

Notice that we stabilize the problem by enriching the test space with respect to the trial space in the alternating direction manner. Now, in the first sub-step we have $M^y = \tilde{M}^y$, $M^z = \tilde{M}^z$ and $M^x \neq \tilde{M}^x$, and $M^{yT} = M^y$, $M^{zT} = M^z$. Now, in the first sub-step we have

$$
\begin{pmatrix} G & B \\ B^T & 0 \end{pmatrix} = \begin{pmatrix} [\tilde{M}^x + \tilde{K}^x] & [M^x + \frac{\tau}{2}(K^x + G^x)] \\ [M^x + \frac{\tau}{2}(K^x + G^x)]^T & 0 \end{pmatrix} \otimes M^y \otimes M^z \tag{29}
$$

in the second sub-step

$$
\begin{pmatrix} G & B \\ B^T & 0 \end{pmatrix} = M^x \otimes \begin{pmatrix} [\tilde{M}^y + \tilde{K}^y] & [M^y + \frac{\tau}{2}(K^y + G^y)] \\ [M^y + \frac{\tau}{2}(K^y + G^y)]^T & 0 \end{pmatrix} \otimes M^z \tag{30}
$$

and in the third sub-step

$$\begin{pmatrix} G & B \\ B^T & 0 \end{pmatrix} = M^x \otimes M^y \otimes \begin{pmatrix} [\tilde{M}^z + \tilde{K}^z] & [M^z + \frac{\tau}{2}(K^z + G^z)] \\ [M^z + \frac{\tau}{2}(K^z + G^z)]^T & 0 \end{pmatrix} \quad (31)$$

All these matrices are the Kronecker products of three multi-diagonal sub-matrices, and they can be factorized in a linear $\mathcal{O}(N)$ computational cost.

4   Numerical Results

4.1   Manufactured Solution Problem

In order to verify the order and accuracy of the Douglas-Gunn time-integration schemes with IGA-FEM discretization and the direction splitting solver, we construct a time-dependent advection-diffusion problem with manufactured solution.

$$\frac{du}{dt} - \nabla \cdot (\alpha \nabla u) + \beta \cdot \nabla u = f,$$

with $\alpha = 10^{-2}$, $\beta = (1, 0, 0)$, with zero Dirichlet boundary conditions solved on a square $[0, 1]^3$ domain. We setup the forcing function $f(x, y, z; t)$ in such a way that it delivers the manufactured solution of the form $u_{exact}(x, y, z; t) = \sin(\pi x) \sin(\pi y) \sin(\pi z) \sin(\pi t)$ on a time interval $[0, 2]$.

We solve the problem with residual minimization method on $32 \times 32 \times 32$ mesh with different time steps, as presented in Fig. 1, using the Douglas-Gunn time integration scheme and the direction splitting solver using the Kronecker product structure of the matrices.

We compute the error between the exact solution u_{exact} and the numerical solution u_h. We present the comparisons with different time step size τ. We compute relative error $\|u_{\text{exact}}(t) - u_{\text{h}}(t)\|_{L^2} / \|u_{\text{exact}}(t)\|_{L^2} \cdot 100\%$ and plot it in Fig. 1. The horizontal lines represent the time step size selected for the entire simulation, and the vertical lines present the numerical error with respect to the known exact solution.

The Douglas-Gunn scheme is of the second order accurate, down to the accuracy of 10^{-5}.

4.2   Pollution Propagation Simulations

In this section, we describe the numerical simulation of three-dimensional model advection-diffusion problem over a 3D cube shape domain with dimensions $5000 \times 5000 \times 5000$ m.

$$\frac{du}{dt} - \nabla \cdot (\alpha \nabla u) - \beta \cdot \nabla u = f, \quad (32)$$

In our equation we have the diffusion coefficients $\alpha = (50, 50, 0.5)$. We utilize tensor products of 1D B-splines along the x, y, and z. We apply the alternating direction implicit solver with three intermediate time steps. The velocity field is $\beta = (\beta^x(t), \beta^y(t), \beta^z(t)) = (\cos a(t), \sin a(t), v(t))$ where $a(t) = \frac{\pi}{3}(\sin(s) + \frac{1}{2}\sin(2.3s)) + \frac{3}{8}\pi$, $v(t) = \frac{1}{3}\sin(s)$, $s = \frac{t}{150}$. The source is given by $f(p) = (r-1)^2(r+1)^2$, where $r = \min(1, (|p-p_0|/25)^2)$, p represents the distance from the source, and p_0 is the location of the source $p_0 = (3,3,2)$. The initial state is defined as the constant concentration of the order of 10^{-6} in the entire domain (numerical zero).

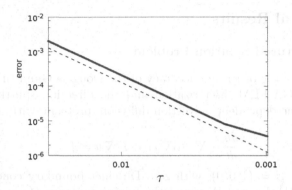

Fig. 1. Numerical error for Douglas-Gunn time integration scheme on $32 \times 32 \times 32$ mesh for different time steps.

The physical meaning of this setup is the following. We model the propagation of the pollutant generated by a single source modeled by the f function, distributed by the wind blowing with changing directions, modeled by β function, and the diffusion phenomena modeled by the coefficients α The computational domain unit is meter $[m]$, the wind velocity β is given in meters per second $[\frac{m}{s}]$, and the diffusion coefficient α is given in square meters per second $[\frac{m^2}{s}]$. The units for the solution are then kilograms per cube meter $[\frac{kg}{m^3}]$. We expect from the numerical results to observe the propagation of the pollutant as distributed by the wind and the diffusion process.

Our first numerical results concern the computational mesh with a size of $50 \times 50 \times 50$ elements with quadratic B-splines. We employ standard Galerkin formulation here with direction splitting, without the residual minimization method. We perform 300 time steps of the numerical simulation. The snapshots presented in Fig. 2 represent time steps 100, 200 and 300. We observe unexpected "oscillations" and "reflections". Since the simulation is supposed to model the propagation of the pollutant from a chimney by means of the advection (wind) and diffusion phenomena, the oscillations and reflections on the boundary are not expected there. Both these phenomena appear and disappear during the entire simulation; they do not cause a blowup of the entire simulations, just unexpected local behavior.

To improve the spatial stability of the simulation, we add now the residual minimization method on top of the Galerkin setup. Thus, the second simulation was performed again on the mesh size with $50 \times 50 \times 50$ elements, with quadratic B-splines for trial and cubic B-Splines for test. The snapshots from the numerical results are presented in Fig. 2. We perform 300 time steps, and we present the snapshots in time steps 100, 200, and 300.

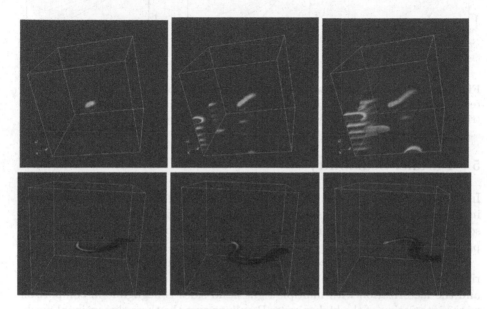

Fig. 2. Top panel: simulation with Galerkin method. Snapshots from the time steps 10, 20, and 30 of with quadratic C1 B-splines over $50 \times 50 \times 50$ mesh, without stabilization. **Bottom panel:** simulation with residual minimization method. Snapshots from the time steps 100, 200 and 300 of the first problem simulation with quadratic C1 B-splines for trial and cubic C2 B-splines for test over $50 \times 50 \times 50$ mesh.

We use the implicit extension of the parallel code [14] for shared memory Linux cluster nodes. The total simulation time was 100 min on a laptop with i7 6700Q processor 2.6 GHz (8 cores with HT) and 16B or RAM. We emphasize that ADI is not an iterative solver. It is just a linear $\mathcal{O}(N)$ computational cost solver that performs Gaussian elimination for matrices having Kronecker product structure. Thus, the solution obtained by the solver is exact (up to the round-off errors). In this sense, we do not present the iterations or convergence of the ADI solver, since it is executed once per each time step. In other words, we can perform 300 Gaussian elimination, each with 1,000,000 unknowns, with the high accuracy resulting from the ADI direct solver (only round-off errors are involved), on a laptop with eight cores, with the implicit method, within 1.5 h.

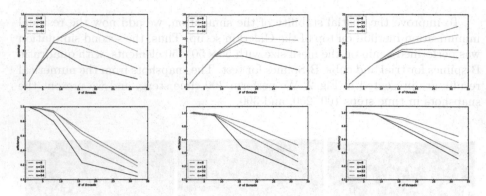

Fig. 3. Speedup (top line) and efficiency (bottom line) for $p = 1$ for trial, $p = 2$ for testing (first column), $p = 2$ for trial, $p = 3$ for testing (second column), and $p = 3$ for trial, $p = 4$ for testing (third column).

5   Parallel Scalability

Implementation of the ideas described in the preceding sections has been created in C++ and parallelized using our code for IGA-FEM simulations with ADI solver [14], extended to the implicit method. We use the GALOIS framework for parallelization [21–24].

The parallelization concerns mainly the algorithm of the integration of the right-hand-side vector. The cost of generation of the one-dimensional matrices and the factorization with multiple-right-hand sides is negligible in comparison to the integration of the higher-order B-splines over the three-dimensional mesh. The parallel implementation in GALOIS involves the usage of *Galois::for_each*, *Galois::Runtime::LL::SimpleLock*.

```
1.  for each element E = [ξ_{l_x}, ξ_{l_x+1}] × [ξ_{l_y}, ξ_{l_y+1}] × [ξ_{l_z}, ξ_{l_z+1}] in parallel do
2.      U^{loc} ← 0
3.      for each quadrature point ξ = (X_{k_x}, X_{k_y}, X_{k_z}) do
4.          x ← Ψ_E(ξ), W ← w_{k_x} w_{k_y} w_{k_z}, u, Du ← 0
5.          for I ∈ I(E) do
6.              u ← u + U_I^{(t)} B_I(ξ), Du ← Du + U_I^{(t)} ∇B_I(ξ)
7.          endfor
8.          for I ∈ I(E) do
9.              v ← B_I(ξ), Dv ← ∇B_I(ξ), U_I^{loc} ← U_I^{loc} + W |E| (uv + Δt F(u, Du, v, Dv))
10.         endfor
11.     endfor
12.  endfor
13.  synchronized
14.     for I ∈ I(E) do
15.         U_I^{(t+1)} ← U_I^{(t+1)} + U_I^{loc}
16.     endfor
17.  end
```

The speedup and efficiency of the code are presented in Fig. 3. We can draw the following conclusions from the presented plots:

- For linear B-splines for trial and quadratic B-splines for testing and large grids $32 \times 32 \times 32$ and $64 \times 64 \times 64$ the speedup grows up to 16 cores. It is around 10–11 for 16 cores. The corresponding efficiency for 16 cores is around 0.7. Then, for 32 cores the speedup went down since for more than 20 cores used the hyperthreading is utilized.
- For quadratic B-splines for trial and cubic B-splines and large grids $32 \times 32 \times 32$ and $64 \times 64 \times 64$ the speedup grows up to 16 cores. It is around 12–14 for 16 cores. The corresponding efficiency for 16 cores is around 0.8–0.9. Then, for 32 cores and $32 \times 32 \times 32$ mesh the speedup grows up to 17, and for $64 \times 64 \times 64$ mesh is decreases slightly since for more than 20 cores the hyperthreading is used.
- For cubic B-splines for trial and quartic B-splines for testing and large grids $32 \times 32 \times 32$ and $64 \times 64 \times 64$ the speedup grows up to 32 cores. It is around 15 for 16 cores (near perfect speedup) and around 20 for 32 cores, where we use the hyperthreading (more than 20 cores). The corresponding efficiency for 16 cores is around 0.9–1.0. Then, for 32 cores the efficiency decreases slightly down to 0.6–0.7.
- Increasing the mesh size increases the parallel scalability up to $32 \times 32 \times 32$ mesh. Larger mesh, $64 \times 64 \times 64$ performs slightly worse than $32 \times 32 \times 32$ mesh.
- The most interesting observation is that while increasing the B-splines order we observe the improvement of the parallel scalability. This is important from the point of view of the stabilization with the residual minimization method. The order of B-splines in the test space is increased to enforce the stabilization, and when we increase the order to obtain the stabilization, we also improve the parallel scalability.

6 Conclusions

We introduced an isogeometric finite element method for an implicit simulations of the advection-diffusion problem with Douglas-Gunn time-integration scheme that results in a Kronecker product structure of the matrix in every time step. The application of B-spline basis functions for the approximation of the numerical solutions results in a smooth, higher order approximation of the solution. It also enables for the residual minimization stabilization with a linear computational cost $\mathcal{O}(N)$ of the direct solver. The method has been verified on a three-dimensional advection-diffusion problem. Our future work will involve the extension of the model to more complicated equations and geometries. In particular, we plan to use the isogeometric alternating direction implicit solver for tumor growth simulations in two- and three-dimensions [13,17]. Our equations can also be extended to model a pollution problem, with different chemical components, propagating and reacting together through space in time, as described in [18].

Acknowledgments. The work of Maciej Paszyński, Marcin Łoś and visit of Judit Muñoz-Matute at AGH have been supported by National Science Centre, Poland grant no. 2017/26/M/ST1/00281. The J. Tinsley Oden Faculty Fellowship Research Program at the Institute for Computational Engineering and Sciences (ICES) of the University of Texas at Austin has supported the visit of Maciej Paszyński to ICES.

References

1. Peaceman, D.W., Rachford Jr., H.H.: The numerical solution of parabolic and elliptic differential equations. J. Soc. Ind. Appl. Math. **3**, 28–41 (1955)
2. Douglas, J., Gunn, J.E.: A general formulation of alternating direction methods. Numer. Math. **6**(1), 428–453 (1964)
3. Birkhoff, G., Varga, R.S., Young, D.: Alternating direction implicit methods. Adv. Comput. **3**, 189–273 (1962)
4. Cottrell, J.A., Hughes, T.J.R., Bazilevs, Y.: Isogeometric Analysis: Towards Unification of CAD and FEA. Wiley, Hoboken (2009)
5. Douglas, J., Rachford, H.: On the numerical solution of heat conduction problems in two and three space variables. Trans. Am. Math. Soc. **82**, 421–439 (1956)
6. Gao, L., Calo, V.M.: Fast isogeometric solvers for explicit dynamics. Comput. Methods Appl. Mech. Eng. **274**(1), 19–41 (2014)
7. Gao, L., Calo, V.M.: Preconditioners based on the alternating-direction-implicit algorithm for the 2D steady-state diffusion equation with orthotropic heterogeneous coefficients. J. Comput. Appl. Math. **273**(1), 274–295 (2015)
8. Guermond, J.L., Minev, P.: A new class of fractional step techniques for the incompressible Navier-Stokes equations using direction splitting. C.R. Math. **348**(9–10), 581–585 (2010)
9. Guermond, J.L., Minev, P., Shen, J.: An overview of projection methods for incompressible flows. Comput. Methods Appl. Mech. Eng. **195**, 6011–6054 (2006)
10. Keating, J., Minev, P.: A fast algorithm for direct simulation of particulate flows using conforming grids. J. Comput. Phys. **255**, 486–501 (2013)
11. Gurgul, G., Woźniak, M., Łoś, M., Szeliga, D., Paszyński, M.: Open source JAVA implementation of the parallel multi-thread alternating direction isogeometric L2 projections solver for material science simulations. Comput. Methods Mater. Sci. **17**, 1–11 (2017)
12. Łoś, M., Woźniak, M., Paszyński, M., Dalcin, L., Calo, V.M.: Dynamics with matrices possessing Kronecker product structure. Procedia Comput. Sci. **51**, 286–295 (2015)
13. Łoś, M., Paszyński, M., Kłusek, A., Dzwinel, W.: Application of fast isogeometric L2 projection solver for tumor growth simulations. Comput. Methods Appl. Mech. Eng. **316**, 1257–1269 (2017)
14. Łoś, M., Woźniak, M., Paszyński, M., Lenharth, A., Pingali, K.: IGA-ADS: isogeometric analysis FEM using ADS solver. Comput. Phys. Commun. **217**, 99–116 (2017)
15. Łoś, M., Paszyński, M.: Applications of alternating direction solver for simulations of time-dependent problems. Comput. Sci. **18**(2), 117–128 (2017)
16. Woźniak, M., Łoś, M., Paszyński, M., Dalcin, L., Calo, V.M.: Parallel fast isogeometric solvers for explicit dynamics. Comput. Inform. **36**(2), 423–448 (2017)
17. Łoś, M., Kłusek, A., Hassam, M.A., Pingali, K., Dzwinel, W., Paszyński, M.: Parallel fast isogeometric L2 projection solver with GALOIS system for 3D tumor growth simulations. Comput. Methods Appl. Mech. Eng. **343**, 1–22 (2019)

18. Oliver, A., Montero, G., Montenegro, R., Rodríguez, E., Escobar, J.M., Pérez-Foguet, A.: Adaptive finite element simulation of stack pollutant emissions over complex terrain. Energy **49**, 47–60 (2013)

19. Wachspress, E.L., Habetler, G.: An alternating-direction-implicit iteration technique. J. Soc. Ind. Appl. Math. **8**, 403–423 (1960)

20. Łoś, M., Muñoz-Matute, J., Muga, I., Paszyński, M.: Isogeometric residual minimization method (iGRM) with direction splitting for non-stationary advection-diffusion problems. Comput. Math. Appl. (2019, in press)

21. Pingali, K., et al.: The tao of parallelism in algorithms. SIGPLAN Not. **46**(6), 12–25 (2011)

22. Hassaan, M.A., Burtscher, M., Pingali, K.: Ordered vs. unordered: a comparison of parallelism and work-efficiency in irregular algorithms. In: Proceedings of the 16th ACM Symposium on Principles and Practice of Parallel Programming, PPoPP 2011 (2011)

23. Lenharth, A., Nguyen, D., Pingali, K.: Priority queues are not good concurrent priority schedulers. In: Träff, J.L., Hunold, S., Versaci, F. (eds.) Euro-Par 2015. LNCS, vol. 9233, pp. 209–221. Springer, Heidelberg (2015). https://doi.org/10. 1007/978-3-662-48096-0_17

24. Kulkarni, M., Pingali, K., Walter, B., Ramanarayanan, G., Bala, K., Chew, L.P.: Optimistic parallelism requires abstractions. ACM SIGPLAN Not. **42**(6), 211–222 (2007)

25. Demkowicz, L.: Babuśka <=> Brezzi, ICES-Report 0608, 2006, The University of Texas at Austin, USA. https://www.ices.utexas.edu/media/reports/2006/0608. pdf

26. Babuśka, I.: Error bounds for finite element method. Numer. Math. **16**, 322–333 (1971)

27. Brezzi, F.: On the existence, uniqueness and approximation of saddle-point problems arising from Lagrange multiplier. ESAIM: Mathematical Modelling and Numerical Analysis - Modélisation Mathématique et Analyse Numérique 8.R2, pp. 129–151 (1974)

28. Hughes, T.J.R., Scovazzi, G., Tezduyar, T.E.: Stabilized methods for compressible flows. J. Sci. Comput. **43**(3), 343–368 (2010)

29. Franca, L.P., Frey, S.L., Hughes, T.J.R.: Stabilized finite element methods: I. Application to the advective-diffusive model. Comput. Methods Appl. Mech. Eng. **95**(2), 253–276 (1992)

30. Franca, L.P., Frey, S.L.: Stabilized finite element methods: II. The incompressible Navier-Stokes equations. Comput. Methods Appl. Mech. Eng. **99**(2–3), 209–233 (1992)

31. Brezzi, F., Bristeau, M.-O., Franca, L.P., Mallet, M., Rogé, G.: A relationship between stabilized finite element methods and the Galerkin method with bubble functions. Comput. Methods Appl. Mech. Eng. **96**(1), 117–129 (1992)

32. Demkowicz, L., Gopalakrishnan, J.: Recent developments in discontinuous Galerkin finite element methods for partial differential equations. In: Feng, X., Karakashian, O., Xing, Y. (eds.) An Overview of the DPG Method. IMA Volumes in Mathematics and its Applications, vol. 157, pp. 149–180 (2014)

33. Paszyński, M., Pardo, D., Calo, V.M.: Direct solvers performance on h-adapted grids. Comput. Math. Appl. **70**(3), 282–295 (2015)

34. Chan, J., Evans, J.A.: A Minimum-residual finite element method for the convection-diffusion equations. ICES-Report 13-12 (2013)
35. Broersen, D., Dahmen, W., Stevenson, R.P.: On the stability of DPG formulations of transport equations. Math. Comput. **87**, 1051–1082 (2018)
36. Broersen, D., Stevenson, R.: A robust Petrov-Galerkin discretisation of convection-diffusion equations. Comput. Math. Appl. **68**(11), 1605–1618 (2014)

Numerical Simulation of Heat Transfer in an Enclosure with Time-Periodic Heat Generation Using Finite-Difference Method

Igor Miroshnichenko[1(✉)] and Mikhail Sheremet[2]

[1] Regional Scientific and Educational Mathematical Centre,
Tomsk State University, 634050 Tomsk, Russia
miroshnichenko@mail.tsu.ru
[2] Laboratory on Convective Heat and Mass Transfer,
Tomsk State University, 634050 Tomsk, Russia
sheremet@math.tsu.ru

Abstract. This paper reports a numerical investigation of highly coupled system of partial differential equations, simulating the fluid flow and heat transfer in a large-scale enclosure with time-periodic heat generation. The bottom wall of the enclosure is insulated, and heat exchange with the environment is modeled at other external boundaries. The heater with time-periodic heat generation is located at the bottom of the enclosure. The internal surfaces of both the heater and walls are assumed to be gray. Air is the working fluid and the Rayleigh number is 10^9. To solve the governing equations with dimensionless vorticity – stream function – temperature variables, the finite difference method has been used. The developed model has been validated through a comparison with data of other authors. The effect of surface emissivity and periodic heat generation on Nusselt numbers and both stream function and temperature distributions has been investigated. The results showed that the influence of the thermal radiation on total thermal transmission increases with surface emissivity of walls and heater surfaces. The present numerical method can be applied in several engineering problems, such as designing passive cooling systems and the simulation of heat transfer in building constructions.

Keywords: Finite-difference method · Heat transfer · Stream function

1 Introduction

Many industrial processes are associated with convective heat transfer in large-scale enclosures. One significant application from an energy-saving point of view is an effective optimization of energy consumption for heating buildings. The main and relevant topics for construction and design of buildings are to provide both energy efficiency and thermal comforts for inhabitants. The efficient optimization and thorough design of buildings require modern experimental and numerical approaches. In specialized literature one can find a large number of articles relating to convective heat transfer inside a large-scale enclosure due to its applications in buildings [1–3].

© Springer Nature Switzerland AG 2020
V. V. Krzhizhanovskaya et al. (Eds.): ICCS 2020, LNCS 12143, pp. 149–162, 2020.
https://doi.org/10.1007/978-3-030-50436-6_11

Several studies on free convection inside enclosures have been focused on attic space, where the free convection mechanism is sensitive. A comprehensive review of thermal transmission in attic-shaped spaces has been presented by Saha and Khan [4]. Their findings have indicated that most works have been performed in the laminar and transition regimes. At the same time, as practice shows, the airflow in the real attic must be fully turbulent [5, 6]. Therefore, numerical analysis of hydrodynamics and thermal transmission inside attic (or other large areas) should be carried out in this regime. The review of Das et al. [7] has summarized the studies on natural convection in different (non-square) shapes of enclosures and enclosures with wavy and curved walls. A number of works showed that change of both the angle and aspect ratio of the parallelogrammic and triangular cavities had a big effect on the flow fields. The influence of the different parameters such as the Darcy number, Prandtl number, Rayleigh number, volume fraction of the nanoparticles and irreversibility distribution ratios has also been analyzed. These results provide helpful insight into possible strategies to enhance convective heat transfer inside complicated enclosures. Analysis of the effect of a room heater (location, size and power) on heat transfer and air flow is high important problem. Experimental and numerical simulation of turbulent modes of natural convection in a large-scale cavity with a small heat source has been carried out by Zhang et al. [8]. The central plane velocity fields above the plate have been measured using PIV method and temperature measurements have been obtained using thermocouples. The numerical modeling was conducted by Fluent software using various models of turbulence. The analysis of inertia and buoyancy force discussed in this paper helps to better understand the nature of flow inside the enclosure.

It should be noted that radiative heat transfer plays an important role in thermal transmission inside enclosures. As practice shows, radiation can have a big effect both on hydrodynamic characteristics and heat transfer even at relatively low temperature differences [9–12]. A detailed review of works in the field of turbulent free convection without and with radiation in rectangular cavities was conducted by Miroshnichenko and Sheremet [13]. Inclination angle and shape of the cavities, thermal boundary and initial conditions, various radiative properties and heat source location have also been investigated. Sharma et al. [14] have investigated the interaction effects between turbulent thermogravitational convective heat transport and thermal radiation in bottom heated cavities. The aspect ratio in this study varies from 0.5 to 2 and Ra based on cavity width varies from 10^8 to 10^{12}. They have found a correlation using aspect ratio and Rayleigh number for determining the mean convective Nusselt number. Kogawa et al. [15] have studied the radiation impact on turbulent modes of free convective energy transport inside closed volume. They have analyzed four different radiation models (surface radiation, non-radiation, gas radiation and combined radiation) to understand the radiation influence of the gas and from the borders. Their results have shown that the effect of gas radiation on radiative energy transport is insignificant, while the surface radiation effect is dominant.

To accurately predict convective-radiative heat transfer, various numerical approaches, namely, finite difference and finite volume methods, finite element and lattice Boltzmann methods are used [16–18]. The impact of non-uniform and uniform heating of inclined walls on free convection in an isosceles triangular cavity has been studied by Basak et al. [19] using the finite element method. They have considered two various cases of thermal boundary conditions, viz., non-uniformly and uniformly heating of two inclined walls. The main goal of that study was to analyze the flow and temperature fields with detailed analysis of heat transfer estimates for free convection in triangular cavity. Their findings indicated that geometry does not have much effect on flow structure at small Prandtl numbers.

To the best of our knowledge, investigation of the impact of time periodic heat generation on convective flows in an enclosure having internal heating has not received due attention. The aim of this work is to simulate unsteady turbulent energy transport in a large-scale enclosure with a heater using the finite difference method. Analysis of other heat transfer modes (radiation and conduction inside the heater and solid walls) noticeably affects natural convection and essentially complicates the mathematical model. Moreover, we conducted a comprehensive study of the effect of time-periodic heat generation inside the enclosure.

2   Mathematical and Physical Models

The geometry of the problem as shown in Fig. 1 is a large-scale enclosure bounded by massive walls of finite conductivity with a local heat-generating element.

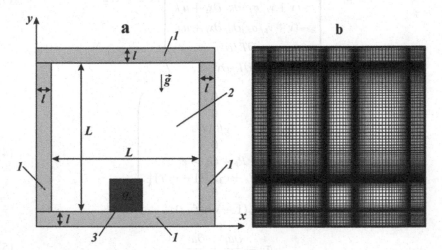

Fig. 1. The domain of interest (a): 1 - solid walls, 2 - air, 3 - heat source of constant temperature, computational domain and grid (b).

The heat-generating element is located on the bottom wall and it has time-periodic heat generation. Time-periodic heat generation was determined as $q = q_v(1 - sin(f \cdot t))$. The solid walls of the enclosure are of finite thermal conductivity λ_w and finite thickness l. The external border of the lower wall is adiabatic. No-slip conditions are accepted for cavity walls. The heat exchange with the outside is simulated at external surfaces. All internal walls of the large-scale enclosure are diffuse, opaque and gray emitters. The considered air flow is turbulent in nature. The properties of air are supposed to be permanent except for the density where the Boussinesq approximation is considered.

Taking into account the above assumptions the Reynolds-averaged Navier-Stokes equations can be written in the following form [20, 21]:

$$\frac{\partial G}{\partial t} + \frac{\partial S_1}{\partial x_1} + \frac{\partial S_2}{\partial x_2} = R \tag{1}$$

$$G = \begin{pmatrix} u_1 \\ u_2 \\ T \\ k \\ \varepsilon \\ T \\ T \end{pmatrix} \tag{2}$$

$$S_i = \begin{pmatrix} u_i \\ p\delta_{i1} - (v+v_t)\sigma_{i1} + u_i u_1 \\ p\delta_{i2} - (v+v_t)\sigma_{i2} + u_i u_2 \\ -(\alpha+\alpha_t)\partial T/\partial x_i + u_i T \\ -(v+v_t/\sigma_k)\partial k/\partial x_i + u_i k \\ -(v+v_t/\sigma_\varepsilon)\partial \varepsilon/\partial x_i + u_i \varepsilon \\ \alpha_w \partial T/\partial x_i \\ s\alpha_{hs}\partial T/\partial x_i \end{pmatrix}, \forall\, i = 1,2 \tag{3}$$

$$R = \begin{pmatrix} 0 \\ 0 \\ g\beta\Delta T \\ 0 \\ P_k + G_k - \varepsilon \\ (c_{1\varepsilon}(P_k + c_{3\varepsilon}G_k) - c_{2\varepsilon}\varepsilon)\frac{\varepsilon}{k} \\ 0 \\ q_v(1 - \sin(f \cdot t)) \end{pmatrix} \tag{4}$$

$$\sigma_{ij} = \frac{\partial u_i}{\partial x_j} + \frac{\partial u_j}{\partial x_i}, \tag{5}$$

where G_k defines the generation or dissipation of turbulent kinetic energy due to buoyancy, P_k describes the production of k and present terms are expressed as

$$P_k = v_t\left[2\left(\frac{\partial u_1}{\partial x_1}\right)^2 + 2\left(\frac{\partial u_2}{\partial x_2}\right)^2 + \left(\frac{\partial u_1}{\partial x_2} + \frac{\partial u_2}{\partial x_1}\right)^2\right], \quad G_k = -\frac{g\beta v_t}{\mathrm{Pr}_t}\frac{\partial T}{\partial x_2}. \tag{6}$$

Here x_1, x_2 are the physical coordinates; T is the temperature; u_1, u_2 are the velocity components in the projection on the x_1 and x_2 axes, respectively; ε is the dissipation rate of the kinetic energy of turbulence; α_w is the coefficient of thermal diffusivity of the solid wall material; v is the coefficient of kinematic viscosity; v_t is the coefficient of turbulent viscosity; g is the gravitational acceleration; k is the kinetic energy of turbulence; α_t is the coefficient of turbulent thermal diffusivity; β is the temperature coefficient of volume expansion; t is the time; L is the characteristic size of the cavity (Fig. 1). The Kolmogorov–Prandtl formula $v_t = c_\mu k^2 / \varepsilon$ was used to calculate turbulent viscosity, q_v is the volume density of heat flux.

The governing Eqs. (1)–(5) can be written in a slightly different form, which eliminates the need to search for a pressure field. The form of equations, where the variables are the vorticity (ω) and stream function (ψ), allows us to reduce the number of differential equations, thereby lead to a decrease in the time needed for calculations. In the present paper, a numerical study of the convective-radiative thermal transmission has been carried out in dimensionless variables. The scales of temperature, velocity, distance, time, kinetic energy of turbulence, dissipation rate of kinetic energy of turbulence, stream function and vorticity are chosen as $\Delta T = T_{hs} - T^e$, $\sqrt{g\beta\Delta TL}$, L,

$\sqrt{g\beta\Delta T/L}$, $g\beta\Delta TL$, $\sqrt{g^3\beta^3(\Delta T)^3 L}$, $\sqrt{g\beta\Delta TL^3}$, $\sqrt{L/g\beta\Delta T}$, respectively.

With the aim of detailed study of the both energy and momentum transport near the solid walls, a non-uniform grid has been introduced using a special algebraic coordinate transformation [20, 21]:

$$\xi = a + \frac{b-a}{2}\left\{1 + \mathrm{tg}\left[\frac{\pi\kappa}{b-a}\left(X - \frac{a+b}{2}\right)\right] \Big/ \mathrm{tg}\left[\frac{\pi}{2}\kappa\right]\right\},$$

$$\eta = a + \frac{b-a}{2}\left\{1 + \mathrm{tg}\left[\frac{\pi\kappa}{b-a}\left(Y - \frac{a+b}{2}\right)\right] \Big/ \mathrm{tg}\left[\frac{\pi}{2}\kappa\right]\right\}.$$

where κ is a compaction parameter and a, b are the geometrical characteristics.

Derivatives of the first and second orders in spatial coordinates are:

$$\frac{\partial\xi}{\partial X} = \frac{\pi\kappa}{2\cdot\mathrm{tg}\left\{\frac{\pi\kappa}{2}\right\}\cos^2\left\{\frac{\pi\kappa}{2}(2X-1)\right\}}, \quad \frac{\partial\eta}{\partial Y} = \frac{\pi\kappa}{2\cdot\mathrm{tg}\left\{\frac{\pi\kappa}{2}\right\}\cos^2\left\{\frac{\pi\kappa}{2}(2Y-1)\right\}},$$

$$\frac{\partial^2\xi}{\partial X^2} = \frac{(\pi\kappa)^2}{\mathrm{tg}\left\{\frac{\pi\kappa}{2}\right\}} \frac{\sin\left\{\frac{\pi\kappa}{2}(2X-1)\right\}}{\cos^3\left\{\frac{\pi\kappa}{2}(2X-1)\right\}}, \quad \frac{\partial^2\eta}{\partial Y^2} = \frac{(\pi\kappa)^2}{\mathrm{tg}\left\{\frac{\pi\kappa}{2}\right\}} \frac{\sin\left\{\frac{\pi\kappa}{2}(2Y-1)\right\}}{\cos^3\left\{\frac{\pi\kappa}{2}(2Y-1)\right\}}.$$

The air flow and heat transfer were determined by subsequent characteristics: the Prandtl number (Pr), the Ostrogradsky number (Os) and the Rayleigh number (Ra). These parameters are:

$$Pr = \frac{v}{\alpha_{air}}, \quad Os = \frac{q_v L^2}{\lambda_{hs}\Delta T}, \quad Ra = \frac{g\beta\Delta TL^3}{v\alpha_{air}}.$$

It should be noted that the volumetric heat generation from the heater is described by the Ostrogradsky number. Taking into account the algebraic coordinate transformation noted above boundary and initial conditions are considered in the following form:

At $\tau = 0$

$$Psi(\xi, \eta, 0) = \Omega(\xi, \eta, 0) = K(\xi, \eta, 0) = E(\xi, \eta, 0) = 0, \ \Theta(\xi, \eta, 0) = 0.5$$

At $\tau > 0$

at the boundary $\eta = 0$: $\frac{\partial \Theta}{\partial \eta} = 0$;

at the boundaries $\xi = 0$ and $\xi = 1 + 2l/L$: $\frac{\partial \xi}{\partial X} \frac{\partial \Theta}{\partial \xi} = Bi \cdot \Theta$ (the heat exchange with an external environment is simulated);

at the boundary $\eta = 1 + 2l/L$: $\frac{\partial \eta}{\partial Y} \frac{\partial \Theta}{\partial \eta} = Bi \cdot \Theta$ (the heat exchange with an outside is simulated);

at the heater surface: $\frac{\partial \Theta_{hs}}{\partial \tilde{n}} = \frac{\lambda_{air}}{\lambda_{hs}} \frac{\partial \Theta_{air}}{\partial \tilde{n}} - N_{rad}Q_{rad}$;

at the internal surfaces of the walls-air boundary, parallel to the axis $O\xi$:

$$\Psi = 0, \frac{\partial \Psi}{\partial \eta} = 0, \ \Theta_w = \Theta_{air}, \ \lambda_{w,air} \frac{\partial \eta}{\partial Y} \frac{\partial \Theta_w}{\partial \eta} = \frac{\partial \eta}{\partial Y} \frac{\partial \Theta_{air}}{\partial \eta} - N_{rad}Q_{rad};$$

at the internal surfaces of the walls-air boundary, parallel to the axis $O\eta$:

$$\Psi = 0, \frac{\partial \Psi}{\partial \xi} = 0, \ \Theta_w = \Theta_{air}, \ \lambda_{w,air} \frac{\partial \xi}{\partial X} \frac{\partial \Theta_w}{\partial \xi} = \frac{\partial \xi}{\partial X} \frac{\partial \Theta_{air}}{\partial \xi} - N_{rad}Q_{rad}.$$

The boundary conditions for the turbulent characteristics were presented in detail previously in [21]. It is necessary to understand the impact of radiative mechanism of energy transfer on the hydrodynamics and thermal transmission in the analyzed domain. The non-dimensional net radiative thermal flux Q_{rad} is defined as [21]:

$$Q_{rad,k} = R_k - \sum_{i=1}^{N} F_{k-i}R_i,$$

$$R_k = (1 - \tilde{\varepsilon}_k) \sum_{i=1}^{N} F_{k-i}R_i + \tilde{\varepsilon}_k(1 - \zeta)^4 \left(\Theta_k + 0.5 \frac{1+\zeta}{1-\zeta} \right)^4.$$

Here $Bi = hL/\lambda_w$ is the Biot number, σ is the Stefan-Boltzmann constant, λ_{air} is the air thermal conductivity, $N_{rad} = \sigma T_{hs}^4 L/[\lambda_{air}(T_{hs} - T^e)]$ is the radiation number, F_{k-i} is the view factor from kth unit to the ith unit of the chamber, $\alpha_{i,j} = \alpha_i/\alpha_j$ is the thermal diffusivity ratio, $\lambda_{i,j} = \lambda_i/\lambda_j$ is the thermal conductivity ratio, $\zeta = T^e/T_{hs}$ is the temperature parameter, $Q_{rad,k}$ is the net radiative thermal flux (dimensionless), R_k is the dimensionless radiosity of the kth unit of a chamber, $\tilde{\varepsilon}$ is the surface emissivity.

To solve the set of governing equations the finite-difference method is used. The system (1)–(5) is a combination of elliptic and parabolic equations. So, the second-order accurate central differences scheme was used to describe the elliptical equation. The successive over relaxation method was used to solve the difference equation. To solve the parabolic equations the locally one-dimensional scheme was adopted. The diffusion terms in parabolic equations were discretized using central differences scheme of the second-order accuracy, whereas accurate upwind difference scheme was applied to discretize the convective terms. First-order scheme was employed for the transient term. The resulting systems of linear equations were worked out by tridiagonal matrix algorithm (Thomas method).

Next, we consider in detail the solution of elliptic equations for the stream function. Taking into account the mesh transformation, a spatio-temporal uniform grid was constructed:

$$\xi_i = ih_\xi, \quad \eta_j = jh_\eta, \quad \tau_n = n\tau,$$

here h_ξ, h_η – space steps; τ – time step; $i = \overline{0, N_\xi}$; $j = \overline{0, N_\eta}$; $n = \overline{0, M}$.

Denote $f(\xi_i, \eta_j, \tau_n) = f_{i,j}^n$, to approximate the derivatives of the first and second orders, the central differences are used:

$$\frac{\partial f}{\partial \xi} \approx \frac{f_{i+1,j} - f_{i-1,j}}{2h_\xi}, \quad \frac{\partial^2 f}{\partial \xi^2} \approx \frac{f_{i+1,j} - 2f_{i,j} + f_{i-1,j}}{h_\xi^2},$$

$$\frac{\partial f}{\partial \eta} \approx \frac{f_{i,j+1} - f_{i,j-1}}{2h_\eta}, \quad \frac{\partial^2 f}{\partial \eta^2} \approx \frac{f_{i,j+1} - 2f_{i,j} + f_{i,j-1}}{h_\eta^2}.$$

The approximation relation for the time derivative is as follows:

$$\frac{\partial f}{\partial \tau} \approx \frac{f_{i,j}^{n+1} - f_{i,j}^n}{\tau}.$$

Poisson's equation for the stream function taking into account the coordinate transformation is defined as:

$$\frac{d^2\xi}{dX^2}\frac{\partial \Psi}{\partial \xi} + \left(\frac{d\xi}{dX}\right)^2\frac{\partial^2\Psi}{\partial \xi^2} + \frac{d^2\eta}{dY^2}\frac{\partial \Psi}{\partial \eta} + \left(\frac{d\eta}{dY}\right)^2\frac{\partial^2\Psi}{\partial \eta^2} = -\Omega.$$

One approach to solving elliptic equations is to use central differences to approximate second-order derivatives. As a result, we obtain the following discrete equation:

$$\frac{d^2\xi}{dX^2}\frac{\Psi^k_{i+1,j} - \Psi^{k+1}_{i-1,j}}{2h_\xi} + \left(\frac{d\xi}{dX}\right)^2\frac{\Psi^k_{i+1,j} - 2\Psi^{k+1}_{i,j} + \Psi^{k+1}_{i-1,j}}{h_\xi^2} + \frac{d^2\eta}{dY^2}\frac{\Psi^k_{i,j+1} - \Psi^{k+1}_{i,j-1}}{2h_\eta}$$

$$+ \left(\frac{d\eta}{dY}\right)^2\frac{\Psi^k_{i,j+1} - 2\Psi^{k+1}_{i,j} + \Psi^{k+1}_{i,j-1}}{h_\eta^2} = -\Omega_{i,j},$$

here k is a number of iterations.

Denote

$$\frac{d\xi}{dX} = a, \quad \frac{d^2\xi}{dX^2} = b, \quad \frac{d\eta}{dY} = c, \quad \frac{d^2\eta}{dY^2} = d.$$

Further, the resulting system of linear algebraic equations is solved by successive over relaxation method:

$$\begin{cases} \widehat{\Psi}^{k+1}_{i,j} = \dfrac{h_\eta^2 h_\xi b\left(\Psi^k_{i+1,j} - \Psi^{k+1}_{i-1,j}\right)}{4\left(c^2 h_\xi^2 + a^2 h_\eta^2\right)} + \dfrac{h_\eta^2\left(\Psi^k_{i+1,j} + \Psi^{k+1}_{i-1,j}\right)}{2\left(c^2 h_\xi^2 + a^2 h_\eta^2\right)} + \dfrac{h_\xi^2 h_\eta d\left(\Psi^k_{i,j+1} - \Psi^{k+1}_{i,j-1}\right)}{4\left(c^2 h_\xi^2 + a^2 h_\eta^2\right)} \\ \quad + \dfrac{h_\xi^2\left(\Psi^k_{i,j+1} + \Psi^{k+1}_{i,j-1}\right)}{2\left(c^2 h_\xi^2 + a^2 h_\eta^2\right)} + \dfrac{h_\xi^2 h_\eta^2\Omega_{i,j}}{2\left(c^2 h_\xi^2 + a^2 h_\eta^2\right)}, \\ \Psi^{k+1}_{i,j} = \Psi^k_{i,j} + \varpi\left(\widehat{\Psi}^{k+1}_{i,j} - \Psi^k_{i,j}\right). \end{cases}$$

The relaxation parameter ϖ was chosen experimentally from the results of many numerical experiments. A thorough criterion for stream function variable is used to receive converged solutions at each time step. The convergence condition $\left|\Psi^{k+1}_{ij} - \Psi^k_{ij}\right| < 10^{-6}$ must be satisfied by variable Ψ^k_{ij} at any grid point (i, j), here k is a given iteration parameter. This difference scheme is unconditionally stable and has an approximation order $O\left(h_\xi^2 + h_\eta^2\right)$.

Table 1. Variation of the average Nusselt numbers

Ra	[22]	[23]	[24]	Present data
10^7	16.79	16.523	–	17.13
10^8	30.506	30.225	28.78	33.06
10^9	57.35	–	62.0	60.54

In order to verify the numerical algorithm, the developed computational code has been validated successfully using various numerical results. In the case of natural convection inside the differentially heated enclosure, the developed computational code has been validated successfully using the numerical results (variation of the average Nusselt number) of Dixit and Babu [22], Zhuo and Zhong [23] and Le Quere [24] (Table 1).

The numerical simulation has been carried out on coarser and finer meshes, in order to check the grid independence. These computational tests with various grids allowed choosing the suitable grid size selection without compromising both accuracy and CPU time. The mesh sensitivity investigation was examined for $\tilde{\varepsilon} = 0$, Ra = 10^9. Three different meshes of 60×60, 120×120 and 156×156 were applied to verify the grid independence. The fluid flow rate, average convective Nusselt number at the heater surface and average heater temperature are shown in Table 2. The obtained results have shown that the 120×120 grid provided a good accuracy. For example, the maximum difference in terms of the average convective Nusselt number between case of 120×120 points and case of 156×156 points is less than 5.6%. In this connection, the non-uniform 120×120 grid has been selected for analysis.

Table 2. Grid independence study for Ra = 10^9, $\tilde{\varepsilon} = 0, f = 0$.

Grid size	Nu_{conv}	$|\psi|_{max}$	Θ_{hs}
60×60	48.78	0.031	0.764
120×120	61.54	0.028	0.783
156×156	64.974	0.027	0.786

3 Results and Discussion

Numerical simulation is reported for the following values of key parameters: $Pr = 0.7$, Ra = 10^9, $h/L = 0.1$, $\zeta = 0.82$, $Os = 1$, $N_{rad} = 245.36$, $0 \le \tilde{\varepsilon} \le 1$. The main attention was paid to the impact of both surface emissivity and time-periodic heat generation on distributions of both integral parameters (average radiative and convective Nusselt numbers at the heater surface) and local parameters (streamlines and isotherms). The geometry (Fig. 1) is selected with an eye to simulate thermal transmission and air motion within the room with a heater. The present results are received using one Intel Core i7 processor of 3.30 GHz with 16 GB memory RAM.

Figure 2 shows the distribution of isolines of the stream function and temperature as a function of the surface emissivity ($\tilde{\varepsilon} = 0.3$ $- a$, $\tilde{\varepsilon} = 0.9$ $- b$) at $f = 0.001\pi$. Two low-intensity symmetric convective cells determining clockwise (right) and counterclockwise (left) circulations are formed inside the large-scale enclosure. The formation of the present vortex structures is caused by cooling of the analyzed area due to the heat exchange with the environment (since the ambient temperature is lower than the initial temperature inside the enclosure), as well as the influence of the heater which located on the bottom wall of the enclosure. The presence of an upward flow of warm air in the central part of the cavity reflects the formation of a thermal plume. The bottom solid wall is a zone least affected by the environment, this is due to heat dissipation from the heat source. With an increase in the surface emissivity, the intensity of convective flow decreases, which is confirmed by the characteristic decrease in the maximum value of the stream function in the core of the convective cell $|\Psi|_{max}^{\tilde{\varepsilon}=0.9} = 0.0255 < |\Psi|_{max}^{\tilde{\varepsilon}=0.3} = 0.0269$. Just to clarify again, thermal transmission is caused by both the effect of time-periodic heat generation from the heater and the cooling of the enclosure owing to the heat exchange with an environment. The radiation significantly affects the distribution of temperature inside the cavity. In general, an increment of surface emissivity allows increasing the mean temperature inside the air-filled enclosure.

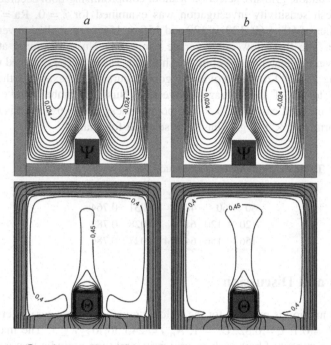

Fig. 2. Isotherms Θ and streamlines Ψ for different values of surface emissivity at $f = 0.001\pi$: $\tilde{\varepsilon} = 0.3 \ - a, \tilde{\varepsilon} = 0.9 \ - b$

The impact of the surface emissivity of heater and walls surfaces, as well as dimensionless time on the average convective and radiative Nusselt numbers (at the surface of the heater) has been analyzed numerically in Fig. 3. It should be noted that if the boundary conditions are not isothermal (for example, volumetric heat generation from the heater) then achieving the steady state conditions is extremely difficult. Time-periodic heat generation was determined as $q = q_v(1 - sin(f\tau))$. When surface emissivity values increase, the temperature gradient at the heater surface is reduced and natural convection is also weakened. So, with an increase in $\tilde{\varepsilon}$, a characteristic decrease in the intensity of convective heat transfer has been observed (at $\tau = 5000$, the average convective Nusselt number decreases by 8.1% when changing of $\tilde{\varepsilon}$ from 0.3 to 0.9), while a growth of the average radiative Nusselt number with increasing surface emissivity values has been observed (at $\tau = 5000$, a change of $\tilde{\varepsilon}$ from 0.3 to 0.9 leads to an increase in the average radiative Nusselt number up to 3.47 times). Although profiles are time-periodic, the peaks and valleys are increasing with time. This is due to the fact that heat conduction equation inside the heater has an internal heat generation term.

More detailed influence of the values of surface emissivity on the temperature profiles at the middle cross-Sect. 10 = 0.6 is depicted in Fig. 4. A rise of surface emissivity of internal surfaces affects a slight decreasing temperature directly within the heater. This fact also leads to a growth of the mean temperature within the large-scale enclosure. It should be noted that the increase in surface emissivity is manifested in a noticeable decrease in temperature inside the heater.

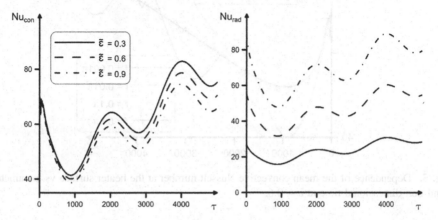

Fig. 3. Variation of the average convective (Nu_{con}) and radiative (Nu_{rad}) Nusselt numbers at the heat source surface with the dimensionless time and the surface emissivity $\tilde{\varepsilon}$ at $f = 0.001\pi$.

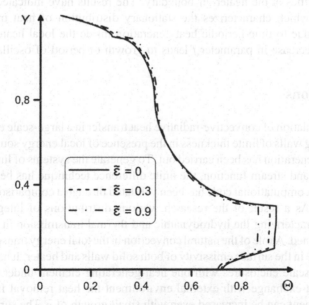

Fig. 4. Temperature profiles at $X = 0.6$ for different values of surface emissivity

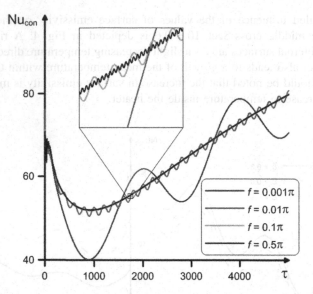

Fig. 5. Dependence of the mean convective Nusselt number at the heater surface vs. parameter f and non-dimensional time at $\tilde{\varepsilon} = 0.6$.

In order to study the influence of time-periodic heat generation on energy transport, the convective heat exchange from the heater was estimated by the average convective Nusselt number. Figure 5 presents the average convective Nusselt number for various surface emissivities at the heater-air boundary. The results have indicated that there is no time point which characterizes the stationary distribution of heat transfer coefficients. This is due to time-periodic heat generation inside the local heater. It is worth noting that a decrease in parameter f leads to growth of period of oscillation.

4 Conclusions

Numerical simulation of convective-radiative heat transfer in a large-scale enclosure with heat-conducting walls of finite thickness in the presence of local energy source with time-periodic heat generation has been carried out. To generate the systems of linear equations using vorticity and stream function, the finite difference technique has been employed. The developed computational code has been validated through a comparison with data of other authors. As a result of the research, various distributions of integral and local parameters characterizing the hydrodynamic and thermal transmission in the enclosure have been obtained. A role of the natural convection in the total energy transport decreases with an increase in the surface emissivity of both solid walls and heater. It has been shown that in a large-scale enclosures with one heat-generating element under conditions of convective heat exchange with external environment the heat removal from the heat-generation element can be increased even with small growth of $\tilde{\varepsilon}$. The surface radiation has a significant effect on the total heat exchange and can reach more than 50% of the total heat flux, particularly if the heater and wall surfaces have high emissivity.

Acknowledgements. This work was supported by the Russian Science Foundation (Project No. 19-79-00296).

References

1. Li, X., Tu, J.: Evaluation of the eddy viscosity turbulence models for the simulation of convection–radiation coupled heat transfer in indoor environment. Energy Build. **184**, 8–18 (2019)
2. Rahimi, M., Sabernaeemi, A.: Experimental study of radiation and free convection in an enclosure with a radiant ceiling heating system. Energy Build. **42**, 2077–2082 (2010)
3. Muresan, C., Menezo, C., Bennacer, R., Vaillon, R.: Numerical simulation of a vertical solar collector integrated in a building frame: radiation and turbulent natural convection coupling. Heat Transfer Eng. **27**, 29–42 (2006)
4. Saha, S.C., Khan, M.M.K.: A review of natural convection and heat transfer in attic-shaped space. Energy Build. **43**, 2564–2571 (2011)
5. Ridouane, E., Campo, A., Hasnaoui, M.: Turbulent natural convection in an air-filled isosceles triangular enclosure. Int. J. Heat Fluid Flow **27**, 476–489 (2006)
6. Ghasemi, M., Fathabadi, M., Shadaram, A.: Numerical analysis of turbulent natural convection heat transfer inside a triangular-shaped enclosure utilizing computational fluid dynamic code. Int. J. Numer. Methods Heat Fluid Flow **18**, 14–23 (2008)
7. Das, D., Roy, M., Basak, T.: Studies on natural convection within enclosures of various (non-square) shapes – a review. Int. J. Heat Mass Transfer **106**, 356–406 (2017)
8. Zhang, X., Su, G., Yu, J., Yao, Z., He, F.: PIV measurement and simulation of turbulent thermal free convection over a small heat source in a large enclosed cavity. Build. Environ. **90**, 105–113 (2015)
9. Velusamy, K., Sundararajan, T., Seetharamu, K.N.: Interaction effects between surface radiation and turbulent natural convection in square and rectangular enclosures. J. Heat Transfer **123**, 1062–1070 (2001)
10. Shati, A.K.A., Blakey, S.G., Beck, S.B.M.: A dimensionless solution to radiation and turbulent natural convection in square and rectangular enclosures. J. Eng. Sci. Technol. **7**, 257–279 (2012)
11. Vivek, V., Sharma, A.K., Balaji, C.: Interaction effects between laminar natural convection and surface radiation in tilted square and shallow enclosures. Int. J. Thermal Sci. **60**, 70–84 (2012)
12. Martyushev, S.G., Miroshnichenko, I.V., Sheremet, M.A.: Numerical analysis of spatial unsteady regimes of conjugate convective-radiative heat transfer in a closed volume with an energy source. J. Eng. Phys. Thermophys. **87**, 124–134 (2014)
13. Miroshnichenko, I.V., Sheremet, M.A.: Turbulent natural convection heat transfer in rectangular enclosures using experimental and numerical approaches: a review. Renew. Sustain. Energy Rev. **82**, 40–59 (2018)
14. Sharma, A.K., Velusamy, K., Balaji, C., Venkateshan, S.P.: Conjugate turbulent natural convection with surface radiation in air-filled rectangular enclosures. Heat Mass Transfer **50**, 625–639 (2007)
15. Kogawa, T., Okajima, J., Sakurai, A., Komiya, A., Maruyama, S.: Influence of radiation effect on turbulent natural convection in cubic cavity at normal temperature atmospheric gas. Int. J. Heat Mass Transfer **104**, 456–466 (2017)
16. Wang, H., Xin, S., Le Quere, P.: Numerical study of natural convection-surface radiation coupling in air-filled square cavities. C.R. Mecanique **334**, 48–57 (2006)

17. Alvarado, R., Xaman, J., Hinojosa, J., Alvarez, G.: Interaction between natural convection and surface thermal radiation in tilted slender cavities. Int. J. Thermal Sciences **47**, 355–368 (2008)
18. Nia, M.F., Nassab, S.A.G., Ansari, A.B.: Transient combined natural convection and radiation in a double space cavity with conducting walls. Int. J. Therm. Sci. **128**, 94–104 (2018)
19. Basak, T., Roy, S., Krishna Babu, S., Balakrishnan, A.R.: Finite element analysis of natural convection flow in a isosceles triangular enclosure due to uniform and non-uniform heating at the side walls. Int. J. Heat Mass Transf. **51**, 4496–4505 (2008)
20. Miroshnichenko, I.V., Sheremet, M.A.: Effect of thermal conductivity and emissivity of solid walls on time-dependent turbulent Conjugate convective-radiative heat transfer. J. Appl. Comput. Mech. **5**(2), 207–216 (2019)
21. Miroshnichenko, I.V., Sheremet, M.A.: Numerical simulation of turbulent natural convection combined with surface thermal radiation in a square cavity. Int. J. Numer. Methods Heat Fluid Flow **25**, 1600–1618 (2015)
22. Dixit, H.N., Babu, V.: Simulation of high rayleigh number natural convection in a square cavity using the Lattice Boltzmann Method. Int. J. Heat Mass Transf. **49**, 727–739 (2006)
23. Zhuo, C., Zhong, C.: LES-based filter-matrix lattice boltzmann model for simulating turbulent natural convection in a square cavity. Int. J. Heat Fluid Flow **42**, 10–22 (2013)
24. Le Quéré, P.: Accurate solutions to the square thermally driven cavity at High Rayleigh number. Comput. Fluids **20**, 29–41 (1991)

Development of an Object-Oriented Programming Tool Based on FEM for Numerical Simulation of Mineral-Slurry Transport

Sergio Peralta[1], Jhon Cordova[1], Cesar Celis[1], and Danmer Maza[2(✉)]

[1] Pontificia Universidad Católica del Perú, Lima, Peru
{sergio.peralta,jcordovaa}@pucp.pe,
ccelis@pucp.edu.pe
[2] Pontifícia Universidade Católica do Rio de Janeiro, Rio de Janeiro, Brazil
danmer@lmmp.mec.puc-rio.br

Abstract. The early stages of the development of a finite element method (FEM) based computational tool for numerically simulating mineral-slurry transport involving both Newtonian and non-Newtonian flows are described in this work. The rationale behind the conception, design and implementation of the referred object-oriented programming tool is thus initially highlighted. A particular emphasis is put on several architectural aspects accounted for and object class hierarchies defined during the development of the tool. Next one of the main modules composing the tool under development is further described. Finally, as a means of illustration, the use of the FEM based tool for simulating two-dimensional laminar flows is discussed. More specifically, canonical configurations widely studied in the past are firstly accounted for. A more practical application involving the simulation of a mineral-slurry handling device is then studied using the power-law rheological model. The results from the simulations carried out highlight the usefulness of the tool for realistically predicting the associated flow behavior. The FEM based tool discussed in this work will be used in future for carrying out high-fidelity numerical simulations of turbulent multiphase flows including fluid-particle interactions.

Keywords: Computational fluid dynamics · Finite element method · Object-oriented programming · Non-Newtonian fluid · Mineral-slurry transport

1 Introduction

Mining companies continuously set aside large amounts of resources to optimize their mineral extraction processes. This occurs because energy costs and greenhouse gas emissions can be significantly reduced by optimizing such processes [1]. Notice that mineral processing often involves transporting mixtures of water and/or chemical solutions carrying ground rocks known as mineral-slurries [2]. Mineral-slurry transport involves indeed multiphase flows featuring solid, liquid and gas phases. Consequently, in order to optimize mineral-slurry transporting systems, the associated complex flows need to be carefully characterized. Based on their rheological behavior, mineral-slurries may

© Springer Nature Switzerland AG 2020
V. V. Krzhizhanovskaya et al. (Eds.): ICCS 2020, LNCS 12143, pp. 163–177, 2020.
https://doi.org/10.1007/978-3-030-50436-6_12

exhibit Newtonian or non-Newtonian rheological properties [3]. Particle size, slurry density, slurry viscosity, mass flow rate, and friction losses are thus the main design factors when designing or selecting transport slurry systems [2]. In practice both experimental or numerical techniques can be used to study and optimize mineral extraction processes [4]. Carrying out experimental tests is nevertheless relatively expensive. In contrast, numerical modeling allows performing numerical experiments at lower costs than laboratory or full-scale experiments.

Different numerical methods can be used to numerically solve the diverse multiphase flows present in practical applications. The finite element method (FEM) [5] constitutes one of the referred numerical methods. Industrial device designing and/or operating conditions optimization are some of the main FEM practical applications. In mining engineering FEM has been used for various purposes in the past [6–9]. Notice that there are several issues to be accounted for when properly developing a FEM-based computational tool. In order to obtain reliable numerical results in the shortest possible computational time for instance, FEM models need to be coded such to maximize the computational resources involved. A suitable programming paradigm should be also firstly selected before implementing a computational tool based on FEM. Object-oriented programming (OOP) has been preferred in the past for carrying out the implementation of some FEM solvers [10]. Currently there are available several commercial and open-source computational packages for FEM related applications [11–14]. Even though some of the aforementioned computational packages can be used for simulating mineral-slurry transport, it has been decided to develop a new computational tool. The main reason behind this important decision is that the new tool will include besides FEM based solvers several other numerical methods suitable for grinding processes modelling. In general, the referred new computational package is expected to be flexible enough, reducing the issues present when coupling different numerical approaches and properly modeling complex flows.

In this work, the early stages of the development of a FEM based computational tool for numerically simulating mineral-slurry transport involving both Newtonian and non-Newtonian fluids are described. Accordingly, Sect. 2 describes the rationale behind the conception, design and implementation of the referred FEM based tool. A particular emphasis is put on several architectural aspects accounted for and object class hierarchies defined during its development. One of the main modules composing the tool under development is further described in Sect. 3, including the rheological models implemented so far in the FEM based tool under development. In Sect. 4 in turn, as a means of illustration, the use of the tool for simulating two-dimensional laminar flows is discussed. More specifically, canonical configurations widely studied in the past are firstly accounted for [15–17]. A more practical application involving the simulation of a mineral-slurry flow device is then studied using a non-Newtonian fluid model. Finally, Sect. 5 summarizes the main conclusions drawn from the results obtained here.

2 FEM Tool Development

2.1 Development Context

The FEM tool discussed here constitutes one of the modules of a larger computational package under development called CFLOWSS (Complex FLOWS Solver) [18]. As its name indicates CFLOWSS aims to numerically solve different complex flows including mineral-slurries. Currently several Eulerian and Lagrangian approaches are being implemented in CFLOWSS. The feasibility of easily carrying out one-way or two-way coupling between different numerical approaches represents therefore one of the key features of CFLOWSS. As highlighted in Fig. 1, some of the Eulerian numerical techniques initially accounted for include finite volume methods (FVM) [19] and finite element methods (FEM) [5]. Notice that it was decided to implement a FEM module in CFLOWSS due to its proven applicability for non-Newtonian flows numerical modelling [20]. Lagrangian approaches include in turn discrete elements methods (DEM) [21] and smoothed particle hydrodynamics (SPH) [22]. Other numerical techniques such as spectral methods (SM) [23] and particle finite element methods (PFEM) [24] will be considered as well in future. Notice that the multiphysics involved when modeling complex flows justifies the inclusion of the numerical techniques indicated above.

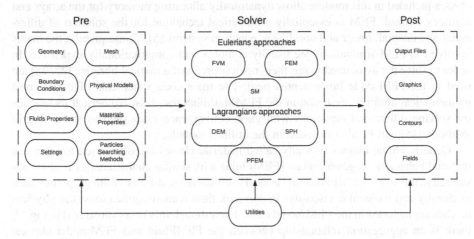

Fig. 1. Main modules composing CFLOWSS (Complex FLOWS Solver).

CFLOWSS is implemented using both an OOP paradigm and C++ as the main programming language. The development of the CFLOWSS modules, models and numerical algorithms is continuous in order to improve the accuracy of the flow modeling processes undertaken with its aid. Three main modules compose CFLOWSS, (i) Pre, (ii) Solver and (iii) Post (Fig. 1). Geometry, mesh, physical models and other settings required for starting a simulation case are defined in the CFLOWSS Pre module. The CFLOWSS Solver module includes in turn the Eulerian and Lagrangian approaches available for the numerical simulation of complex flows. One of these numerical techniques (FEM) and its associated models and modules will be further discussed in the following sections. Finally, the CFLOWSS Post module mainly deals with the creation of results-containing output files able to be understood by well-known post-processing tools such as ParaView [25].

2.2 FEM Module Class Hierarchy

Details of the implementation of the FEM related modules in the CFLOWSS computational tool are provided in this section. Notice that the OOP paradigm used here favor the FEM implementation because FEM is essentially a modular method [26]. Many objects and abstract data have been thus extracted from the FEM formulation. As expected the FEM code implementation is based on many OOP principles such as abstraction, encapsulation, inheritance, composition and polymorphism [10]. For the sake of brevity however, only the most relevant FEM class hierarchy (Fig. 2) is discussed here. All classes shown in Fig. 2 were implemented within the CFLOWSS modules described above. In Fig. 2 the relationship between classes is represented by different line patterns. Solid lines represent inheritance. Dashed lines indicate in turn composition. Finally dash-dotted lines imply data transfer between friend classes.

The CFLOWSS utility module shown in Fig. 1 and Fig. 2 includes the main utilities required to perform a FEM simulation. For instance, the FEMArray and FEMMatrix classes included in this module allow dynamically allocating memory for the arrays and matrices utilized. FEM is essentially a numerical technique for the solution of differential and integral, linear and non-linear, equation systems [5]. Consequently, the maths involved in FEM simulations are usually complex and computationally expensive. In order to solve the associated maths then, open source mathematical libraries [27, 28] are used in the FEM code implemented here. The main code statements of the referred mathematical libraries are found in the FEMMathlibraries class. Numerical techniques for solving differential equations with good convergence rates such as the multigrid methods [15] will be also included in the utilities module.

Generic virtual classes typically called material classes are often used to represent material behaviors. A generic class (FEMFluid) with similar characteristics is also considered in this work. FEMFluid class is used to characterize different fluid properties such as density and molecular viscosity. In complex flow related applications, the physical models are included in the FEMmodel class. Even though this is not illustrated in Fig. 2, there is an aggregation relationship between the FEMFluid and FEMmodel classes. FEMnonNewtonianModel is a generic class that has no physical meaning, which include information and procedures common to all non-Newtonian derived models. Two well-known non-Newtonian models, power law [16] and Carreau-Yasuda [17], have been implemented so far for simulating mineral-slurry transport. For instance, the FEMPowerLaw class is used to estimate the shear rate and the power-law apparent viscosity in a non-Newtonian power-law fluid application. The FEMTruncatedPowerLaw class involves an improvement in the power-law model achieved by using a four-parameter model [17]. In order to generate a FEM mesh, it is necessary to have a properly defined geometry and information about each element node. FEMMesh class results thus from a composition of FEMGeometry and FEMNode classes. Notice that FEMGeometry class allows setting the geometric configuration accounted for in a FEM simulation, whereas FEMNode class is used to store nodal information. FEMBoundaryConditions class is used in turn to set the governing equations boundary conditions (BC). Configurable BC include Dirichlet and Neumman BC [5]. Finally, FEMBasicFunctions allows configuring the FEM interpolation functions on each element.

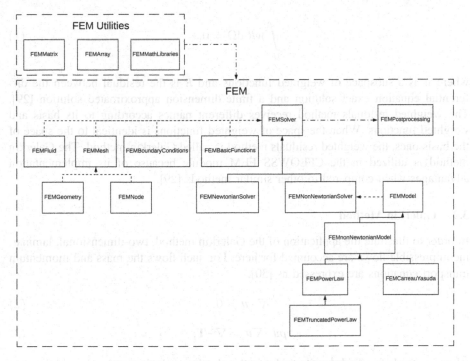

Fig. 2. FEM module class hierarchy in CFLOWSS.

The main class of the CFLOWSS FEM module is the FEMSolver class. This class is composed of FEMFluid, FEMMesh, FEMBoundaryConditions and FEMBasicFunctions classes. That is, the FEMSolver class instantiates objects of these classes and accesses through these objects all methods included in such classes. Furthermore, the FEMSolver class has also access to all mathematical algorithms related classes required to solve any FEM simulation. Non-linear equations systems for instance are linearized and then solved with numerical methods such as the Newton-Raphson one [5]. Both FEMNewtonianSolver and FEMNonNewtonianSolver classes are inherited from the FEMSolver one (Fig. 2). The FEMnonNewtonianModel generic class composes the latter class. Finally, results related data coming out from the FEMSolver class are sent to the FEMPostprocessing class for further post-processing.

3 FEM Formulation

3.1 Finite Element Method Approach

In the CFLOWSS FEM module under discussion here, the particular approach used for solving the flow governing equations is the weighted residuals method [5]. In general, for a domain Ω, the weighted residuals formulation implies solving a equation of the form,

$$\int_{\Omega} wR \, \mathrm{d}\Omega = 0, \tag{1}$$

where w is a subspace of weighted functions and R is the residual between the differential equation exact solution and a finite-dimension approximated solution [29]. The weighted residuals method receives different names according to its basis and weighted functions. When the space of weighted functions is identical to the space of the basis ones, the weighted residuals method is called Galerkin method. The Galerkin method is utilized in the CFLOWSS FEM module because of its implementation advantages when compared to other similar methods [29].

3.2 Galerkin Method

In order to illustrate the application of the Galerkin method, two-dimensional, laminar, incompressible flows are accounted for here. For such flows the mass and momentum transport equations are expressed as [30],

$$\nabla \cdot \boldsymbol{u} = 0, \tag{2}$$

$$\rho \boldsymbol{u} \cdot \nabla \boldsymbol{u} = \nabla \cdot \mathbf{T}, \tag{3}$$

where ρ stands for fluid density and $\boldsymbol{u}$ is the absolute velocity vector. In addition, for Newtonian fluids, the total stress tensor $\mathbf{T}$ is computed according to,

$$\mathbf{T} = -p\mathbf{I} + \mu \left[\nabla \boldsymbol{u} + (\nabla \boldsymbol{u})^T \right]. \tag{4}$$

In Eq. (4), p represents the absolute pressure and μ the fluid dynamic viscosity.

The application of the Galerkin method to a flow governed by the transport equations highlighted above leads to x and y-axis momentum equations of the form,

$$R_{mx}^i = \int_{\Omega} \rho \psi_i \left[u \frac{\partial u}{\partial x} + v \frac{\partial u}{\partial y} \right] + \frac{\partial \psi_i}{\partial x} \left[-p + 2\mu \frac{\partial u}{\partial x} \right]$$
$$+ \frac{\partial \psi_i}{\partial y} \left[\mu \left(\frac{\partial u}{\partial y} + \frac{\partial v}{\partial x} \right) \right] \mathrm{d}\Omega - \int_{\Gamma} \psi_i f_x \mathrm{d}\Gamma = 0 \ i = 1, \ldots, N, \tag{5}$$

$$R_{my}^i = \int_{\Omega} \rho \psi_i \left[u \frac{\partial v}{\partial x} + v \frac{\partial v}{\partial y} \right] + \frac{\partial \psi_i}{\partial y} \left[-p + 2\mu \frac{\partial v}{\partial y} \right]$$
$$+ \frac{\partial \psi_i}{\partial x} \left[\mu \left(\frac{\partial u}{\partial y} + \frac{\partial v}{\partial x} \right) \right] \mathrm{d}\Omega - \int_{\Gamma} \psi_i f_y \mathrm{d}\Gamma = 0 \ i = 1, \ldots, N. \tag{6}$$

In these equations, ψ_i are the weighted functions related to the momentum equations, which are equivalent to the basis ones. Biquadratic functions have been taken into account for the basis functions implementation in the CFLOWSS FEM module. N here is the number of algebraic momentum equations. Notice that the line integrals $\int_\Gamma \psi_i f \mathrm{d}\Gamma$ appearing in Eqs. (5) and (6) allow imposing boundary conditions on the domain Ω. The mass transport equation reads in turn as follows,

$$R_c^i = \int_\Omega \left(\frac{\partial u}{\partial x} + \frac{\partial v}{\partial y} \right) \chi_i \mathrm{d}\Omega = 0 \; i = 1, \ldots, M, \tag{7}$$

where χ_i are the weighted functions related to the continuity equations. Linear functions have been accounted for in this case. M represents here the number of algebraic equations related to continuity. In order to complement the FEM modeling detailed here, some rheological models implemented in the CFLOWSS FEM based tool for simulating non-Newtonian flows are described in the following section.

3.3   Rheological Modeling of Non-Newtonian Flows

Mineral slurries rheological properties vary according to its solids concentration. When the mineral slurries' solids concentration increases, the associated apparent viscosity η is no longer constant but depends on shear stress τ and/or shear rate $\dot{\gamma}$ [3]. For non-Newtonian flows related applications, the dynamic viscosity μ appearing in Eqs. (5) and (6) has to be replaced with a term of the form $\eta(\dot{\gamma})$. In this work, the $\dot{\gamma}$ magnitude is calculated locally using the second invariant of the rate-of-strain tensor [16]. The associated components of the rate-of-strain tensor are estimated in turn from the local velocity field [31]. Three typical models commonly used to compute the $\eta(\dot{\gamma})$ term are included in the CFLOWSS FEM based tool.

Power-Law Model
The simplest representation of the apparent dynamic viscosity of a non-Newtonian fluid is given by the power-law model [16],

$$\eta(\dot{\gamma}) = K\dot{\gamma}^{n-1}, \tag{8}$$

where K and n are the empirically obtained consistency and flow behavior indexes, respectively. Notice that this model may present some issues at relatively low and high shear rates, where additional model parameters are required to properly characterize the non-Newtonian fluids' behavior [17]. The power-law model can be used then with caution in non-Newtonian flow applications such as mineral slurries [3].

Carreau-Yasuda Model
The Carreau-Yasuda model is a five-parameter model used when there are large deviations from the power-law one at relatively low and high shear rates [16]. Non-Newtonian rheological properties of flows carrying solid particles such as mineral

slurries may be described indeed by the Carreau-Yasuda model [3]. In this model, the apparent dynamic viscosity is estimated from,

$$\eta(\dot{\gamma}) = \mu_\infty + (\mu_0 - \mu_\infty)(1 + (\lambda\dot{\gamma})^a)^{\frac{n-1}{a}}. \tag{9}$$

Where μ_0 is the zero-shear rate viscosity, μ_∞ the infinite-shear-rate viscosity, λ the fluid relaxation time and a is a dimensionless parameter describing the width of the transition region between the zero-shear-rate region and the power-law one [32].

Truncated Power-Law Model

The four-parameter truncated power-law model represents in turn an improvement on the pure power-law one for all shear rate range [17]. The apparent dynamic viscosity is computed in this case as,

$$\eta(\dot{\gamma}) = \begin{cases} \mu_0, & \dot{\gamma} < \dot{\gamma}_1, \\ K\dot{\gamma}^{n-1}, & \dot{\gamma}_1 < \dot{\gamma} < \dot{\gamma}_2, \\ \mu_\infty, & \dot{\gamma}_2 < \dot{\gamma}. \end{cases} \tag{10}$$

In this model, the shear rates at which the low- and high-viscosity cut-offs are introduced are estimated from, respectively,

$$\dot{\gamma}_1 = (\frac{K}{\mu_0})^{\frac{1}{1-n}}, \qquad \dot{\gamma}_2 = (\frac{K}{\mu_\infty})^{\frac{1}{1-n}}. \tag{11}$$

Some results obtained from the use of the CFLOWSS FEM based tool accounting for these different rheological models are described in the following section.

4   Results and Discussion

The main results obtained from the simulation of two-dimensional laminar flows using the CFLOWSS FEM based tool are discussed in this section. Canonical configurations widely studied in the past are firstly accounted for, followed by a practical application involving the simulation of a mineral-slurry flow device and non-Newtonian flows.

4.1   Newtonian Laminar Flow Between Infinite Parallel Plates

The fully-developed two-dimensional Newtonian laminar incompressible steady-state flow between two infinite parallel plates is firstly considered here. The associated geometry accounted for is schematically shown in Fig. 3(a). The analytical solution characterizing this particular flow was obtained from Wang's work [33].

Velocity fields computed using thirty different meshes (ranging from 2 to 1800 elements) were initially compared with their corresponding analytical solution and the obtained average root-mean-square error (RMSE) was about 4.9E−08. For illustrative purposes, velocity profiles computed using a 1250-elements mesh and analytically are compared in Fig. 3(b). As noticed from this plot, the numerical and analytical results agree relatively well.

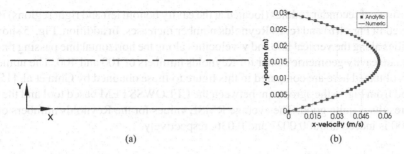

(a)   (b)

Fig. 3. (a) Geometry accounted for when computing Newtonian laminar flows between parallel plates. (b) x-velocity profiles obtained analytically and using the CFLOWSS FEM based tool.

4.2 Lid-Driven Square Cavity

The lid-driven cavity configuration has been widely utilized in the past as a standard verification case for new CFD codes [34]. This configuration has been also used here for verifying the CFLOWSS FEM based tool accounting for a Newtonian flow. The numerical results discussed in this section have been compared with those obtained by Ghia et al. [15] at Reynolds numbers of 100 and 400. Accordingly, Fig. 4(a) shows RMSE based mesh independency results involving the u and v velocity components along the vertical and horizontal lines passing through the cavity geometric center. These results indicate in particular that when the mesh has more than 200 elements, the RMSE values remain almost constant and less than 5%.

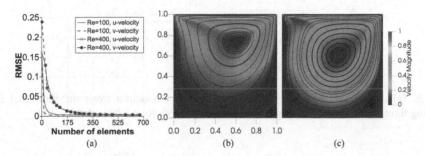

(a)   (b)   (c)

Fig. 4. (a) RMSE vs number of mesh elements. Lid-driven cavity velocity contours and streamline patterns computed by the CFLOWSS FEM based tool for (b) Re = 100 and (c) Re = 400.

The lid-driven cavity velocity contours and streamline patterns computed by the CFLOWSS FEM based tool for Reynolds numbers of 100 and 400 are shown in Fig. 4(b) and (c), respectively. A 676-elements mesh with an aspect ratio equal to one was utilized for obtaining these results, which are comparable to those obtained by Ghia et al. [15]. In particular, the streamline patterns included in these plots reveal that the main eddy center moves towards the cavity geometric center as Reynolds number increases [34].

The growth of secondary eddies (located at the cavity bottom left and right regions) is also observed in Fig. 4(b) and (c) as Reynolds number increases. In addition, Fig. 5 shows u-velocities along the vertical line and v-velocities along the horizontal line passing through the square cavity geometric center for Reynolds numbers of 100 and 400. The numerical results obtained here are compared in this figure to those obtained by Ghia et al. [15]. As noticed from Fig. 5 the agreement between the CFLOWSS FEM based tool and the Ghia et al. results is quite good. The average RMSE values for the Reynolds numbers of 100 and 400 is indeed about 0.0027 and 0.036, respectively.

4.3   Non-Newtonian Laminar Flow Between Infinite Parallel Plates

For the same considerations described in Sect. 4.1, in order to verify the CFLOWSS FEM based tool capabilities for dealing with non-linear fluid viscosities, a non-Newtonian flow passing through two infinite parallel plates has been simulated. The non-Newtonian flow velocity profiles computed then using the Carreau-Yasuda, power-law and truncated power-law models are shown in Fig. 6. When possible, the numerical results obtained here are compared with their respective analytical solutions [16, 17]. Notice that for the Carreau-Yasuda model there is no analytical solution available [32].

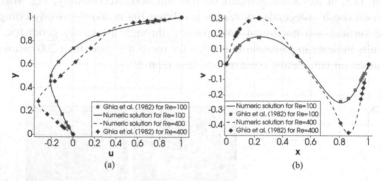

Fig. 5. (a) u-velocities along the vertical line and (b) v-velocities along the horizontal line passing through the square cavity geometric center for both Re = 100 and Re = 400.

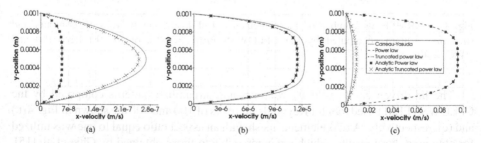

Fig. 6. Non-Newtonian flow velocity profiles obtained using Carreau-Yasuda, power-law and truncated power-law models at different pressure gradients. (a) 1 Pa/m, (b) 5 Pa/m, (c) 75 Pa/m.

In addition, the fluid properties and geometric configuration simulated in this case were extracted from [17], accounting for a density value of 1 kg/m³. The main results indicate that, accounting for a 50 × 25 elements mesh, RMSE values of 2.4E−23 and 3.7E−13 in Fig. 6(a), 5.6E−09 and 2.7E−10 in Fig. 6(b), and 2.6E−07 and 6.6E−7 in Fig. 6(c) characterize the truncated power-law and power-law models, respectively. Moreover, the influence of the pressure gradient values on the velocity profiles associated with each rheological model included in Fig. 6 is similar to that described in literature [17].

4.4 Slurry Receiving Chamber

The receiving chamber of a slurry distribution box [35] has been simulated as well using the CFLOWSS FEM based tool as an example of the engineering situations where this tool can be applied. It is worth noticing that turbulent models for multiphase flows including fluid-particle interactions will be implemented later on in the CFLOWSS FEM based tool so they are not currently available yet. Even so, currently this tool can produce reliable results for some operating conditions characterizing mineral slurry-handling devices. For instance, when the mineral-slurry both features relatively low flow velocities and contains large amounts of solid particles, the mineral-slurry transport may be considered as a laminar non-Newtonian flow [36]. The referred situation can be modeled using the current version of the CFLOWSS FEM based tool under discussion here. In particular, different flow patterns, velocity fields and geometric configurations associated with the associated slurry-handling devices can be assessed using it. Accordingly, a particular geometric configuration of a slurry receiving chamber has been accounted for in this work as shown in Fig. 7(a). Notice from this figure that the vertical distance of the mineral-slurry exit measured from the device bottom is denoted by Y. This vertical distance has been varied in the numerical simulations carried out here for flow pattern analyses. All results discussed in this section have been obtained accounting for a non-Newtonian power-law model, and mineral-slurry related parameters reading as follows, (i) $\rho = 1370$ kg/m³, (ii) $\mu = 1.6$ Pa.s and (iii) $n = 0.4$ [16].

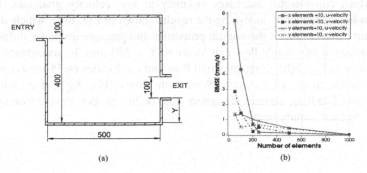

(a)   (b)

Fig. 7. (a) Slurry receiving chamber geometric configuration (units in mm). (b) RMSE versus number of mesh elements.

A mesh-independency assessment has been initially carried out as shown in Fig. 7 (b), accounting for a 1000 elements mesh as the reference mesh for computing the RMSE values. Several numerical simulations were so performed gradually increasing the number of mesh elements along one axis and keeping constant this number along the other one, i.e., varying the mesh elements aspect ratio. For instance, the square-symbols curves in Fig. 7(b) were obtained from simulations where the number of mesh elements along the x-axis was constant and equal to 10. Similarly, the ×-symbols curves correspond to simulations where the number of mesh elements along the y-axis was equal to 10. Notice that the RMSE values of the two velocity components were accounted for in this initial verification stage. The results shown in Fig. 7(b) indicate in particular when the mesh has more than 500 elements, the RMSE values (lower than 0.5 mm/s) no longer vary significantly.

The main results obtained from the numerical simulations of the slurry receiving chamber accounted for here are summarized in Fig. 8. In particular, Fig. 8(a) to (c) and Fig. 8(d) to (f) show, respectively, the streamlines and velocity contours characterizing pure Newtonian (n = 1.0) and non-Newtonian mineral-slurry (n = 0.4) flows passing through the receiving chamber when varying the flow exit location. For obtaining these results, a constant velocity profile at the receiving chamber entry of 75 mm/s was initially imposed. The referred results highlight in particular that both velocity fields and eddy sizes are influenced by the fluid's rheological properties and flow exit locations. When increasing Y indeed, progressively larger eddies are formed in the chamber bottom left corner and those in the top right one almost disappear. Properly locating these eddies is important because sedimentation is fostered in eddies-containing regions.

Pressure drops as a function of the flow exit location (Y value), the flow entry velocity and the fluid rheological properties were also studied as illustrated in Fig. 8(g). This plot shows firstly that for the three imposed entry velocities, the highest pressure drops are obtained at the flow exit relative locations of 0 and 1, Y = 0 and Y = 400 mm, respectively. In addition, for relatively low flow entry velocities ($\approx$25 mm/s), compared to the Newtonian case (n = 1), the mineral-slurry flow (n = 0.4) presents higher pressure drops. This finding comes from the use of the power-law model, which considerably increases viscosity at low velocity gradients. This last aspect can be confirmed by analyzing the results shown in Fig. 8(h), which shows the relationship between fluid's rheological properties and pressure drops for different flow entry velocities at one single flow exit location, Y = 250 mm. In accordance with the results shown in Fig. 8(h), at relative high flow entry velocities ($\approx$75 mm/s), relatively higher pressure drops characterize Newtonian flows (Fig. 8(g)). Following these results, slurry-handling devices designers can define proper device configurations according to their requirements.

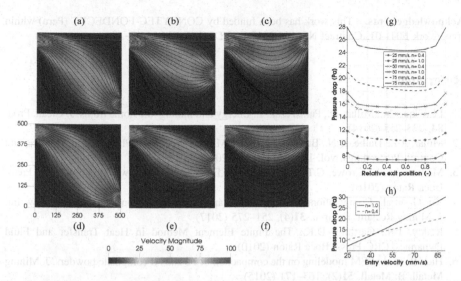

Fig. 8. Velocity contours and streamline patterns characterizing Newtonian (n = 1.0) ((a) to (c)) and mineral-slurry (n = 0.4) flows passing through the receiving chamber for exit positions Y of 0 mm ((a), (d)), 200 mm ((b), (e)) and 400 mm ((c), (f)). (g) Pressure drop versus relative exit position for different imposed inlet velocities and rheological properties. (h) Pressure drop versus receiving chamber entry velocity for Newtonian and mineral-slurry flows at Y = 250 mm.

5 Conclusions

In this work, the early stages of the development of a FEM based computational tool for numerically simulating mineral-slurries transport involving both Newtonian and non-Newtonian flows were described. The rationale behind the conception, design and implementation of the referred FEM based tool was thus initially described. Several architectural aspects accounted for and object class hierarchies defined during its development were detailed next. The particular approach used for solving the flow governing equations, i.e., the Galerkin method, a variation of the weighted residuals one, was emphasized as well, along with the non-Newtonian flow models implemented in the referred tool for dealing with mineral-slurries rheological aspects. As a means of illustration, the FEM based computational tool discussed in this work was used for studying two-dimensional, laminar, incompressible flows. Canonical configurations studied in the past were firstly accounted for, followed by a practical application involving the simulation of a mineral-slurry handling device and non-Newtonian flows. When possible, the numerical results obtained here were compared with analytical solutions and data available in literature and the corresponding agreement was relatively good. It is concluded therefore that the FEM based tool is useful for realistically predicting the associated flow behavior. Once fully developed, the computational tool discussed in this work will be used for carrying out high-fidelity numerical simulations of turbulent multiphase flows including fluid-particle interactions.

Acknowledgements. This work has been funded by CONCYTEC-FONDECYT (Peru) within framework E041-01, Contract No. 155-2018-FONDECYT-BM-IADT-AV.

References

1. Levesque, M., Millar, D., Paraszczak, J.: Energy and mining–the home truths. J. Clean. Prod. **84**, 233–255 (2014)
2. Mular, A.L., Halbe, D.N., Barratt, D.J. (eds.): Mineral Processing Plant Design, Practice, and Control: Proceedings, vol. 1. SME, Denver (2002)
3. Michaelides, E., Crowe, C.T., Schwarzkopf, J.D.: Multiphase Flow Handbook. CRC Press, Boca Raton (2016)
4. Xu, G., et al.: Computational fluid dynamics applied to mining engineering: a review. Int. J. Mining Reclam. Environ. **31**(4), 251–275 (2017)
5. Reddy, J.N., Gartling, D.K.: The Finite Element Method in Heat Transfer and Fluid Dynamics. CRC Press, Boca Raton (2010)
6. Han, P., et al.: FEM modeling on the compaction of Fe and Al composite powders. J. Mining Metall. B: Metall. **51**(2), 163–171 (2015)
7. Pater, Z., Bartnicki, J., Kazanecki, J.: 3D finite elements method (FEM) analysis of basic process parameters in rotary piercing mill. Metalurgija **51**(4), 501–504 (2012)
8. Kovačević, D., et al.: Optimal finite elements method (FEM) model for the jib structure of a waterway dredger. Metalurgija-Zagreb **51**(1), 113 (2012)
9. Kocich, R., Kliber, J., Herout, M.: Finite elements method (FEM) simulation based prediction of deformation and temperature at rolling of tubes on a pilgrim mill. Metalurgija **48**(4), 267–271 (2009)
10. Bittencourt, M.L.: Using C++ templates to implement finite element classes. Eng. Comput. **17**(7), 775–788 (2000)
11. COMSOL Multiphysics. https://www.comsol.com
12. ANSYS. https://www.ansys.com
13. Deal.II. https://www.dealii.org/
14. ELMER. https://www.csc.fi/web/elmer
15. Ghia, U.K.N.G., Ghia, K.N., Shin, C.T.: High-resolutions for incompressible flow using the Navier-Stokes equations and a multigrid method. J. Comput. Phys. **48**(3), 387–411 (1982)
16. Chhabra, R.P., Richardson, J.F.: Non-Newtonian Flow and Applied Rheology: Engineering Applications. Butterworth-Heinemann, Oxford (2011)
17. Lavrov, A.: Flow of truncated power-law fluid between parallel walls for hydraulic fracturing applications. J. Non-Newtonian Fluid Mech. **223**, 141–146 (2015)
18. Angeles, L., Celis, C.: Assessment of neighbor particles searching methods for discrete element method (DEM) based simulations. In: VI International Conference on Particle-Based Methods – Fundamentals and Applications (2019)
19. Versteeg, H.K., Malalasekera, W.: An Introduction to Computational Fluid Dynamics: The Finite, vol. Method. Pearson Education, London (2007)
20. Maza, D., Carvalho, M.S.: Trailing edge formation during slot coating of rectangular patches. J. Coat. Technol. Res. **14**(5), 1003–1013 (2017). https://doi.org/10.1007/s11998-017-9962-1
21. Matuttis, H.-G., Chen, J.: Understanding the Discrete Element Method: Simulation of Non-Spherical Particles for Granular and Multi-Body Systems. Wiley, Hoboken (2014)
22. Violeau, D.: Fluid Mechanics and the SPH Method: Theory and Applications. Oxford University Press, Oxford (2012)

23. Canuto, C., et al.: Spectral Methods: Evolution to Complex Geometries and Applications to Fluid Dynamics. Springer, Heidelberg (2007). https://doi.org/10.1007/978-3-540-30728-0
24. Oñate, E., Owen, R. (eds.): Particle-Based Methods: Fundamentals and Applications, vol. 25. Springer, Dordrecht (2011). https://doi.org/10.1007/978-94-007-0735-1
25. ParaView. https://www.paraview.org/
26. Kumar, S.: Object-oriented finite element programming for engineering analysis in C++. JSW 5(7), 689–696 (2010)
27. SUNDIALS. https://computing.llnl.gov/projects/sundials
28. GSL. https://www.gnu.org/software/gsl/
29. Jacob, F., Ted, B.: A First Course in Finite Elements. Wiley, Hoboken (2007)
30. Cengel, Y.A.: Fluid Mechanics. Tata McGraw-Hill Education, New York (2010)
31. Gabbanelli, S., Drazer, G., Koplik, J.: Lattice Boltzmann method for non-Newtonian (power-law) fluids. Phys. Rev. E 72(4), 046312 (2005)
32. Wang, C.-H., Ho, J.-R.: A lattice Boltzmann approach for the non-Newtonian effect in the blood flow. Comput. Math. Appl. 62(1), 75–86 (2011)
33. Wang, C.Y.: Exact solutions of the steady-state Navier-Stokes equations. Ann. Rev. Fluid Mech. 23(1), 159–177 (1991)
34. Shankar, P.N., Deshpande, M.D.: Fluid mechanics in the driven cavity. Ann. Rev. Fluid Mech. 32(1), 93–136 (2000)
35. Adriasola, J.M., Janssen, R.H.A.: Best practices for design of slurry flow distributions (2018)
36. Abulnaga, B.E.: Slurry Systems Handbook. McGraw-Hill, New York (2002)

Descending Flight Simulation of Tiltrotor Aircraft at Different Descent Rates

Ayato Takii[1](✉), Masashi Yamakawa[1], and Shinichi Asao[2]

[1] Kyoto Institute of Technology, Matsugasaki,
Sakyo-ku, Kyoto 606-8585, Japan
kpp_fsl_ta@yahoo.co.jp
[2] College of Industrial Technology, 1-27-1, Amagasaki, Hyogo 661-0047, Japan

Abstract. Helicopters and tiltrotor aircrafts are known to fall into an unstable state called vortex ring state when they descend rapidly. This paper presents a six degrees of freedom descending flight simulation of a tiltrotor aircraft represented by the V-22 Osprey, considering the interaction between fluid and a rigid body. That is, an aircraft affects the surrounding flow field by rotating the rotors, and flies with the generated force as thrust. Similarly, an orientation of the airframe is controlled by aerodynamic force which is generated by manipulating the shape. This numerical analysis is a complicated moving boundary problem involving motion of an air-frame or rotation of rotors. As a numerical approach, the Moving Computational Domain (MCD) method in combination with the multi-axis sliding mesh approach is adopted. In the MCD method, the whole computational domain moves with objects in the domain. At this time, fluid flow around the objects is generated by the movement of the boundaries. In addition, this method removes computational space restrictions, allowing an aircraft to move freely within the computational space regardless of a size of a computational grid. The multi-axis sliding mesh approach allows rotating bodies to be placed in a computational grid. Using the above approach, the flight simulation at two different descent rates is performed to reveal a behavior of a tiltrotor aircraft and a state of the surrounding flow field.

Keywords: Computational fluid dynamics · Tiltrotor · Flight simulation · Vortex ring state

1 Introduction

It is known that rotorcrafts such as a helicopter or a tiltrotor fell into unstable state called vortex ring state (VRS) when it descends at high descent rate or at an angle close to vertical descent. At this time, lift generated by the rotor disc is significantly reduced. This phenomenon occurs when the descent speed approaches the rotor wake speed and airflow recirculate through the rotor disk. VRS has also caused several rotorcraft crashes. VRS accidents have also been reported on a tiltrotor V-22 Osprey which has recently attracted attention. This was due to the pilot descending beyond the recommended descent range [1]. The behavior of rotors in VRS has long been known to aerodynamic experts, and a considerable number of studies have been reported [2]. Many researches have been investigating the VRS characteristics of the rotor by

© Springer Nature Switzerland AG 2020
V. V. Krzhizhanovskaya et al. (Eds.): ICCS 2020, LNCS 12143, pp. 178–190, 2020.
https://doi.org/10.1007/978-3-030-50436-6_13

experimental tests and numerical simulations [3, 4]. However, most of these studies focused on flight tests or wind tunnel tests in steady conditions. Vortex ring state is a complex phenomenon with large unsteady airflow. Hence practical investigations can also be useful to fully understand aerodynamics and develop accurate prediction methods. Researches on unsteady simulation of helicopters have been conducted [5, 6]. However, few studies on tiltrotor aircraft have been seen.

The purpose of this study is to compute unsteady flow around a tiltrotor aircraft when it descends and to simulate the motion of the aircraft in such airflow. The six degrees of freedom descending flight simulation is performed on a tiltrotor aircraft represented by V-22 Osprey, considering an interaction between fluid and a rigid body. By performing numerical computations at different descent rates, effects on motion of an aircraft and complex fluid phenomena caused by aircraft movement are analyzed. To perform descent flight simulations, coupled computation is conducted that integrate both flight dynamics which deals with aircraft movement and fluid dynamics which deals with fluid flow around aircraft. In other words, the aircraft is treated as a rigid body and flies under force generated by interaction with the surrounding fluid flow. To solve such a complicated moving boundary problem, a combination of the MCD method based on the unstructured moving-grid finite-volume method and the multi-axis sliding mesh method was adopted as numerical approach.

2 Numerical Approach

2.1 Governing Equation

To solve the flow field around a tiltrotor aircraft, three-dimensional Euler equations are used as governing equations. The equations in the conservation form are written as follows:

$$\frac{\partial q}{\partial t} + \frac{\partial E}{\partial x} + \frac{\partial F}{\partial y} + \frac{\partial G}{\partial z} = 0, \tag{1}$$

$$q = \begin{bmatrix} \rho \\ \rho u \\ \rho v \\ \rho w \\ e \end{bmatrix}, E = \begin{bmatrix} \rho u \\ \rho u^2 + p \\ \rho uv \\ \rho uw \\ u(e+p) \end{bmatrix}, F = \begin{bmatrix} \rho v \\ \rho uv \\ \rho v^2 + p \\ \rho vw \\ v(e+p) \end{bmatrix}, G = \begin{bmatrix} \rho w \\ \rho uw \\ \rho vw \\ \rho w^2 + p \\ w(e+p) \end{bmatrix}, \tag{2}$$

where q is the conserved quantity vector, E, F, G are the inviscid flux vectors. As unknowns, ρ is the density, u, v, w are the x, y, z components of the velocity vector and e is the total energy per unit volume. The working fluid assumed to be perfect gas, the pressure p is defined as follows:

$$p = (\gamma - 1)\left[e - \frac{1}{2}\rho\left(u^2 + v^2 + w^2\right)\right], \tag{3}$$

where γ is the specific heat ratio. In this paper, γ is taken as 1.4, and the initial conditions of density, pressure, velocity components in the x, y and z directions are given by $\rho = 1.0$, $p = 1.0/\gamma$, $u = 0.0$, $v = 0.0$, $w = 0.0$, respectively.

2.2   Unstructured Moving-Grid Finite-Volume Method

The unstructured moving-grid finite-volume method [7] is used to perform computation involving movement and deformation of grids. In the method, fluxes are evaluated on a control volume in the space-time unified domain (x, y, z, t) so that the geometric conservation law (GCL) [8] is satisfied. By applying the divergence theorem to space-time unified control volume, the three-dimensional Euler equations are deformed as follows:

$$\int_{\Omega} \left(\frac{\partial q}{\partial t} + \frac{\partial E}{\partial x} + \frac{\partial F}{\partial y} + \frac{\partial G}{\partial z} \right) d\Omega = \int_{\partial \Omega} (E, F, G, q) \cdot \tilde{n} dV$$

$$= \sum_{l=1}^{6} (E\tilde{n}_x + F\tilde{n}_y + G\tilde{n}_z + q\tilde{n}_t)_l = 0, \tag{4}$$

where Ω is the control volume, $\tilde{n} = (\tilde{n}_x, \tilde{n}_y, \tilde{n}_z, \tilde{n}_t)$ is the outward unit normal vector on $\partial \Omega$ and l indicates faces of the control volume.

2.3   Moving Computational Domain Method

In this paper, the flow field around a tiltrotor aircraft that occurs when the aircraft descends is computed. In the traditional approach, an aircraft is placed in a uniform flow and a flow around that is computed. However, it cannot be applied to this simulation involving free movement of an aircraft. Such an analysis needs to be treated as a moving boundary problem. The moving computational domain (MCD) method [9–14] is a technique for computing fluid flow around moving object. In this method, whole computational domain moves with an object which is inside the domain as shown in Fig. 1. At this time, fluid flow around an object is generated by movement of boundary surfaces. Specifically, it is caused by the boundary conditions given to the moving boundary. In addition, this method removes computational space limitations. Therefore, an aircraft can move freely in the calculation space regardless of the size of computational grid. Combining with the multi-axis sliding mesh approach [15], this method is applied not only to flight of an aircraft, but also to rotation of two rotors. The computational procedure is based on the unstructured moving-grid finite-volume method described in the last subsection. Flow variables are defined at the center of cells in unstructured mesh. The flux vectors are evaluated using the Roe flux difference splitting scheme [16] with MUSCL scheme. Gradient limiter is Hishida's limiter of van Leer type [17]. The two-stage Runge-Kutta method is used to solve local time stepping.

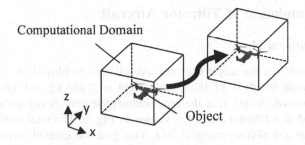

Fig. 1. Conceptual figure of the MCD method

2.4 Coupled Computation

In this paper, position and orientation of a tiltrotor aircraft is determined by coupled computation by considering the interaction between the body and surrounding fluid flow. An aircraft is considered as rigid body in three-dimensional space. Therefore, the motion of an aircraft is determined by solving the equations of motion of six degrees of freedom, consisting of three degrees of freedom translation and rotation. Newton's law of motion and Euler's rotation equation are used for translation and rotation, respectively. Each equation of motion is as follows:

$$m\frac{d^2r}{dt^2} = F, \tag{5}$$

$$I\frac{d\omega}{dt} + \omega \times I\omega = T, \tag{6}$$

where m is the mass of the aircraft, r is the position vector of the center of the aircraft, F is the force vector, I is the inertia tensor (written in matrix form), ω is the angular velocity vector and T is the torque vector around the center. The torque and force are calculated by integrating the pressure on surface of the aircraft. These equations are discretized as follows:

$$mv^{n+1} = mv^n + F^n\Delta t, \tag{7}$$

$$r^{n+1} = r^n + 0.5(v^{n+1} + v^n)\Delta t, \tag{8}$$

$$I\omega^{n+1} = I\omega^n + (T^n - \omega^n \times I\omega^n)\Delta t, \tag{9}$$

where a subscript n is the time level and Δt is the time step.

3 Flight Simulation of Tiltrotor Aircraft

3.1 Computational Mesh

V-22 Osprey which is the major tilt-rotor aircraft was employed as a computational model. The overall length is 17.48 m, the weight is 21545 kg and the rotor speed is 397 rpm. The overall length L is the representative length. A computational domain and surface grid of a tiltrotor aircraft is shown in Fig. 2. The total number of cells is 3297890 and the size of computational 30 L. This grid was created by using MEGG3D [18, 19]. In this computation, the computational domain is divided into 7 domains to use multi-axis sliding mesh approach. Figure 3 shows domain decomposition. Computational grids for each domain are shown from Figs. 4, 5, 6 and 7.

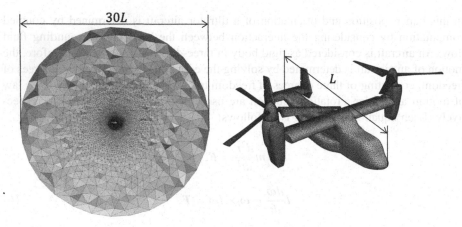

Fig. 2. Computational domain and surface grid

Fig. 3. Domain decomposition

Domain1   Domain2

Fig. 4. Mesh around rotors

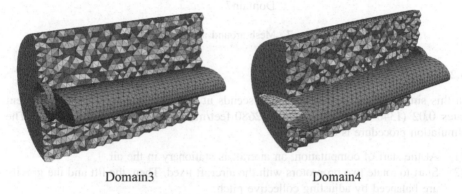

Domain3   Domain4

Fig. 5. Mesh around wings

In this simulation, the aircraft converts into ... rises 0.02 (Fig. ...) 0.050 feet ... the simulation procedure is as follows:

(1) Assume start of computation, an aircraft is stationary in the air.
(2) Start to rotate the rotors with the aircraft fixed. The lift and the weight are balanced by adjusting collective pitch.
(3) Unlock aircraft having ... descend, adjusting ... to give pitch to maintain the constant descent rate.

The predictive motion of the aircraft is controlled by tilting two propeller disks, forward and ... In addition, ... controlled by increasing ... and (Fig. ...

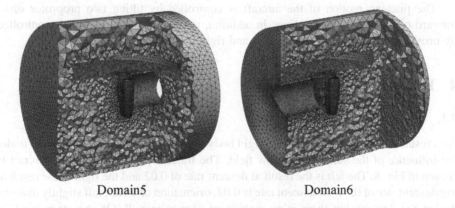

Domain5   Domain6

Fig. 6. Mesh around nacelles

... rigid body ... the influence of the ... gravity field. The ... shown in Fig. 8. The left is the resultant descent rate of 0.02 and the right is the result descent rate ... and descent rate is 0.02, ... during the descent, but there is an attitude of change ... overall. On the other hand, ... descent rate 0.04, ... equilibrium of the aircraft is largely disturbed ... time elapses. These changes of pitch angle of an aircraft about their own axes indicated by Fig. 9. As shown in the figure, in both cases, pitch angle sum to change amount = 300, in the case of K = 0.02, no further change, seen after some time ... aircraft is initially disturbed. On the other hand, in the case of K = 0.04, pitch angle have ...

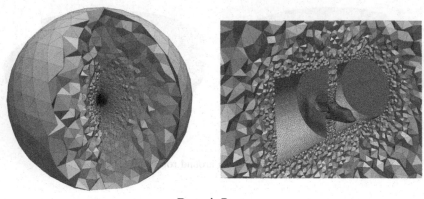

Domain7

Fig. 7. Mesh around a fuselage

3.2 Flight Conditions

In this simulation, a tiltrotor aircraft descends at different descent rates. The descent rates 0.02 (1340 feet/min) and 0.04 (2680 feet/min) are used for this analysis. The simulation procedure is as follows:

(1) At the start of computation, an aircraft is stationary in the air.
(2) Start to rotate two proprotors with the aircraft fixed. Then, the lift and the gravity are balanced by adjusting collective pitch.
(3) Unlock aircraft having been fixed and make an aircraft descend, adjusting collective pitch to maintain the constant descent rate.

The pitching motion of the aircraft is controlled by tilting two proprotor disks forward or backward at same time. In addition, the rolling of the aircraft is controlled by providing a difference between left and right collective pitches.

4 Results

4.1 Movement of Tiltrotor Aircraft

As a result of computation with fluid-rigid body interaction, an aircraft descended under the influence of the surrounding flow field. The trajectory of descent of an aircraft is shown in Fig. 8. The left is the result at descent rate of 0.02 and the right is the result at the descent rate of 0.04. At descent rate is 0.02, orientation of an aircraft slightly disturbs during the descent, but there is no significant change overall. On the other hand, at descent rate 0.04, it is confirmed that orientation of an aircraft is largely disturbed as time elapses. Time changes of pitch angle of an aircraft about these two cases is indicated by Fig. 9. As shown in the figure, in both cases, pitch angles start to change around $t = 300$. In the case of $V_z = 0.02$, no further change is seen after orientation of an aircraft is initially disturbed. On the other hand, in the case of $V_z = 0.04$, pitch angle of an aircraft gradually increases, and is expected to increase in the future. From the above, it was confirmed that orientation of an aircraft was more disturbed as the descent rate increased.

Fig. 8. Trajectory of an aircraft movement at descent rate 0.02 (left) and 0.04 (right)

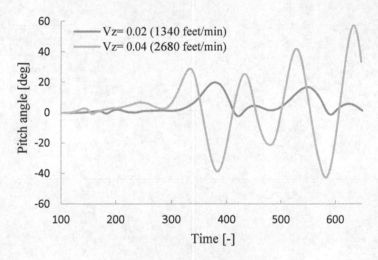

Fig. 9. Pitch angle of an aircraft at each descent rate V_z

4.2 Flow Fields

Figure 10 is isosurfaces ($V = 0.1$) of magnitude of the flow velocity around the tiltrotor aircraft during descent at descent rate 0.02. It illustrates a state from the start of descent. The figure at the upper left shows the state immediately after the descent started, where the blade pitch has been reduced and acceleration has started to the target descent rate. At this time, the iso-surface of the magnitude value of the velocity is in the state of clinging around the rotors. In this situation, you can see that the lift loses to gravity and the aircraft descends. The right figure next to that shows a little advanced time level.

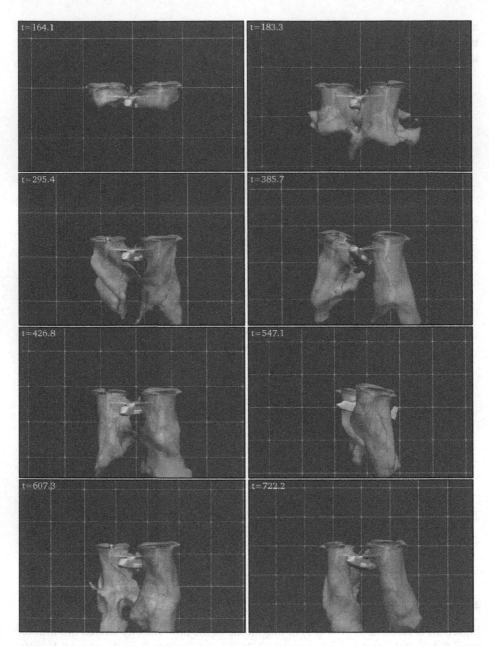

Fig. 10. Isosurfaces of magnitude of flow velocity ($V = 0.1$) at descent rate 0.02 (about 1340 feet/min)

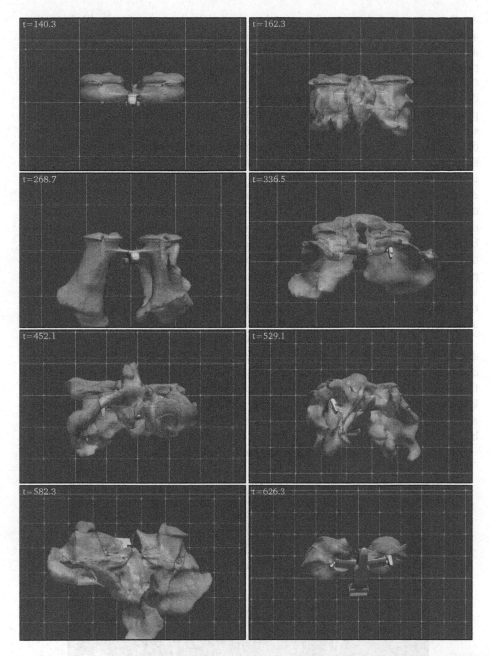

Fig. 11. Isosurfaces of magnitude of flow velocity ($V = 0.1$) at descent rate 0.04 (about 2680 feet/min)

As the velocity of the aircraft approaches the target descent rate, the blade pitch is adjusted so that lift and gravity balance and maintain the constant speed of the body. As a result, the downwash blowing down from the rotors appeared again. Therefore, at this point, it can be considered that the aircraft obtained the lift as a reaction force against the downwash. After that, the body descends at an almost constant speed. During the descent, it is constantly confirmed that downwash is blown down from left and right proprotors. This indicates that the aircraft flies with proper lift. Next, Fig. 11 is iso-surfces ($V = 0.1$) of magnitude of the flow velocity around an aircraft during descent at descent rate 0.04, illustrating a state from the start of descent. As in the previous case, the upper left figure shows the state just after the aircraft has started to descent. The isosurface has its state clinging around the rotors due to the acceleration to the target descent rate. Then, for a while after reaching the target descent rate, you can see the downwash that is blown down from the left and right rotors (about $t = 270$). However, unlike previous case, an aircraft begins to pitch and the surrounding airflow begins to be greatly disturbed after a while. At this point, the downwash which has been blown down from left and right proprotors disappeared. Without the downwash, the pro-protors is not able to get enough lift. Moreover, once this turbulence has occurred, vicious circle between large changes in the attitude and turbulence of the fluid around the aircraft continue to repeat, increasing the instability of the body. As a result, the aircraft loses control to maintain its flight attitude.

Figure 12 and Fig. 13 show visualization of flow velocity vectors in a typical state during descent at each descent rate. At descent rate 0.02, it is confirmed that air flows downward from proprotors. On the other hand, at descent rate 0.04, it is confirmed that flow velocity vector is swirling at tip of proprotor. This is a vortex ring state, also known as power settling. In this situation, airflows that should be blown down from rotors is immediately recirculated to the rotors and results in a decrease in lift provided by that. Therefore, it is considered that an orientation control by proprotors does not work sufficiently and the aircraft becomes more unstable.

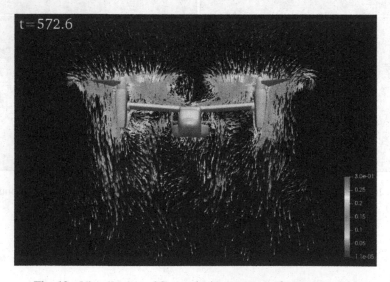

Fig. 12. Visualization of flow velocity vectors at descent rate 0.02

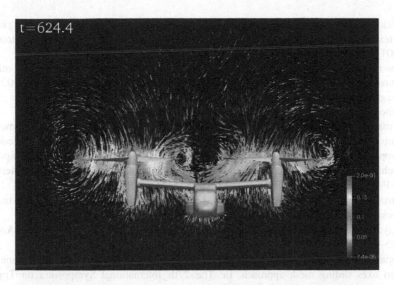

Fig. 13. Visualization of flow velocity vectors at descent rate 0.04

5   Conclusions

Six degrees of freedom descending flight simulation was conducted for a tiltrotor aircraft represented by V-22 Osprey, considering interaction of fluid and rigid body. To solve complicated moving boundary problem, a combination of the MCD method based on the unstructured moving-grid finite-volume method and the multi-axis sliding mesh method was adopted as numerical approach. Descending simulation is computed in two cases: descent rate 0.02 and 0.04. The results of movement of descending aircraft was presented by trajectory and pitch angle. Next, the state of flow fields around the aircraft was presented in a form of visualization of an isosurface of flow velocity and flow velocity vectors. At descent rate 0.02, the aircraft continued a stable descent flight throughout. At descent rate 0.04, it is confirmed that orientation of an aircraft is largely disturbed as time elapses. At that time, it was found from the flow velocity vectors that the aircraft has fallen into vortex ring state. This flight simulation qualitatively showed that a tiltrotor aircraft would become unstable and fall into vortex ring state during flight at high descent rate.

Acknowledgements. This publication was subsidized by JKA through its promotion funds from KEIRIN RACE.

References

1. Gertler, J.J.: V-22 osprey tilt-rotor aircraft: background and issues for congress. In: Library of Congress Washington DC Congressional Research Service, p. 57 (2009)
2. Johnson, W.: Model for Vortex Ring State Influence on Rotorcraft Flight Dynamics. NASA Ames Research Center, P76, Moffett Field, CA, United States (2005)

3. Betzina, M.D.: Tiltrotor descent aerodynamics: a small-scale experimental investigation of vortex ring state. In: American Helicopter Society 57th Annual Forum, Washington, D.C. (2001)
4. Brown, R.E., Leishman, J.G., Newman, S.J., Perry, FJ.: Blade twist effects on rotor behaviour in the vortex ring state. In: 28th European rotorcraft forum, Bristol (2002)
5. Grzegorczyk, K.: Analysis of influence of helicopter descent velocity changes on the phenomena of vortex ring state. Adv. Sci. Technol. Res. J. **7**(17), 35–41 (2013)
6. Surmacz, K., Ruchała, P., Stryczniewicz, W.: Wind tunnel tests of the development and demise of Vortex Ring State of the rotor. In: Advances in Mechanics: Theoretical, Computational and Interdisciplinary Issues, Proceedings of the 3rd Polish Congress of Mechanics (PCM) and 21st International Conference on Computer Methods in Mechanics (CMM), Gdansk, 8–11 September 2015, pp. 551–554. CRC Press, Leiden (2016)
7. Yamakawa, M., Matusno, K.: Unstructured moving-grid finite-volume method for unsteady shocked flows. J. Comput. Fluids Eng. **10**(1), 24–30 (2005)
8. Obayashi, S.: Freestream capturing for moving coordinates in three dimensions. AIAA J. **30**, 1125–1128 (1992)
9. Yamakawa, M., Chikaguchi, S., Asao, S.: Numerical simulation of tilt-rotor plane using multi axes sliding mesh approach. In: The 27th International Symposium on Transport Phenomena, Honolulu (2016)
10. Watanabe, K., Matsuno, K.: Moving computational domain method and its application to flow around a high-speed car passing through a hairpin curve. J. Comput. Sci. Technol. **3**(2), 449–459 (2009)
11. Yamakawa, M., Mitsunari, N., Asao, S.: Numerical simulation of rotation of intermeshing rotors using added and eliminated mesh method. Procedia Comput. Sci. **108**, 1883–1892 (2017)
12. Asao, S., et al.: Simulations of a falling sphere with concentration in an infinite long pipe using a new moving mesh system. Appl. Thermal Eng. **72**, 29–33 (2014)
13. Asao, S., et al.: Parallel computations of incompressible flow around falling spheres in a long pipe using moving computational domain method. Comput. Fluids **88**, 850–856 (2013)
14. Yamakawa, M., Takekawa, D., Matsuno, K., Asao, S.: Numerical simulation for a flow around body ejection using an axisymmetric unstructured moving grid method. Comput. Thermal Sci. **4**(3), 217–223 (2012)
15. Takii, A., Yamakawa, M., Asao, S., Tajiri, K.: Six degrees of freedom numerical simulation of tilt-rotor plane. In: Rodrigues, J.M.F., et al. (eds.) ICCS 2019. LNCS, vol. 11536, pp. 506–519. Springer, Cham (2019). https://doi.org/10.1007/978-3-030-22734-0_37
16. Roe, P.L.: Approximate riemann solvers parameter vectors and difference schemes. J. Comput. Phys. **43**, 357–372 (1981)
17. Hishida, M., Hashimoto, A., Murakami, K., Aoyama, T.: A new slope limiter for fast unstructured CFD solver FaSTAR. In: Proceedings of 42nd Fluid Dynamics Conference/Aerospace Numerical Simulation Symposium, pp. 1–10 (2010). (in Japanese)
18. Ito, Y., Nakahashi, K.: Surface triangulation for polygonal models based on CAD data. Int. J. Numer. Methods Fluids **39**(1), 75–96 (2002)
19. Ito, Y.: Challenges in unstructured mesh generation for practical and efficient computational fluid dynamics simulations. Comput. Fluids **85**, 47–52 (2013)

The Quantization Algorithm Impact in Hydrological Applications: Preliminary Results

Alessio De Rango[1]($\boxtimes$)(iD), Luca Furnari[1](iD), Donato D'Ambrosio[2](iD), Alfonso Senatore[1](iD), Salvatore Straface[1](iD), and Giuseppe Mendicino[1](iD)

[1] DIAm, University of Calabria, Rende, Italy
{alessio.derango,luca.furnari,alfonso.senatore,salvatore.straface,
giuseppe.mendicino}@unical.it
[2] DeMACS, University of Calabria, Rende, Italy
donato.dambrosio@unical.it

Abstract. A computationally efficient surface to groundwater coupled hydrological model is being developed based on the Extended Cellular Automata (XCA) formalism. The three-dimensional unsaturated flow model was the first to be designed and implemented in OpenCAL. Here, the response of the model with respect to small variations of the quantization threshold has been assessed, which is the OpenCAL's quantization algorithm's parameter used for evaluating cell's steady state condition. An unsaturated flow test case was considered where the elapsed times of both the non quantized execution and the execution run by setting the quantization threshold to zero (with respect to the moist content variable) were already evaluated. The model response has been assessed in terms of both accuracy and computational performance in the case of an MPI/OpenMP hybrid execution. Results have pointed out that a very good tradeoff between accuracy and computational performance can be achieved, allowing for a considerable speed-up of the model against a very limited loss of precision.

Keywords: Extended Cellular Automata · Unsaturated flow modelling · Computational efficiency

1 Introduction

Due to climate change, water management has become a key factor for sustainable development scenarios [1,18]. The development of increasingly efficient hydrological models is an essential element in the study of the water cycle dynamics. Many models have been proposed to predict these phenomena (cfr. [27]). Most of them use a physical approach based on partial differential equations (PDEs) to describe the real phenomenon. Nevertheless, since their analytical solution is often unknown, an approximate solution is obtained by applying a numerical method such as Finite Differences, Finite Elements, or Finite Volumes (cf. [6,14]). In most cases, a time-explicit recurrence relation is obtained,

© Springer Nature Switzerland AG 2020
V. V. Krzhizhanovskaya et al. (Eds.): ICCS 2020, LNCS 12143, pp. 191–204, 2020.
https://doi.org/10.1007/978-3-030-50436-6_14

which expresses the next state of the generic (discrete) element of the system as a function of the current states of a limited number of neighboring elements. For this reason, the most explicit schemes can be formalized in terms of Cellular Automata (CA) [25], which is one of the best-known and most widely used decentralized discrete parallel calculation models (cf. [4,5,8,9,16,20,22,26,29]). Nevertheless, a kind of global control over the simulation is often useful to make the modeling of certain processes more straightforward. Therefore, both decentralized transitions and steering operations can be adopted to model complex systems (cfr. [10]).

Different software systems were proposed to model complex systems formalized in terms of Cellular Automata. Among them, *Camelot* was developed as a (commercial) integrated development environment (IDE) based on the CARPET C-like MPI-based language [2]. Another example is *libAuToti* [30], which represented a C++ free application program interface (API) based on MPI. Unfortunately, both systems are no longer developed. However, the *OpenCAL* (Open Computing Abstraction Layer) open source software library [7] has been recently proposed as an alternative to the aforementioned software systems. In particular, it permits to exploit heterogeneous parallel computational devices on clusters of interconnected workstations. Both CPUs and GPUs, as well as other many-core accelerators, are supported. Despite its recent release, OpenCAL was already applied to the simulation of different physical phenomena, including a debris flow evolving on real topographic surface, as well as to the implementation of graphics convolutional filters, fractal generation algorithms. In addition, a particle system based on the Discrete Element Method was also implemented and preliminarily tested [12,13].

In [11] we developed a preliminary unsaturated model, $XCA - Flow$, based on the discretization of Darcy's law, by adopting an explicit Finite Difference scheme to obtain a discrete formulation [17,23]. We then implemented it by using OpenCAL, and therefore applied the model to simulate a three-dimensional test case. Specifically, giving rise to a topologically connected phenomenon, it was simulated for assessing the computational advantage that the OpenCAL's *quantization algorithm* proved to be able to provide in similar cases [7]. The *quantization threshold* used to characterize the stationary cells was set to zero, meaning that the reference hydraulic head difference had to overcome zero in order to *activate* the cell. The highest speed-up (with respect to the parallel non quantized simulation) of about 4.4 was achieved by considering a hybrid distributed/shared memory execution on a dual socket Intel Xeon-based workstation, using 2 MPI processes and 16 OpenMP threads, for a total of 32 computing threads. In particular, the fastest execution of the non quantized simulation required about 535 s using 32 threads on a single MPI process, while the quantized simulation lasted 122 s only. This good result suggested us to further investigate the OpenCAL's quantization algorithm in order to understand if a good tradeoff between accuracy and computational performance was possible in the case non-zero quantization thresholds were considered.

The paper is organized as follows. Section 2 briefly outlines the $XCA - Flow$ unsaturated flow model, while Sect. 3 briefly presents the OpenCAL architecture and its quantization algorithm. Section 4 illustrates the case study, together with the outcomes and the accuracy evaluation, while Sect. 5 assesses the simulations outcomes and presents the computational performances achieved. Finally, Sect. 6 concludes the paper with a general discussion envisaging possible future developments.

2   The $XCA - Flow$ Flow Model

The $XCA - Flow$ was developed for addressing several typical hydrological problems, from parcel/hillslope to large basin scales, benefiting from the computational advantages allowed by the XCA-based approach. The direct discrete formulation of the Richards' equation where the $XCA - Flow$ model relies on is thoroughly described by Mendicino et al. [23], who already used the XCA formalism for developing an unsaturated flow model using the CAMELot environment. Here, for the sake of clarity and completeness, the main equations of the model are summarized.

The Richards' equation is a non linear degenerate elliptic-parabolic partial differential equation [19] describing double-phases flow in porous media, it is given by combining the mass conservation equation with the momentum conservation equation represented by the Darcys' law. Considering the pressure head ψ as the dependent variable, the Richards' equation for an isotropic porous medium is written as [3]:

$$C_c(\psi)\frac{\partial\psi}{\partial t} - \nabla\big[K(\psi)\nabla\psi\big] - \frac{\partial K(\psi)}{\partial z} = 0 \tag{1}$$

where the pressure head ψ [m] is related to the hydraulic head h [m] by the equation $\psi = h - z$, being z [m] the elevation, $C_c(\psi)$ is the specific ritention capacity [m^{-1}], given by the relation $C_c(\psi) = d\theta/dh$, where in turns θ is the moisture content [m^3 m^{-3}] and $K(\psi)$ is the hydraulic conductivity [m s^{-1}].

The $XCA - Flow$ model solving Eq. 1 is formally defined as:

$$XCA - Flow = \langle(D, S, I), X, Q, (\Sigma, \Phi), \Gamma, (T, t), \omega, \tau_{\mathbb{N}}\rangle$$

where:

- $D = [0, n_r - 1] \times [0, n_c - 1] \times [0, n_s - 1] \subset \mathbb{Z}^3$ is the three-dimensional discrete computational domain, with n_r, n_c, and n_s the number of rows, columns, and slices, respectively.
- $S = [0, n_r \cdot \Delta s] \times [0, n_c \cdot \Delta s] \times [0, n_s \cdot \Delta s] \subset \mathbb{R}^3$ is the continuum three-dimensional realm corresponding to D, subdivided in cubic cells of side Δs. The function μ defines the mapping to D as:

$$\mu : D \to \mathbb{R}^3$$

$$(\iota_1, \iota_2, \iota_3) \mapsto (\iota_1 \cdot \Delta s, \iota_2 \cdot \Delta s, (n_s - 1 - \iota_3) \cdot \Delta s)$$

- $I = \{I_\rho, I_\beta, I_\tau \subseteq D\}$ is the set of domain interfaces, where
  - $I_\rho \subseteq \pi_{s=0}$ is the boundary region, belonging to the top surface $\pi_{s=0}$, that is affected by rain;
  - I_β the remaining domain boundary region over which no input rain is considered;
  - $I_\tau = D \setminus I_\rho \cup I_\beta$ the set of cells belonging to the inner domain, where the system evolution occurs.
- $X = \{(0,0,0), (-1,0,0), (0,-1,0), (0,1,0), (1,0,0), (0,0,-1), (0,0,1)\}$ is the von Neumann neighborhood.
- $Q = Q_\theta \times Q_K \times Q_h \times Q_{C_c} \times Q_{conv}$ is the set of states for the cell, where:
  - Q_θ is the set of values representing the soil moisture content;
  - Q_K is the set of values representing the unsaturated hydraulic conductivity;
  - Q_h is the set of values representing the hydraulic total head;
  - Q_{C_c} is the set of values representing the specific retention capacity;
  - Q_{conv} is the set of values representing the temporal step size which guarantees the numerical convergence;
- $\Sigma = \{\sigma_\rho, \sigma_\beta, \sigma_\tau\}$ is the set of local transition functions or kernels. In particular:
  - $\sigma_\rho : Q_h \times Q_{C_c} \to Q_h$ accounts for input rain. Specifically, if h, C_c, r_{ir}, Δt, and Δs^2 denote hydraulic head, specific retention capacity, rain intensity rate, time interval corresponding to a transition step, and surface of the cell side, respectively, σ_ρ updates the hydraulic head within the cell as:

$$h' = h + \frac{r_{ir} \cdot \Delta t}{\Delta s^2 \cdot C_c}$$

  - $\sigma_\beta : Q_h \to Q_h$ sets the boundary condition on the boundary cells that are not affected by rain. The Neumann boundary conditions, which fix the water flow to a constant value, for instance $h' = h$ represent a no flow condition; the Dirichlet boundary conditions, which fix the hydraulic head to a constant value, can be adopted.
  - $\sigma_\tau : Q^{|X|} \to Q$ corresponds to the discrete time explicit resolution of the Eq. 1 and can be written as:

$$h'_c = h_c + \frac{\Delta t \left[\sum_{\alpha=1}^6 \overline{K}_\alpha (h_\alpha - h_c) \right]}{\Delta s^2 C_c}$$

where $\overline{K}_\alpha$ is the average hydraulic conductivity between the current cell c and the cell in the α neighbor calculated using:

$$\overline{K}_\alpha = \frac{2\Delta s^3}{\frac{\Delta s^3}{K_\alpha} + \frac{\Delta s^3}{K_c}}$$

also the substates q_θ, q_K and q_{c_c} are updated according to the constitutive equations between ψ, θ and K proposed by [24, 31].

Finally the substate q_{conv} is updated adopting the Courant-Friedrichs-Lewy condition to achieve numerical convergence:

$$\Delta t \leq \frac{\Delta s^2 C_c}{\sum_{\alpha=1}^{6} \overline{K}_\alpha}$$

Optionally, the σ_τ kernel can add/remove cells to/from the set of active cells A in case the *quantization* algorithm is exploited in the OpenCAL implementation of $XCA - Flow$. In this case, a gradient threshold must be set on one or more cell substates that, if exceeded/not-exceeded, produces the cell activation/deactivation.

- $\Phi = \{\phi_\rho, \phi_\beta, \phi_\tau\}$ is the set of functions applying the local transition of Σ to the non-local domains defined by I. In particular:

  - $\phi_\rho : Q^{|I_\rho|} \rightarrow Q^{|I_\rho|}$ applies the σ_ρ local transition to the I_ρ interface, to account for input rain;
  - $\phi_\beta : Q^{|I_\beta|} \rightarrow Q^{|I_\beta|}$ applies the σ_β local transition to the I_β interface, to account for boundary conditions;
  - $\phi_\tau : Q^{|I_\tau^+|} \rightarrow Q^{|I_\tau|}$ applies the σ_τ local transition to the I_τ interface. Here, I_τ^+ denotes the union of I_τ and the set of cells belonging to $D \setminus I_\tau$ that are needed to guarantee a complete neighborhood to each boundary cell of I_τ. According to the Cellular Automata definition, only states of cells in I_τ are updated.

- $\Gamma = \{\gamma_t \mid \gamma_t : Q_{conv}^{|D|} \rightarrow \mathbb{R}\}$ is the set of global functions, where γ_t evaluates a reduction over the Q_{conv} substate in order to evaluate the physical time corresponding to a state transition of the automaton. Specifically, if Δt denotes the time step size, we have:

$$\Delta t = \min_{\iota \in D} q_{conv_\iota}$$

- $T = \phi_\rho \circ \phi_\beta \circ \phi_\tau \circ \gamma_t$ is the function determining the automaton global transition. It is obtained by preliminary applying the elementary processes to the related interfaces, and then the reduction function needed to evaluate the time interval corresponding to a computational step.
- $\Delta t \in \mathbb{R}^+$ is the quantity corresponding to the physical time interval simulated by a state transition of the automaton. It is evaluated step by step by considering the γ_t reduction function.
- $\omega : \mathbb{R} \rightarrow \{false, true\}$ is the termination criterion, based on the simulation elapsed time. When the prefixed simulated time interval is complete, ω returns $false$ and the simulation terminates.
- $\tau_{\mathbb{N}} : \mathbb{N} \times C \rightarrow C$ is the XCA *control unit*. At step $n = 0$, the XCA is in the initial configuration C_0. $\tau_{\mathbb{N}}$ is then applied at discrete steps, by producing a sequence of configurations $C_1, C_2, \cdots$, until the ω termination criterion is satisfied.

3   A Brief Overview of the OpenCAL Library and Its Quantization Algorithm

OpenCAL (Open Computation Abstraction Layer) is an open source parallel C/C++ software library based on the Extended Cellular Automata (XCA) computational paradigm [15], which exposes as a Domain Specific Language (DSL). As a consequence, being the XCA formalism quite general, OpenCAL also provides support to other computational models like Cellular Automata, Finite Difference and, more in general, to structured grid-based methods.

Once the simulation model has been properly defined using the XCA formalism, OpenCAL allows for its straightforward implementation. As a matter of fact, the spatial computational domain, the variables representing the cell's state (substates), and the neighborhood can be easily defined, as well as the local state transition function and possible global operators. Moreover, the initial conditions of the system, and a termination criterion to the simulation can be defined with the minimum effort. Eventually, OpenCAL provides embedded optimization algorithms and allows for a fine grained control over the simulation [7].

One of the most important advantages of using OpenCAL is that, once the serial implementation of a model has been completed, different parallel versions can be easily obtained, including those for multi- and many-core shared memory devices, as well as for distributed memory systems, thanks to the adoption of parallel underlying APIs like OpenMP, OpenCL, and MPI. Furthermore, the embedded optimizations adopted in the serial version are transparently translated in the different parallel execution contexts.

Among the above cited optimizations, the quantization algorithm can provide a significant speed-up in the simulation of topologically connected phenomena. Chosen a model's substate and a stationary condition threshold for the cell, the algorithm allows to skip stationary cells, meaning that not only the transition function is not computed, but also that stationary cells are completely skipped, i.e., they are not visited by the loops spanning the domain. To this end, the set of *active* cells, A, must be preliminary defined, usually at the initialization stage. Only cells belonging to A are processed. The set A must therefore be kept updated during the simulation by referring to specific add/remove API functions. Similarly to the case of the substates, even the active cell's data structure needs to be updated before applying the next transition. Further details regarding the quantization algorithm can be found in [7].

4   Description of the Test Case

The case of study investigated in this work is a three-dimensional test based on a real experiment conducted in the Jornada Test Site near Las Cruces, New Mexico, by the University of Arizona and reproduced using two-dimensional numerical modeling by [28] and [21]. It concerned a heterogeneous terrain with very dry initial conditions, having three main horizontal soil layers and a fourth soil type inserted in the deepest of them (Fig. 1). The main hydraulic properties of the four soil types are reported in Table 1. The original two-dimensional test case domain was 8.00 m long and 6.50 m deep, with a uniform rain rate r_{ir} of 0.02 m d^{-1} for a length of 2.25 m along the left side of the upper boundary layer (Fig. 1). Along the other boundaries, Neumann's conditions (no flow) were imposed.

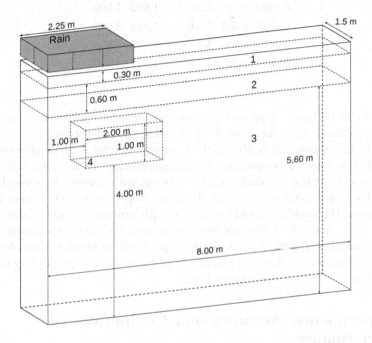

Fig. 1. Cross-section of the computational domain. It shows the four different zones, the first is located on the top of the domain and extends vertically for 0.30 m, the second extends for 0.60 m, the third for the remaining 5.60 m. The fourth zone is inserted and surrounded by the third zone. The grey zone on the top of the domain highlights the surface area directly affected by the rain.

As in [11], the original domain was extended along the third dimension in order to obtain a 1.50 m wide three-dimensional grid of cubic cells with edge length $\Delta s = 0.05$ m. Accordingly, a grid of $n_r = 160$ rows, $n_c = 30$ columns, and $n_s = 130$ slices was obtained. The input rain was accordingly extended to the third dimension (Fig. 1). The system initial conditions were defined by imposing a constant total head value of -7.3 m all over the domain, coinciding with the test case already considered in [11]. This configuration permitted to exploit the OpenCAL's quantization algorithm i) by initializing the set of non-stationary cells, A, to the domain interface affected by rain, and ii) by activating a stationary cell c each time the $\Delta h_c > \tau$ condition was satisfied, where Δh_c and τ are the hydraulic head difference between the central (already active) cell and the neighboring cell c, and τ the fixed activation threshold, respectively. Since

Table 1. Properties of the four different soil types. Each row indicates a different zone from the first to the fourth, each column represents a soil property. θ_r is the residual soil moisture content, θ_s is the satutared soil moisture content, α [m^{-1}] and n are soil parameters adopted in the Van Genuchten model, K_s [m d^{-1}] is the saturated hydraulic conductivity.

Zone	θ_r	θ_s	α (m^{-1})	n	K_s (m d^{-1})
1	0.1020	0.368	3.34	1.982	7.909
2	0.0985	0.351	3.63	1.632	4.699
3	0.0859	0.325	3.45	1.573	4.150
4	0.0859	0.325	3.45	1.573	41.50

h-form of the Richards' equation [3] was considered to solve the unsaturated flow, the hydraulic head has been used to define the activation condition. Note that, each time a new cell is activated, the model activates its neighboring cells as well. This is needed to guarantee mass conservation. In fact, if a cell c losses some mass due to the hydraulic head condition with respect to a neighboring cell c_n, this latter must be active at the same step to receive the mass from c. In particular, the problem could occur at the phenomenon's propagation front.

As in [11], a total of 30 days were simulated for each experiment, which required a total of 25,986 computational steps, with an average time step $\Delta t = 99.75$ s. The outcome is shown in Fig. 2. The phenomenon propagates up to about 3 m depth, not affecting the deepest layers.

5   Experiments, Accuracy and Computational Performance

Different simulations of the test case described in Sect. 4 were performed by varying the hydraulic head quantization threshold, with the purpose to assess both the accuracy and the computational performance with respect to the non quantized (reference) simulation. The values adopted for the quantization threshold are listed in Table 2. Note that, the set of experiments here considered extends the study described in [11], where the only $\tau = 0$ threshold was taken into account.

Figure 3 shows the differences of the variable ψ between the reference and the $\tau = 0.1, 0.25, 0.5, 1$ quantized simulations. The error, which is limited to about -3.3 m, increases (in absolute value) with the quantization threshold. In particular, the differences are close to zero in most parts of the domain, with the major errors localized near the phenomenon propagation front (cf. Figs. 2 and 3).

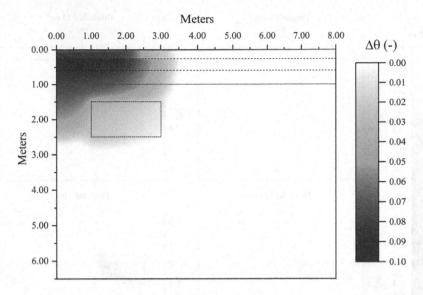

Fig. 2. The soil moisture content variation $\Delta\theta$ [-] on a vertical section at the final step of the 30-days reference simulation. Black dotted lines indicate the different zones, the red line shows the domain decomposition between the two nodes considered.

Regarding the computational aspects, the OpenMP/MPI OpenCAL component was used to implement the XCA-Flow model, and a workstation running Arch Linux, equipped whit two Intel 8 core (16 threads) Xeon E5-2650 sockets, was used to run the simulations. In particular, the experiments were performed by referring to a 2 MPI processes/16 OpenMP threads parallel configuration. Specifically, the 3D domain was decomposed along the third (vertical) dimension over the 2 MPI processes, resulting in two sub-domains of 20 and 110 slices, respectively (cf. Fig. 2). Note that, the adopted parallel set up is the same of [11], which permitted to achieve the best performance on the considered hardware by executing the reference simulation in 729.15 s. The results achieved are summarized in Table 2 in terms of elapsed times. Figure 4 shows both the mean square error of the ψ variable with respect to the reference simulation and the corresponding elapsed times of the simulations executed by considering the different quantization thresholds listed in Table 2. The error is close to zero up to $\tau = 0.1$, afterward it increases up to the value of about $2.6 \cdot 10^{-3}$ m^2. Differently, the elapsed times decrease with the quantization threshold, from a value of about 300 s to a value of less than 100 s.

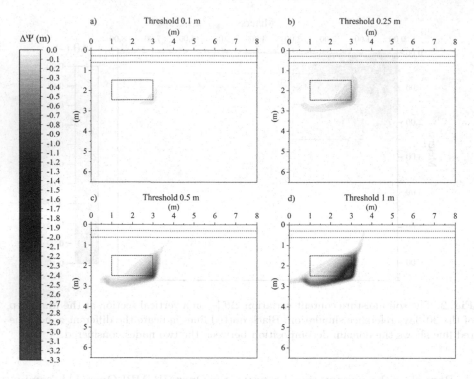

Fig. 3. Vertical sections showing the differences of the $\Delta\psi$ variable [m] between the final configurations of the reference simulation and the simulations performed by considering the τ a) 0.1 m, b) 0.25 m, c) 0.5 m, and d) 1 m quantization thresholds.

Table 2. Threshold values τ [m] adopted in this work with the corresponding mean square error [m²], based on the ψ variable calculated only on the active cells, and elapsed time [s].

Threshold (m)	Mean square error (m²)	Elapsed time (s)
0	0	291.77
0.001	1.74e−10	183.47
0.01	2.92e−08	155.69
0.1	6.23e−06	137.23
0.25	6.46e−05	122.89
0.5	4.36e−04	109.47
1.0	2.59e−03	99.15

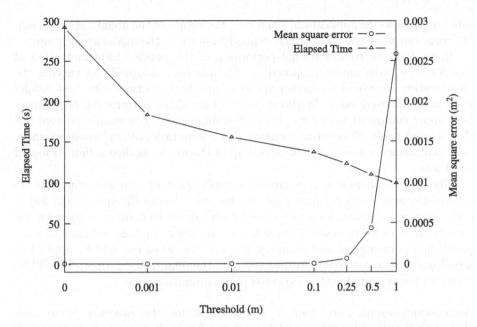

Fig. 4. Mean square error [m²], based on the ψ variable calculated only on the active cells, and elapsed time [s] achieved using different quantization threshold. A further simulation with no threshold has been executed with an elapsed time of 729.15 s.

6 Conclusions

In this paper, the implementation of an unsaturated water flow model is presented based on the OpenCAL library, which allows exploiting the XCA formalism. Specifically, the impact of the quantization algorithm is presented, which allows reducing the computational effort at the expense of the accuracy level.

A three-dimensional test case was considered to assess computational performance and accuracy, which simulates the infiltration in the first soil layers produced by a uniform rain rate. Small variations of the quantization threshold τ were assessed in order to activate the non-stationary cells. In particular, τ refers to the hydraulic head variation between adjacent cells.

From an accuracy assessment point of view, it was observed that the quantization algorithm produces a delay in the propagation front since the non-active cells, which are not taken into account during the computation, become active when the quantization threshold is reached. As a consequence, there is a time cumulative error near the propagation front, which is tightly related to the chosen quantization threshold. In particular, most of the thresholds used for this specific case study, on a 30 days simulation, generate a low relative error. The greatest error of about 3.3 m has been obtained by adopting a threshold $\tau = 1$ m, which could not be tolerable for some applications. Although, the other thresholds $\tau \in [0.001, 0.5]$ m generate smaller errors. The acceptability of the results achieved, also in these cases, relies on the specific application. Therefore, the choice of the quantization threshold turns out to be a key factor and depends mainly on the model application.

Regarding the computational performance, the parallel implementation of the $XCA - Flow$ model proposed in [11] has been adopted, by varying the quantization threshold to further speeding-up the execution. The best results have been achieved using the threshold $\tau = 1\,\text{m}$, which permitted a 66% execution boost compared to the $\tau = 0$ parallel simulation. The results pointed out that a good trade off between accuracy and computational performance can be achieved, allowing a considerable speed-up of the model against a limited loss of precision.

The future outlook will regard the application of dynamic quantization thresholds, which will permit to reduce the error during the simulation. Moreover, further hydrological variables will be considered to define the quantization thresholds in order to assess if there is any further advantage in terms of computational performance and accuracy. Finally, $XCA - Flow$ will be applied to a real basin scale case study. In this context, the quantization algorithm will be a crucial factor to reduce the expected high computing time.

Acknowledgement. Luca Furnari acknowledges for the program "POR Calabria FSE/FESR 2014/2020 - Mobilità internazionale di Dottorandi e Assegni di ricerca/Ricercatori di Tipo A" Actions 10.5.6 and 10.5.12.

References

1. Abd-Elaty, I., Sallam, G.A., Straface, S., Scozzari, A.: Effects of climate change on the design of subsurface drainage systems in coastal aquifers in arid/semi-arid regions: case study of the nile delta. Sci. Total Environ. **672**, 283–295 (2019)
2. Cannataro, M., Di Gregorio, S., Rongo, R., Spataro, W., Spezzano, G., Talia, D.: A parallel cellular automata environment on multicomputers for computational science. Parallel Comput. **21**(5), 803–823 (1995)
3. Celia, M.A., Bouloutas, E.T., Zarba, R.L.: A general mass-conservative numerical solution for the unsaturated flow equation. Water Resour. Res. **26**(7), 1483–1496 (1990)
4. Cervarolo, G., Mendicino, G., Senatore, A.: A coupled ecohydrological-three-dimensional unsaturated flow model describing energy, H2O and CO2 fluxes. Ecohydrology **3**(2), 205–225 (2010)
5. Cervarolo, G., Mendicino, G., Senatore, A.: Coupled vegetation and soil moisture dynamics modeling in heterogeneous and sloping terrains. Vadose Zone J. **10**, 206–225 (2011)
6. Chang, K.S., Song, C.J.: Interactive vortex shedding from a pair of circular cylinders in a transverse arrangement. Int. J. Numer. Meth. Fluids **11**(3), 317–329 (1990)
7. D'Ambrosio, D., et al.: The open computing abstraction layer for parallel complex systems modeling on many-core systems. J. Parallel Distrib. Comput. **121**, 53–70 (2018)
8. D'Ambrosio, D., De Rango, A., Rongo, R.: Opencal simulation of the 1992 tessina landslide. In: 30th European Modeling and Simulation Symposium, EMSS 2018, pp. 210–217 (2018)

 9. D'Ambrosio, D., De Rango, A., Spataro, D., Rongo, R., Spataro, W.: Applications of the OpenCAL scientific library in the context of CFD: applications to debris flows. In: 2017 IEEE 14th International Conference on Networking, Sensing and Control (ICNSC), pp. 720–725 (2017)
10. Dattilo, G., Spezzano, G.: Simulation of a cellular landslide model with CAMELOT on high performance computers. Parallel Comput. **29**(10), 1403–1418 (2003)
11. De Rango, A., et al.: OpenCAL system extension and application to the three-dimensional Richards equation for unsaturated flow. Comput. Math. Appl. (Article in Press). https://doi.org/10.1016/j.camwa.2020.05.017
12. De Rango, A., Napoli, P., D'Ambrosio, D., Spataro, W., Di Renzo, A., Di Maio, F.: Structured Grid-Based Parallel Simulation of a Simple DEM Model on Heterogeneous Systems, pp. 588–595 (2018)
13. De Rango, A., Spataro, D., Spataro, W., D'Ambrosio, D.: A first multi-GPU/multi-node implementation of the open computing abstraction layer. J. Comput. Sci. **32**, 115–124 (2019)
14. Deng, X., Min, Y., Mao, M., Liu, H., Tu, G., Zhang, H.: Further studies on geometric conservation law and applications to high-order finite difference schemes with stationary grids. J. Comput. Phys. **239**, 90–111 (2013)
15. Di Gregorio, S., Serra, R.: An empirical method for modelling and simulating some complex macroscopic phenomena by cellular automata. Future Gener. Comput. Syst. **16**, 259–271 (1999)
16. Filippone, G., D'Ambrosio, D., Marocco, D., Spataro, W.: Morphological coevolution for fluid dynamical-related risk mitigation. ACM Trans. Model. Comput. Simul. (TOMACS) **26**(3), 18 (2016)
17. Folino, G., Mendicino, G., Senatore, A., Spezzano, G., Straface, S.: A model based on cellular automata for the parallel simulation of 3D unsaturated flow. Parallel Comput. **32**(5), 357–376 (2006)
18. Kundzewicz, Z., et al.: The implications of projected climate change for freshwater resources and their management. Hydrol. Sci. J. **53**(1), 3–10 (2008)
19. List, F., Radu, F.A.: A study on iterative methods for solving richards' equation. Comput. Geosci. **20**(2), 341–353 (2016)
20. Lucà, F., D'Ambrosio, D., Robustelli, G., Rongo, R., Spataro, W.: Integrating geomorphology, statistic and numerical simulations for landslide invasion hazard scenarios mapping: an example in the Sorrento Peninsula (Italy). Comput. Geosci. **67**(1811), 163–172 (2014)
21. McCord, J.T., Goodrich, M.T.: Benchmark testing and independent verification of the VS2DT computer code. Technical report, Sandia National Labs (1994)
22. Mendicino, G., Pedace, J., Senatore, A.: Stability of an overland flow scheme in the framework of a fully coupled eco-hydrological model based on the Macroscopic Cellular Automata approach. Commun. Nonlinear Sci. Numer. Simul. **21**(1–3), 128–146 (2015)
23. Mendicino, G., Senatore, A., Spezzano, G., Straface, S.: Three-dimensional unsaturated flow modeling using cellular automata. Water Resour. Res. **42**(11), 2332–2335 (2006)
24. Mualem, Y.: A new model for predicting the hydraulic conductivity of unsaturated porous media. Water Resour. Res. **12**(3), 513–522 (1976)
25. von Neumann, J.: Theory of Self-Reproducing Automata. University of Illinois Press, Champaign (1966). Edited by Arthur W. Burks
26. Senatore, A., et al.: Accelerating a three-dimensional eco-hydrological cellular automaton on GPGPU with OpenCL. In: AIP Conference Proceedings, vol. 1776, p. 080003 (2016)

27. Senatore, A., Mendicino, G., Smiatek, G., Kunstmann, H.: Regional climate change projections and hydrological impact analysis for a Mediterranean basin in Southern Italy. J. Hydrol. **399**(1), 70–92 (2011)
28. Smyth, J., Yabusaki, S., Gee, G.: Infiltration evaluation methodology-letter report 3: selected tests of infiltration using two-dimensional numerical models. Pacific Northwest Laboratory, Richland (1989)
29. Spataro, D., D'Ambrosio, D., Filippone, G., Rongo, R., Spataro, W., Marocco, D.: The new SCIARA-fv3 numerical model and acceleration by GPGPU strategies. Int. J. High Perform. Comput. Appl. **31**(2), 163–176 (2017)
30. Spingola, G., D'Ambrosio, D., Spataro, W., Rongo, R., Zito, G.: Modeling complex natural phenomena with the libAuToti cellular automata library: an example of application to lava flows simulation. In: PDPTA - International Conference on Parallel and Distributed Processing Techniques and Applications, pp. 277–283 (2008)
31. Van Genuchten, M.T.: Calculating the unsaturated hydraulic conductivity with a new closed-form analytical model. Researh Reprot - Water Resources Program, Department of Civil Engineering, Princeton University, Princeton (1978)

An Expanded Mixed Finite Element Method for Space Fractional Darcy Flow in Porous Media

Huangxin Chen[1,2] and Shuyu Sun[2($\boxtimes$)]

[1] School of Mathematical Sciences and Fujian Provincial Key Laboratory on
Mathematical Modeling and High Performance Scientific Computing,
Xiamen University, Fujian 361005, China
chx@xmu.edu.cn

[2] Physical Science and Engineering Division (PSE), King Abdullah University
of Science and Technology (KAUST), Thuwal 23955-6900, Saudi Arabia
shuyu.sun@kaust.edu.sa

Abstract. In this paper an expanded mixed formulation is introduced
to solve the two dimensional space fractional Darcy flow in porous media.
By introducing an auxiliary vector, we derive a new mixed formulation
and the well-possedness of the formulation can be established. Then
the locally mass-conservative expanded mixed finite element method is
applied for the solution. Numerical results are shown to verify the efficiency of the proposed algorithm.

Keywords: Space fractional Darcy flow · Extended mixed finite
element method · Well-posedness

1 Introduction

Fractional partial differential equations (PDE) have been explored as an important tool to develop more accurate mathematical models to describe complex
anomalous systems such as phase transitions, anomalous diffusions. In this paper,
we focus on the modeling and simulation of flow in porous media. In particular,
when considering modeling of flow in fractured porous media, one may consider
different fractional time derivatives in matrix and fracture regions due to the
different memory properties. For instance, Caputo [3] apply the time fractional
PDE to model the flow in fractured porous media. However, there still exists the
fact that there is a steady state for flow in fractured porous media. Thus, we
consider the steady state space fractional PDE for the modeling and simulation
of flow in porous media.

The work of Huangxin Chen was supported by the NSF of China (Grant No. 11771363,
91630204, 51661135011), the Fundamental Research Funds for the Central Universities
(Grant No. 20720180003). The work of Shuyu Sun was supported by King Abdullah
University of Science and Technology (KAUST) through the grant BAS/1/1351-01-01.

© Springer Nature Switzerland AG 2020
V. V. Krzhizhanovskaya et al. (Eds.): ICCS 2020, LNCS 12143, pp. 205–216, 2020.
https://doi.org/10.1007/978-3-030-50436-6_15

Until now, a number of articles referring to the space fractional PDE have appeared in literature (see [13] and the reference therein). Most of the works concern on the fractional Laplacian equation in unbounded or bounded domain, and there exist different kinds of definitions of fractional Laplacian, such as spectral/Fourier definition, singular integral representation, via the standard Laplacian (elliptic extension), directional representation, et al. Lots of numerical methods have been developed for the space fractional PDE, for instance, the finite different method, the finite volume method, the spectral method, et al (see [9,12,14–16,19] and the references therein). In particular, the finite element methods have been firstly developed and analyzed by Ervin, Roop for the space fractional PDE, and then by other authors in a series of works [2,5–7,18]. In these works, standard Galerkin finite element methods are always applied for the fractional Laplacian equation. However, when the standard Galerkin finite element methods are applied to the flow equation, the mass conservation can not be retained.

For the space fractional Darcy flow, J. H. He [10] firstly studied seepage flow in porous media and used fractional derivatives to describe the fractional Darcy's law behavior. In [10], the permeability can only be assumed to be diagonal and the PDE system violates the principle of Galilean invariance. Some numerical methods have been developed for such kind of equation, e.g., see [2]. In this paper, we will apply the fractional gradient operator defined by Meerschaert et al. [17] to write down the space fractional Darcy flow for the two dimensional problem which obeys the principle of Galilean invariance. In order to develop the locally mass-conservative finite element method for the fractional Darcy flow, Chen and Wang [4] proposed a new mixed finite element method for a one-dimensional fractional Darcy flow. In this work, we extend the locally mass-conservative mixed finite element method to the two dimensional problems which can be easily extended to three dimensional problems. By introducing a new auxiliary vector, we can obtain the new expanded mixed formulation for the fractional Darcy flow and the well-possedness of the new formulation can be well established.

The rest of the paper is organized as follows. In Sect. 2, we introduce the mathematical model for the two dimensional space fractional Darcy flow in porous media. Then we introduce the expanded mixed formulation and establish its well-posedness in Sect. 3, and show the expanded mixed finite element method and the detailed implementation in Sect. 4. Some numerical results are given in Sect. 5 to verify the efficiency of the proposed algorithm. Finally we provide a conclusion in Sect. 6.

2   Preliminary

In this section we will follow [7,17] to recall the definitions of the directional integral, the directional derivative operators, the fractional gradient operator, and then introduce the space fractional Darcy's law which obeys the principle of Galilean invariance. We use the standard notations and definitions for Sobolev spaces (cf. [1]) throughout the paper. Since our work focus on the two dimensional problem, the following definitions are given for the functions in $\mathcal{R}^2$.

Definition 1 *(cf. [7]). Let $\mu > 0$ and $\theta \in \mathcal{R}$. The μ-th order fractional integral in the direction $\boldsymbol{\theta} = (\cos\theta, \sin\theta)$ is defined by*

$$D_\theta^{-\mu} v(x,y) = \int_0^\infty \frac{\xi^{\mu-1}}{\Gamma(\xi)} v(x - \xi\cos\theta, y - \xi\sin\theta) d\xi,$$

where $\Gamma(\cdot, \cdot)$ is a Gamma function.

Definition 2 *(cf. [7]). Let n be a positive integer. The n-th order derivative in the direction of $\boldsymbol{\theta} = (\cos\theta, \sin\theta)$ is given by*

$$D_\theta^n v(x,y) = (\cos\theta \frac{\partial}{\partial x} + \sin\theta \frac{\partial}{\partial y})^n v(x,y).$$

Definition 3 *(cf. [7]). Let $\mu > 0, \theta \in \mathcal{R}$. Let n be an integer such that $n - 1 \leq \mu < n$, and define $\sigma = n - \mu$. Then the μ-th order directional derivative in the direction of $\boldsymbol{\theta} = (\cos\theta, \sin\theta)$ is defined by*

$$D_\theta^\mu v(x,y) = D_\theta^{-\sigma} D_\theta^n v(x,y).$$

Definition 4 *(cf. [17]). Let $\alpha \in (0,1)$. The fractional gradient operator with respect to the measure M is defined by*

$$\nabla_M^\alpha v = \int_0^{2\pi} \boldsymbol{\theta} D_\theta^\alpha v M(\theta) d\theta,$$

where $\boldsymbol{\theta} = (\cos\theta, \sin\theta)$ is a unit vector, D_θ^α is the Riemann-Liouville fractional directional derivative and $M(\theta)$ is a positive (probability) density function satisfying $\int_0^{2\pi} M(\theta) d\theta = 1$.

By the Lemma 5.6 in [7], we have $D_\theta^\alpha v = D_\theta^{\alpha-1} D_\theta^1 v = D_\theta^{\alpha-1}(\boldsymbol{\theta} \cdot \nabla v)$. Thus, we write down the steady state space fractional Darcy flow as follows:

$$\nabla \cdot \boldsymbol{u} = f \qquad \text{in } \Omega, \tag{2.1a}$$

$$\boldsymbol{u} = -\boldsymbol{K}\nabla_M^\alpha p \qquad \text{in } \Omega, \tag{2.1b}$$

$$p = 0 \qquad \text{in } \mathcal{R}^2 \setminus \Omega, \tag{2.1c}$$

where $\boldsymbol{u}, p, f$ are fluid velocity, pressure and source term, $\boldsymbol{K}$ is a bounded symmetric and positive definite permeability tensor, and $\nabla_M^\alpha p = \int_0^{2\pi} \boldsymbol{\theta} D_\theta^{\alpha-1}(\boldsymbol{\theta} \cdot \nabla p) M(\theta) d\theta$. For brevity, we consider the fractional Darcy flow in a bounded domain with homogeneous boundary condition for pressure and assume the pressure to be zero outside the domain. For the problem with non-homogeneous boundary conditions, some addition techniques such as the lifting approach and other strategies introduced in [13] can be further applied to solve the problem.

For the one dimensional space fractional diffusion equation, the regularity of solution was obtained in [8,11]. In this work, we consider the two dimensional model and assume the regularity of solution for the system (2.1) as $p \in H_0^{\frac{\alpha+1}{2}}(\Omega)$. The gravity effect can also be considered in the space fractional Darcy law as $\boldsymbol{u} = -\boldsymbol{K}(\nabla_M^\alpha p + \rho \boldsymbol{g})$, and the above model can also be extended to the three dimensional problem.

3   Expanded Mixed Formulation

In this section we present a mass-conservative mixed formulation for the fractional Darcy flow (2.1) and establish the well-posedness of the weak formulation. Firstly we define the notations for some Sobolev spaces as follows:

$$\boldsymbol{V} := H(\mathrm{div}, \Omega) \cap [H^{\frac{1-\alpha}{2}}(\Omega)]^2, \quad \boldsymbol{H} := [H^{\frac{\alpha-1}{2}}(\Omega)]^2, \quad Q := L^2(\Omega).$$

In order to propose a well-posed mixed formulation for the space fractional Darcy flow, we introduce a new auxiliary vector $\boldsymbol{w} = \nabla p \in \boldsymbol{H}$. Now we present the expanded mixed formulation for (2.1) as follows: Find $(\boldsymbol{u}, \boldsymbol{w}, p) \in \boldsymbol{V} \times \boldsymbol{H} \times Q$, such that

$$(\boldsymbol{w}, \boldsymbol{\eta}) + (p, \nabla \cdot \boldsymbol{\eta}) = 0, \tag{3.1a}$$

$$(\boldsymbol{u}, \boldsymbol{v}) + (\boldsymbol{K} \int_0^{2\pi} \boldsymbol{\theta} D_\theta^{\alpha-1}(\boldsymbol{\theta} \cdot \boldsymbol{w}) M(\theta) d\theta, \boldsymbol{v}) = 0, \tag{3.1b}$$

$$(\nabla \cdot \boldsymbol{u}, q) = (f, q), \tag{3.1c}$$

for any $(\boldsymbol{\eta}, \boldsymbol{v}, q) \in \boldsymbol{V} \times \boldsymbol{H} \times Q$. Since we assume that $\boldsymbol{V}$ and $\boldsymbol{H}$ are dual spaces, $\boldsymbol{w}, \boldsymbol{u} \in \boldsymbol{V}$ and $\boldsymbol{\eta}, \boldsymbol{v} \in \boldsymbol{H}$, we can see that the inner products for $(\boldsymbol{w}, \boldsymbol{\eta})$ and $(\boldsymbol{u}, \boldsymbol{v})$ in (3.1) are well defined.

Now we define $\boldsymbol{U} := \boldsymbol{H} \times Q$ and let

$$a(\boldsymbol{\tau}, \boldsymbol{\chi}) := (\boldsymbol{K} \int_0^{2\pi} \boldsymbol{\theta} D_\theta^{\alpha-1}(\boldsymbol{\theta} \cdot \boldsymbol{w}) M(\theta) d\theta, \boldsymbol{v}),$$

$$b(\boldsymbol{\chi}, \boldsymbol{u}) := (\boldsymbol{u}, \boldsymbol{v}) + (\nabla \cdot \boldsymbol{u}, q),$$

$$b(\boldsymbol{\tau}, \boldsymbol{\eta}) := (\boldsymbol{w}, \boldsymbol{\eta}) + (p, \nabla \cdot \boldsymbol{\eta}),$$

for $\boldsymbol{\tau} = (\boldsymbol{w}, p) \in \boldsymbol{U}, \boldsymbol{\chi} = (\boldsymbol{v}, q) \in \boldsymbol{U}$. Then the expanded mixed formulation (3.1) can be equivalently rewritten as follows: For any $(\boldsymbol{\chi}, \boldsymbol{\eta}) \in \boldsymbol{U} \times \boldsymbol{V}$, find $(\boldsymbol{\tau}, \boldsymbol{u}) \in \boldsymbol{U} \times \boldsymbol{V}$ such that

$$a(\boldsymbol{\tau}, \boldsymbol{\chi}) + b(\boldsymbol{\chi}, \boldsymbol{u}) = (f, q), \tag{3.2a}$$

$$b(\boldsymbol{\tau}, \boldsymbol{\eta}) = 0. \tag{3.2b}$$

In the following, we will aim to prove the well-posedness of the mixed system (3.2). We define $\boldsymbol{Z} = \{\boldsymbol{\chi} \in \boldsymbol{U} : b(\boldsymbol{\chi}, \boldsymbol{\eta}) = 0, \forall \boldsymbol{\eta} \in \boldsymbol{V}\}$. We start the proof from the following key lemma which can be proved by the similar technique in [4]. We denote by C with or without subscript a positive constant. These constants can take on different values in different occurrences.

Lemma 1. *Let $\chi = (v, q) \in Z$. We have that $\chi \in Z$ if and only if $q \in H_0^{\frac{\alpha+1}{2}}(\Omega)$ and $v = \nabla q \in [H^{\frac{\alpha-1}{2}}(\Omega)]^2$.*

Proof. If $q \in H_0^{\frac{\alpha+1}{2}}(\Omega)$ and $v = \nabla q \in [H^{\frac{\alpha-1}{2}}(\Omega)]^2$, we have

$$(v, \eta) = (\nabla q, \eta) = -(q, \nabla \cdot \eta) + \int_{\partial\Omega} q\eta \cdot n, \quad \forall \eta \in V.$$

Since $q \in H_0^{\frac{\alpha+1}{2}}(\Omega)$, we have $(v, \eta) + (q, \nabla \cdot \eta) = 0$, i.e., $\chi = (v, q) \in Z$.

Now we let $\chi = (v, q) \in Z$. Firstly, we let $\eta = (1, 0)^T$ or $(0, 1)^T$ in $b(\chi, \eta)$ and we get $v \in [L^1(\Omega)]^2$. Then we let $\eta \in \mathcal{D}(\Omega)$ where $\mathcal{D}(\Omega)$ denotes the set of all functions $\phi \in C^\infty(\Omega)$ which vanish outside a compact subset of Ω. Then we have

$$(v, \eta) = -(q, \nabla \cdot \eta) = (\nabla q, \eta), \quad \forall \eta \in [\mathcal{D}(\Omega)]^2,$$

which yields that $v = \nabla q \in [L^1(\Omega)]^2$ and $q \in W^{1,1}(\Omega)$. By the density and Sobolev imbedding theories, we have that for any $q \in W^{1,1}(\Omega)$, there exist $q_\epsilon \in C^1(\Omega)$ and a constant $C > 0$ such that

$$\|q - q_\epsilon\|_{W^{1,1}(\Omega)} \le \epsilon, \quad \|q - q_\epsilon\|_{L^\infty(\Omega)} \le C\|q - q_\epsilon\|_{W^{1,1}(\Omega)} \le C\epsilon.$$

Now, for any $\eta \in V$, we have

$$(q, \nabla \cdot \eta) = (q - q_\epsilon, \nabla \cdot \eta) + (q_\epsilon, \nabla \cdot \eta)$$

$$= (q - q_\epsilon, \nabla \cdot \eta) - (\nabla q_\epsilon, \eta) + \int_{\partial\Omega} q_\epsilon \eta \cdot n$$

$$= -(\nabla q, \eta) + \int_{\partial\Omega} q\eta \cdot n$$

$$+ (q - q_\epsilon, \nabla \cdot \eta) - (\nabla q_\epsilon - \nabla q, \eta) + \int_{\partial\Omega} (q_\epsilon - q)\eta \cdot n. \tag{3.3}$$

By the density argument and the imbedding theory, we can see that the last three terms on the right-hand side of the last equality in (3.3) become zero as $\epsilon \to 0$. Thus combining the above derivation and $v = \nabla q$, we have

$$(q, \nabla \cdot \eta) = -(\nabla q, \eta) + \int_{\partial\Omega} q\eta \cdot n = -(v, \eta) + \int_{\partial\Omega} q\eta \cdot n, \quad \forall \eta \in V. \tag{3.4}$$

Since $\chi = (v, q) \in Z$, we have $(q, \nabla \cdot \eta) + (v, \eta) = 0$, which together with (3.4) yields that $q = 0$ on $\partial\Omega$. Thus we have $q \in W_0^{1,1}(\Omega)$.

By the definition of fractional gradient operator, we have

$$\nabla_M^{\frac{\alpha+1}{2}} q = \int_0^{2\pi} \theta D_\theta^{\frac{\alpha+1}{2}} q M(\theta) d\theta = \int_0^{2\pi} \theta D_\theta^{\frac{\alpha-1}{2}} \theta \cdot \nabla q M(\theta) d\theta$$

$$= \int_0^{2\pi} \theta D_\theta^{\frac{\alpha-1}{2}} \theta \cdot v M(\theta) d\theta \in L^2(\Omega).$$

Thus we obtain $q \in H^{\frac{\alpha+1}{2}}(\Omega)$ which together with $q = 0$ on $\partial\Omega$ yields $q \in H_0^{\frac{\alpha+1}{2}}(\Omega)$ and $v = \nabla q \in [H^{\frac{\alpha-1}{2}}(\Omega)]^2$. Now we conclude the proof. $\square$

Next we introduce another two important tools to prove the well-posedness of (3.2).

Lemma 2 *(cf. [7]). Let $\mu > 0$. For each $\theta \in [0, 2\pi)$, there holds*

$$(D_\theta^\mu q, D_{\theta+\pi}^\mu q)_{L^2(\Omega)} = \cos(\pi\mu)\|D_\theta^\mu q\|_{L^2(\Omega)}^2, \quad \forall q \in H_0^\mu(\Omega) \text{ and } q = 0 \text{ in } \mathcal{R}^2 \setminus \Omega.$$

Lemma 3 *(cf. [7]). For any $q \in H_0^\mu(\Omega)$ and $q = 0$ in $\mathcal{R}^2 \setminus \Omega$, there holds the fractional Poincaré-Friedrichs inequality as follows:*

$$\int_0^{2\pi} \|D_\theta^\mu q\|_{L^2(\Omega)}^2 M(\theta)d\theta \geq C\|q\|_{L^2(\Omega)}^2.$$

Now for any $\chi = (v, q) \in U$, we denote

$$|v|_{\frac{\alpha-1}{2},M}^2 = \int_0^{2\pi} \|D_\theta^{\frac{\alpha-1}{2}} \boldsymbol{\theta} \cdot v\|_{L^2(\Omega)}^2 M(\theta)d\theta$$

and

$$\|\chi\|_U^2 := \|q\|_{L^2(\Omega)}^2 + |v|_{\frac{\alpha-1}{2},M}^2.$$

In the following, in order to show the proof in brevity, we assume the permeability tensor $\boldsymbol{K} = K\boldsymbol{I}$ with a positive constant $K > 0$ and $\boldsymbol{I}$ is identity matrix.

Lemma 4. *For any $\chi \in Z$, we have*

$$a(\chi, \chi) \geq C\|\chi\|_U^2.$$

Proof. By Lemma 1, for any $\chi = (v, q) \in Z$, we have $v = \nabla q$ and $q \in H_0^{\frac{\alpha+1}{2}}(\Omega)$. Then by the Theorems 2.1–2.2 in [7], the fact $D_\theta^1 = -D_{\theta+\pi}^1$ and the Lemma 2, we have

$$a(\chi, \chi) = (K \int_0^{2\pi} \boldsymbol{\theta} D_\theta^{\alpha-1}(\boldsymbol{\theta} \cdot v)M(\theta)d\theta, v)$$

$$= (K \int_0^{2\pi} D_\theta^{\alpha-1}(\boldsymbol{\theta} \cdot \nabla q)M(\theta)d\theta, \boldsymbol{\theta} \cdot \nabla q)$$

$$= \int_0^{2\pi} (K D_\theta^{\frac{\alpha-1}{2}}(\boldsymbol{\theta} \cdot \nabla q), D_{\theta+\pi}^{\frac{\alpha-1}{2}}(\boldsymbol{\theta} \cdot \nabla q))M(\theta)d\theta$$

$$= -\int_0^{2\pi} (K D_\theta^{\frac{\alpha+1}{2}} q, D_{\theta+\pi}^{\frac{\alpha+1}{2}} q)M(\theta)d\theta$$

$$= \sin\frac{\pi\alpha}{2} \int_0^{2\pi} \|\sqrt{K} D_\theta^{\frac{\alpha+1}{2}} q\|_{L^2(\Omega)}^2 M(\theta)d\theta.$$

We note that $D_\theta^{\frac{\alpha+1}{2}} q = D_\theta^{\frac{\alpha-1}{2}} \boldsymbol{\theta} \cdot v$. Then we can get the desired estimate by the above equality and the fractional Poincaré-Friedrichs inequality in the Lemma 3. □

Lemma 5. *There holds*

$$\inf_{\eta \in V} \sup_{\chi \in U} \frac{b(\chi, \eta)}{\|\chi\|_U \|\eta\|_V} \geq C.$$

Proof. For any $\eta \in V$, we let $\chi_0 = (\eta, \nabla \cdot \eta)$, then we have

$$\sup_{\chi \in U} \frac{b(\chi, \eta)}{\|\chi\|_U} \geq \frac{b(\chi_0, \eta)}{\|\chi_0\|_U} = \frac{(\eta, \eta) + (\nabla \cdot \eta, \nabla \cdot \eta)}{(\|\eta\|_{\frac{\alpha-1}{2}, M}^2 + \|\nabla \cdot \eta\|_{L^2(\Omega)}^2)^{1/2}}.$$

By the imbedding theory, we easily have

$$\sup_{\chi \in U} \frac{b(\chi, \eta)}{\|\chi\|_U} \geq C \|\eta\|_V,$$

which directly yields the desired estimate. $\qquad\square$

By the Lemmas 4–5 and the Babuška-Brezzi theory, we finally obtain the following theorem to state the well-posedness of (3.2), and this also indicates the well-posedness of the mixed system (3.1).

Theorem 1. *There exists a unique solution (τ, u) for the mixed system (3.2).*

4 Expanded Mixed Finite Element Method and Its Implementation

In this section we will introduce the expanded mixed finite element method for the mixed formulation (3.1) and show the details of implementation. Let $\mathcal{T}_h$ be the quasi-uniform structured or unstructured mesh on Ω. We define

$$\begin{aligned}
V_h &= \{\eta_h \in V : \eta_h|_T \in RT_0(T), \ \forall T \in \mathcal{T}_h\}, \\
H_h &= \{v_h \in H : v_h|_T \in [P_0(T)]^2, \ \forall T \in \mathcal{T}_h\}, \\
Q_h &= \{q_h \in Q : q_h|_T \in P_0(T), \ \forall T \in \mathcal{T}_h\}.
\end{aligned}$$

We remark that the high order mixed finite element spaces can also be used in $V_h \times H_h \times Q_h$ if the solution is smooth enough. For the problem with a low regularity solution, the approximation based on the low order mixed finite element space is advised. We utilize V_h, H_h, Q_h as the approximate spaces for V, H, Q. The expanded mixed finite element method for the space fractional Darcy flow (2.1) is defined as: Find $(u_h, w_h, p_h) \in V_h \times H_h \times Q_h$ such that

$$(w_h, \eta_h) + (p_h, \nabla \cdot \eta_h) = 0, \tag{4.1a}$$

$$(u_h, v_h) + (K \int_0^{2\pi} \theta D_\theta^{\alpha-1}(\theta \cdot w_h) M(\theta) d\theta, v_h) = 0, \tag{4.1b}$$

$$(\nabla \cdot u_h, q_h) = (f, q_h), \tag{4.1c}$$

for any $(\boldsymbol{\eta}_h, \boldsymbol{v}_h, q_h) \in \boldsymbol{V}_h \times \boldsymbol{H}_h \times Q_h$.

We note that the key step in implementation of (4.1) is how to discretize the following term:

$$T_f = (\boldsymbol{K} \int_0^{2\pi} \boldsymbol{\theta} D_\theta^{\alpha-1}(\boldsymbol{\theta} \cdot \boldsymbol{w}_h) M(\theta) d\theta, \boldsymbol{v}_h).$$

We assume that $M(\theta)$ has the discrete form

$$M(\theta) = \sum_{k=1}^{L} \omega_k \delta(\theta - \theta_k), \quad \sum_{k=1}^{L} \omega_k = 1,$$

where δ is the Dirac delta function. For brevity, we assume $\boldsymbol{K} = K\boldsymbol{I}$, then

$$T_f = \sum_{k=1}^{L} \omega_k K(D_{\theta_k}^{\alpha-1}(\boldsymbol{\theta}_k \cdot \boldsymbol{w}_h), \boldsymbol{\theta}_k \cdot \boldsymbol{v}_h).$$

Since $\boldsymbol{\theta}_k = (\cos\theta_k, \sin\theta_k)$ and $\boldsymbol{w}_h, \boldsymbol{v}_h \in \boldsymbol{H}_h$ are piecewise constant vector functions with the basis functions $(0,1)^T$ and $(1,0)^T$, the key implementation of T_f lies in the computation of

$$\hat{T}_f^\theta = (D_\theta^{\alpha-1} \chi_{K'}, \chi_K), \quad K', K \in \mathcal{T}_h,$$

where χ_S is an indicator function for a set S in $\mathcal{R}^2$.

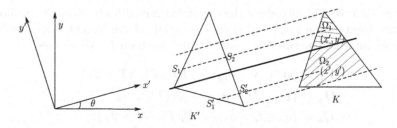

Fig. 1. Illustration for computing $\hat{T}_f^\theta$ on the unstructured mesh.

If $\mathcal{T}_h$ is an unstructured mesh with a triangular partition of Ω, we can compute $\hat{T}_f^\theta$ as follows: Let $\nu = 1 - \alpha$, $\Omega_1, \Omega_2 \in K$, we have

$$\hat{T}_f^\theta = (D_\theta^{\alpha-1} \chi_{K'}, \chi_K)_K$$
$$= \int_{\Omega_1} \left(\frac{(x'-S_1)^\nu}{\Gamma(\nu+1)} - \frac{(x'-S_2)^\nu}{\Gamma(\nu+1)}\right) dx' dy' + \int_{\Omega_2} \left(\frac{(x'-S_1')^\nu}{\Gamma(\nu+1)} - \frac{(x'-S_2')^\nu}{\Gamma(\nu+1)}\right) dx' dy',$$

where S_1, S_2, S_1', S_2' are the coordinates in the rotating coordinate system (see Fig. 1).

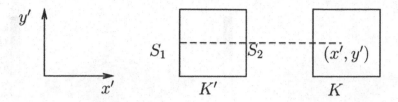

Fig. 2. Illustration for computing $\hat{T}_f^\theta$ on the structured mesh.

The implementation of $\hat{T}_f^\theta$ can also be similarly implemented on the structured mesh. For simplicity, we implement (4.1) on the structured mesh with

$$M(\theta) = \sum_{k=1}^{4} \frac{1}{4}\delta(\theta - \theta_k), \text{ where } \theta_k = \frac{\pi}{2}(k-1).$$

Then, when $\theta = 0, \pi/2, \pi, 3\pi/2$, we have (see Fig. 2)

$$\hat{T}_f^\theta = (D_\theta^{\alpha-1}\chi_{K'}, \chi_K)_K = \int_K (\frac{(x'-S_1)^\nu}{\Gamma(\nu+1)} - \frac{(x'-S_2)^\nu}{\Gamma(\nu+1)})dx'dy'.$$

5   Numerical Experiments

In this section we show some numerical results to verify the efficiency of the expanded mixed finite element method. In the following examples, we assume the porous medium is isotropic and $\boldsymbol{K} = \boldsymbol{I}$. We implement the proposed algorithm on the structured mesh with $M(\theta) = \sum_{k=1}^{4} \frac{1}{4}\delta(\theta - \theta_k)$, where $\theta_k = \frac{\pi}{2}(k-1)$.

Example 1. In this example we test the steady state space fractional Darcy flow (2.1) on a unit square domain with 81×81 grid. We let $f = 1$ and $f = -1$ in four grid cells of $\mathcal{T}_h$ respectively and let $f = 0$ in other region.

We choose $\alpha = 0.1, 0.5, 0.7, 0.9$ to test the algorithm. From Fig. 3 we can clearly see how the parameter α influences the solution.

Example 2. We remark that the expanded mixed FEM can also be applied to solve the following space fractional transport in porous media:

$$\frac{\partial c}{\partial t} + \nabla \cdot (\boldsymbol{u}c + \boldsymbol{D}(\boldsymbol{u})\nabla_M^\alpha c) = f \qquad \text{in } \Omega,$$

$$c = 0 \qquad \text{in } \mathcal{R}^2 \setminus \Omega,$$

$$c(x,0) = c_0(x) \qquad \text{in } \Omega.$$

In this example, we test the space fractional transport on $[0,4] \times [0,1]$ with 128×32 grid. We assume the velocity $\boldsymbol{u} = (0.1, 0)^T$. We denote $r^2 = (x - 1/8)^2 + (y-1/2)^2$ and let $f = 0$, $c_0(\boldsymbol{x}) = 10^5 \times 2^{1-1/(1-5r^2/4)}$ in the local region $r^2 \le 10^{-3}$ and $c_0(\boldsymbol{x}) = 0$ in other region.

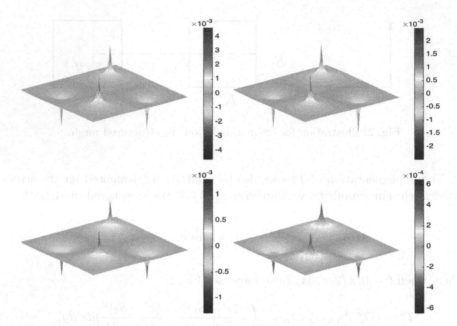

Fig. 3. With $K = 1$. Top-Left: solution of pressure ($\alpha = 0.1$). Top-Right: solution of pressure ($\alpha = 0.5$). Bottom-Left: solution of pressure ($\alpha = 0.7$). Bottom-Right: solution of pressure ($\alpha = 0.9$).

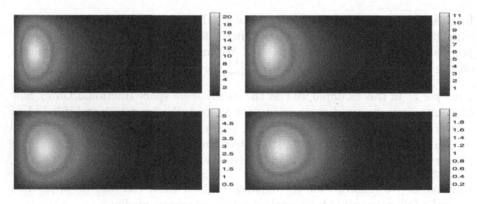

Fig. 4. The concentration solutions with different α at $T = 1.2$ with time step $\delta t = 0.01$. Top-Left: $\alpha = 0.1$. Top-Right: $\alpha = 0.5$. Bottom-Left: $\alpha = 0.7$. Bottom-Right: $\alpha = 0.9$.

We test this example by choosing $\alpha = 0.1, 0.5, 0.7, 0.9$ and compute the solutions with time step $\delta t = 0.01$ until the time $T = 1.2$. From Fig. 4 we can also clearly see how the parameter α influences the distribution of concentration at a fixed time. We can see that the diffusion effect is more obvious when the parameter α becomes large.

6   Conclusion

In this paper we discuss an expanded mixed finite element method for the solution of the two dimensional space fractional Darcy flow in porous media. The locally mass-conservation can be retained by this mixed scheme. The well-posedness of the expanded mixed formulation is proved and the implementation of the algorithm is given in details. Numerical results are shown to verify the efficiency of this mixed scheme. The mixed scheme for the space fractional Darcy flow with non-homogeneous boundary conditions will be investigated in the future work.

References

1. Adams, R.: Sobolev Spaces. Academic Press, New York (1975)
2. Bu, W., Tang, Y., Yang, J.: Galerkin finite element method for two-dimensional Riesz space fractional diffusion equations. J. Comput. Phys. **276**, 26–38 (2014)
3. Caputo, M.: Models of flux in porous media with memory. Water Resour. Res. **36**, 693–705 (2000)
4. Chen, H., Wang, H.: Numerical simulation for conservative fractional diffusion equations by an expanded mixed formulation. J. Comput. Appl. Math. **296**, 480–498 (2016)
5. Ervin, V.J., Roop, J.P.: Variational formulation for the stationary fractional advection dispersion equation. Numer. Methods Partial Differ. Equ. **22**, 558–576 (2005)
6. Ervin, V.J., Heuer, N., Roop, J.P.: Numerical approximation of a time dependent, nonlinear, space-fractional diffusion equation. SIAM J. Numer. Anal. **45**, 572–591 (2007)
7. Ervin, V.J., Roop, J.P.: Variational solution of fractional advection dispersion equations on bounded domains in R^d. Numer. Methods Partial Differ. Equ. **23**, 256–281 (2007)
8. Ervin, V.J., Heuer, N., Roop, J.P.: Regularity of the solution to 1-D fractional order diffusion equations. Math. Comput. **87**, 2273–2294 (2018)
9. Gao, G., Sun, Z.: A compact finite difference scheme for the fractional sub-diffusion equations. J. Comput. Phys. **230**, 586–595 (2011)
10. He, J.H.: Approximate analytical solution for seepage flow with fractional derivatives in porous media. Comput. Methods Appl. Mech. Eng. **167**, 57–68 (1998)
11. Jin, B., Lazarov, R., Pasciak, J., Rundell, W.: Variational formulation of problems involving fractional order differential operators. Math. Comput. **84**, 2665–2700 (2015)
12. Li, C., Zeng, F., Liu, F.: Spectral approximations to the fractional integral and derivative. Fract. Calc. Appl. Anal. **15**, 383–406 (2012)
13. Lischke, A., et al.: What is the fractional Laplacian? A comparative review with new results. J. Comput. Phys. **404**, 109009 (2020)
14. Liu, F., Zhuang, P., Turner, I., Burrage, K., Anh, V.: A new fractional finite volume method for solving the fractional diffusion equation. Appl. Math. Model. **38**, 3871–3878 (2014)
15. Meerschaert, M.M., Tadjeran, C.: Finite difference approximations for fractional advection-dispersion flow equations. J. Comput. Appl. Math. **172**, 65–77 (2004)
16. Meerschaert, M.M., Scheffler, H.P., Tadjeran, C.: Finite difference methods for two dimensional fractional dispersion equation. J. Comput. Phys. **211**, 249–261 (2006)

17. Meerschaert, M.M., Mortensen, J., Wheatcraft, S.W.: Fractional vector calculus for fractional advection-dispersion. Phys. A **367**, 181–190 (2006)

18. Roop, J.P.: Computational aspects of FEM approximation of fractional advection dispersion equations on bounded domains in R^2. J. Comput. Appl. Math. **193**, 243–268 (2006)

19. Wang, H., Basu, T.S.: A fast finite difference method for two-dimensional space-fractional diffusion equations. SIAM J. Sci. Comput. **34**, 2444–2458 (2012)

Prediction of the Free Jet Noise Using Quasi-gas Dynamic Equations and Acoustic Analogy

Andrey Epikhin[1,2](✉) ⬤ and Matvey Kraposhin[1,3] ⬤

[1] Ivannikov Institute for System Programming of the RAS,
Moscow 109004, Russia
andrey.epikhin@bk.ru, m.kraposhin@ispras.ru
[2] Bauman Moscow State Technical University, Moscow 105005, Russia
[3] Keldysh Institute of Applied Mathematics of the RAS,
Moscow 125047, Russia

Abstract. The paper is focused on the numerical simulation of acoustic properties of the free jets from circle nozzle at low and moderate Reynolds numbers. The near-field of compressible jet flow is calculated using developed regularized (quasi-gas dynamic) algorithms solver QGDFoam. Acoustic noise is computed for jets with M = 0.9, Re = 3600 and M = 2.1, Re = 70000 parameters. The acoustic pressure in far field is predicted using the Ffowcs Williams and Hawkings analogy implemented in the libAcoustics library based on the OpenFOAM software package. The determined properties of the flow and acoustic fields are compared with experimental data. The flow structures are characterized by the development of the Kelvin-Helmholtz instability waves, which lead to energy outflux in the radial direction. Their further growth is accompanied by the formation of large and small-scale eddies leading to the generation of acoustic noise. The results showed that for selected jets the highest levels of generated noise is obtained at angles around 30° which agrees well with experimental data.

Keywords: Aeroacoustics · Jet · Noise · Regularized equations · OpenFOAM

1 Introduction

The main cause of acoustic noise from high-speed turbulent jets is hydrodynamic instability. Depending on the scale, frequency, area of occurrence of these instabilities and their structures it is possible to classify such parameters of a supersonic jet noise source as radiation direction, intensity, spectrum frequency [1, 2]. The following noise sources are allocated for free supersonic jets [3–7]: large-scale turbulent structures; small-scale turbulence; broadband noise caused by the interaction of the shock-wave structure of the non-isobaric jet with hydrodynamic instability (Mach waves); narrowband noise caused by resonant regimes between the shock-wave structure of the jet and hydrodynamic instability (screech tone). The papers [8] and [9] present a generalized (for sub- and supersonic cases) law of scaling the noise level from average jet parameters and present empirical dependencies for prediction of turbulent mixing noise

© Springer Nature Switzerland AG 2020
V. V. Krzhizhanovskaya et al. (Eds.): ICCS 2020, LNCS 12143, pp. 217–227, 2020.
https://doi.org/10.1007/978-3-030-50436-6_16

spectra from supersonic jets. At present, there is a need for additional experimental and numerical studies to improve the existing methodology for predicting acoustic noise from trans- and supersonic jets and the mechanisms of their generation. This will ultimately help in the design of aircraft and rocket engines and reduce the overall noise generated. However, conducting experimental research is very difficult and expensive. In turn, the significant increase in computational power leads to the fact that the numerical prediction of acoustic noise from trans- and supersonic jets becomes relevant and important stage of research.

The qualitative simulation of the interaction of high-speed flow, hydrodynamic instability and external environment is critical for the tasks related to the study of acoustic noise of trans- and supersonic jets. Important while modeling the mixing processes in the near field is the exact resolution of formation of hydrodynamic instability and their interaction with the main flow. The numerical methods used in such cases must have the ability to resolve both the high-speed flows described by the Euler equations and the viscous flows at significantly subsonic velocities. However, a known problem of many of the existing explicit methods based on the approximate methods of the Godunov type (local Lax-Friedrichs (Rusanov), HLL, HLLC, etc.) implemented in computing packages is the limited area of applicability of these algorithms at the values of the Mach number less than 1. In turn, well-proven methods such as SIMPLE or PISO are unappropriated for highly compressible flows. An alternative to the approaches described above is the use of hybrid approaches [10] and algorithms based on regularized or quasi-gas dynamic (QGD) equations [11, 12]. The regularized gas dynamic equations are analogous to the Navier-Stokes equations and are used to describe flows in various tasks. QGD-algorithms are characterized by uniformity of approximating expressions, ease of use and physically determined by the numerical viscosity. Given the complexity of the processes occurring at the outflow of high-speed turbulent jets, it is appropriate to apply integral analogies to the solution of the Ffowcs Williams and Hawkings equation to solve the problem of noise prediction in the far field [13, 14].

The various approaches and algorithms described above for solving gas dynamic equations, and the integral method for determining acoustic pressure in the far field are implemented in the OpenFOAM open source package. In works [15–17] authors conduct research of various methods of numerical modeling implemented in Open-FOAM package and formulate recommendations for calculation of jet flows at small Reynolds numbers. So in researches [15, 17] it is established that the resolution of a computational grid to simulate the process of formation and propagation of instabilities in space should be not less than 30–40 cells per diameter (CPD). In order to expand the application scope of the developed QGDFoam solver, the validation of the QGD algorithms on the tasks of trans- and supersonic jets flow of perfect viscous gas and the acoustic noise generated by them at small and moderate Reynolds numbers (Re = 3600 and 70000) were carried out. The flow field and acoustic properties from jets for the selected conditions are well studied [18–20] and can be used to validate numerical methods and schemes.

2 Governing Equations and Methods

2.1 Quasi-gas Dynamic Equations

Regularized or quasi-gas dynamic system of equations includes equations of balance of mass, momentum and energy [11, 21]:

$$\frac{\partial \rho}{\partial t} + \nabla \cdot \vec{j}_m = 0, \tag{1}$$

$$\frac{\partial \rho \vec{U}}{\partial t} + \nabla \cdot \left(\vec{j}_m \otimes \vec{U} \right) + \nabla p = \nabla \cdot \Pi, \tag{2}$$

$$\frac{\partial E}{\partial t} + \nabla \cdot \left(\frac{\vec{j}_m}{\rho} (E + p) \right) + \nabla \cdot \vec{q} = \nabla \cdot \left(\Pi \vec{U} \right), \tag{3}$$

$$\vec{j}_m = \rho \left(\vec{U} - \vec{w} \right), \tag{4}$$

$$\Pi = \Pi_{NS} + \Pi_{QGD}, \tag{5}$$

$$\vec{q} = \vec{q}_{NS} + \vec{q}_{QGD}, \tag{6}$$

where ρ – density; $\vec{U}$ – velocity; $\vec{j}_m$ - mass flux density; p – pressure; E – total energy per unit volume; Π - viscous stress tensor; $\vec{q}$ – heat flux; Π_{NS}, $\vec{q}_{NS}$.- classical Navier-Stokes viscous stress tensor and heat flux; $\vec{w}$, Π_{QGD}, $\vec{q}_{QGD}$ - additional dissipative terms [11, 21] in the corresponding equations with coefficient, which is denoted as τ.

As an extension of the classical system of Navier-Stokes equations, the QGD system contains additional constituents that are proportional to a small coefficient τ having the time dimension. When parameter τ tends to zero, QGD system of equations reduces to Navier-Stokes system. In dimensionless form, the value of τ is proportional to the Knudsen number. For compressible gases, τ is too small to use its direct value because it does not provide the required stability of the numerical algorithm. In this case, the role of the free path in the numerical algorithm can perform the step of the computational grid:

$$\tau = \alpha_{QGD} \frac{\Delta_h}{a}, \tag{7}$$

where α_{QGD} is a small constant in the range from 0 to 1, which is the tuning parameter of the numerical QGD algorithm; Δ_h - computational grid step; a is the sound velocity. When solving problems with high values of Mach and Reynolds numbers, the introduced dissipation with the help of τ-supplements is insufficient, and therefore additional viscosity is introduced into the system as a coefficient in the viscous stress tensor:

$$\mu \rightarrow \mu + Sc_{QGD} p \tau, \tag{8}$$

where Sc_{QGD} is a scheme parameter that ensures its stability at high values of local Mach number. Values of α_{QGD} and Sc_{QGD} are selected as the trade-off between stability of the numerical algorithm and numerical diffusion. The theoretically justified guide on how to select values of coefficients is presented in [22, 23]. The general rule for the adjustment of these coefficients consists in gradual decrease from the baseline configuration: α_{QGD} = 0.4–0.5 and Sc_{QGD} = 1. According to the definition (7), (8) the decrease of α_{QGD} could be considered as the mesh refinement. According to the previous research [15] the recommended value of α_{QGD} and Sc_{QGD} are 0.15 and 0, respectively.

The system of Eqs. (1)–(3) is implemented in QGDFoam solver based on Open-FOAM package [24, 25]. More details about the code, including the governing equations can be found in papers [21, 24].

2.2 Acoustic Analogy

One of the main approaches used to model acoustic wave propagation is the solution of the Ffowcs Williams and Hawkings equation. There are various integral formulations for solving this equation. In this paper we used the Farassat 1A formulation [26, 27], implemented in the libAcoustics library developed by the authors [28, 29]. More details about the governing equations of this formulation can be found in paper [27]. This analogy is used to determine the far field noise generated by an acoustic source moving in a gas. This formulation is widely used to calculate the acoustic pressure from the rotor or jet flows. The following formula was used to calculate the sound pressure level (SPL):

$$SPL(dB) = 20log_{10}\left(p_{rms}/p_{ref}\right), \tag{9}$$

where p_{rms} is the root mean square sound pressure, p_{ref} is reference sound pressure.

The developed library based on OpenFOAM package is publicly available and can be compiled independent of any modules of the main package and the type of solvers.

3 Numerical Setup

The problem of trans- and supersonic jets outflow with the following initial data is considered: M = 0.9, Re = 3600; M = 2.1, Re = 70000. The flow parameters and geometry match an experimental study conducted by Stromberg et al. [18] and Troutt et al. [20] and results will be compared to their data whenever possible.

The computational domain was a rectangular parallelepiped in which the outlet boundary is removed by 100D, the side boundary by 20D, where D is the diameter of the nozzle exit. The inlet boundary corresponded to the exit of the round nozzle and coincides with the beginning of coordinates (see Fig. 1). In addition, the computational grid has been refined on 30D downstream. Based on the recommendations presented in [17], a calculation grid with 40 CPD and a total number of cells of 33 million is constructed. Figure 2 shows a fragment of the grid near the nozzle exit.

The numerical simulation was performed using QGDFoam solver with the following setting parameters defined in [17]: α_{QGD} = 0.15, Sc_{QGD} = 0. Adams-Bashforth

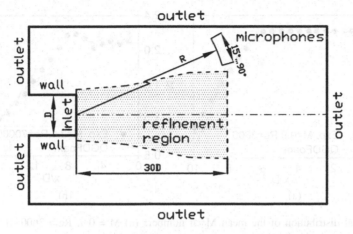

Fig. 1. Computational domain and boundary conditions.

Fig. 2. Fragment of the computational grid near the nozzle exit.

(backward) scheme was used for time derivatives approximation. The turbulence model has not been applied for jet at small Reynolds number. For the case M = 2.1 and Re = 70000, the LES approach with Smagorinsky sub-grid model was used.

The libAcoustics library with Farassat 1A formulation was used to predict acoustic pressure in the far field. Virtual microphones were located at a distance of R = 30D and 40D for jets with low and moderate Reynolds number respectively, the angle of microphones position θ was set from 15 to 90° to the jet axis.

4 Results and Discussion

4.1 Near Field Calculation

Figure 3 presents the axial distribution of the Mach number averaged over time in comparison with experimental data. Figures 4 and 5 show the flow field of the jets under study.

The flow field at low Reynolds number is laminar for considerable part of the potential core region. The potential core length of the jet at moderate Reynolds number

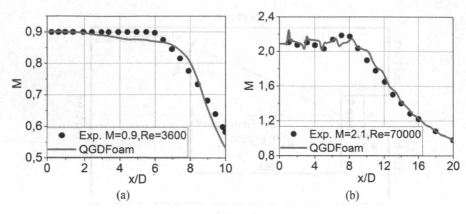

(a)   (b)

Fig. 3. Axial distribution of the mean Mach number: (a) M = 0.9, Re = 3600; (b) M = 2.1, Re = 70000.

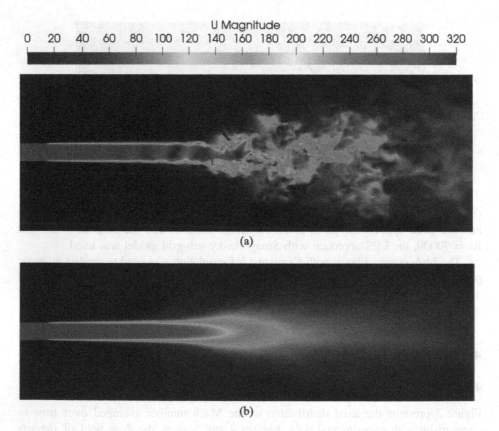

(a)

(b)

Fig. 4. The jet velocity distribution at M = 0.9, Re = 3600: (a) instantaneous; (b) mean.

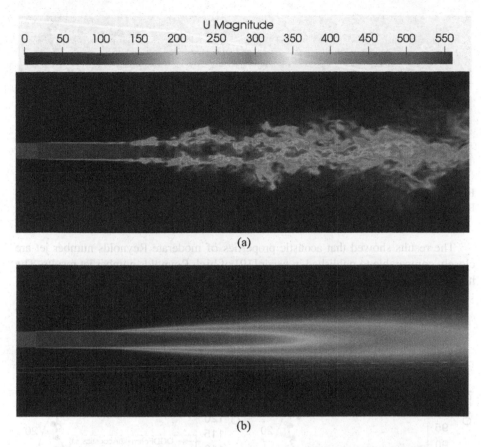

Fig. 5. The jet velocity distribution at M = 2.1, Re = 70000: (a) instantaneous; (b) mean.

is between 8 and 10 diameters and characterized by weak shock cell structure. The sonic point in this jet is reached between 18 and 20 diameters downstream of the nozzle exit.

4.2 Far Field Noise Prediction

Based on the recommendations [13, 14] on the choice of the form of the control surface to calculate the acoustic pressure from jet flows, two open surfaces were constructed, which are schematically depicted in Fig. 6 (s1 is green, s2 is red). The control surfaces start 0.1D downstream of the inflow boundary and extend to 35D along the streamwise direction. Sound pressure level for two tasks are calculated in terms of a reference pressure scaled to the ambient pressure: $p_{ref} = 2 \cdot 10^{-5} \cdot (p_c/p_a)$, where p_c is the pressure in the test chamber, p_a is standard atmospheric pressure.

The OASPL directivity distribution for low and moderate Reynolds numbers jets are presented in Fig. 7.

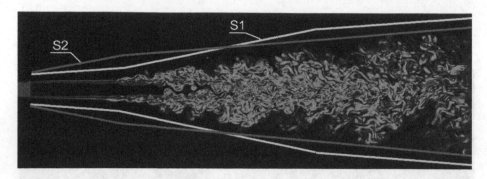

Fig. 6. Schematic showing the two open control surfaces surrounding the jet at moderate Reynolds number (Divergence of velocity contours are shown). (Color figure online)

The results showed that acoustic properties of moderate Reynolds number jet are closely comparable to published in paper [19] of high Reynolds number jet results. The highest levels of generated noise occur at angles around 30°.

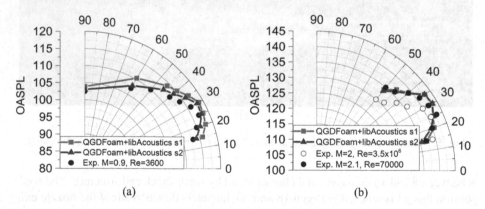

Fig. 7. Sound pressure level directivity distributions for QGDFoam solver with different control surfaces: (a) M = 0.9, Re = 3600; (b) M = 2.1, Re = 70000.

The effect of Reynolds number on the acoustic frequency spectrum at the point corresponding maximum acoustic noise are presented on Fig. 8. At the low Reynolds number acoustic spectrum is dominated by a narrow band of frequencies. At the moderate Reynolds number, it is quite full. The broad peak is around $St = f \cdot D/U = 0.22$ that coincides approximately with the low Reynolds condition.

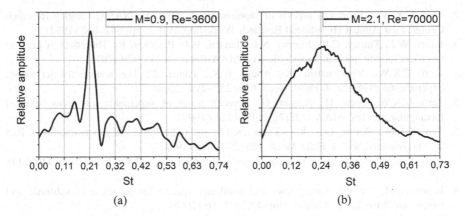

Fig. 8. Acoustic spectrum in the maximum noise direction ($\theta = 30°$): (a) M = 0.9, Re = 3600; (b) M = 2.1, Re = 70000.

5 Conclusion

The numerical experiments have allowed to investigate the applicability of the developed approach based on implementation of the QGD equations using Open-FOAM package. The developed solver QGDFoam has been coupled with libAcoustics library. Validation of QGDFoam solver was performed on the problems of trans- and supersonic jets of perfect viscous gas and acoustic noise generated by them at low and moderate Reynolds numbers. The calculated gas-dynamic parameters of near-field and acoustic pressure in the far field correspond well with the experimental data.

The analysis of the received data has allowed to expand applicability of QGD algorithms for numerical simulation of the free jet flow at moderate Reynolds numbers. Thus, the resolution of the calculated grid should be at least 40 CPD, and the regularization parameters are $Sc_{QGD} = 0$, $\alpha_{QGD} = 0.15$.

It was found that the jet at moderate Reynolds number both the flow and acoustic properties are considerably more complex than the at low Reynolds number conditions. The major of acoustic noise propagation direction of the investigated jets occur at angles around 30° to the jet axis. The noise spectrum of jet at the low Reynolds number is a narrow band of frequencies in turn at moderate Reynolds number it is quite full.

Acknowledgments. This work was supported by the RSF under the grant No. 19-11-00169. The results of the work are obtained using computational resources of MCC NRC «Kurchatov Institute», http://computing.nrcki.ru.

References

1. Crighton, D.G.: Orderly structure as a source of jet exhaust noise: survey lecture. In: Fiedler, H. (ed.) Structure and Mechanisms of Turbulence II. Lecture Notes in Physics, vol. 76. Springer, Heidelberg (1978). https://doi.org/10.1007/BFb0012619

2. Tam, C.K.W.: Theoretical aspects of supersonic jets noise. In: NASA, Langley Research Center, First Annual High-Speed Research Workshop, Part 2, pp. 647–662 (1992)
3. Baars, W.J., Tinney, C.E., Murray, N.E., Jansen, B.J., Panickar, P.: The effect of heat on turbulent mixing noise in supersonic jets. AIAA Paper, 2011-1029 (2011)
4. Tam, C.K.W., Viswanathan, K., Ahuja, K.K., Panda, J.: The sources of jet noise: experimental evidence. J. Fluid Mech. **615**, 253–292 (2008)
5. Tam, C.K.W., Shen, H., Raman, G.: Screech tones of supersonic jets from bevelled rectangular nozzles. AIAA J. **35**(7), 1119–1125 (1997)
6. Tam, C.K.W., Burton, D.E.: Sound generated by instability waves of supersonic flows. Part 2. Axisymmetric jets. J. Fluid Mech. **138**, 273–295 (1984)
7. Tam, C.K.W.: Mach wave radiation from high-speed jets. AIAA J. **47**(10), 2440–2448 (1984)
8. Kandula, M.: On the scaling laws and similarity spectra for jet noise in subsonic and supersonic flow. Int. J. Acoust. Vibr. **13**(1), 3–16 (2008)
9. Tam, C.K.W., Golebiowski, M., Seiner, J.M.: On the two components of turbulent mixing noise from supersonic jets. AIAA Paper, 96-1716 (1996)
10. Kraposhin, M.V., Banholzer, M., Pfitzner, M., Marchevsky, I.K.: A hybrid pressure-based solver for nonideal single-phase fluid flows at all speeds. Int. J. Numer. Methods Fluids **88** (2), 79–99 (2018)
11. Elizarova, T.G.: Quasi-Gas Dynamic Equations. Springer, Berlin (2009). https://doi.org/10.1007/978-3-642-00292-2
12. Chetverushkin, B.N.: Kinetic schemes and quasi-gas-dynamic system of equations. Russ. J. Numer. Anal. Math. Model. **20**(4), 337–351 (2005)
13. Uzun, A., Lyrintzis, A.S., Blaisdell, G.A.: Coupling of integral acoustics methods with LES for jet noise prediction. AIAA Paper, 4982-5001 (2004)
14. Shur, M., Spalart, P., Strelets, M.: Noise prediction for increasingly complex jets. Part I: methods and tests. Int. J. Aeroacoustics **4**, 213–246 (2005)
15. Kraposhin, M.V., Epikhin, A.S., Elizarova, T.G., Vatutin, K.A.: Simulation of transonic low-Reynolds jets using quasi-gas dynamics equations. J. Phys: Conf. Ser. **1382**(1), 012019 (2019)
16. Al-Zoubi, A., Beilke, J., Korchagova, V.N., Strizhak, S.V., Kraposhin, M.V.: Comparison of the performance of open-source and commercial CFD packages for simulating supersonic compressible jet flows. In: Proceedings - 2018 Ivannikov Memorial Workshop (IVMEM 2018), pp. 61–65 (2019)
17. Epikhin, A., Kraposhin, M., Vatutin, K.: The numerical simulation of compressible jet at low Reynolds number using OpenFOAM. In: E3S Web of Conferences, vol. 128 (2019)
18. Stromberg, J.L., McLaughlin, D.K., Troutt, T.R.: Flow field and acoustic properties of a Mach number 0.9 jet at a low Reynolds number. J. Sound Vibr. **72**(2), 159–176 (1980)
19. McLaughlin, K., Seiner, J.M., Liu, C.H.: On noise generated by large scale instabilities in supersonic jets. AIAA Paper, 80-0964 (1980)
20. Trouttt, T.R., McLaughlin, D.K.: Experiments on the flow and acoustic properties of a moderate-Reynolds-number supersonic jet. J. Fluid Mech. **116**, 123–156 (1982)
21. Kraposhin, M.V., Ryazanov, D.A., Smirnova, E.V., Elizarova, T.G., Istomina, M.A.: Development of OpenFOAM solver for compressible viscous flows simulation using quasi-gas dynamic equations. In: Proceedings - 2017 Ivannikov ISPRAS Open Conference (ISPRAS 2017), pp. 117–123 (2018)
22. Zlotnik, A.A., Lomonosov, T.A.: Conditions for L2-dissipativity of linearized explicit difference schemes with regularization for 1D barotropic gas dynamics equations. Comput. Math. Math. Phys. **59**, 452–464 (2019)

23. Zlotnik, A.A., Lomonosov, T.A.: L2-dissipativity of the linearized explicit finite-difference scheme with a kinetic regularization for 2D and 3D gas dynamics system of equations. Appl. Math. Lett. **103**, 106198 (2020)
24. Kraposhin, M.V., Smirnova, E.V., Elizarova, T.G., Istomina, M.A.: Development of a new OpenFOAM solver using regularized gas dynamic equations. Comput. Fluids **166**, 163–175 (2018)
25. QGDFoam solver. https://github.com/unicfdlab/QGDsolver
26. Brentner, K.S., Farassat, F.: An analytical comparison of the acoustic analogy and Kirchhoff formulations for moving surfaces. AIAA J. **36**, 1379–1386 (1998)
27. Brès, G.A., Pérot, F., Freed, D.: A Ffowcs Williams–Hawkings solver for lattice Boltzmann based computational aeroacoustics. AIAA Paper, 2010-3711 (2010)
28. Epikhin, A., Evdokimov, I., Kraposhin, M., Kalugin, M., Strijhak, S.: Development of a dynamic library for computational aeroacoustics applications using the OpenFOAM open source package. Procedia Comput. Sci. **66**, 150–157 (2015)
29. libAcoustics library. https://github.com/unicfdlab/libAcoustics

Simulation Based Exploration of Bacterial Cross Talk Between Spatially Separated Colonies in a Multispecies Biofilm Community

Pavel Zarva and Hermann J. Eberl[(✉)]

Department of Mathematics and Statistics, University of Guelph,
Guelph, ON N1G 2W1, Canada
{pzarva,heberl}@uoguelph.ca

Abstract. We present a simple mesoscopic model for bacterial cross-talk between growing biofilm colonies. The simulation setup mimics a novel microfludic biofilm growth reactor which allows a 2D description. The model is a stiff quasilinear system of diffusion-reaction equations with simultaneously a super-diffusion singularity and a degeneracy (as in the porous medium equation) that leads to the formation of sharp interfaces with finite speed of propagation and gradient blow up. We use a finite volume method with arithmetic flux averaging, and a time adaptive stiff time integrator. We find that signal and nutrient transport between colonies can greatly control and limit biofilm response to induction signals, leading to spatially heterogeneous biofilm behavior.

Keywords: Biofilm · Cross talk · Degenerate diffusion · Mathematical model · Reaction-diffusion · Simulation · Super-diffusion

1 Introduction

Bacterial biofilms are microbial aggregates on immersed, biotic or abiotic, surfaces or interfaces. Biofilm communities consist usually of several colonies of same or different species. They form wherever environmental conditions permit microbial growth [18]. In the initial (reversible) stage of biofilm formation cells attach to the surface, called substratum in the biofilm literature, and begin the production of extracellular polymeric substances (EPS), a protective layer in which they are themselves embedded and multiply. The colonies grow and eventually may merge into larger assemblages. In these aggregates, diffusion gradients develop that lead to variations in local growth conditions for cells in dependence of their location in the biofilm. Under some circumstances this can lead to the formation of microniches, which allows for example for anaerobic pockets in otherwise aerobic biofilms. Biofilm based processes are engineered for example in wastewater

Supported by Natural Sciences and Engineering Research Council of Canada (NSERC): RGPIN-2019-05003, RTI-2016-00080.

treatment, soil remediation, or also in biofuel production. On the other hand, in a medical or industrial context, biofilm growth is often unwanted and leads to detrimental consequences, such as biofouling and biocorrosion of equipment, public health risks, or bacterial infections [27]. Biofilm borne infections are more difficult to eradicate than other bacterial infections, for example due to the mechanical and chemical protection that the encasing EPS layer offers to the cells.

Quorum sensing is a mechanism that bacteria use to coordinate gene expression in groups. Cells produce a signal (also called autoinducers) at a basal rate. When the signal concentration passes an induction threshold, the cells undergo a change in gene expression [12]. Initially, as the name suggests, this was thought to be a communication mechanism to assess group strength. However, it is understood now that in spatially structured populations and in non completely mixed environments this induction process is also affected by transport processes acting on the signal [16]. For example it was suggested based on theoretical and computational studies that in hydrodynamic environments larger upstream colonies can play a major role in induction of smaller downstream colonies [10], which also has been confirmed experimentally since [17]. In certain biofilms, quorum sensing is used as a stress response mechanism to control virulence factors, such as elastases, pyocyanin, cyanide and exotoxins [1]. It was suggested that biofilm formation itself, i.e. the onset of EPS production, in some species is controlled by a quorum sensing mechanism [2]. Based on computational studies, it was argued [11] that a quorum sensing induction mechanism can explain a switch from an early stages mode of biofilm growth, during which resources are primarily invested in increasing population size, to a mode of growth during which resources are invested in protection by producing EPS.

Cross-talk refers to bacteria of one species or strain responding to signals produced by another one [3,4]. In gram negative bacteria N-acyl homoserone lactones (AHL) are an important group of quorum sensing signalling molecules. Although some strains have been reported to produce up to 20 different AHLs [3], the number of known AHLs is limited [24]. Therefore, some signal molecules are used by, or recognised by several strains or species, leading to interspecies communication. This concept of cross-talk challenges the notion that quorum sensing is purely an autoinduction mechanism.

Mathematical models of biofilms have been used since the 1980s, specifically in wastewater engineering. Originally, these were based on the assumption of stratification parallel to the substratum. At the end of the 1990s, after it became appreciated that biofilms are not flat layers many different models for the growth of spatially heterogeneous biofilms were proposed, utilising agent based techniques, cellular automata, or partial differential equations. The models in the latter groups can be subdivided into such that view a biofilm as a mechanical object, and those that consider them a spatially structured population. Experimentally, biofilms have been characterised as both. One model that can be motivated from both angles is a density dependent diffusion reaction framework [6,20]. This model framework has been extended to describe several biofilm processes, including some that involve quorum sensing [7,8,10,11,13–15,26]. In our current study we will extend this framework to include a simple example of bacterial cross-talk. An earlier example of a model of cross talk in biofilms can be found in [24], in which sender and receiver bacteria were in close proximity, in which it was assumed that biofilm colonies do not grow in time, and in which bacteria received nutrients from an agar layer on which they were grown. In our study the focus is on the role of diffusion of signals and nutrients between colonies that are spaced apart. One of the big, unsolved problems in multi-dimensional biofilm modeling is the inherently multi-scale nature of biofilm processes. Often one is interested in mesocopic features, i.e. in processes on the actual colony scale. To simulate such processes, requires a good set of boundary conditions that describe how the domain of simulation (e.g. a small open section of a much larger flow channel) is connected with the greater physical environment. This problem in simulations is in essence the same problem experimental studies suffer from, for example because modern microscopes can only resolve small sections of an experimental reactor. In this paper we introduce a new and simple computational setup, which circumvents this problem, mimicking a new microfluidic growth chamber for biofilm experiments that was recently introduced in [22].

2   Mathematical Model

2.1   Biological Processes and Governing Equations

We consider a dual species biofilm community consisting of two species of bacteria. Both species compete for a single, shared, growth limiting carbon substrate, which is subject to Fickian diffusion. Substrate uptake follows Monod kinetics, i.e. the uptake rate is proportional to the local substrate concentration when limited, and approximately constant where and when substrate is abundant. Substrate uptake translates into production of new biomass. We also account for cell death at a constant rate. Biofilm colonies do not expand spatially if locally space is available to accommodate newly produced cells. When the maximum physically attainable biofilm density is approached, colonies start spreading. We model this by the density dependent diffusion mechanism that was introduced in [6] and that can be derived both from the viewpoint of a biofilm as a mechanical object and as a spatially structured microbial populations [20]. Experimentally, biofilms have been characterised as both.

We assumed that the bacteria of one species ('sender') produce a signal molecule at a basal rate. If the local signal concentration passes an induction threshold, cells up-regulate and the rate of signal production increases by one order of magnitude. We do not explicitly distinguish between up- and down-regulated cell fractions, but make this distinction implicitly in terms of local signal concentration, and whether it is above or below the induction threshold value. Upregulated cells are assumed to have different growth behavior than down-regulated cells. More specifically, in this exploratory study, we assume that their substrate uptake rate increases, in order to invest resources into protection mechanisms, which we do not model explicitly. The autoinducer signal is transported in the domain by Fickian diffusion and degrades abiotically. The second species ('receiver') responds to signals and upregulates, i.e. increases substrate uptake if the signal surpasses induction threshold, but it does not produce the signal.

The model is cast in terms of the dependent variables M_1, volume fraction of sender species; M_2, volume fraction of receiver species; C, concentration of carbon substrate; S concentration of signal. The independent variables are time $t > 0$ and location $x \in \Omega \subset \mathbb{R}^d$. The model reads

$$\partial_t M_1 = \nabla \left[D(M_1) \nabla M_1 \right] + \mu_1 f(C; \kappa_1) M_1 - \delta_1 M_1 \tag{1}$$

$$\partial_t M_2 = \nabla \left[D(M_2) \nabla M_2 \right] + \mu_2 f(C; \kappa_2) M_2 - \delta_2 M_2 \tag{2}$$

$$\partial_t C = D_C \Delta C - \sum_{i=1}^{2} \nu_i \Big(1 + \sigma_i h(S; \tau_i) \Big) f(C; \kappa_i) M_i \tag{3}$$

$$\partial_t S = D_S \Delta S + f(C; \kappa_1) \Big(\beta h(S, \tau_1) + \alpha \Big) M_1 - \gamma S_1 \tag{4}$$

In this model the coefficient function f is the Monod function that describes the dependence of the growth process on the substrate concentration C. This function saturates $f \approx 1$ if $C \gg \kappa$ and f is proportional to C as $C \ll \kappa$. Function h is a Hill function that describes the induction process, i.e. the transition from a down-regulated state with $h \approx 0$ when $S \ll \tau_i$, to an up-regulated state with $h \approx 1$ if $S \gg \tau$. We have

$$f(C; \kappa) = \frac{C}{\kappa + C}, \qquad h(S; \tau) = \frac{S^n}{\tau^n + S^n}, \qquad n \approx 2.2 \sim 2.5. \tag{5}$$

The function $D(M)$ describes spatial movement of biomass in dependence of local density,

$$D(M) = D_M \frac{M^a}{(1 - M)^a}. \tag{6}$$

It combines two nonlinear diffusion effects: (i) for $M \to 0$ it behaves like $D(M) \sim M^a$, i.e we have a porous medium degeneracy with $D(0) = 0$ that implies that the biofilm/water interfaces moves with finite speed, and (ii) for $M \to 1$ it behaves like $D(M) \sim (1 - M)^{-a}$, i.e. we have a super-diffusion singularity, which ensures that biomass expands spatially as $M \to 1$ and ensures $M < 1$ even if biomass continues to grow as long as substrates are available. Both effects are necessary to describe biofilm formation. Note that the biomass motility coefficient D_M is much smaller than the diffusion coefficients $D_{C,S}$ of substrate and signal. We remark that a single species model formulation is used here for both biomass species independently, although mutli-species formulations (with or without cross-diffusion) are available [13, 20]. This is justified because in our simulation setup, as described below, both species will be separated throughout, and no merging of colonies of different kinds will take place. In the (local) absence of one species, the dual-species models reduce to the single species models.

In this model formulation, all parameters are non-negative. See Table 1 for their definition and default values that will be used in the simulations.

Table 1. Model parameters and their default values used in the simulations.

Parameter	Symbol	Value
Growth rate of species i	μ_i	1.0
Cell lysis rate for species i	δ_i	0.1
Substrate uptake rate of down-regulated cells of species	ν_i	1.0×10^4
Half saturation concentration for species i	κ_i	0.1
Factor of increased substrate uptake of up-regulated cells of species i	σ_i	1.0
Induction threshold for species i	τ_i	1.0
Basal signal production rate	α	4500
Increased signal production rate of upregulated cells	β	45000
Abiotic signal degradation rate	γ	0
Dimerisation exponent for signal production	n	2.5
Diffusion coefficient of carbon substrate	D_C	33
Diffusion coefficient of signal	D_S	16.5
Biomass motility coefficient	D_M	1.0×10^{-6}
Biomass diffusion nonlinearity exponent	a	4
Growth chamber length/width	L	0.5
External substrate concentration	C_0	1.0

2.2 Physical Configuration and Initial and Boundary Conditions

The simulation setup in our study is inspired by a micro-fluidic device that was introduced in [22] and is based on [23]. It consists of a rectangular growth chamber of extensions $2L \times L \times H$, which is subdivided by a membrane into two sections of size $L \times L \times H$. This membrane is permeable to dissolved substrates C and S, but not to bacteria. Along the lateral boundaries opposite of the internal membrane, both growth chambers are connected to tangential flow channels by like membranes. In the flow channels the substrate concentration is kept at a bulk concentration $C = C_0$, and the signal concentration is kept at nil. Thus, via diffusion across these external membranes, substrate is supplied to the growth channels, and signals are removed. The other boundaries of the growth chamber are impermeable to substrates. The height of the growth chamber H is comparable to the length scale of bacteria, and $H \ll L$. This allows us to consider biofilm colonies as two-dimensional, spreading across the substratum but having constant height. The system is described then by a two-dimensional rectangular domain $\Omega = [0, 2L] \times [0, L]$. The boundary conditions are

$$\partial_n M_{1,2}|_{\partial\Omega} = 0, \tag{7}$$

$$C|_{x_1 \in \{0,2L\}} = C_0, \ S|_{x_1 \in \{0,2L\}} = 0, \ \partial_n C|_{x_2 \in \{0,L\}} = 0, \ \partial_n S|_{x_2 \in \{0,L\}} = 0. \tag{8}$$

Although the experimental device on which our simulation setup is based has a membrane separating both sections of the growth chamber, we do not specify internal boundary conditions. The purpose of the membrane is to allow the substrate and the signal, C and S, to freely diffuse between both parts of the domain, but to prevent the colony bacteria from crossing over. Due to the porous medium degeneracy $D(0) = 0$ in (6), initial data with compact support imply solutions with compact support. Thus a colony that initially is separated away from the inner boundary will at most reach it after some finite time T. Our simulations are always terminate before that occurs.

The initial conditions for the dissolved concentrations in our simulations are

$$S|_{t=0} \equiv 0, \quad C|_{t=0} = C_0. \tag{9}$$

For the biomass we introduce two subdomains $\Omega_1(0) \subset [0, L - \eta] \times [0, L] \subset \Omega$ and $\Omega_2(0) \subset [L + \eta, 2L] \times [0, L] \subset \Omega$, $0 < \eta \ll L$, such that

$$M_i|_{t=0} = \begin{cases} m_{oi} \geq 0, \, x \in \Omega_i(0), \\ 0 \qquad\quad x \notin \Omega_i(0) \end{cases}, \quad i = 1, 2. \tag{10}$$

Here $\Omega_i(0)$ consists of possibly several not connected compact regions that are very small compared to the domain Ω. Mostly we will use these to be spherical or constructed from overlapping spherical regions.

2.3   Numerical Implementation

Besides the stiffness introduced by the disparity of time scales in the reaction terms, the model features two effects that make its numerical simulation challenging: the super-diffusion singularity as $M \to 1$ leads to blow-up of the diffusion coefficient, whereas the degeneracy at $M = 0$ introduces a sharp interface that propagates with finite speed and along which the biomass gradient blows up. The model is first discretised in space, then in time, following the approach laid out in [13,14]: For spatial discretisation a standard finite volume method is used on a uniform rectangular grid of size $2m \times m$. Integrating (1) over each grid cell and using the Divergence Theorem gives

$$\frac{d}{dt} \int_{v_{i,j}} M_1 dx dy = \int_{\partial v_{i,j}} J_n ds + \int_{v_{i,j}} R_1 dx dy, \quad i = 1, ..., 2m, j = 1, ..., m, \tag{11}$$

where v_{ij} denotes the domain of the cell with grid index (i, j), $J_n := D(M_1)\partial_n M_1$ denotes the outward normal flux across the grid cell boundary, and R_1 stands for the reaction terms in (1). To evaluate the area integrals in (11), we evaluate the dependent variables at the grid cell center,

$$M_{1;i,j}(t) := M_1(t, x_i, y_j) \approx M_1\big(t, (i - 1/2)\,\Delta x, (j - 1/2)\,\Delta x\big)$$

with $\Delta x = H/m$ and approximate the integral by the midpoint rule. The line integral in (11) is evaluated by considering every edge of the grid cell separately using the midpoint rule. To this end, the diffusion coefficient $D(M_1)$ in the midpoint of the cell edge is approximated by arithmetic averaging from the neighboring grid cell center points, and the derivative of M_1 across the cell edge by a central finite difference. It was shown previously that this approximates biofilm interface propagation well [5,14]. We get then for the biomass density in the grid cell center the ordinary differential equation

$$\frac{d}{dt} M_{1;i,j} = \frac{1}{\Delta x} \left(J_{i+\frac{1}{2},j} + J_{i-\frac{1}{2},j} + J_{i,j+\frac{1}{2}} + J_{i,j-\frac{1}{2}} \right) + R_{1;i,j}. \tag{12}$$

For the fluxes we have

$$J_{i+\frac{1}{2},j} = \begin{cases} \frac{1}{2\Delta x}(D(M_{1;i+1,j})+D(M_{1;i,j}))(M_{1;i+1,j} - M_{1;i,j}) & \text{for} \quad i < 2m, \\ 0 & \text{for} \quad i = 2m, \end{cases}$$

$$J_{i-\frac{1}{2},j} = \begin{cases} 0 & \text{for} \quad i = 1, \\ \frac{1}{2\Delta x}(D(M_{1;i,j}) + D(M_{1;i-1,j}))(M_{1;i-1,j} - M_{1;i,j}) & \text{for} \quad i > 1, \end{cases}$$

$$J_{i,j+\frac{1}{2}} = \begin{cases} \frac{1}{2\Delta x}(D(M_{1;i,j+1}) + D(M_{1;i,j}))(M_{1;i,j+1} - M_{1;i,j}) & \text{for} \quad j < m, \\ 0 & \text{for} \quad j = m, \end{cases}$$

$$J_{i,j-\frac{1}{2}} = \begin{cases} 0 & \text{for} \quad j = 1, \\ \frac{1}{2\Delta x}(D(M_{1;i,j-1}) + D(M_{1;i,j}))(M_{1;i,j-1} - M_{1;i,j}), & \text{for} \quad j > 1. \end{cases}$$

The spatial discretization of the equations for M_2, S, C follows the same principle, replacing the Neumann boundary conditions along the lateral boundaries by Dirichlet conditions for S and C. After spatial discretization and lexiographical ordering we obtain a system of $8m^2$ differential equations of the form

$$\begin{cases} \frac{d\mathbf{M_1}}{dt} = \mathcal{D}(\mathbf{M_1})\mathbf{M_1} + \mathcal{R}_1(\mathbf{C})\mathbf{M_1} \\ \frac{d\mathbf{M_2}}{dt} = \mathcal{D}(\mathbf{M_2})\mathbf{M_2} + \mathcal{R}_2(\mathbf{C})\mathbf{M_2} \\ \frac{d\mathbf{C}}{dt} = \mathcal{D}_C\mathbf{C} - \mathcal{R}_{1,C}(\mathbf{C,S})\mathbf{M_1} - \mathcal{R}_{2,C}(\mathbf{C,S})\mathbf{M_2} + \mathbf{b}_C \\ \frac{d\mathbf{S}}{dt} = \mathcal{D}_S\mathbf{S} + \mathcal{R}_S(\mathbf{C,S})\mathbf{M_1} - \gamma\mathcal{I}\mathbf{S} + \mathbf{b}_S \end{cases} \tag{13}$$

where vectors $\mathbf{M_1}, \mathbf{M_2}, \mathbf{C}, \mathbf{S}$ contain the grid approximations of the dependent variables, $\mathcal{I}$ is the identity matrix in $\mathbb{R}^{2m^2 \times 2m^2}$. The matrices $\mathcal{D}$ and $\mathcal{D}_{N,S}$ contain the spatial derivative terms. They are symmetric, and weakly diagonally dominant with non-positive main diagonals and non-negative off-diagonals. The matrices $\mathcal{R}_1, \mathcal{R}_2, \mathcal{R}_{1,C}, \mathcal{R}_{2,C}, \mathcal{R}_S$ are diagonal and contain the reaction terms; vectors $\mathbf{b}_{CS}$ contain contributions of the Dirichlet boundary conditions.

The semi-discrete system (13) is a sparse lattice ODE that satisfies a Lipschitz condition and preserves non-negativity.We use the embedded Rosenbrock-Wanner Method ROS3PRL [21]. The Jacobian linear systems are solved using a Jacobi-preconditioned BiCGSTAB method. Most of the code was implemented in-house using Fortran (with Intel, GNU, and Portland Group compilers) and prepared for execution on shared memory computers using OpenMP. For the BiCGSTAB algorithm we used an OpenMP parallelised version of the implementation in Sparskit [19,25]. This uses reverse communication and external inner products and sparse matrix vector products. Both the diffusion operators of the model and the Jacobian matrices are organised in sparse diagonal format. Simulations reported here have been carried out on Intel Xeon based Lenovo P520 and P710 workstations, and on two heterogeneous clusters of the Compute Canada network (SHARCNET's Graham, and WestGrid's cedar). ParaView was used for visual postprocessing.

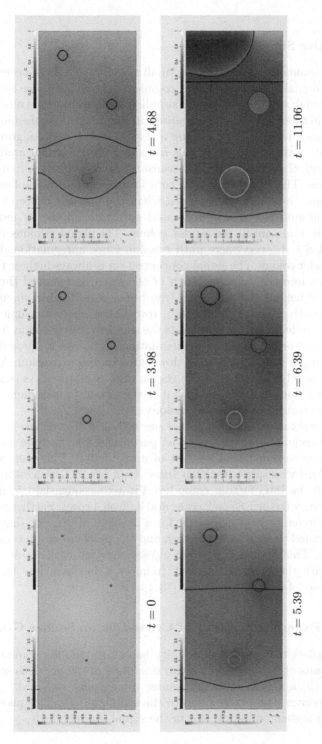

Fig. 1. Typical simulation of the model as described in Sect. 3.1. Shown are for different simulation times: the nutrient concentration C in the background (greyscale), and biomass densities $M_{1,2}$ and signal concentration S as contour lines. The colony in the left half of the domain is the sender colony, the two colonies in the right half are the receiver colonies. The contour line at which S is at induction threshold, $S = 1$, is emphasised by a thicker blue contour line.

3 Results

3.1 An Illustrative Simulation

For illustration, we simulate a system with a small sender colony in the center of the right half of the domain, and two receiver colonies in the left half, one is closer to the boundary through which nutrients are supplied and autoinducers removed, one closer to the center line. The results are shown in Fig. 1. Initially, the biomass in all colonies increases; they start expanding at $t \approx 1$. As the colonies grow the substrate concentration there decreases. This induces steeper nutrient gradients at the boundaries and, thus, increases mass transfer into the system, promoting growth of all colonies. The close a colony is to a lateral boundary, the faster it expands. This becomes especially pronounced when the nutrient concentration becomes small, at around $t \approx 6$. The increased substrate uptake by induced cells leads to a local minimum in the sender colony. The signal remains below induction threshold $S < 1$, everywhere until $t \approx 4.5$. Once upregulation starts, it is rapid in the sender colony. The signal concentration is always highest there. From here it diffuses into the domain. Much of the signal is removed through the lateral boundary behind the sender colony, but a considerable amount is transported into the other half of the domain, toward the receiver colonies. At $t = 5.39$ the signal at induction threshold levels the smaller receiver colony, which is quickly encompassed in the region of up-regulation $S > 1$. As it upregulates, its nutrient consumption increases, slowing down growth and expansion. At $t = 6.39$ this colony is entirely contained in the region of upregulation. As signal is continuously removed from the environment, it never exceeds a maximum value of $S \approx 4.2$, which is attained in the sender colony.

This simulation was carried out on a grid of size 400×200. To validate parallel performance of the code, a speed-up test was performed for the time interval $3.90 < t < 4.69$, which encompasses the onset of upregulation. This was carried out on a Lenovo P520 worksation with an Intel Xeon E5-1660 processor, and code compiled using the Intel fortran compiler. Vis-a-vis sequential execution, we found acceleration by factor 1.85 on 2 cores and 3.05 on 4 cores. Further speed-up beyond this was found to be insignificant for a problem of this size (e.g. 3.46 on 8 cores). We repeated this with the gfortran and the portland group complers with similar results. This finding is in good agreement with earlier results for similar problems with the same or different (namely a Jacobian free but not error controlled) time integration strategies [13, 14, 19].

3.2 Simulation Experiment: Effect of Cross-Talk on Sender Growth

To investigate the effect that cross-talk activity has on the biofilm community, we compare the results of (i) the default setup described in the previous section with two variants: (ii) a system with the sender colony only, $m_{o2} = 0$; (iii) a system where the receiver does not respond to the signal, $\sigma_2 = 0$. This allows us to study how in our system cross-talk affects the signal producer.

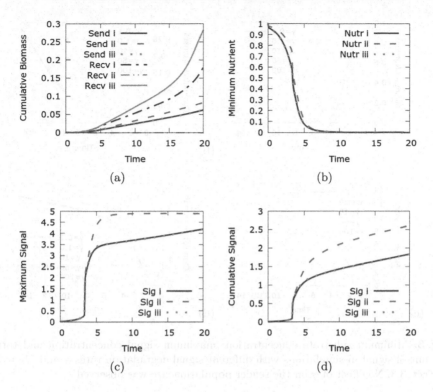

Fig. 2. Total biomass, maximum signal concentration, total amount of signal, and minimum substrate concentration in simulations of cases (i)–(iii) of Sect. 3.2.

We report in Fig. 2 the following quantities of interest, which are all functions of time: total mass of sender and receiver species, minimum substrate concentration in the system, maximum signal concentration, and total amount of signal. In the absence of receiver colonies, case (ii), the nutrient availability for senders increases due to a lack of competition. This leads to a faster growth of senders, thus more signals are produced. Also the minimum nutrient concentration remains slightly above the case with three colonies present. If the receivers do not respond to the signal, case (iii), they do not increase nutrient uptake when the signal concentration enters a range of induction, $S > 1$. Thus the available nutrients are more efficiently converted into new biomass, leading to a growth advantage *vis-a-vis* the base line scenario of case (i). This increased growth of receivers does not translate into a reduction in nutrients available to senders, and thus the signal concentrations in case (iii) is almost identical to case (i).

3.3 Simulation Experiment: The Role of Signal Degradation

The model contains in (4) abiotic signal degradation at rate γ. In the simulations in Sects. 3.1, 3.2 this term was turned off, $\gamma = 0$. This parameter will be key in exploring in future studies quorum quenching strategies in cross-talk systems, i.e. attempts to suppress quorum sensing by lowering signals, which often have

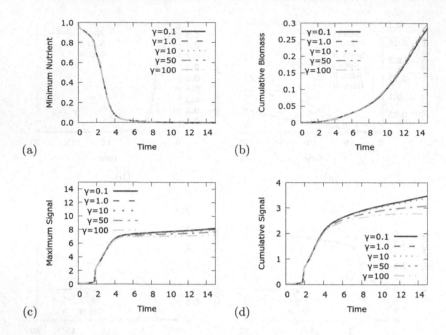

Fig. 3. Minimum substrate concentration, maximum signal concentration and total amount of signal in simulations with different signal degradation rates γ and $D_S = 8$, cf. Sect. 3.3. No effect of γ on the sender population size was observed.

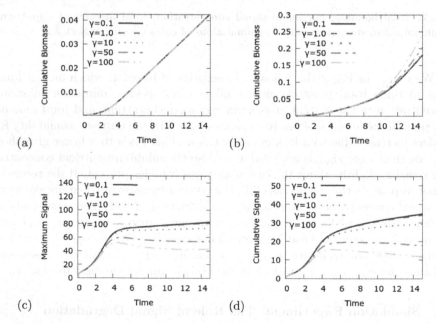

Fig. 4. Total biomass, maximum signal concentration and total amount of signal in simulations with varying signal degradation rates γ, and $D_S = 0.8$, cf. Sect. 3.3. No effect of γ on the minimum substrate concentration was observed.

been suggested as adjuvant strategies to antibiotics in the treatment of biofilm borne infections [9]. Increasing γ will lower the signal concentration and thus delay (or even prevent) upregulation, which leads in our formulation to an increased substrate concentration, thus to faster biomass growth, hence more signal being produced, etc. The net effect of these processes pulling in opposite direction is difficult to predict, and likely depends on the actual parameter values.

We vary γ in the range $\gamma = 0.1 \sim 100$. The signal diffusion coefficient was set to $D_S = 8$. In Fig. 3 we present the same quantities of interests that we have reported in Sect. 3.2. Differences in signal concentration as γ varies become noticeable only after the onset of induction, at $t \approx 3$. However, differences in signal concentration in a range clearly above induction threshold, $S > \tau$, have only little effect, since there $h(S, \tau) \approx 1$. Accordingly, in our simulations, γ has almost now influence on the population and cross-talk dynamics.

To explore this further, we carried out additional simulations. Although the diffusion coefficients are difficult to control experimentally, we re-ran the simulation with $D_S = 80$, a value that is likely too large to be practically relevant. Increased diffusion implies a faster removal of autoinducer signals. Also in these simulations one could observe the effect of γ on the signal concentration, but due to their fast washout, the system was never induced, such that $h(S, \tau) \approx 0$. Therefore, also in this regime signal degradation did not affect population and cross-talk behaviour (data not shown). On the other hand, for a substantially lower signal diffusion concentration, $D_S = 0.8$ (too low to be practically relevant), autoinducers are not removed fast enough and can accumulate, cf Fig. 4. Due to the slow transport in the system, reactions dominates over diffusion, so that the signal concentration in the receiver region remains small, making more substrate available for growth of receiver population. This suggests that the efficacy of signal degradation depends strongly on the mass transport conditions in the system, which often are not easy to control, e.g. in a medical context.

4   Conclusion

In the course of this study the following lessons were learned:

1. The dynamics and efficacy of bacterial cross-talk, like quorum sensing itself, is greatly affected by transport processes in the environment and not a mere consequence of population size. Such transport effects can introduce spatial heterogeneity in population behaviour. It is conceivable to have in one biofilm community receiver colonies of the same species that are entirely induced, and others that remain down-regulated.

2. In our simulations the only effect of upregulation was an increased consumption of nutrients. We did not track explicitly virulence factors into which these additional resources are invested. In more specific applications, accounting for the effect of up-regulation in more detail, the question will arise whether the purpose of cross-talk between species is to benefit the sender over the recipient or *vice versa*, or whether the relationship is symbiotic. This will also be a question about the nature of the underlying autoinduction system, which will

possibly introduce additional positive or negative feedback. Further investigation will be warranted, requiring a model extension and to account for additional Fickian or degenerate diffusion-reaction equations, thus increasing computational complexity.

3. When developing quorum quenching strategies to control cross-talk it will be paramount that those need to take into account also physical conditions of the system, in particular as they pertain to signal transport, and not only biological and biochemical aspects.

4. The simulation setup that we introduced here was inspired by a recently proposed experimental biofilm growth reactor. It allows us to reduce the influence of physical unrealistic boundary condition effects, to which most multi-dimensional biofilm simulation studies are exposed. Moreover, it allows to use a 2D description as a natural setting for such simulations. This seems a promising setup for further multi-dimensional biofilm simulation studies going forward.

Acknowledgement. We thank WestGrid, SHARCNET and Compute Canada for access to compute resources used for some simulations that we reported here.

References

1. Barr, H.L., Halliday, N., Cámara, M., Barrett, D.A., Williams, P., Forrester, D.L., et al.: Pseudomonas aeruginosa quorum sensing molecules correlate with clinical status in cystic fibrosis. Eur. Respir. J. **46**, 1046–1054 (2015)
2. Davies, D.G.: The involvement of cell-to-cell signals in the development of a bacterial biofilm. Science **280**, 295–298 (1998)
3. Doberva, M., Stien, D., Sorres, J., Hue, N., Sanchez-Ferandin, S., Eparvier, V., et al.: Large diversity and original structures of acyl-homoserine lactones in strain MOLA 401, a marine rhodobacteraceae bacterium. Front. Microbiol. **8**, 1152 (2017)
4. Dulla, G.F.J., Lindow, S.E.: Acyl-homoserine lactone-mediated cross talk among epiphytic bacteria modulates behavior of Pseudomonas syringae on leaves. ISME J. **3**, 825–834 (2009)
5. Eberl, H.J., Demaret, L.: A finite difference scheme for a degenerated diffusion equation arising in microbial ecology. Electron. J. Differ. Equ. **15**, 77–95 (2007)
6. Eberl, H.J., Parker, D.F., van Loosdrecht, M.C.M.: A new deterministic spatio-temporal continuum model for biofilm development. J. Theor. Med. **3**(3), 161–175 (2001)
7. Emerenini, B., Hense, B.A., Kuttler, C., Eberl, H.J.: A mathematical model of quorum sensing induced biofilm detachment. PlosOne **10**(7), e0132385 (2015)
8. Emerenini, B.O., Sonner, S., Eberl, H.J.: Mathematical analysis of a quorum sensing induced biofilm dispersal model. Math. Biosc. Eng. **14**(3), 625–653 (2017)
9. Fong, J., Zhang, C., Yang, R., Boo, Z.Z., Tan, S.K., Nielsen, T.E., et al.: Combination therapy strategy of quorum quenching enzyme and quorum sensing inhibitor in suppressing multiple quorum sensing pathways of P. aeruginosa. Sci. Rep. **8**(1), 1155 (2018)
10. Frederick, M.R., Kuttler, C., Hense, B.A., Müller, J., Eberl, H.J.: A mathematical model of quorum sensing in patchy biofilm communities with slow background flow. Can. Appl. Math. Quart. **18**(3), 267–298 (2010)

11. Frederick, M.R., Kuttler, C., Hense, B.A., Eberl, H.J.: A mathematical model of quorum sensing regulated EPS production in biofilms. Theor. Biol. Med. Model. **8**, 8 (2011)

12. Fuqua, W., Winans, S., Greenberg, E.: Quorum sensing in bacteria: the LuxR-LuxI family of cell density-responsive transcriptional regulators. J. Bacteriol. **176**, 269–275 (1994)

13. Ghasemi, M., Eberl, H.J.: Extension of a regularization based time-adaptive numerical method for a degenerate diffusion-reaction biofilm growth model to systems involving quorum sensing. Procedia Comput. Sci. **108**, 1893–1902 (2017)

14. Ghasemi, M., Eberl, H.J.: Time adaptive numerical solution of a highly degenerate diffusion-reaction biofilm model based on regularisation. J. Sci. Comput. **74**, 1060–1090 (2018)

15. Ghasemi, M., Hense, B.A., Kuttler, C., Eberl, H.J.: Simulation based exploration of quorum sensing triggered resistance of biofilms to antibiotics. Bull. Math. Biol. **80**(7), 1736–1775 (2018)

16. Hense, B.A., Kuttler, C., Mïler, J., Rothballer, M., Hartmann, A., Kreft, J.-U.: Does efficiency sensing unify diffusion and quorum sensing? Nat. Rev. Microbiol. **5**, 230–239 (2007)

17. Kim, M.K., Ingremeau, F., Zhao, A., Bassler, B.L., Stone, H.A.: Local and global consequences of flow on bacterial quorum sensing. Nat. Microbiol. **1**, 15005 (2016)

18. Lewandowski, Z., Beyenal, H.: Fundamentals of Biofilm Research. CRC Press, Boca Raton (2007)

19. Muhammad, N., Eberl, H.J.: OpenMP parallelization of a mickens time-integration scheme for a mixed-culture biofilm model and its performance on multi-core and multi-processor computers. In: Mewhort, D.J.K., Cann, N.M., Slater, G.W., Naughton, T.J. (eds.) HPCS 2009. LNCS, vol. 5976, pp. 180–195. Springer, Heidelberg (2010). https://doi.org/10.1007/978-3-642-12659-8_14

20. Rahman, K.A., Sudarsan, R., Eberl, H.J.: A mixed culture biofilm model with cross-diffusion. Bull. Math. Biol. **77**(11), 2086–2124 (2015)

21. Rang, J.: Improved traditional Rosenbrock-Wanner methods for stiff ODEs and DAEs. J. Comput. Appl. Math. **286**, 128–144 (2015)

22. Shankles, P.G.: Interfacing to biological systems using microfluidics. Ph.D. dissertation, University of Tennessee, Knoxville (2018)

23. Shankles, P.G., Timm, A.C., Doktycz, M.J., Retterer, S.T.: Fabrication of nanoporous membranes for tuning microbial interactions and biochemical reactions. J. Vac. Sci. Tech. B. **33**(6), 06FM031–06M031-8 (2015)

24. Silva, K.P.T., Chellamuthu, P., Boedicker, J.Q.: Quantifying the strength of quorum sensing crosstalk within microbial communities. PLoS Comput. Biol. **13**(1), e1005809 (2017)

25. Saad, Y.: SPARSKIT: a basic tool for sparse matrix computations (1994). http://www.users.cs.umn.edu/saad/software/SPARSKIT/sparskit.htm

26. Sonner, S., Efendiev, M.A., Eberl, H.J.: On the well-posedness of a mathematical model of quorum-sensing in patchy biofilm communities. Math. Meth. Appl. Sci. **34**(13), 1667–1684 (2011)

27. Stewart, P.S., Costerton, J.E.: Antibiotic resistance of bacteria in biofilms. Lancet **358**, 135–38 (2001)

Massively Parallel Stencil Strategies for Radiation Transport Moment Model Simulations

Marco Berghoff[1]([✉])(iD), Martin Frank[1](iD), and Benjamin Seibold[2](iD)

[1] Steinbuch Centre for Computing, Karlsruhe Institute of Technology,
Karlsruhe, Germany
{marco.berghoff,martin.frank}@kit.edu
[2] Department of Mathematics, Temple University, Philadelphia, PA 19122, USA
seibold@temple.edu

Abstract. The radiation transport equation is a mesoscopic equation in high dimensional phase space. Moment methods approximate it via a system of partial differential equations in traditional space-time. One challenge is the high computational intensity due to large vector sizes (1 600 components for P39) in each spatial grid point. In this work, we extend the calculable domain size in 3D simulations considerably, by implementing the StaRMAP methodology within the massively parallel HPC framework NAStJA, which is designed to use current supercomputers efficiently. We apply several optimization techniques, including a new memory layout and explicit SIMD vectorization. We showcase a simulation with 200 billion degrees of freedom, and argue how the implementations can be extended and used in many scientific domains.

Keywords: Radiation transport · Moment methods · Stencil code · Massively parallel

1 Introduction

The accurate computation of radiation transport is a key ingredient in many application problems, including astrophysics [28,35,44], nuclear engineering [8,9,32], climate science [24], nuclear medicine [20], and engineering [29]. A key challenge for solving the (energy-independent) radiation transport equation (RTE) (1) is that it is a mesoscopic equation in a phase space of dimension higher than the physical space coordinates. Moment methods provide a way to approximate the RTE via a system of macroscopic partial differential equations (PDEs) defined in traditional space-time. Here we consider the P_N method [7], which is based on an expansion of the solution of (1) in Spherical Harmonics. It can be interpreted as a moment method or, equivalently, as a spectral semi-discretization in the angular variable. Advantages of the P_N method over angular discretizations by collocation (discrete ordinates, S_N) [30] is that it preserves rotational invariance. A drawback, particular in comparison to nonlinear

© Springer Nature Switzerland AG 2020
V. V. Krzhizhanovskaya et al. (Eds.): ICCS 2020, LNCS 12143, pp. 242–256, 2020.
https://doi.org/10.1007/978-3-030-50436-6_18

moment methods [1,18,23,31,40,42], are spurious oscillations ("wave effects") due to Gibbs phenomena. To keep these at bay, it is crucial that the P_N method be implemented in a flexible fashion that preserves efficiency and scalability and allows large values of N.

Studies and applications of the P_N methods include [7,25,26,34]. An important tool for benchmarking and research on linear moment methods is the StaRMAP project [38], developed by two authors of this contribution. Based on a staggered grid stencil approach (see Sect. 2.1), the StaRMAP approach is implemented as an efficiently vectorized open-source MATLAB code [36]. The software's straightforward usability and flexibility have made it a popular research tool, used in particular in numerous dissertations, and it has been extended to other moment models (filtered [16] and simplified [33]), and applied in radiotherapy simulations [12,19]. For 2D problems, the vectorized MATLAB implementation allows for serial or shared memory (MATLAB's automatic usage of multiple cores) parallel execution speeds that are on par with comparable implementations of the methodology in C++. The purpose of this paper is to demonstrate that the same StaRMAP methodology also extends to large-scale, massively parallel computations and yields excellent scalability properties.

While S_N solvers for radiation transport are important production codes and major drivers for method development on supercomputers (one example is DENOVO [10], which is one of the most time-consuming codes that run in production mode on the Oak Ridge Leadership Computing Facility [27]), we are aware of only one work [14] that considers massively parallel implementations for moment models.

The enabler to transfer StaRMAP to current high-performance computing (HPC) systems is the open-source NAStJA framework [2,5], co-developed by one author of this contribution. NAStJA is a massively parallel framework for stencil-based algorithms on block-structured grids. The framework has been shown to efficiently scale up to more than ten thousand threads [2] and run simulations in several areas, using the phase-field method for water droplets [4], the phase-field crystal model for crystal–melt interfaces [15] and cellular Potts models for tissue growth and cancer simulations [6] with millions of grid points.

2 Model

The radiation transport equation (RTE) [8]

$$
\partial_t \psi(t, x, \Omega) + \Omega \cdot \nabla_x \psi(t, x, \Omega) + \Sigma_t(t, x)\psi(t, x, \Omega)
$$
$$
= \int_{S^2} \Sigma_s(t, x, \Omega \cdot \Omega')\psi(t, x, \Omega')\, d\Omega' + q(t, x, \Omega), \tag{1}
$$

equipped with initial data $\psi(0, x, \Omega)$ and suitable boundary conditions, describes the evolution of the density ψ of particles undergoing scattering and absorption in a medium (units are chosen so that the speed of light $c = 1$). The phase space consists of time $t > 0$, position $x \in \mathbb{R}^3$, and flight direction $\Omega \in S^2$.

The medium is characterized by the cross-section Σ_t (see below) and scattering kernel Σ_s. Equation (1) stands representative for more general radiation problems, including electron and ion radiation [11] and energy-dependence [21].

Moment methods approximate (1) by a system of macroscopic equations. In 1D slab geometry, expand the Ω-dependence of ψ in a Fourier series, $\psi(t, x, \mu) = \sum_{\ell=0}^{\infty} \psi_\ell(t, x) \frac{2\ell+1}{2} P_\ell(\mu)$, where μ is the cosine of the angle between Ω and x-axis, and P_ℓ are the Legendre polynomials. Testing (1) with P_ℓ and integrating yields equations for the Fourier coefficients $\psi_\ell = \int_{-1}^{1} \psi P_\ell \, d\mu$ as

$$\partial_t \psi_\ell + \partial_x \int_{-1}^{1} \mu P_\ell \psi \, d\mu + \Sigma_{t\ell} \psi_\ell = q_\ell \qquad \text{for } \ell = 0, 1, \dots , \qquad (2)$$

where $\Sigma_{t\ell} = \Sigma_t - \Sigma_{s\ell} = \Sigma_a + \Sigma_{s0} - \Sigma_{s\ell}$ and $\Sigma_{s\ell} = 2\pi \int_{-1}^{1} P_\ell(\mu) \Sigma_s(\mu) \, d\mu$. Using the three-term recursion for Legendre polynomials, relation (2) becomes

$$\partial_t \psi_\ell + \partial_x \left(\frac{\ell+1}{2\ell+1} \psi_{\ell+1} + \frac{\ell}{2\ell+1} \psi_{\ell-1} \right) + \Sigma_{t\ell} \psi_\ell = q_\ell.$$

These equations can be assembled into an infinite system $\partial_t \boldsymbol{u} + M \cdot \partial_x \boldsymbol{u} + C \cdot \boldsymbol{u} = \boldsymbol{q}$, where $\boldsymbol{u} = (\psi_0, \psi_1, \dots)^T$ is the vector of moments, M is a tri-diagonal matrix with zero diagonal, and $C = \text{diag}(\Sigma_{t0}, \Sigma_{t1}, \dots)$ is diagonal. The slab-geometry P_N equations are now obtained by omitting the dependence of ψ_N on ψ_{N+1} (alternative interpretations in [13, 22, 37]).

In 2D and 3D, there are multiple equivalent ways to define the P_N equations (cf. [7, 38]). StaRMAP is based on the symmetric construction using the moments $\psi_\ell^m(t, x) = \int_{S^2} \overline{Y_\ell^m(\Omega)} \psi(t, x, \Omega) \, d\Omega$, with the complex spherical harmonics $Y_\ell^m(\mu, \varphi) = (-1)^m \sqrt{\frac{2\ell+1}{4\pi} \frac{(\ell-m)!}{(\ell+m)!}} \, e^{im\varphi} P_\ell^m(\mu)$, where $\ell \geq 0$ is the moment order, and $-\ell \leq m \leq \ell$ the tensor components. Appropriate substitutions [38] lead to real-valued P_N equations. In 3D the moment system becomes

$$\partial_t \boldsymbol{u} + M_x \cdot \partial_x \boldsymbol{u} + M_y \cdot \partial_y \boldsymbol{u} + M_z \cdot \partial_z \boldsymbol{u} + C \cdot \boldsymbol{u} = \boldsymbol{q} , \qquad (3)$$

where the symmetric system matrices M_x, M_y, M_z are sparse and possess a very special pattern of nonzero entries (see [36, 38]). That coupling structure between unknowns (same in 2D and 1D) enables elegant and effective staggered grid discretizations upon which StaRMAP is based.

2.1 Numerical Methodology

We consider the moment system (3) in a rectangular computational domain $(0, L_x) \times (0, L_y) \times (0, L_z)$ with periodic boundary conditions (see below). The domain is divided into $n_x \times n_y \times n_z$ rectangular equi-sized cells of size $\Delta x \times \Delta y \times \Delta z$. The center points of these cells lie on the base grid

$$G_{111} = \left\{ \left(\left(i - \tfrac{1}{2}\right) \Delta x, \left(j - \tfrac{1}{2}\right) \Delta y, \left(k - \tfrac{1}{2}\right) \Delta z \right) \mid (1, 1, 1) \leq (i, j, k) \leq (n_x, n_y, n_z) \right\}.$$

The first component of $\boldsymbol{u}$ (the zeroth moment, which is the physically meaningful radiative intensity) is always placed on G_{111}. The other components of $\boldsymbol{u}$ are then placed on the 7 other staggered grids $G_{211} = \{(i\Delta x, (j-1/2)\,\Delta y, (k-1/2)\,\Delta z)\}$, $G_{121} = \{((i-1/2)\,\Delta x, j\Delta y, (k-1/2)\,\Delta z)\}, \ldots, G_{222} = \{(i\Delta x, j\Delta y, k\Delta z)\}$, following the fundamental principle that an x-derivative of a component in (3) that lives on a $(1, \bullet, \bullet)$ grid updates a component that lives on the corresponding $(2, \bullet, \bullet)$ grid. Likewise, x-derivatives of components on $(2, \bullet, \bullet)$ grids update information on the $(1, \bullet, \bullet)$ grids; and analogously for y- and z-derivative. A key result, proved in [38], is that this placement is, in fact, always possible.

Due to this construction, all spatial derivatives can be approximated via simple second-order half-grid centered finite difference stencils: two x-adjacent values, for instance living on the $(1, 1, 1)$ grid, generate the approximation

$$\partial_x u(i\Delta x, (j-\tfrac{1}{2})\Delta y, (k-\tfrac{1}{2})\Delta z) = \frac{u_{(i+\frac{1}{2}, j-\frac{1}{2}, k-\frac{1}{2})} - u_{(i-\frac{1}{2}, j-\frac{1}{2}, k-\frac{1}{2})}}{\Delta x} + O(\Delta x^2)$$

on the $(2, 1, 1)$ grid. We now call the G_{111}, G_{221}, G_{122}, and G_{212} grids "even", and the G_{211}, G_{121}, G_{112}, and G_{222} grids "odd".

The time-stepping of (3) is conducted via bootstrapping between the even and the odd grid variables. This is efficiently possible because of the approximate spatial derivatives of the even/odd grids update *only* the components that live on the odd/even grids. Those derivative components on the dual grids are considered "frozen" during a time-update of the other variables, leading to the decoupled update ODEs

$$\begin{cases} \partial_t \boldsymbol{u}^{\mathrm{e}} + C^{\mathrm{e}} \cdot \boldsymbol{u}^{\mathrm{e}} = \boldsymbol{q}^{\mathrm{e}} - (M_x^{\mathrm{eo}} \cdot D_x + M_y^{\mathrm{eo}} \cdot D_y + M_z^{\mathrm{eo}} \cdot D_z)\boldsymbol{u}^{\mathrm{o}} \\ \partial_t \boldsymbol{u}^{\mathrm{o}} + C^{\mathrm{o}} \cdot \boldsymbol{u}^{\mathrm{o}} = \boldsymbol{q}^{\mathrm{o}} - (M_x^{\mathrm{oe}} \cdot D_x + M_y^{\mathrm{oe}} \cdot D_y + M_z^{\mathrm{oe}} \cdot D_z)\boldsymbol{u}^{\mathrm{e}} \end{cases} \tag{4}$$

for the vector of even moments $\boldsymbol{u}^{\mathrm{e}}$ and the vector of odd moments $\boldsymbol{u}^{\mathrm{o}}$. In (4), the right-hand sides are constant in time (due to the freezing of the dual variables, as well as the source $\boldsymbol{q}$). Moreover, because C^{e} and C^{o} are diagonal, the equations in (4) decouple further into scalar ODEs of the form

$$\partial_t u_k(\boldsymbol{x}, t) + \bar{c}_k(\boldsymbol{x}) u_k(\boldsymbol{x}, t) = \bar{r}_k(\boldsymbol{x}),$$

whose exact solution is

$$u_k(\boldsymbol{x}, t + \Delta t) = u_k(\boldsymbol{x}, t) + \Delta t\, (\bar{r}_k(\boldsymbol{x}) - \bar{c}_k(\boldsymbol{x}) u_k(\boldsymbol{x}, t))\, E(-\bar{c}_k(\boldsymbol{x})\Delta t). \tag{5}$$

Here $\boldsymbol{x} = (x, y, z)$ is the spatial coordinate, and $E(c) = (\exp(c) - 1)/c$ (see [38] for a robust implementation of this function). To achieve second order in time, one full time-step (from t to $t + \Delta t$) is now conducted via a Strang splitting

$$\boldsymbol{u}(\boldsymbol{x}, t + \Delta t) = S^{\mathrm{o}}_{\frac{1}{2}\Delta t} \circ S^{\mathrm{e}}_{\Delta t} \circ S^{\mathrm{o}}_{\frac{1}{2}\Delta t} \boldsymbol{u}(\boldsymbol{x}, t), \tag{6}$$

where $S^{\mathrm{o}}_{\frac{1}{2}\Delta t}$ is the half-step update operator for the odd variables, and $S_{\Delta t}{}^{\mathrm{e}}$ the full-step update operator for the even variables, both defined via (5).

The convergence of this method, given that $\Delta t < \min\{\Delta x, \Delta y, \Delta z\}/3$, has been proven in [38]. Stability is generally given even for larger time-steps if scattering is present.

3   Implementation

In the following section, we present our implementation of the StaRMAP model and applied optimizations that are required to run on current HPC systems efficiently. NAStJA-StaRMAP v1.0 [3] is published under the Mozilla Public License 2.0 and the source-code is available at https://gitlab.com/nastja/starmap.

3.1   The NAStJA Framework

The StaRMAP methodology described above was implemented using the open-source NAStJA framework[1]. The framework was initially developed to explore non-collective communication strategies for simulations with a large number of MPI ranks, as will be used in exascale computing. It was developed in such a way that many multi-physics applications based on stencil algorithms can be efficiently implemented in a parallel way. The entire domain is build of blocks in a block-structured grid. These blocks are distributed over the MPI ranks. Inside each block, regular grids are allocated for the data fields. The blocks are extended with halo layers that hold a copy of the data from the neighboring blocks. This concept is flexible, so it can adaptively create blocks where the computing area moves. The regular structure within the blocks allows high-efficiency compute kernels, called sweeps. Every process holds information only about local and adjacent blocks. The framework is entirely written in modern C++ and makes use of template metaprogramming to achieve excellent performance without losing flexibility and usability. Sweeps and other actions are registered and executed by the processes in each time-step for their blocks so that functionality can be easily extended. Besides, sweeps can be replaced by optimized sweeps, making it easy to compare the optimized version with the initial one.

3.2   Optimizations

Starting with the 3D version of the MATLAB code of StaRMAP, the goal of this work was to develop a highly optimized and highly parallel code for future real-time simulations of radiation transports.

Basic Implementation. The first step was to port the MATLAB code to C++ into the NAStJA framework. Here we decide to use spatial coordinates (x, y, z) as the underlying memory layout. At each coordinate, the vector of moments u is stored. The sub-grids G_{111} to G_{222} are only considered during the calculation and are not saved separately. This means that the grid points on G_{111} and all staggered grid points are stored at the non-staggered (x, y, z)-coordinates. Thus it can be achieved that data that are needed for the update is close to each other in the memory. As for Eq. (4) described, all even components are used to calculate the odd components and vice versa. This layout also allows the usage of a relatively small stencil. The D3C7 stencil, which reads for three dimensions, the central data point and the first six direct neighbors, is sufficient.

---

[1] The MPL-2.0 source-code is available at https://gitlab.com/nastja/nastja.

For parallelization, we use NAStJA's block distribution and halo exchange mechanisms. The halo is one layer that holds a copy of the u vectors from the neighboring blocks. Since a D3C7 stencil is used, it is sufficient to exchange the six first neighboring sides. Figure 1 left shows the grid points in NAStJA's cells and the halo layer. For the implemented periodic boundary condition, we use this halo exchange to copy NAStJA-cells from one side to the opposite side, even if only half of the moments are needed to calculate the central differences.

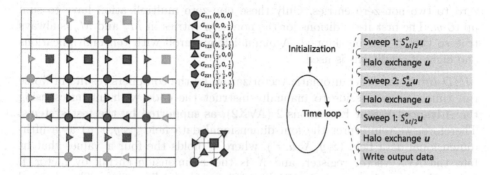

Fig. 1. Left: Staggered grids for the first z-layer. The odd coordinates are blue triangles and the even coordinates are marked by red shapes. The NAStJA-cells are blue squares. The base (111) grid is denoted by the black lines. The grid points with dotted border are the halo layer or the periodic boundary copy, the light grid points are not used. Center: 3D NAStJA-cell with the base grid point G_{111} (red circle) and the seven staggered grid points. Right: Action and sweep setup in NAStJA. (Color figure online)

For the calculation of the four substeps in Eq. (6), two different sweeps are implemented, each sweep swipes over the spatial domain in z, y, x order. The updates of each u component for each cell is calculated as followed. Beginning with the first substep, sweep S^o calculates $d_x u, d_y u, d_z u$ of the even components as central differences, laying on the odd components. Then, the update of the odd components using this currently calculated $d_x u, d_y u, d_z u$ is calculated. After the halo layer exchange, sweep S^e calculates the second substep. Therefore, first, the $d_x u, d_y u, d_z u$ of the odd components are calculated, followed by the update of the even components. A second halo layer exchange proceeds before the sweep S^o is called again to complete with the third substep. The time-step is finalized by a third halo layer exchange and an optional output. Figure 1 right shows the whole sweep setup of one time-step in the NAStJA framework.

The time-independent parameter values as q, $\bar{c}_k(x)$, and $E(-\bar{c}_k(x)\Delta t/2)$ are stored in an extra field on the non-staggered coordinates. Here, $\bar{c}_k(x)$ for $k \geq 1$ are identical. Their values on the staggered grid positions are interpolated.

Reorder Components. For optimization purposes, the calculation sweeps can easily exchange in NAStJA. Two new calculation sweeps are added for each of the following optimization steps. The computational instructions for the finite differences of the components on one sub-grid are the same, as well as the inter-polated parameter values. Components of the vector u are reordered, in that way that components of individual sub-grids are stored sequentially in memory. First, the even then, the odd sub-grid components follow, namely G_{111}, G_{221}, G_{212}, G_{122}, G_{211}, G_{121}, G_{112}, and G_{222}.

Unroll Multiplications. The calculation of $w = M_x \cdot d_x u + M_y \cdot d_y u + M_z \cdot d_z u$ is optimized by manually unroll and skipping multiplication. The Matrices M_x and M_y have in each row one to four non-zero entries while the Matrix M_z has zero to two non-zero entries. Only these non-zero multiplication have to sum up to w. The first if-conditions for the non-zero entries in M_x and M_y is always true so that it can be skipped. A manual loop-unroll with ten multiplications and eight if-conditions is used.

SIMD Intrinsics. The automatic vectorization by the compilers results in worse run times. So we decide to manually instruct the code with intrinsics using the Advanced Vector Extensions 2 (AVX2), as supported by the test systems. Therefore, we reinterpret the four-dimensional data field (z, y, x, u) as a fifth-dimensional data field (z, y, X, u, x'), where x' holds the four x values that fit into the AVX vector register, and X is the x-dimension shrink by factor 4. Currently, we only support multiples of 4 for the x-dimension. The changed calculation sweeps allow calculating four neighbored values at once. The fact that the studied number of moments are multiples of 4 ensures that all the memory access are aligned. With this data layout, we keep the data very local and can still benefit from the vectorization.

4   HPC System

To perform the scaling test, we use a single node (kasper) and the high-performance computing systems ForHLR II, located at Karlsruhe Institute of Technology (fh2). The single node has two quad-core Intel Xeon processors E5-2623 v3 with Haswell architecture running at a base frequency of 3 GHz (2.7 GHz AVX), and have 4×256 KB of level 2 cache, and 10 MB of shared level 3 cache. The node has 54 GB main memory.

The ForHLR II has 1152 20-way Intel Xeon compute nodes [39]. Each of these nodes contains two deca-core Intel Xeon processors E5-2660 v3 with Haswell architecture running at a base frequency of 2.6 GHz (2.2 GHz AVX), and have 10×256 KB of level 2 cache, and 25 MB of shared level 3 cache. Each node has 64 GB main memory, and an FDR adapter to connect to the InfiniBand 4X EDR interconnect. In total, 256 nodes can be used, which are connected by a quasi fat-tree topology, with a bandwidth ratio of 10:11 between the switches and leaf switches. The leaf switches connect 23 nodes. The implementation of Open MPI in version 3.1 is used.

5   Results and Discussion

In this section, we present and discuss single core performance results as well as scaling experiments run on a high-performance computing system.

The presented performance results are measured in MLCUP/s, which stands for "million lattice cell component updates per second". This unit takes into account that the amount of data depends on the number of lattice cells and the number of moments.

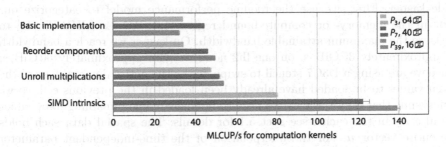

Fig. 2. Performance of the various optimization variants of the calculation sweeps running on a single core. The block size ▨ was chosen so that the number of the total components is approximately equal for all number of moments $M_\bullet$. The marks denote the average of three runs. The error bars indicate the minimum and maximum.

5.1 Performance Results

Single Core Performance. The starting point of our HPC implementation was a serial MATLAB code. A primary design goal of StaRMAP is to provide a general-purpose code with several different functions. In this application, we focus on specific cases, but let the number of moments be a parameter. A simple re-implementation in the NAStJA framework yields the same speed as the MATLAB code but has the potential to run in parallel and thus exceed the MATLAB implementation.

Figure 2 shows the performance of the optimization describes in Sect. 3.2. The measurements based on the total calculation sweep time per time-step, i.e., two sweep S^o + sweep S^e. In all the following simulations, we use cubic blocks, such that a block size of 40 refers to a cubic block with an edge length of 40 lattice cells without the halo. In legends, we write 40 ▨ . The speedup from the basic implementation to the reorder components version is small for P_3 and P_7 but significant for P_{39} (+54%). The number of components on each subgrid is small for the first both but large for P_{39}, so the overhead of the loops over all components becomes negligible. Unrolling brings an additional speedup of 38% for P_3, 14% for P_7, and 9% for P_{39}. Vectorization has the smallest effect for P_3 (+70%). For P_7 we gain +138% and +160% for P_{39}.

The combination of all optimizations results in a total speedup of factor 2.36, 2.77, 4.35 for P_3, P_7, P_{39}, respectively. This optimization enables us to simulate sufficiently large domains in a reasonable time to obtain physically meaningful results. Note, these results run with a single thread, so the full L3 cache is used.

Since the relative speedup does not indicate the utilization of a high-performance computing system, we have additionally analyzed the absolute performance of our code. In the following, we will concentrate on the single-node performance of our optimized code.

We show the performance analysis of the calculation sweeps on the single node kasper. First, we use the roofline performance model to categorize our code in the memory- or compute-bound region [43]. We use LIKWID [41] to measure the maximum attainable bandwidth. On kasper we reach a bandwidth of approximately 35 GiB/s, on one fh2 node we gain approximately 50 GiB/s. Since we are using a D3C7 stencil to swipe across the entire domain, four of the seven values to be loaded have already been loaded in the previous cell, so we can assume that only three values need to be loaded. The remaining data values are already in the cache, see Sect. 5.1 for details. The spatial data each holds the entire vector u. For the interpolation of the time-independent parameter data, 130 Byte are not located in the cache and have to be loaded for on lattice update. The sweeps have to load 24 Byte per vector component. Remember that we need three sweeps to process one time-step, so an average of 94.5 Byte for P_3 are loaded per lattice component update, 77.6 Byte, 72.2 Byte for P_7, P_{39}, respectively. If we only consider the speak-performance on fh2 of 50 GiB/s · 72.2 Bytes/LCUP = 3 785 MLCUP/s and 2 527 MLCUP/s on kasper. That is far away from what we measured—an indication that we are operating on the compute-bound side. Counting the floating-point operations for one time-step, we get $392 + 40v_e + 50v_o$ FLOP, where v_e is the number of even and v_o the number of odd vector components.

So an average of 68.3 FLOP for P_3 are used per lattice component update, 50.5 FLOP, 45.1 FLOP for P_7, P_{39}, respectively. This results in an arithmetic intensity on the lower bound of 0.72 FLOP/Byte to 0.62 FLOP/Byte for P_3, P_{39}, respectively. The Haswell CPU in kasper has an AVX base frequency of 2.7 GHz [17] and can perform 16 floating-point operations with double precision per cycle. This results in 43.2 GFLOP/s per core. The achieved 139.1 MLCUP/s per core corresponds to 6.3 GFLOP/s and so to 15% peak-performance.

Cache Effects. Even if the analysis in the previous section shows that our application is compute-bound, it is worth taking a look at the cache behavior. Running large blocks will result in an excellent parallel scaling because of the computational time increase by $O(n^3)$ and the communication data only by $O(n^2)$.

To discover the cache behavior, we run 20 single jobs in parallel on one node on the fh2, this simulates the 2.5 MiB L3 cache per core. Figure 3 show the performance for different block sizes. For P_7, a maximum block size of 13 fits into the L2 cache, here the largest performance can be seen. At a block size of 35, the performance drops, which can be explained by the fact that with a maximum block size of 40, three layers fit into the L3 cache. For P_3, a block size of 20 still fits into the L2 cache, so here is the peak, up to a block size of 80 the performance remains almost constant after dropping firstly, this is the size where the three layers fit into the L3 cache. The maximum for P_{39} is at a block size of 5, here the three layers fit into the L3 cache. We have not tested a smaller block size, because of the overhead of loops becomes too big. We will use the marked block sizes for the scaling analysis in the following sections. The block size of 20 was chosen so that all three moment orders can be compared here.

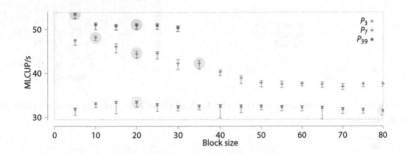

Fig. 3. Performance of the calculation sweeps for different block sizes.

Scaling Results. To examine the parallel scalability of our application, we consider weak scaling for different block sizes. During one run, each process gets a block of the same size. So we gain accurate scaling data that does not depend on any cache effects described in Sect. 5.1. First, we look at one node of the fh2, and then at the performance across multiple nodes, with each node running 20 processes at the 20 cores. We use up to 256 nodes, which are 5 120 cores. Figure 4(a) shows P_7-runs with different block sizes, where the MPI processes distributed equally over the two sockets. All three block sizes show similar, well-scaling behavior. Moreover, the whole node does not reach the bandwidth limit of 3 785 MLCUP/s, which confirms that the application is on the compute-bound side.

Before conducting scaling experiments, we evaluate the various parts of the application. Therefore, we show the amount of used calculation and communication time in Fig. 4(b). The calculation time for one time-step consists of the time used by two sweeps S^o and one sweep S^e. The communication time sums up the time used for the three halo exchanges. A high communication effort of about 50% is necessary. This proportion rarely changes for different vector lengths.

Figure 5 shows the parallel scalability of the application for different vector lengths and block sizes. The results of runs with one node are used as the basis for the efficiency calculations. In (a) three regimes are identifiable, P_3, 80 ▥ and P_{39}, 20 ▥ are more expensive and take a long time. P_7, 35 ▥ is in the middle, and the remainder takes only a short average time per time-step. As expected, this is also reflected in the efficiency in (b). The expensive tasks scale slightly better with approximately 80% efficiency on 256 nodes, 5120 cores. The shorter tasks still have approximately 60% efficiency. From one to two nodes, there is a drop in some jobs; the required inter-node MPI communication can explain this.

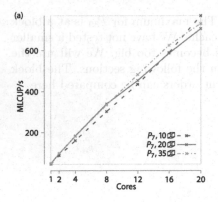
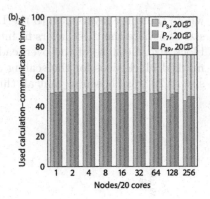

Fig. 4. (a) Single Node scaling on fh2. (b) Calculation time (dark) versus communication time (light).

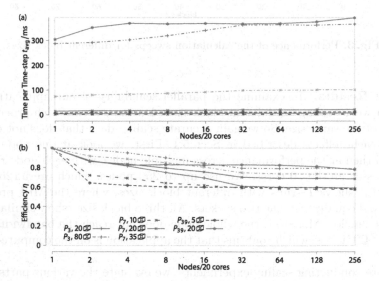

Fig. 5. MPI scaling (a) average time per time-step and (b) efficiency on fh2 for up to 5 120 cores.

From 32 nodes, the efficiency of all sizes is almost constant. This is because a maximum of 23 nodes is connected to one switch, i.e., the jobs must communicate via an additional switch layer. For runs on two to 16 nodes, the job scheduler can distribute the job to nodes connected to one switch but does not have to.

5.2   Simulation Results

With the parallelizability and scalability of the methodology and implementation established, we now showcase its applicability in a representative test example. We consider a cube geometry that resembles radiation transport (albeit with simplified physics) in a nuclear reactor vessel, consisting of a reactor core with fuel rods, each 1 cm (5 grid-points) thick, surrounded by water (inner box in Fig. 6, and concrete (outer box). The non-dimensional material parameters are: source $q_0 = 2$, absorption $\Sigma_a^w = 10$, $\Sigma_a^c = 5$, scattering $\Sigma_s = 1$. The spatial resolution of the rod geometry and surrounding has a grid size of 500 ⌀ , which we compute on up to 2000 cores via moment resolutions P_3, P_7, P_{19}, P_{29}, and P_{39}, depicted in Fig. 6 right. As one can see by comparing P_N, $N \geq 19$, the P_{19} simulation is well-resolved.

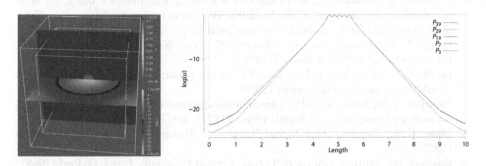

Fig. 6. Left: Rod geometry surrounded by water and concrete. The vertical slice shows u and the plane $\log_{10}(u)$. Right: Plot of the intensity $\log_{10}(u)$ over the section.

6   Conclusion

We have developed and evaluated a massively parallel simulation code for radiation transport based on a moment model, which runs efficiently on current HPC systems. With this code, we show that large domain sizes are now available. Therefore, an HPC implementation is of crucial importance. Starting from the reference implementation of StaRMAP in MATLAB, we have developed a new, highly optimized implementation that can efficiently run on modern HPC systems. We have applied optimizations at various levels to the highly complex stencil code, including explicit SIMD vectorization. Systematic performance engineering at the node-level resulted in a speedup factor of 4.35 compared to the original code and 15% of peak performance at the node-level. Besides, we have shown excellent scaling results for our code.

Acknowledgments. This work was performed on the supercomputer ForHLR funded by the Ministry of Science, Research and the Arts Baden-Württemberg and by the Federal Ministry of Education and Research. B. Seibold wishes to acknowledge support by the National Science Foundation through grant DMS–1719640.

References

1. Anile, A.M., Pennisi, S., Sammartino, M.: A thermodynamical approach to Eddington factors. J. Math. Phys. **32**, 544–550 (1991)
2. Berghoff, M., Kondov, I., Hötzer, J.: Massively parallel stencil code solver with autonomous adaptive block distribution. IEEE Trans. Parallel Distrib. Syst. **29**, 2282–2296 (2018)
3. Berghoff, M., Frank, M., Seibold, B.: StaRMAP - A NAStJA Application (2020). https://doi.org/10.5281/zenodo.3741415
4. Berghoff, M., Kondov, I.: Non-collective scalable global network based on local communications. In: 2018 IEEE/ACM 9th Workshop on Latest Advances in Scalable Algorithms for Large-Scale Systems (scalA), pp. 25–32. IEEE (2018)
5. Berghoff, M., Rosenbauer, J., Pfisterer, N.: The NAStJA Framework (2020). https://doi.org/10.5281/zenodo.3740079
6. Berghoff, M., Rosenbauer, J., Schug, A.: Massively parallel large-scale multi-model simulation of tumor development (2019)
7. Brunner, T.A., Holloway, J.P.: Two-dimensional time dependent Riemann solvers for neutron transport. J. Comput. Phys. **210**(1), 386–399 (2005)
8. Case, K.M., Zweifel, P.F.: Linear Transport Theory. Addison-Wesley, Boston (1967)
9. Davison, B.: Neutron Transport Theory. Oxford University Press, Oxford (1958)
10. Evans, T.M., Stafford, A.S., Slaybaugh, R.N., Clarno, K.T.: Denovo: a new three-dimensional parallel discrete ordinates code in SCALE. Nuclear Technol. **171**(2), 171–200 (2010). https://doi.org/10.13182/NT171-171
11. Frank, M., Herty, M., Schäfer, M.: Optimal treatment planning in radiotherapy based on Boltzmann transport calculations. Math. Mod. Meth. Appl. Sci. **18**, 573–592 (2008)
12. Frank, M., Küpper, K., Seibold, B.: StaRMAP – a second order staggered grid method for radiative transfer: application in radiotherapy. In: Sundar, S. (ed.) Advances in PDE Modeling and Computation, pp. 69–79. Ane Books Pvt. Ltd. (2014)
13. Frank, M., Seibold, B.: Optimal prediction for radiative transfer: a new perspective on moment closure. Kinet. Relat. Models **4**(3), 717–733 (2011). https://doi.org/10.3934/krm.2011.4.717
14. Garrett, C.K., Hauck, C., Hill, J.: Optimization and large scale computation of an entropy-based moment closure. J. Comput. Phys. **302**, 573–590 (2015). https://doi.org/10.1016/j.jcp.2015.09.008
15. Guerdane, M., Berghoff, M.: Crystal-melt interface mobility in bcc Fe: linking molecular dynamics to phase-field and phase-field crystal modeling. Phys. Rev. B **97**(14), 144105 (2018)
16. Hauck, C.D., McClarren, R.G.: Positive P_N closures. SIAM J. Sci. Comput. **32**(5), 2603–2626 (2010)
17. Intel Corporation: Intel Xeon Processor E5 v3 product family: specification update. Technical report 330785-011, Intel Corporation (2017)
18. Kershaw, D.S.: Flux limiting nature's own way. Technical report UCRL-78378, Lawrence Livermore National Laboratory (1976)
19. Küpper, K.: Models, Numerical Methods, and Uncertainty Quantification for Radiation Therapy. Dissertation, Department of Mathematics, RWTH Aachen University (2016)

20. Larsen, E.W.: Tutorial: the nature of transport calculations used in radiation oncology. Transp. Theory Stat. Phys. **26**, 739 (1997)
21. Larsen, E.W., Miften, M.M., Fraass, B.A., Bruinvis, I.A.D.: Electron dose calculations using the method of moments. Med. Phys. **24**, 111–125 (1997)
22. Larsen, E.W., Morel, J.E., McGhee, J.M.: Asymptotic derivation of the multigroup P_1 and simplified P_N equations with anisotropic scattering. Nucl. Sci. Eng. **123**, 328–342 (1996)
23. Levermore, C.D.: Relating Eddington factors to flux limiters. J. Quant. Spectrosc. Radiat. Transfer **31**, 149–160 (1984)
24. Marshak, A., Davis, A.: 3D Radiative Transfer in Cloudy Atmospheres. Springer, Heidelberg (2005). https://doi.org/10.1007/3-540-28519-9
25. McClarren, R.G., Evans, T.M., Lowrie, R.B., Densmore, J.D.: Semi-implicit time integration for P_N thermal radiative transfer. J. Comput. Phys. **227**(16), 7561–7586 (2008)
26. McClarren, R.G., Holloway, J.P., Brunner, T.A.: On solutions to the P_n equations for thermal radiative transfer. J. Comput. Phys. **227**(3), 2864–2885 (2008)
27. Messer, O.B., D'Azevedo, E., Hill, J., Joubert, W., Berrill, M., Zimmer, C.: MiniApps derived from production HPC applications using multiple programing models. Int. J. High Perform. Comput. Appl. **32**(4), 582–593 (2018). https://doi.org/10.1177/1094342016668241
28. Mihalas, D., Weibel-Mihalas, B.: Foundations of Radiation Hydrodynamics. Dover (1999)
29. Modest, M.F.: Radiative Heat Transfer, 2nd edn. Academic Press (1993)
30. Morel, J.E., Wareing, T.A., Lowrie, R.B., Parsons, D.K.: Analysis of ray-effect mitigation techniques. Nuclear Sci. Eng. **144**, 1–22 (2003)
31. Müller, I., Ruggeri, T.: Rational Extended Thermodynamics, 2nd edn. Springer, New York (1993). https://doi.org/10.1007/978-1-4612-2210-1
32. Murray, R.L.: Nuclear Reactor Physics. Prentice Hall (1957)
33. Olbrant, E., Larsen, E.W., Frank, M., Seibold, B.: Asymptotic derivation and numerical investigation of time-dependent simplified P_N equations. J. Comput. Phys. **238**, 315–336 (2013)
34. Olson, G.L.: Second-order time evolution of P_N equations for radiation transport. J. Comput. Phys. **228**(8), 3072–3083 (2009)
35. Pomraning, G.C.: The Equations of Radiation Hydrodynamics. Pergamon Press (1973)
36. Seibold, B., Frank, M.: StaRMAP code. http://www.math.temple.edu/~seibold/research/starmap
37. Seibold, B., Frank, M.: Optimal prediction for moment models: crescendo diffusion and reordered equations. Contin. Mech. Thermodyn. **21**(6), 511–527 (2009). https://doi.org/10.1007/s00161-009-0111-7
38. Seibold, B., Frank, M.: StaRMAP - a second order staggered grid method for spherical harmonics moment equations of radiative transfer. ACM Trans. Math. Softw. **41**(1), 4:1–4:28 (2014)
39. Steinbuch Centre for Computing: Forschungshochleistungsrechner ForHLR II. https://www.scc.kit.edu/dienste/forhlr2.php
40. Su, B.: Variable Eddington factors and flux limiters in radiative transfer. Nucl. Sci. Eng. **137**, 281–297 (2001)
41. Treibig, J., Hager, G., Wellein, G.: LIKWID: a lightweight performance-oriented tool suite for x86 multicore environments. In: Proceedings of PSTI 2010, the First International Workshop on Parallel Software Tools and Tool Infrastructures, San Diego, CA (2010)

42. Turpault, R., Frank, M., Dubroca, B., Klar, A.: Multigroup half space moment approximations to the radiative heat transfer equations. J. Comput. Phys. **198**, 363–371 (2004)
43. Williams, S., Waterman, A., Patterson, D.: Roofline: an insightful visual performance model for multicore architectures. Commun. ACM **52**(4), 65–76 (2009)
44. Zeldovich, Y., Raizer, Y.P.: Physics of Shock Waves and High Temperature Hydrodynamic Phenomena. Academic Press (1966)

Hybrid Mixed Methods Applied to Miscible Displacements with Adverse Mobility Ratio

Iury Igreja[✉] [iD] and Gabriel de Miranda[iD]

Computer Science Department and Computational Modeling Graduate Program,
Federal University of Juiz de Fora, Juiz de Fora, MG, Brazil
{iuryigreja,gabriel.miranda}@ice.ufjf.br

Abstract. We propose stable and locally conservative hybrid mixed
finite element methods to approximate the Darcy system and convection-
diffusion problem, presented in a mixed form, to solve miscible displacements considering convective flows with adverse mobility ratio. The stability of the proposed formulations is achieved due to the choice of non-
conforming Raviart-Thomas spaces combined to upwind scheme for the
convection-dominated regimes, where the continuity conditions, between
the elements, are weakly enforced by the introduction of Lagrange multipliers. Thus, the primal variables of both systems can be condensed
in the element level leading a positive-definite global problem involving
only the degrees of freedom associated with the multipliers. This approach, compared to the classical conforming Raviart-Thomas, present
a reduction of the computational cost because, in both problems, the
Lagrange multiplier is associated with a scalar field. In this context, a
staggered algorithm is employed to decouple the Darcy problem from
the convection-diffusion mixed system. However, both formulations are
solved at the same time step, and the time discretization adopted for
the convection-diffusion problem is the implicit backward Euler method.
Numerical results show optimal convergence rates for all variables and
the capacity to capture the formation and the propagation of the viscous fingering, as can be seen in the comparisons of the simulations of
the Hele-Shaw cell with experimental results of the literature.

Keywords: Raviart-Thomas spaces · Hybrid mixed methods · Locally
conservative methods · Upwind stabilization · Adverse mobility ratio ·
Hele-Shaw

1 Introduction

The miscible displacement of a higher viscosity fluid in a porous medium raises
considerable attention to a variety of applications like hydrology, blood motion

This study was financed in part by the Coordenação de Aperfeiçoamento de Pessoal
de Nível Superior - Brasil (CAPES) - Finance Code 001. This work was also partially
supported by CNPq and UFJF.

in vessels, industrial processes involving filtration such as sugar refining, carbon sink, oil recovery and groundwater exploration in porous media [6,14,30]. During the transport of an injected fluid in a reservoir, it is common the formation of fingers and channels in the flow, and this can happen even in a homogeneous media if the injected fluid is less viscous than the resident fluid. Injection of a fluid with different viscosity of the reservoir-resident fluid usually produces complex concentration patterns along with displacement [5,6,18,28].

In the numerical point of view, since approximations are adopted to simulate the fluids displacement with different viscosities using meshes of elements with the cell diameter smaller than the length of a formed finger, this can lead to inaccurate representations of the interfaces of these fronts, due to the tendency that mixtures, with a higher mobility ratio M, have to generate highly branched fractal structures [5,14,29]. The mobility ratio relates the viscosity of the resident fluid ν_{res} to the viscosity of the injected fluid ν_{inj}

$$M = \nu_{res}/\nu_{inj}, \tag{1}$$

for $M > 1$, or adverse mobility ratio, the injected fluid is less viscous than resident fluid, and in this case, the experiments predict that the displacement front is physically unstable where small perturbations can be forming multiple viscous fingerings [1,18,26,27]. This unstable behavior is also true mathematically, as can be seen in [27] and in references therein. Therefore, this problem requires the employ of robust numerical methods that accurately solve this phenomenon.

The Partial Differential Equations (PDE) that govern the phenomenon of the displacement of the fluid mixtures consist of a system formed by Darcy's problem and Transport problem. Some successful numerical approximations employing finite element methods to solve the Darcy problem can be found in [2,8,11,16, 19,25], for the Transport equation can be seen in [9,10,12,17,20,23] and for the Darcy-Transport coupled system we mention the works [13,15,17,21,31].

The objective of this work is to propose an equivalent finite element method for Darcy and Transport problems. In this sense, we rewrite the Transport equation in a mixed form in terms of diffusive flux and concentration. Thus, we have two mixed systems. However, the use of finite element methods for mixed problems requires compatibility between the approximation spaces [3,7]. Stable formulations have proven successful as can be seen in [8,25]. The Raviart-Thomas spaces [25], referenced here by $\mathcal{RT}_k$, was developed for the mixed problems, like Darcy problem, is be able to simulate problems in heterogeneous medium with stability, mass conservation and optimal convergence rates. However, Arnold and Brezzi in [2] proposed an hybridization, employing Lagrange multipliers associated to the scalar field, that gives rise to a positive-definite system obtaining the same accuracy of the conforming Raviart-Thomas spaces but with fewer unknowns. An extension of this approach for the mixed convection-diffusion problems are developed and analyzed by Egger and Schöberl [12], where a formulation is proposed by the combination of upwind techniques used in discontinuous Galerkin methods for hyperbolic problems with conservative discretizations of mixed methods for elliptic problems.

In this context, we propose an equivalent locally conservative numerical method for the Darcy and the convection-diffusion problems employing the Raviart-Thomas spaces in a non-conforming way in both formulations. In this case, the continuity is weakly imposed via Lagrange multipliers associated with the scalar field defined on the edges/faces of the elements. Thus, all interest variables (velocity, pressure, diffusive flux and concentration) degrees-of-freedom are condensed in the element level leading to a global problem involving only the degrees-of-freedom of the multiplier. After solving the global problem, the approximate solution of the multiplier is plugged into the local problems to recover the discontinuous approximation of the primal variables. This approach significantly reduces the computational cost compared to the use of conforming Raviart-Thomas spaces.

The methodology to solve this system is based on a staggered algorithm to decouple the hydrodynamics from the hyperbolic system, resulting in a scheme that uses a locally conservative hybrid mixed finite element method to approximate both problems. The two problems are solved in the same time scale, applying an implicit backward finite difference scheme in time to approximate the Transport equation. Moreover, the spurious oscillations characteristics of the convection-dominated regimes are mitigated through of an upwind scheme associated with the multiplier and concentration values, evaluated on the edges of the elements [12]. Numerical simulations are presented and demonstrate optimal convergence rates for all variables and a good capture of the viscous fingering instabilities on the interface between the miscible fluids compared to experimental simulations in Hele-Shaw cells with rectilinear flows.

This paper is organized as follows: In Sect. 2, we present the Darcy-transport model problem. In Sect. 3, notations and definitions required to present the hybrid methods are described. The stable and equivalent mixed hybrid method for the Darcy and transport is presented in Sect. 4. Section 5 is devoted to convergence study and numerical simulations, comparing the approximate solution with experimental results in Hele-Shaw cell, considering adverse mobility ratio and high Péclet number. And finally, in Sect. 6, we present the concluding remarks of this work.

2 Model Problem

The model problem is described by a PDE system composed by two subsystems, the Darcy problem, considering a incompressible flow and neglecting gravitational forces, and the Transport equation. Thus, let $\Omega \subset \mathbb{R}^d$ be the domain, with $d = 2, 3$, and the boundary $\partial\Omega = \partial\Omega_N \cup \partial\Omega_D$ in time interval $(0, T]$, the problem can be written as follows

Given the concentration c and the functions $\overline{p}$ and f, find the pair $[\mathbf{u}, p]$, such that:

$$\begin{aligned}
\mathbf{u} &= -\mathbb{K}\nabla p \text{ in } \quad \Omega, \\
\operatorname{div}(\mathbf{u}) &= f \text{ in } \quad \Omega, \\
\mathbf{u} \cdot \mathbf{n} &= 0 \quad \text{ on } \quad \partial\Omega_N, \\
p &= \overline{p} \quad \text{ on } \quad \partial\Omega_D.
\end{aligned} \tag{2}$$

Given the Darcy velocity $\mathbf{u}$, *porosity* ϕ *and the functions* g, c_0 *and* $\bar{c}$, *find* c, *such that:*

$$\begin{aligned}
\phi\frac{\partial c}{\partial t} + \mathbf{u} \cdot \nabla c - \mathrm{div}(\mathbb{D}\nabla c) &= g \text{ in } & \Omega \times (0, T], \\
c(\mathbf{x}, 0) &= c_0(\mathbf{x}) & \text{in } \Omega, \\
\mathbb{D}\nabla c \cdot \mathbf{n} &= 0 & \text{on } \partial\Omega_N \times (0, T], \\
c &= \bar{c} & \text{on } \partial\Omega_D \times (0, T].
\end{aligned} \tag{3}$$

where the variables of interest, p, $\mathbf{u}$, and c are respectively the hydrostatic pressure, the Darcy's velocity and the fluid concentration. The coefficients of the equations are $\phi = \phi(\mathbf{x})$ the effective porosity of the medium, $\mathbb{K} = \mathbb{K}(c, \mathbf{x}) = \frac{\mathbb{G}}{\nu}$ the hydraulic conductivity tensor, a proportionality coefficient that takes into account the characteristics of the medium, including size, distribution, form, and arrangement of the particles, and the viscosity of the fluids. Thus, $\mathbb{G} = \mathbb{G}(\mathbf{x})$ and $\nu = \nu(c)$ denotes the medium permeability and the fluid viscosity, respectively. Finally, $\mathbb{D} = \mathbb{D}(\mathbf{u})$ is the dispersion tensor that can be defined as

$$\mathbb{D}(\mathbf{u}) = \alpha_m \mathbb{I} + \|\mathbf{u}\| \left[\alpha_l \mathbb{E}(\mathbf{u}) + \alpha_t(\mathbb{I} - \mathbb{E}(\mathbf{u}))\right], \quad \mathbb{E}(\mathbf{u}) = \frac{\mathbf{u} \otimes \mathbf{u}}{\|\mathbf{u}\|^2},$$

with $\|\mathbf{u}\|^2 = \sum_{i=1}^{d} u_i^2$, $\otimes$ denoting the tensorial product, $\mathbb{I}$ the identity tensor, α_m being a molecular diffusion coefficient, α_l the longitudinal dispersion and α_t the transverse dispersion. In miscible displacement of a fluid through another in a reservoir, the dispersion is physically more important than the molecular diffusion [15, 24]. Thus, we assume the following properties $0 < \alpha_m \le \alpha_l$, $\alpha_l \ge \alpha_t > 0$ and $0 < \phi \le 1$.

With the gravitational effect neglected, besides the mobility ratio, another dimensionless parameter determines the behavior of the model is the Péclet number, $P_e = \|\mathbf{u}\|L/\|\mathbb{D}\|$, where L is the channel length. For miscible displacements in an petroleum reservoir, the viscosity of the fluid mixture may depend on the concentration of the injected fluid through a nonlinear function, the quarter-power viscosity law [30]

$$\nu(c) = \nu_{res}[1 - c + M^{\frac{1}{4}}c]^{-4}, \quad c \in [0, 1] \tag{4}$$

where M defined in Eq. (1) denotes the mobility ratio. From the Eq. (4) we can observe that, for $M \neq 1$ the subsystems (2) and (3) become tightly coupled.

In order to generate equivalence between transport problem (3) and Darcy problem (2), we rewrite the transport problem in a mixed form including the diffusive flux $\boldsymbol{\sigma} = -\mathbb{D}\nabla c$, which gives rise to the following problem

$$\begin{aligned}
\boldsymbol{\sigma} + \mathbb{D}\nabla c &= 0 & \text{on } \Omega \times (0, T], \\
\phi\frac{\partial c}{\partial t} + \mathbf{u} \cdot \nabla c + \mathrm{div}(\boldsymbol{\sigma}) &= g \text{ on } \Omega \times (0, T],
\end{aligned} \tag{5}$$

with boundary and initial conditions given in (3).

3   Notations and Definitions

To introduce the equivalent stable hybrid formulation for Darcy (2) and Transport (3) systems, we first recall some notations and definitions. Therefore, let $H^m(\Omega)$ the usual Sobolev space equipped with norm $\|\cdot\|_{m,\Omega} = \|\cdot\|_m$ and seminorm $|\cdot|_{m,\Omega} = |\cdot|_m$, with $m \geq 0$. For $m = 0$, we present $L^2(\Omega) = H^0(\Omega)$ as the space of square integrable functions and $H_0^1(\Omega)$ the subspace of functions in $H^1(\Omega)$ with zero trace on $\partial\Omega$. In additional, we also define the Hilbert space associated to the divergence operator

$$H(\text{div}, \Omega) = H(\text{div}) = \{\mathbf{w} \in [L^2(\Omega)]^d, \text{div}\mathbf{w} \in L^2(\Omega)\},$$

with norm

$$\|\mathbf{w}\|_{H(\text{div})}^2 = \|\mathbf{w}\|_0^2 + \|\mathbf{w}\|_0^2.$$

Let $\mathcal{T}_h$ be a regular finite element partition of the domain Ω, defined by

$$\mathcal{T}_h = \{K\} := \text{an union of all elements } K$$

and let

$$\mathcal{E}_h = \{e; e \text{ is an edge/face of } K \text{ for all } K \in \mathcal{T}_h\}$$

denotes the set of all edges/faces e of all elements K,

$$\mathcal{E}_h^0 = \{e \in \mathcal{E}_h; e \text{ is an interior edge/face}\}$$

is the set of interior egdes/faces, and

$$\mathcal{E}_h^\partial = \{e \in \mathcal{E}_h; e \subset \partial\Omega\}$$

the set of edges/faces of $\mathcal{E}_h$ on the boundary of Ω. We assume that the domain Ω is a polygonal and $\mathcal{T}_h$ is a regular partition of Ω. Thus, there exists $c > 0$ such that $h \leq c h_e$, where h_e is the diameter of the edge/face $e \in \partial K$ and h, the mesh parameter, is the element diameter. For each element K we associate a unit outward normal vector $\mathbf{n}_K$.

The $\mathcal{RT}_k$ spaces [25] are constructed by mapping polynomials defined on the reference element $\hat{K} = [-1, 1]^2$ to each element K of the mesh $\mathcal{T}_h$. We denote by $F_K : \hat{K} \to K$ the invertible, bilinear map of the two domains in $\mathbb{R}^d$, $d = 2, 3$. A scalar-valued function $\hat{\varphi}$ on $\hat{K}$ transforms to a function $\varphi = P_K^0 \hat{\varphi}$ on K by the composition

$$\varphi(\boldsymbol{x}) = (P_K^0 \hat{\varphi})(\boldsymbol{x}) = \hat{\varphi}(\hat{\boldsymbol{x}}), \tag{6}$$

with $\boldsymbol{x} = F_K(\hat{\boldsymbol{x}})$. A vector-valued function $\hat{\boldsymbol{\varphi}}$ on $\hat{K}$ transforms to a function $\boldsymbol{\varphi} = P_K^1 \hat{\boldsymbol{\varphi}}$ on K via the Piola transform

$$\boldsymbol{\varphi}(\boldsymbol{x}) = (P_K^1 \hat{\boldsymbol{\varphi}})(\boldsymbol{x}) = \frac{1}{J_K(\hat{\boldsymbol{x}})}[\mathrm{D}F_K(\hat{\boldsymbol{x}})]\hat{\boldsymbol{\varphi}}(\hat{\boldsymbol{x}}), \tag{7}$$

where $\mathbf{D}\boldsymbol{F}_K(\hat{\boldsymbol{x}})$ is the Jacobian matrix of the mapping $\boldsymbol{F}_K$ and $J_K(\hat{\boldsymbol{x}})$ is its determinant.

The (discontinuous) $\mathcal{RT}_k$ space of index k, $\mathcal{U}_h^k \times \mathcal{P}_h^k$, is defined to be

$$\mathcal{U}_h^k = \left\{ \mathbf{v}_h \in [L^2(\Omega)]^2; \mathbf{v}_h|_K \in P_K^1(\mathbb{Q}_{k+1,k}(\hat{K}) \times \mathbb{Q}_{k,k+1}(\hat{K})), \forall K \in \mathcal{T}_h \right\}, \quad (8)$$

$$\mathcal{P}_h^k = \left\{ q_h \in L^2(\Omega); q_h|_K \in P_K^0(\mathbb{Q}_k(\hat{K})), \forall K \in \mathcal{T}_h \right\}, \quad (9)$$

where $\mathbb{Q}_{i,j}$ denotes the set of polynomial functions of degree up to i in x and up to j in y. We also define the following sets of functions on the mesh skeleton:

$$\mathcal{M}_h^k = \left\{ \mu \in L^2(\mathcal{E}_h); \ \mu|_e \in p_k(e), \ \forall e \in \mathcal{E}_h^0, \ \mu|_e = \overline{\mu}, \forall e \in \mathcal{E}_h^\partial \cap \partial\Omega_D \right\}, \quad (10)$$

$$\bar{\mathcal{M}}_h^k = \left\{ \mu \in L^2(\mathcal{E}_h); \ \mu|_e \in p_k(e), \ \forall e \in \mathcal{E}_h^0, \ \mu|_e = 0, \forall e \in \mathcal{E}_h^\partial \cap \partial\Omega_D \right\}, \quad (11)$$

where $p_k(e)$ denotes the set of polynomial functions of degree up to k on e, and $\overline{\mu}$ is the Dirichlet boundary condition function associated to $\overline{p}$ in the Darcy problem and $\overline{c}$ in the transport equation.

4   Equivalent Hybrid Mixed Method

From the definitions presented in the previous section, we can write the following hybrid mixed formulation for the Darcy and Transport systems. For this, we define the product spaces $\mathbf{V}_h^k = \mathcal{U}_h^k \times \mathcal{P}_h^k \times \mathcal{M}_h^k$ and $\bar{\mathbf{V}}_h^k = \mathcal{U}_h^k \times \mathcal{P}_h^k \times \bar{\mathcal{M}}_h^k$ and the variable sets

$$\mathbf{X}_h^D = [\mathbf{u}_h, p_h, \lambda_h^p] \in \mathbf{V}_h^k \qquad \text{and} \qquad \mathbf{X}_h^T = [\boldsymbol{\sigma}, c_h, \lambda_h^c] \in \mathbf{V}_h^k$$

concerning the variables related to the Darcy problem (2) and the variables related to transport problem respectively, we can show the following semi-discrete formulation to the Darcy problem

Given c_h, find $\mathbf{X}_h^D \in \mathbf{V}_h^k$, such that

$$A(\mathbf{X}_h^D, \mathbf{Y}_h) = F(\mathbf{Y}_h), \qquad \forall \mathbf{Y}_h \in \bar{\mathbf{V}}_h^k. \quad (12)$$

On the other hand, the transport problem can be formulated by

Given $\mathbf{u}_h$, find $\mathbf{X}_h^T \in \mathbf{V}_h^k$, such that $\mathbf{Y}_h \in \bar{\mathbf{V}}_h^k$

$$\phi\frac{\partial c_h}{\partial t} + B(\mathbf{X}_h^T, \mathbf{Y}_h) = G(\mathbf{Y}_h), \qquad \forall \mathbf{Y}_h \in \bar{\mathbf{V}}_h^k. \quad (13)$$

In this context, we define the following generalized bilinear form for Darcy and transport problems

$$A(\mathbf{X}_h, \mathbf{Y}_h) = \sum_{K \in \mathcal{T}_h} \left[\int_K \mathbb{C}\mathbf{w}_h \cdot \mathbf{v}_h dx - \int_K s_h \nabla \cdot \mathbf{v}_h dx + \int_{\partial K} \lambda_h (\mathbf{w}_h \cdot \mathbf{n}_K) ds \right.$$

$$\left. + \int_{\partial K} \mu_h (\mathbf{w}_h \cdot \mathbf{n}_K) ds - \int_K q_h \nabla \cdot \mathbf{w}_h dx \right], \quad (14)$$

where $\mathbf{X}_h = [\mathbf{w}_h, s_h, \lambda_h]$ and $\mathbf{Y}_h = [\mathbf{v}_h, q_h, \mu_h]$. Taking $\mathbf{X}_h = \mathbf{X}_h^D$ and $\mathbb{C} = \mathbb{K}^{-1}$ this form can be adapted to Darcy's problem and to $\mathbf{X}_h = \mathbf{X}_h^T$ and $\mathbb{C} = \mathbb{D}^{-1}$ can be adapted to the transport problem. The source term of (12) and (13) is obtained directly from the multiplication of the respective functions f and g by q_h with a negative sign.

The convective term of the transport equation is stabilized by an upwind strategy proposed by [12], which can be defined as

$$A_{conv}(\mathbf{X}_h^T, \mathbf{Y}_h) = \int_K c_h \mathbf{u}_h \cdot \nabla q_h \mathrm{dx} + \int_{\partial K} \mathbf{u}_h \cdot \mathbf{n}_K \{\lambda_h^c/c_h\}(\mu_h - q_h)ds \quad (15)$$

with

$$\{\lambda/c\} := \begin{cases} \lambda, & e \subset \partial K^{\mathrm{in}} \\ c, & e \subset \partial K^{\mathrm{out}} \end{cases},$$

where the element outflow boundary is defined as $\partial K^{\mathrm{out}} := \{e \in \partial K : \mathbf{u} \cdot \mathbf{n}_K > 0\}$ and the element inflow boundary as $\partial K^{\mathrm{in}} = \partial K \backslash \partial K^{\mathrm{out}}$.

Therefore, the form $A(\cdot, \cdot)$ of the formulation (12) is given by the bilinear form $A(\cdot, \cdot)$ defined in (14) taking $\mathbf{X}_h = \mathbf{X}_h^D$ and $\mathbb{C} = \mathbb{K}^{-1}$. On the other hand, the compact bilinear form $B(\cdot, \cdot)$ of the formulation (13) is defined as

$$B(\mathbf{X}_h^T, \mathbf{Y}_h) = A\left(\mathbf{X}_h^T, \mathbf{Y}_h\right) + A_{conv}\left(\mathbf{X}_h^T, \mathbf{Y}_h\right), \quad (16)$$

with $\mathbb{C} = \mathbb{D}^{-1}$.

It is important to emphasize that according to the numerical analysis of the presented hybrid formulations, using Raviart-Thomas spaces, *a priori* error estimates in the $L^2(\Omega)$-norm ensures optimal convergence rates for the velocity, pressure and concentration. For more details see [2, 12].

4.1 Time Discretization and Resolution Algorithm

Setting the time step $\Delta t > 0$, such that $N = T/\Delta t$ and $t_n = n\Delta t$ with $n = 0, 1, 2, ..., N$ and the time interval $I_h = \{0 = t_0, t_1, ..., t_N = T\}$ which defines a partition of $I = (0, T]$, we can discretize the time derivative term of the semi-discrete formulation (13) employing implicit backward finite difference scheme as follows

Given $\mathbf{u}_h^{n+1}$ and c_h^0, find $\mathbf{X}_h^{T,n+1} \in \mathbf{V}_h^k \times I_h$, such that for all $\mathbf{Y}_h \in \bar{\mathbf{V}}_h^k$

$$\phi \frac{c_h^{n+1} - c_h^n}{\Delta t} + B(\mathbf{X}_h^{T,n+1}, \mathbf{Y}_h) = G(\mathbf{Y}_h), \quad \mathrm{com} \quad n = 0, 1, 2, ..., N \quad (17)$$

The resolution methodology is focused on decoupling Darcy and transport problems. Thus, given the initial condition c_h^0, the viscosity is evaluated using the Eq. (4) and used to solve the Darcy problem (12), once the velocity $\mathbf{u}_h$ is computed the transport problem is solved using the formulation (17). This resolution algorithm is repeated until it reaches the final time T.

To reduce the computational cost of the problem resolution at each time step, the static condensation technique is employed. Thus, element-level problems are condensed in favor of the Lagrange multiplier, generating a global system with degrees of freedom associated with the multiplier only, which in this case is a scalar. Also, the new system of equations is positive-definite, allowing for using simpler and more robust solvers. After a global system resolution, the variables u_h, p_h, σ_h and c_h are retrieved in the element level [15,16]. In this work, the deal.II library [4] is used to solve these problems.

5   Numerical Results

In this section, the proposed hybrid method (12)–(17) is tested through convergence studies and numerical simulations that are compared with experimental results presented by Malhotra et al. [18] in adverse mobility rate scenarios.

5.1   Convergence Study

Here, we study the convergence rates of the proposed hybrid mixed formulations in a domain $\Omega = [0,1]^2$ and in the time interval $I = [0, 0.2]$, adopting the sources $f = 0$ e $g = 2\pi^2 \sin(\pi(x + y - 2t))$ and the parameters $\mathbb{G} = \mathbb{I}$, $\nu(c) = 0.2c - 0.5$, $\alpha_{mol} = 0.01$, $\alpha_l = \alpha_t = 0$ and $\phi = 1.0$, is possible to derive the following analytical solution [22]

$$ \mathbf{u} = [1,1]^T, \quad p = \frac{2}{10\pi}\cos(\pi(x+y-2t)) + \frac{1}{2}(x+y), \quad c = \sin(\pi(x+y-2t)). \quad (18) $$

The initial and boundary conditions are determined by the evaluation of the exact solution on time $t = 0$ and on the boundary of the domain $\partial\Omega_D$, respectively.

For the h-convergence study, we adopt meshes with $4, 16, 64, 256, 1024, 4096$ quadrilateral elements with the same polynomial order $k = 0, 1, 2$ for the Lagrange multiplier and the primal variables (velocity, pressure, diffusive flux and concentration) and the time step $\Delta t = h^{k+1}$ to reduce the effects of the error associated to the time discretization. Hence, the spatial error governs the overall error. The results can be seen in Table 1, where it is possible to observe optimal convergence rates for all variables, i.e., order $k + 1$.

5.2   Numerical Simulations in Hele-Shaw Cells

The following results are performed in a Hele-Shaw cell where the flow channel is 84 cm long and 5 cm wide, as described in the work of Malhotra et al. [18]. In this experimental work, Malhotra and collaborators injecting water with dye into glycerol solutions to quantify the growth of the mixing zone in miscible viscous fingering.

In this example, dyed water was injected into 79% glycerol solution at a rate of 4.69 ml/min with a mobility ratio $M = 50$ and a Péclet number $P_e = 10234$ gives rise to the profiles presented in the left side of the Fig. 1. In this context, adopting the same experimental data in a mesh of 1500×100 elements with

Table 1. h-Convergence study from the approximations c_h, $\mathbf{u}_h$ and p_h.

Order	Cells	$\|c - c_h\|_0$		$\|\mathbf{u} - \mathbf{u}_h\|_0$		$\|p - p_h\|_0$	
		Error	Rate	Error	Rate	Error	Rate
0	4	7.894e−1	–	5.147e−1	–	1.158e−1	–
	16	5.852e−1	0.43	4.399e−1	0.23	5.187e−2	1.16
	64	3.704e−1	0.66	2.708e−1	0.70	2.865e−2	0.86
	256	2.162e−1	0.78	1.627e−1	0.74	1.584e−2	0.86
	1024	1.077e−1	1.01	9.175e−2	0.83	7.949e−3	0.99
	4096	5.351e−2	1.01	4.892e−2	0.91	3.987e−3	1.00
1	4	5.273e−1	–	4.579e−1	–	2.920e−2	–
	16	1.834e−1	1.52	1.543e−1	1.57	1.120e−2	1.38
	64	4.454e−2	2.04	4.624e−2	1.74	2.801e−3	2.00
	256	1.156e−2	1.95	1.223e−2	1.92	7.422e−4	1.92
	1024	2.874e−3	2.01	3.092e−3	1.98	1.861e−4	2.00
	4096	7.173e−4	2.00	7.754e−4	2.00	4.685e−5	1.99
2	4	3.120e−1	–	2.612e−1	–	1.849e−2	–
	16	4.340e−2	2.85	4.589e−2	2.51	2.791e−3	2.73
	64	5.654e−3	2.94	6.119e−3	2.91	3.709e−4	2.91
	256	7.062e−4	3.00	7.705e−4	2.99	4.658e−5	2.99
	1024	8.975e−5	2.98	1.005e−4	2.94	6.157e−6	2.92

Fig. 1. Hele-Shaw cell experiment (left) developed by Malhotra et al. [18] compared with the approximate solution (right) at time $t = 22, 48, 63, 68, 80, 90$ s, adopting 1500×100 elements, $M = 50$, $\mathrm{P_e} = 10234$ and time step $\Delta t = 0.1$ s.

polynomial order $k = 1$, time step $\Delta t = 0.1\,\mathrm{s}$, $\mathbb{G} = \mathbb{I}$, $\alpha_{mol} = 1.53 \times 10^{-7}\,\mathrm{m}^2/\mathrm{s}$, $\alpha_l = \alpha_{mol}$ and $\alpha_t = 1.53 \times 10^{-8}\,\mathrm{m}^2/\mathrm{s}$ and $\phi = 1.0$, we develop numerical simulations to compare with experimental results at time $t = 22, 48, 63, 68, 80, 90\,\mathrm{s}$, as can be seen in the right side of the Fig. 1.

As can be seen in Fig. 1, the numerical results show a good performance of the proposed method in capturing both the formation of viscous fingers and the propagation of the water front. Moreover, the upwind scheme employed is capable of stabilizing the convection-dominated regime caused by the reduction of the resident fluid viscosity adopted in the experiment that generates a Péclet number of the 10^4 order.

6   Conclusions

In this work, we proposed an equivalent stable hybrid mixed method adopting non-conforming Raviart-Thomas spaces for the Darcy and the Transport systems to solve miscible displacements with adverse mobility ratio. For the convection-diffusion equation, an upwind scheme was employed to stabilizing the convection-dominated regimes, and the implicit backward Euler approach was used to the time discretization. The continuity was weakly imposed by the Lagrange multiplier associated with the pressure field for the Darcy problem and the concentration for the Transport mixed problem. This approach gives rise to a positive-definite global matrix with reduced computational cost compared to classical conforming Raviart-Thomas formulations. To solve this coupled problem, we employed a staggered approach to decoupling the systems using the same time scale for both problems. The numerical studies confirmed optimal convergence rates for velocity, pressure and concentration. In addition, simulations adopting an adverse mobility ratio, comparing the approximate solution with experimental results in Hele-Shaw cell, demonstrated that the proposed hybrid formulations are capable of capturing the formation and propagation of the viscous fingering even in cases of high Péclet number.

References

1. Aftosmis, M.J.: Upwind method for simulation of viscous flow on adaptively refined meshes. AIAA J. **32**(2), 268–277 (1994). https://doi.org/10.2514/3.11981
2. Arnold, D.N., Brezzi, F.: Mixed and nonconforming finite element methods: implementation, postprocessing and error estimates. ESAIM: M2AN **19**(1), 7–32 (1985). https://doi.org/10.1051/m2an/1985190100071
3. Babuška, I.: Error-bounds for finite element method. Numer. Math. **16**(4), 322–333 (1971). https://doi.org/10.1007/BF02165003
4. Bangerth, W., Hartmann, R., Kanschat, G.: deal.II - a general-purpose object-oriented finite element library. ACM Trans. Math. Softw. **33**(4) (2007). https://doi.org/10.1145/1268776.1268779
5. Bettelheim, E., Agam, O., Zabrodin, A., Wiegmann, P.: Singularities of the Hele-Shaw flow and shock waves in dispersive media. Phys. Rev. Lett. **95**, 244504 (2005). https://doi.org/10.1103/PhysRevLett.95.244504

6. Bischofberger, I., Ramachandran, R., Nagel, S.R.: Fingering versus stability in the limit of zero interfacial tension. Nat. Commun. **5**, 5265 (2014). https://doi.org/10.1038/ncomms6265

7. Brezzi, F.: On the existence, uniqueness and approximation of saddle-point problems arising from Lagrange multipliers. Revue Française d'Automatique Informatique et Recherche Opérationnelle, Séries Rouge **8**(R-2), 129–151 (1974)

8. Brezzi, F., Douglas, J., Marini, D.: Two families of mixed finite elements for second order elliptic problems. Numer. Math. **47**(2), 217–235 (1985). https://doi.org/10.1007/BF01389710

9. Brooks, A., Hughes, T.: Streamline-upwind/Petrov-Galerkin methods for advection dominated flows, vol. 2, pp. 283–292. Calgary University, Calgary, Canada, January 1980

10. Cockburn, B., Dong, B., Guzmán, J., Restelli, M., Sacco, R.: A hybridizable discontinuous Galerkin method for steady-state convection-diffusion-reaction problems. SIAM J. Sci. Comput. **31**(5), 3827–3846 (2009). https://doi.org/10.1137/080728810

11. Correa, M., Loula, A.: Unconditionally stable mixed finite element methods for Darcy flow. Comput. Methods Appl. Mech. Eng. **197**(17–18), 1525–1540 (2008). https://doi.org/10.1016/j.cma.2007.11.025

12. Egger, H., Schöberl, J.: A hybrid mixed discontinuous Galerkin finite-element method for convection-diffusion problems. IMA J. Numer. Anal. **30**(4), 1206–1234 (2009). https://doi.org/10.1093/imanum/drn083

13. Fabien, M.S., Knepley, M., Riviere, B.: A high order hybridizable discontinuous Galerkin method for incompressible miscible displacement in heterogeneous media. Results Appl. Math., 100089 (2020). https://doi.org/10.1016/j.rinam.2019.100089

14. Homsy, G.M.: Viscous fingering in porous media. Ann. Rev. Fluid Mech. **19**(1), 271–311 (1987). https://doi.org/10.1146/annurev.fl.19.010187.001415

15. Igreja, I.: A new approach to solve the stokes-darcy-transport system applying stabilized finite element methods. In: Rodrigues, J.M.F., et al. (eds.) ICCS 2019. LNCS, vol. 11539, pp. 524–537. Springer, Cham (2019). https://doi.org/10.1007/978-3-030-22747-0_39

16. Igreja, I., Loula, A.F.: A stabilized hybrid mixed DGFEM naturally coupling Stokes-Darcy flows. Comput. Methods Appl. Mech. Eng. **339**, 739–768 (2018). https://doi.org/10.1016/j.cma.2018.05.026

17. Lee, S., Wheeler, M.F.: Adaptive enriched Galerkin methods for miscible displacement problems with entropy residual stabilization. J. Comput. Phys. **331**, 19–37 (2017). https://doi.org/10.1016/j.jcp.2016.10.072

18. Malhotra, S., Sharma, M.M., Lehman, E.R.: Experimental study of the growth of mixing zone in miscible viscous fingering. Phys. Fluids **27**(1), 014105 (2015). https://doi.org/10.1063/1.4905581

19. Masud, A., Hughes, T.J.: A stabilized mixed finite element method for Darcy flow. Comput. Methods Appl. Mech. Eng. **191**(39–40), 4341–4370 (2002). https://doi.org/10.1016/S0045-7825(02)00371-7

20. Nguyen, N., Peraire, J., Cockburn, B.: An implicit high-order hybridizable discontinuous galerkin method for linear convection-diffusion equations. J. Comput. Phys. **228**(9), 3232–3254 (2009). https://doi.org/10.1016/j.jcp.2009.01.030

21. Núñez, Y., Faria, C., Loula, A., Malta, S.: Um método híbrido de elementos finitos aplicado a deslocamentos miscíveis em meios porosos heterogêneos. Revista Internacional de Métodos Numéricos para Cálculo y Diseño en Ingenieria **33**(1), 45–51 (2017). https://doi.org/10.1016/j.rimni.2015.10.002

22. Ohlberger, M.: Convergence of a mixed finite element: finite volume method for the two phase flow in porous media. East West J. Numer. Math. **5**, 183–210 (1997)
23. Oikawa, I.: Hybridized discontinuous Galerkin method for convection–diffusion problems. Jpn. J. Ind. Appl. Math., 1–20 (2014). https://doi.org/10.1007/s13160-014-0137-5
24. Peaceman, D.W.: Fundamentals of Numerical Reservoir Simulation, vol. 6. Elsevier, Amsterdam (2000)
25. Raviart, P.A., Thomas, J.M.: A mixed finite element method for 2-nd order elliptic problems. In: Galligani, I., Magenes, E. (eds.) Mathematical Aspects of Finite Element Methods. LNM, vol. 606, pp. 292–315. Springer, Heidelberg (1977). https://doi.org/10.1007/BFb0064470
26. Rouy, E., Tourin, A.: A viscosity solutions approach to shape-from-shading. SIAM J. Numer. Anal. **29**(3), 867–884 (1992). https://doi.org/10.1137/0729053
27. Russell, T., Wheeler, M., Chiang, C.: Large-scale simulation of miscible displacement by mixed and characteristic finite element methods, pp. 85–107. SIAM, Philadelphia, January 1986
28. Saffman, P.G., Taylor, G.I.: The penetration of a fluid into a porous medium or Hele-Shaw cell containing a more viscous liquid. Proc. Roy. Soc. Lond. Ser. A. Math. Phys. Sci. **245**(1242), 312–329 (1958). https://doi.org/10.1098/rspa.1958.0085
29. Sander, L., Ramanlal, P., Ben-Jacob, E.: Diffusion-limited aggregation as a deterministic growth process. Phys. Rev. A **32**(5), 3160 (1985). https://doi.org/10.1103/PhysRevA.32.3160
30. Settari, A., Price, H., Dupont, T., et al.: Development and application of variational methods for simulation of miscible displacement in porous media. Soc. Petrol. Eng. J. **17**(03), 228–246 (1977). https://doi.org/10.2118/5721-PA
31. Zhang, J., Zhu, J., Zhang, R., Yang, D., Loula, A.F.: A combined discontinuous Galerkin finite element method for miscible displacement problem. J. Comput. Appl. Math. **309**, 44–55 (2017). https://doi.org/10.1016/j.cam.2016.06.021

Smart Systems: Bringing Together Computer Vision, Sensor Networks and Machine Learning

Learn More from Context: Joint Modeling of Local and Global Attention for Aspect Sentiment Classification

Siyuan Wang[1,2](✉) [ID], Peng Liu[1,2], Jinqiao Shi[1,3], Xuebin Wang[1,2], Can Zhao[1,2], and Zelin Yin[1,2]

[1] Institute of Information Engineering, Chinese Academy of Sciences, Beijing, China
{wangsiyuan,pengliu1995,wangxuebin,zhaocan,yinzelin}@iie.ac.cn
[2] School of Cyber Security, University of Chinese Academy of Sciences, Beijing, China
[3] Beijing University of Posts and Telecommunications, Beijing, China
shijinqiao@bupt.edu.cn

Abstract. Aspect sentiment classification identifies the sentiment polarity of the target that appears in a sentence. The key point of aspect sentiment classification is to capture valuable information from sentence. Existing methods have acknowledged the importance of the relationship between the target and the sentence. However, these approaches only focus on the local information of the target, such as the positional relationship and the semantic similarity between the words in a sentence and the target. Moreover, the global information of the interaction of words in sentence and their influence on the final prediction of sentiment polarity are ignored in related works. To tackle this issue, the present paper proposes Joint Modeling of Local and Global Attention (LGAJM), with the following two aspects: (1) the study develops a position-based attention network concentrating on the local information of semantic similarity and position information of the target. (2) In order to fetch global information, such as context information and interaction between words in sentences, the self-attention network is introduced. Besides, a BiGRU-based gating mechanism is proposed to weight the outputs of these two attention networks. The model is evaluated on two datasets: laptop and restaurant from SemEval 2014. Experimental results demonstrate the high effectiveness of the proposed method in aspect sentiment classification.

Keywords: Aspect sentiment classification · Attention · Gating · CNN

1 Introduction

Aspect sentiment classification, with its inherent challenges and wide applications, has been an important task in natural language processing (NLP) and draws wide attention both by industry and academia. Aspect sentiment classification aims to identify the sentiment polarity of targets (positive, neutral,

© Springer Nature Switzerland AG 2020
V. V. Krzhizhanovskaya et al. (Eds.): ICCS 2020, LNCS 12143, pp. 271–284, 2020.
https://doi.org/10.1007/978-3-030-50436-6_20

negative) that appear in a given sentence. For example, the second sentence in Figs. 1, the sentiment polarity of the target *"menu"* should be identified as negative while it would be positive polarity for *"dishes"*. Therefor, a critical demand is to extract valuable information and precisely recognize sentiment polarity in aspect level sentiment classification.

The existing methods dealing with aspect sentiment classification can be classified into two categories: feature-based methods, such as support vector machine [5], are labor-intensive and highly depend on the quality of features. Neural-network-based methods, such as Target-Dependent LSTM (TD-LSTM) and Target-Connection LSTM (TC-LSTM) [13], can learn features without feature engineering and have been widely used in the fields of NLP. However, these neural-network-based methods cannot effectively identify the importance of words in a sentence. Consequently, the attention mechanism is introduced to promote the neural-network-based methods. The attention mechanism is used to help the model to pay more attention to essential words in a sentence. For example, AE-LSTM [17], ATAE-LSTM [17] and IAN [7] are all designed with attention networks to improve their model for aspect sentiment classification.

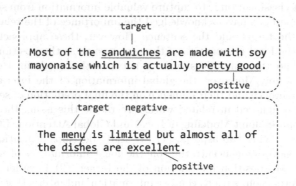

Fig. 1. Examples for aspect-level sentiment classification

These existing methods only considered local information about the target. On the other hand, this paper argues that global information plays an important role in aspect sentiment classification. For instance, considering the first sentence in the Fig. 1, words near the target usually get more attention in previous methods, but the sentiment polarity of aspect *"sandwiches"* is determined by opinion words *"pretty good"* which are far away from the aspect and get little attention. Therefore, for the sake of approving the accuracy of the model, global information should be considered in the model.

LGAJM, proposed by the present study, consists of a word representation module, multi-attention module, BiGRU-based gating module, and CNN classifier module and aims at tracking the aforementioned problems. In LGAJM, the word representation module generates different representations of individual sentence words by the target. The multi-attention module contains two attention networks: local attention network and global attention network. These two attention networks can capture not only local information about the aspect but also global information about the sentence, thereby improving classification performance. The BiGRU-based gating module is used to weight the importance of two outputs of the multi-attention module and combine them. The CNN classifier is used to extract n-gram features of the sentence and make sentiment polarity prediction. Experimental results and comparisons on benchmark SemEval2014 datasets have shown the superior performance of the proposed method.

The contribution of this paper can be summarized as follows:

- LGAJM employs the local attention network to model semantic information and position relevance to extract local relevant information. LGAJM also utilizes the global attention network to model the interaction of sentence words and obtain global contextual information.
- The study propose a BiGRU-based gating mechanism that effectively combines the local attention network and global attention network.
- Experimental results on two benchmark datasets illustrate that the proposed model significantly outperforms the comparative baselines.

2   Related Work

Conventional methods for aspect sentiment classification are rule-based methods and statistic-based methods. Statistic-based methods, such as SVM [5] and MaxEnt-LDA [19], build features between the target and sentence, and then predict the sentiment polarity of target. However, statistic-based methods are highly dependent on the quality of feature engineering work and laborious job.

Recently studies on neural-network-based methods can automatically encode original features as continuous and low-dimensional vectors without feature engineering. Some LSTM-based methods, such as TD-LSTM [13] and TC-LSTM [13] make prediction based on the relationship between sentence and aspect. However, these models ignore which words are more important in a sentence.

The Attention mechanism is taken advantage of assigning different weights to words of different importance in sentences. For example, AE-LSTM and ATAE-LSTM [17] calculate the attention weights with a standard attention mechanism. However, these methods ignore the position information. EAM [4] and PBAN [3] fist use the position information in their model. Position information is used to select keywords and ignore other unimportant words in EAM [4]. In PBAN [3] position information is used to combine position embedding with word embedding as the inputs. But when position information makes a significant influence in a model, global information is hardly considered in the same model.

In an effective attention mechanism, both local position information and global information should require more attention. Multi-head attention [15] has been recognized as an effective mechanism to advance the existing attention functions [6,8,9] and showed its effect on many tasks such as [12,16]. It can catch global information by carrying different features of the target and sentence in different subspaces.

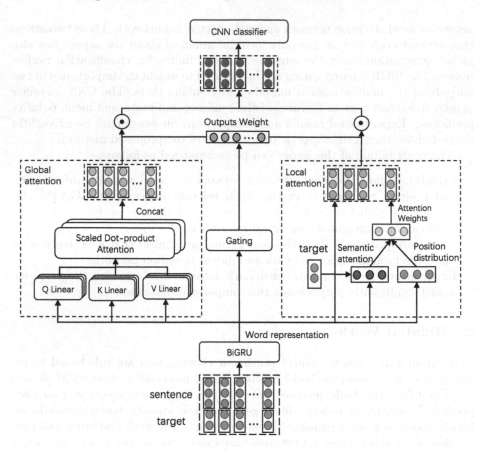

Fig. 2. Architecture of our model.

3   Model Description

This section gives a brief description of the architecture of the model in this paper. Our model consists of four modules: a word representation module, mutli-attention module, BiGRU-based gating module and CNN classifier module, as shown in Fig. 2. Given a sentence $sen = (w_1, w_2, w_3, ..., w_n)$ containing n words and target entries $a = (a_1, a_2, a_3..., a_m)$ consisting of m words which is a sub-sequence of sentence sen, our model aims to identify the sentiment polarity P of the target. The sentiment polarity of the target belongs to the set $P = \{$"positive", "negative", "neutral"$\}$.

3.1   Word Representation Module

To get high dimensional embedding with abundant information for each word, we employ the pre-trained word vector GloVe [10] to map each word. Define

$L \in \mathbb{R}^{d \times |V|}$ as embedding layer from Glove, where d is the dimension of vectors and $|V|$ is the size of the vocabulary set V. Each sentence word and each target word are required from Glove. Sentence words embeddings are defined as $(e_{w_1}, e_{w_2}, e_{w_3}, ..., e_{w_n})$ and target words embeddings are defined as $(e_{a_1}, e_{a_2}, e_{a_3}, ..., e_{a_n})$.

As targets may have different effects on the final representations of the words in sentence, a sentence should be represented differently based on aspect words. Firstly, a vector a_{avg} is calculated to represent the aspect. Then we generate different representations of each context word by individual target representation. The vector s of words in sentence is defined as follows:

$$a_{avg} = \frac{1}{n} \sum_{i=1}^{n} e_{a_i} \qquad (1)$$

$$s_i = [e_{w_i}; a_{avg}] \qquad (2)$$

where i indicates the ith word representation, e_{a_i} and e_{w_i} are the ith word embeddings of target and sentence, the dimension of s_i is $2d$. Equation 1 outputs a vector a_{avg} calculated by averagely pooling all word embeddings of target representing the aspect. So s_i is the representation of ith words in sentence under the corresponding target. s_i is generated by concatenating ith sentence word embedding and a_{avg}.

BLSTM models the context dependency with the forward LSTM and the backward LSTM. BLSTM has been widely adopted to NLP tasks [2]. The forward LSTM handles the sentence from left to right, and the backward LSTM handles it from right to left. By BLSTM, model can obtain the bidirectional hidden state of each word. However, to simplify our model, we choose bidirectional GRU which is similar to BLSTM but has fewer parameters and lower computational complexity. LGAJM employs a BiGRU to accumulate the context information for each word of the input sentence.

At each time step t, the new words representation $\overrightarrow{h_t}$ which contains the past information is computed. The update process at time t is as follows:

$$z_t = \sigma(W_z s_t + U_z \overrightarrow{h_{t-1}}) \qquad (3)$$

$$r_t = \sigma(W_t s_t + U_z \overrightarrow{h_{t-1}}) \qquad (4)$$

$$g_t = tanh(W s_t + U(r_t \circ \overrightarrow{h_{t-1}})) \qquad (5)$$

$$\overrightarrow{h_t} = (1 - z_t) \circ \overrightarrow{h_{t-1}} + z_t \circ g_t \qquad (6)$$

where σ and $tanh$ are sigmoid and hyperbolic tangent functions, s_t is the representation of word at time t, $W_z, W_t, W \in \mathbb{R}^{d_h \times 2d}$ and $U_z, U_t, U \in \mathbb{R}^{d_h \times d_h}$ are trainable parameters where d_h is the size of $\overrightarrow{h_t}$, z and r simulate binary switches that control to update the information from the current input or not and forget the information in the memory cells or not, respectively.

Then the other representation containing the future information $\overleftarrow{h_t}$ is obtained in the same way, and we concatenate $\overrightarrow{h_t}$ and $\overleftarrow{h_t}$ to generate a new representation h_t for each word:

$$h_t = [\overrightarrow{h_t}; \overleftarrow{h_t}] \tag{7}$$

3.2 Attention Module

We have got the contextual representation of each word in sentence. In this section, we employ two attention networks. One is position-based local attention network to capture the local information of target. The other is the self-attention-based global attention network to obtain the global contextual information and model the word interaction of the whole sentence.

Local Attention Network: In terms of the local information of target, there is no doubt that semantic relationship between the words of sentence and target is a meaningful information. The most relevant opinion words of target are found to be near the target by surveying the datasets. So our model adapts the position-based local attention mechanism to capture semantic information and position relevance. In this attention network, the attention weight is obtained based on the local position information and the semantic relationship between sentence and target.

Firstly, we gain the distance of d_i between the ith sentence words and target words:

$$d_i = \begin{cases} |i - a_s| & i < a_s \\ 0 & a_s \leq i \geq a_e \\ |i - a_e| & i > a_e \end{cases} \tag{8}$$

where i is ith word index in the sentence, a_s is the begin index of the target and a_e is the end index of the target.

Then we map d_i to a value between 0 and 1, in the same interval as on part of attention weight:

$$\tilde{d_i} = 1 - \frac{d_i}{n} \tag{9}$$

where n is the length of sentence.

Another part of attention k_i is calculated by model input representation h_i and target representation a_{avg}. This part is expected to consider semantic information. The calculation process is as follows:

$$k_i = tanh(h_i W^T a_{avg}) \tag{10}$$

where $tanh$ is the hyperbolic tangent functions.

$\tilde{d_i}$ and k_i are multiplied and do normalization to get the final attention weight $\widetilde{att_i^p}$:

$$att_i^p = \tilde{d_i} k_i \tag{11}$$

$$\widetilde{att}_i^p = \frac{e^{att_i^p}}{\sum e^{att_i^p}} \tag{12}$$

The ith output m_i^l of local attention network is:

$$m_i^l = \widetilde{att}_i^p h_i \tag{13}$$

where h_i is the contextual representation of ith word.

Global Attention Network: When target relevant opinion words distribute in a long sentence, previous attention mechanism based on local information can't perform well. However, the self-attention mechanism allows the model to jointly attend to information from different representation subspaces at different positions and has been proven its effectiveness in the field of machine translation [9]. Therefore the self-attention is utilized to obtain a new representation of the whole sentence, so that our model is capable to extract the global context information from the sentence.

The original self-attention network adds position information into word embeddings which changes the original representation. Expanded dimension of word representation with position embedding is aimed at fitting this task. The original word representation h_i is processed as follows:

$$\tilde{h}_i = [h_i; pos_i] \tag{14}$$

where position embedding pos_i is obtained by looking up a position embedding matrix $\mathbb{L} \in \mathbb{R}^{d \times V}$ which is according to the distance between the word and target. Position embedding matrix is randomly initialized, and updated during the training process. Then query matrix Q_i^j, key matrix K_i^j, value matrix V_i^j are constructed by new words representation the $\tilde{h}_i$:

$$Q_i^j, K_i^j, V_i^j = \tilde{h}_i W_j^Q, \tilde{h}_i W_j^K, \tilde{h}_i W_j^V \tag{15}$$

where W_j^Q, W_j^K, W_j^V are trainable parameters and j represents the word representation at the jth semantic subspace.

The ith output of self-attention is calculated as follows:

$$head_j = softmax(\frac{Q_i^j K_i^{j\,T}}{\sqrt{d_k}}) V_i^j \tag{16}$$

$$m_i^g = Concat(head_1, ..., head_h) W^O \tag{17}$$

where d_k is the dimension of K, h is the count of head, i represents the index of input in global attention network and m_i^g is the final representation of word in global attention network.

3.3   BiGRU-Based Gating Mechanism

As attention module has two outputs, $(m_1^l, m_2^l, m_3^l, ..., m_n^l)$ are the outputs of position-based attention network corresponding to the local information, and

$(m_1^g, m_2^g, m_3^g, ..., m_n^g)$ are the output of multi-head attention network related to the global information. To effectively combining these outputs, we introduce the gating mechanism. The gating mechanism has proven to be useful for the RNN, and it is used to control the information flowing in a network. Gating mechanism is expected to learn an appropriate weight of two outputs in the process of model training and combine two outputs to the final word representation. As the importance of word global and local representation is related to the contextual information, the final weight is expected to determined based on it. So we make that the representation of the ith word cross the BiGRU network to obtain the weight of the output of local attention module and global attention module.

A *sigmoid* activate function which scales a scalar to 0–1 is used to convert ith word representation to a weight f_i:

$$f_i = sigmoid(h_i V^T) \tag{18}$$

where $V^T \in \mathbb{R}^{d_h \times 1}$ is trainable parameter, d_h is the dimension of h_t.

So the weight of the ith output of local attention is f_i and the weight of the ith output of global attention is $1 - f_i$. The final representation m_i of the ith word is as follows:

$$m_i = f_i * m_i^l + (1 - f_i) * m_i^g \tag{19}$$

3.4   CNN Layer

CNN is introduced to extract n-gram features of sentence. To produce multiple features for CNN, the convolutional layer and the max-pooling layer capture the most crucial features:

$$c_i = \text{relu}(W_{conv}[m_{i-1}, m_i, m_{i+1}] + b_{conv}) \tag{20}$$
$$z = \text{maxpooling}(c_1, c_2, ..., c_n) \tag{21}$$

where $i \in [1, n]$, W_{conv} and b_{conv} are parameters of the convolutional kernel and z is the final representation of the sentence with the target.

Finally, z is fed into a softmax layer to predict sentiment polarity y_i of the target:

$$y_i = softmax(W_s z + b_s) \tag{22}$$

where W_s, b_s are the weight matrix and bias, and y_i is the predicting polarity of the target.

So the training object is minimizing cross-entropy function:

$$loss = -\sum log(y_i pol_i) \tag{23}$$

where pol_i is the real sentiment polarity of target.

4 Experiments

4.1 Experimental Setting

Dataset Settings: Experiments are conducted on two datasets: Restaurant and Laptop from SemEval2014 task4 [11]. These datasets contain lots of user reviews and each review associated with a list of opinion targets and corresponding sentiment polarities. After preprocessing, statistics of the final datasets are given in Table 1.

Hyper-parameters: 20% of the training data is selected randomly as the development set to tune hyperparameters. The network is trained for 200 epochs and select the best model according to the performance on the development set. The final hyperparameters are selected when they produce the highest accuracies on the development set. All word embedding are initialized by the pre-trained Glove and the dimension of the word embedding and target embedding are set to 300. The dimension of the BiGRU hidden vectors is 100. The convolutional kernel size is 3. The weight matrices and bias are given the initial value by sampling from the uniform distribution $\bigcup(0.01, 0.01)$. Dropout is applied on the input word embeddings for purpose of avoiding overfitting. Adam with learning rate set to 0.0005 is used for optimizing the models.

Table 1. Statistic results of Laptop and Restaurant.

	Set	Total	Positive	Negative	Neutral
Laptop	Train	2328	994	870	464
	Test	638	341	128	169
Restaurant	Train	3608	2164	807	637
	Test	1120	728	196	196

4.2 Compared Methods

To evaluate the effectiveness of our method, LGAJM is compared with the following methods:

- **SVM:** SVM is a traditional support vector machine based model with extensive feature engineering [5].
- **TD-LSTM:** TD-LSTM uses two LSTMs to model the left and right contexts of the target separately, then predicts the sentiment polarity of target based on concatenated context representations [13].
- **ATAE-LSTM:** ATAE-LSTM learns attention embeddings and combines them with the LSTM hidden states to predict the sentiment polarity of target [18].

- **MemNet:** MemNet employs multiple attention layers over the word embeddings and bases on the output of the last layer predicting the sentiment polarity of target. And it introduces the position information [14].
- **IAN:** IAN adapts two LSTMs to model context and target separately, then interactively get the final representation of context and aspect [7].
- **RAM:** RAM employs multi-hop attention on the hidden states of a positon-weighted bidirectional LSTM to extract the relevant information of target, and nonlinearly combine the result from memory with a GRU [1].
- **PBAN:** PBAN incorporates position embedding and word embedding, then use bidirectional attention mechanism between target and sentence to get the final representation of context and target [3].

To investigate the effectiveness of our model, the variants of the LGAJM model are listed as follows: Local attention+CNN(LAM), Global attention+CNN(GAM). The previous two models are set to verify the effectiveness of two attention mechanisms and prove that two attention mechanisms complement each other well. These two models make up our ablation experiments.

4.3 Result Analysis

Table 2. Average accuracies and Macro-F1 scores over 5 runs with random initializations. The best results are in bold. The marker * indicates that our full model LGAJM is significantly better than TD-LSTM, ATAE-LSTM, MemNet, IAN, RAM, and PBAN with $p < 0.05$ based on one-tailed unpaired t-test.

	Models	LAPTOP		REST	
		Acc.	Macro-F1	Acc.	Macro-F1
Baselines	SVM [5]	70.49	NA	80.16	NA
	ATAE-LSTM [18]	68.70	NA	77.20	NA
	MemNet [14]	72.21	NA	80.95	NA
	IAN [7]	72.10	NA	78.60	NA
	PBAN [3]	74.12	NA	81.16	NA
	TD-LSTM [13]	71.83	68.43	78.00	66.73
	RAM [1]	74.49	**71.35***	80.23	70.80
Ablated models	GAM	73.35	69.51	79.65	70.75
	LAM	74.14	70.16	80.0	70.80
Full model	LGAJM	**74.92***	**71.35***	**81.16***	**71.73***

The accuracy and Macro-F1 of our proposed model is evaluated and compare with others basic model on Laptop and Restaurant datasets. As shown by the result in Table 2, the final reported numbers are obtained as the best value over 5 runs with random initialization. We can get the following observations:

(1) LGAJM achieves the best performance among all methods on both accuracy and macro-F1 for all datasets. This proves the effectiveness of our method.
(2) LAM performs better than SVM, IAN, TD-LSTM, and ATAE-LSTM which not follow position information on both Laptop and Restaurant datasets, which suggests the effectiveness of local position information. GAM also outperforms than SVM, IAN, TD-LSTM, MemNet and ATAE-LSTM on Laptop datasets and outperforms than IAN, TD-LSTM, and ATAE-LSTM on Restaurant datasets, which verifies global information can help final prediction. The reason for LAM outperforming than GAM is that more related opinion words of target are near the target words. So usually local position information can play a better role.
(3) In conclusion, LGAJM performs best on all evaluating indicators and all datasets because LGAJM concentrates on not only local position information but also global information. Two attention networks outputs are combined by BiGRU-based gating mechanism which shows local information and global information is complemented and BiGRU-based gating can combine these effectively.

Analysis of Local Attention. The relevant opinion words of target are usually found near the target. The proposed model thus introduces position information to model a position-based attention network that can pay more attention to the adjacency words of the target. As shown in Table 2, experimental results show that the LAM model usually performs better than base methods. To further illustrate the contribution of the local attention mechanism, some sentences are picked from the datasets tested by the ablated model LAM. Figure 3 demonstrates the different distribution of attention. For the sentence in Fig. 3, the target wordw are *"menu"* and *"dishes"* and the opinion words are *"limited"* and

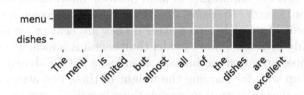

Fig. 3. LAM ablation model's attention distribution

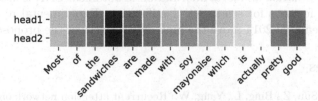

Fig. 4. GAM ablation model's target word attention distribution

"excellent". These opinion words all near the target. The local attention network can precisely assign a significantly higher weight to the real opinion word *"limited"* and *"excellent"* than to other words separately. Experimental results show the advantage of the local position especially when there is a short distance between opinion words and target words.

Analysis of Global Attention. The global attention mechanism is required to capture the global information, especially when the relevant opinion words of the target are distributed in a long sentence. Similarly to previous experiments, the ablation experiments are also set up and examples are picked to check the effectiveness of the global attention mechanism. Figure 4 illustrates the difference. There is a long-distance between the opinion word *"good"* and target words, and as expected, the ablated model LAM assigns low weight to the related opinion word *"good"*. However, the ablated model GAM assigns obviously higher weight to the word *"good"*. Experimental results show the ability of self-attention to capture global information as well as the effectiveness of global information.

Analysis of Gating Mechanism. In the proposed model, the gating mechanism is chosen to control information flow in the model and combine the two outputs of the attention module to ensure that the model considers both local and global information. When the LAM model assigns low weight to the opinion word *"good"* and GAM model assigns higher weight, the gating mechanism resets high weight to the output of GAM. The existing model can thus make the correct prediction. Results show that two attention mechanisms are complementary and the BiGRU-based-gating mechanism demonstrates its strengths.

5   Conclusion

This paper verified the importance of local position information and global context information. In order to ensure that the proposed attention network focuses on relevant words of the target no matter if they are near or far away from the aspect, the study presents a novel model that contains a local attention mechanism and a global attention mechanism. Besides, a BiGRU-based gating mechanism is come up with for learning the weight of these two attention networks' outputs and it adjusts the word representations based on context information of sentence words. The final representation of each word in sentence incorporating the global information and local information is obtained. CNN is used to extract local n-gram features for sentiment classification. Experimental results on two datasets of SemEval 2014 prove the high effectiveness of the proposed model.

References

1. Chen, P., Sun, Z., Bing, L., Yang, W.: Recurrent attention network on memory for aspect sentiment analysis. In: EMNLP, pp. 452–461. Association for Computational Linguistics (2017)

2. Cho, K., et al.: Learning phrase representations using RNN encoder-decoder for statistical machine translation. In: Proceedings of the 2014 Conference on Empirical Methods in Natural Language Processing (EMNLP), Doha, Qatar, October 2014, pp. 1724–1734. Association for Computational Linguistics (2014)

3. Gu, S., Zhang, L., Hou, Y., Song, Y.: A position-aware bidirectional attention network for aspect-level sentiment analysis. In: Proceedings of the 27th International Conference on Computational Linguistics (COLING 2018), Santa Fe, New Mexico, USA, 20–26 August 2018, pp. 774–784 (2018)

4. He, R., Lee, W.S., Ng, H.T., Dahlmeier, D.: Effective attention modeling for aspect-level sentiment classification. In: Proceedings of the 27th International Conference on Computational Linguistics (COLING 2018), Santa Fe, New Mexico, USA, 20–26 August 2018, pp. 1121–1131 (2018)

5. Kiritchenko, S., Zhu, X., Cherry, C., Mohammad, S.: NRC-Canada-2014: detecting aspects and sentiment in customer reviews. In: SemEval@COLING, pp. 437–442. The Association for Computer Linguistics (2014)

6. Li, J., Tu, Z., Yang, B., Lyu, M.R., Zhang, T.: Multi-head attention with disagreement regularization. In: Proceedings of the 2018 Conference on Empirical Methods in Natural Language Processing, Brussels, Belgium, 31 October–4 November 2018, pp. 2897–2903 (2018)

7. Ma, D., Li, S., Zhang, X., Wang, H.: Interactive attention networks for aspect-level sentiment classification. In: IJCAI, pp. 4068–4074. ijcai.org (2017)

8. Meng, F., Zhang, J.: DTMT: a novel deep transition architecture for neural machine translation. In: The Thirty-Third AAAI Conference on Artificial Intelligence (AAAI 2019), The Thirty-First Innovative Applications of Artificial Intelligence Conference (IAAI 2019), The Ninth AAAI Symposium on Educational Advances in Artificial Intelligence (EAAI 2019), Honolulu, Hawaii, USA, 27 January–1 February 2019, pp. 224–231 (2019)

9. Ott, M., Edunov, S., Grangier, D., Auli, M.: Scaling neural machine translation. In: Proceedings of the Third Conference on Machine Translation: Research Papers (WMT 2018), Belgium, Brussels, 31 October–1 November 2018, pp. 1–9 (2018)

10. Pennington, J., Socher, R., Manning, C.D.: Glove: global vectors for word representation. In: EMNLP, pp. 1532–1543. ACL (2014)

11. Pontiki, M., Galanis, D., Pavlopoulos, J., Papageorgiou, H., Androutsopoulos, I., Manandhar, S.: SemEval-2014 task 4: aspect based sentiment analysis. In: SemEval@COLING, pp. 27–35. The Association for Computer Linguistics (2014)

12. Tan, Z., Wang, M., Xie, J., Chen, Y., Shi, X.: Deep semantic role labeling with self-attention. In: Proceedings of the Thirty-Second AAAI Conference on Artificial Intelligence (AAAI 2018), the 30th innovative Applications of Artificial Intelligence (IAAI 2018), and the 8th AAAI Symposium on Educational Advances in Artificial Intelligence (EAAI 2018), New Orleans, Louisiana, USA, 2–7 February 2018, pp. 4929–4936 (2018)

13. Tang, D., Qin, B., Feng, X., Liu, T.: Effective LSTMS for target-dependent sentiment classification. In: COLING, pp. 3298–3307. ACL (2016)

14. Tang, D., Qin, B., Liu, T.: Aspect level sentiment classification with deep memory network. In: EMNLP, pp. 214–224. The Association for Computational Linguistics (2016)

15. Vaswani, A., et al.: Attention is all you need. In: Advances in Neural Information Processing Systems 30: Annual Conference on Neural Information Processing Systems 2017, 4–9 December 2017, Long Beach, CA, USA, pp. 5998–6008 (2017)

16. Verga, P., Strubell, E., McCallum, A.: Simultaneously self-attending to all mentions for full-abstract biological relation extraction. In: Proceedings of the 2018 Conference of the North American Chapter of the Association for Computational Linguistics: Human Language Technologies (NAACL-HLT 2018), New Orleans, Louisiana, USA, 1–6 June 2018, Volume 1 (Long Papers), pp. 872–884 (2018)
17. Wang, Y., Huang, M., Zhu, X., Zhao, L.: Attention-based LSTM for aspect-level sentiment classification. In: Proceedings of the 2016 Conference on Empirical Methods in Natural Language Processing (EMNLP 2016), Austin, Texas, USA, 1–4 November 2016, pp. 606–615 (2016)
18. Wang, Y., Huang, M., Zhu, X., Zhao, L.: Attention-based LSTM for aspect-level sentiment classification. In: EMNLP, pp. 606–615. The Association for Computational Linguistics (2016)
19. Zhao, W.X., Jiang, J., Yan, H., Li, X.: Jointly modeling aspects and opinions with a MaxEnt-LDA hybrid. In: Proceedings of the 2010 Conference on Empirical Methods in Natural Language Processing (EMNLP 2010), 9–11 October 2010, MIT Stata Center, Massachusetts, USA, A meeting of SIGDAT, a Special Interest Group of the ACL, pp. 56–65 (2010)

ArtPDGAN: Creating Artistic Pencil Drawing with Key Map Using Generative Adversarial Networks

SuChang Li[1], Kan Li[1(✉)], Ilyes Kacher[2], Yuichiro Taira[3], Bungo Yanatori[3], and Imari Sato[4]

[1] Beijing Institute of Technology, Beijing, China
likan@bit.edu.cn
[2] Qwant Research, Paris, France
[3] Tokyo University of the Arts, Tokyo, Japan
[4] National Institute of Informatics, Tokyo, Japan

Abstract. A lot of researches focus on image transfer using deep learning, especially with generative adversarial networks (GANs). However, no existing methods can produce high quality artistic pencil drawings. First, artists do not convert all the details of the photos into the drawings. Instead, artists tend to use strategies to magnify some special parts of the items and cut others down. Second, the elements in artistic drawings may not be located precisely. What's more, the lines may not relate to the features of the items strictly. To address above challenges, we propose ArtPDGAN, a novel GAN based framework that combines an image-to-image network to generate key map. And then, we use the key map as an important part of input to generate artistic pencil drawings. The key map can show the key parts of the items to guide the generator. We use a paired and unaligned artistic drawing dataset containing high-resolution photos of items and corresponding professional artistic pencil drawings to train ArtPDGAN. Results of our experiments show that the proposed framework performs excellently against existing methods in terms of similarity to artist's work and user evaluations.

Keywords: Generative adversarial networks · Deep learning · Artistic pencil drawing

1 Introduction

Pencil drawing is one of the most appreciated technique in quick sketching or finely-worked depiction. Researchers are quite interested in pencil drawings because it is a combination of observation, analysis and experience of the authors. So study pencil drawing can be help to the progress of artificial intelligence. And it usually takes several hours to finish a fine drawing (Fig. 1), even for an experienced artist with professional training, which attracts people work on the pencil drawing generation algorithms. In former methods, pencil drawing generation

© Springer Nature Switzerland AG 2020
V. V. Krzhizhanovskaya et al. (Eds.): ICCS 2020, LNCS 12143, pp. 285–298, 2020.
https://doi.org/10.1007/978-3-030-50436-6_21

was split into two components, the structure map that define region boundaries, and the tone map that reflects differences in the amount of light falling on a region as well as its intensity or tone and even texture [20]. However, we learned from artists that artistic pencil drawing should be able to capture the characteristics of the items and emphasize them. We give the images which key part have been labeled a name called key map. The key maps labeled by artists are also shown in Fig. 1.

Designing an algorithm or a framework which can study from artistic drawings and automatically transform an input photo into high-quality artistic drawings is highly desired. It can be used in many areas such as animation and advertisement. In particular, the development of deep learning which uses networks to perform image style transfer was also proposed [5]. Recently, generative adversarial network (GAN) [8] based style transfer methods (e.g. [1,2,11,30]) with datasets of (paired or unpaired) photos and stylized images have achieved abundant good results.

Based the knowledge of artists, generating artistic pencil drawings are quite different with pencil styles studied in previous work [17,20]. The differences can be summarized into three aspects. First, the artists will not convert all the details of the photos directly into their drawings, they will find the most important regions to magnify and simplify other parts at the same time. Second, artists will not locate the elements in pencil drawings precisely, which makes it a challenge for the methods based on similarity or correspondence (e.g. Pix2Pix [11]). Finally, artists put lines in pencil drawings that are not directly related to the basic vision features in the view or photograph of the items. Therefore, even state-of-the-art image style transfer algorithms (e.g. [5,11,16,18,22,30]) often fail to produce vivid and realistic artistic pencil drawings.

To address the above challenges, we propose ArtPDGAN, a novel GAN based architecture which combines with an image-to-image network for transforming photos to high-quality artistic pencil drawings. ArtPDGAN can generate key maps for the original photos and use the key maps to synthesis the artistic pencil drawings. To learn key region for different object shapes effectively, our architecture involves a specialized image-to-image network to capture key map.

The main contributions of this work are summarized as follows:

- We propose a GAN based framework for artistic pencil drawing generation, which combines with a specialized image-to-image network to generate high-quality and expressive artistic drawings from real photos.
- In order to imitate artists better, we also propose a key map dedicated to the artists' emphasizing parts. This make our model more imitation of the artist than previous works. To our knowledge, it is the first one to apply the key map in artistic style transfer, which is an idea based on the knowledge of artists.
- The experiments demonstrate the ability of our model to synthesize artistic pencil drawings which are more close to the artists' drawings than the state-of-the-art methods.

Fig. 1. Examples of artistic pencil drawing and key map

2 Related Work

Pencil drawing generation has been widely studied in sketch extraction and deep learning style transfer. In this section, we will summarize related work in these two aspects respectively.

2.1 Sketch Extraction

Traditional edge extraction methods like [3,7,14] usually deal with the edge extraction problem with fuzzy mathematics as well as other algebraic algorithms. However, the edges are not as natural as human-made ones even though they are easy to calculate because of their discontinuity. Works such as [26,29] are also based on neural network. However, their results are still quite different with artistic drawings.

2.2 Style Transfer Using Neural Networks

Large numbers of approaches have been proposed to learn style transfer from examples, because it is too hard to describe the styles semantically. The Image Analogy approach [9] requires the input and the output are strictly aligned because it is designed for pairs training. Liao et al. [18] proposed a method called Deep Image Analogy which also requires the example images and target image to have similar content, if they are not aligned with each other. By finding semantically meaningful correspondences between two input images, they first compute correspondences between feature maps extracted by a network, and then finish visual attribute transfer. Deep Image Analogy works well on photo-to-style transfer problem, but when being applied to our artistic pencil drawing style, the subjects in the generated images look fuzzy due to the light texture. Due to the difficulty of gathering aligned and paired data, advanced neural style transfer methods (e.g. [11,25,31]) can be hardly used in common style transfer applications.

There are also methods (e.g. [10,13,15,19,30]) which can learn mappings from unpaired data. Most of them are designed with cycle-consistency theorem. However, these methods do not work well on capturing the tone texture.

Gatys et al. [5] is a milestone for neural style transfer research. They use the Gram matrices [4] in a VGG network to capture the content and style representations of images. The stylization process is achieved by minimizing the distances between the style features of the content image and the style image. This method works well on oil painting style transfer and get good results for famous artists' styles like Van Gogh. But it is not suitable for pencil drawing style transfer because it takes style as a kind of texture while style in pencil drawing is too little to be easily captured. The perceptual loss which based on high-level features was proposed by Johnson et al. [12], and became one of the best loss in image style transfer problem. For the same reason as [6], their texture-based loss function is not suitable for our style. In addition to aforementioned limitations for artistic pencil drawing style transfer, most existing methods require the style image to be close to the content image. And they do not apply the key map to improve the generated results.

Our models are different with the works above. We use an image-to-image network to generate key map, then combine it with structure map and tone map as the input to our generator. Our ArtPDGAN is able to use the features from these three kinds of maps and generate artistic pencil drawings.

3   Model

Our approach is based on the knowledge learned from artists that pencil drawings can be separated into three main components: structure map, tone map and key map. Each of them delineate different parts of the pencil drawings. The structure map is used to detect object boundaries as well as other boundaries in the scene. The cross hatching and other tonal technique for lighting, texture, and materials should be shown in tone map. The key map which our method replied on can help to find the key parts of the items. Our method combines all the maps above to do the style transformation.

With the dataset provided by artists, our model is trained to learn to translate from input maps to artistic pencil drawings. Our training process is based on the idea of paired image-to-image translation frameworks, though our model does not need the aligned data. Since the artists will abstract and transform the photos, the aligned training data are not helpful to artistic characteristics. We use two GAN-based models in our framework, one is used for generating the artistic pencil drawings and the other is for the key map generation. These two models are trained together because we want the whole framework can learn how to imitate the artists. Figure 2 shows the main modules of the proposed framework. We generate the training maps by using abstraction filters and functions on pencil drawings to estimates edges and tones. These filters can produce similar abstract maps from pencil drawings as from real photographs. Hence, at test time, we use the same abstraction filters on the input photograph, to produce input maps which are in the same domain as the training inputs.

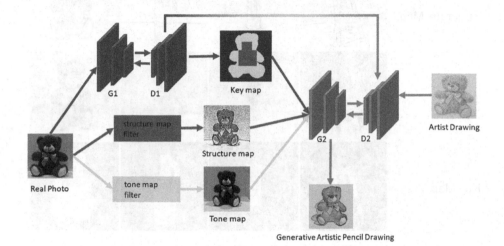

Fig. 2. The structure of the ArtPDGAN

3.1   Structure Map and Tone Map

For generating structure map, we use the adaptive filter [28], which can mark material edges of images, even the highly-textured ones. To extract the tone map, we apply a simple mapping on the pixels of photos to generate a smoothing output as the tone extraction. The mapping can be formulated as

$$Tone\ map = \lfloor original\ image/9 \rfloor \tag{1}$$

This formula is based on the experience of the artists. The artists only use a few channels of tone compared with real photos. So we directly compress the values of the pixels. Examples of structure map and tone map are shown in Fig. 3. The tone maps are actually invisible and we managed to show them as images.

3.2   Key Map

The goal of our key map is to help the generator network learn the artistic characteristics. Our model mainly based on the fact that artists will use their experience to find the key parts and emphasize them. Our model can learn from

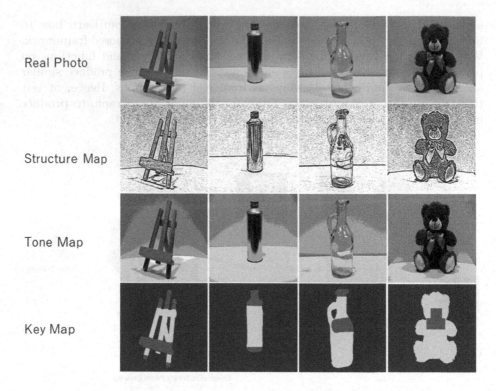

Fig. 3. Examples of structure map, tone map and key map

the labeled key map and find the key parts by itself as the artists do. Examples of key map used by our model are shown in Fig. 3. The key map network aims at extracting key regions from the photos. The important parts in the key maps are marked in red. As it's impossible to learn the key map for every photo at test time, we apply a specialized image-to-image network based GAN model to preprocess inputs and generate the key map at test time. Our training key map pair is labeled by artist. And the 300 training pairs contains almost all common shapes. The image-to-image network contains the Generator G_1 and the Discriminator D_1, and use the loss as [11]

$$L = arg \min_{G_1} \max_{D_1} L_{cGAN}(G_1, D_1) + \lambda L_{L1}(G_1) \tag{2}$$

where

$$L_{cGAN} = E_{x,y}[log\, D_1(x,y)] + E_{x,z}[log(1 - D_1(G_1(x,z)))] \tag{3}$$

During the training stage, the model G_1 and D_1 will work together to learn how to find the key map by using aligned and paired data created from key map labeled by artists and real photos. Then the whole framework learns to apply the emphasizing according to the key maps. At test-time, the model will generate the key map using real photos only.

3.3   Pencil Drawing Network

The G_2 and D_2 are the Generator and Discriminator of pencil drawing network (Fig. 2). We first use the key map network to generate the key maps. Then we encode key map and the real photos as feature maps and concatenated with the features of the structure map and the tone map. The pencil drawing network will use the features to generate artistic pencil drawings. Due to the lines of artistic pencil drawings are not related to the real photo precisely, there always some artifacts around the lines and make the results look fuzzy. To address this issue, we apply feature matching based on the idea of [21]. We extract the features from the Discriminator D_2 and regulate them to match the features from the Discriminator D_1 of the key map network.

3.4   Loss Function

As tried in the previous works [17], loss functions do not perform well alone, our loss function is combined by three loss functions. However, we do not use the classical pixel-based reconstruction loss L_{rec}. The reason is described above, as the artists will not directly change every detail of the real photos into drawings, the L_{rec} will lead the model to a less artistic results. Our loss function can be described as

$$L_{all} = \alpha * L_{adv} + \beta * L_{fea} + \gamma * L_{per} \tag{4}$$

Adversarial Loss. As the traditional conditional GAN, we use the discriminator network D_2 to discriminate the real samples from the pencil drawings and generated results. And the goal of the generator G_2 works on opposite, trying to generate images which cannot be judged from the real ones by the discriminator D_2. This can be achieved by using an adversarial loss:

$$L_{adv} = \min_{G_2} \max_{D_2} P_Y [log D_2(y)] + P_X[log(1 - D_2(G_2(x)))] \tag{5}$$

In which P_Y and P_X represent the distributions of pencil drawing samples y and their generated samples x.

Feature Match Loss. To match features extract from the Discriminator of image-to-image network D_1 and the Discriminator of pencil drawing network D_2, we use formula introduced by [21].

$$L_{fea} = \|f(D_1(x)) - f(D_2(x))\|_2^2 \tag{6}$$

In which $f(x)$ denote activations on an intermediate layer of the discriminator.

Perceptual Loss. The perceptual loss [12] was performed well in minimizing the feature differences. It makes the results sharper than traditional reconstruction loss L_{rec}. So we also apply it to help the generated samples look more close to the artistic pencil drawing.

$$L_{per} = \sum_{i=1}^{4} \|\Phi_i(G_2(x)) - \Phi_i(y)\|_2^2 \tag{7}$$

where x, y are the input and the pencil drawing from artists, G is the translation model, and Φ stands for the VGG-19 [23] network up to the $ReLU\_i\_1$ layer.

4 Experiments

We implemented ArtPDGAN using PyTorch [24] and execute experiments on a computer with an NVIDIA Titan X GPU. Our model only need 200 epochs' training, and the training time is about 6 h. The generator G takes color photos as input and output the gray drawings whose size is 512 * 512. So the numbers of input and output channels are 3 and 1, respectively. In all our experiments, the parameters in Eq. 4 are fixed at $\alpha = 1.0$, $\beta = 0.5$, $\gamma = 0.5$. In order to guarantee the fairness, all the evaluation results showed in this section are based on the generated results of test data, and all the images are resized to 256 * 256.

The generated results are shown in the Fig. 4. We use 100 images in user study 1 and 80 images in user study 2. We finally collect the feedback from 108 users of totally 9920 scores.

4.1 Ablation Study in ArtPDGAN

We perform an ablation study on our unique factor, the key map. As it shown in Fig. 5, the user study between ArtPDGAN and ArtPDGAN without key map show that the key map is critical to our ArtPDGAN and help to produce high-quality results of artistic pencil drawings.

4.2 Comparison with State-of-the-art

We compare ArtPDGAN with three state-of-the-art style transfer methods: Gatys, CycleGAN and Im2Pencil.

Gatys, Pix2Pix and CycleGAN are classical and famous style transfer models of different types of data. Im2Pencil is accepted by CVPR 2019, which stands for the latest approach. Our dataset is made up of paired and unaligned data, which do not satisfy the conditions of the Pix2Pix model. So we choose CycleGAN as the comparative method because it can be applied in the unpaired and unaligned dataset. Qualitative results of comparison with Gatys, CycleGAN and Im2Pencil are shown in Fig. 5, Fig. 6, Fig. 7, Fig. 8 and Table 1. Figure 8 shows the results of one of the user studies, which rates the similarity between the results of different algorithms and the artists' works. In the user study, we divided the users into two groups, an inexperienced user group and an experienced user group, based on their drawing experience. The reason is that users with different drawing experience may have different focuses while looking at images. Experienced users my easily to find out the key map like the artist, while inexperienced users may pay more attention on realistic details of the object in the drawings.

Gatys' method takes one content image and one style image as input by default, so we use the exact drawing in the training set as the style and content image to model the target style for a fair comparison. Im2pencil provides many

Real
Photo   Gatys   IM2pencil   CycleGAN   ArtPDGAN   Artist
Drawing

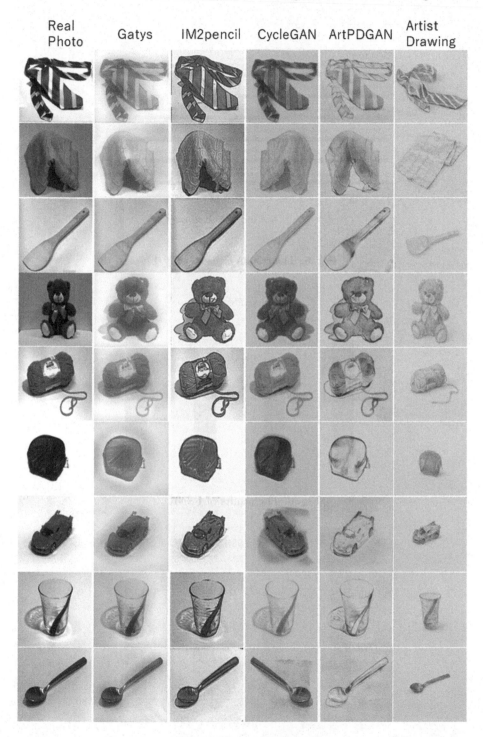

Fig. 4. The generated results

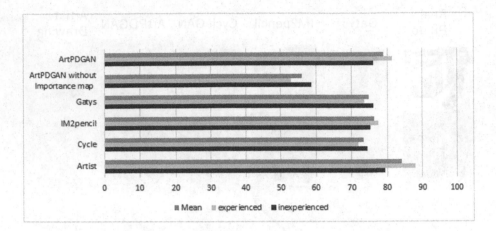

Fig. 5. The results of user study 1

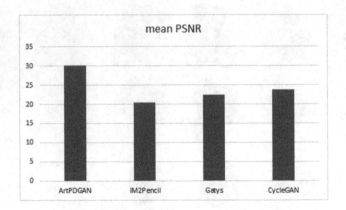

Fig. 6. The results of PSNR

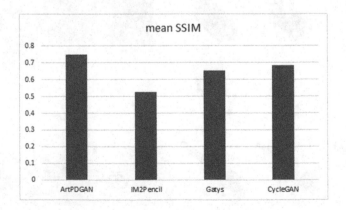

Fig. 7. The results of SSIM

Table 1. Experiment results

Model	Mean PSNR	Max PSNR	Mean SSIM	Max SSIM
ArtPDGAN	30.0945	34.5141	0.7502	0.8368
Im2Pencil	18.5237	24.1215	0.5256	0.7295
Gatys	22.3423	26.6630	0.6521	0.7226
CycleGAN	23.7886	31.7173	0.6838	0.7801

styles to choose from, we just use the fundamental style in all the experiment. As Fig. 5 shows, Gatys' method generates good results for artistic pencil stylization. The CycleGAN gets good results among PSNR, SSIM and user study, while the Im2pencil only gets very good manual score. And our ArtPDGAN is always the best one among these methods.

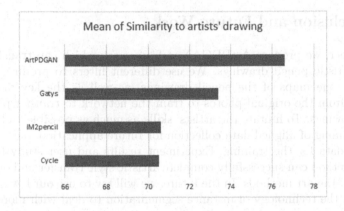

Fig. 8. The results of user study 2

We can find that Gatys's results lose some features and have two or more kinds of styles of regions. And the results contain many artifacts. The reason behind is that the method regards style as texture information in the Gram matrix, which cannot capture our artistic pencil style which only has little texture. And the artistic deformation caused the imprecise content loss. CycleGAN also cannot imitate the artistic drawings well. As shown in Fig. 8, CycleGAN's results do not look like an artist's drawing. CycleGAN is unable to preserve important features because it uses the cycle-consistency to constrain network. The cycle-consistency which uses only unsupervised information, is less accurate than a supervised method and leads to problems that it's hard for the loss to accurately recover all the details of domains. Im2pencil generates results that preserve some aspect of artistic drawings, but they also have many artifacts. The structural lines, making the stylized result unlike the artistic drawings, and the xDOG filter [27] makes the tone maps look too close to black-and-white photos.

The main reason of these problems is that they did not use the real pictures drawing by artists.

In comparison, our method captures the key regions accurately and generates high-quality results with artist's drawing style. Moreover, our results are very close to the drawings drawn by the artists than other methods. For quantitative evaluation, we compare our ArtPDGAN with artists' works, Gatys, CycleGAN and Im2Pencil using the user study, which are a widely used in GAN evaluation. We also measure the similarity between generated artistic pencil drawings and real artistic pencil drawings use the PSNR, SSIM and the user study at the same time. The comparison results are presented in Table 1. The scores show that our method has a higher PSNR value and SSIM value, which means that our drawings is closer to the artistic pencil drawings than Im2Pencil, CycleGAN and Gtays. The same as the results of user investigation. In other words, these results indicate ArtPDGAN captures better artistic pencil drawings distribution than comparative methods.

5   Conclusion and Future Work

In this paper, we propose ArtPDGAN, a framework which can transform photos into artistic pencil drawings. We use different filters to produce structure maps and tone maps of the pencil drawings, as well as the key maps which generated from the original photos to train the network to transfer photos into artistic drawings. To imitate the artists' skills as much as possible and avoid the time consuming of aligned data collection for future application, our model uses unaligned data for the training. Experiment results and user study both show that our method can successfully complete artistic style transfer and outperform the state-of-the-art methods. In the future, we will try to use our key maps combined with the technology of instance segmentation to deal with more complex photos.

Acknowledgment. This research was supported by Beijing Natural Science Foundation (No. L181010, 4172054), and National Basic Research Program of China (No. 2013CB329605).

References

1. Azadi, S., Fisher, M., Kim, V.G., Wang, Z., Shechtman, E., Darrell, T.: Multicontent GAN for few-shot font style transfer. arXiv Computer Vision and Pattern Recognition (2017)
2. Chen, Y., Lai, Y., Liu, Y.: CartoonGAN: generative adversarial networks for photo cartoonization, pp. 9465–9474 (2018)
3. Gao, W., Zhang, X., Yang, L., Liu, H.: An improved Sobel edge detection, vol. 5, pp. 67–71 (2010)
4. Gatys, L.A., Ecker, A.S., Bethge, M.: Texture synthesis using convolutional neural networks. arXiv Computer Vision and Pattern Recognition (2015)

5. Gatys, L.A., Ecker, A.S., Bethge, M.: Image style transfer using convolutional neural networks, pp. 2414–2423 (2016)
6. Gatys, L.A., Ecker, A.S., Bethge, M., Hertzmann, A., Shechtman, E.: Controlling perceptual factors in neural style transfer, pp. 3730–3738 (2017)
7. Gonzalez, C.I., Melin, P., Castro, J.R., Mendoza, O., Castillo, O.: An improved Sobel edge detection method based on generalized type-2 fuzzy logic. Soft. Comput. **20**(2), 773–784 (2016)
8. Goodfellow, I., et al.: Generative adversarial nets, pp. 2672–2680 (2014)
9. Hertzmann, A., Jacobs, C.E., Oliver, N., Curless, B., Salesin, D.: Image analogies, pp. 327–340 (2001)
10. Huang, X., Liu, M., Belongie, S., Kautz, J.: Multimodal unsupervised image-to-image translation. arXiv Computer Vision and Pattern Recognition (2018)
11. Isola, P., Zhu, J., Zhou, T., Efros, A.A.: Image-to-image translation with conditional adversarial networks. arXiv Computer Vision and Pattern Recognition (2016)
12. Johnson, J., Alahi, A., Feifei, L.: Perceptual losses for real-time style transfer and super-resolution. arXiv Computer Vision and Pattern Recognition (2016)
13. Kim, T., Cha, M., Kim, H., Lee, J.K., Kim, J.: Learning to discover cross-domain relations with generative adversarial networks, pp. 1857–1865 (2017)
14. Kumar, S., Saxena, R., Singh, K.: Fractional Fourier transform and fractional-order calculus-based image edge detection. Circ. Syst. Sig. Process. **36**(4), 1493–1513 (2017)
15. Lee, H., Tseng, H., Huang, J., Singh, M., Yang, M.: Diverse image-to-image translation via disentangled representations, pp. 36–52 (2018)
16. Li, C., Wand, M.: Combining Markov random fields and convolutional neural networks for image synthesis. arXiv Computer Vision and Pattern Recognition (2016)
17. Li, Y., Fang, C., Hertzmann, A., Shechtman, E., Yang, M.: Im2Pencil: controllable pencil illustration from photographs, pp. 1525–1534 (2019)
18. Liao, J., Yao, Y., Yuan, L., Hua, G., Kang, S.B.: Visual attribute transfer through deep image analogy. ACM Trans. Graph. **36**(4), 120 (2017)
19. Liu, M., Breuel, T.M., Kautz, J.: Unsupervised image-to-image translation networks. arXiv Computer Vision and Pattern Recognition (2017)
20. Lu, C., Xu, L., Jia, J.: Combining sketch and tone for pencil drawing production, pp. 65–73 (2012)
21. Salimans, T., Goodfellow, I., Zaremba, W., Cheung, V., Radford, A., Chen, X.: Improved techniques for training GANs. arXiv Learning (2016)
22. Shih, Y., Paris, S., Barnes, C., Freeman, W.T., Durand, F.: Style transfer for headshot portraits. In: International Conference on Computer Graphics and Interactive Techniques, vol. 33, no. 4, p. 148 (2014)
23. Simonyan, K., Zisserman, A.: Very deep convolutional networks for large-scale image recognition. arXiv Computer Vision and Pattern Recognition (2014)
24. Steiner, B., et al.: PyTorch: an imperative style, high-performance deep learning library (2019)
25. Wang, T., Liu, M., Zhu, J., Tao, A., Kautz, J., Catanzaro, B.: High-resolution image synthesis and semantic manipulation with conditional GANs, pp. 8798–8807 (2018)
26. Wei, Y., Wang, Z., Xu, M.: Road structure refined cnn for road extraction in aerial image. IEEE Geosci. Remote Sens. Lett. **14**(5), 709–713 (2017)
27. Winnemoller, H.: XDoG: advanced image stylization with extended difference-of-Gaussians, pp. 147–156 (2011)

28. Wittenmark, B.: Adaptive filter theory. Automatica **29**(2), 567–568 (1993)
29. Xie, S., Tu, Z.: Holistically-nested edge detection. arXiv Computer Vision and Pattern Recognition (2015)
30. Zhu, J., Park, T., Isola, P., Efros, A.A.: Unpaired image-to-image translation using cycle-consistent adversarial networks, pp. 2242–2251 (2017)
31. Zhu, J., et al.: Toward multimodal image-to-image translation. arXiv Computer Vision and Pattern Recognition (2017)

Interactive Travel Aid for the Visually Impaired: from Depth Maps to Sonic Patterns and Verbal Messages

Piotr Skulimowski(✉) and Pawel Strumillo

Institute of Electronics, Lodz University of Technology, Lodz, Poland
piotr.skulimowski@p.lodz.pl

Abstract. This paper presents user trials of a prototype micro-navigation aid for the visually impaired. The main advantage of the system is its small form factor. The device consists of a Structure Sensor depth camera, a smartphone, a remote controller and a pair of headphones. An original feature of the system is its interactivity. The user can activate different space scanning modes and different sound presentation schemes for 3D scenes on demand. The results of the trials are documented by timeline logs recording the activation of different interactive modes. The aim of the first trial was to test system capability for aiding the visually impaired to avoid obstacles. The second tested system efficiency at detecting open spaces. The two visually impaired testers performed the trials successfully, although the times required to complete the tasks seem rather long. Nevertheless, the trials show the potential usefulness of the system as a navigational aid and have enabled us to introduce numerous improvements to the tested prototype.

Keywords: Assistive technology · Electronics travel aids · Visual impairment · Depth maps · Scene segmentation · Sonification

1 Introduction

Building an electronic travel aid (ETA) for the visually impaired, whether dedicated to micro- or macro-navigation tasks, has proven itself to be a difficult interdisciplinary challenge. Research efforts to build an out-of-laboratory ETA device date back to 1889, when the first attempts were undertaken by a Polish scientist named Kazimierz Noiszewski. Noiszewski built the "Electroftalm", a device that converted visual signals to auditory stimulation and utilized the photoelectric properties of Selenium cells [1]. However, the device was heavy and inconvenient to use, so did not find any practical applications. Today, there are small form factor devices available that enable reconstruction of the 3D structure of the environment, such as structural light and stereovision. Such techniques have been applied in practice, e.g. in prototype electronic travel aids (ETAs) assisting the blind in mobility [2]. However, analysis of the operation of ETA solutions, and the fact that they have not been accepted or widely used by blind people, points to the need to develop more efficient, user-centered methods of presenting information about the three-dimensional environment to the blind.

© Springer Nature Switzerland AG 2020
V. V. Krzhizhanovskaya et al. (Eds.): ICCS 2020, LNCS 12143, pp. 299–311, 2020.
https://doi.org/10.1007/978-3-030-50436-6_22

Major barriers faced by blind people are spatial orientation and independent mobility. This is because the visually impaired cannot built accurate and dynamic internal cognitive models of the environment using their non-visual senses. The primary navigational aid for the blind is a white cane, which increases the range of haptic perception of the environment but only to a range of about 1.5 m. The authors of a very recent study [3] conducted a state of the art survey of wearable assistive devices for the blind and visually impaired. They investigated more than 70 different solutions, of which 29 used scene depth reconstructed from time of flight (ToF) cameras or stereovision. Based on this review, it can be concluded that the existing solutions do not enable the visually impaired to use interactive techniques to explore the environment.

In this work, we describe novel non-visual techniques which engage the senses of touch and hearing to give the visually impaired an impression of their environment. The novelty of the proposed solution is the interactive mode of audio-haptic presentation, whereby the user can flexibly control the amount of incoming information. The blind user can acquire a capacity for selective attention, similar to that for the sense of vision. This work is a continuation of our earlier studies, reported in [4], where we proposed a scheme for sonification of U-depth maps (i.e. histograms of the consecutive columns of the depth map) of the environment. The sound was generated as a sum of sinusoidal sounds, each one dedicated to the predefined depth range. The user interactively selected the depth map region for sonification by using touch gestures on the mobile device screen.

2   Related Work

Recently, there have been many scientific efforts aimed at building assistive devices for the visually impaired. Comprehensive reviews of research into technological aids for the visually impaired are given in [3, 5]. Electronic travel aids are devices or systems that aid mobility and navigation for the visually impaired [6]. These personal systems convert data about the environment into auditory or haptic stimulation [7]. Such use of healthy sensory modality to substitute a lost sense is termed sensory substitution [8]. Some approaches attempt to use sound to encode spatial data about the environment [9]. Others use haptic or tactile displays to present the geometric layout of the environment [10]. Text-to-speech (TTS) technologies have also been developed, which enable the visually impaired to access the written word. Notable examples are the optophone and Kurtzweil's "reading machine" [11]. Simpler technologies convert visual images into haptic or auditory modalities. Audio description is a method of explaining graphics or images in the form of verbal narration, commenting on the visual information. However, this requires human participation in interpreting the imaged scene content. Other solutions are based on automatic conversion of visual data into auditory signals [6] or haptic stimulation [12].

Unfortunately, few of the technical solutions to aid the visually impaired present non-visual information about the environment in an interactive way [13]. By interactivity, we mean functionality that would enable the blind user to take over control of the incoming stream of non-visual information presented by the device or computer application about the environment. Only a few interactive assistive approaches for the visually impaired are reported in literature. For example, in [14] line elements of images were sonified, while in [15] Nintendo's Wiimote controller was utilized to provide the user with combined sonic and haptic feedback. The image was presegmented and the detected regions were assigned descriptors that were further mapped to sounds. However, the mapping method was not explained. A multilevel sonification approach has been proposed in [16], whereby low-level image features such as edges, colors, and textures are sonified in parallel, and machine learning algorithms are applied to recognize and verbally describe objects to the user. Finally, a system was reported in [17], which integrates a geographic information system of a building with computer vision for navigating a blind individual. Its advantage is that only one camera, e.g. smartphone camera can be used without any need for any additional hardware.

3 Materials and Methods

3.1 System Hardware Platform

Our ETA device consists of a depth camera connected to a smartphone with Android OS, a pair of headphones and a remote controller. Using the remote controller, the blind user can set non-visual presentation parameters and the operating mode of the device. Figure 1 shows the system mounted on a mannequin head.

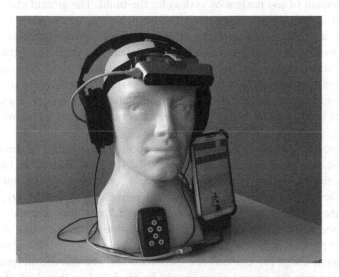

Fig. 1. Components of the ETA system, consisting of a depth camera, an Android OS smartphone, a pair of headphones and a remote controller.

The depth camera is a Structure Sensor (SS). It is a lightweight device (65 g) that uses an infrared structured light pattern projector and a low-range infrared CMOS camera to reconstruct 3D scene geometry, with a range of depth reconstruction of 0.4–5 m.

The SS generates depth images with a rate of 30 fps with 640 × 480 pixel spatial resolution (or 60 fps and 320 × 240 pixels), both with depth relative accuracy of less 1%. We selected the 60 fps mode. The horizontal field of view (FoV) of the SS is 58° and the vertical FoV is 45°. The SS device is mounted on the head using rubber bands (see Fig. 1). A special rig is used to set various camera tilt angles. The chosen camera does not work outdoor in a direct sunlight, but it can be replaced in the future with a new class of depth sensors equipped with stereo infrared cameras and laser projectors. Although, wearing a headset in not very comfortable, it allows to scan the scene by moving user's head. The Android OS smartphone is a Samsung Note 3 equipped with a Qualcomm Snapdragon 800 processor. The image processing software is written in Java and uses OpenNI2 drivers. The remote controller is a small form factor device to be held in the user's hand, with a set of convex keys that "click" when pressed. The controller communicates with the smartphone via a Bluetooth link.

3.2 Image Processing

Our approach to designing an ETA is that the blind person should construct a higher level understanding of the environment from low-level features such as edges, fragments of planes or open space devoid of obstacles. This approach is strongly supported by earlier studies on visual rehabilitation and by in-depth understanding of the neural mechanisms behind nonvisual multisensory presentation of space [8, 18].

Reliable estimation of the position of the ground plane in a scene is a very important element of any navigation system for the blind. The ground plane should not be identified as an obstacle and should not be presented to the user. However, the accuracy of its detection determines other important system capabilities, such as identification of small but important obstacles or defects in the ground. Many automatic methods of detecting ground planes have been proposed [19, 20], especially for vehicle-related applications. These solutions, however, fail in the case of scenes with complicated structures, such as where the ground plane is visible only in a small part of the image, or in imaging systems where 6DoF (Degrees of Freedom) camera motion is enabled.

In the research described in this work, we applied the method of determining the ground plane described in our previous study [4]. This method is based on the assumptions that the system is placed on the user's forehead and the distance from the camera to the ground is within the presumed range. Knowing the orientation and location of the ground plane versus the camera optical axis, it is possible to determine the area of the plane in the image even if it is visible in only a small region of the image. In the present study, we limit the search area of the ground plane to the bottom part of the depth image [4]. Our earlier experiments showed that the tilt (pitch angle) of the camera towards the ground plane should be slightly less than half of the vertical field of view of the camera, which is 22°. This makes it possible to observe both the ground plane and the objects located slightly above the user's head. With these assumptions, the distance to the obstacle should be defined in a new coordinate system, which is related to the ground plane camera O' rather than to the camera coordinate system O (see Fig. 2).

Let the ground plane equation $Ax + By + Cz + D = 0$ and the origin of the coordinate system associated with the ground O' be the point that is the projection of the origin of the coordinate system associated with the camera O on the ground plane. The new origin of the O' coordinate system in the O coordinate system has coordinates $[x_F, y_F, z_F]$, which are determined by assuming that the origin of the camera coordinate system is the closest point to the ground plane.

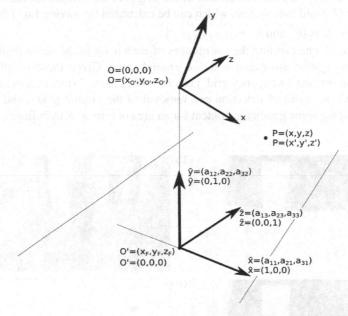

Fig. 2. Geometric configuration of the camera coordinate system (O) and the ground related coordinate system (O'). Coordinate systems O, O', point coordinates P, and the basis vectors are defined in the camera coordinate system (top line) and in the ground related coordinate system (bottom line), respectively.

The basis unit vectors of the O' coordinate system in the camera coordinate system O are $\hat{x} = (a_{11}, a_{21}, a_{31})$, $\hat{y} = (a_{12}, a_{22}, a_{32})$, $\hat{z} = (a_{13}, a_{23}, a_{33})$. The basis vector $\hat{y}$ has the same direction as the normal vector to the ground plane $[A, B, C]$. The basis vector $\hat{z} = [x_F, y_F, z_F]$ is oriented along the line for which the scene depth is defined. Finally, $\hat{x}$ can be computed from the outer product $\hat{x} = \hat{z} \times \hat{y}$. It can be proven [21] that any point (x', y', z') defined in ground-related coordinate system O' has the following (x, y, z) coordinates in the camera coordinate system O:

$$
\begin{bmatrix} x \\ y \\ z \end{bmatrix} = \begin{bmatrix} x_F \\ y_F \\ z_F \end{bmatrix} + \begin{bmatrix} a_{11} & a_{12} & a_{13} \\ a_{21} & a_{22} & a_{23} \\ a_{31} & a_{32} & a_{33} \end{bmatrix} \begin{bmatrix} x' \\ y' \\ z' \end{bmatrix} \tag{1}
$$

Similarly, an arbitrary point (x, y, z) in the camera coordinate system O has (x', y', z') coordinates in the new O' ground-related coordinate system:

$$\begin{bmatrix} x' \\ y' \\ z' \end{bmatrix} = \begin{bmatrix} x_{O'} \\ y_{O'} \\ z_{O'} \end{bmatrix} + \begin{bmatrix} a_{11} & a_{21} & a_{31} \\ a_{12} & a_{22} & a_{32} \\ a_{13} & a_{23} & a_{33} \end{bmatrix} \begin{bmatrix} x \\ y \\ z \end{bmatrix} \tag{2}$$

where $[x_{O'}, y_{O'}, z_{O'}]^T$ are the coordinates of the origin of the camera coordinate system given in the O' coordinate system, which can be calculated by solving Eq. 1 for the pair of points $x = [0, 0, 0]^T$ and $x' = [x_{O'}, y_{O'}, z_{O'}]^T$.

Using Eq. 2, one can find the coordinates of each point in the scene defined in the O' coordinate system associated with the ground plane. Given these coordinates, an obstacle map (termed "occupancy grid") can be built. Figure 3 shows an example of an indoor scene, the result of detection and removal of the ground plane, and the occupancy grid of the scene geometric content for an area of 6 m × 5 m in front of the user.

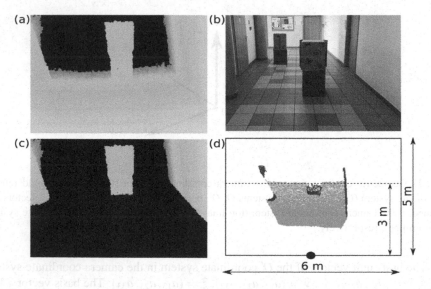

Fig. 3. Depth map captured by the camera shown in pseudo-colors (a), the corresponding indoor 3D scene image (b), the depth map with ground plane removed (c), and the occupancy grid (top view) generated for the depth map (d) (the green region represents the detected ground plane, brown regions represent detected obstacles, and the black dot denotes the position of the user in the scene). (Color figure online)

3.3 Presentation Methods

Two obstacle presentation methods are proposed: one based on parking sensors, the other on horizontal space scanning in the entire angular range of the SS depth camera (see Fig. 4). In the first method, which is the default mode, obstacles are detected in front of the system user in two adjacent cuboids, each measuring the selected depth range, the user's height (estimated automatically by the system), and half the width of the user's shoulders (Fig. 4a). The user is informed by audio signals resembling a car parking sensor the nearest obstacle(s) in each cuboid. Sounds with a constant frequency of 457 Hz are generated for 30 ms. The interval between successive sounds dT in (milliseconds) depends on the distance to an obstacle according to the heuristic formula:

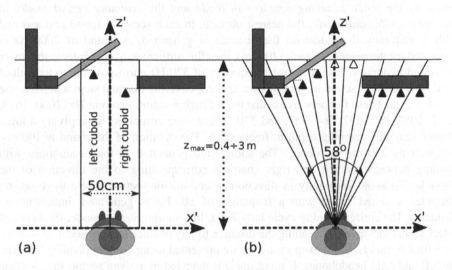

Fig. 4. Comparison of the proposed sound modes for presenting obstacles: a) in the *default mode* obstacles are detected in front of the user only, b) the *scanning mode* presents the obstacle by horizontal scanning of the of the field of view. The black and white small triangles indicate depth values communicated to the user. Black triangles indicate sectors in which scene objects are within system depth range, whereas white triangles denote sectors in which scene objects are further than 3 m away from the camera.

$$dT = 0.2d + \frac{d^2}{40000} \qquad (3)$$

where d is given in millimeters. The sound volume depends on the distance: the closer the obstacle, the louder the sound. In the absence of obstacles a low, quiet and cyclic sound with a frequency of 300 Hz is generated, confirming that the system is operating correctly.

If the user is not moving (such inactivity is detected by analyzing signals form the accelerometers integrated into the mobile device carried by the user) then the "verbal mode" of the system is automatically activated. In this mode, the system informs the user about the nearest distance (in meters) to any obstacle by short verbal commands, e.g. "zero three" means 0.3 m, "one two" means 1.2 m. There is a distinction between distances from the obstacles in each cuboid: if the difference between the distances is less than 10 cm then the smallest distance value is read centrally in the headphones. If the difference is larger than 10 cm, then two values are read sequentially, for the left and right cuboid separately. The voice commands are panned slightly left or right for the corresponding cuboid. To avoid a cluttered sequence of verbal cues, the next distance is communicated to the user after not less than 2 s.

The second mode of system operation, *scanning mode*, is activated on demand by pressing the appropriate button on the remote control. In this presentation mode, the entire space in the depth camera field of view is divided into 9 vertical sectors, each with an angular range of approx. 6.5°. The scanning mode features two options (selected by the user): *scanning sonification mode* and the *scanning verbal mode*. In *scanning sonification mode*, the nearest obstacle in each sector is found and a sound with a frequency dependent on the distance is generated. A sound of 2000 Hz is generated for the nearest obstacles (0.4 m), then the further away the obstacle the lower the sound frequency, with the lowest frequency of 370 Hz corresponding to a depth of 3 m. Nine different sound frequencies are applied, covering in equal steps a depth range of 0.4 ÷ 3 m. These frequencies assume the following values given in Hz: 2000, 1619, 1312, 1062, 860, 697, 564, 457, and 370. These were computed by applying a logarithmic scale for mapping depth to frequencies. The duration of the sound is 100 ms, followed by 150 ms of silence. The sounds are generated in the headphones with panning between the left and right channels corresponding to the direction of the obstacle. To avoid ambiguity in direction detection, for sectors where there are no obstacles a muted sound with a frequency of 300 Hz is generated, indicating no obstacles. The entire scanning cycle lasts 2.4 s. In *scanning verbal mode*, the user can select verbal messages for reading the distance to any obstacles in consecutive sectors. As with the sounds, the spoken messages are presented using volume panning between the left and right headphones. If no obstacle is detected in a given sector, the message "empty" is spoken.

4   Mobility Tests of the System and Discussion

Mobility tests of a prototype version of the system were conducted with two blind people who were also asked for their opinions and ideas for improvements. The tests were approved by the Bioethics Commission at the Medical University of Lodz, Poland. Prior to the mobility trails, the users were acquainted with the system features, given time to test the system, and asked to share their impression of its usability. Table 1 summarizes user comments and steps taken to improve system functionality.

The mobility tests of the system were carried out with participation of two visually impaired individuals, a woman aged 34 (Tester 1) and a man aged 37 (Tester 2). Tester 1 belongs to visual impairment category 4 as defined by the World Health Organization [22], whereas Tester 2 belongs to category 3. Tester 1 is aided by a guide dog and Tester 2 uses a white cane as a primary mobility aid. During the trials, the testers did not use their primary travel aids.

Two indoor mobility testing scenarios were carried out:

1) a walk down a corridor with cardboard boxes positioned at random locations to evaluate the ability of the system to detect obstacles and aid the user to maintain the direction of travel along the corridor,
2) a walk down an empty corridor and finding a door opening, to evaluate the ability of the system to locate and present off-path empty spaces.

Table 1. Main user comments addressing system usability features and modifications of the system in response to user comments

No.	User comment	System modification
1	Increased speed of response to changes in obstacle location	In the *default mode* the system response to changing position of obstacle(s) was increased to a rate close to 60 fps
2	Frequencies used in the *scanning sonification mode* are too high and considered annoying	Frequencies reduced to the range of 370 Hz ÷ 2000 Hz
3	100% left/right panning of the sound in the *default mode* is misleading because the perceived obstacle(s) were well off the collision track	Sound panning was shifted to a more central location to position the sounds within the virtual cuboids in front of the users (see Fig. 4a)
4	No user interaction for the presentation mode. Interaction control of system functions and parameters was strongly advised by the users	Interactive presentation modes added on demand, i.e. *scanning sonification mode* and *scanning verbal mode*. This involved adding a remote control
5	The system depth detection range of 5 m is too large. Distant obstacles are detected which are irrelevant to the current mobility task of the blind user	The default obstacle detection range was reduced to 3 m (approx. twice the range of a white cane)

Figure 5 shows photographs taken during mobility testing scenarios 1) and 2), which were carried out indoors on the university campus. The testers were informed that they could freely use the different operating modes of the system, should the need arise. We implemented system capability to record time-stamped logs, noting which system operating modes were activated during the testing trials. A video recorded during one of the trials containing on-line segmentation results of the scene and the generated sound messages is available at: https://youtu.be/dAYiw-OhjEw.

4.1 Trial 1: A Walk Down a Corridor with Cardboard Box Obstacles

The task of the user in this trial was to walk along a corridor approx. 3 m wide and avoid two obstacles randomly positioned on the walking path before stopping in front of a wall (see Fig. 5). The length of the path was approx. 12 m. The users kept the smartphone securely in their pocket and held the remote controller that served to activate the different operating modes of the system in their hand.

The data recorded by the smartphone application during this test is illustrated in Fig. 6. The upper panel shows the result for Tester 1 and the bottom panel the results for Tester 2. Note that Tester 1 needed T = 87 s to walk the path, whereas Tester 2 walked the same path in less than T = 70 s. Tester 1 frequently stopped to listen to voiced messages giving information on the distance to the obstacles (indicated by arrows on a timeline). Tester 1 preferred this presentation mode to the "parking sensor" mode, which is the default. Also note that Tester 1 did not switch on any of the scanning modes. Tester 2, on the other hand, used the scanning verbal mode twice and the scanning sonification mode 7 times.

Fig. 5. The two proposed testing scenarios: a walk down a corridor with cardboard boxes simulating obstacles (a), finding a door opening on an empty corridor (b).

Moreover, he stopped 12 times during trail, while Tester 1 stopped 27 times, which is one stop every 3 s (on average) while walking. These times seem long (in particular for Tester 1). However, we observed that both testers were excessively testing the capabilities of the system, rather than concentrating on achieving the task quickly (the results were obtained for one preliminary trial).

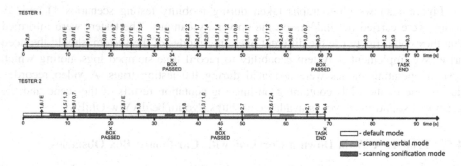

Fig. 6. Timelines showing how Testers 1 and 2 performed the task of walking along a corridor with obstacles. Vertical arrows indicate when the testers stopped and voice messages informed them of the distance to obstacle(s) in meters (E – denotes empty space). The crosses below the timelines indicate when the testers avoided the obstacles (no obstacle collisions were noted).

4.2 Trial 2: A Walk Down an Empty Corridor and Finding a Door Opening

The aim of the second test scenario was to evaluate how efficiently the system can aid the blind user in finding an open space, in this case an opened door. The task was to walk along an empty corridor and find a door opening. The users were told on which side of the corridor they should expect the door opening, but were not given any indication as to the location of the door along the corridor. The distance to the door opening from the starting point was in fact approx. 10 m.

The timelines showing how the testers performed in this trial are shown in Fig. 7. Note again that Tester 2 completed the task in a much shorter time (T = 40 s) than Tester 1 (T = 85 s). As in the first trial, Tester 1 was using mainly the default system mode and only twice switched on the scanning verbal mode. Tester 1 preferred to stop and by head turns obtain confirmation of her position relative to the corridor wall. This strategy was not very efficient, because it did not exploit the scanning modes designed to solve this type of task. Note also that Tester 1 touched the wall in spite of having been warned that she was approaching an obstacle (see the upper panel in Fig. 7). Tester 2, on the other hand, showed much greater skill in efficiently using the system. In the second part of the test, he frequently used the scanning sonification mode to search for the door opening (see the lower panel in Fig. 7) and stopped only three times while walking along the path.

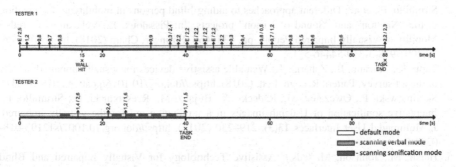

Fig. 7. Timelines showing how Testers 1 and 2 performed the task of walking along an empty corridor and locating a door opening (explanations of the symbols are given in the caption in Fig. 6).

5 Conclusions

In this study, we presented user trials of a prototype ETA system aimed at aiding the visually impaired to complete micro-navigation tasks. The system has a small form factor and lightweight. An important feature of the system is its interactivity. The user can activate different space scanning methods and sound presentation modes on demand (see Fig. 4). Both verbal messages and sonification schemes for presenting 3D scenes are available. An important property of the tested prototype is that all computations related to processing depth images (e.g. ground plane detection) and auditory presentations are performed on-line, on the smartphone platform. Initial trials of the system, although carried out with participation of only two visually impaired users, have shown the potential viability of the system as a micro-navigation aid. User feedback about the system was used to modify the sound presentation parameters and improve system functionality (see Table 1).

Comparing to our previous work [4] and added value is the way of interacting with the remote control, which is much more convenient than using the touch screen of the phone. Moreover, presenting data based on the 3D scene information instead of its U-depth representation is more detailed and more intuitive according to the comments of the trial blind participants.

Finally, we are aware that more rigorous trials are needed with a larger group of visually impaired users. In particular the testers should be given more time to attain automatism for efficient use of the system. Further research should also test how helpful the system is in combination with the users' primary travel aids (a white cane or a guide dog).

References

1. Starkiewicz, W., Kuliszewski, T.: Progress report on the elektroftalm mobility aid. In: Proceedings of the Rotterdam Mobility Research Conference, pp. 27–38. American Foundation for the Blind, New York (1965)
2. Strumillo, P., et al.: Different approaches to aiding blind persons in mobility and navigation in the "Naviton" and "Sound of Vision" projects. In: Pissaloux, E., Velázquez, R. (eds.) Mobility of Visually Impaired People, pp. 435–468. Springer, Cham (2018). https://doi.org/10.1007/978-3-319-54446-5_15
3. Tapu, R., Mocanu, B., Zaharia, T.: Wearable assistive devices for visually impaired: a state of the art survey. Pattern Recogn. Lett. (2018). https://doi.org/10.1016/j.patrec.2018.10.0311
4. Skulimowski, P., Owczarek, M., Radecki, A., Bujacz, M., Rzeszotarski, D., Strumillo, P.: Interactive sonification of U-depth images in a navigation aid for the visually impaired. J. Multimodal User Interfaces 13(3), 219–230 (2018). https://doi.org/10.1007/s12193-018-0281-3
5. Hersh, M., Johnson, M. (eds.): Assistive Technology for Visually Impaired and Blind People. Springer, London (2008). https://doi.org/10.1007/978-1-84628-867-8
6. Bujacz, M., Strumillo, P.: Sonification: review of auditory display solutions in electronic travel aids for the blind. Arch. Acoust. 41(3), 401–414 (2016)
7. Dakopoulus, D., Bourbakis, N.G.: Wearable obstacle avoidance electronic travel aids for blind: a survey. IEEE Trans. Man Cybern. Part C Appl. Rev. Syst. 40(1), 25–35 (2010)
8. Maidenbaum, S., Abboud, S., Amedi, A.: Sensory substitution: closing the gap between basic research and widespread practical visual rehabilitation. Neurosci. Biobehav. Rev. 41, 3–15 (2014)
9. Dobrucki, A., Plaskota, P., Pruchnicki, P., Pec, M., Bujacz, M., Strumillo, P.: Measurement system for personalized head-related transfer functions and its verification by virtual source localization trials with visually impaired and sighted individuals. J. Audio Eng. Soc. 58(9), 724–738 (2010)
10. Brock, A., Jouffrais, Ch.: Interactive audio-tactile maps for visually impaired people. ACM SIGACCESS Access. Comput. (ACM Digit. Libr.) 113, 3–12 (2015). Association for Computing Machinery (ACM)

11. Kurzweil, R.: The Kurzweil reading machine: a technical overview. Science, Technology and the Handicapped (1976)
12. Pissaloux, E., Maingreaud, F., Velazquez, R., Fontaine, E.: Concept of the walking cognitive assistance: experimental validation. ASME Int. J. Adv. Model. C **67**, 75–86 (2007)
13. Pissaloux, E.E., Velazquez, R. (eds.): Mobility in Visually Impaired People - Fundamentals and ICT Assistive Technologies. Springer, Cham (2018). https://doi.org/10.1007/978-3-319-54446-5
14. Ramloll, R., Yu.W., Brewster, S., Riedel, B., Burton, M., Dimihgen, R.: Constructing sonified haptic line graphs for the blind student: first steps. In Proceedings of the 4th International ACM Conference on Assistive Technologies, pp. 17–25 (2000)
15. O'Neill, Ch., Ng, K.: Hearing images: interactive sonification interface for images. In: EVA 2008 London Conference 22–24 July, pp. 188–195 (2008)
16. Banf, M., Blanz, V.: A modular computer vision sonification model for the visually impaired. In: Proceedings of International Conference of Auditory Display (2012)
17. Serrão, M., et al.: Computer vision and GIS for the navigation of blind persons in buildings. Univ. Access Inf. Soc. **14**(1), 67–80 (2015)
18. Zhang, M., Weisser, V.D., Stilla, R., Prather, S., Sathian, K.: Multisensory cortical processing of object shape and its relation to mental imagery. Cogn. Affect. Behav. Neurosci. **4**(2), 251–259 (2004)
19. Kwon, J., Dragon, R., Gool, L.V.: Joint tracking and ground plane estimation. IEEE Signal Process. Lett. **23**(11), 1514–1517 (2016)
20. Lin, Y., Guo, F., Li, S.: Road obstacle detection in stereo vision based on UV-disparity. J. Inf. Comput. Sci. **11**(4), 1137–1144 (2014)
21. Stark, M.: Analytical geometry. Mathematical Monographies 26. Mathematical Institute, Polish Academy of Sciences (1951). (in Polish)
22. WHO webpage. International classification of diseases for mortality and morbidity statistics. http://www.who.int/classifications/icd/en/. Accessed 04 Jan 2020

Ontology-Driven Edge Computing

Konstantin Ryabinin[✉][iD] and Svetlana Chuprina[iD]

Perm State University, Bukireva Street 15, 614990 Perm, Russia
kostya.ryabinin@gmail.com, chuprinas@inbox.ru

Abstract. The paper is devoted to new aspects of ontology-based approach to control the behavior of Edge Computing devices. Despite the
ontology-driven solutions are widely used to develop adaptive mechanisms to the specifics of the Internet of Things (IoT) and ubiquitous
computing ecosystems, the problem of creating withal full-fledged, easy
to handle and efficient ontology-driven Edge Computing still remains
unsolved. We propose the new approach to utilize ontology reasoning
mechanism right on the extreme resource-constrained Edge devices, not
in the Fog or Cloud. Thanks to this, on-the-fly modifying of device functions, as well as ad-hoc monitoring of intermediate data processed by
the device and interoperability within the IoT are enabled and become
more intelligent. Moreover, the smart leverage of on-demand automated
transformation of Machine-to-Machine to Human-Centric IoT becomes
possible. We demonstrate the practical usefulness of our solution by the
implementation of ontology-driven Smart Home edge device that helps
locating the lost things.

Keywords: Ontology engineering · Internet of Things · Edge
Computing · Ubiquitous computing · Human-machine interaction ·
Human-Centric IoT · Firmware generation

1 Introduction

The development and deployment of a ubiquitous computing environment faces
many challenges related to the communicability, reconfigurability and context
awareness of individual data processing nodes within this environment. There
are a lot of ways to tackle these challenges by introducing intermediate layers in
the sensor and actuator network, where certain groups of devices are controlled
by hubs. Hubs are normally quite powerful nodes capable of complex algorithms
execution including different steering and data aggregation techniques based
on Machine Learning or Semantic Web. Such nodes are often denoted as "Fog
Nodes", and the corresponding set of computation approaches is called "Fog
Computing" [11].

The reported study is supported by Ministry of Science and Higher Education of the
Russian Federation, State Assignment No. FSNF-2020-0017 (Research Project of Perm
State University, 2020–2022).

© Springer Nature Switzerland AG 2020
V. V. Krzhizhanovskaya et al. (Eds.): ICCS 2020, LNCS 12143, pp. 312–325, 2020.
https://doi.org/10.1007/978-3-030-50436-6_23

The responsibility of Fog Nodes includes tracking of the connected sensors and actuators, performing data preprocessing and aggregation, maintaining the temporal data storage and data transfer, as well as providing needed machine-to-machine (M2M) and human-machine interfaces (HMIs). To make the entire network more flexible, hubs should be aware of their usage context [21], which enables smart control over the potentially transient end-point devices. This, however, restricts the use-cases to the star-like network topologies, where hubs become mandatory links inside the data processing chains and can run into potential bottlenecks both in stability and performance.

One of the ways to improve the stability and performance within ubiquitous computing environment is based on the "Edge Computing" [5] concept, whereby the end-point devices (sensors and actuators themselves, so-called "Edge Nodes") leverage their own computing capabilities to partially offload the hubs. But traditionally the Edge Nodes, being an essential part of data processing chains, are unaware of their role in the entire system, and provide neither configuration nor monitoring interface past the hub. As a result, the flexibility, transparency and reconfigurability of each particular sensor subnet (up to the nearest hub) is still an issue: every change or access for monitoring in the sensor network involves the nearest Fog Node, that still has to enumerate and track all the Edge Nodes.

We propose to push the intelligent capabilities to the edge of the sensor/actuator network making the Edge Nodes as smart, as possible. For this, we utilize ontology engineering methods and means, which allow us to generate the firmware for the corresponding devices making their functioning be driven by the reusable formal knowledge representation model. Previously we successfully applied ontology engineering methods to automate firmware and middleware generation for the devices within the ecosystem of the Internet of Things (IoT) [14]. The current work is devoted to Edge Computing, entirely governed by ontologies. In this case, changes in the firmware of edge devices are completely avoided in the favor of changing the functioning model stored as an ontology.

Compared to the traditional firmware-based approach, ontology-driven Edge Computing ensures the semantic protocols, which allow on-the-fly reconfiguration of edge devices, inspecting their roles and functioning patterns at runtime (by means of self-documentation capabilities), transparently monitoring all the stages of their work (including the monitoring of the intermediate data and their transformations, should it be required). Moreover, if required, ontologies enable the flexible interconnection within the sensor network and the autonomous functioning of intelligent Edge nodes, when no network facilities are available. All this in turn, appears to be a powerful advantage to build IoT ecosystem upon, since the ontology-driven access allows to establish ad-hoc human-machine interaction sessions with any IoT device on demand. Thereby the human-it-the-loop scenario is supported as a key for Human-Centric IoT [2].

Our goal was to make a step towards hardware implementation of task ontologies by organizing the ontology-driven functioning of the light Edge devices based on very resource-constrained microcontroller units (MCUs) like ESP8266 (80 KiB RAM, 80 MHz CPU), ATmega328 (2 KiB RAM, 16 MHz CPU) or even ATtiny45 (256 B RAM, 8 MHz CPU).

We implemented the suggested approach within SciVi Smart system [14] (https://scivi.tools) and tested it by developing a practically useful IoT device.

2   Key Contributions

In this paper, we present an approach to organize Edge Computing governed by ontologies. The main idea is to replace the traditional firmware of edge device by the ontology reasoner, while the underlying extensible ontology describes the schematics, role and functioning principles of this device.

The following key results of the conducted research are presented:

1. The unified ontology-driven approach to perform configurable and inspectable computing on Edge devices within sensor/actuator network in a unified intelligent way. This allows not only to improve the M2M interconnections, but also to automate the transforming M2M IoT into Human-Centric IoT by means of ad-hoc monitoring and steering Edge devices.
2. The ontology "cognitive compression" method, whereby all the redundant information is trimmed, yet the essential structure of ontology remains retrievable and preserves its semantic power. Removing the excessive ontology nodes and relations, using the topological sorting for the remaining ones and applying multilevel structure layout to describe data flow chains in observable and concise form we managed to fit them in the RAM of tiny MCUs.
3. Software to generate extra-lightweight configurable reasoner for ontologies compressed by the method (2). This reasoner is written in C++ (some parts are written manually, but some are generated automatically according to the user's preferences) and can run on a wide range of MCUs, including ESP8266, STM32, ATmega and even ATtiny.
4. Practically useful implementation of the suggested approach by creating ontology-driven Smart Home Edge device that can help to find lost or forgotten things.

3   Related Work

Ontology engineering appears to be a powerful methodology to leverage interoperability of heterogeneous devices within IoT ecosystems [16], as it brings advanced semantics and context-awareness to the ubiquitous computing [10,19]. The emerging convergence of IoT and cognitive technologies is denoted as Semantic Web of Things (SWoT) [7]. Traditionally, only M2M interaction is considered, and ontologies are mainly used to describe the properties and capabilities of the devices to automate their communication and coordination for the sake of cooperation [3,7,16,21]. However, it is nowadays obvious that the evolution of IoT is not limited by M2M, but also requires technologies to facilitate human-in-the-loop scenario. This direction of IoT is called Human-Centric one [2] and demands both semantic technologies and advanced software and hardware HMIs, including tangible ones [6].

Ubiquitous computing environment traditionally includes 3 main levels of computing: Cloud [11], Fog [11] and Edge [5]. All the levels are characterized by the distributed data processing, but the computation power and energy consumption decrease dramatically form the Cloud to the Edge, so each level requires its own approaches both in hardware and in software. In this paper, we focus on the Edge level.

As stated by K. Sahlmann et al., the key feature required by Edge devices is their adaptiveness that can be reached by using ontology-driven solutions [17]. But the main obstacles to the implementation of such solutions are storage capacity and computing power of the Edge nodes [9,12,16,18,20]. The traditional ontology reasoners are well performed on the desktop computers and capable to run on smartphones [4,8], but still cannot even be started on MCUs [18] despite rapid evolution of Edge hardware. Main problems of Edge devices, which hinder straightforward use of ontology-driven techniques, are the following [9,12,16,18,20]:

1. Low RAM capacity. It often appears to be impossible to arrange all the needed data structures in the Edge device memory, not even the ontology itself.
2. Low CPU frequency. Even if the ontology and related supplementary data fit in the device memory, reasoning process takes then too much time and appears to be practically useless as it is incompatible with the real-time tasks (while Edge Computing often requires real-time operation).
3. Low power. As the reasoning requires a lot of computations and has considerable memory footprint, it is an energy consuming process. But Edge devices are often autonomous, so should consume as less power as possible, or their power sources will drain too fast.

X. Su et al. tackle these problems by introducing so-called "Entity Notation": the concise format for knowledge representation and network package payload encoding [20]. According to the evaluations provided, this format enables drastic reduction of the ontology size compared with standard OWL or RDF notation (compression ratio compared to RDF is about 20 times), while the computational burden and energy consumption decrease as well. This format enables efficient interconnection for MCUs and can be encoded/decoded in a straightforward way, but it still cannot ensure the fitting of entire ontology into the device RAM for the full-fledged reasoning.

Another promising way to fit ontologies to the Edge device memory is proposed by Sahlmann et al. [16]. But the role of ontologies is limited to describing the capabilities of Edge devices without affecting their behavior. In the later work K. Sahlmann et al. proposed ontology-driven Edge device virtualization [17]. This appears to be a step further towards ontology-driven Edge Computing, however the reasoning process takes place on the hub (Fog) devices, which have much more computational power compared to Edge devices.

H. Dibowski et al. presented a very elaborate work about the usage of ontologies to describe the device capabilities, but also to retrieve devices, select their operation mode, parametrize them, and automatically evaluate their interoperability [3]. However, the reasoning takes place on the Fog node as well (authors used a quite powerful PC as an aggregator for a large amount of IoT devices).

C. Seitz et al. suggest embedding an ontology-based expert system to automate diagnosis of different anomalies within IoT and robotics systems [18]. The software solution proposed by Seitz et al. performs well on the resource constrained embedded devices running under Linux-based operating system. However, the memory consumption of this software reaches several megabytes, so it is suitable for microcomputers (like e.g. Raspberry Pi), and not for MCUs.

Very impressive results were obtained by H. Abdulrab et al., who developed so-called "Ontology Mediators" – the semantic integration components for ubiquitous computing environment [1]. This is a special model-driven middleware to enable transparent interconnection of different devices, including the Edge ones (running under ECOS). The interconnection based on Ontology Mediators comprises device communication as well as fusion and smart transformation of sensors' data. In fact, this work is the closest one to what we want to achieve, since Ontology Mediators are governed by the knowledge model and the target hardware are very resource-constrained. However, this solution foremost addresses the M2M communication, while the human-in-the-loop scenario is not elaborated.

The main distinctive features of our approach are the following:

1. We suggest full-fledged ontology-driven Edge Computing solution, assuming the behavior of Edge Computing devices (e.g. sensing, data processing, actuation and communication) is fully controlled by task ontologies. Thereby we a make a step towards hardware implementation of ontologies.
2. We target not only microcomputers like Raspberry Pi, but also extremely resource-constrained MCUs (e.g. ESP8266 and even ATtiny45).
3. We focus not only on M2M interconnection, but also on human-in-the-loop scenario (utilizing the semantic power of ontologies to leverage the creation of ad-hoc HMI with Edge devices to enable on-demand device monitoring and steering) and on autonomous functioning of Edge devices (when no network and correspondingly no external steering is available).

4   Proposed Solution

4.1   Background

In our previous research we used ontology-driven solutions to automate the calibration and monitoring of IoT sensor-based devices by means of Smart system SciVi scientific visualization tools [15], as well as to create custom hardware HMI based on IoT technologies [14]. SciVi provides the high-level user interface to describe the behavior and interconnection of custom IoT devices and of middleware to glue together different parts of IoT ecosystem in a graphical form using data flow diagrams (DFDs). Moreover, SciVi enables the automatic firmware or middleware generation for IoT devices according to these DFDs.

Data processing operations, visualization features and code generation mechanism of SciVi are governed by the set of underlying ontologies organized as a repository. The detailed description of this process can be found in [13–15].

But until now the firmware for IoT devices generated by SciVi was completely imperative: it implements algorithm described by DFD in a traditional sequential fashion.

In the present work we extend this mechanism by the new module that enables generation of embedded reasoners capable of running direct on the IoT Edge devices and turning them into ontology-driven ones.

4.2 Proposed Ontology-Driven Edge Computing Pipeline

The main idea of ontology-driven Edge Computing is schematically presented in the Fig. 1 and described below.

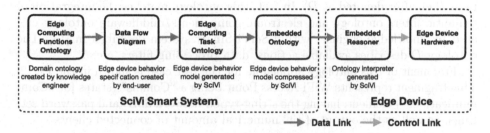

Fig. 1. Ontology-driven Edge Computing (arrows represent data links).

We propose the following lifecycle of the ontology-driven Edge device. First, the embedded reasoner should be generated and installed on MCU. This step is normally performed just once, as a single reasoner covers wide variety of tasks the MCU can be used for. Then, Edge device should be assembled by attaching required sensors and actuators to the MCU. This step can be repeated whenever the device is upgraded. After that, the user should describe the device behavior as DFD within SciVi. This step can be repeated whenever the device role in the computation process should be changed. DFD is automatically transformed into the task ontology that is cognitively compressed and stored in the concise binary format we call EON (Embedded or Edge ONtology). EON-encoded ontology is transferred to Edge device using wired or wireless connection. Afterwards, the reasoner on the device' side starts working, so the device performs described actions. Custom monitoring and steering of the device is allowed in SciVi via special queries to the embedded reasoner. This step is normally repeated multiple times during the device usage, facilitating the human-in-the-loop scenario.

To test the proposed approach, we created relatively simple but useful IoT device that is "a reusable pager for the silent things". There are a lot of things (for example, suitcases, glasses, wallets, keychains, etc.) you may want to find in a unified way by a ring signal, just like calling a lost cell phone. The pager we created is removable/reusable; can be attached to the "silent thing" and "called" anytime: it maintains a WiFi access point and buzzes whenever someone connects. The pager also has a "night mode": when the illumination is low, the frequency of buzzer signal decreases, so it is perceived quieter. Thanks to the ontology-based functioning, the base frequency of buzzing can be altered according to the particular thing the pager is attached to, so the user will be able to distinguish, which one is "responding" to the call. We used the following hardware: ESP8266 MCU with a built-in 802.11 (WiFi) capable communication module as a core, passive buzzer HW-508 to play sound and photoresistor VT90N2 to measure illumination strength. These components are very common and cheap, which make it easy to reproduce the results described in the paper.

4.3   Edge Computing Functions Ontology

First of all, we propose an extensible domain ontology of Edge Computing functions (hereafter denoted as D). In fact, this ontology matches the integration of semantic filters ontology and electronic components/middleware ontology created by knowledge engineer within SciVi Smart system described in [14]. The ontology D describes available actions, data processing filters, etc.

Fragment of this ontology applicable for the test case is shown in the Fig. 2. This fragment represents WiFi Access Point being a "Comms" (states for "Communications") element, having the string-typed network SSID and password settings, as well as the output of the numerical amount of connected clients.

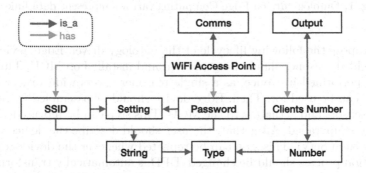

Fig. 2. Fragment of Edge Computing Functions ontology.

4.4   Data Flow Diagram

Based on the ontology D, SciVi automatically generates a set of instruments enabling end-user to compose DFD and thereby visually describe the behavior of Edge device. DFD describing the behavior of the test case Edge device is shown in the Fig. 3.

Each node represents the individual data processing stage and links represent the data transfer. It is worth noting, that each node has a corresponding ontology fragment under the hoot, like the one shown in the Fig. 2 (which matches the "Wifi Access Point" node). The diagram defines simple algorithm of Edge device

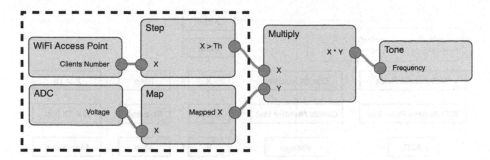

Fig. 3. Data flow diagram representing the behavior of Edge device.

behavior: the number of connected clients, obtained from the Wifi Access Point is tested against the threshold 1 ("Step" node), and multiplied ("Multiply" node) by the value obtained from the ADC (the photoresistor is connected to) and mapped to [100, 1000] ("Map" node). The actual quotients like threshold value and mapping segment borders are tuned in the settings window, that opens by clicking the corresponding node, so they are not depicted in the figure. The multiplication result is then used as a frequency of tone ("Tone" node) played back by the device' buzzer. The part of DFD highlighted by the dotted line will be discussed in the next section.

In real-world, users can define more complex DFDs and, if the provided set of nodes is not enough for them, new ones can be easily added by changing the ontology D instead of modifying the SciVi core source code.

4.5   Edge Computing Task Ontology

The user-defined DFD is than used by SciVi Smart system to automatically create the task ontology (hereafter denoted as T). In fact, this ontology is the representation of DFD in an ontological form.

Each node of T can be computable, this means, it can have an attribute defining a specific function that should be evaluated during the reasoning process to determine the node's value. The nodes can be chained by use_for relations, which denote the data flow.

It must be noted, that only the ontology T should be stored and handled on the Edge device side. The ontology D can be stored elsewhere, for example, on the Fog device or even in the Cloud.

Fragment of task ontology T, generated according to the above mentioned DFD and controlling the device is shown in the Fig. 4. This fragment corresponds to the part of DFD highlighted with the dotted line in the Fig. 3. Nodes filled gray and relations depicted by dotted arrows are not stored on Edge device, whereby nodes filled white and yellow, and the relations depicted by solid lines are stored on Edge device. The difference in storing the white and yellow nodes is discussed in the next section.

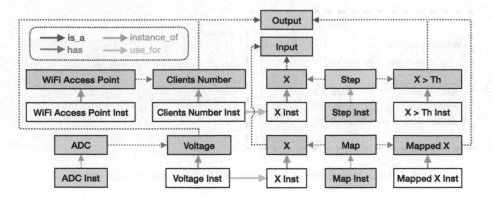

Fig. 4. Fragment of task ontology describing the behavior of Edge device.

4.6 Cognitive Compression and Embedded Ontology

Task ontology T should be uploaded to the target device to govern its functioning. However, it should first be compressed to fit the Edge device memory.

As stated in [18], size of the ontology is the most significant parameter that should be optimized when it comes to reasoning on the Edge. Speaking about MCU, capacity of the memory is often restricted to less than 1 KiB (e.g. popular MCU ATtiny45 possesses 256 B of RAM and 256 B of EEPROM). Traditional formats like OWL or RDF are not even close to be handled on such hardware [20].

Searching through the available information resources, we were unable to find any format that would enable fitting a solid meaningful ontology into less than 256 B. So, we developed our own representation format called EON. It is a concise binary format highly optimized for Edge Computing tasks. The EON format is created special for task ontologies (T) trimming all the redundant information and assuming that, if required, this information can be unambiguously restored with a help of the domain ontology (D). We call this trimming "cognitive compression", and this is an essential part of EON.

Encoding and decoding of ontologies representation in EON format can be expressed in a formal way as follows: $T_{EON} = \sigma(\kappa(T, D)), T = \kappa^{-1}(\sigma^{-1}(T_{EON}), D)$, where σ denotes serialization into EON binary form; κ – cognitive compression; σ^{-1} – deserialization to traditional form e.g. OWL; κ^{-1} – cognitive decompression. T_{EON} denotes task ontology stored in EON format, while T and D are represented in some traditional format e.g. in OWL.

The function σ takes an ontology as a set of nodes, their attributes and relations, and dumps it to a sequence of bytes. This sequence is chunk-based and contains only two chunks (preceded by length in bytes): relations chunk and attributes chunk. It must be noted, that to reduce the size of the byte sequence, we do not store the names of nodes or relations inside it, just integer identifiers (IDs). The names are actually not needed for reasoning on Edge device, so they are trimmed by the function κ.

Relations chunk stores triplets similar to RDF, but represented by integer IDs only. Each triplet is packed into 16 or 24 bit depending on the size of ID. In the current form EON is limited by 64 nodes inside T (after cognitive compression), 65536 nodes inside D and 9 relation types (the possible bitwise layouts are discussed below). These limits appear to be sufficient for computing tasks performed on the low-performance Edge devices. In case of more complex tasks (performed on correspondingly more powerful devices) we can extend EON relation triplets to 32 bit, which will enable handling significantly larger ontologies.

Attributes chunk stores key-value pairs, where key is the integer ID of node and value is its attribute. Normally, attributes are the functions that should be evaluated to calculate the result associated with the node in terms of data flow. Functions may contain operators' ID, nodes' IDs and constants. The table of available operators is tied to the reasoner and is discussed in the upcoming section. If the function contains node ID, the corresponding result of this node is substituted during the reasoning process. Constants can be of different data type. Right now, numerical types (signed and unsigned integers up to 32 bit and 32 bit float) and null-terminated ASCII strings are supported.

Functions are stored in the postfix notation to reduce the entry size. Each function component and each attribute have a distinctive bit marker. Thereby, no length values are stored to reduce the size of entire sequence.

Task ontology T describes the data flow inside Edge device and communication protocols with the other computing nodes in the network. In the data flow, order of operations is crucial. As the operations are encoded to the ontology nodes' attributes, we suggest storing their order implicitly as the order of nodes. As seen in the Fig. 4, we apply topological sorting to the nodes, so the data flow represented by use_for relations is aligned from left to right, building a multilevel structure layout of ontology graph representation. This layout is both human- and machine-readable, incorporating the knowledge about the operations order. When dumped to EON format, this partial order of nodes is preserved, so the reasoner can evaluate the nodes one by one, and any time the next node requires the data from the previous ones, these data are already computed. This reduces both the size of result ontology and the computation time.

Next, when preparing T to dumping to EON, we trim all the redundant knowledge that can be unambiguously restored by traversing the domain ontology D and contains no descriptions of actions essential for Edge device functioning. The trimmed nodes are filled gray in the Fig. 4, the trimmed relations are shown as dotted arrows. Currently we remove all relations except instance_of (because it denotes the connection of T and D and thereby is needed to restore the trimmed knowledge) and use_for (because it directly denotes the data flow). It must be noted, that EON, as an ontology representation format, can support much more relations, letting space for future modifying (e.g. generalizing) the cognitive compression algorithm.

The names of nodes and relations are also trimmed (replaced by the IDs), so the cognitive decompression is possible only if the ontology D is presented. However, as long as the cognitive decompression takes place in SciVi that runs on the Fog Node or in Cloud, access to D is not an issue. To reduce the size of resulting byte sequence, we enumerate the nodes filled white in the Fig. 4 starting from 1, but to make the decompression possible, we keep the IDs from the ontology D for the nodes filled yellow. So, for each `instance_of` relation, the source node is the one from T (so its ID is just a number of node in the multilevel structure layout of T), and the destination node is denoted by its ID in D.

To represent the IDs in the EON byte sequence, following layouts are used (s denotes bit of source node ID, r – bit of relation ID, d – bit of destination node ID):

1. s s s s s s 0 r r r d d d d d d
   (regular relation for the nodes inside T).
2. s s s s s s 1 0 d d d d d d d d
   (`instance_of` relation joining T and D, if D has no more than 256 entities).
3. s s s s s s 1 1 d d d d d d d d d d d d d d d d
   (`instance_of` relation joining T and D, if D has from 256 to 65536 entities).

Last, but not least, we optimize the storage of constants by finding the smallest type including particular value (e.g. the value 1.0 is treated as 8 bit integer, while 1.1 – as 32 bit float).

In our test case, the ontology, which fragment is shown in the Fig. 4, took only 80 bytes after cognitive compression and dumping to EON (this appears to be more than 800 times smaller than storing it in OWL format).

4.7   Embedded Reasoner

The aim of embedded reasoner is to ensure the evaluation of data flow represented by the ontology T in the EON format. The architecture of the reasoner is shown in the Fig. 5, where data links depict the transfer of data between the reasoner modules, and control links represent the transfer of commands.

To enable portability across different MCUs, the reasoner is implemented in C++. Currently, we have tested it on ESP8266 (within WeMos D1 mini platform), ATmega328 (within Arduino platform) and ATtiny45 (as a standalone MCU). The compilation process is automated within SciVi Smart system [14].

The **Functions Module** steers the hardware and provides the table of available operators, which enables the MCU's instruction set to be used within ontologies. The code of this module is generated automatically by SciVi according to the user's choice of libraries that should be supported in the particular reasoner. This configurability allows reducing the size of reasoner's binary representation (e.g. ATtiny45 has only 4 KiB flash memory to store the code).

The **Communication Module** allows sending and receiving data and commands to/from SciVi. Basically, 4 commands are supported: `Upload` (to save new ontology T on the MCU), `Download` (to retrieve T from MCU), `Get` (to receive the particular evaluated result of given node within T) and `Set` (to change the evaluated result of given node by the new one, that was evaluated on the side of

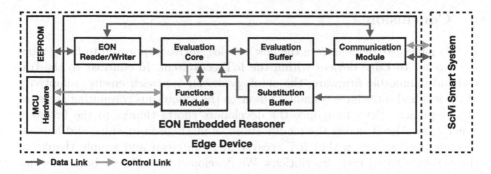

Fig. 5. Architecture of EON embedded reasoner.

SciVi). Communication takes part either via WebSocket over WiFi (in case of ESP8266), via modified set of AT-commands via RS-232 serial port (in case of ATmega328) or via USI SPI (in case of ATtiny45).

The ontology T is stored in the MCU's EEPROM. The reading and writing of EON format is ensured by the **EON Reader/Writer**. The **Evaluation Core** performs traversing the EON-encoded ontology and calls operators provided by **Function Module** whenever needed. Evaluation results are stored in the **Evaluation Buffer**; the results obtained from SciVi (via **Communication Module**) are stored in the **Substitution Buffer**. **Substitution Buffer** allows steering the MCU externally by overriding the results evaluated by the **Evaluation Core**. This in turn enables building ad-hoc user interface with Edge device and opens a gate for the on-demand transformation of M2M functioning into the Human-Centric one.

When the ontology T encoded to EON format and uploaded to Edge device, its embedded reasoner starts executing the tasks described. This is an autonomous process (suitable for M2M functioning scenario), however the end-user can anytime connect to the device using SciVi Smart system to monitor the data flow or even to take control over it. For this, ontology T is retrieved back to SciVi, decoded from EON with help of the ontology D and transformed into the DFD shown to the user. After that, the user can add visualization nodes to this diagram utilizing the visual analytics capabilities of SciVi [15], or change the data links to modify the flow itself. Data for monitoring are retrieved in real-time (using `Get` command discussed above), and the modifications of the data flow take effect immediately (using `Set` command).

The changes in data flow are kept as long as the communication session with Edge device persists. By disconnecting, all the changes are automatically reverted by the embedded reasoner (by wiping the **Substitution Buffer**), so Edge device restores its initial functioning. If the user wants to store the changes, updated DFD should be dumped to task ontology and committed to Edge device.

5   Conclusion

In this paper, we demonstrate the new ontology-driven approach to bring more intelligence to Edge devices within the IoT ecosystem. In contrast to the traditional immutable firmware, the ontology-driven approach ensures adaptivity of devices and introduces semantic power to the ubiquitous computing systems. Our approach allows to reduce the developers efforts thanks to the firmware generation. Also it makes the modifications of machine-to-machine and human-machine interactions within IoT ecosystem more clean and simple thanks to the ontology-based task descriptions. We developed the efficient binary format EON and cognitive compression algorithm to represent task ontologies on Edge devices' side, implemented the set of instruments to automate the creation of corresponding ontologies and the lightweight configurable reasoner using the tools of our Smart system SciVi.

We tested our approach by creating a simple yet useful Edge device that helps to locate lost things. SciVi Smart system is an open source project available on https://github.com/scivi-tools. The subproject discussed in this paper is available on https://github.com/scivi-tools/scivi.eon.

Our approach can be refined by adding the task ontology verification stage to find out whether the particular ontology from the repository is suitable for the device of certain hardware configuration. Also, the embedded reasoner efficiency should be studied in terms of CPU overhead and energy consumption.

We plan to use this approach to create function-reach hardware HMI for complex visual analytics tasks by studying multiparametric modeling of users' communication processes in the Internet social services [14].

References

1. Abdulrab, H., Babkin, E., Kozyrev, O.: Semantically enriched integration framework for ubiquitous computing environment. In: Babkin, E. (ed.) Ubiquitous Computing, pp. 177–196. IntechOpen, London (2011). https://doi.org/10.5772/15262. chap. 9
2. Calderon, M., Delgadillo, S., Garcia-Macias, A.: A more human-centric internet of things with temporal and spatial context. Proc. Comput. Sci. **83**, 553–559 (2016). https://doi.org/10.1016/j.procs.2016.04.263
3. Dibowski, H., Kabitzsch, K.: Ontology-based device descriptions and device repository for building automation devices. EURASIP J. Embed. Syst. **2011**(1), 1–17 (2011). https://doi.org/10.1155/2011/623461
4. Guclu, I., Li, Y.-F., Pan, J.Z., Kollingbaum, M.J.: Predicting energy consumption of ontology reasoning over mobile devices. In: Groth, P., et al. (eds.) ISWC 2016. LNCS, vol. 9981, pp. 289–304. Springer, Cham (2016). https://doi.org/10.1007/978-3-319-46523-4_18
5. Hamilton, E.: What is edge computing: the network edge explained (2018). https://www.cloudwards.net/what-is-edge-computing/. Accessed 08 Jan 2020
6. Ishii, H.: Tangible bits: beyond pixels. In: Proceedings of the 2nd International Conference on Tangible and Embedded Interaction, pp. XV-XXV (2008). https://doi.org/10.1145/1347390.1347392

7. Jara, A.J., Olivieri, A.C., Bocchi, Y., Jung, M., Kastner, W., Skarmeta, A.F.: Semantic web of things: an analysis of the application semantics for the IoT moving towards the IoT convergence. Int. J. Web Grid Serv. **10**(2/3), 244–272 (2014). https://doi.org/10.1504/IJWGS.2014.060260

8. Koopmann, P., Hähnel, M., Turhan, A.-Y.: Energy-efficiency of OWL reasoners— frequency matters. In: Wang, Z., Turhan, A.-Y., Wang, K., Zhang, X. (eds.) JIST 2017. LNCS, vol. 10675, pp. 86–101. Springer, Cham (2017). https://doi.org/10.1007/978-3-319-70682-5_6

9. Li, P.: Semantic Reasoning on the Edge of Internet of Things. Ph.D. thesis, University of Oulu, master's thesis, Degree Programme in Computer Science and Engineering (2016)

10. Pardo, E., Espes, D., Le-Parc, P.: A framework for anomaly diagnosis in smart homes based on ontology. Proc. Comput. Sci. **83**, 545–552 (2016). https://doi.org/10.1016/j.procs.2016.04.255

11. Pisani, F., Borin, E.: Fog vs. cloud computing: should i stay or should i go? In: Proceedings of the Workshop on INTelligent Embedded Systems Architectures and Applications, pp. 27–32 (2018). https://doi.org/10.1145/3285017.3285026

12. Ruta, M., Scioscia, F., Sciascio, E.D.: A mobile matchmaker for resource discovery in the ubiquitous semantic web. In: 2015 IEEE International Conference on Mobile Services, pp. 336–343 (2015). https://doi.org/10.1109/MobServ.2015.76

13. Ryabinin, K., Chuprina, S.: High-level toolset for comprehensive visual data analysis and model validation. Proc. Comput. Sci. **108**, 2090–2099 (2017). https://doi.org/10.1016/j.procs.2017.05.050

14. Ryabinin, K., Chuprina, S., Belousov, K.: Ontology-driven automation of iot-based human-machine interfaces development. In: Rodrigues, J.M.F., et al. (eds.) ICCS 2019. LNCS, vol. 11540, pp. 110–124. Springer, Cham (2019). https://doi.org/10.1007/978-3-030-22750-0_9

15. Ryabinin, K., Chuprina, S., Kolesnik, M.: Calibration and monitoring of IoT devices by means of embedded scientific visualization tools. In: Shi, Y., et al. (eds.) ICCS 2018. LNCS, vol. 10861, pp. 655–668. Springer, Cham (2018). https://doi.org/10.1007/978-3-319-93701-4_52

16. Sahlmann, K., Scheffler, T., Schnor, B.: Ontology-driven device descriptions for IoT network management. In: 2018 Global Internet of Things Summit, GIoTS (2018). https://doi.org/10.1109/GIOTS.2018.8534569

17. Sahlmann, K., Schwotzer, T.: Ontology-based virtual IoT devices for edge computing. In: Proceedings of the 8th International Conference on the Internet of Things (2018). https://doi.org/10.1145/3277593.3277597

18. Seitz, C., Schönfelder, R.: Rule-based OWL reasoning for specific embedded devices. In: Aroyo, L., et al. (eds.) ISWC 2011. LNCS, vol. 7032, pp. 237–252. Springer, Heidelberg (2011). https://doi.org/10.1007/978-3-642-25093-4_16

19. de Souza, W.L., Prado, A.F., Forte, M., Cirilo, C.E.: Content adaptation in ubiquitous computing. In: Babkin, E. (ed.) Ubiquitous Computing, pp. 67–94. IntechOpen, London (2011). https://doi.org/10.5772/15940. chap. 4

20. Su, X., Riekki, J., Haverinen, J.: Entity notation: enabling knowledge representations for resource-constrained sensors. Pers. Ubiquit. Comput. **16**, 819–834 (2012). https://doi.org/10.1007/s00779-011-0453-6

21. Venkatesh, J., Chan, C., Rosing, T.: An Ontology-Driven Context Engine for the Internet of Things. Technical report CS2015-1009, Department of Computer Science and Engineering, University of California, San Diego (2015). https://csetechrep.ucsd.edu/Dienst/UI/2.0/Describe/ncstrl.ucsd_cse/CS2015-1009. Accessed 08 Jan 2020

Combined Metrics for Quality Assessment of 3D Printed Surfaces for Aesthetic Purposes: Towards Higher Accordance with Subjective Evaluations

Jarosław Fastowicz⊙, Piotr Lech⊙, and Krzysztof Okarma(✉)⊙

Department of Signal Processing and Multimedia Engineering, Faculty of Electrical
Engineering, West Pomeranian University of Technology in Szczecin,
Sikorskiego 37, 70-313 Szczecin, Poland
{jfastowicz,piotr.lech,okarma}@zut.edu.pl

Abstract. Objective quality assessment for 3D printing purposes may
be considered as one of the most useful applications of machine vision
in smart monitoring related to the development of the Industry 4.0 solutions. During recent years several approaches have been proposed, assuming observing the side surfaces, mainly based on the analysis of the regularity of visible patterns, which represent the consecutive printed layers. These methods, based on the use of general purpose image quality
assessment (IQA) metrics, Hough transform, entropy and texture analysis, make it possible to classify the printed samples, independently of
the filament's colour, into low and high quality classes, with the use of
photos or 3D scans of the side surfaces.

The next step of research, investigated in this paper, is the combination of various proposed approaches to develop a combined metric,
possibly highly correlated with subjective opinions. Since the correlation of single metrics developed mainly for classification is relatively low,
their combination makes it possible to achieve much better results, verified using an original, newly developed database containing 107 captured
images and 3D scans of the 3D printed surfaces with various colours and
local distortions caused by external factors, together with Mean Opinion
Scores (MOS) gathered from independent observers. Obtained results are
promising and may be a starting point for further research towards the
optimisation of the newly developed metrics for the automatic assessment of the 3D printed surfaces, mainly for aesthetic purposes.

Keywords: Surface quality assessment · 3D printing · Image entropy ·
Depth maps · Image analysis · Additive manufacturing

1 Introduction

Automatic objective quality assessment of 3D printed surfaces is currently one of
the most dynamically developing areas of image analysis for emerging applications.

V. V. Krzhizhanovskaya et al. (Eds.): ICCS 2020, LNCS 12143, pp. 326–339, 2020.
https://doi.org/10.1007/978-3-030-50436-6_24

Observing a rapid growth of popularity of 3D printing (additive manufacturing), as well as the availability of affordable high quality cameras, a natural direction of an extensive research is the application of image analysis methods for smart monitoring of the 3D printing process. The goal of such methods is to make it possible not only to control the progress of the 3D printing procedure but also to prevent the occurrence of some minor errors or even abort the manufacturing process in the case of poor quality of obtained objects.

Vision based assessment of 3D printed surfaces is a natural extension of research activities related to *in situ* monitoring of the manufacturing process and non-destructive evaluation (NDE), which have been reported recently, e.g. with the use of Optical Coherence Tomography (OCT) for selective laser sintering [12], spatially resolved acoustic spectroscopy [13] or using top view stereo cameras to obtain the cloud of 3D points further compared with the model [14].

An interesting approach to video based detection of defects during the 3D printing process has been presented by Straub [30], where five cameras with Raspberry Pi units connected using Ethernet cables have been used to capture the images of the manufactured object. Although two major types of issues - lack of filament causing the "dry printing" and premature job termination - have been detected properly, the system's high sensitivity to changes of environmental conditions and camera motions has been a major problem in this approach.

Some other approaches to imaging in quality assessment of 3D prints utilize "process signatures" used for fused deposition of ceramic materials [7,8], as well as the analysis of "road paths" for identification of under- and over-filling comparing them to the predefined models [5].

Another interesting approach is the non-destructive evaluation based on ultrasonic imaging and X-rays [34] as well as using electromagnetic methods [2], however applicable mainly for off-line quality assessment of previously manufactured objects. Some of the recently presented methods require the comparison with the model of the printed objects [18], whereas some other attempts are based on previous time-consuming training [3,4,31] or additional filtering [27]. Nevertheless, most of the proposed approaches utilizing machine vision are used for process monitoring fault detection rather than quality assessment of the manufactured objects [6]. One of recent examples [36] is the use of fringe projector with the analysis of small subregions with the use of local point features. An interesting application for multi-material 3D printing has also been presented in MultiFab project [28], together with automated positioning system utilizing the data obtained from the precisely calibrated OCT 3D scanner.

Nevertheless, the main goal of our research is not only the monitoring of the progress of production and the state of the printing device, but primarily an automatic smart quality assessment of the manufactured object during the printing process, which is usually relatively long. Such possibility would be useful for saving the filament, energy and time in case of detection of too low quality, making it possible to stop the manufacturing process and warn the user immediately.

Since in some cases some minor issues may be corrected after the manufacturing process, or even during printing in some devices, the classification of the printed surfaces into two classes representing high and low quality 3D prints may be insufficient. Despite of encouraging classification results obtained during recent years, an approach to quality assessment using a continuous scale would be more demanding. Considering the methodology typically used in general purpose image quality assessment (IQA), where the objective quality metrics should be highly correlated with subjective opinions, typically expressed as Mean Opinion Score (MOS) values or Differential MOS, a unique dedicated database of 3D printed samples together with subjective scores has been prepared making it possible to verify existing methods. Additionally, a novel approach to surface quality assessment of 3D prints based on the combination of different methods optimized towards high correlation with subjective scores has been proposed and verified in the paper, leading to satisfactory results.

2   Methods of Surface Quality Assessment Based on Classification

Automatic quality assessment of 3D printed surfaces based on the analysis of images captured by side view cameras can be conducted using different approaches, including texture analysis, adaptation of general purpose IQA methods, image entropy, detection of patterns based on Hough transform or the use of descriptors for gradient analysis based on Histogram of Oriented Gradients (HOG). Nevertheless, some of the above mentioned approaches can be applied for the assumed colour of the filament and should be additionally tuned for each colour, hence their practical usefulness may be limited. The main purpose of all these methods is related to the classification of the observed surfaces into two major groups representing high and low quality samples, although in some experiments additional "moderately low" and "moderately high" quality samples have been distinguished.

The application of texture analysis is based on the assumption that the statistical distribution of colours of the neighbouring pixels should be similar for the whole image. Hence, analysing the chosen Haralick features, calculated using the Grey-Level Co-occurrence Matrix (GLCM), smaller homogeneity may be observed for lower quality 3D prints [10,19]. Nevertheless, the proposed methods require time-consuming computations of several GLCMs for various offsets, used during further analysis of changes of Haralick features. In view of the necessary computational efforts and an average accuracy, this approach has not been further investigated. This decision results also from the experiments conducted for the whole developed database, leading to worse results in comparison to the other methods.

On the other hand, it can be assumed that, due to the regularity of the patterns representing the consecutive printed layers, observed by a side located camera, the image entropy should be significantly lower for high quality 3D prints. Dividing the image into $N \times N$ fragments, the local entropy values should also be small and similar to each other if there are no visible artifacts, caused e.g. by the lack of filament or being the result of overfilling. Hence, the variance of the local entropy should also be low for high quality surfaces. Since the image entropy is also strongly dependent on the filament's colour, the combination of the local entropy and its variance calculated for HSV and RGB colour spaces has been proposed in the paper [23], leading to colour independent method of quality assessment. Further improvement of classification results, verified for a larger database of the flat 3D printed samples, has been obtained due to the use of entropy based method applied for the depth maps obtained by a 3D scanner [9].

Another possibility of quality evaluation of the 3D printed surfaces is related to the use of some of the general purpose IQA metrics. As the most universal widely used metrics, such as e.g. Structural Similarity (SSIM) [32] or Feature Similarity (FSIM) [35] belong to the group of full-reference methods, which require the knowledge of the original undistorted image, their direct application would require the comparison with the model of the printed surface. To overcome this issue, the division of the image into blocks has been proposed making it possible to calculate the mutual similarities between the image fragments [24]. In the presence of geometrical artifacts the mutual similarities for the image fragment containing the distortions decreases noticeably. A similar approach may be considered for the calculations of correlation, also with the use of the Monte Carlo method [25] to decrease the amount of computations.

Analysing the structure of the 3D printed flat surface with well visible layers, being the result of placing the melted filament over the already hardened polymer, one may expect a high number of straight lines, which should be easily extracted using the Hough transform with appropriate parameters [11]. Nevertheless, as verified experimentally, its direct application may be troublesome, especially for some brighter filaments. In spite of this, due to the additional application of histogram equalization using the well-known CLAHE method, as well as the random choice of the analysed image regions, a relatively high classification accuracy (about 0.8) may be achieved [11].

Another investigated approach is the application of the HOG features [17] calculated locally for various orientations. Since for high quality surface the luminance changes should be well predictable and the horizontal changes should be much smaller than the dominating vertical ones, assuming that the sample is not rotated, the analysis of directional gradients may be a useful tool for the assumed quality evaluation. A high accuracy of classification, independently of the colour of the filament, may be achieved using the signed orientations and 4 bins for the calculation of the HOG features, assuming the final classification using the standard deviation of the HOG features [17].

Although all the previously proposed approaches presented above have been developed for the classification purposes, it has been assumed that they may be additionally verified by means of the database containing the subjective quality scores collected in perceptual experiments. Such results, obtained after the analysis of the opinions provided by human observers, may be useful for the optimization purposes to ensure high correlation of the developed metrics with subjective evaluation, similarly as in general purpose IQA.

3 The 3D Prints Database

The database containing 107 images of the 3D printed flat surfaces together with their depth maps and subjective evaluation results, expressed as Mean Opinion Score (MOS) values, has been prepared with the use of three 3D printers: Prusa i3, RepRap Pro Ormerod 3 and da Vinci 1.0 Pro 3-in-1. All the samples have been prepared using the most popular Fused Deposition Modelling (FDM) technology from 9 different colour ABS (Acrylonitrile Butadiene Styrene) filaments. In comparison with another popular thermoplastic polymer, namely Polyactic Acid (PLA), this material is more abrasion resistant but requires higher working temperature, as its melting point is about 200 °C. It is also lightweight and has good mechanical properties, however its fumes emitted during the printing process may be toxic [1,29].

Since the quality of the manufactured objects are dependent on many conditions, including the quality of materials used for the construction of the 3D printer and the quality of the filament, regardless of some independent factors, the presence of some typical distortions has been forced by changes of temperature, filament's delivery speed or configuration parameters of the stepper motors. All the obtained samples containing various amount of distortions caused mainly by over- and under-filling, including the presence of cracks, have been independently assessed by 92 human observers using the typical scale from 1 (very poor) to 5 (very good). Additionally the obtained MOS values have been compared with the previously utilized expert opinions to confirm the correctness of the obtained results. Some sample images together with MOS values, are presented in Fig. 1.

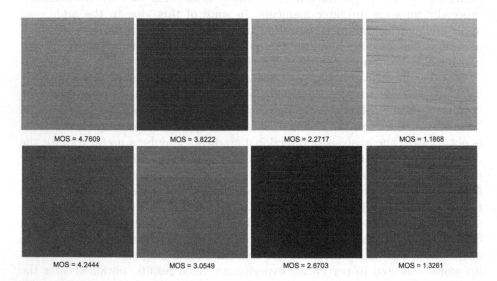

Fig. 1. Sample representative images of the 3D printed flat surfaces with their average subjective quality scores.

The images have been acquired using Sony DSC-HX100V camera with an automatic white balance, 5 mm focal length and the exposure time 1/125 s without flash, preventing a fixed distance. A distributed illumination has been used to prevent strong reflections using three lamps. The depth maps have been obtained as the 1928×1928 pixels 16-bit greyscale images, being the result of the normalization of the STL files representing the 3D models obtained from the 3D point clouds. They have been achieved as the result of the 3D scanning process using the ATOS 3D scanner manufactured by GOM company with the use of fringe pattern perpendicular to the visible layers on the printed surface [9].

The assumption of an automatic quality evaluation of 3D printed surfaces discussed in this paper is its high accordance with subjective opinions, similarly as in general purpose IQA methods, and therefore the proposed approach should be considered as useful mainly for aesthetic purposes rather than e.g. evaluation of mechanical properties. Such extension would require the analysis of the 3D structure of the manufactured object, acquired e.g. using terahertz methods, and is planned as a part of further research. Another possible extension of the database, planned in future work, may be an addition of images of the non-planar objects, where the entropy based methods may be the most suitable.

4   Idea of the Combined Metric

One of the main goals of the general purpose IQA is to obtain the possibly highest correlation between the objective and subjective quality scores. Unfortunately, single metrics, such as SSIM [32] or much better FSIM [35], usually require the additional non-linear mapping recommended by the Visual Quality Experts Group (VQEG) due to the some specific properties of the Human Visual System (HVS). Since various IQA databases are used for the verification and optimization of newly proposed metrics, the parameters of the logistic function typically used for such mapping may vary for different datasets.

As different general purpose metrics utilize various kinds of image informations, the idea of combined/hybrid metrics has been proposed by the combination of three different metrics using their weighted product [20], leading to a significant increase of the Pearson's Linear Correlation Coefficient (PLCC) for raw quality scores without the necessity of non-linear mapping. Such idea has been extended by the replacement of some metrics by newer ones [21,26], as well as its application for multiply distorted images [22] and recently by the use of no-reference metrics [15].

The general form of the combined metric with exponent weights analysed in the paper can be expressed as

$$Q_{combined} = \prod_{i=1}^{K} \text{Metric}_i^{\text{weight}_i}, \qquad (1)$$

where K is the number of weighted metrics (originally $K = 3$ [20,21]).

As the component metrics, further subjected to optimization of their weights, all the previously examined methods of quality evaluation of the 3D printed surfaces, have been used, particularly those described in Sect. 2. For the additional verification of the proposed approach, two rank-order correlation coefficients have been calculated, similarly as typically used in general purpose IQA. Nevertheless, in image quality assessment both these coefficients, namely Spearman Rank Order Correlation Coefficient (SROCC) and Kendall Rank Order Correlation Coefficient (KROCC), are considered as the measures of the prediction monotonicity, whereas PLCC measures the prediction accuracy. Sperman's ρ is defined as:

$$\rho = 1 - \frac{6 \cdot \sum d_i^2}{n \cdot (n^2 - 1)} , \tag{2}$$

where n is the number of images and d_i is the difference between the position of the i-th image in two sequences ordered according to subjective and objective scores, respectively.

Kendall's τ coefficient is defined as:

$$\tau = \frac{n_c - n_d}{0.5 \cdot n \cdot (n - 1)} , \tag{3}$$

where n_c and n_d are the numbers of concordant and discordant, being the positions of two images in the same two sequences sorted according the subjective and objective quality scores, respectively.

Both rank-order coefficients are independent of the differences of the perceived and measured quality, since only the order of the sorted images is considered regardless of the "quality distances" between them, and therefore they do not require any non-linear mapping functions which would not influence the monotonicity of the sequences of the quality scores.

5   Analysis of Experimental Verification

To verify the possible increase of the correlation of the objective metrics with subjective evaluations due to the application of the combined metrics, all correlation coefficients have been calculated firstly for the single methods proposed in previous papers. Analysing the obtained results, presented in Table 1, the best results may be achieved using the methods based on the entropy of the depth map as well as the mutual Feature Similarity calculations. An interesting observation is that for the HOG based metrics much better PLCC values may be obtained for the kurtosis of HOG values but rank-order correlations are higher for standard deviation of HOG originally proposed in [17]. Nevertheless, there is no single method with the PLCC and SROCC exceeding 0.7 and Kendall's τ is slightly higher than 0.5 only for FSIM based metrics.

Considering the results of verification presented in Table 1, additionally illustrated by the scatter plots presented in Fig. 2, all further experiments have

Table 1. Correlation coefficients between the single objective metrics and subjective quality scores obtained for the developed database.

Method	PLCC	SROCC	KROCC
SSIM - 4 blocks [24]	0.3048	0.3017	0.1938
SSIM - 16 blocks [24]	0.3996	0.4012	0.2746
FSIM - 4 blocks [24]	0.6780	0.6826	0.5114
FSIM - 16 blocks [24]	0.6756	0.6865	0.5195
Colour independent entropy [23]	0.4816	0.4920	0.3480
Global entropy of depth map	0.4820	0.4694	0.3283
Sum of local entropy of depth map	0.6936	0.6674	0.4814
Mixed entropy of depth map + CLAHE [9]	0.6603	0.6547	0.4606
Hough + CLAHE [11]	0.1619	0.3147	0.2365
Standard deviation of HOG [17]	0.4294	0.6053	0.4850
Kurtosis of HOG	0.5075	0.5177	0.3874

Table 2. Correlation coefficients between the optimized combined metrics and subjective quality scores obtained for the developed database.

Combined methods	PLCC	SROCC	KROCC
$FSIM_4$ and $E_{localdepth}$	0.7575	0.7249	0.5492
$FSIM_4$, $E_{localdepth}$ and Hough + CLAHE [11]	0.8166	0.7960	0.6110
$FSIM_4$, $E_{localdepth}$, Hough + CLAHE [11] and kurtosis of HOG	**0.8353**	0.8215	0.6403
$FSIM_4$, EMV_{hue256} [23], Hough + CLAHE [11] and kurtosis of HOG	0.8332	**0.8332**	**0.6448**

started with the optimization of the combined metric based on FSIM and local entropy of depth map. Assuming the combined metric based on formula (1) expressed as

$$Q_{comb2} = FSIM_4^{\alpha} \cdot E_{localdepth}^{\beta} ,\qquad(4)$$

where $FSIM_4$ is the average mutual Feature Similarity assumed for the division of the image into 4 blocks, $E_{localdepth}$ is the average local entropy of the depth map assuming its division into 16 blocks as proposed in [9] and the weighting coefficients α and β have been subjected to optimization leading to the increase of the PLCC value to 0.7575 (for $\alpha = 1.6$ and $\beta = -1.2$) as presented in Table 2.

During further experiments some other metrics presented above have been included in the general formula of the combined metric (1) with optimized exponential weights leading to the results presented in Table 2. As can be observed,

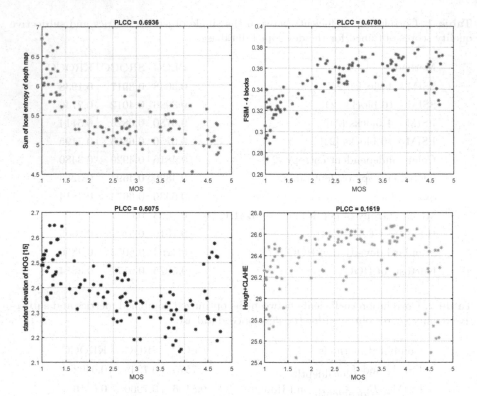

Fig. 2. Scatter plots obtained using selected single metrics and MOS values for 107 samples from the developed database.

the best results have been achieved for the combination of four metrics with the following respective weighting coefficients: $\alpha = 2.9$, $\beta = -1.6$, $\gamma = -14$ and $\delta = -1.8$, where two latter weighting coefficients should be applied for the metric proposed in [11] and kurtosis of HOG features, respectively. Replacing the metric based on the entropy of depth maps by the product of the average local image entropy calculated for the hue component in HSV colour space and its variance, assuming the division of the image into 256 regions [23], makes it possible ot increase the SROCC and KROCC values with slightly worse Pearson's correlation. The optimized coefficients have been obtained by the unconstrained non-linear optimization using the MATLAB *fminsearch* function, based on simplex search method, additionally verified using some gradient-based methods.

To illustrate the advantages of the proposed approach, the scatter plots illustrating the relationships between the subjective and objective metrics for 107 samples included in the developed database are presented in Figs. 2 and 3. Observing these plots, higher linearity of the relation between the MOS and proposed combined metrics can be easily noticed.

Since the calculations of all metrics for a single 1600×1600 pixels image takes less than 2 seconds in MATLAB environment, installed on a PC with

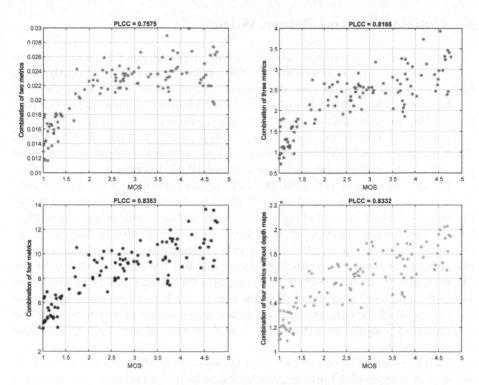

Fig. 3. Scatter plots obtained using proposed combined metrics and MOS values for 107 samples from the developed database.

Intel i7 processor clocked at 2.8 GHz and 16 GB of RAM, the proposed approach should be fast enough also for *in situ* quality monitoring of the 3D prints during a relatively slow typical manufacturing process, even with the use of hardware solutions with lower computational efficiency. Due to the independence of computations performed for each of the individual metrics, some parallelization possibilities of calculations may be considered as well.

Although the verification of the proposed methods has been conducted off-line, the only limitation of the presented approach for on-line applications is related to the necessity of acquisition of depth maps in addition to images captured by side located cameras. In the case of removing the element based on entropy of the depth map for the optimized weights of three other metrics, the PLCC decreases to 0.7877 with SROCC = 0.7854 and KROCC = 0.5898. Nevertheless, similar solutions based on projection on fringe patterns have also been considered by some other researchers [36]. An alternative solution is the use of entropy based method analysed in [23] for images captured by camera calculated in HSV colour space, leading to even better rank-order correlations, as shown in Table 2.

6   Conclusions and Future Work

Application of the proposed combined metrics makes it possible to increase the correlation with subjective evaluation of 3D printed surfaces significantly, from below 0.7 obtained for the best single metric to over 0.83 achieved for the best combination of four methods with optimized weighting coefficients. In comparison to the use of combined metrics for general purpose IQA [15,20,21,26], the increase of the correlation coefficients is much larger, partially due to a high diversity of the combined metrics, which utilize different methods, such as Feature Similarity, entropy, Hough transform and HOG descriptors.

The development of the database of 3D prints containing the results of subjective evaluation opens some new possibilities for the development of even better metrics, optimized in view of correlation with aesthetic evaluations. Nevertheless, an interesting direction of our future research may also be the extension of the database by the results of some other non-destructive evaluation methods, e.g. using terahertz technology, to obtain full information related to the 3D structure of the manufactured objects, also for off-line quality inspection in view of mechanical properties. Another issue, which is worth investigating, is the extension of the dataset towards further development of methods useful for evaluation of non-planar surfaces, e.g. based on entropy and mutual similarity of image regions.

An interesting overview of various approaches to quality control in seven different technologies of 3D printing can be found in [16], whereas some other open challenges are specified in [33], where it has been stated that *"the development of 3D printing technologies is still underway, meaning that there are multiple alternatives without an absolute rule for choosing among them"*. In view of these needs, the proposed approach to combination of multiple methods of surface quality assessment can be considered as one of the potentially useful solutions for emerging applications related to video based quality control in 3D printing. Such methodology may also be further adapted for some other 3D printing methods and materials than the most popular Fused Deposition Modelling with the use of thermoplastic polymer filaments, such as PLA or ABS.

References

1. Azimi, P., Zhao, D., Pouzet, C., Crain, N.E., Stephens, B.: Emissions of ultra-fine particles and volatile organic compounds from commercially available desktop three-dimensional printers with multiple filaments. Environ. Sci. Technol. **50**(3), 1260–1268 (2016)
2. Busch, S.F., Weidenbach, M., Fey, M., Schäfer, F., Probst, T., Koch, M.: Optical properties of 3D printable plastics in the THz regime and their application for 3D printed THz optics. J. Infrared Millimeter Terahertz Waves **35**(12), 993–997 (2014). https://doi.org/10.1007/s10762-014-0113-9
3. Chauhan, V., Surgenor, B.: A comparative study of machine vision based methods for fault detection in an automated assembly machine. Proc. Manuf. **1**, 416–428 (2015). https://doi.org/10.1016/j.promfg.2015.09.051

4. Chauhan, V., Surgenor, B.: Fault detection and classification in automated assembly machines using machine vision. Int. J. Adv. Manuf. Technol. **90**(9), 2491–2512 (2016). https://doi.org/10.1007/s00170-016-9581-5
5. Cheng, Y., Jafari, M.A.: Vision-based online process control in manufacturing applications. IEEE Trans. Autom. Sci. Eng. **5**(1), 140–153 (2008). https://doi.org/10.1109/TASE.2007.912058
6. Delli, U., Chang, S.: Automated process monitoring in 3D printing using supervised machine learning. Proc. Manuf. **26**, 865–870 (2018). https://doi.org/10.1016/j.promfg.2018.07.111
7. Fang, T., Jafari, M.A., Bakhadyrov, I., Safari, A., Danforth, S., Langrana, N.: Online defect detection in layered manufacturing using process signature. In: Proceedings of IEEE International Conference on Systems, Man and Cybernetics, San Diego, CA, USA, vol. 5, pp. 4373–4378 (1998). https://doi.org/10.1109/ICSMC.1998.727536
8. Fang, T., Jafari, M.A., Danforth, S.C., Safari, A.: Signature analysis and defect detection in layered manufacturing of ceramic sensors and actuators. Mach. Vis. Appl. **15**(2), 63–75 (2003). https://doi.org/10.1007/s00138-002-0074-1
9. Fastowicz, J., Grudziński, M., Tecław, M., Okarma, K.: Objective 3D printed surface quality assessment based on entropy of depth maps. Entropy **21**(1), 97 (2019). https://doi.org/10.3390/e21010097
10. Fastowicz, J., Okarma, K.: Texture based quality assessment of 3D prints for different lighting conditions. In: Chmielewski, L.J., Datta, A., Kozera, R., Wojciechowski, K. (eds.) ICCVG 2016. LNCS, vol. 9972, pp. 17–28. Springer, Cham (2016). https://doi.org/10.1007/978-3-319-46418-3_2
11. Fastowicz, J., Okarma, K.: Quality assessment of photographed 3D printed flat surfaces using Hough transform and histogram equalization. J. Univ. Compu. Sci. **25**(6), 701–717 (2019). http://www.jucs.org/jucs_25_6/quality_assessment_of_photographed
12. Gardner, M.R., et al.: In situ process monitoring in selective laser sintering using optical coherence tomography. Opt. Eng. **57**, 041407 (2018). https://doi.org/10.1117/1.OE.57.4.041407
13. Hirsch, M., et al.: Assessing the capability of in-situ nondestructive analysis during layer based additive manufacture. Addit. Manuf. **13**, 135–142 (2017). https://doi.org/10.1016/j.addma.2016.10.004
14. Holzmond, O., Li, X.: In situ real time defect detection of 3D printed parts. Addit. Manuf. **17**, 135–142 (2017). https://doi.org/10.1016/j.addma.2017.08.003
15. Ieremeiev, O., Lukin, V., Ponomarenko, N., Egiazarian, K.: Combined no-reference IQA metric and its performance analysis. Electron. Imaging **2019**(11), 260-1–260-7 (2019). https://doi.org/10.2352/ISSN.2470-1173.2019.11.IPAS-260
16. Kim, H., Lin, Y., Tseng, T.L.B.: A review on quality control in additive manufacturing. Rapid Prototyping J. **24**(3), 645–669 (2018). https://doi.org/10.1108/RPJ-03-2017-0048
17. Lech, P., Fastowicz, J., Okarma, K.: Quality evaluation of 3D printed surfaces based on HOG features. In: Chmielewski, L.J., Kozera, R., Orłowski, A., Wojciechowski, K., Bruckstein, A.M., Petkov, N. (eds.) ICCVG 2018. LNCS, vol. 11114, pp. 199–208. Springer, Cham (2018). https://doi.org/10.1007/978-3-030-00692-1_18
18. Makagonov, N.G., Blinova, E.M., Bezukladnikov, I.I.: Development of visual inspection systems for 3D printing. In: 2017 IEEE Conference of Russian Young Researchers in Electrical and Electronic Engineering, EIConRus, pp. 1463–1465, February 2017. https://doi.org/10.1109/EIConRus.2017.7910849

19. Okarma, K., Fastowicz, J.: No-reference quality assessment of 3D prints based on the GLCM analysis. In: Proceedings of the 2016 21st International Conference on Methods and Models in Automation and Robotics, MMAR, pp. 788–793 (2016). https://doi.org/10.1109/MMAR.2016.7575237

20. Okarma, K.: Combined full-reference image quality metric linearly correlated with subjective assessment. In: Rutkowski, L., Scherer, R., Tadeusiewicz, R., Zadeh, L.A., Zurada, J.M. (eds.) ICAISC 2010. LNCS (LNAI), vol. 6113, pp. 539–546. Springer, Heidelberg (2010). https://doi.org/10.1007/978-3-642-13208-7_67

21. Okarma, K.: Combined image similarity index. Opt. Rev. **19**(5), 349–354 (2012). https://doi.org/10.1007/s10043-012-0055-1

22. Okarma, K.: Quality assessment of images with multiple distortions using combined metrics. Elektronika Ir Elektrotechnika **20**(6), 128–131 (2014). https://doi.org/10.5755/j01.eee.20.6.7284

23. Okarma, K., Fastowicz, J.: Color independent quality assessment of 3D printed surfaces based on image entropy. In: Kurzynski, M., Wozniak, M., Burduk, R. (eds.) CORES 2017. AISC, vol. 578, pp. 308–315. Springer, Cham (2018). https://doi.org/10.1007/978-3-319-59162-9_32

24. Okarma, K., Fastowicz, J.: Adaptation of full-reference image quality assessment methods for automatic visual evaluation of the surface quality of 3D prints. Elektronika Ir Elektrotechnika **25**(5), 57–62 (2019). https://doi.org/10.5755/j01.eie.25.5.24357

25. Okarma, K., Lech, P.: Monte Carlo based algorithm for fast preliminary video analysis. In: Bubak, M., van Albada, G.D., Dongarra, J., Sloot, P.M.A. (eds.) ICCS 2008. LNCS, vol. 5101, pp. 790–799. Springer, Heidelberg (2008). https://doi.org/10.1007/978-3-540-69384-0_84

26. Oszust, M.: Decision fusion for image quality assessment using an optimization approach. IEEE Signal Process. Lett. **23**(1), 65–69 (2016). https://doi.org/10.1109/LSP.2015.2500819

27. Scime, L., Beuth, J.: Anomaly detection and classification in a laser powder bed additive manufacturing process using a trained computer vision algorithm. Addit. Manuf. **19**, 114–126 (2018). https://doi.org/10.1016/j.addma.2017.11.009

28. Sitthi-Amorn, P., et al.: MultiFab: a machine vision assisted platform for multi-material 3D printing. ACM Trans. Graph. **34**(4), 129:1–129:11 (2015). https://doi.org/10.1145/2766962

29. Stephens, B., Azimi, P., Orch, Z.E., Ramos, T.: Ultrafine particle emissions from desktop 3D printers. Atmos. Environ. **79**, 334–339 (2013)

30. Straub, J.: Initial work on the characterization of additive manufacturing (3D printing) using software image analysis. Machines **3**(2), 55–71 (2015). https://doi.org/10.3390/machines3020055

31. Tourloukis, G., Stoyanov, S., Tilford, T., Bailey, C.: Data driven approach to quality assessment of 3D printed electronic products. In: Proceedings of the 38th International Spring Seminar on Electronics Technology, ISSE, pp. 300–305 (2015)

32. Wang, Z., Bovik, A., Sheikh, H., Simoncelli, E.: Image quality assessment: from error visibility to structural similarity. IEEE Trans. Image Process. **13**(4), 600–612 (2004). https://doi.org/10.1109/TIP.2003.819861

33. Wu, H., Chen, T.: Quality control issues in 3D-printing manufacturing: a review. Rapid Prototyping J. **24**(3), 607–614 (2018). https://doi.org/10.1108/RPJ-02-2017-0031

34. Zeltmann, S.E., Gupta, N., Tsoutsos, N.G., Maniatakos, M., Rajendran, J., Karri, R.: Manufacturing and security challenges in 3D printing. JOM **68**(7), 1872–1881 (2016). https://doi.org/10.1007/s11837-016-1937-7

35. Zhang, L., Zhang, L., Mou, X., Zhang, D.: FSIM: A feature similarity index for image quality assessment. IEEE Trans. Image Process. **20**(8), 2378–2386 (2011). https://doi.org/10.1109/TIP.2011.2109730
36. Zhao, X., Lian, Q., He, Z., Zhang, S.: Region-based online flaw detection of 3D printing via fringe projection. Meas. Sci. Technol. **31**(3), 035011 (2019). https://doi.org/10.1088/1361-6501/ab524b

Path Markup Language for Indoor Navigation

Yang Cai$^{(\boxtimes)}$, Florian Alber, and Sean Hackett

Carnegie Mellon University, 4720 Forbes Ave., Pittsburgh, USA
ycai@cmu.edu, {falber, shackett}@andrew.cmu.edu

Abstract. Indoor navigation is critical in many tasks such as firefighting, emergency medical response, and SWAT response, where GPS signals are not available. Prevailing approaches such as beacons, radio signal triangulation, SLAM, and IMU methods are either expensive or impractical in extreme conditions, e.g. poor visibility and sensory drifting. In this study, we develop a path markup language for pre-planning routes and interacting with the user on a mobile device for real-time indoor navigation. The interactive map is annotated with walkable paths and landmarks that can be used for inertial motion sensor-based navigation. The wall-following and landmark-checking algorithms help to cancel drifting errors along the way. Our preliminary experiments show that the approach is affordable and efficient to generate annotated building floor path maps and it is feasible to use the map for indoor navigation in real-time on a mobile device with motion sensors. The method can be applied to intelligent helmets and mobile phones, including potential applications of first responders, tour guide for buildings, and assistance for visually impaired users.

Keywords: Indoor navigation · SLAM · IMU · EMS · Firefighting · Markup language · Map

1 Introduction

Indoor navigation is a growing routine in modern urban lives. People need to navigate through large infrastructures such as airports, hospitals, museums, schools, shopping malls, subways, and factories. It is critical for extreme cases such as firefighting, emergency medical responses, and SWAT team responses. It is logical to consider indoor navigation like a GPS navigator, including display of the floor map, updating current position on the map, and updating the landmarks nearby. Unfortunately, GPS signals are normally weak or available in buildings. It is rather difficult to sense the location of the user without infrastructure-dependent sensory systems. Furthermore, most available floor plans contain either too much irrelevant information such as toilet seats and furniture in the room, or too little navigational information, e.g. no connection to the user's location nor user's orientation. First responders often have a brief look at the paper drawings of a building and try to memorize it before entering. To make the matter worse, first responders often encounter the buildings with heavy smoke and poor visibility, where prevailing vision-based navigation methods would fail because we virtually have to navigate in a dark environment.

Who navigates well in the dark? The answer is blind people. In the 1970's, there was a blackout in New York City. Many people had to walk to their home without any

© Springer Nature Switzerland AG 2020
V. V. Krzhizhanovskaya et al. (Eds.): ICCS 2020, LNCS 12143, pp. 340–352, 2020.
https://doi.org/10.1007/978-3-030-50436-6_25

light. Some blind people volunteered to guide sighted people home. In this study, we want to explore a markup language to annotate building's floor plans so that the user can navigate through the building with inertial motion sensors without infrastructure-dependent beacons.

2 Related Studies

Similar to a GPS navigation system, we need a digitally annotated map that works with positioning systems. There are many geographically tagging markup languages and geocodes. For example, Keyhole Markup Language (KML) enables geographically tagging buildings, streets, and events with GPS coordinates [1]. KML has been adopted by Google Maps and Google Earth and it is a part of the Open Geospatial Consortium (OGC) [2]. OpenStreetMap [3] is an open source for millions of footprints of buildings, contributed by users. In addition to the GPS coordinates, there is the Military Grid Reference System (MGRS) [4] standardized by NATO militaries for locating points on Earth. MGRS is a multi-resolution geocode system that consists of a grid zone designator, followed by the 100,000-m^2 identifier and then the numerical location with a pair of easting and northing values. The longer digits, the more accuracy. For example, 4QFJ 123 678 has a precision level of 100 m and 4QFJ 12345 67890 can reach a precision level of 1 m. Geocode can be also embedded into images, for example, GeoTiff image format turns pixels into GPS coordinates with its metadata [5]; and HDF (Hierarchical Data Format) standardizes NASA Earth Observation System (EOS) data products, including multiple spectrum satellite sensory data and multiple object data files [6].

Outdoor map markup languages and geocode provide starting points and context for indoor navigation. A few indoor geocode and markup languages are extensions of outdoor ones. OGC's CityGML [7] is an open data model and XML-based format for the storage and exchange of virtual 3D city models. The aim of the development of CityML is to reach an international standardization for the basic entities, attributes, and relations of a 3D city model. Some schematic descriptions are relevant to emergency responses such as RoofSurfaceType, WallSurfaceType, FloorSurfaceType, and GroundSurfaceType [9]. The three dimensional descriptions about the buildings, bridges, streets, and grounds provide important information about the elevation or depth of city objects and benefit to indoor navigation. In 2014, OGC further released IndoorGML in 2014 [8], which is a data model and exchange format for the representation of the indoor navigation aspects. IndoorGML provides a topographic and semantic model of indoor space that connects to related standards like CityGML, Open Floor Plan [9], and OpenStreetMap. IndoorGML also describes multiple layers of indoor components including topographic space, WiFi sensor space, and RFID sensor space. All of the geocode, standards and exchange formats enable data sharing and emergency services [10]. In addition, OGC also released Augmented Reality Markup Language 2.0 (ARML) [11] to allow users to describe virtual objects in an augmented reality (AR) scene with their appearances and their location related to the real world. ARML 2.0 also defines a script language to dynamically modify the AR scene based on user behavior and user input, e.g. turning head or raising a hand, etc.

In summary, the OGC has moved from 2D geocode to 3D geocode, and moved from cities to building, from outdoor to indoor, and from schematic world to virtual reality, and augmented reality worlds. Perhaps, the most valuable outcome of the OGC geocode and standards lies in its potential of crowdsourcing for massive geocoded data about buildings, interiors, and floor plans. However, despite a broad spectrum of indoor facility markup languages, the existing methods are too abstract and complicated; and there are many gaps to fill in order to be useful in indoor navigation. For example, how to convert a floor plan drawing in PDF format into a geocoded floor plan? How to use the floor plan interactively in real-time indoor navigation?

Localization is the most critical component in indoor navigation. We need to know where the user is and which direction the user is heading in the building. Traditional Dead Reckoning, or Inertial Motion Unit (IMU) sensor-based approaches can work in totally dark environments, but they have notorious accumulative drifting problems over a period of time [12]. Recent Renaissance of IMU sensor-based methods are enhanced for better accuracy by placing IMU sensors on shoes [13] or fusing with other sensors [14]. The prevailing approach is the beacon-based localization, including ultrasound [15], LoRa [16], and WiFi [17]. Installing and calibrating beacons in a building are expensive and there are wall-attenuation problems [18]. There are growing technologies of infrastructure-free localization by mobile beacons [19] or collaborative positioning [20]. Rapidly growing mobile robotics technologies bring new dimensions to indoor navigation. Simultaneous Localization and Mapping (SLAM) algorithm [21] has been popular for 3D modeling from motion, tracking and mapping at the same time. Single RGB camera-based visual SLAM can generate the motion trajectory in a relative 3D space. Stereo and RGB-Depth camera-based SLAM yield absolute 3D coordinates of the trajectory. Visual SLAM is computationally expensive and it often fails in poor lighting, smoky, or feature-less environments such as a painted white wall. Some RGB-D sensor-based SLAM incorporate IMU sensors for more accurate localization results. LiDAR-based SLAM can work in the dark by tracking the 3D point clouds but its cost is very expensive [22]. Thermal IR cameras can also be used for SLAM but its images are rather low-resolution and it's expensive as well [23].

In summary, there is no silver bullet in indoor localization. The technologies for large-scale localization in normal environments or extreme conditions such as fire and smoke are not mature yet. There are gaps between the available technologies and applications. For example, there is little connection between the geographic markup languages and indoor navigation technologies. Many sensors need pre-calibration. Some of them such as magnetic field sensors need to be calibrated each time of usage. Self-calibration methods have implemented, for example, DJI drones use a motor to rotate the magnetic sensor before taking off [24]. A few novel concepts might pave the way for affordable and practical indoor localization, for example, the mobile device for helping visually impaired users to navigate indoors [25]. The assistive technology is affordable and interactive with a wall-following function. In nature, there are also other modalities for navigation based on smell intensity, lighting, sound, magnetic field, and simply tactile sensing [26]. Biomimicry teaches us to look into novel sensors and fusion algorithms, for example, the one-dimensional LiDAR and IMU sensor for first-person view imaging [14].

3 Path Markup Language

Although OGC's IndoorGML includes the IndoorNavigation module, it only provides standards and a high-level framework, rather than functional indoor navigation solutions. In this study, we propose the Path Markup Language (PML) as a data model and schema specifically for indoor navigation pre-incident planning and real-time navigation guidance. Currently, PML contains the following geocode elements: footprint, floor plan, path, landmark, and waypoint. These objects can be expressed in XML schema.

Footprint is the boundary of a building in a polygon. It can be extracted directly from Google Earth manually, or with machine vision. The coordinate points can also be downloaded from Google Maps or OpenStreetMap but it is not guaranteed because it depends on user online contributions. The coordinates are normally in GPS format and the sequence is counter-clockwise. The XME schema of Footprint is following:

```
<pml:Footprint>
        <pml:Polygon>
                <pml:coordinates>0,0 100,0 100,100 0,100 0,0
                </coordinates>
        </pml:Polygon>
</pml:Footprint>
```

The floor plan is a hierarchical structure of anchor points, footprint, rooms, paths, landmarks, and waypoints. It takes at least 3 anchor points to scale and align a floor plan to a georeferenced map such as Google Maps. Normally floor plan drawings are CAD drawings without any georeference. In PML, we overlay the floor plan to Google Maps by scaling, rotating and translating to extract the GPS coordinates directly.

Path is a critical element in indoor navigation. We assume a building is not an empty stadium. Instead, it contains walls, hallways, and barriers. We assume that humans and robots can only walk on the paths without breaking walls or barriers. This assumption helps to reduce the IMU-based localization drifting through walls. Instead, the estimated trajectory will be along the Paths. In the PML, a Path is omnidirectional and it is a sequence of line segments with widths.

```
<pml:Path>
        <pml:Name> "Hallway" </pml:Name>
        <pml:Width> 1 </pml:Width>
                <pml:Line>
                        <pml:coordinates>0,0 10,0 10,10 0,10
                        </coordinates>
                </pml:Line>
</pml:Path>
```

Landmark is also a critical element in PML. Here we assume the IMU-based indoor navigation system has the sensory drifting problem and there are landmarks along the paths. When the user approaches the landmark nearby, the navigation system will send a confirmation request. Once the landmark is acknowledged, the localization track starts over again and the drift is canceled before it is accumulated further. A Landmark can be labeled with a symbol, for example, "E" as elevator and "S" as stairs. It also can be displayed with an icon.

```
<pml:Landmark>
        <pml:Name> "E" </pml:Name>
        <pml:Icon>elevatorIcon.pmg </pml:Icon>
                <pml:coordinates>5,5
                </coordinates>
</pml:Landmark>
```

Finally, Waypoint is the location of the user including heading and coordinates. It will be updated in real-time to display the current position and orientation of the user. Waypoints can be stored and displayed as a digital pheromone along the Paths. The pheromone trace can be turned off (0), without decay (1), or with decay (2).

```
<pml:Waypoint>
        <pml:Name> "Me" </pml:Name>
        <pml:Icon>RedArrow.pmg </pml:Icon>
        <pml:coordinates>7,9</coordinates>
        <pml:heading>245</pml:heading>
        <pml:trace>2</pml:trace>
</pml:Waypoint>
```

4 Mobile System Architecture

To implement PML, we aim to combine geographic markup language with real-time indoor navigation algorithms into a simple and affordable working system. The system contains two modules like displayed in Fig. 1: Map Generation and Navigation Guide. For the Map Generation module, a map can be generated for any building with a floor plan and that can be GPS tagged using downloaded or online Google Maps. The floor plan of the building is required in order to map the building's indoor features. This floor plan is imported as an image and overlaid with the footprint of the building from Google Maps which provides the GPS coordinates for navigation within the floor plan. The overlaid floor plan can be scaled and rotated to match the building's footprint on Google Maps. The map is then annotated with important features including the rooms, hallways, flights of stairs, elevators, doorways, etc. The annotated map is then exported as a csv file for example, which can then be used by the navigation guide application.

The Navigation Guide module utilises the generated map as a bounded region to navigate within. Tracking begins at the entrance to a building where Android Localisation (GPS, mobile network, etc.) is still accurate and can be used as a true starting point. This starting position can then be confirmed or be set manually if localistion is not accurate e.g. starting inside the building. The use of the Android Step Detection and Android IMU are then used to track the movement and orientation of the user through the buildings walkable paths. The wall-following algorithm reduces drifting by bounding the tracked path within the walkable paths and hugging corners. The landmark-checking is a manual approach to correct drifting when the user approaches a landmark. After reaching the landmark the user can confirm this and the user position will be updated to this landmark.

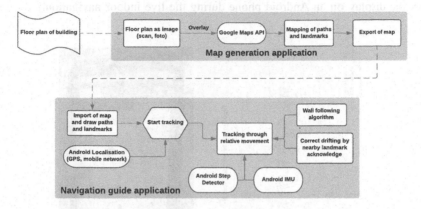

Fig. 1. System architecture

The pseudo code for map generation is as follows:

```
load Google Maps API;
user imports floor plan image as overlay;
transform image to fit GPS footprint on Google Maps;
lock overlay image based on true GPS coordinates;
start mapping based on overlayed floor plan:
        draw paths;
        set landmarks;
export map;
```

PML also includes real-time human-computer interaction interfaces. The pseudo Code for Indoor Navigation:

```
import map;
draw simple map of paths and landmarks;
set starting point based on Android Localisation;
start tracking:
        get step event from Android Step Detector;
        get imu data to calculate direction of movement;
        Wall-Following Algorithm;
        if user confirms landmark:
                update position and correct drift;
```

5 Map Annotation

For the navigation app to have a floor plan with a walkable area and identifiable features, a geocode-annotated map must be supplied. After an image of the building's floor plan is overlayed with the Google Maps building footprint, all annotations that are added will be tagged with GPS coordinates. The path of walkable areas can be added as polygons and a number of landmarks including stairs, doors, elevators, corners can be tagged on the map with a corresponding icon shown in Fig. 2. The left image of Fig. 3 shows the annotated floor plan with paths and landmarks. The right images on Fig. 3 shows the display on an Android phone during the live indoor navigation.

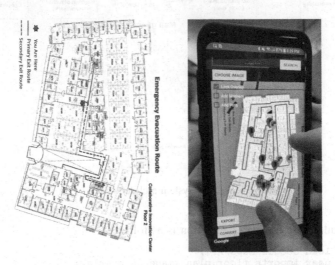

Fig. 2. The floor plan (left) and overlaid floor plan on top of Google Maps building (right)

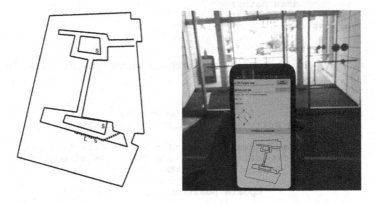

Fig. 3. Generated navigation map with footprint, paths and landmarks (left) and the display on an Android phone (right)

6   Inertial Sensory Fusion for Steps and Orientation

The Navigation Guide application currently uses the built-in sensors of a standard Android phone. For this approach data from the IMU and from the magnetic field sensor is used to detect the relative movement inside a building. The first challenge is to detect the movement of the user, which can be defined for a person walking over the steps taken. This is a simple approach to track the person and enables it already to test our navigation concept. For later use it would be necessary to detect the size of the steps or calibrate the application for every user and its own step size. Additionally, for a real usage scenario it's necessary to update this movement detection to a more complex one, with which different moving styles can be tracked. Especially for the firefighter scenario, where a variety of walking, crawling, shuffle walking and other movement styles are frequently used. To detect the steps for a walking scenario on an Android phone, it's possible to detect the steps over a simple state machine based on the peaks in the acceleration data or to use the already built-in Step Detector in the Android SDK as described in [27]. During the first trial runs it was noticeable that the already built-in feature can detect steps very accurately. The Step Detector analyzes the acceleration data of the phone and based on that it detects a step movement, which triggers an event. This event can be used to account for the step and track the movement of the person.

With the detection of the movement of the user it is now important to detect where the user is heading. For that the approach is to use the magnetic field sensor and detect the direction of the movement, with the limitation that external magnetic fields can disturb the detection. In this application the data of the magnetic field sensor gets read out and it gets filtered by a lowpass filter in the form of an exponentially weighted moving average like:

$$C[i] = \alpha \cdot B[i] + (1 - \alpha) \cdot C[i - 1] \tag{1}$$

where, $B[i]$ and $C[i]$ are input and output on the discrete time-domain data with $i \in N_0$ and α as the corresponding weight variable.

Those filtered values are a relative measurement from the phone and now it's necessary to define the orientation of the phone to calculate the right directions. This is already possible with built-in functions of the Android SDK. First a so-called rotation matrix can be calculated, which transforms the magnetic field measurement based on the gravity measurement from the device coordinate system in a global coordinate system like described in [27]. The rotation matrix can then be used to calculate the orientation of the phone as Azimuth, Pitch and Roll. For our movement the Azimuth is especially important, because it describes the rotation around the gravity axis as the angle between the facing direction of the user and the direction to the magnetic north pole. For that reason the Azimuth can directly be used to define the orientation of the movement of the user.

This angle could be inaccurate if only one measurement of the direction of the movement gets evaluated. For that reason the phone constantly measures the orientation and whenever a step is taken, we can average the measuring values during the step and use the average angle of the direction to place the next step. The new location of the step at $x[i]$ and $y[i]$ coordinates can be calculated over the current position and the average Azimuth φ_{avg} to:

$$x[i] = x[i - 1] + \Delta x \cdot \sin(\varphi_{avg}), \qquad y[i] = y[i - 1] + \Delta y \cdot \cos(\varphi_{avg}) \tag{2}$$

with the scaled step size Δx and Δy. Those scaled step sizes result from the scaling of the step size to the geological coordinates of the steps, which converts the feet per step to latitude and longitude per step. With that it's possible to define the next step and the direction of the movement.

7 Wall-Following Algorithm

Our major assumption is that the user only walks, crawls, or runs along the predefined paths. The user won't walk through a wall, for example. This assumption helps the indoor navigation algorithm to reduce the impact of the IMU sensory integral drifting errors. The wall-following algorithm estimates the user's position by the measurement data from the accelerometers and magnetic sensors. Due to the drifting error, the estimated position may drift away from the annotated path, for example, pass through the wall in the hallway. Therefore, we need to detect the collision between the estimated user location and the boundary of the path, e.g. a wall.

A collision occurs whenever the next step would be outside of a walkable path and the line of the step intersects with the path outline. For that reason it's possible to check for collisions after every new calculated step, if it would be outside an allowed path. The implementation of this check iterates over the path outline polygons and checks if there is an intersection with the connection line between the last step to the new one. After iterating over the path outline polygons, the result directly shows if a collision for this new step would occur. When the collision is detected, the algorithm corrects the trajectory and updates the user's position *along the border of the path*. Figure 4 illustrates the wall following method.

Additionally, it could be possible that at a path crossing the tracking takes a wrong turn and follows the wrong path like displayed in the right illustration of Fig. 4. If the direction of those two paths diverge, then the tracking would sooner or later head into a wall. If we now account for the steps it would make in that direction and solve those conflicts with the above described collision detection, so it could be possible that those counted steps would reach over to the other path. If that is the case, we can cut this corner and move the position to the other path and go from there. With that we lose accuracy, but we can undo a potentially fatal error of the tracking approach.

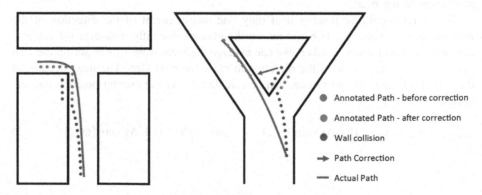

Fig. 4. Wall following (left) and corner cutting (right)

8 Indoor Navigation Experiments

A preliminary lab experiment has been conducted at an office building on the first floor including hallways, elevators and stairs. The length of the building is about 200 m. The footprint and floor plan are available from Google Maps. We also obtained the scanned floor plan with details of rooms. After aligning the scanned floor plan with Google Maps, we obtained the geocode coordinates of the floor plan. We then annotated landmarks on the floor plan with elevators, stairs, and doors. Figure 5 through Fig. 6 shows the results of four tests in the building. Our tests show that the wall-following algorithm indeed corrected the IMU drifting errors and put the trajectories back to the path. We found that corners on the floor plan might be helpful to be additional landmarks. The tests also show weaknesses of the algorithm to be improved, for example, the starting point of indoor navigation. We need to start tracking the location waypoint before entering the building when the GPS signal is available. We also found the collision detection algorithm may get stuck at a certain point when the walking angle is perpendicular to the wall. Besides, if the landmarks are too far apart, or the drifting error is too large, then the navigation might fail. Nevertheless, our initial experiments prove that this simple and affordable approach is feasible in a realistic building environment and have a reasonable accuracy within 1–1.5 m, which is acceptable to many humanitarian rescue and recovery tasks.

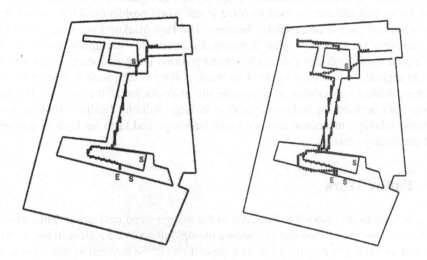

Fig. 5. Indoor navigation experiment at the office building (part 1)

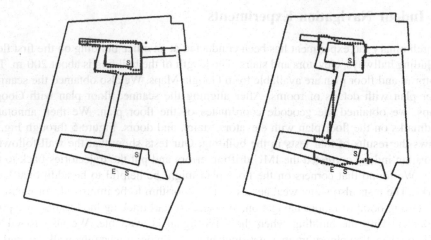

Fig. 6. Indoor navigation experiment at the office building (part 2)

9   Conclusions

In this paper, we proposed a novel Path Markup Language for indoor navigation applications. We have shown that the mapping and navigation can be integrated into a modular system and can be used to solve a real world problem tackling indoor navigation without the use of expensive beacons. The Path Markup Language can be used to intuitively create a floor plan featuring landmarks for navigation purposes. The navigation application can then track a users position using pedometers and wall following algorithms to reduce errors. Landmark polling is then used to reset this error, allowing human supervision to overcome the inaccuracies of the system. The technology can be used for indoor navigation in large building facilities such as malls, airports, subways, museums, schools, office buildings, and factories for tour guidance, and emergency services.

10   Future Work

Several features are planned to be added to the mapping and navigation applications to further enhance their functionality. Adding multiple floors to a building in the mapping app and the ability to transition between these floors in the navigation app. This would be achieved through a push notification when in the location of stairs/elevator landmark, e.g. "Up 1 floor", "Down 1 floor". With these features a 3D view of the building's floors and the user's current position could be added.

Like satellite navigation systems in vehicles do not display a full road map, only a small section that the car is currently in, and that rotates based on the orientation of the car, this feature would improve the view and intuition of the navigation app.

Automatic sensor calibration is necessary for the future work, including walking stride calibration, magnetic sensor calibration, as well as altimeter calibration. The more sensors we throw in the system, the more calibration we need. Automatic calibration can be implemented with sensory fusion, e.g. calibrating stride with laser distance measurement and calibrating altimeter with satellite signals while the system is outside of the building.

In addition, we are developing the indoor navigation on a helmet for first responders to view the navigational information from a projected heads-up display (HUD). This would free up their hands for emergency services and enhance the augmented reality experience in harsh environments, for example, see-through smoke and walls.

Acknowledgement. This work was performed under the financial assistance award 70NANB17H173 from the U.S. Department of Commerce, National Institute of Standards and Technology, PSCR Division and PSIA Program. This project is also funded in part by Carnegie Mellon University's Mobility21 National University Transportation Center, which is sponsored by the US Department of Transportation. The authors are grateful to the NIST PSCR Program Manager Jeb Benson for his comments and suggestions about the technical development of the hyper-reality helmet system.

References

1. Keyhole Markup Language (KML). https://developers.google.com/kml
2. Open Geospatial Consortium (OGC). https://www.opengeospatial.org/standards/kml/
3. OpenStreetMap. https://www.openstreetmap.org/#map=5/38.007/-95.844
4. Military Grid Reference System. https://en.wikipedia.org/wiki/Military_Grid_Reference_System
5. GeoTiff, WikiPedia. https://en.wikipedia.org/wiki/GeoTIFF
6. HDF. https://nsidc.org/data/hdfeos/intro.html
7. CityGML. https://www.opengeospatial.org/standards/citygml
8. IndoorGML. http://www.indoorgml.net/
9. Schema of OGC CityML. http://schemas.opengis.net/citygml/building/2.0/building.xsd
10. OGC Hosts Indoor Location and Floor Plan Standards Forum. https://www.opengeospatial.org/pressroom/pressreleases/1122
11. ARML. https://www.opengeospatial.org/standards/arml
12. Dead Reckoning. https://en.wikipedia.org/wiki/Dead_reckoning
13. Xiao, Z., Wen, H., Markham, A., Trigoni, N.: Robust Indoor Positioning with Lifelong Learning. https://www.cs.ox.ac.uk/files/9047/Xiao%20et%20al.%202015.pdf
14. Cai, Y., Hackett, S., Alber, F.: Heads-Up LiDAR imaging, to appear on IS&T, Electronic Imaging Conference, 20 January 2020
15. Lin, Q., An, Z., Yang, L.: Rebooting ultrasonic positioning systems for ultrasound-incapable smart devices. https://arxiv.org/pdf/1812.02349.pdf
16. Indoor positioning via LoRaWAN, indoornavigation.com. https://www.indoornavigation.com/wiki-en/indoor-positioning-via-lorawan
17. WiFi positioning system, WikiPedia. https://en.wikipedia.org/wiki/Wi-Fi_positioning_system

18. Zafari, F., Gkelias, A., Leung, K.K.: A survey of indoor localization systems and technologies. arXiv: https://arxiv.org/pdf/1709.01015.pdf
19. Wang, Q., Lou, H., Men, A., Zhao, F.: An infrastructure-free indoor localization algorithm on smartphone. Sensors **18**(10), 3317. https://www.researchgate.net/publication/328067193_ An_Infrastructure-Free_Indoor_Localization_Algorithm_for_Smartphones
20. Noh, Y., Yamaguchi, H., Lee, U.: Infrastructure-free collaborative indoor positioning schema for time-critical team operations. IEEE Trans. SMC. **48**(3) (2018). https://ieeexplore. ieee.org/abstract/document/7747408
21. SLAM, WikiPedia. https://en.wikipedia.org/wiki/Simultaneous_localization_and_mapping
22. Mathworks. Implement SLAM with Lidar Scans (2020). https://www.mathworks.com/help/ nav/ug/implement-simultaneous-localization-and-mapping-with-lidar-scans.html
23. Shin, Y.S., Kim, A.: Sparse depth enhanced direct thermal-infrared SLAM beyond the visible spectrum. arXiv:1902.10892
24. DJI Mavic Pro manual. https://dl.djicdn.com/downloads/mavic/20171219/Mavic%20Pro% 20User%20Manual%20V2.0.pdf
25. Sato, D., Oh, U., Naito, K., Takagi, H., Kitani, K., Asakawa, C.: NavCog3: an evaluation of a smartphone-based blind indoor navigation assistant with semantic features in a large-scale environment. In: ASSET 2017, Baltimore, MD, USA, 29 October–1 November 2017. https://www.ri.cmu.edu/wp-content/uploads/2018/01/p270-sato.pdf
26. Barrie, D.: Supernavigators: Exploring the Wonders of How Animals Find Their Way. The Experiment, LLC, New York (2019)
27. Google, Android Developer API references. https://developer.android.com/reference

Smart Fire Alarm System with Person Detection and Thermal Camera

Yibing Ma[1](✉), Xuetao Feng[1], Jile Jiao[1], Zhongdong Peng[1],
Shenger Qian[1], Hui Xue[1], and Hua Li[2]

[1] Alibaba Group, Beijing, China
{yibing.myb, xuetao.fxt, jile.jjl, zhongdong.pzd,
shenger.qse, hui.xueh}@alibaba-inc.com
[2] Institute of Computing Technology, Chinese Academy of Sciences, Beijing,
China
lihua@ict.ac.cn

Abstract. Fire alarm is crucial for safety of life and property in many scenes. A good fire alarm system should be small-sized, low-cost and effective to prevent fire accidents from happening. In this paper we introduce a smart fire alarm system used in kitchen as a representative scenario. The system captures both thermal and optical videos for temperature monitoring and person detection, which are further used to predict potential fire accident and avoid false alarm. Thermal videos are used to record the temperature change in region-of-interests, for example, cookware. YOLOv3-tiny algorithm is modified for person detection and can be iteratively improved with the hard examples gathered by the system. To implement the system on an edge device instead of a server, we propose a high-efficiency neural network inference computing framework called TuringNN. Comprehensive rules enable the system to appropriately respond to different situations. The proposed system has been proved effective in both experiments and numerous cases in complex practical applications.

Keywords: Fire alarm system · Thermal camera · Person detection · TuringNN

1 Introduction

Fire prevention is crucial for the safety of life and property in residential and public scenes such as restaurants, hotels and factories. According to statistics [1, 2], cooking is the leading cause of hotel fires (46%) and one out of every 12 hotels reports a structural fire per year in US. A fire accident usually happens when the cookware is kept heated on the stove without supervision. Although fire regulations have been made to prevent fire accidents, they are sometimes ignored or disobeyed. The earlier the accidents can be detected, the easier they can be dealt with and the fewer damages can be caused. Therefore, it is necessary to build a smart alarm system for fire prevention. The system needs two essential modules, potential fire sensing and proper disposal strategy.

V. V. Krzhizhanovskaya et al. (Eds.): ICCS 2020, LNCS 12143, pp. 353–366, 2020.
https://doi.org/10.1007/978-3-030-50436-6_26

There are various methods for fire sensing such as by detecting smoke, carbon monoxide or flame. However, these methods work only when the fire has broken out. In contrast, monitoring the temperature of combustibles makes it possible to prevent fire before it occurs. There are three necessary ingredients for most fires, i.e. fuel, oxygen and heat. It is feasible to predict potential fire accidents when the temperature rises too quickly or near the ignition point. Temperature measurement sites in application scenario, e.g. stoves in kitchen, are usually numerous and irregularly distributed. Consequently, it is more practical to deploy remote and wide-ranged thermal cameras than single-point thermometers to monitor temperature.

Nevertheless, merely monitoring temperature may lead to the problem of false alarm when someone is working in the scene, for example a chef in the kitchen. As a correction, it is necessary to judge the presence of persons. The task can be fulfilled by person detection. With the development of computer vision and deep learning, person detection has been widely researched and various algorithms have been proposed. Surveillance cameras have become common in public places, providing clear and real-time videos. But there are still challenges for person detection in different sites. For instance, chefs in kitchen usually wear white suits with masks and tall hats, which are rare in public datasets and difficult to detect for existing models.

Moreover, the system should be reliable, low-cost and easy to deploy. Transmitting all the videos to one computing server is a waste of network bandwidth. Instead, person detection should be implemented on distributed mobile processors on an edge device and only detection results are sent to communication server. To ensure the speed of person detection, a device-specific deep learning acceleration framework is necessary.

In this paper, we propose a smart fire alarm system with bi-spectrum camera and embedded Industry Personal Computer (IPC). The system is applied in kitchens and the diagram is shown in Fig. 1. The camera has an optical lens and a thermal lens to capture visible videos and monitor the temperature of cookware respectively. The processor performs person detection with YOLOv3-tiny [3] model. The algorithm is modified to deal with special appearance and occlusions of chefs in kitchen, and the inconsistency of aspect ratio between different training datasets. To implement the system on a mobile processor, we propose a high-performance neural network inference computing framework called TuringNN. Person detection and temperature analysis are completed in real time on the IPC. When the temperature reaches preset thresholds and no person is detected, staff in charge are alarmed with buzzers in the kitchen, phone calls or instant messages from mobile applications. Meanwhile, the IPC reports an alarm along with current temperature, video frames and detection results to the server. In severe cases when temperature gets too high or rises too fast, the IPC sends a signal to controller to shut off the stove. As the system runs, incorrect person detection cases are collected to improve the model, and disposal rules can be updated according to feedback. The system has been applied for kitchen fire alarm in Freshhema [4], a combination of fresh food supermarket and restaurant with over 130 chain stores all over China. By adjusting detection model and disposal rules, we can apply the system to various application situations.

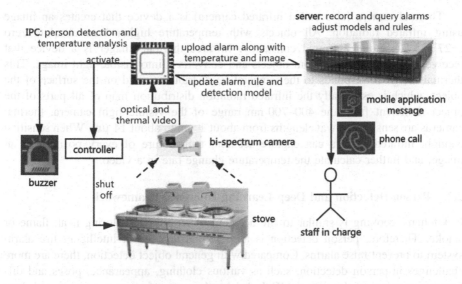

Fig. 1. The diagram of our system.

The remainder of the paper is organized as follows. Section 2 reviews the related work on fire alarm systems and person detection. Section 3 presents the details of the proposed system. Section 4 introduces the experiments and shows the respective results. Section 5 concludes the whole paper and gives some directions for future work.

2   Related Works

2.1   Fire Alarm System and Thermal Camera

Fire safety solutions have been widely researched for years. A fire alarm system has a number of devices working together to detect fire and warn people through visual and audio appliances when smoke, flame, carbon monoxide or other emergencies occur. These alarms may be activated automatically from smoke detectors and heat detectors or be activated via manual fire alarm activation devices like manual call points. Siemens [5] proposed a solution of fire detection in kitchens with fire detector which consists of smoke, carbon monoxide and flame detectors. However, this solution only works after fire has happened. Moreover, although Siemens claims that selectable parameter settings make the detectors immune to deceptive phenomena, they are not well adaptive to environment changes. Cheng et al. [6] proposed a fire safety device for stove-top burner. It senses the temperature of the cooking utensil and automatically shuts off the flow of electricity or gas to the burner when the temperature of the cooking utensil exceeds threshold. However, the device does not consider the presence of persons, which may cause false alarms. There are some other fire alarm solutions such as [19–21], which are similar to [5] and [6]. The alarms are made after the fire happens, which lead to delayed handling.

Thermal camera (also called infrared camera) is a device that creates an image using infrared radiation. All objects with temperature higher than absolute zero (−273.15 °C) radiate infrared constantly. Infrared thermal imager is a device that receives infrared radiation from objects and converts it into visible-light image. This thermal image corresponds to the temperature distribution field on the surface of the object, which is essentially the infrared radiation distribution map of all parts of the object. Different from the 400–700 nm range of the visible light camera, thermal cameras are sensitive to wavelengths from about 1 μm to about 14 μm. When sensitive enough, infrared camera can capture accurate temperature of every position in an image, and further calculate the temperature change rate in a video.

2.2   Person Detection and Deep Learning Inference Framework

In kitchens, cooking is similar to fire accident when only monitoring heat, flame or smoke. Therefore, person detection is an essential task in an intelligent fire alarm system to prevent false alarms. Compared with general object detection, there are more challenges in person detection, such as various clothing, appearance, poses and different viewpoints of the camera. With the development of computer vision and deep learning, various object detection algorithms are proposed such as Faster R-CNN [7], SSD [8] and YOLO [9]. Although new algorithms are constantly proposed and better performances have been achieved, they usually require lots of computing resources and long inference time. In application scenarios, deep learning models should be economical and efficient on mobile platforms with limited computing power.

Numerous techniques have been proposed to accelerate deep learning models on low-power devices. For example, model quantization refers to the process of reducing the number of bits that represent a number [22]. In the context of deep learning, the predominant numerical format used for research and for deployment has so far been 32-bit floating point. However, the pursuit for reduced bandwidth and computing requirements of deep learning models has driven research into using lower-precision numerical formats. It has been proved that weights and activations can be represented using 8-bit integers without incurring significant loss in accuracy. By combining various acceleration techniques, researchers have built deep learning inference frameworks for mobile devices, such as TensorFlow Lite [11] and NCNN [12]. The frameworks make deep learning models faster and smaller while maintain or even improve the performance by combining multiple optimization methods. Despite the convenience of public frameworks, there are still adaption problems for specific devices and network structures, and there is still space for further optimization.

3   The Proposed System

The workflow of our system is demonstrated in Fig. 2. It consists of four modules, (a) temperature measurement, (b) person detection, (c) TuringNN and (d) disposal rules.

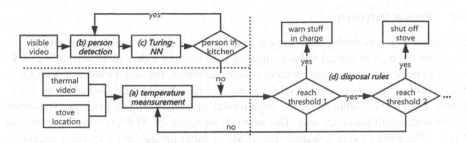

Fig. 2. The workflow of our system.

3.1   Temperature Measurement

The temperature is measured with thermal videos. First of all, the locations of stoves are manually annotated in thermal videos, as shown in Fig. 3 right. The temperature of target locations is sampled with an adjustable time interval. The heating curve is fit with least square method according to the discrete temperature data points in observation period. The slope of the curve, namely current temperature ascent rate R, is calculated as

$$R = \frac{n \sum_{i=1}^{n} it y_i - \sum_{i=1}^{n} it \sum_{i=1}^{n} y_i}{n \sum_{i=1}^{n} (it)^2 - \left(\sum_{i=1}^{n} it \right)^2} \tag{1}$$

where the last n temperature data points are used for calculation and n is set to 20 in our system, t denotes the temperature check time interval and is set according to current temperature as shown in Table 1, y_i represents the values of the n temperature data points. If the difference between the current temperature and the previous temperature is more than 20 °C, the current temperature is ignored as a noise.

Fig. 3. Optical video (left) and thermal video (right) of the same site. The temperature measurement locations are annotated as red boxes. (Color figure online)

3.2 Person Detection

Person detection is deployed on optical video because the color and texture of persons are more obvious in optical video compared with thermal video, as shown in Fig. 3. In order to achieve accurate and efficient person detection, the small model of YOLOv3-tiny is utilized in our system. YOLOv3-tiny is pruned from YOLOv3 [3] with fewer layers and scales. Compared to the large model of YOLOv3, YOLOv3-tiny is 5 times faster with small accuracy loss. The network structure of YOLOv3-tiny is shown in Fig. 4. The model extracts feature map (FM) of input image with a 13-layer backbone network. YOLOv3-tiny predicts bounding boxes at two different scales (with stride 8 and 16) with the idea of Feature Pyramid Network for Object Detection (FPN) [10]. Two detection headers, which are separately built on the top of two feature maps with different scales, are responsible for detecting objects with different sizes. As shown in Fig. 4(b), each grid in the detection header is assigned with K different anchors, and thus predicts K detections that consist of 4 bounding box offsets, 1 objectiveness and C class predictions. In our case, K is set to 3 and C is 1 to only detect person. The final result tensor of detection header has a shape of $N \times N \times (K(4 + 1 + C))$ where $N \times N$ denotes the spatial size of the last convolutional feature map. Three modifications are proposed to make it more suitable for our system.

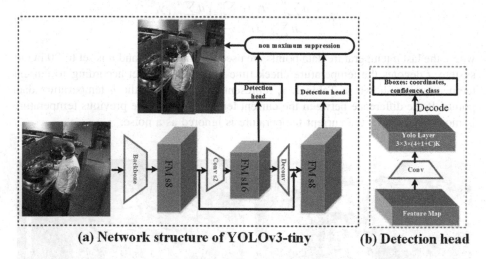

(a) Network structure of YOLOv3-tiny   (b) Detection head

Fig. 4. The network structure of YOLOv3-tiny. K denotes the number of anchors in each scale and is set to 3 in YOLOv3-tiny.

Wide augmentation, which means data augmentation while keeping aspect ratio. To deal with the difference of person appearance between public datasets and kitchen and to guarantee the generalization ability of the model, we train it with both public datasets and application dataset of surveillance videos. As shown in Fig. 5(a) and Fig. 5(b), public datasets have various aspect ratios while surveillance videos have fixed aspect ratio of 16:9. When training the model, if we directly resize all images into the same aspect ratio, some ground truths will suffer from severe distortion, as illustrated in Fig. 5(c). Therefore, we keep the aspect ratio of images unchanged by adding borders to the image in data augmentation, as demonstrated Fig. 5(d).

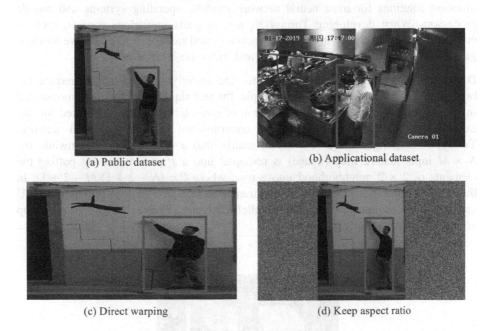

(a) Public dataset

(b) Applicational dataset

(c) Direct warping

(d) Keep aspect ratio

Fig. 5. Different ways to change image size.

Ignore Label. The original YOLO distinguishes positive and negative objects with an Intersection-Over-Union (IOU) threshold. However, the ground truth is ambiguous in some cases such as crowd, heavy occlusion and blur. To deal with this issue, we introduce "ignore" label in annotation. The object is annotated with ignore label if it is a crowd of persons, a part of a person or when we are not sure if the object is a person or not. In training, if the IOU between a box proposal and any ground truth box with ignore label is larger than threshold, the proposal is ignored in loss calculation. In this way the noise of unclear targets is avoided.

Iterative Improvement. With the operation of the system, the incorrect person detection instances are accumulated along with the false fire alarms. The instances can be used as hard examples (images that cause the model make errors) to improve the detection model. The new model is finetuned from current model with new instances as well as original training data, which effectively avoid overfitting. Each time the model is updated, the accumulation of hard examples and the next iteration becomes slower. Finally, the model turns stable and accurate.

3.3 Neural Network Inference Acceleration with TuringNN

TuringNN is a novel deep learning inference framework optimized for mobile heterogeneous computing on Android, iOS, Linux and Windows devices. It has optimization solutions for most neural network models, operating systems and mobile processors. When developing TuringNN, we set goals in various aspects, such as customization, resource-saving, high performance and easy-using. To achieve all these goals, comprehensive technical solutions and tricks are considered.

Data Rearrangement. Rearrange data for the underlying hardware to reduce the bottleneck of memory reading. For example, the technique of Im2col is demonstrated in Fig. 6. A straightforward implementation of convolution is not well suited for fast execution on CPU. When dealing with convolutional layers in a neural network, TuringNN rearranges the 2×2 kernel matrix into a 4×1 array. Meanwhile the $N \times M$ input feature map (image) is reshaped into a $P \times 4$ matrix by putting the elements of 2×2 neighborhood into a row, where $P = (N - 2 + 1)(M - 2 + 1)$. In this way, the operation of convolution is translated to matrix multiplication. The CPU can then take advantage of special parallel hardware (SSE or MMX) to speed up convolution substantially.

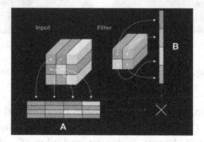

Fig. 6. Im2col-based implementation of convolution.

Quantitative Calculation. 8-bit data is calculated for the underlying hardware to increase data multiplication and throughput. The quantitative calculation chart is shown in Fig. 7. Firstly, TuringNN runs a model of 32-bit floating point (FP32) with a calibration dataset and records the maximum and minimum of the feature map in each layer. During inference, the weights and input in each layer are converted into 8-bit integers (INT8) with the recorded maximum and minimum, and then the output is turned back to FP32. The quantization and dequantization formula are noted in Eq. (2), where F and Q are the FP32 and INT8 values, while F_{max} and F_{min} are the maximum and minimum FP32 values of each layers. As shown in Fig. 7 left, the dequantization of the previous operation and the quantization of the next operation will cancel each other out. Therefore, TuringNN combines the dequantization and quantization before each operation when there are multiple quantified operations in succession, as illustrated in Fig. 7 right.

$$Q = \frac{255(F - F_{min})}{F_{max} - F_{min}} - 128, \quad F = \frac{(Q + 128)(F_{max} - F_{min})}{255} + F_{min} \quad (2)$$

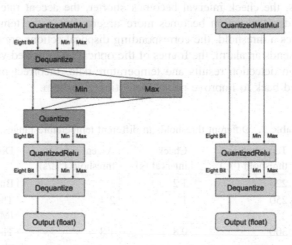

Fig. 7. The quantitative calculation charts. Dequantization are canceled out by quantization of the next layer.

Multithreading. TuringNN makes full use of the ability of multithreading. When the compiler supports OpenMP, we set the concurrent number according to the hardware device, and the concurrent time-consuming cycle will use multi-core to complete the operation. On Apple's platform, we use Grand Central Dispatch (GCD) to dispatch the time-consuming cycle, and the system dynamically allocates the concurrent number, so as to obtain better performance than OpenMP. On Android and Linux platform, using thread pool for multithreading management and control can make more efficient use of multi-core and large-small-core hardware resources.

Memory Layout. Different data arrangement formats are designed for different operators. For example, operators such as convolution and pool run on the memory layout of batch number/channel/height/width (NCHW), while operators such as post and split use the memory layout of NHWC. When the memory layouts of adjacent operators are inconsistent, conversion operators are inserted.

Fusing. When transforming models, we merge layers that do not need to exist independently to reduce the amount of calculation. For example, by merging batch normalization into convolution, the cost of batch normalization is saved while the operation result is not affected.

3.4 Disposal Rule and Devices

The disposal rules of industrial personal computer (IPC) in our system are demonstrated in Fig. 2. The optical lens and thermal lens focus on the same view of the kitchen. Person detection is continuously run in optical videos. If no person is detected, current temperature of cookware is measured from thermal videos. When food is fried, the oil temperature rises fast and then slows down. According to this rule, different temperature thresholds are set for temperature analysis, as listed in Table 1. As the temperature rises, the check interval becomes shorter, the accent rate threshold gets smaller and the disposal action becomes more urgent. When the temperature or the ascent rate reaches a threshold, the corresponding disposal actions are triggered. Each time the system sends an alarm, the frames of the optical and infrared videos are saved along with person detection results and temperature data. Incorrect person detection results will be fed back to improve person detection algorithm.

Table 1. Different thresholds in different temperature ranges.

Temperature range (°C)	Temperature threshold (°C)	Check interval (s)	Ascent rate threshold (°C/s)	Disposal actions
180–220	220	1.2	3	Buzzer
220–250	250	1	2.4	The above and IM app
250–300	300	0.8	1.8	The above and phone call
Above 300	350	0.6	1.5	The above and shut off stove

The bi-spectrum camera in our system (Fig. 8) is DH-TPC-BF2221-T manufactured by Dahua Technology [13]. It has an optical lens and a thermal lens to capture optical videos and monitor temperature. The thermal lens has many advantages such as fast response (latency less than 40 ms), wide temperature measurement range (−20 °C to 550 °C) and small error (less than 5 °C). The type of IPC (Fig. 9) is UNO-2372G produced by Advantech [14]. It has Intel ® atom Baytrial E3845 quad core processor up to 1.91 GHZ and 4 GB DDR3L memory.

Fig. 8. The bi-spectrum camera in our system.

Fig. 9. The IPC in our system.

4   Experiments and Analyses

To guarantee the generalization of the person detection model, we build the training and testing dataset with our application dataset and multiple public datasets including Pascal VOC (only person) [15], COCO2017 (only person) [16], KITTI [17] and CityPersons [18]. There are 60000 and 80000 images in application dataset and public datasets respectively, and 20% of images in each dataset are randomly chosen as test set and the remainder for training. The results of the following experiments are based on the test set.

The results of different person detection algorithms are listed in Table 2. Compared with original YOLOv3-tiny, the model with ignore label improves the precisions of both public and application datasets, showing its general effectiveness. As a contrast, wide augmentation significantly improves the performance on application dataset while not affects the performance on public datasets. The reason is that wide augmentation adapts the model to the highly concentrated aspect ratio of the application dataset (16:9) with input size of 384×224 instead of the original 288×288. By combining ignore label and wide augmentation (IL $\times$ WA), the model achieves better scores on application dataset, showing the coordination of the two modifications. The last line indicates the model finetuned with 500 hard examples. It further improves the precision on application dataset, proving the effectiveness of iterative model improvement.

Table 2. Performances of different person detection models. AP stands for average precision [23].

Models	Input size	Public datasets			Application dataset		
		AP	AP50	AP75	AP	AP50	AP75
Original Yolov3-tiny	288×288	0.28	0.632	0.252	0.237	0.514	0.176
Ignore label	288×288	0.365	0.724	0.339	0.319	0.613	0.262
Wide augmentation	384×224	0.361	0.708	0.326	0.355	0.684	0.281
IL + WA	384×224	**0.363**	**0.709**	**0.327**	0.362	0.718	0.306
500 hard examples	384×224	0.361	0.706	0.326	**0.367**	**0.739**	**0.309**

We speed up the person detection model above with different model acceleration frameworks, and the inference time is listed in Table 3. Darknet [3] is the original framework, and all the models have the same precision. The inference time is calculated with the same processor of our system. It is clear that all the acceleration frameworks have large speedup over original Darknet. Moreover, TuringNN shows the advantage in efficiency over the other open-source frameworks. It is 17.4% faster than NCNN [12], 48.5% faster than Tensorflow Lite [11] and 370% faster than Darknet, indicating the effectiveness of the optimizations in TuringNN.

Table 3. Speed of different model optimization frameworks.

Framework	Inference time (ms)	Frame per second
Darknet	620	1.61
Tensorflow Lite	196	5.1
NCNN	155	6.45
TuringNN	**132**	**7.58**

In summary, the features of existing fire alarm systems and the proposed system are compared in Table 4. It is obvious that our system has advantage over existing ones in accuracy, speed, cost and so on. The proposed fire alarm system has been applied for kitchen safety in Freshhema for over 6 months. An alarm is defined as effective if the oil temperature has risen into the ranges in Table 1 and nobody is near the stove, i.e. both the temperature measurement and person detection are correct. For each store, the system makes over 100 effective alarms per month. The alarmed cases do not necessarily lead to fire accident, but the warnings do reduce potential fire risks and improve the safety awareness of the staff.

Table 4. Comparison of existing fire alarm systems and the proposed system.

	Existing fire alarm systems	The proposed system
Sensor	Flame, smoke, carbon monoxide detectors	Regular and thermal bi-spectrum camera
Timeliness	After fire happens	Before fire happens
False alarm prevention	None	Person detection
Warning mode	Buzzer, phone call to fire station, delayed response	Buzzer, IM app, phone call to staff, shut off stove, quick response
System upgrade	Device replacement, difficult	Model or parameter change, easy

5   Conclusion

In this paper we propose a smart kitchen fire alarm system. The proposed system prevents fire accidents by monitoring the temperature of cookware with thermal cameras. Moreover, the system achieves quick response and avoids false alarm via person detection. By using YOLOv3-tiny algorithm and model quantification framework TuringNN, we are able to implement the system on small-scale low-cost edge devices for practical situations. Experiments demonstrate the effectiveness of our system. Future research will focus on person detection on thermal videos and improving the precision and recall of the system with the detections of both optical and thermal videos.

References

1. National Fire Protection Association: Structure Fires in Hotel and Motels. Report, September 2015
2. U.S. Fire Administration: Hotel and Motel Fires (2014–2016). Topical Fire Report Series, vol. 19, no. 4 (2018)
3. Redmon, J., Farhadi, A.: YOLOv3: an incremental improvement (2018). arXiv preprint arXiv:1804.02767. https://pjreddie.com/darknet/yolo/
4. Freshhema. (Chinese). https://www.freshhema.com/. Accessed 05 Jan 2020. Introduction in English. https://equalocean.com/retail/20190324-freshhema-future-of-alibaba. Accessed 05 Jan 2020
5. Siemens Switzerland Ltd.: Fire detection in kitchens. https://www.downloads.siemens.com/download-center/Download.aspx?pos=download&fct=getasset&id1=A6V10430602. Accessed 05 Jan 2020
6. Cheng, Y., Cheng, L.: Fire Safety Device for Stove-top Burner. Patent, US5945017A, United Sates (1999)
7. Ren, S., He, K., GirShick, R., Sun, J.: Faster R-CNN: towards real-time object detection with region proposal networks. In: Advances in Neural Information Processing Systems (NIPS) (2015)
8. Liu, W., et al.: SSD: single shot multibox detector. In: Leibe, B., Matas, J., Sebe, N., Welling, M. (eds.) ECCV 2016. LNCS, vol. 9905, pp. 21–37. Springer, Cham (2016). https://doi.org/10.1007/978-3-319-46448-0_2
9. Redmon, J., Divvala, S., Girshick, R., Farhadi A.: You only look once: unified, real-time object detection. arXiv:1506.02640 (2015)
10. Lin, T., Dollár, P., Girshick, R., He, K., Hariharan, B., Belongie, S.: Feature pyramid networks for object detection. arXiv:1612.03144 (2016)
11. TensorFlow Lite. https://www.tensorflow.org/lite. Accessed 05 Jan 2020
12. NCNN Github. https://github.com/Tencent/ncnn. Accessed 05 Jan 2020
13. Dahua Technology. https://us.dahuasecurity.com/. Accessed 05 Jan 2020
14. Advantech. https://buy.advantech.com/?country=United%20States. Accessed 05 Jan 2020
15. Everingham, M., Van Gool, L., Williams, C.K.I., Winn, J., Zisserman, A.: The PASCAL visual object classes challenge 2012 (VOC2012) results
16. Lin, T.-Y., et al.: Microsoft COCO: common objects in context. In: Fleet, D., Pajdla, T., Schiele, B., Tuytelaars, T. (eds.) ECCV 2014. LNCS, vol. 8693, pp. 740–755. Springer, Cham (2014). https://doi.org/10.1007/978-3-319-10602-1_48

17. Geiger, A., Lenz, P., Urtasun, R.: Are we ready for autonomous driving? The kitti vision benchmark suite. In: CVPR (2012)
18. Zhang, S., Benenson, R., Schiele, B.: CityPersons: a diverse dataset for pedestrian detection. In: 2017 IEEE Conference on Computer Vision and Pattern Recognition (CVPR), pp. 4457–4465 (2017)
19. Koorsen Fire & Security, Inc.: Installation, service, and inspection for a range of systems. https://www.koorsen.com/products-services/fire-protection/fire-alarm-systems/. Accessed 09 Apr 2020
20. Nearby Engineers Inc.: Fire Suppression Systems for Commercial Kitchens. https://www.ny-engineers.com/blog/fire-suppression-systems-for-commercial-kitchens. Accessed 09 Apr 2020
21. Encore Fire Protection Inc.: Protecting Your Kitchen: Restaurant Fire Safety Systems. http://blog.encorefireprotection.com/blog/protecting-your-kitchen-restaurant-fire-safety-systems. Accessed 09 Apr 2020
22. Krishnamoorthi, R.: Quantizing deep convolutional networks for efficient inference: A whitepaper. arXiv:1806.08342 (2018)
23. Lin, T.-Y., et al.: Microsoft COCO: common objects in context. arXiv:1405.0312 (2014)

Data Mining for Big Dataset-Related Thermal Analysis of High Performance Computing (HPC) Data Center

Davide De Chiara[1], Marta Chinnici[2(✉)], and Ah-Lian Kor[3]

[1] ENEA-ICT Division, C.R. Portici Piazzale Enrico Fermi 1, 80055 Naples, Italy
davide.dechiara@enea.it
[2] ENEA-ICT Division, C.R Casaccia Via Anguillarese 301, 00123 Rome, Italy
marta.chinnici@enea.it
[3] Leeds Beckett University, Leeds, UK
a.kor@leedsbeckett.ac.uk

Abstract. Greening of Data Centers could be achieved through energy savings in two significant areas, namely: compute systems and cooling systems. A reliable cooling system is necessary to produce a persistent flow of cold air to cool the servers due to increasing computational load demand. Servers' dissipated heat effects a strain on the cooling systems. Consequently, it is necessary to identify hotspots that frequently occur in the server zones. This is facilitated through the application of data mining techniques to an available big dataset for thermal characteristics of High-Performance Computing ENEA Data Center, namely Cresco 6. This work presents an algorithm that clusters hotspots with the goal of reducing a data centre's large thermal-gradient due to uneven distribution of server dissipated waste heat followed by increasing cooling effectiveness.

Keywords: Data Center · HPC · Data mining · Big data · Thermal · Hotspot · Cooling · Thermal management

1 Introduction

A large proportion of worldwide generated electricity is through hydrocarbon combustion. Consequently, this causes a rise in carbon emission and other Green House Gasses (GHG) in the environment, contributing to global warming. Data Center (DC) worldwide were estimated to have consumed between 203 to 271 billion kWh of electricity in the year 2010 [1] and in 2017, US based DCs alone used up more than 90 billion kilowatt-hours of electricity [14]. According to [2], unless appropriate steps are taken to reduce energy consumption and go-green, global DC share of carbon emission is estimated to rise from 307 million tons in 2007 to 358 million tons in 2020. Servers in DCs consume energy that is proportional to allocated computing loads, and unfortunately, approximately 98% of the energy input is being dissipated as waste heat energy. Cooling systems are deployed to maintain the temperature of the computing servers at the vendor specified temperature for consistent and reliable performance. Koomey [1] emphasises that a DC energy input is primarily consumed by cooling and compute systems (comprising servers in chassis and racks). Thus, these two systems

© Springer Nature Switzerland AG 2020
V. V. Krzhizhanovskaya et al. (Eds.): ICCS 2020, LNCS 12143, pp. 367–381, 2020.
https://doi.org/10.1007/978-3-030-50436-6_27

have been critical targets for energy savings. Computing-load processing entails jobs and tasks management. On the other hand, DC cooling encompasses the installation of cooling systems and effective hot/cold aisle configurations. Thermal mismanagement in a DC could be the primary contributor to IT infrastructure inefficiency due to thermal degradation. Server microprocessors are the primary energy consumers and waste heat dissipators [4]. Generally, existing DC air-cooling systems are not sufficiently efficient to cope with the vast amount of waste heat generated by high performance-oriented microprocessors. Thus, it is necessary to disperse dissipated waste heat so that there will be an even distribution of waste heat within a premise to avoid overheating. Undeniably, a more effective energy savings strategy is necessary to reduce energy consumed by a cooling system and yet efficient in cooling the servers (in the compute system). One known technique is thermal-aware scheduling where a computational workload scheduling is based on waste heat. Thermal-aware schedulers adopt different thermal-aware approaches (e.g. system-level for work placements [16]; execute 'hot' jobs on 'cold' compute nodes; predictive model for job schedule selection [17]; ranked node queue based on thermal characteristics of rack layouts and optimisation (e.g. optimal setpoints for workload distribution and supply temperature of the cooling system). Heat modelling provides a model that links server energy consumption and their associated waste heat. Thermal-aware monitoring acts as a thermal-eye for the scheduling process and entails recording and evaluation of heat distribution within DCs. Thermal profiling is based on useful monitoring information on workload-related heat emission and is useful to predict the DC heat distribution. In this paper, our analysis explores the relationship between thermal-aware scheduling and computer workload scheduling. This is followed by selecting an efficient solution to evenly distribute heat within a DC to avoid hotspots and cold spots. In this work, a data mining technique is chosen for hotspots detection and thermal profiling for preventive measures. The novel contribution of the research presented in this paper is the use of real thermal characteristics big dataset for ENEA High Performance Computing (HPC) CRESCO6 compute nodes. Analysis conducted are as follows: hotspots localisation; users categorisations based on submitted jobs to CRESCO6 cluster; compute nodes categorisation based on thermal behaviour of internal and surrounding air temperatures due to workload related waste heat dissipation. This analysis aims to minimise employ thermal gradient within a DC IT room through the consideration of the following: different granularity levels of thermal data; energy consumption of calculation nodes; IT room ambient temperature. An unsupervised learning technique has been employed to identify hotspots due to the variability of thermal data and uncertainties in defining temperature thresholds. This analysis phase involves the determination of optimal workload distribution to cluster nodes. Available thermal characteristics (i.e. exhaust temperature, CPUs temperatures) are inputs to the clustering algorithm. Subsequently, a series of clustering results are intersected to unravel nodes (identified by IDs) that frequently fall into high-temperature areas. The paper is organised as follows: Sect. 1 – Introduction; Sect. 2 – Background: Related Work; Sect. 3 – Methodology; Sect. 4 – Results and Discussion; Sect. 5 – Conclusions and Future Work.

2 Background: Related Work

In the context of High Performance Computing Data Center (HPC-DC), it is essential to satisfy service level agreements with minimal energy consumption. This will involve the following: DC efficient operations and management within recommended IT room requirements, specifications, and standards; energy efficiency and effective cooling systems; optimised IT equipment utilisation. DC energy efficiency has been a long standing challenge due to multi-faceted factors that affect DC energy efficiency and adding to the complexity, is the trade-off between performance in the form of productivity and energy efficiency. Interesting trade-offs between geolocations and DC energy input requirements (e.g. cold geolocations and free air-cooling; hot, sunny geolocations and solar powered renewable energy) are yet to be critically analysed [8]. One of the thermal equipment-related challenge is raising the setpoint of cooling equipment or lowering the speed of CRAC (Computer Room Air Conditioning) fans to save energy, may in the long-term, decrease the IT systems reliability (due to thermal degradation). However, a trade-off solution (between optimal cooling system energy consumption and long-term IT system reliability) is yet to be researched on [8]. Another long-standing challenge is IT resource over-provisioning that causes energy waste due to for idle servers. Relevant research explores optimal allocation of PDUs (Power Distribution Units) for servers, multi-step algorithms for power monitoring, and on-demand provisioning reviewed in [8]. Other related work addresses workload management, network-level issues as optimal routing, Virtual Machines (VM) allocation, and balance between power savings and network QoS (Quality of Service) parameters as well as appropriate metrics for DC energy efficiency evaluation. One standard metric used by a majority of industrial DCs is Power Usage Effectiveness (PUE) proposed by Green Grid Consortium [2]. It shows the ratio of total DC energy utilisation with respect to the energy consumed solely by IT equipment. A plethora of DC energy efficiency metrics evaluate the following: thermal characteristics; ratio of renewable energy use; energy productivity of various IT system components, and etc. There is a pressing need to provide a holistic framework that would thoroughly characterise DCs with a fixed set of metrics and reveal potential pitfalls in their operations [3]. Though some existing research work has made such attempts but to date, we are yet to have a standardised framework [9, 10, 13]. To reiterate, the thermal characteristics of the IT system ought to be the primary focus of an energy efficiency framework because it is the main energy consumer within a DC. Several researches have been conducted to address this issue [12]. Sungkap et al. [11] propose an ambient temperature-aware capping to maximize power efficiency while minimising overheating. Their research includes an analysis of the composition of energy consumed by a cloud-based DC. Their findings for the composition of DC energy consumption are approximately 45% for compute systems; 40% for refrigeration-based air conditioning; remaining 15% for storage and power distribution systems. This implies that approximately half of the DC energy is consumed by non-computing devices. In [6], Wang and colleagues present an analytical model that describes DC resources with heat transfer properties and workloads with thermal features. Thermal modelling and temperature estimation from thermal sensors ought to consider the emergence of server hotspots and thermal solicitation due to the increase in inlet air temperature, inappropriate positioning of a rack or even inadequate room ventilation. Such phenomena are unravelled by thermal-aware location analysis. The thermal-aware server provisioning approach to minimise the total DC energy consumption calculates the value of energy by considering the maximum working temperature of the servers. This approach should consider the fact that any rise in the inlet temperature rise may cause the servers to reach the maximum temperature resulting in thermal stress, thermal degradation, and severe damage in the long run. Typical different identified types of thermal-aware scheduling are reactive, proactive and mixed. However, there is no reference to heat-modelling or thermal-monitoring and profiling. Kong and colleagues [4] highlight important concepts of thermal-aware profiling, thermal-aware monitoring, and thermal-aware scheduling. Thermal-aware techniques are linked to the minimisation of waste heat production, heat convection around server cores, task migrations, and thermal-gradient across the microprocessor chip and microprocessor power consumption. Dynamic thermal management (DTM) techniques in microprocessors encompasses the following: Dynamic Voltage and Frequency Scaling (DVFS), Clock gating, task migration, and Operating System (OS) based DTM and scheduling. In [5], Parolini and colleagues propose a heat model; provide a brief overview of power and thermal efficiency from microprocessor micro-level to DC macro-level. To reiterate, it is essential for DC energy efficiency to address thermal awareness in order to better understand the relationship between both the thermal and the IT aspects of workload management. In this paper, the authors incorporate thermal-aware scheduling, heat modelling, thermal aware monitoring and thermal profiling using a big thermal characteristic dataset of a HPC-Data Center. This research involves measurement, quantification, and analysis of compute nodes and refrigerating machines. The aim of the analysis is to uncover underlying causes that causes temperatures rise that leads to the emergence of thermal hotspots. Overall, effective DC management requires energy use monitoring, particularly, energy input, IT energy consumption, monitoring of supply air temperature and humidity at room level (i.e. granularity level 0 in the context of this research), monitoring of air temperature at a higher granularity level (i.e. at Computer Room Air Conditioning/Computer Room Air Handler (CRAC/CRAH) unit level, granularity level 1). Measurements taken are further analysed to reveal extent of energy use and economisation opportunities for the improvement of DC energy efficiency level (granularity level 2). DC energy efficiency metrics will not be discussed in this paper. However, the discussion in the subsequent section primarily focuses on thermal guidelines from American Society of Heating, Refrigerating And AC Engineers (ASHRAE) [7].

3 Methodology

To reiterate, our research goal is to reduce DC wide thermal-gradient, hotspots and maximise cooling effects. This entails the identification of individual server nodes that frequently occur in the hotspot zones through the implementation of a clustering algorithm on the workload management platform. The big thermal characteristics dataset of ENEA Portici CRESCO6 computing cluster is employed for the analysis.

It has 24 measured values (or features) for each single calculation node (see Table 1) and comprises measurements for the period from May 2018 to January 2020. Briefly, the cluster CRESCO6 is a High-Performance Computing System (HPC) consisting of 434 calculation nodes with a total of 20832 cores. It is based on Lenovo Think System SD530 platform, an ultra-dense and economical two-socket server in a 0.5 U rack form factor inserted in a 2U four-mode enclosure. Each node is equipped with 2 Intel Xeon Platinum 8160 CPUs (each with 24 cores) and a clock frequency of 2.1 GHz; a RAM size of 192 GB, corresponding to 4 GB/core. A low-latency Intel Omni-Path 100 Series Single-port PCIe 3.0 x16 HFA network interface. The nodes are interconnected by an Intel Omni-Path network with 21 Intel Edge switches 100 series of 48 ports each, bandwidth equal to 100 GB/s, and latency equal to 100 ns. The connections between the nodes have 2 tier 2:1 no-blocking tapered fat-tree topology. The power consumption massive computing workloads amount to a maximum of 190 kW.

Table 1. Thermal dataset – description of features.

Node name	Server ID, integer from 1 to 434
Timestamp	timestamp of a measurement
System, CPU, Memory Power	Node instantaneous power, memory power, CPU power use in three corresponding columns (expressed in Watt)
Fan 1a, Fan1b, ..., Fan 5a, Fan 5b	Speed of a cooling fan installed inside the node (expressed in RPM - revolutions per minute)
System, CPU, Memory, I/O utilisation	Percentage values of CPU, RAM memories and I/O utilisation
Inlet, CPU1, CPU2, Exhaust temperature	Temperature at the front, inside (on CPU1 and CPU2) and at the rear of every single node (expressed in Celsius)
SysAirFlow	Speed of air traversing the node expressed in CFM (cubic foot per minute)
DC energy	Meter of total energy used by the node, updated at corresponding timestamp and expressed in kWh

3.1 Energy Saving Approach

This work incorporates thermal-aware scheduling, heat modelling, and thermal monitoring followed by subsequent user profiling based on "waste heat production" point of view. Thermal-aware DC scheduling is designed based on data analytics conducted on real data obtained from running cluster nodes in a real physical DC. For the purpose of this work, approximately 20 months' worth of data has been collected. Data collected are related to: relevant parameters for each node (e.g. inlet air temperature, internal temperature of each node, energy consumption of CPU, RAM, memory, etc...); environmental parameters (e.g. air temperatures and humidity in both the hot and cold aisles); cooling system related parameters (e.g. fan speed); and finally, individual users who submit their jobs to cluster node. This research focuses on the effect of dynamic workload assignment on energy consumption and performance of both the computing and cooling systems. The constraint is that each arrived job must be assigned irrevocably to a particular server without any information about impending incoming jobs. Once the job has been assigned, no pre-emption or migration is allowed, which a rule is typically adhered to for HPC applications due to high data reallocation incurred costs. In this research, we particularly explore an optimised mapping of nodes that have to be physically and statically placed in advance to one of the available rack slots in the DC. This will form a matrix comprising computing units with specific characteristics and certain resource availability level at a given time t. The goal is to create a list of candidate nodes to deliver "calculation performance" required by a user's job. When choosing the candidate nodes, the job-scheduler will evaluate the suitability of the thermally cooler nodes (which at the instant t) based on their capability to satisfy the calculation requested by a user (in order to satisfy user's SLA). To enhance the job scheduler decision making, it is essential to know in advance, the type of job a user will submit to a node(s) for computation. Such insight is provided by several years' worth of historical data and advanced data analytics using machine learning algorithms. Through Platform Load Sharing Facility (LSF) accounting data we code user profiles into 4 macro-categories:

1. *CPU_intensive*
2. *MEMORY_intensive*
3. *CPU&MEMORY_intensive*
4. *CPU&MEMORY_not intensive*

This behavioural categorisation provides an opportunity to save energy and better allocate tasks to cluster nodes to reduce overall node temperatures. Additionally, when job allocation is evenly distributed, thermal hotspots and cold spots could be avoided. The temperatures of the calculation nodes could be evened out, thus, resulting in a more even distribution of heat across the cluster.

3.2 Users and Workload Understanding Profiled log

Based on thermal data, it is necessary to better understand in-depth what users do and how they manage to solicit the calculation nodes for their jobs. The three main objectives of understanding users' behaviour are as follows: Identify parameters based on the diversity of submitted jobs for user profiling; Analyse the predictability of various resources (e.g. CPU, Memory, I/O) and identify their time-based usage patterns; Build predictive models for estimating future CPU and memory usage based on historical data carried out in the LSF platform. Abstraction of behavioural patterns in the job submission and its associated resource consumption is necessary to predict future resource requirements. This is exceptionally vital for dynamic resource provisioning in a DC. User profile is created based on submitted job-related information and to reiterate, the 4 macro categories of user profiles are: 1) CPU-intensive, 2) disk-intensive, 3) both CPU and MEMORY-intensive, or 4) neither CPU-nor MEMORY-intensive. A crosstab of the accounting data (provided by the LSF platform) and resource consumption data help guide the calculation of relevant thresholds that code jobs into several distinct utilisation categories. For instance, if the CPU load is high (e.g., larger than 90%) during almost 60% of the job running time for an application, then the job can be labelled as a CPU-intensive one. The goal is for the job-scheduler to optimise task scheduling when a job with the same AppID (i.e. the same type of job) or same username is re-submitted to a cluster. In case of a match with the previous AppID or username, relevant utilisation stats from the profiled log are retrieved. Based on the utilisation patterns, this particular user/application will be placed into one of the 4 previously discussed categories. Once a job is categorised, a thermally suitable node is selected to satisfy the task calculation requirements. A task with high CPU and memory requirement will not be immediately processed until the node temperature is well under a safe temperature threshold. Node temperature refers to the difference between the node's outlet exhaust air and inlet air temperatures (note: this generally corresponds to the air temperature in the aisles cooled by the air conditioners).

3.3 Real-Time Workload Management Based on Thermal Awareness: Cluster Evaluation

It is necessary to have a snapshot of relevant thermal parameters (e.g. temperatures of each component in the calculation nodes) for each cluster to facilitate efficient job allocation by the job-scheduler. Generally, a snapshot is obtained through direct interrogation of the nodes and installed sensors in their vicinity, or inside the calculation nodes. For each individual node, the temperatures of the CPUs, memories, instantaneous energy consumption and peed of the cooling fans are evaluated undeniably, the highly prioritised parameter is the difference between the node's inlet and exhaust air temperatures. If there is a marked difference, it is evident that the node is very busy (with jobs that require a lot of CPU or memory-related resource consumption). Therefore, for each calculation node, relevant data is monitored in real time, and subsequently, virtually stored in a matrix that represents the state of the entire cluster. Each matrix cell represents the states of a node (represented by relevant parameters). For new job allocation, the scheduling algorithm will choose a node based on its states depicted in the matrix (e.g. recency or Euclidean distance). Through this, generated waste heat is evenly distributed over the entire "matrix" of calculation nodes so that hotspots could be significantly reduced. Additionally, a user profile is an equally important criterion for resource allocation. This is due to the fact that user profiles provide insights into user consumption patterns and the type of submitted jobs and their associated parameters. For example, if we know that a user will perform CPU-intensive jobs for 24 h, we will allocate the job in a "cell" (calculation node) or a group of cells (when the number of resources requires many calculation nodes) that are physically well distributed or with antipodal locations. This selection strategy aims to evenly spread out the high-density nodes followed by the necessary cooling needs. This will help minimise DC hotspots and ascertain efficient cooling with reduction in cooling-related energy consumption.

4 Results and Discussion

As previously discussed, we have created user profiles based on submitted job-related information. Undeniably, these profiles are dynamic because they are constantly revised based on user resource consumption behaviour. For example, a user may have been classified as "CPU intensive" for a certain time period. However, if the user's submitted jobs are no longer CPU intensive, then the user will be re-categorised. The deployment of the thermal-aware job scheduler generally aims to reduce the overall CPU/memory temperatures, and outlet temperatures of cluster nodes. The following design principles guide the design and implementation of the job: 1) Job categories - assign an incoming job to one of these 4 categories: CPU-intensive, memory-intensive, neither CPU nor memory-intensive, and both CPU and memory-intensive tasks; 2) Utilisation monitoring - monitoring CPU and memory utilisation while making scheduling decisions; 3) Redline temperature control - ensure operating CPUs and memory under threshold temperatures; 4) Average temperatures maintenance - monitor average CPU and memory temperatures in a node and manage an average exhaust air temperature across a cluster. To reiterate, user profile categorisation is facilitated by maintaining a log profile of both CPU and memory utilisation for every job (with an associated user) processed in the cluster. A log file contains the following user-related information: (1) user ID; (2) Application identification; (3) the number of submitted jobs; (4) CPU utilisation; (5) memory utilisation.

4.1 Data Analysis: Results and Discussion

A list of important thermal management-related terms is as follows: 1) CPU-intensive - applications that is computation intensive (i.e. requires a lot of processing time); 2) Memory-intensive-a significant portion of these applications require RAM processing and disk operations; 3) Maximum (redline) temperature - the maximum operating temperature specified by a device manufacturer or a system administrator; 4) Inlet air temperature - the temperature of the air flowing into a data node (i.e. temperature of the air sucked in from the front of the node); 5) Exhaust air temperature - the temperature of the air coming out from a node (the temperature of the air extracted from the rear of the node). By applying these evaluation criteria, we have built an automated procedure that provides insight into the 4 user associated categories (based on present and historical data). Obviously, the algorithm always makes a comparison between a job just submitted by a user and the time series (if any) of the same user. If the application launched or the type of submitted job remains the same, then the user will be grouped into one of the 4 categories (based on a supervised learning algorithm) During each job execution, the temperature variations of the CPUs and memories are recorded at pre-established time intervals. Finally, it continuously refines the user behaviour based on the average length of time the user uses for the job. This will provide a more accurate user (and job) profile because it provides reliable information on the type of job processed in a calculation node and its total processing time. The job scheduler will exploit such information for better job placement within an ideal array of calculation nodes in the cluster. A preliminary study is conducted. To provide insight into the functioning of the clusters. For 8 months, we have observed the power consumption

(Fig. 1) and temperature (Fig. 2) profiles of the nodes with workloads. We have depicted energy consumed by the various server components (CPU, memory, other) in Fig. 3 and presented a graph that highlights the difference in energy consumption between idle and active nodes (Fig. 4).

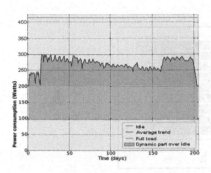

Fig. 1. The representative shape of Power profile portion on average for all available nodes. Power consumption dataset for a subset of 200 days.

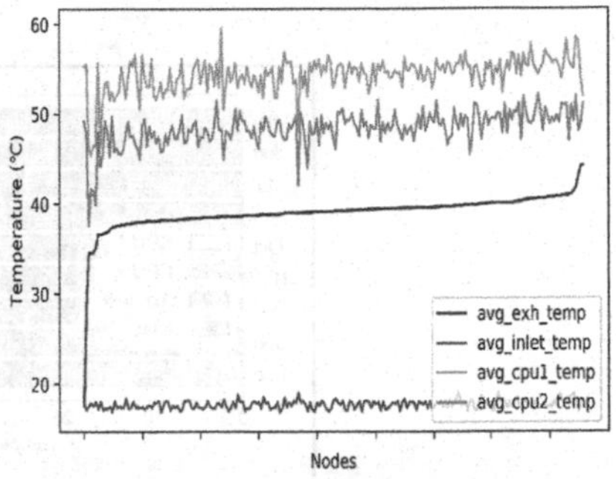

Fig. 2. Temperature profiles (subset of 1 month) on average for all available nodes. Nodes are sorted in the order of exhaust air temperature increase.

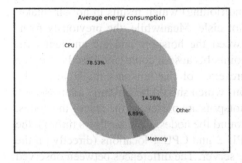

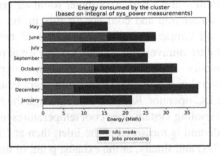

Fig. 3. Average energy partioning for all nodes of cluster CRESCO6

Fig. 4. Energy consumption in idle and active nodes (subset of 8 months)

It is observed that for each node, an increase in load effects an increase in temperature difference between inlet and exhaust air for that particular node. Figure 5 depicts the average observed inlet air temperature (blue segment, and in the cold aisle), and exhaust air temperature at their rear side (amaranth segment, in the hot aisle). Note the temperature measurements are also taken two CPUs adjacent to every node. The setpoints of the cooling system are approximately 18 °C at the output and 24 °C at the input of the cooling system – as respectively shown in Fig. 5 as blue and red vertical lines. However, it appears that the lower setpoint is variable (supply air at 15–18 °C) while the higher setpoint varies from 24–26 °C. As observed from the graph, the cold aisle maintains the setpoint temperature at the inlet of the node, which affirms the efficient design of the cold aisle (i.e. due to the use of plastic panels to isolating the cold aisle from other spaces in the IT room). However, the exhaust air temperature has registered on average, 10 °C higher level than the hot aisle setpoint. Notably, exhaust temperature sensors are directly located at the rear of the node (i.e. in the hottest parts of the hot aisle).

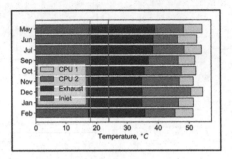

Fig. 5. Temperature observed on average in all nodes during 9 months with vertical lines corresponding to cold and hot aisle setpoints. (Color figure online)

Therefore, it is observed that hotspots are immediately located at the back of server racks, while the hot aisle air is cooled down to the 24–26 °C. This is due the cooling system at the CRAC (computer room air conditioning) which results in hot air intake, air circulation and cold-hot air mix in the hot aisle. Meanwhile, the previously mentioned temperature difference of 10 °C between the hotspots and the ambient temperature unravels the cooling system weak points because it could not directly cool the hotspots. In the long term, the constant presence of the hotspots might affect the servers' performance (i.e. thermal degradation) which should be carefully addressed by the DC operator. Remarkably, although the hotspots are present at the rear of the nodes, the cooling system does cool temperatures around the nodes. Cold air flows through the node and is measured at the inlet, then at CPU 2 and CPU 1 locations (directly on the CPUs) and finally, at the exhaust point of the server. The differences between observed temperature ranges in these locations are averaged for all the nodes. An investigation on the observed temperature distribution contributes to the overall understanding of the thermal characteristics, as it provides an overview of the prevailing temperatures shown in Fig. 5 and Fig. 6. For every type of thermal sensors, the temperature values are recorded as an integer number, so the percentage of occurrences of each value is calculated. The inlet air temperature is registered around 18 °C in the majority of cases and has risen up to 28 °C in around 0.0001% of cases. It could be concluded that the cold aisle temperature remains around the 15–18 °C setpoint for most of the monitored period. Ranges of the exhaust temperature and those of CPUs 1 and 2 are in the range 20–60 °C with most frequently monitored values in the intervals of 18–50 °C. Although these observations might incur measurement errors, they reveal severs that are at risks of frequent overheating when benchmarked with manufacturer's recommendation data sheets.

Additionally, this study focuses on variation between subsequent thermal measurements with the aim of exploring temperature stability around the nodes. All temperature types have distinct peaks of zero variation which decreases symmetrically and assumes a Gaussian distribution. It could be concluded that temperature tends to be stable in the majority of monitored cases. However, the graphs for exhaust and CPUs 1 and 2 temperature variation (Fig. 6 reveal that less than 0.001% of the recorded measurements show an amplitude of air temperature changes of 20 °C or more occurring at corresponding locations.

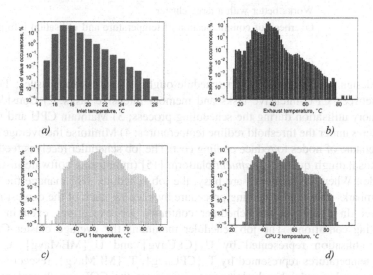

a)   b)

c)   d)

Fig. 6. Distribution of monitored temperature values taken for all nodes and months.

Sudden infrequent temperature fluctuations are less dangerous compared to prolonged periods of constantly high temperatures. Nevertheless, further investigation is needed to uncover causes of abrupt temperature changes so that appropriate measures could be undertaken by DC operators to maintain prolonged periods of constantly favourable conditions. We propose a scheduler upgrade which aims to optimise CPU and memories-related resource allocation, as well as exhaust air temperatures without relying on profile information. Prescribed targets for the proposed job scheduler are shown in Table 2.

Table 2. Schema with prefixed target for improved job scheduler.

	Proposed job scheduler
Strategy	Schedules job based on utilisation and temperature information gathered at run-time
Job assignment	Assigns a job to the coolest node in a cluster at any point in time
Job scheduling	Schedules a job on the coolest node in a cluster
Temperature control	Maintains uniform temperate across a cluster
Node activity	At least 50% are active nodes at any given time in a cluster
Pros	Works better with a large cluster
Cons	Overhead of communication of temperature and utilisation information

The design of the proposed job schedule ought to address four issues: 1) Differentiate between CPU-intensive tasks and memory-intensive tasks; 2) Consider CPU and memory utilisation during the scheduling process; 3) Maintain CPU and memory temperatures under the threshold redline temperatures; 4) Minimise the average exhaust air temperature of nodes to reduce cooling cost. The job scheduler receives feedback of node status through queried *Confluent* platform [15] (monitoring software installed on each node). When all the nodes are busy, the job scheduler will manage the temperatures, embarks on a load balancing procedure by keeping track of the coolest nodes in the cluster. In doing so, the scheduler continues job executions even in hot yet undamaging conditions. The job scheduler maintains the average cluster CPU and memory utilisation represented by U_{CPUavg} and U_{MEMavg}, CPU and memory temperatures represented by T_{CPUavg}, T_{MEMavg}, respectively. The goal of our enhanced job scheduler is to maximise the COP (coefficient of performance). Below are the 7 constraints (at nodes level) for our enhanced scheduler:

1. *check constraint T_{CPU}^{i} (instant i) $< T_{CPUAvg}$*
2. *otherwise, check constraint $T_{Mem}^{i} < T_{Memavg}$*
3. *$T_{Memavg} < T_{MemMax}$ and $T_{CPUavg} < T_{CPUMax}$*
4. *$T_{out}^{i} \leqslant (\sum_{i=1}^{N} Tout)/N$*
5. *Each job is assigned to utmost one node*
6. *Minimise response time of job*

With the first and second constraints are satisfied, ensure that the memory and CPU temperatures remain below the threshold temperatures. If a cluster's nodes exceed the redline threshold, then optimise the temperature by assigning jobs to the coolest node in the cluster. The third constraint specifies that if the average temperature of memory or CPU rises above the maximum temperature, then the scheduler should stop scheduling tasks as it might encounter hardware failures. The fourth constraint states that the exhaust air temperature of a node should be the same or less than the average exhaust air temperature of the cluster (taking into consideration N number of nodes). The fifth constraint ensures that a node gets utmost one job at a single point in time. The last point aims at reducing the completion time of a job to achieve optimal performance.

The following is the description of our algorithm:

```
****matrix of node with position r-ow and c-olumn****
Cluster= matrix[r,c]
user=getUSERfromSubmittedJob_in_LSF
Jobtype= getJobProfile(user)

****push the values of utilization and temperature for cpu and memory into matrix*****
        for (i=0; i=number_of_node;i++) do
                nodename = getnodeName(i)
                U_icpu = getCPU_Utilization(nodename)
                U_imemory = getMEMORY_Utilization(nodename)
                T_icpu = getCPU_Temperature(nodename)
                T_imemory = getMEMORY_Temperature(nodename)
        End for

*************if a user is not profiled ***************
        if Jobtype= null then
        ***********try to understand job type at run time***********
                if (Ucpu <= U_threshold_cpu) && (Umemory <= U_threshold_memory) then
                        Jobtype=easyJob
                else if (Ucpu>U_threshold_cpu) && (Umemory < U_threshold_memory) then
                        Jobtype=CPUintensiveJob
                else if (Ucpu<U_threshold_cpu) && (Umemory > U_threshold_memory) then
                        Jobtype=MEMORYintensiveJob
                else
                        Jobtype=CPU&MEMORYintensiveJob
                end if
        end if

******** I try to find the candidate nodes for each type of job***********
avgTempCluster= avgTemp(Cluster)
minT_nodename= getTempNodename(minTemp(Cluster))
maxT_nodename=getTempNodename(maxTemp(Cluster))

***********intervals of temperatures for candidate nodes*************
bestCPUIntensiveNode=getNode (minT_nodename, minT_nodename+25%))
bestMEMORYIntensiveNode= getNode(minT_nodename+50%, minT_nodename+75%)
bestCPU&MEMORYIntensiveNode= getNode(minT_nodename+25%, minT_nodename+50%)
bestEasyJob= getNode(maxT_nodename, maxT_nodename-25% )

*****************job assignments*************************
        if Jobtype= CPUintensiveJob then
                assignJob (bestCPUIntensiveNode)
        else if Jobtype= MEMORYintensiveJob then
                assignJob (bestMemoryIntensiveNode)
        else if Jobtype= CPU&MEMORYintensiveJob then
                assignJob(bestCPU&MEMORYIntensiveNode)
        else
                assignJob(bestEasyJob)
        end if
```

The algorithm feeds into the node matrix by considering the physical arrangement of every single node inside the racks. Firstly, obtain the profile of the user who puts in a resource request for resources. This is done by retrieving the user's profile from a list of stored profiles. The algorithm is executed for all the nodes to appreciate resource utilisation level and temperature profiles each node. If the user profile does not exist, then when a user executes a job for the first time, the algorithm calculates a profile instantaneously. All the indicated threshold values are operating values calculated for each cluster configuration and are periodically recalculated and revised according to the use of the cluster nodes. Subsequently, some temperature calculations are made from the current state of the cluster (through a snapshot of thermal profile). Finally, the last step is to assign the job to the node based on the expected type of job. Through this, the algorithm helps avert the emergence of hotspots and cold spots by uniformly distributing the jobs in the cluster.

5 Conclusions and Future Work

In order to support sustainable development goals, energy efficiency ought to be the ultimate goal for a DC with a sizeable high-performance computing facility. To reiterate, this work primarily focuses on two of major aspects: IT equipment energy productivity and thermal characteristics of an IT room and its infrastructure. The findings of this research are based on the analysis of available monitored thermal characteristics-related data for CRESCO6. These findings feed into recommendations for enhanced thermal design and load management. In this research, clustering performed on big datasets for CRESCO6 IT room temperature measurements, has grouped nodes into clusters based on their thermal ranges followed by uncovering the clusters they frequently subsume during the observation period. Additionally, a data mining algorithm has been employed to locate the hotspots and approximately 8% of the nodes have been frequently placed in the hot range category (thus labelled as hotspots). Several measures to mitigate risks associated with the issue of hotspots have been recommended: more efficient directional cooling, load management, and continuous monitoring of the IT room thermal conditions. This research brings about two positive effects in terms of DC energy efficiency. Firstly, being a thermal design pitfall, hotspots pose as a risk of local overheating and servers thermal degradation due to prolonged exposure to high temperatures. Undeniably, information of hotspots localisation could facilitate better thermal management of the IT room where waste heat is evenly distributed. Thus, it ought to be the focus of enhanced thermal management in the future. Secondly, we discussed ways to avert hotspots through thermal-aware resource allocation (i.e. select the coolest node for a new incoming job), and selection of nodes (for a particular job) that are physically distributed throughout the IT room.

References

1. Koomey, J.: Growth in Data Center Electricity use 2005 to 2010, pp. 1–24. Analytics Press (2011). https://doi.org/10.1088/1748-9326/3/3/034008
2. Greenpeace: How Dirty Is Your Data? A Look at the Energy Choices That Power Cloud Computing (2011)
3. Reddy, V.D., et al.: Metrics for sustainable data centers. IEEE Trans. Sustain. Comput. 2(3), 290–303 (2017)
4. Kong, J., Chung, S.W., Skadron, K.: Recent thermal management techniques for microprocessors. ACM Comput. Surv. 44(3) (2012). https://doi.org/10.1145/2187671. 2187675

5. Parolini, L., Sinopoli, B., Krogh, B.H., Wang, Z.: A cyber-physical systems approach to data center modeling and control for energy efficiency. Proc. IEEE **100**, 254–268 (2012). https://doi.org/10.1109/JPROC.2011.2161244

6. Wang, L., et al.: Thermal aware workload placement with task-temperature profiles in a datacenter. J. Supercomput. **2012**(61), 780–803 (2012). https://doi.org/10.1007/s11227-011-0635-z

7. ASHRAE Technical Committee 9.9: Thermal Guidelines for Data Processing Environments – Expanded Data Center Classes and Usage Guidance (2011)

8. Jin, X., Zhang, F., Vasilakos, A.V., Liu, Z.: Green Data centers: a survey, perspectives, and future directions. arXiv, vol. 1608, no. 00687 (2016)

9. Chinnici, M., Capozzoli, A., Serale, G.: Measuring energy efficiency in data centers. In: Pervasive Computing Next Generation Platforms for Intelligent Data Collection, Chap. 10, pp. 299–351 (2016)

10. Chinnici, M., et al: Data center, a cyber-physical system: improving energy efficiency through the power management. In: 2017 IEEE 15th International Conference on Dependable, Autonomic and Secure Computing, 15th International Conference on (DASC/PiCom/DataCom/CyberSciTech), pp. 269–272 (2017)

11. Yeo, S., et al.: ATAC: ambient temperature-aware capping for power efficient datacenters. In: Proceedings of the 5th ACM Symposium on Cloud Computing, Seattle, WA, USA (2014). https://doi.org/10.1145/2670979.2670966

12. Capozzoli, A., et al.: Thermal metrics for data centers: a critical review. Energy Procedia **62**, 391–400 (2014)

13. Capozzoli, A., Chinnici, M., Perino, M., Serale, G.: Review on performance metrics for energy efficiency in data center: the role of thermal management. In: Klingert, S., Chinnici, M., Rey Porto, M. (eds.) E2DC 2014. LNCS, vol. 8945, pp. 135–151. Springer, Cham (2015). https://doi.org/10.1007/978-3-319-15786-3_9

14. Vxchnge site. https://www.vxchnge.com/. Accessed 27 Mar 2020

15. Confluent site. https://docs.confluent.io/platform.html. Accessed 27 Mar 2020

16. Van Damme, T., De Persis, C., Tesi, P.: Optimized thermal-aware job scheduling and control of data centers. IEEE Trans. Control Syst. Technol. **27**(2), 760–771 (2019). https://doi.org/10.1109/TCST.2017.2783366

17. Varsamopoulos, G., Banerjee, A., Gupta, S.K.S.: Energy efficiency of thermal-aware job scheduling algorithms under various cooling models. In: Ranka, S., et al. (eds.) IC3 2009. CCIS, vol. 40, pp. 568–580. Springer, Heidelberg (2009). https://doi.org/10.1007/978-3-642-03547-0_54

A Comparison of Multiple Objective Algorithms in the Context of a Dial a Ride Problem

Pedro M. M. Guerreiro[1], Pedro J. S. Cardoso[1,2]([✉]),
and Hortênsio C. L. Fernandes[3]

[1] ISE, Universidade do Algarve, Faro, Portugal
{pmguerre,pcardoso}@ualg.pt
[2] LARSyS, Universidade do Algarve, Faro, Portugal
[3] Yellowfish Travel, Lda, Albufeira, Portugal
hortensio@yellowfishtransfers.com

Abstract. In their operations private chauffeur companies have to solve variations of the multiple objective dial a ride problem. The number and type of restrictions make the problem extremely intricate and, when manually done, requires specialized people with a deep knowledge of the *modus operandi* of the company and of the environment in which the procedure takes place. Nevertheless, the scheduling can be automated through mean of computational methods, allowing to deliver solutions faster and, possible, optimized. In this context, this paper compares six algorithms applied to solving a multiple objective dial a ride problem, using data from a company mainly working in the Algarve, Portugal. The achieved results show that ϵ-MOEA overcomes the other algorithms tested, namely the NSGA-II, NSGA-III, ϵ-NSGA-II, SPEA2, and PESA2.

Keywords: DARP · Evolutionary algorithms · Multiple objective optimization · Private drivers scheduling

1 Introduction

One of the problems addressed by private chauffeur companies is the design of optimal routes and schedules to pick up, drive, and deliver clients to drop off locations previously scheduled. The optimization has to attend to a large number of parameters and restrictions, e.g.: capacity demands (number of persons, volume of the luggage, large/special sport items etc.) or service restrictions (transportation of children, disabled people, drivers availability, drivers working

---

This work was supported by the Portuguese Foundation for Science and Technology (FCT), project LARSyS - FCT Plurianual funding 2020–2023. We also thank Yellowfish Travel, Lda for the valuable data, participation in the problem definition and solution development.

V. V. Krzhizhanovskaya et al. (Eds.): ICCS 2020, LNCS 12143, pp. 382–396, 2020.
https://doi.org/10.1007/978-3-030-50436-6_28

hours, vehicles availability, etc.). The problem is even more intricate and dynamic when issues like delayed services, absence of workers, traffic congestion, vehicle breakdowns or service cancellations (just to give some examples) are taken into account.

In its general form, the described problem is a variation of the dial a ride problem (DARP), where clients obligatorily provide the pick up and drop off locations and time, and other parameters such as number of passengers and their ages or luggage volume (e.g., see [6,26] for formal definition of the problem). DARP consists in, given a characterized fleet of vehicles, designing (optimal) routes and schedules to satisfy clients requests of pick up and deliver in certain time windows, not necessarily using all available vehicles. DARP is known to be NP-hard and has been addressed in many ways including exact and meta-heuristic solutions [14]. For example, to solve a route planning problem at a senior activity center, a heterogeneous DARP with configurable vehicle capacity was solved by Qu and Bard [27], using a branch and price and cut algorithm. A branch and cut algorithm capable of solving small to medium-size instances was developed by Cordeau [5] along with a mixed-integer programming formulation. In [29] a pick up and delivery problem with time windows and a DARP tested a branch and cut algorithms with families of inequalities on several instance sets of the problems. In [31] an ant colony optimization algorithm was applied to minimize the fleet size required to solve a DARP. The study in [9] aimed to develop and test different genetic algorithms to find appropriate encodings and configurations, specifically for the case with time windows. Parallel implementations were also studied as the tabu search variants described and compared in [1] (applied to a static DARP). A two-stage hybrid meta-heuristic method using ant colony optimization and tabu search for the vehicle routing problem with constraints of simultaneous pick up and delivery, and time windows was presented by Lai and Tong [21]. Another parallel approach embedded with a multi-start heuristic for solving the vehicle routing problem with simultaneous pick up and delivery was proposed in [30]. Berbeglia, Cordeau and Laporte presented an hybrid algorithm for the dynamic DARP [2], which combines an exact constraint programming algorithm and a tabu search heuristic.

The DARP was formulated with distinct objectives such as the minimization of the routes' duration, ride time, waiting time, number of vehicles used, operational cost, etc. (for a more complete list see [7]). In the criteria context, the problem has been addressed in several multiple objective (MO) formulations. In the MO case, several possibly antagonistic objectives are evaluated producing a set of solutions over a partially ordered objective space [12]. For instance, Chevrier et al. [4] addressed the problem of lack of transport service in sparsely inhabited areas as a demand responsive transport problem, comparing the non-dominated sorting genetic algorithm II (NSGA-II), the strength pareto evolutionary algorithm 2 (SPEA2), and the indicator based evolutionary algorithm (IBEA). In the same work, improvements using an iterative local search (ILS), added in the mutation operator, are used to select the best approach in a solution capable of producing answers in a short period of time. Another application of NSGA-II to a multiple objective variation of the DARP problem considering disruptive scenarios (e.g., accidents with the transporting vehicles, vehicle breakdown, and traffic jams) was presented by Issaoui *et al.* [19]. A bi-objective formulation of DARP is solved using NSGA-II in [16]. The formulation consists in the determination of routes to be performed by a fleet of vehicles available to serve geographically dispersed customers. For more in-depth surveys on the subject, please refer to the works of Cordeau and Laporte [6] and Ho *et al.* [18].

This paper compares six multiple objective evolutionary algorithms (MOEA) applied to solving an instance of the multiple objective dial a ride problem (MO-DARP), namely: Multi-Objective Evolutionary Algorithm based on ϵ-dominance (ϵ-MOEA), Non-dominated Sorting Genetic Algorithm II (NSGA-II), NSGA-III, ϵ-dominance Non-dominated Sorting Genetic Algorithm II (ϵ-NSGAII), improved Strength Pareto Evolutionary Algorithm (SPEA2) and Pareto Envelope-based Selection Algorithm 2 (PESA2). Data used to produce the experimental results was retrieved from the working flow of a private chauffeur company with a heterogeneous fleet, which mainly works in the Algarve, Portugal. The objectives were discussed with the company and set to be: (a) the minimization of total distance traveled by the vehicles to serve all requests, (b) the minimization of the drivers' wages difference, and (c) the minimization of the total number of empty seats while satisfying all requests. Results show that the algorithms can return a very large set of valid solution but, for the tested instances and in general, ϵ-MOEA overcomes the other algorithms in indicators such as hypervolume, generational distance, and inverted generational distance.

The added value of this paper is the comparison of the computational performance of the six mentioned state-of-the-art algorithms over a set of DARP instances with different sizes, allowing us to propose ϵ-MOEA as a good starting point (since, e.g., computational time was not a major concern) to solve the private chauffeur company's problem.

The remaining document is structured as follows. The next section deals with the problem's formulation and multiple objective optimization issues. Section 3 briefly describes the algorithmic solutions tested and discusses the problem's codification and chosen operators. The next Section, Sect. 4, presents and discusses results obtained with the experimented methods. The last section presents a conclusion and future work.

2   Problem Formulation

The MO-DARP addressed in this paper was formulated in [24] and took up again in [15]. The problem is formulated over a directed network where nodes are pick up, delivery or shift start/end locations, and edges have associated to them distance and traversing time. A request is characterized by a pick up node, a drop off node, the number and type of passengers (load), and pick up (or drop off) time. Driver and vehicles are connected as there are as many depots as drivers, since it is considered that a vehicle starts and ends at the driver's homes (start/end location). Each vehicle has a capacity and is allocated to one driver. Drivers' working time window and maximum workload are also parameters of the model. The formulation allows for a vehicle to be shared by more than one driver, given that the driver's time windows do not overlap with each other, and the end point of one shift coincides with the start point of the next shift. In the algorithm, these cases are solved considering dummy-duplicated cars with the same characteristics of the original ones.

Regarding scheduling issues, the model takes into consideration operational costs, the number of unused vehicle seats while serving a request, and the cumulative salary of the driver regarding the drivers' average salary (part of the salary of the drivers depends on the number and type of services). Therefore, for a given set of services, the goal of the problem is expressed by a multiple objective function which comprises (a) the minimization of total distance made by the vehicles to serve all services, (b) the minimization of the drivers' wages difference, and (c) the minimization of the total number of empty seats while satisfying all requests. Being obvious that the first objective minimizes the operational costs, the third one also has the same effect as optimizing the number of seats to the request will reduce the use of larger vehicles, which usually have associated higher costs (e.g., larger consumption) and might be less comfortable for the passengers. The second objective reduces the overload of some of the drivers/vehicles which might represent a security issue, promoting, at the same time, more balanced wages among drivers. Most requests solving issues are more or less common sense as, all requests are satisfied by a single vehicle, the vehicle which picks up the customer is the same that drops him off, a vehicle starts and ends its daily service at its depot, vehicles' capacity restrictions for each service are verified, none of the requests of a vehicle overlap in time, and workers only works within the defined time window.

As these different objectives are conflicting, when a solution is better in one objective is normally worse in other(s), e.g., a solution will not improve/decrease the total distance traveled by the vehicles to serve all services (objective (a)) without increasing at least one of the other two objectives (drivers' wages difference (b) or total number of empty seats while satisfying all requests (c)). This means that the problem will not have one best solution, but rather a collection of trade-off solutions, each better in one objective, but worse in other(s). If no decision formula is known (e.g., priority on the objectives), a human decision maker should be presented with this collection of "best solutions", to decide which one should be used, based on the company's politics.

To determine these "best solutions", the Pareto dominance is used: suppose an objective function $F = (f_1, f_2, \ldots, f_m)$ such that $f_i : \Omega \to R$ ($i \in \{1, 2, \ldots, m\}$, Ω is the solutions/search space, and R is an ordered set), and, without loss generality, all objectives of our MO problem are to be minimized. The solution $X \in \Omega$ dominates solution $Y \in \Omega$, $X \prec Y$, if the following pair of conditions are verified: $f_i(X) \leq f_i(Y)$ for all $i \in \{1, 2, \ldots, m\}$ and exists $j \in \{1, 2, \ldots, m\}$ such that $f_j(X) < f_j(Y)$. The solution of the MO problem is the set of solutions which are not dominated by any other solution in the search space, i.e., $\mathcal{P} = \{X \in \Omega | \nexists_{Y \in \Omega} : Y \prec X\}$, called Pareto set. The set of images of the solutions in the Pareto set is called Pareto front, $\mathcal{F} = \{y \in R^m | y = F(X), X \in \mathcal{P}\}$.

However, it is not always possible to compute the Pareto set as, for instance, it can be very large or there are no "efficient" algorithms to compute it. In this case, decision makers might be satisfied when a good approximation set is returned. To measure the quality of approximation set, many indicators can be found in literature [3,28,33]. Desirable features of those indicators include convergence (toward the Pareto front), spread (extent of the front, i.e., distance between the front's extreme solutions), distribution (evenness of the front, i.e., uniformity of the objective distances between pairs of "adjacent" solutions), and Pareto compliance (the ranking established by the indicator does not contradict Pareto optimality, i.e., a front ranked better by one metric must be Pareto preferred over the ones which are ranked worst).

In this work, we will compare the results of the algorithms using the hypervolume (HV), the generational distance (GD) and the inverted generational distance (IGD) indicators. The (i) hypervolume, HV, uses a reference point to measure the size of the objective space covered by the approximation front [35]. HV considers accuracy, diversity and cardinality, being the only known unary metric with this capability. In our case, the hypervolume ratio is computed has the ratio between the hypervolume of the approximation front and the hypervolume of a reference front (i.e., best know approximation set, computed as the non-dominated elements of all known solutions) will be used. The (ii) generational distance, GD, of an approximation front A is computed as the sum of the distance between each solution in A to the closest objective vector in a reference front, averaged over the size of A. The reference front can be the Pareto front or the best known approximation of it. GD only estimates the convergence of the approximation set toward the reference front, and the results can be biased toward approximation sets of poor quality having large cardinality. The (iii) inverted generational distance, IGD, which compared with GD, reverses the order of the fronts considered as input, i.e., IGD is computed as the sum of the distances between the elements of the reference front to the closest element the approximation front, averaged over the size of the reference front. This makes IGD less sensitive to the size of the approximation set and intuitively measures more effectively convergence, spread, and distribution. It should be mentioned that, while the hypervolume is Pareto compliant, GD and IGD are not.

3 Algorithmic Solutions

This section presents the adopted solution encoding, general genetic algorithms (GA) flow and briefly introduces the six algorithms used to solve the MO-DARP problem presented in Sect. 2.

3.1   Solution Encoding and General GA

The solution of the MO-DARP problem corresponds to the assignment of drivers/vehicles to services. E.g., considering a set of services (S_1, S_2, S_3, S_4), ordered for instance by pick up time, a solution could be something like $(1, 2, 1, 3)$ meaning that services S_1 and S_3 are to be served by driver/vehicle 1, S_2 is to be served by driver/vehicle 2, and service S_4 for by driver 3.

Given this representation using vectors of integers, the basic implementation of the based-GA is quite straightforward and can be summarized as follows [14]: (i) initialize the population by randomly attributing drivers/vehicles to services. (ii) Use the objective function values and number of violated restrictions to compute the fitness of each individual. (iii) Select parents supported on their rank and apply the crossover operator with probability, p_c. Figure 1 – left exemplifies a k-point crossover operation [32] where Parent 1 and Parent 2 are combined to generate Offspring 1 and Offspring 2. In the example, the operator randomly chooses $\kappa = 3$ (a parameter of the algorithm) cutting points which is followed by a swap between the blocks defined by those cutting points. (iv) In the next step, offspring suffer a mutation with probability p_m. In our case, the mutation operator randomly changes the vehicle/driver assigned to a service, e.g., see Fig. 1 – right. (v) As an optional step, not applied in our experiments, we can use a local search operator and/or solution correction. For instance, the local search can apply "neighbor" operator to improve offspring, generally improving the convergence in terms of fitness and time/number of iterations/number of objective function evaluations. Solution correction can also improve the algorithm performance by applying operations to ensure that solutions going to the next generation are feasible. (vi) Finally, the offspring are evaluated for their fitness (and possibly crowding distance, distance to reference points, etc.) and between parents and offspring, a new population, with the same size of the original one, is returned. This population moves to the next generation, returning to step (iii), or if the stopping criteria are met, the feasible non-dominated solutions in the last population are returned as the proposed solution.

As seen, GA evolve a population of solutions through a number of generations, using two main operators: crossover and mutation. In short, the first operator passes the parents lineament to the offspring and the second implements diversity on the population, preventing it from becoming trapped in some local minima and avoiding the stagnation of the population. Given a population and its offspring, in a greedy manner, the best individuals will be part of the next generation, maintaining a fixed population size (N). If everything goes well, the new population has moved towards the optimal solution, while maintaining its diversity.

One crucial difference between a generic (single objective) GA and a multiple objective GA, is the selection of the individuals that move to the next generation, as in a multiple objective problem, generally there is not a single best solution for the problem and therefore the selection of elements moving to the next generation is done using the Pareto dominance.

Another thing worth mention is that, as the representation used allows for invalid solutions (e.g., a vehicle has two services that overlap in time), these are treated in the algorithm by implementing a restriction-violation policy, meaning that each solution will have assigned to it the number of restrictions it violates, and solutions that do not violate any restriction will always be considered "better" than those that violate any restriction, this means that the former will always be moved to the next generation (regardless of their fitness values) prior to any of the latter, and if any of these last will survive, the ones with less violations are preferred.

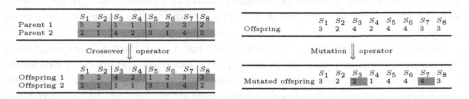

Fig. 1. Solution representation using vectors of integers. Example of the crossover (left) and mutation (right) operators.

3.2   Algorithms Overview

NSGA-II. As the name suggests, Non-dominated Sorting Genetic Algorithm II, NSGA-II, proposed by Deb *et. al.* [11], is a genetic algorithm specially adapted for the optimization of multiple objective problems. NSGA-II uses Pareto dominance to rank solutions to their respective "layer" in the approximation set, i.e., non-dominated solutions are in rank 1, solutions dominated by first rank solutions are in rank 2, solutions dominated by solutions in rank 1 and 2 are placed in rank 3, and so on. The individuals that pass to the next generation are consecutively selected from the lowest rank layers. When not all individuals in a given rank can pass to the next generation, NSGA-II uses the notion of crowding distance to maintain diversity, so that the solutions discarded are those in a more crowded space.

NSGA-III. Another algorithm is NSGA-III [10], characterized by the authors as "a reference-point-based many-objective evolutionary algorithm following NSGA-II framework that emphasizes population members that are non-dominated, yet close to a set of supplied reference points". In this algorithm, the maintenance of diversity among population is aided by supplying and adaptively updating a number of well-spread reference points (r), replacing the crowding distance operator. Equal to NSGA-II, "lowest" layers' elements are added while not surpassing the population's size. When adding a layer that surpasses the population size then some elements are selected taking into consideration a set of reference points which can either be predefined in a structured manner or supplied (preferentially) by the user. Worth mentioning that the population size is not a parameter, but is determined automatically using the number of reference points.

PESA2. Proposed by Corne *et al.* [8] the Pareto Envelope-based Selection Algorithm 2, PESA2, uses a selection technique in which the unit of selection are hyper-boxes located in the objective space. The selective fitness is made over the hyper-boxes populated by at least one individual of the current approximation set. The selected individual is then randomly chosen from the elected hyper-boxes. The method was proven to return solutions which are better spread along the Pareto frontier than individual based selection tested methods.

SPEA2. Proposed by Zitzler, Laumanns, and Thiele [34], the improved Strength Pareto Evolutionary Algorithm, SPEA2, starts by initializing the population with a set of candidate solutions. Then, the best solutions are stored in an archive, insulated from the population. Non-dominated individuals are featured by combining dominance count and dominance rank methods. In other words, for each individual the number of dominated individuals and the number of dominating individuals in the population allows to compute a fitness value. In each iteration, non-dominated individuals from the union of the archive and the current population are updated, maintaining archive's size, either by including dominated individual from the current pool or removing individual from the archive, basing the decision on the nearest neighbor Euclidean distance. The mating pool used to generate the next population is filled by the individuals of the archive and the offspring are then generated by a set of variation operators.

ε-MOEA. Deb, Mohan and Mishra [13] proposed a steady-state Multi-Objective Evolutionary Algorithm, ε-MOEA, based on the ε-dominance, concept introduced in [22]. This algorithm uses a two co-evolving populations: an EA population, and an archive population. The former starts with an random population, and the later gets the non-dominated solutions from the former. In each generation, two solutions, one from each population, are mated to create offspring, which are then checked for inclusion in both population and archive, using the ε-dominance concept. Worth mention that, while the population has a fixed number of individuals, meaning that an individual must be removed in order to incorporated an offspring in the population, the archive can grow as large as needed, to incorporate all non-dominated solution. A detailed explanation can be found in [13].

ε-NSGA-II. Kollat and Reed [20] have extended the NSGA-II algorithm, by adding ε-dominance [22], archiving and adaptive population sizing. The ε-dominance allows the user to specify the precision with which they want to quantify each objective. The ε-NSGA-II uses a series of "connected runs", and, as the search progresses, the population size is adapted based on the number of ε-non-dominated solutions found and stored in the archive, which are then used to direct the search using an injection scheme, where 25% of the subsequent population are composed by ε-non-dominated solutions from the archive, and the other 75% will be generated randomly.

As seen, the last two algorithms use the ϵ-dominance concept to maintain the diversity of the population. Basically, this means that the search space is divided in a number of grids (or hyper-boxes) of ϵ-size, and in each grid (or hyper-box) only one solution in allowed (see [22] for a detailed explanation), guaranteeing this way the inexistence of any area for the solutions to converge.

Table 1. Tuned algorithms' parameters.

Algorithms	κ	p_c	p_m	ϵ	r	N
NSGA-II	4	0.9	0.01	–	–	100
NSGA-III	4	0.9	0.01	–	12	–
SPEA2	4	0.9	0.01	–	–	100
PESA2	4	0.9	0.01	–	–	100
ϵ-NSGA-II	4	0.9	0.01	0.001	–	100
ϵ-MOEA	4	0.9	0.01	0.001	–	100

4   Experimental Results

Starting by a brief characterization of the private chauffeur company's testing data, this section compares the algorithms presented in Sect. 3. So, data used in this section is part of the dataset of the services requested between January 2012 and December 2019 to a private chauffeur company of Algarve. Data is characterized by a seasonal variation where the summer months, in particular July, August, and September, have the largest number of services. The number of services per hour also varies during a tipical day. As a curiosity, the peak number of services in a single hour was reached in a Saturday of September 2019 with 70 services requested between 8 and 9 a.m. On the other other hand, the peak number of services for a single day was also reached in September 2019 with 552 requests.

The code was developed is Python, using the Platypus framework for evolutionary computing, which includes optimization algorithms and analysis tools for MO optimization [17]. To compute the distance between the different service locations, the Open Source Routing Machine (OSRM) [23] was used, which is a C++ routing engine for shortest paths in road networks, supported on Open Street Maps cartography [25]. Implementations were run on an Intel i7-5820K at 3.3 GHz, 32 GB RAM, running Microsoft Windows 10. To simplify the presentation, algorithms' parameters were previously tuned varying their values over large predetermined sets, being the final values summarized in Table 1. The stopping criteria was set as 50,000 evaluations of the objective function and the algorithms were run 10 times for each instance. Furthermore, given the large number of days/problem instances available and in order to represent the seasonality of this type of service, 4 scenarios with different degrees of complexity were selected, namely, by order of scenario, with 40, 99, 300 and 526 requests, and 14, 25, 52 and 70 vehicles, respectively. As a note, the vehicle set used in each scenario was limited to the vehicles used by the company on the selected day.

Table 2 shows the achieved mean values (standard deviation values inside parenthesis) for HV, GD, and IGD indicators. In bold are the best achieved indicator's values. Required to calculate the indicators' values, the reference front for each scenario was computed as the best know approximation front, i.e., it was computed has the non-dominated elements of all known solutions in the objective space. From the table's analysis and over the introduced arrangement, it is deductible that ϵ-MOEA achieves the best results. Only on Scenario 4, NSGA-II had best HV but with a difference of 0.002 units. This results allows us to believe that ϵ-MOEA has better convergence, spread, and distribution when compared with the others, again over this set of instances. We should also point out that in general, algorithms achieved a population of non-dominated solutions in every case, i.e., all solutions/population elements belong to their approximation set. This fact allowed us to avoid taking measures relative to feasibility of the solution (e.g., implementing post optimization/feasibility operators), maintaining the purity of the algorithms.

Table 2. Hypervolume, generational distance, and inverted generational distance mean (standard deviation) values for the 4 scenarios.

Indicator	Scenario	NSGA-II	NSGA-III	SPEA2	PESA2	ϵ-NSGA-II	ϵ-MOEA
HV	1	0.896 (.02)	0.898 (.02)	0.907 (.01)	0.905 (.01)	0.918 (.02)	**0.934** (.02)
	2	0.819 (.03)	0.859 (.02)	0.831 (.02)	0.826 (.04)	0.798 (.02)	**0.881** (.03)
	3	0.654 (.07)	0.738 (.03)	0.626 (.04)	0.699 (.06)	0.540 (.07)	**0.784** (.06)
	4	0.593 (.06)	**0.757** (.05)	0.605 (.06)	0.695 (.07)	0.478 (.05)	0.755 (.08)
GD	1	0.006 (.00)	0.004 (.00)	0.004 (.00)	0.005 (.00)	0.002 (.00)	**0.001** (.00)
	2	0.012 (.00)	0.006 (.00)	0.004 (.00)	0.008 (.00)	0.003 (.00)	**0.002** (.00)
	3	0.022 (.01)	0.014 (.00)	0.010 (.00)	0.015 (.00)	0.014 (.00)	**0.004** (.00)
	4	0.031 (.01)	0.017 (.00)	0.018 (.01)	0.020 (.01)	0.028 (.00)	**0.010** (.00)
IGD	1	0.078 (.01)	0.070 (.01)	0.066 (.01)	0.067 (.00)	0.051 (.01)	**0.040** (.01)
	2	0.122 (.02)	0.078 (.01)	0.087 (.01)	0.098 (.02)	0.114 (.02)	**0.057** (.01)
	3	0.236 (.05)	0.146 (.02)	0.199 (.03)	0.167 (.04)	0.272 (.06)	**0.106** (.02)
	4	0.350 (.08)	0.178 (.04)	0.272 (.05)	0.213 (.05)	0.407 (.05)	**0.160** (.05)

To better illustrate the difference in the solutions, only extreme solutions (those with the minimum values in each objective) will be examined next. This does not mean that a decision maker will select on of those. Figure 2 shows three typical scheduling solutions for Scenario 1 obtained with ϵ-MOEA, where each line represents a vehicle's scheduling (red bars for service time and blue bars the connection time, i.e., moving without passengers). The plots shows the extreme solutions for each objective, namely: (top) minimal total distance, (middle) minimal difference in drivers' wages, and (bottom) minimal number of empty seats. Argumentably, although expectable, it is possible to assert that the top solution (with a total of 2745 Km) has less transit without passengers (blue bars) than the others (with 3088 Km and 2900 Km, respectively). Easy to observe is the fact that, the bottom solution has 3 vehicles without any service, as these were vehicle with more seats and were not scheduled. This last solution shows that is conceivable for the company to diminish the number of large vehicles, at least in the low season.

As an example of a larger instance, Fig. 3 presents the scheduling for Scenario 4 considering the minimal distance (extreme) solution, obtained with ϵ-MOEA. The presented solution accounts for 38987 Km and 924 empty seats. If the number of empty seats was to be minimized (solution is not presented), the number of empty seats extreme solution accounted for 41660 Km and 844 empty seats. As conclusion, these solutions require a decision maker which, in accordance with the company policy, must decide, for instance, if saving 2673 Km is better or worse than using larger vehicles (with, probably, higher fuel consumption). Or simply choose a more "balanced" solution between the ones returned by the elected algorithm.

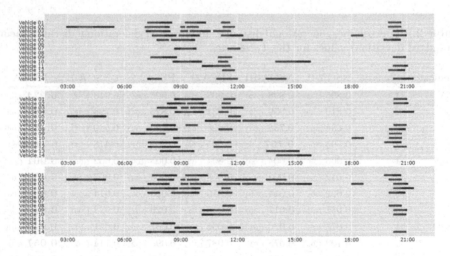

Fig. 2. Examples of scheduling obtained with ϵ-MOEA for scenario 1, considering the objective-extreme: (top) total distance, (middle) drivers' wages difference, and (bottom) number of empty seats.

Fig. 3. Examples of scheduling obtained with ε-MOEA for Scenario 4, considering the objective-extreme total distance

5 Conclusion

This paper studies algorithmic solutions for the scheduling operation of a private chauffeur company. The problem was previously formalized as a multiple objective dial a ride problem, with three objectives: (a) minimization of total distance traveled by the vehicles to serve all requests, (b) minimization of the drivers' wages difference, and (c) minimization of the total number of empty seats while satisfying all requests. Six algorithms (ε-MOEA, NSGA-II, NSGA-III, ε-NSGA-II, SPEA2, and PESA2) were compared using hypervolume, generational distance, and inverted generational distance indicators over 4 different size instances of the problem. Over this arrangement, ε-MOEA could be considered the best solution as it presented the best indicator values, except for one indicator in one scenario.

As future work an analysis on the usage of this and other algorithms online (or quasi-online) is due. Another not yet studied issue is the problem's dynamic, where new requests or vehicles are inserted/removed during or after optimization process finished.

References

1. Attanasio, A., Cordeau, J.F., Ghiani, G., Laporte, G.: Parallel tabu search heuristics for the dynamic multi-vehicle dial-a-ride problem. Parallel Comput. **30**(3), 377–387 (2004). https://doi.org/10.1016/j.parco.2003.12.001
2. Berbeglia, G., Cordeau, J.F., Laporte, G.: A hybrid tabu search and constraint programming algorithm for the dynamic dial-a-ride problem. INFORMS J. Comput. **24**(3), 343–355 (2012). https://doi.org/10.1287/ijoc.1110.0454
3. Bezerra, L.C.T., López-Ibáñez, M., Stützle, T.: An empirical assessment of the properties of inverted generational distance on multi- and many-objective optimization. In: Trautmann, H., et al. (eds.) EMO 2017. LNCS, vol. 10173, pp. 31–45. Springer, Cham (2017). https://doi.org/10.1007/978-3-319-54157-0_3
4. Chevrier, R., Liefooghe, A., Jourdan, L., Dhaenens, C.: Solving a dial-a-ride problem with a hybrid evolutionary multi-objective approach: Application to demand responsive transport. Appl. Soft Comput. **12**(4), 1247–1258 (2012). https://doi.org/10.1016/j.asoc.2011.12.014
5. Cordeau, J.F.: A branch-and-cut algorithm for the dial-a-ride problem. Oper. Res. **54**(3), 573–586 (2006). https://doi.org/10.1287/opre.1060.0283
6. Cordeau, J.F., Laporte, G.: The dial-a-ride problem: models and algorithms. Ann. Oper. Res. **153**(1), 29–46 (2007). https://doi.org/10.1007/s10479-007-0170-8
7. Cordeau, J.F., Laporte, G.: The dial-a-ride problem (DARP): variants, modeling issues and algorithms. Q. J. Belg. Fr. Ital. Oper. Res. Soc. **1**(2), 89–101 (2003). https://doi.org/10.1007/s10288-002-0009-8
8. Corne, D.W., Jerram, N.R., Knowles, J.D., Oates, M.J.: PESA-II: Region-based selection in evolutionary multiobjective optimization. In: Proceedings of the Genetic and Evolutionary Computation Conference GECCO 2001, pp. 283–290. Morgan Kaufmann Publishers (2001)
9. Cubillos, C., Rodriguez, N., Crawford, B.: A study on genetic algorithms for the DARP problem. In: Mira, J., Álvarez, J.R. (eds.) IWINAC 2007. LNCS, vol. 4527, pp. 498–507. Springer, Heidelberg (2007). https://doi.org/10.1007/978-3-540-73053-8_50
10. Deb, K., Jain, H.: An evolutionary many-objective optimization algorithm using reference-point-based nondominated sorting approach, part I: solving problems with box constraints. IEEE Trans. Evol. Comput. **18**(4), 577–601 (2014). https://doi.org/10.1109/TEVC.2013.2281535
11. Deb, K., Pratap, A., Agarwal, S., Meyarivan, T.: A fast and elitist multiobjective genetic algorithm: NSGA-II. IEEE Trans. Evol. Comput. **6**(2), 182–197 (2002). https://doi.org/10.1109/4235.996017
12. Deb, K.: Multi-objective Optimization Using Evolutionary Algorithms. Wiley, Hoboken (2001)
13. Deb, K., Mohan, M., Mishra, S.: A fast multi-objective evolutionary algorithm for finding well-spread pareto-optimal solutions. Technical report (2003). http://www.iitk.ac.in/kangal/papers/k2003003.pdf
14. Gendreau, M., Potvin, J.Y.: Handbook of Metaheuristics, vol. 2. Springer, New York (2010). https://doi.org/10.1007/978-1-4419-1665-5
15. Guerreiro, P.M.M., Cardoso, P.J.S., Fernandes, H.C.L.: Applying NSGA-II to a multiple objective dial a ride problem. In: Rodrigues, J.M.F., et al. (eds.) ICCS 2019. LNCS, vol. 11540, pp. 55–69. Springer, Cham (2019). https://doi.org/10.1007/978-3-030-22750-0_5

16. Haddadene, S.A., Labadie, N., Prodhon, C.: NSGA-II enhanced with a local search for the vehicle routing problem with time windows and synchronization constraints. IFAC Papers Line **49**(12), 1198–1203 (2016). https://doi.org/10.1016/j.ifacol.2016.07.671

17. Hadka, D.: Platypus - Multiobjective Optimization in Python (2015). https://platypus.readthedocs.io. Accessed 08 Feb 2019

18. Ho, S.C., Szeto, W., Kuo, Y.H., Leung, J.M., Petering, M., Tou, T.W.: A survey of dial-a-ride problems: literature review and recent developments. Transp. Res. Part B Methodol. **111**, 395–421 (2018). https://doi.org/10.1016/j.trb.2018.02.001

19. Issaoui, B., Khelifi, L., Zidi, I., Zidi, K., Ghédira, K.: A contribution to the resolution of stochastic dynamic dial a ride problem with NSGA-II. In: 13th International Conference on Hybrid Intelligent Systems, HIS 2013, pp. 54–59 (2013)

20. Kollat, J.B., Reed, P.M.: The value of online adaptive search: a performance comparison of NSGAII, ε-NSGAII and ε-MOEA. In: Coello Coello, C.A., Hernández Aguirre, A., Zitzler, E. (eds.) EMO 2005. LNCS, vol. 3410, pp. 386–398. Springer, Heidelberg (2005). https://doi.org/10.1007/978-3-540-31880-4_27

21. Lai, M., Tong, X.: A metaheuristic method for vehicle routing problem based on improved ant colony optimization and tabu search. J. Ind. Manage. Optim. **8**(2), 469–484 (2012). https://doi.org/10.3934/jimo.2012.8.469

22. Laumanns, M., Thiele, L., Deb, K., Zitzler, E.: Combining convergence and diversity in evolutionary multiobjective optimization. Evol. Comput. **10**(3), 263–282 (2002). https://doi.org/10.1162/106365602760234108, http://www.mitpressjournals.org/doi/10.1162/106365602760234108

23. Luxen, D., Vetter, C.: Real-time routing with openstreetmap data. In: Proceedings of the 19th ACM SIGSPATIAL International Conference on Advances in Geographic Information Systems, GIS 2011, New York, NY, USA, pp. 513–516. ACM (2011). https://doi.org/10.1145/2093973.2094062

24. Morais, A.C., Torres, L., Dias, T.G., Cardoso, P.J.S., Fernandes, H.: A combined data mining and tabu search approach for single customer dial-a-ride problem. In: 7th International Conference on Metaheuristics and Nature Inspired Computing, Marrakech, Morocco, pp. 121–123, October 2018

25. OpenStreetMap contributors: Planet dump (2017). https://planet.osm.org, https://www.openstreetmap.org

26. Parragh, S.N., Doerner, K.F., Hartl, R.F.: A survey on pickup and delivery problems. J. für Betriebswirtschaft **58**(1), 21–51 (2008). https://doi.org/10.1007/s11301-008-0033-7

27. Qu, Y., Bard, J.F.: A branch-and-price-and-cut algorithm for heterogeneous pickup and delivery problems with configurable vehicle capacity. Transp. Sci. **49**(2), 254–270 (2014). https://doi.org/10.1287/trsc.2014.0524

28. Riquelme, N., Lücken, C.V., Baran, B.: Performance metrics in multi-objective optimization. In: 2015 Latin American Computing Conference, CLEI, pp. 1–11, October 2015. https://doi.org/10.1109/CLEI.2015.7360024

29. Ropke, S., Cordeau, J.F., Laporte, G.: Models and branch-and-cut algorithms for pickup and delivery problems with time windows. Netw. Int. J. **49**(4), 258–272 (2007). https://doi.org/10.1002/net.20177

30. Subramanian, A., Drummond, L., Bentes, C., Ochi, L., Farias, R.: A parallel heuristic for the vehicle routing problem with simultaneous pickup and delivery. Comput. Oper. Res. **37**(11), 1899–1911 (2010). https://doi.org/10.1016/j.cor.2009.10.011

31. Tripathy, T., Nagavarapu, S.C., Azizian, K., Ramasamy Pandi, R., Dauwels, J.: Solving dial-a-ride problems using multiple ant colony system with fleet size minimisation. In: Chao, F., Schockaert, S., Zhang, Q. (eds.) UKCI 2017. AISC, vol. 650, pp. 325–336. Springer, Cham (2018). https://doi.org/10.1007/978-3-319-66939-7_28
32. Umbarkar, A., Sheth, P.: Crossover operators in genetic algorithms: a review. ICTACT J. Soft Comput. **06**(01), 1083–1092 (2015). https://doi.org/10.21917/ijsc.2015.0150
33. Zitzler, E., Thiele, L., Laumanns, M., Fonseca, C.M., da Fonseca, V.G.: Performance assessment of multiobjective optimizers: an analysis and review. IEEE Trans. Evol. Comput. **7**(2), 117–132 (2003). https://doi.org/10.1109/TEVC.2003.810758
34. Zitzler, E., Laumanns, M., Thiele, L.: SPEA2: Improving the strength pareto evolutionary algorithm (2001). https://doi.org/10.3929/ETHZ-A-004284029, http://hdl.handle.net/20.500.11850/145755
35. Zitzler, E., Thiele, L.: Multiobjective evolutionary algorithms: a comparative case study and the strength Pareto approach. IEEE Trans. Evol. Comput. **3**(4), 257–271 (1999). https://doi.org/10.1109/4235.797969

Software Engineering for Computational Science

Lessons Learned in a Decade of Research Software Engineering GPU Applications

Ben van Werkhoven[1]([✉]) [iD], Willem Jan Palenstijn[2] [iD], and Alessio Sclocco[1] [iD]

[1] Netherlands eScience Center, Amsterdam, The Netherlands
{b.vanwerkhoven,a.sclocco}@esciencecenter.nl
[2] Centrum Wiskunde & Informatica (CWI), Amsterdam, The Netherlands
w.j.palenstijn@cwi.nl

Abstract. After years of using Graphics Processing Units (GPUs) to accelerate scientific applications in fields as varied as tomography, computer vision, climate modeling, digital forensics, geospatial databases, particle physics, radio astronomy, and localization microscopy, we noticed a number of technical, socio-technical, and non-technical challenges that Research Software Engineers (RSEs) may run into. While some of these challenges, such as managing different programming languages within a project, or having to deal with different memory spaces, are common to all software projects involving GPUs, others are more typical of scientific software projects. Among these challenges we include changing resolutions or scales, maintaining an application over time and making it sustainable, and evaluating both the obtained results and the achieved performance.

Keywords: Software engineering · Research software engineering · GPU Computing · Research software

1 Introduction

Since the introduction of programmable units in Graphics Processing Units (GPUs) scientists have been using GPUs because of their raw compute power and high energy efficiency [20]. Today, many of the top500 supercomputers are equipped with GPUs [27] and GPUs are a driving force behind the recent surge in machine learning [1,4,5]. However, developing GPU applications requires computations to be parallelized using specialized programming languages, and to achieve high performance requires to understand the underlying hardware [22]. As such, many GPU applications have been developed by Research Software Engineers (RSEs) [2] that have specialized in this field. This paper presents an overview of our experiences and of the lessons learned from developing GPU applications for scientific research in a wide range of domains, with different computational requirements, and a variety of programming languages.

Developing GPU applications requires to make software architectural choices that will be costly to change once implemented, concerning device memory management, host and device code integration across programming languages, and

© Springer Nature Switzerland AG 2020
V. V. Krzhizhanovskaya et al. (Eds.): ICCS 2020, LNCS 12143, pp. 399–412, 2020.
https://doi.org/10.1007/978-3-030-50436-6_29

synchronizing and maintaining multiple versions of computational kernels. This is due to the fact that the commonly used GPU programming systems do not allow developers to easily explore the whole design space. In addition, GPU kernels exhibit large design spaces with different ways to map computations to threads and thread blocks, different layouts to use in specialized memories, specific hardware features to exploit, and values to select for thread block dimensions, work per thread, and loop unrolling factors. As such, auto-tuning is often necessary to achieve optimal and portable performance.

This paper also presents challenges and lessons learned specific to GPU *research software*. Research software is defined by Hettrick et al. [11] as software that is used to generate, process or analyze results that are intended to appear in scientific publications. Research software is often developed using short-lived grant-based research funding, and therefore its development is focused on new features instead of reliability and maintainability [7]. Finally, several surveys have shown that research software is in large part developed by scientists who lack a formal education or even interest in software engineering best practices [10, 11]. The experiences we share in this paper are based on both short-lived collaborations between RSEs and scientists and on collaborations where an RSE was embedded in a research group for a longer period of time.

Advancements in hardware, compilers, and testing frameworks present new opportunities to apply software engineering best practices to GPU applications. However, it is important to realize that GPU research software is often based on existing software developed by scientists with no formal training in software engineering. The software sustainability of GPU research software remains an open challenge as GPU programming remains a specialized field and RSEs are often only involved during short-lived collaborative projects.

The reason to move code to the GPU is often to target larger, more complex problems, which may require the development of new methods to operate at higher resolutions or unprecedented problem scales. Evaluating the results of these applications often requires carefully constructed test cases and expert knowledge from the original developers. Designing fair and reproducible performance experiments that involve applications using different languages, compilers, and hardware is a difficult task. This has led to the publication of controversial performance comparisons, and currently hinders RSEs in publishing about their work.

The rest of this paper is organized as follows. Section 2 lists the applications that we use as case studies in this paper. Section 3 presents the challenges and lessons learned that apply to GPU programming in any context, whereas Sect. 4 focuses on those specific to a research software engineering context. Finally, Sect. 5 summarizes our conclusions.

2   Case Studies

In this section, we introduce some of the scientific GPU applications that we developed, which we use as case studies throughout the paper. These applications span a wide range of scientific domains and host programming languages, as well as target hardware platforms, ranging from GPU-enabled supercomputers to workstations and desktop computers equipped with one or more consumer grade GPUs. The remainder of this section contains a brief description of each application, and highlights some of the challenges encountered; a summary of the characteristics of said applications is provided in Table 1.

Table 1. Overview of the GPU application case studies.

Application name	Scientific domain	Language	Main bottleneck	Target hardware	Existing GPU code
2D & 3D SMLM	Microscopy	MATLAB	Compute	Desktop/server	No
3D GDE	Geospatial databases	C++	Latency	Server	No
AMBER	Radio astronomy	C++	Communication	GPU cluster	No
ASTRA toolbox	Tomography	MATLAB Python	Compute	Desktop/server	No
CISI	Digital forensics	Java	Compute	Desktop	No
KM3NeT L0-Trigger	Particle physics	Python	Compute	GPU cluster	No
Parallel-Horus	Computer vision	C	Compute	GPU cluster	No
POP	Climate modeling	Fortran90	Communication	Supercomputer	No
SAGECAL	Radio astronomy	C++	Compute	GPU cluster	Yes

2D single molecule localization microscopy (SMLM) [12] is a MAT-LAB application, implementing a template-free particle fusion algorithm based on an all-to-all registration, which provides robustness against individual misregistrations and underlabeling. The method combines many different observations into a single super-resolution reconstruction, working directly with localizations rather than pixelated images. The application is very compute intensive as it uses several quadratic algorithms for computing registration scores on the GPU.

3D Geospatial Data Explorer (GDE) [8] is a spatial database management system which provides in-situ data access, spatial operations, and interactive data visualization for large, dense LiDAR data sets. The system was designed to handle a LiDAR scan of the Netherlands of 640 billion points, combined with cadastral information. The main GPU kernel implements a point-in-polygon algorithm for selecting data points within a selected shape. To optimize the latency of this operation, host to device transfers are overlapped with execution on the GPU, and auto-tuning is used extensively.

AMBER [24], the Apertif Monitor for Bursts Encountered in Real-time, is a fully auto-tuned radio-astronomical pipeline for detecting Fast Radio Bursts and other single pulse transients, developed in C++ and OpenCL. All the main computational components of AMBER are accelerated and run on the GPU, and the pipeline is deployed and used in production at the Westerbork radio telescope [29]. The main technical challenge developing this application has been processing, in real-time, the large amount of data produced by the telescope, while the main non-technical challenge has been to properly validate the application's results.

ASTRA Toolbox [19] is a toolbox of high-performance GPU primitives for 2D and 3D tomography aimed at researchers and algorithm developers. The basic forward and backward projection operations are GPU-accelerated, and callable via a C++-interface from MATLAB and Python to enable building new algorithms. The main challenges were the trade-off between flexibility and performance, making effective use of memory caching, and the initial lack of reference code.

Common image source identification (CISI) [31] is a digital forensics tool that clusters a collection of digital photos based on whether they were acquired using the same image sensor. The GPU application implements a pipeline that extracts image noise patterns, as well as two algorithms for computing similarity scores. This application was developed in collaboration with the Netherlands Forensics Institute, where most of the software development happens in Java, as such the host code for the GPU application is written in Java as well.

KM3NeT L0-Trigger is a GPU pipeline designed to process unfiltered data from the KM3NeT neutrino telescope [14]. Most of the existing processing happens on filtered, so called L1 data. The main issue in the design and implementation of this pipeline was the fact that the algorithms that operate on the L1 data were not suitable for processing L0 data. As such, we had to design a new pipeline that correlates L0 hits, clusters them, and classifies whether a neutrino event has occurred.

Parallel Ocean Program (POP) [33] is an ocean general circulation model. POP is a large Fortran 90 code that has been in development and use for a long time. While the code does not contain any particular computational hotspots, we have ported the equation of state and vertical mixing computations to the GPU. The main challenges were overlapping computation with CPU-GPU communication and host language integration. The kernels have a relatively low arithmetic intensity and not much opportunity for code optimizations.

Parallel-Horus [34] is an image processing library that automatically parallelizes image algebra operations. One of the more challenging aspects in integrating GPU kernels was to deal with the separate GPU memory. For most applications that use Parallel-Horus, 2D convolution is the most time consuming operation, for which we have implemented a highly-optimized and auto-tuned GPU kernel [32].

SAGECal [26] is a radio interferometric calibration package supporting a wide range of source models. SAGECal is intended to run on various platforms from GPU clusters to low-end energy-efficient GPUs. We have significantly improved the performance of the original GPU code by removing the use of dynamic parallelism and changing how the problem is mapped to threads and thread blocks. Using intrinsics to accelerate the computation of trigonometric functions also significantly improved performance.

3 GPU Programming Challenges

This section lists the challenges encountered, and the lessons learned, while building the applications described in Sect. 2. Many of these originate from the fact that the GPU application is based on existing software.

3.1 Dealing with Separate Device Memory

The fact that GPUs have separate device memory introduces much complexity. First of all, one needs to decide how to manage the GPU memory itself. Allocating and freeing memory can be costly operations and as such memory allocations are preferably reused over time. Furthermore, the GPU programmer needs to consider what policy to implement for when device memory is full, or when it can be more efficient to forgo using device memory and stream data directly from host memory instead. There are two main perspectives to consider here, depending on whether a library or an application is being developed.

From a library perspective, the application using the library either manages device memory or trusts the library to handle it. In case the library acts as a drop-in replacement for an existing non-GPU library, the library needs to manage device memory and automatically transfer data. For example, in the Parallel-Horus image processing library it is not known in what order the operations will be called and when data needs to be present in host or device memory. In this case, we extended the state-machine used inside the library to keep track of distributed memory to also keep track of host or device copies of the data. This allows the library to insert memory transfers only when necessary and as such many, otherwise redundant, data transfers can be eliminated.

When developing a GPU application, the programmer has a complete view of what needs to be executed on the GPU and in what order. In particular for applications that require low latency GPU operations, it can be important for performance to overlap the data transfers with computations on the GPU. We have used performance modeling to estimate the performance impact of various overlapping techniques [35].

For applications that require frequent, asynchronous, or high-throughput data transfers between host and device memory, it may be necessary to change the way in which host memory is allocated, as these allocations will need to be pinned and page-aligned. This could require extensive changes to the original application and may prove difficult for applications in object-oriented languages such as Java, MATLAB, or modern Fortran.

3.2 Host Language Integration

In many cases, only a relatively small part of an existing application is ported to the GPU, and as such a large part remains in the original programming language. In general, there are two ways of integrating GPU code into applications that are not written in C/C++. Either you write the host code in C/C++ and use some foreign function interface to call that host function from the original application, or you use the language bindings that are available for the OpenCL or CUDA driver APIs in the original program's language. We now briefly discuss how GPU code might be integrated in several different languages that we have used.

C++. While OpenCL and CUDA have been designed for C and C++, their runtime APIs were initially not object oriented. OpenCL now has an official C++ header, but for CUDA a high-level object-oriented runtime API is still lacking. As such, even for C++ there are separately developed API wrappers to create a modern C++ API for CUDA programming [21].

MATLAB offers three different ways to execute GPU code into applications. Firstly, through arrays that can be declared using a GPU-enabled array type. Point-wise arithmetic operations on the array are lazily evaluated and compiled into CUDA kernels on first use. Secondly, PTX-compiled CUDA kernels can be loaded as functions into MATLAB. Finally, the MEX interface allows to call C/C++ functions from MATLAB, which could in turn call GPU kernels.

If you have complex kernels that do more than point-wise operations, you are limited to option 2 and 3. If you also need fine-grained control over GPU memory, for example to reuse GPU memory allocations across kernel invocations, or to control exactly when data is transferred between host and device memory, you have to write the host code in C/C++ and use the MEX interface.

Fortran has fewer options. There is CudaFortran offered by the PGI compiler, which requires the GPU kernels to be written as CudaFortran subroutines. When working on the Parallel Ocean Program, we have written C functions around our kernels and some of the CUDA runtime API functions to interface the kernels written as CUDA/C code from Fortran 90.

Java has a number of different wrappers available for integrating GPU code. The direct language bindings for OpenCL and CUDA, JOCL and JCuda, are most commonly used, but object-oriented programming that adheres to common Java conventions requires additional wrappers. As most Java GPU libraries rely on open source contributions, it is often hard for these projects to keep up with the latest features.

Python has a great number of options, ranging from array, data frame, or tensor libraries with GPU support to language bindings for OpenCL and CUDA through PyOpenCL and PyCuda. The latter are not direct mappings of the C APIs and as such allow GPU programming in a way common to Python programmers. However, both projects rely on open source contributions to be kept up to date. For example, initial basic C++ support in the kernel language already existed at the time of CUDA 1.x, but PyCuda still requires kernels to have C linkage.

3.3 Optimizing Code

One of the most time-consuming aspects of software engineering GPU applications is performance debugging and code optimization. With *code optimization* we mean the process of making changes in the kernel code with the aim to improve kernel performance. Applying code optimization is often necessary to achieve high performance.

The roofline model [36] is an often used tool for analyzing whether the performance of a code is bandwidth or compute bound. While the original roofline model was not specifically developed for GPUs, it has proved an incredibly useful tool for GPU programming. Many adaptations of the roofline model have been introduced, for example a cache-aware roofline model [13], a quantitative model for GPU performance estimation [15], or one that includes PCIe data transfers [35]. While performance models help to understand the bottlenecks in the code, it is still very hard to fully explain the performance at run time.

In general, it is important to understand that moving data around is much more expensive than computing on it. As such, many of the code optimizations that are applied to GPU code concern improving data access patterns, exploiting specific memories as caches, and reusing data. Code optimizations that focus specifically on improving the computations, without changing data access patterns are in our experience quite rare.

In the end, the most frequently applied code optimization is to vary the amount of work per thread and thread block. This technique is referred to under many different names, including thread-block-merge [37], thread coarsening [38], or 1xN Tiling [23]. The reason that this particular code optimization is so effective is because it improves upon many different aspects of the code, including reducing the number of redundant instructions across threads and thread blocks, improving the data access pattern to maximize locality, and improving the amount of reuse of data and/or intermediate computations.

However, in the end the GPU programmer is left with a large design space of how to parallelize the computation and map that onto threads, how many threads to use in each thread block dimension, how much work to give to each thread and thread block, and so on. These parameters are hard to predict and the optimal configuration can be very hard to find by hand [25,30]. As such, we have made extensive use of auto-tuning techniques in many different applications to ensure optimal and portable performance of our applications across different GPUs and problem dimensions.

4   Research Software Challenges

This section presents the challenges and lessons learned with regard to the research software engineering aspects of developing GPU research software.

4.1   Software Engineering Practices

It is important to realize that GPU research software is often based on existing software developed by scientists with no formal training in software engineering [11]. In addition, research software is often developed within projects of fixed duration with no resources for long-term maintenance [7]. The result is that research software often does not follow software engineering best practices, for example having no, little, or outdated documentation, no or only few automated tests, and code optimizations that might no longer be relevant.

Testing. Despite the large number of frameworks that seek to simplify GPU application development, a majority of GPU applications is still developed using low-level languages such as CUDA and OpenCL. This is in part due to the tension between increased abstraction levels and performance requirements that demand developer control over all aspects of the hardware. Another practical reason is that many tools and frameworks are developed as academic works, e.g. part of a PhD project, with no guarantee of sustained maintenance, making it a risk to use such a framework as a basis for a new project.

Unfortunately, not much CUDA and OpenCL code that is used in publications is tested. For a long time there were no testing frameworks that actively supported testing of GPU code. That is why we have extended Kernel Tuner [30] to support testing GPU code from Python.

Dealing with Existing Optimizations. Several of the applications we have worked on included optimizations whose need had faded over time, as compilers improved or new hardware architectures were released. This phenomenon can also be observed in GPU applications.

For example, in the early days of CUDA programming increasing the work per thread required manually unrolling the inner-most loop, avoiding the use of arrays because of memory inefficiencies. As it is often necessary to vary and tune the amount of work assigned to each thread, many applications contained multiple different versions of their GPU kernels with much code duplication as a result. Other developers used custom written code generators to avoid code duplication, resulting in code that is much harder for others to understand. Fortunately, with modern GPUs and compilers, it is possible to vary the work per thread dynamically.

Another example is from the Parallel Ocean Program that has been in development for a long time. A lot of the code we worked with was developed in the late nineties. Computer architectures were very different at the time, where compute was expensive and memory bandwidth was quite cheap. In modern processors, and in particular GPUs, this is quite the opposite, which means that some of the patterns in the original code could now be seen as unnecessary optimizations.

In general, it is very important to realize that the code that you develop may well outlive the hardware you are currently developing for. Trusting the compiler to handle the simple cases of code optimization helps to improve the maintainability and portability of the code. As of CUDA 7.0 and OpenCL 2.x, C++11 is supported in the kernel programming languages, improving readability and maintainability of GPU code. In addition, templates can be used to avoid code duplication.

Selecting the Right Starting Point. Selecting the right starting point for developing GPU kernels is crucial. The first implementation for a GPU kernel is commonly called the *naive* version, which should be simple enough to allow the GPU to exploit massive data-parallelism. The name, however, suggests that there is something wrong with this code, which is definitely not the case. The naive code is a crucial part of the GPU application. Even if it is not executed in production, it should be stored along with the code, because it is necessary for understanding and testing the optimized and tunable versions of the code.

4.2   Targeting Problems at New Scales

An important question that is easy to overlook is whether the existing research software, on which the GPU code will be based, is actually capable of solving the problem at hand. This may sound obvious, but it is important to realize that moving to the GPU is often motivated by the desire to increase in scale and complexity. The research group is developing GPU code because they want to address larger, more computationally demanding problems at some larger scale or at a higher resolution.

At different scales it is often even required to use different algorithms. Not just from a performance perspective, but also because the original code was developed to solve a problem at a different scale, and might simply not produce meaningful results when executed on a larger problem. Secondly, rounding errors and other numerical inaccuracies may accumulate in a stronger fashion when the problem size changes.

For example, ocean models at coarse resolutions often use parameterizations to represent subgrid physical processes that are too small to be resolved at the current scale. However, at higher resolutions, we can do without these parameterizations because the parameterized physical processes are already resolved. As such, using the same algorithm without modification at both resolutions would not produce correct results [6].

We have also observed this when working on the L0-trigger pipeline for the KM3NeT neutrino telescope, where the algorithms that were in place operated on the pre-filtered L1 hits. However, using these algorithms to correlate L0 hits led to nearly all hits being correlated, as such the trigger pipeline did not work at all for the unfiltered L0 hits. In the end, we had to develop entirely new correlation criteria and community detection algorithms for the GPU pipeline.

4.3   Software Sustainability

Lago et al. [16] define *software sustainability* as "the capacity to endure and preserve the function of a system over an extended period of time". In contrast to narrower definitions that only consider software sustainability as a composite measure of software quality attributes [28], the definition by Lago et al. also captures social aspects that are crucial for research software projects to endure.

Translating and parallelizing existing research software for the GPU during a short-lived collaboration creates a risk with regard to software sustainability. Programming GPU applications remains a specialized field that requires advanced technical knowledge and specialized programming languages. The GPU applications that we have developed were often part of temporary collaborations between different research groups. Once the project is finished, who is going to maintain the newly developed GPU code?

Without significant changes to the way in which scientific projects and software development for research is funded, there is no obvious solution that we are aware of. We can only recommend to think about software sustainability from the start. Measures that could be taken are involving the original developers of the code in the GPU development process, as well as documenting why and how the GPU code differs from the existing code.

However, involving the original developers in the development of a GPU version is not always possible. For example, in one project we could not get security clearance to access the internal version control servers, forcing the two groups to develop on separate repositories. Another issue that can occur is that some developers take offense when someone else starts to make significant changes to their code. This could be because accelerating their code can be interpreted as a sign that the original developers did something wrong, or simply because the original code had been extensively validated and any significant changes to the code may require to redo the validation.

Another take on the sustainability problem is to use a domain-specific language (DSL) to introduce a separation of concerns between what needs to be computed and how the computations are mapped to the target hardware platform. This approach is, for example, currently being researched in the weather and climate modeling community in Europe [17].

4.4   Evaluating Results

Evaluating the output quality is one of the main challenges in developing GPU research software. When code is ported to the GPU, the output will not be bit-for-bit the same as it was on the CPU. There are many reasons for this.

Unlike the first programmable GPU architectures, modern GPUs fully implement the IEEE floating-point standard. Other factors do, however, introduce differences in the results. Floating-point arithmetic is not associative and as parallelization changes the order of operations, the results from a parallel algorithm will inherently be different from the sequential version, and will depend on the exact parallelization strategy and parameters chosen. Other reasons include the availability of different instructions (such as fused multiply-add instructions), the use of extended precision (80-bit floating point) in many CPUs, and different compilers that optimize differently.

RSEs of GPU applications will have to live with the fact that the results are not bit-for-bit the same. Whether that is a problem depends on the application. To analyze whether these differences matter for the application requires that you understand what is being computed and why, which requires close collaboration with the scientists that developed the original application.

It can also be difficult to determine whether the differences in results are due to a bug or due to the difference in parallelization and compilers. Testing can help to increase confidence in the GPU application, but we have learned the hard way that testing with randomly generated inputs can give a false sense of correctness. As such, we recommend that the test input is realistic enough to produce realistic output and to not only compare the GPU results with results obtained on the CPU, but also assert that both outputs themselves are correct.

4.5   Evaluating Application Performance

Setting up experiments for performance comparisons of GPU applications, and in particular presenting the performance improvement over earlier versions, can be quite complicated.

In the early days of GPU Computing, many papers reported spectacular performance results. With their paper "Debunking the 100X GPU vs. CPU myth" [18], scientists from Intel made it clear that many of these performance results were based on unfair comparisons. In many cases, the performance of a highly-optimized GPU kernel was compared against an unoptimized, sometimes not even parallelized CPU kernel. Lee et al. [18] demonstrated for a large range of benchmark applications that if both the GPU and the CPU implementations are fully optimized, the performance difference between the two is usually within the range of the theoretical performance difference between the two platforms.

In computer science, the discussion around presenting performance results often assumes that the only question worth answering is: "Which of these two processors is the most efficient for algorithm X?". On the other hand, RSEs often work with scientists from different fields, and thus receive questions such as "How much faster is that MATLAB/Python code that I gave you on the GPU?". This puts the RSEs in a difficult position. You would like to be able to tell the broader scientific community the fact that the application X now runs Y times faster. However, that result on its own is currently very hard to publish, unless you are also willing to spend several more months to also optimize the CPU code and make a 'fair' performance comparison between the two processors.

Not being able to publish about your work could ultimately limit career advancement, because while RSEs are tasked with creating research software, they are often judged by metrics such as scientific publications [3].

When it comes to presenting performance comparisons, it should be absolutely clear what is being compared against what, what hardware is being used by each application, and why that comparison is a realistic and representative use case. It is important that the experimental methodology is sound and is documented so that others may reproduce the results.

In their paper "Where is the data?", Gregg and Hazelwood [9] point out that it is crucial to understand where the input and output data of a GPU kernel is stored. This strongly depends on the rest of the application. It could be that the GPU kernel is part of a pipeline of GPU kernels and that it is safe to assume that the input and output data are present in device memory. However when this is not the case, the data transfer between host and device should be included in performance comparisons.

It is also possible to normalize results in different ways, for example based on hardware purchase costs, energy consumption, theoretical peak performance, person-months spent on code optimization. It all depends on the application whether such normalizations have any relevance, and they do not solve the problem in general. In addition, these normalizations can make it harder to interpret and compare the presented performance results across different publications.

5   Conclusions

GPUs are a very attractive computational platform because they offer high performance and energy efficiency at relatively low cost. However, developing GPU applications can be challenging, in particular in a research software engineering context. For each of the case studies included in this paper, which span a wide range of research domains, programming languages, and target hardware platforms, we have listed what the main challenges were in developing the GPU application. We have grouped the recurring challenges and the lessons learned into two categories, related to either GPU programming in general, or specifically to developing GPU applications as a research software engineer.

In summary, developing GPU applications in general requires fundamental design decisions on how to deal with separate memory spaces, integrate different programming languages, and how to apply code optimizations and auto-tuning.

Developing GPU research software comes with a number of specific challenges, with regard to software engineering best practices, and the quantitative and qualitative evaluation of output results. In general, we recommend to carefully select and if needed rewrite the original application to ensure the starting point is of sufficient code quality and is capable of solving the problem at the scale the GPU application is targeting. When performance comparisons of different applications are of interest to the broader scientific community it is important that RSEs can publish those results, both for the community to take notice of this result and for the RSEs to advance in their academic career. Finally, we note that software sustainability remains an open challenge for GPU research software when RSEs are only involved in the project on a temporary basis.

References

1. Abadi, M., et al.: TensorFlow: large-scale machine learning on heterogeneous distributed systems. arXiv preprint arXiv:1603.04467 (2016)
2. Baxter, R., Hong, N.C., Gorissen, D., Hetherington, J., Todorov, I.: The research software engineer. In: Digital Research Conference, Oxford, pp. 1–3 (2012)
3. Brett, A., et al.: Research Software Engineers: State of the Nation Report 2017, April 2017. https://doi.org/10.5281/zenodo.495360
4. Chen, T., et al.: MXNet: a flexible and efficient machine learning library for heterogeneous distributed systems. arXiv preprint arXiv:1512.01274 (2015)
5. Chetlur, S., et al.: cuDNN: efficient primitives for deep learning. arXiv preprint arXiv:1410.0759 (2014)
6. Gent, P.R.: The Gent-McWilliams parameterization: 20/20 hindsight. Ocean Model. **39**(1–2), 2–9 (2011)
7. Goble, C.: Better software, better research. IEEE Internet Comput. **18**(5), 4–8 (2014)

8. Goncalves, R., et al.: A spatial column-store to triangulate the Netherlands on the fly. In: Proceedings of the 24th ACM SIGSPATIAL International Conference on Advances in Geographic Information Systems, p. 80. ACM (2016)
9. Gregg, C., Hazelwood, K.: Where is the data? Why you cannot debate CPU vs. GPU performance without the answer. In: (IEEE ISPASS) IEEE International Symposium on Performance Analysis of Systems and Software, pp. 134–144. IEEE (2011)
10. Hannay, J.E., MacLeod, C., Singer, J., Langtangen, H.P., Pfahl, D., Wilson, G.: How do scientists develop and use scientific software? In: 2009 ICSE Workshop on Software Engineering for Computational Science and Engineering, pp. 1–8. IEEE (2009)
11. Hettrick, S., et al.: UK research software survey 2014 (2014)
12. Heydarian, H., et al.: Template-free 2D particle fusion in localization microscopy. Nat. Methods 15(10), 781 (2018)
13. Ilic, A., Pratas, F., Sousa, L.: Cache-aware roofline model: upgrading the loft. IEEE Comput. Archit. Lett. 13(1), 21–24 (2013)
14. de Jong, M.: The KM3NeT neutrino telescope. Nucl. Instrum. Methods Phys. Res. Sect. A 623(1), 445–447 (2010). https://doi.org/10.1016/j.nima.2010.03.031. 1st International Conference on Technology and Instrumentation in Particle Physics
15. Konstantinidis, E., Cotronis, Y.: A quantitative roofline model for GPU kernel performance estimation using micro-benchmarks and hardware metric profiling. J. Parallel Distrib. Comput. 107, 37–56 (2017)
16. Lago, P., Koçak, S.A., Crnkovic, I., Penzenstadler, B.: Framing sustainability as a property of software quality. Commun. ACM 58(10), 70–78 (2015)
17. Lawrence, B.N., et al.: Crossing the chasm: how to develop weather and climate models for next generation computers? Geosci. Model Dev. 11(5), 1799–1821 (2018)
18. Lee, V.W., et al.: Debunking the 100x GPU vs. CPU myth: an evaluation of throughput computing on CPU and GPU. ACM SIGARCH Comput. Archit. News 38(3), 451–460 (2010)
19. Palenstijn, W.J., Batenburg, K.J., Sijbers, J.: Performance improvements for iterative electron tomography reconstruction using graphics processing units (GPUs). J. Struct. Biol. 176(2), 250–253 (2011)
20. Portegies Zwart, S.F., Belleman, R.G., Geldof, P.M.: High-performance direct gravitational N-body simulations on graphics processing units. New Astron. 12(8), 641–650 (2007)
21. Rosenberg, E.: CUDA-API-wrappers: Thin C++-flavored wrappers for the CUDA runtime API (2019). https://github.com/eyalroz/cuda-api-wrappers
22. Ryoo, S., Rodrigues, C.I., Baghsorkhi, S.S., Stone, S.S., Kirk, D.B., Hwu, W.M.W.: Optimization principles and application performance evaluation of a multithreaded GPU using CUDA. In: Proceedings of the 13th ACM SIGPLAN Symposium on Principles and Practice of Parallel Programming, pp. 73–82. ACM (2008)
23. Ryoo, S., et al.: Program optimization space pruning for a multithreaded GPU. In: Proceedings of the 6th Annual IEEE/ACM International Symposium on Code Generation and Optimization, pp. 195–204. ACM (2008)
24. Sclocco, A., van Leeuwen, J., Bal, H.E., van Nieuwpoort, R.V.: A real-time radio transient pipeline for ARTS. In: 2015 IEEE Global Conference on Signal and Information Processing (GlobalSIP), pp. 468–472, December 2015. https://doi.org/10.1109/GlobalSIP.2015.7418239

25. Sclocco, A., Bal, H.E., Hessels, J., Van Leeuwen, J., Van Nieuwpoort, R.V.: Auto-tuning dedispersion for many-core accelerators. In: 2014 IEEE 28th International Parallel and Distributed Processing Symposium, pp. 952–961. IEEE (2014)
26. Spreeuw, H., van Werkhoven, B., Diblen, F., Yatawatta, S.: GPU acceleration of the SAGECal calibration package for the SKA. In: Proceedings of the Astronomical Data Analysis Software and Systems XXVIII (ADASS 2019) (2019)
27. Top500 (2020). http://www.top500.org
28. Venters, C., et al.: The blind men and the elephant: towards an empirical evaluation framework for software sustainability. J. Open Res. Softw. **2**(1), e8 (2014)
29. Verheijen, M.A.W., Oosterloo, T.A., van Cappellen, W.A., Bakker, L., Ivashina, M.V., van der Hulst, J.M.: Apertif, a focal plane array for the WSRT. In: AIP Conference Proceedings, vol. 1035, no. 1, pp. 265–271 (2008). https://doi.org/10.1063/1.2973599
30. van Werkhoven, B.: Kernel Tuner: a search-optimizing GPU code auto-tuner. Future Gener. Comput. Syst. **90**, 347–358 (2019)
31. van Werkhoven, B., Hijma, P., Jacobs, C.J., Maassen, J., Geradts, Z.J., Bal, H.E.: A jungle computing approach to common image source identification in large collections of images. Digit. Invest. **27**, 3–16 (2018)
32. van Werkhoven, B., Maassen, J., Bal, H.E., Seinstra, F.J.: Optimizing convolution operations on GPUs using adaptive tiling. Future Gener. Comput. Syst. **30**, 14–26 (2014). https://doi.org/10.1016/j.future.2013.09.003
33. van Werkhoven, B., et al.: A distributed computing approach to improve the performance of the Parallel Ocean Program (v2.1). Geosci. Model Dev. **7**(1), 267–281 (2014)
34. van Werkhoven, B., Maassen, J., Seinstra, F.J.: Towards User Transparent Parallel Multimedia Computing on GPU-Clusters. In: Varbanescu, A.L., Molnos, A., van Nieuwpoort, R. (eds.) ISCA 2010. LNCS, vol. 6161, pp. 28–39. Springer, Heidelberg (2011). https://doi.org/10.1007/978-3-642-24322-6_4
35. van Werkhoven, B., Maassen, J., Seinstra, F.J., Bal, H.E.: Performance models for CPU-GPU data transfers. In: 2014 14th IEEE/ACM International Symposium on Cluster, Cloud and Grid Computing, pp. 11–20. IEEE (2014)
36. Williams, S., Waterman, A., Patterson, D.: Roofline: an insightful visual performance model for multicore architectures. Commun. ACM **52**(4), 65–76 (2009). https://doi.org/10.1145/1498765.1498785
37. Yang, Y., Xiang, P., Kong, J., Zhou, H.: A GPGPU compiler for memory optimization and parallelism management. ACM SIGPLAN Not. **45**, 86–97 (2010)
38. Zoppetti, G.M., Agrawal, G., Pollock, L., Amaral, J.N., Tang, X., Gao, G.: Automatic compiler techniques for thread coarsening for multithreaded architectures. In: Proceedings of the 14th international conference on Supercomputing, pp. 306–315. ACM (2000)

Unit Tests of Scientific Software: A Study on SWMM

Zedong Peng, Xuanyi Lin, and Nan Niu[✉]

Department of Electrical Engineering and Computer Science (EECS),
University of Cincinnati, Cincinnati, OH 45221, USA
{pengzd,linx7}@mail.uc.edu, nan.niu@uc.edu

Abstract. Testing helps assure software quality by executing program
and uncovering bugs. Scientific software developers often find it challenging to carry out systematic and automated testing due to reasons like
inherent model uncertainties and complex floating point computations.
We report in this paper a manual analysis of the unit tests written by
the developers of the Storm Water Management Model (SWMM). The
results show that the 1,458 SWMM tests have a 54.0% code coverage and
a 82.4% user manual coverage. We also observe a "getter-setter-getter"
testing pattern from the SWMM unit tests. Based on these results, we
offer insights to improve test development and coverage.

Keywords: Scientific software · Unit testing · Test oracle · User
manual · Test coverage · Storm Water Management Model (SWMM)

1 Introduction

Scientific software is commonly developed by scientists and engineers to better understand or make predictions about real world phenomena. Without such
software, it would be difficult or impossible for many researchers to do their
work. Scientific software includes both software for end-user researchers (e.g.,
climate scientists and hydrologists) and software that provides infrastructure
support (e.g., message passing and scheduling). Because scientific software needs
to produce trustworthy results and function properly in mission-critical situations, rigorous software engineering practices shall be adopted to assure software
qualities.

Testing, which is important for assessing software qualities, has been
employed extensively in business/IT software. However, developers of scientific
software have found it more difficult to apply some of the traditional software
testing techniques [14]. One chief challenge is the lack of the test oracle. An
oracle in software testing refers to the mechanism for checking whether the program under test produces the expected output when executed using a set of test
cases [2]. Many testing techniques—especially unit testing commonly carried out
in business/IT software development projects—require a suitable oracle to set

© Springer Nature Switzerland AG 2020
V. V. Krzhizhanovskaya et al. (Eds.): ICCS 2020, LNCS 12143, pp. 413–427, 2020.
https://doi.org/10.1007/978-3-030-50436-6_30

up the expectation with which the actual implementation (e.g., sorting inventory items or calculating tax returns) can be compared.

Researchers have therefore proposed different approaches to overcoming the lack of oracle in scientific software testing. For example, a pseudo oracle—an independently developed program that fulfills the same specification as the program under test—has been used in numerical simulation and climate model testing [10, 11]. A pseudo oracle makes the assumption that independently developed reference models will not result in the same failures; however, Brilliant *et al.* [5] reported instances of N-version programming that violated this assumption. Mayer [22] tested image processing applications with statistical oracles by checking the statistical characteristics of test results; yet a statistical oracle cannot decide whether a single test case has passed or failed.

Single test cases are commonly used in unit testing to verify individual program modules, each of which encapsulates some coherent computation (e.g., a procedure, a function, or a method). Although pseudo and statistical oracles are proposed in the research literature, their adoptions seem isolated among scientific software developers. Our ongoing collaborations with the U.S. Environmental Protection Agency's Storm Water Management Model (SWMM) team suggests limited applicability of N-version programming, and hence pseudo oracle especially at the unit testing levels, due to the constrained software development resources. More importantly, the SWMM team has been developing tests, unit tests and other kinds, throughout the project's more than four decades history [32]. To comply with the recent movements toward improving public access to data [31], these tests are released, sometimes together with the source code of SWMM, in GitHub and other repositories. However, little is known about the characteristics of the SWMM tests.

To shorten the knowledge gap, we report in this paper the tests that are publicly available for the SWMM software. We provide a detailed look at who wrote how many tests in what environments, and further analyze the coverage of the unit tests from two angles: how much they correspond to the user manual and to the codebase. The contributions of our work lie in the qualitative characterization and quantitative examination of the tests written and released by the scientific software developers themselves in the context of SWMM. Our results clearly show that oracle *does* exist in scientific software testing, and our coverage analysis reveals concrete ways to improve testing. In what follows, we provide background information and introduce SWMM in Sect. 2. Section 3 presents our search of SWMM tests, Sect. 4 analyzes the test coverage, and finally, Sect. 5 draws some concluding remarks and outlines future work.

2   Background

2.1   Oracle Problem in Testing Scientific Software

Testing is a mainstream approach toward software quality, and involves examining the behavior of a system in order to discover potential faults. Given an input for the system under test, the *oracle problem* refers to the challenge of distinguishing the corresponding desired, correct behavior from observed, potentially incorrect behavior [2]. The oracle of desired and correct behavior of scientific software, however, can be difficult to obtain or may not be readily available. Kanewala and Bieman [14] listed five reasons.

- Some scientific software is written to find answers that are previously unknown; a case in point is the program computing a large graph's shortest path of any arbitrary pair of nodes.
- It is difficult to determine the correct output for software written to test scientific theory that involves complex calculations, e.g., the large, complex simulations are developed to understand climate change [10].
- Due to the inherent uncertainties in models, some scientific programs do not give a single correct answer for a given set of inputs.
- Requirements are unclear or uncertain up-front due to the exploratory nature of the software [15,24].
- Choosing suitable tolerances for an oracle when testing numerical programs is difficult due to the involvement of complex floating point computations.

Barr *et al.* [2] showed that test oracles could be explicitly specified or implicitly derived. In scientific software testing, an emerging technique to alleviate the oracle problem is metamorphic testing [14,30]. For example, Ding and colleagues [8] tested an open-source light scattering simulation performing discrete dipole approximation. Rather than testing the software on each and every input of a diffraction image, Ding *et al.* systematically (or metamorphically) changed the input (e.g., changing the image orientation) and then compared whether the software would meet the expected relation (e.g., scatter and textual pattern should stay the same at any orientation).

While we proposed hierarchical and exploratory ways of conducting metamorphic testing for scientific software [17,18], our work is similar to that of Ding *et al.*'s [8] by gearing toward the entire application instead of checking the software at the unit testing level. This is surprising given that one of the most prevalent stereotypes of metamorphic testing [30] is the trigonometry function, i.e., $\sin(x) = \sin(\pi - x)$, which is targeted at an individual computational unit. Unit tests are especially useful for guarding the developers against programming mistakes and for localizing the errors when they occur. Thus, we are interested in the unit tests written and released by the scientific software developers themselves, and for our current work, the focus is on SWMM.

2.2 Storm Water Management Model (SWMM)

The Storm Water Management Model (SWMM) [32], created by the U.S. Environmental Protection Agency (EPA), is a dynamic rainfall-runoff simulation model that computes runoff quantity and quality from primarily urban areas. The development of SWMM began in 1971 and since then the software has undergone several major upgrades.

Table 1. SWMM tests in six repositories.

Source	Author (#; Role)	# of Tests	Type	Method	Language	SWMM Version
https://github.com/michaeltryby/ swmm-nrtests/tree/master/public/ update-v5111	(N/A; N/A)	8	Numerical regression testing	numpy.allclose	Python, json	5.1.11
https://github.com/ OpenWaterAnalytics/Stormwater- Management-Model/tree/feature- 2dflood/tests/swmm-nrtestsuite/ benchmark/swmm-5112	(N/A; N/A)	27				5.1.12
https://github.com/USEPA/ Stormwater-Management-Model/ tree/develop/tools/nrtest-swmm/ nrtest.swmm	(1; EPA developer)	2				5.1.12
https://drive.google.com/drive/ folders/16gImGSJV7iygX37P- XiRS4WcBK-Y1zvT	(1; EPA developer)	58				5.1.13
https://github.com/michaeltryby/ swmm-nrtests/tree/master/public	(1; EPA developer)	52				5.1.13
https://github.com/ OpenWaterAnalytics/Stormwater- Management-Model/tree/develop/ tests	(3+; EPA developer)	1,458	Unit testing	boost test	C++	5.1.13

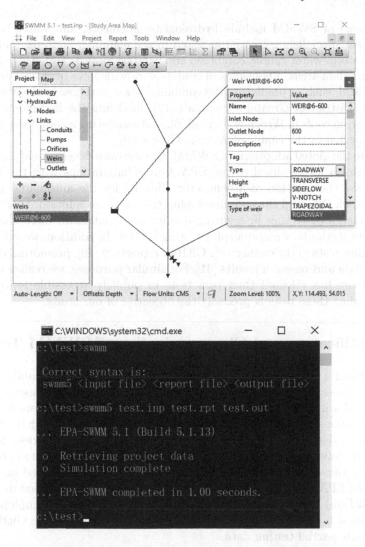

Fig. 1. SWMM running as a Windows application (top) and the computational engine of SWMM running as a console application (bottom).

The most current implementation of the model is version 5.1.13 which was released in 2018. It has modernized both the model's structure and its user interface (UI). The top of Fig. 1 shows a screenshot of SWMM running as a Windows application. The two main parts of SWMM are the computational engine written in C/C++ with about 45,500 lines of code, and the UI written using Embarcadero's Delphi.XE2. Note that the computational engine can be compiled either as a DLL under Windows or as a stand-alone console application under both Windows and Linux. The bottom of Fig. 1 shows that running SWMM in the command line takes three parameters: the input, report, and output files.

The users of SWMM include hydrologists, engineers, and water resources management specialists who are interested in the planning, analysis, and design related to storm water runoff, combined and sanitary sewers, and other drainage systems in urban areas. Thousands of studies worldwide have been carried out by using SWMM, such as predicting the combined sewer overflow in the Metropolitan Sewer District of Greater Cincinnati [12], modeling the hydrologic performance of green roofs in Wrocław, Poland [6], and simulating a combined drainage network located in the center of Athens, Greece [16].

Despite the global adoptions of SWMM, the software development and maintenance remain largely local to the EPA. Our collaborations with the SWMM team involve the creation of a connector allowing for the automated parameter calibration [13], and through developing this software solution, we recognize the importance of testing in assuring quality and contribute hierarchical and exploratory methods of metamorphic testing [17,18]. In addition, we release our metamorphic tests in the connector's GitHub repository [19], promoting the open access to data and research results [31]. For similar purposes, we realize that the SWMM team has released their own tests in publicly accessible repositories. Understanding these tests is precisely the objective of our study.

3 Identification and Characterization of SWMM Tests

We performed a survey analysis of the SWMM tests released in publicly accessible repositories. Our search was informed by the SWMM team members and also involved using known test repositories to find additional ones (snowballing). Table 1 lists the six repositories that we identified, as well as the characteristics of the testing data. The "source" column shows that five repositories are based on GitHub, indicating the adoption of this kind of social coding environments among scientific software developers; however, one source is hosted on Google Drive by an EPA developer, showing that not all tests are embedded or merged in the (GitHub) code branches. Although we cannot claim the completeness of the sources, it is clear that searching only the code repositories like GitHub will result in only partial testing data.

Table 1 shows that three test sets are contributed by individual developers whereas one test set is jointly developed by more than three people. For the other two test sets, we are not certain about the number of authors and their roles, though other GitHub pages may provide inferrable information. The tests that we found can be classified in two categories: numerical regression testing and unit testing. Table 1 also shows that EPA developers adopt Python's numpy.allclose() function to write regression tests. The numpy.allclose() function is used to find if two arrays are element-wise equal within a tolerance, and for SWMM, this type of "allclose" checks whether the output from the newly released code is consistent with that from the previously working code. For regression testing, we count each SWMM input as a test, i.e., a single unit for different code versions to check "allclose". In total, there are 147 regression tests in five repositories.

In contrast, unit testing does not compare different versions of SWMM but focuses on the specific computations of the software. One source of Table 1 contains 1,458 tests written by a group of EPA developers by using the boost environment [7]. In particular, libboost test (version 1.5.4) is used in SWMM, and Boost.Test provides both the interfaces for writing and organizing tests and the controls of their executions. Figure 2 uses a snippet of test_toolkitapi_lid.cpp to explain the three different granularities of SWMM unit tests. At the fine-grained level are the assertions, e.g., line #334 of Fig. 2 asserts "error == ERR_NONE". The value of "error" is obtained from line #333. As shown in Fig. 2, we define a *test* in our study to be one instance that triggers SWMM execution and the associated assertions with that triggering. In Fig. 2, three tests are shown. A group of tests forms a *test case*, e.g., lines #311–616 encapsulate many tests into one BOOST_FIXTURE_TEST_CASE. Finally, each file corresponds to a *test suite* containing one or more test cases. Table 2 lists the seven test suites, and the number of test cases and tests per suite. Averagely speaking, each test suite has 8.7 test cases, and each test case has 23.9 tests.

4   Coverage of SWMM Unit Tests

Having characterized who developed how many SWMM tests in what environments, we turn our attention to the unit tests for quantitative analysis. Our rationales are threefold: (1) a large proportion ($\frac{1,458}{1,605} = 91\%$) of the tests that

```
1     // NOTE: Travis installs libboost test version 1.5.4
      ...
4     #define BOOST_TEST_MODULE "toolkitAPI_lid"
5     #include "test_toolkitapi_lid.hpp"
      ...
60    BOOST_AUTO_TEST_SUITE(test_lid_toolkitapi_fixture)
      ...
310       // Testing for Lid Control Bio Cell parameters get/set
311       BOOST_FIXTURE_TEST_CASE(getset_lidcontrol_BC, FixtureOpenClose_LID)
312       {
          ...
328           // Lid Control
329           // Surface layer get/set check
330           error = swmm_getLidCParam(lid_index, SM_SURFACE, SM_THICKNESS, &db_value);
331           BOOST_REQUIRE(error == ERR_NONE);
332           BOOST_CHECK_SMALL(db_value - 6, 0.0001);
333           error = swmm_setLidCParam(lid_index, SM_SURFACE, SM_THICKNESS, 100);
334           BOOST_REQUIRE(error == ERR_NONE);
335           error = swmm_getLidCParam(lid_index, SM_SURFACE, SM_THICKNESS, &db_value);
336           BOOST_REQUIRE(error == ERR_NONE);
337           BOOST_CHECK_SMALL(db_value - 100, 0.0001);
338
339           error = swmm_getLidCParam(lid_index, SM_SURFACE, SM_VOIDFRAC, &db_value);
          ...
616       }
      ...
2325  BOOST_AUTO_TEST_SUITE_END()
```

Fig. 2. Illustration of SWMM tests and test cases written in the boost environment.

we found are unit tests (2) the tests are intended for the most recent release of SWMM (version 5.1.13), and (3) unit testing requires oracle to be specified which will provide valuable insights into how scientific software developers themselves define test oracles.

Table 2. Test suites, test cases, and tests.

Test suite	Test case	Test
test_output.cpp	14	59
test_swmm.cpp	11	11
test_toolkitapi.cpp	12	128
test_toolkitapi_gage.cpp	1	11
test_toolkitapi_lid.cpp	17	679
test_toolkitapi_lid_results.cpp	5	555
test_toolkitapi_pollut.cpp	1	15
Σ	61	1,458

When unit tests are considered, *coverage* is an important criterion. This is because a program with high test coverage, measured as a percentage, has had more of its source code executed during testing, which suggests it has a lower chance of containing undetected software bugs compared to a program with low test coverage [4]. Practices that lead to higher testing coverage have therefore received much attention. For example, test-driven development (TDD) [23] advocates test-first over the traditional test-last approach, and the studies by Bhat and Nagappan [3] show that the block coverage reached to 79–88% at unit test level in projects employing TDD. While Bhat and Nagappan's studies were carried out at Microsoft, some scientific software demands even higher levels of test coverage. Notably, the European Cooperation for Space Standardization requires a 100% test coverage at software unit level, and Prause *et al.* [27] collected experience from a space software project's developers who stated that 100% coverage is unusual and brings in new risks. Nevertheless, the space software developers acknowledged that 100% coverage is sometimes necessary. Our work analyzes the coverage of SWMM unit tests not only from the source code perspective, but also from the viewpoint of the user manual. Compared to business/IT software, scientific software tends to release authoritative and updated user manual intended for the software system's proper operation. The rest of this section reports the 1,458 unit tests' coverage and discusses our study's limitations.

4.1 SWMM User Manual Coverage

We manually mapped the SWMM unit tests to its version 5.1 user manual [29], and for validation and replication purposes, we share all our analysis data in the institutional digital preservation site Scholar@UC [26]. The 353-page user manual contains 12 chapters and 5 appendices. Our analysis shows that 14, or 82.4%

$(\frac{14}{17})$, are covered by at least one of the 1,458 unit tests. Figure 3 shows the distributions of the unit tests over the 14 user manual chapters/appendices. Because one unit test may correspond to many chapters/appendices, the test total of Fig. 3 is 3,236. The uncovered chapters are: "Printing and Copying" (Chapter 10), "Using Add-In Tools" (Chapter 12), and "Error and Warning Messages" (Appendix E). The error and warning messages are descriptive in nature, and printing, copying, and add-in tools require the devices and/or services external to SWMM. Due to these reasons, it is understandable that no unit tests correspond to these chapters/appendices.

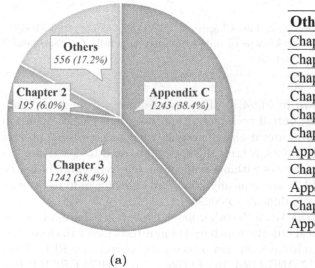

Others	# of Tests
Chapter 5	84 (2.6%)
Chapter 8	84 (2.6%)
Chapter 7	76 (2.3%)
Chapter 11	71 (2.2%)
Chapter 6	69 (2.1%)
Chapter 4	52 (1.6%)
Appendix B	52 (1.6%)
Chapter 9	31 (1.0%)
Appendix D	23 (0.7%)
Chapter 1	8 (0.2%)
Appendix A	6 (0.2%)

(a)   (b)

Fig. 3. (a) Mapping unit tests to user manual chapters/appendices, and (b) Explaining the "Others" part of (a).

Figure 3(a) shows that the unit tests predominantly cover "SWMM's Conceptual Model" (Chapter 3) and "Specialized Property Editors" (Appendix C). The same percentage, 38.4%, of these two parts is not accidental to us. In fact, they share the same subset of the unit tests except for one. We present a detailed look at these parts in Fig. 4. Chapter 3 describes not only the configuration of the SWMM objects (e.g., conduits, pumps, storage units, etc.) but also the LID (low impact development) controls that SWMM allows engineers and planners to represent combinations of green infrastructure practices and to determine their effectiveness in managing runoff. The units presented in §3.2 ("Visual Objects"), §3.3 ("Non-Visual Objects"), and §3.4 ("Computational Methods") thus represent some of the core computations of SWMM. Consequently, unit tests are written for the computations except for the "Introduction" (§3.1) overviewing the Atmosphere, Land Surface, Groundwater, and Transport compartments of SWMM. Surprisingly, more tests are written for the non-visual objects than

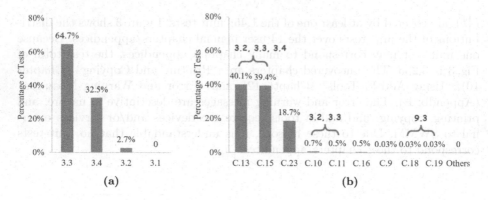

Fig. 4. (a) Breakdowns of unit tests into Chapter 3 ("SWMM's Conceptual Model") of the user manual, and (b) Breakdowns of unit tests into Appendix C ("Specialized Property Editors") of the user manual.

the visual objects, as shown in Fig. 4(a). The visual objects (rain gages, subcatchments, junction nodes, outfall nodes, etc.) are those that can be arranged together to represent a stormwater drainage system, whereas non-visual objects (climatology, transects, pollutants, control rules, etc.) are used to describe additional characteristics and processes within a study area. One reason might be the physical, visual objects (§3.2) are typically combined, making unit tests (e.g., single tests per visual object) difficult to construct.

The non-visual objects (§3.3), on the other hand, express attributes of, or the rules controlling, the physical objects, which makes unit tests easier to construct. For example, two of the multiple-condition orifice gate controls are RULE R2A: "IF NODE 23 DEPTH > 12 AND LINK 165 FLOW > 100 THEN ORIFICE R55 SETTING = 0.5" and RULE R2B: "IF NODE 23 DEPTH > 12 AND LINK 165 FLOW > 200 THEN ORIFICE R55 SETTING = 1.0". For units like RULE R2A and RULE R2B, tests could be written to check whether the orifice setting is correct under different node and link configurations. Under these circumstances, the test oracles are *known* and are given in the user manual (e.g., orifice setting specified in the control rules).

During our manual mappings of the unit tests, we realize the interconnection of the user manual chapters/appendices. One example mentioned earlier is the connection between Chapter 3 and Appendix C. It turns out that such interconnections are not one-to-one, i.e., Appendix C connects to not only Chapter 3 but also to other chapters. In Fig. 4(b), we annotate the interconnections grounded in the SWMM unit tests. For instance, §3.2, §3.3, and §3.4 are linked to §C.10 ("Initial Buildup Editor"), §C.11 ("Land Use Editor"), §C.13 ("LID Control Editor"), and §C.15 ("LID Usage Editor"), indicating the important role of LID plays in SWMM. Although only a very small number of unit tests connects §9.3 ("Time Series Results") with §C.18 ("Time Pattern Editor") and §C.19 ("Time Series Editor"), we posit more tests of this core time-series computation could be developed in a similar way as LID tests (e.g., by using the boost environment illustrated in Fig. 2). A more general speculation that we draw from our analysis is that if some core computation has weak links with the scientific software system's parameters and properties (e.g., Appendix C of the SWMM user manual), then developing unit tests for that computation may require other environments and frameworks like CppTest or CPPUnit; investigating these hypotheses is part of our future research collaborations with the EPA's SWMM team.

4.2   SWMM Codebase Coverage

There are a number of coverage measures commonly used for test-codebase analysis, e.g., Prause *et al.* [27] compared statement coverage to branch coverage in a space software project and showed that branch coverage tended to be lower if not monitored but could be improved in conjunction with statement coverage without much additional effort. For our analysis, we manually mapped the 1,458 unit tests to SWMM's computational engine (about 45,500 lines of code written in C/C++). Like the test-to-user-manual data, we also share our test-to-codebase analysis data in Scholar@UC [26]. At the code file level, the coverage of the 1,458 SWMM unit tests is 54.0%. In line with our user manual analysis results, the code corresponding to the greatest number of unit tests involves runoff and LID, including toposort.c, treatment.c, and runoff.c.

Different from our user manual analysis where we speculated that control rules, such as RULE R2A and RULE R2B, would be among the subjects of unit testing, the actual tests have a strong tendency toward getters and setters. This is illustrated in Fig. 2. Interestingly, we also observe a pattern of "getter-setter-getter" in the tests. In Fig. 2, the test of lines #330–332 first gets swmm_getLidCParam, ensures that there is no error in getting the parameter value (line #331), and compares the value with the oracle (line #332). A minor change is made in the next test where the new "&db_value" is set to be 100, followed by checking if this instance of parameter setter is successful (line #334). The last test in the "getter-setter-getter" sequence immediately gets and checks the parameter value (lines #335–337). Our analysis confirms many instances of this "getter-setter-getter" pattern among the 1,458 unit tests.

It is clear that oracle exists in SWMM unit tests, and as far as the "getter-setter-getter" testing pattern is concerned, two kinds of oracle apply: whether the code crashes (e.g., lines #331, #334, and #336 of Fig. 2) and if the parameter value is close to pre-defined or pre-set value (e.g., lines #332 and #337 of Fig. 2). One advantage of "getter-setter-getter" testing lies in the redundancy of setting a value followed immediately by getting and checking that value, e.g., swmm_setLidCParam with 100 and then instantly checking swmm_getLidCParam against 100. As redundancy improves reliability, this practice also helps with the follow-up getter's test automation. However, a disadvantage here is the selection of the parameter values. In Fig. 2, for example, the oracle of 6 (line #332) may be drawn from SWMM input and/or observational data, but the selection of 100 seems random to us. As a result, the test coverage is low from the parameter value selection perspective, which can limit the bug detection power of the tests.

A post from the SWMM user forum [1] provides a concrete situation of software failure related to specific parameter values. In this post, the user reported that: "The surface depth never even reaches 300 mm in the LID report file" after explicitly setting the parameters of the LID unit (specifically, "storage depth of surface layer" = 300 mm) to achieve the effect [1]. The reply from an EPA developer suggested a solution by changing: "either the infiltration conductivity or the permeability of the Surface, Soil or Pavement layers". Although these layers are part of "LID Controls", and even have their descriptions in §3.3.14 of the SWMM user manual [29], the testing coverage does not seem to reach "storage depth of surface layer" = 300 mm under different value combinations of the Surface, Soil or Pavement layers. We believe the test coverage can be improved when the design of unit tests builds more directly upon the SWMM user manual and when the parameter value selection alters more automatically than currently written in a relatively fixed fashion.

4.3 Threats to Validity

We discuss some of the important aspects of our study that one shall take into account when interpreting our findings. A threat to construct validity is how we define tests. Our work focuses on SWMM unit tests written in the boost environment, as illustrated in Fig. 2. While our units of analysis—tests, test cases, and test suites—are consistent with what boost defines and how the test developers apply boost, the core construct of "tests" may differ if boost evolves or the SWMM developers adopt other test development environments. Within boost itself, for instance, BOOST_AUTO_TEST_CASE may require different ways to define and count tests than BOOST_FIXTURE_TEST_CASE shown in Fig. 2.

An internal validity threat is our manual mapping of the 1,458 SWMM unit tests to the user manual and to the codebase. Due to the lack of traceability information [25, 34, 35] from the SWMM project, our manual effort is necessary in order to understand the coverage of the unit tests. Our current mapping strategy is driven mainly by keywords, i.e., we matched keywords from the tests with the user manual contents and with the functionalities implemented in the code (procedure signatures and header comments). Two researchers independently performed the manual mappings of a randomly chosen 200 tests and achieved a high level of inter-rater agreement (Cohen's $\kappa = 0.77$). We attributed this to the comprehensive documentation of SWMM tests, user manual, and code. The disagreements of the researchers were resolved in a joint meeting, and three researchers performed the mappings for the remaining tests.

Several factors affect our study's external validity. Our results may not generalize to other kinds of SWMM testing (numerical regression tests in particular), to the tests shared internally among the SWMM developers, and to other scientific software with different size, complexity, purposes, and testing practices. As for conclusion validity and reliability, we believe we would obtain the same results if we repeated the study. In fact, we publish all our analysis data in our institution's digital preservation repository [26] to facilitate reproducibility, cross validation, and future expansions.

5   Conclusions

Testing is one of the cornerstones of modern software engineering [9]. Scientific software developers, however, face the oracle challenge when performing testing [14]. In this paper, we report our analysis of the unit tests written and released by the EPA's SWMM developers. For the 1,458 SWMM unit tests that we identified, the file-level code coverage is 54.0% and the user-manual coverage is 82.4%. Our results show that oracle *does* exist in at least two levels: whether the code crashes and if the returned value of a computational unit is close to the expectation. In addition to relying on historical data to define the test oracle [17,18], our study uncovers a new "getter-setter-getter" testing pattern, which helps alleviate the oracle problem by setting a parameter value and then immediately getting and checking it. This practice, though innovative, can be further improved by incorporating the user manual in developing tests and by automating parameter value selection to increase coverage.

Our future work will therefore explore these dimensions while investigating SWMM's numerical regression tests that we identified. While the 82.4% user manual coverage seems high, the 54.0% file-level code coverage begs the question as to whether our current analysis misses some code/functionality that is not user-facing. We will expand our analysis to address this question. We also plan to develop automated tools [20,21,28] for test coverage analysis, and will build initial tooling [33] on the basis of keyword matching drawn from our current operational insights. Our goal is to better support scientists in improving testing practices and software quality.

Acknowledgments. We thank the EPA SWMM team, especially Michelle Simon, Colleen Barr, and Michael Tryby, for the research collaborations. This work is partially supported by the U.S. National Science Foundation (Award CCF-1350487).

References

1. Adei, B., Dickinson, R., Rossman, L.A.: Some observations on LID output. https://www.openswmm.org/Topic/4214/some-observations-on-lid-output. Accessed April 2020
2. Barr, E.T., Harman, M., McMinn, P., Shahbaz, M., Yoo, S.: The oracle problem in software testing: a survey. IEEE Trans. Software Eng. **41**(5), 507–525 (2015)
3. Bhat, T., Nagappan, N.: Evaluating the efficacy of test-driven development: industrial case studies. In: International Symposium on Empirical Software Engineering, pp. 356–363 (2006)
4. Brader, L., Hilliker, H., Wills, A.C.: Chapter 2 Unit Testing: Testing the Inside. Testing for Continuous Delivery with Visual Studio 2012, Microsoft Patterns & Practices (2013)
5. Brilliant, S.S., Knight, J.C., Leveson, N.G.: Analysis of faults in an N-version software experiment. IEEE Trans. Software Eng. **16**(2), 238–247 (1990)
6. Burszta-Adamiak, E., Mrowiec, M.: Modelling of green roofs' hydrologic performance using EPA's SWMM. Water Sci. Technol. **68**(1), 36–42 (2013)

7. Dawes, B., Abrahams, D.: Boost C++ libraries. https://www.boost.org. Accessed April 2020
8. Ding, J., Zhang, D., Hu, X.-H.: An application of metamorphic testing for testing scientific software. In International Workshop on Metamorphic Testing, pp. 37–43 (2016)
9. Dubois, P.F.: Testing scientific programs. Comput. Sci. Eng. 14(4), 69–73 (2012)
10. Easterbrook, S., Johns, T.C.: Engineering the software for understanding climate change. Comput. Sci. Eng. 11(6), 65–74 (2009)
11. Farrell, P.E., Piggott, M.D., Gorman, G.J., Ham, D.A., Wilson, C.R., Bond, T.M.: Automated continuous verification for numerical simulation. Geosci. Model Dev. 4(2), 435–449 (2011)
12. Gudaparthi, H., Johnson, R., Challa, H., Niu, N.: Deep learning for smart sewer systems: assessing nonfunctional requirements. In: International Conference on Software Engineering (SE in Society Track) (2020)
13. Kamble, S., Jin, X., Niu, N., Simon, M.: A novel coupling pattern in computational science and engineering software. In International Workshop on Software Engineering for Science, pp. 9–12 (2017)
14. Kanewala, U., Bieman, J.M.: Testing scientific software: a systematic literature review. Inf. Software Technol. 56(10), 1219–1232 (2014)
15. Khatwani, C., Jin, X., Niu, N., Koshoffer, A., Newman, L., Savolainen, J.: Advancing viewpoint merging in requirements engineering: a theoretical replication and explanatory study. Requirements Eng. 22(3), 317–338 (2017). https://doi.org/10.1007/s00766-017-0271-0
16. Kourtis, I.M., Kopsiaftis, G., Bellos, V., Tsihrintzis, V.A.: Calibration and validation of SWMM model in two urban catchments in Athens, Greece. In: International Conference on Environmental Science and Technology (2017)
17. Lin, X., Simon, M., Niu, N.: Exploratory metamorphic testing for scientific software. Comput. Sci. Eng. 22(2), 78–87 (2020)
18. Lin, X., Simon, M., Niu, N.: Hierarchical metamorphic relations for testing scientific software. In: International Workshop on Software Engineering for Science, pp. 1–8 (2018)
19. Lin, X., Simon, M., Niu, N.: Releasing scientific software in GitHub: a case study on SWMM2PEST. In: International Workshop on Software Engineering for Science, pp. 47–50 (2019)
20. Mahmoud, A., Niu, N.: Supporting requirements to code traceability through refactoring. Requirements Eng. 19(3), 309–329 (2013). https://doi.org/10.1007/s00766-013-0197-0
21. Mahmoud, A., Niu, N.: TraCter: a tool for candidate traceability link clustering. In: International Requirements Engineering Conference, pp. 335–336 (2011)
22. Mayer, J.: On testing image processing applications with statistical methods. In: Software Engineering, pp. 69–78 (2005)
23. Niu, N., Brinkkemper, S., Franch, X., Partanen, J., Savolainen, J.: Requirements engineering and continuous deployment. IEEE Softw. 35(2), 86–90 (2018)
24. Niu, N., Koshoffer, A., Newman, L., Khatwani, C., Samarasinghe, C., Savolainen, J.: Advancing repeated research in requirements engineering: a theoretical replication of viewpoint merging. In: International Requirements Engineering Conference, pp. 186–195 (2016)
25. Niu, N., Wang, W., Gupta, A.: Gray links in the use of requirements traceability. In: International Symposium on Foundations of Software Engineering, pp. 384–395 (2016)

26. Peng, Z., Lin, X., Niu, N.: Data of SWMM Unit Tests. http://dx.doi.org/10.7945/zpdh-7a44. Accessed April 2020
27. Prause, C.R., Werner, J., Hornig, K., Bosecker, S., Kuhrmann, M.: Is 100% test coverage a reasonable requirement? Lessons learned from a space software project. In: Felderer, M., Méndez Fernández, D., Turhan, B., Kalinowski, M., Sarro, F., Winkler, D. (eds.) PROFES 2017. LNCS, vol. 10611, pp. 351–367. Springer, Cham (2017). https://doi.org/10.1007/978-3-319-69926-4_25
28. Reddivari, S., Chen, Z., Niu, N.: ReCVisu: a tool for clustering-based visual exploration of requirements. In: International Requirements Engineering Conference, pp. 327–328 (2012)
29. Rossman, L.A.: Storm Water Management Model User's Manual Version 5.1. https://www.epa.gov/sites/production/files/2019-02/documents/epaswmm5_1_manual_master_8-2-15.pdf. Accessed April 2020
30. Segura, S., Fraser, G., Sánchez, A.B., Cortés, A.R.: A survey on metamorphic testing. IEEE Trans. Software Eng. **42**(9), 805–824 (2016)
31. Sheehan, J.: Federally funded research results are becoming more open and accessible. https://digital.gov/2016/10/28/federally-funded-research-results-are-becoming-more-open-and-accessible/. Accessed April 2020
32. United States Environmental Protection Agency. Storm Water Management Model (SWMM). https://www.epa.gov/water-research/storm-water-management-model-swmm. Accessed April 2020
33. Wang, W., Gupta, A., Niu, N., Xu, L.D., Cheng, J.-R.C., Niu, Z.: Automatically tracing dependability requirements via term-based relevance feedback. IEEE Trans. Ind. Inf. **14**(1), 342–349 (2018)
34. Wang, W., et al.: Complementarity in requirements tracing. IEEE Trans. Cybern. **50**(4), 1395–1404 (2020)
35. Wang, W., Niu, N., Liu, H., Wu, Y.: Tagging in assisted tracing. In: International Symposium on Software and Systems Traceability, pp. 8–14 (2015)

NUMA-Awareness as a Plug-In
for an Eventify-Based Fast
Multipole Method

Laura Morgenstern[1,2]([✉]), David Haensel[1], Andreas Beckmann[1],
and Ivo Kabadshow[1]

[1] Jülich Supercomputing Centre, Forschungszentrum Jülich, 52425 Jülich, Germany
{l.morgenstern,d.haensel,a.beckmann,i.kabadshow}@fz-juelich.de
[2] Faculty of Computer Science, Technische Universität Chemnitz,
09111 Chemnitz, Germany

Abstract. Following the trend towards Exascale, today's supercomputers consist of increasingly complex and heterogeneous compute nodes. To exploit the performance of these systems, research software in HPC needs to keep up with the rapid development of hardware architectures. Since manual tuning of software to each and every architecture is neither sustainable nor viable, we aim to tackle this challenge through appropriate software design. In this article, we aim to improve the performance and sustainability of FMSolvr, a parallel Fast Multipole Method for Molecular Dynamics, by adapting it to Non-Uniform Memory Access architectures in a portable and maintainable way. The parallelization of FMSolvr is based on Eventify, an event-based tasking framework we co-developed with FMSolvr. We describe a layered software architecture that enables the separation of the Fast Multipole Method from its parallelization. The focus of this article is on the development and analysis of a reusable NUMA module that improves performance while keeping both layers separated to preserve maintainability and extensibility. By means of the NUMA module we introduce diverse NUMA-aware data distribution, thread pinning and work stealing policies for FMSolvr. During the performance analysis the modular design of the NUMA module was advantageous since it facilitates combination, interchange and redesign of the developed policies. The performance analysis reveals that the runtime of FMSolvr is reduced by 21% from 1.48 ms to 1.16 ms through these policies.

Keywords: Non-uniform memory access · Multicore programming ·
Software architecture · Fast multipole method

1 Introduction

1.1 Challenge

The trend towards higher clock rates stagnates and heralds the start of the exascale era. The resulting supply of higher core counts leads to the rise of

© Springer Nature Switzerland AG 2020
V. V. Krzhizhanovskaya et al. (Eds.): ICCS 2020, LNCS 12143, pp. 428–441, 2020.
https://doi.org/10.1007/978-3-030-50436-6_31

Non-Uniform Memory Access (NUMA) systems for reasons of scalability. This requires not only the exploitation of fine-grained parallelism, but also the handling of hierarchical memory architectures in a sustainable and thus portable way. We aim to tackle this challenge through suitable software design since manual adjustment of research software to each and every architecture is neither sustainable nor viable for reasons of development time and staff expenses.

Our use case is FMSolvr, a parallel Fast Multipole Method (FMM) for Molecular Dynamics (MD). The parallelization of FMSolvr is based on Eventify, a tailor-made tasking library that allows for the description of fine-grained task graphs through events [7,8]. Eventify and FMSolvr are published as open source under LGPL v2.1 and available at www.fmsolvr.org. In this article, we aim to improve the performance and sustainability of FMSolvr by adapting it to NUMA architectures through the following contributions:

1. A layered software design that separates the algorithm from its parallelization and thus facilitates the development of new features and the support of new hardware architectures.
2. A reusable NUMA module for Eventify that models hierarchical memory architectures in software and enables rapid development of algorithm-dependent NUMA policies.
3. Diverse NUMA-aware data distribution, thread pinning and work stealing policies for FMSolvr based on the NUMA module.
4. A detailed comparative performance analysis of the developed NUMA policies for the FMM on two different NUMA machines.

1.2   State of the Art

MD has become a vital research method in biochemistry, pharmacy and materials science. MD simulations target strong scaling since their problem size is typically small. Thus, the computational effort per compute node is very low and MD applications tend to be latency- and synchronization-critical. To exploit the performance of NUMA systems, MD applications need to adapt to the differences in latency and throughput caused by hierarchical memory.

In this work, we focus on the FMM [6] with computational complexity $\mathcal{O}(N)$. We consider the hierarchical structure of the FMM as a good fit for hierarchical memory architectures. We focus on the analysis of NUMA effects on FMSolvr as a demonstrator for latency- and synchronization-critical applications.

We aim at a NUMA module for FMSolvr since various research works [1,3,4] prove the positive impact of NUMA awareness on performance and scalability of the FMM. Subsequently, we summarize the efforts to support NUMA in current FMM implementations.

ScalFMM [3] is a parallel, C++ FMM-library. The main objectives of its software architecture are maintainability and understandability. A lot of research about task-based and data-driven FMMs is based on ScalFMM. The authors devise the parallel data-flow of the FMM for shared memory architectures in [3] and for distributed memory architectures in [2]. In ScalFMM NUMA policies can be set by the user via the OpenMP environment variables OMP_PROC_BIND and OMP_PLACES. However, to the best of our knowledge, there is no performance analysis regarding these policies for ScalFMM.

KIFMM [15] is a kernel-independent FMM. In [4] NUMA awareness is analyzed dependent on work unit granularity and a speed-up of 4 on 48 cores is gained.

However, none of the considered works provides an implementation and comparison of different NUMA-aware thread pinning and work stealing policies. From our point of view, the implementation and comparison of different policies is of interest since NUMA awareness is dependent on the properties of the input data set, the FMM operators and the hardware. Therefore, the focus of the current work is to provide a software design that enables the rapid development and testing of multiple NUMA policies for the FMM.

2   Fundamentals

2.1   Sustainability

We follow the definition of sustainability provided in [11]:

Definition 1. *Sustainability. A long-living software system is sustainable if it can be cost-efficiently maintained and evolved over its entire life-cycle.*

According to [11] a software system is further on *long-living* if it must be operated for more than 15 years. Due to a relatively stable problem set, a large user base and the great performance optimization efforts, HPC software is typically long-living. This holds e.g. for the molecular dynamics software GROMACS or Coulomb-solvers such as ScaFaCos. Fortran FMSolvr (included in ScaFaCos), the predecessor of C++ FMSolvr, is roughly 20 years old. Hence, sustainability is our main concern regarding C++ FMSolvr.

According to [11], sustainability comprises non-functional requirements such as maintainability, modifiability, portability and evolvability. We add performance and performance portability to the properties of sustainability, to adjust the term to software development in HPC.

2.2   Performance Portability

Regarding performance portability, we follow the definition provided in [14]:

Definition 2. *Performance Portability. For a given set of platforms H, the performance portability Φ of an application a solving problem p is:*

$$\Phi(a, p, H) = \begin{cases} \dfrac{|H|}{\sum_{i \in H} \frac{1}{e_i(a,p)}} & \text{if } i \text{ is supported } \forall i \in H \\ 0 & \text{otherwise} \end{cases}$$

where $e_i(a, p)$ is the performance efficiency of application a solving problem p on platform i.

2.3 Non-uniform Memory Access

NUMA is a shared memory architecture for modern multi-processor and multi-core systems. A NUMA system consists of several NUMA nodes. Each NUMA node is a set of cores together with their local memory. NUMA nodes are connected via a NUMA interconnect such as Intel's Quick Path Interconnect (QPI) or AMD's HyperTransport.

The cores of each NUMA node can access the memory of remote NUMA nodes only by traversing the NUMA interconnect. However, this exhibits notably higher memory access latencies and lower bandwidths than accessing local memory. According to [12], remote memory access latencies are about 50% higher than local memory access latencies. Hence, memory-bound applications have to take data locality into account to exploit the performance and scalability of NUMA architectures.

2.4 Fast Multipole Method

Sequential Algorithm. The fast multipole method for MD is a hierarchical fast summation method (HFSM) for the evaluation of Coulombic interactions in N-body simulations. The FMM computes the Coulomb force $\mathbf{F}_i$ acting on each particle i, the electrostatic field $\mathbf{E}$ and the Coulomb potential Φ in each time step of the simulation. The FMM reduces the computational complexity of classical Coulomb solvers from $\mathcal{O}(N^2)$ to $\mathcal{O}(N)$ by use of hierarchical multipole expansions for the computation of long-range interactions.

The algorithm starts out with a hierarchical space decomposition to group particles. This is done by recursively bisecting the simulation box in each of its three dimensions. We refer to the developing octree as FMM tree. The input data set to create the FMM tree consists of location $\mathbf{x}_i$ and charge q_i of each particle i in the system as well as the algorithmic parameters multipole order p, maximal tree depth d_{max} and well-separateness criterion ws. All three algorithmic parameters influence the time to solution and the precision of the results.

We subsequently introduce the relations between the boxes of the FMM tree referring to [5], since the data dependencies between the steps of the algorithm are based on those:

- **Parent-Child Relation:** We refer to box x as parent box of box y if x and y are directly connected when moving towards the root of the tree.
- **Near Neighbor:** We refer to two boxes as near neighbors if they are at the same refinement level and share a boundary point.
- **Interaction Set:** We refer to the interaction set of box i as the set consisting of the children of the near neighbors of i's parent box which are well separated from i.

Based on the FMM tree, the sequential workflow of the FMM referring to [10] is stated in Algorithm 1, with steps 1. to 5. computing far field interactions and step 6. computing near field interactions.

Algorithm 1. Fast Multipole Method

Input: Positions and charges of particles

Output: Electrostatic field **E**, Coulomb forces **F**, Coulomb potential Φ

0. Create FMM Tree:

Hierarchical space decomposition

1. Particle to Multipole (P2M):

Expansion of particles in each box on the lowest level of the FMM tree into multipole moments ω relative to the center of the box.

2. Multipole to Multipole (M2M):

Accumulative upwards-shift of the multipole moments ω to the centers of the parent boxes.

3. Multipole to Local (M2L):

Translation of the multipole moments ω of the boxes covered by the interaction set of box i into a local moment μ for i.

4. Local to Local (L2L):

Accumulative downwards-shift of the local moments μ to the centers of the child boxes.

5. Local to Particle (L2P):

Transformation of the local moment μ of each box i on the lowest level into far field force for each particle in i.

6. Particle to Particle (P2P):

Evaluation of the near field forces between the particles contained by box i and its near neighbors for each box on the lowest level by computing the direct interactions.

FMSolvr: An Eventify-Based FMM. FMSolvr is a task-parallel implementation of the FMM. According to the tasking approach for FMSolvr with Eventify [7,8], the steps of the sequential algorithm do not have to be computed completely sequentially, but may overlap. Based on the sequential workflow, we differentiate six types of tasks (P2M, M2M, M2L, L2L, L2P and P2P) that span a tree-structured task graph. Since HFSMs such as the FMM are based on a hierarchical decomposition, they exhibit tree-structured, acyclic task graphs and use tree-based data structures. Figure 1 provides a schematic overview of the horizontal and vertical task dependencies of the FMM implied by this parallelization scheme. For reasons of comprehensible illustration, the dependencies are depicted for a binary tree and thus a one-dimensional system, even though the FMM simulates three-dimensional systems and thus works with octrees. In this work, we focus on the tree-based properties of the FMM since these are decisive for its adaption to NUMA architectures. For further details on the functioning of Eventify and its usage for FMSolvr please refer to [7,8].

3   Software Architecture

3.1   Layering: Separation of Algorithm and Parallelization

Figure 2 provides an excerpt of the software architecture of FMSolvr by means of UML. FMSolvr is divided into two main layers: the algorithm layer and the parallelization layer. The algorithm layer encapsulates the mathematical details of the FMM, e.g. exchangeable FMM operators with different time complexities and memory footprints. The parallelization layer hides the details of parallel hardware from the algorithm developer. It contains modules for threading and vectorization and is designed to be extended with specialized modules, e.g. for locking policy, NUMA-, and GPU-support. In this article, we focus on the description of the NUMA module. Hence, Fig. 2 contains only that part of the software architecture which is relevant to NUMA. We continuously work on further decoupling of the parallelization layer as independent parallelization library, namely *Eventify*, to increase its reusability and the reusability of its modules, e.g. the NUMA module described in Sect. 4.

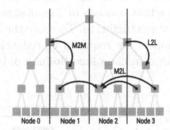

Fig. 1. Exemplary horizontal and vertical data dependencies that lead to inter-node data transfer.

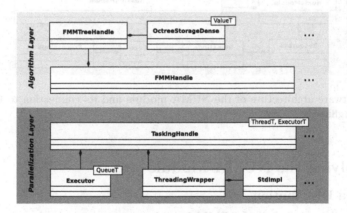

Fig. 2. Layer-based software architecture of FMSolvr. The light gray layer shows an excerpt of the UML-diagram of the algorithm layer that encapsulates the algorithmic details of the FMM. The dark gray layer provides an excerpt of the UML-diagram of the parallelization layer that encapsulates hardware and parallelization details. Both layers are coupled by using the interfaces `FMMHandle` and `TaskingHandle`. (Color figure online)

3.2 NUMA-Module: Modeling NUMA in Software

Figure 3 shows the software architecture of the NUMA module and its integration in the software architecture of FMSolvr. Regarding the integration, we aim to preserve the layering and the according separation of concerns as effectively as possible. However, effective NUMA-awareness requires information from the data layout, which is part of the algorithm layer, and from the hardware, which is part of the parallelization layer. Hence, a slight blurring of both layers is unfortunately unavoidable in this case. After all, we preserve modularization by providing an interface for the algorithm layer, namely NumaAllocator, and an interface for the parallelization layer, namely NumaModule.

Based on the hardware information provided by the NUMA distance table, we model the NUMA architecture of the compute node in software. Hence, a compute node is a Resource that consists of NumaNodes, which consist of Cores, which in turn consist of ProcessingUnits. To reuse the NUMA module for other parallel applications, developers are required to redevelop or adjust the NUMA policies (see Sect. 4) contained in class NumaModule only.

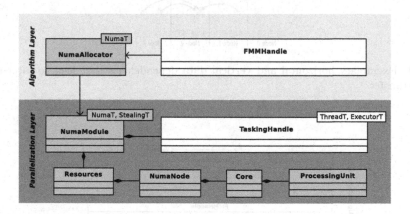

Fig. 3. Software architecture of the NUMA module and its connection to FMMHandle and TaskingHandle.

4   Applying the NUMA-Module

4.1 Data Distribution

As described in Sect. 2.4, the FMM exhibits a tree-structured task graph. As shown in Fig. 1, this task graph and its dedicated data are distributed to NUMA nodes through an equal partitioning of the tasks on each tree level. To assure data locality, a thread and the data it works on are assigned to the same NUMA node. Even though this is an important step to improve data locality, there are still task dependencies that lead to data transfer between NUMA nodes. In order to reduce this inter-node data transfer, we present different thread pinning policies.

4.2 Thread Pinning Policies

Equal Pinning. With *Equal Pinning* (EP), we pursue a classical load balancing approach (cf. *Scatter Principally* [13]). This means that the threads are equally distributed among the NUMA nodes. Hence, this policy is suitable for NUMA systems with homogeneous NUMA nodes only.

Algorithm 2 shows the pseudocode for NUMA-aware thread pinning via policy EP. Let t be the number of threads, n be the number of NUMA nodes and tpn be the number of threads per NUMA node.

The determined number of threads is mapped to the cores of each NUMA node in an ascending order of physical core-ids. This means that threads with successive logical ids are pinned to neighboring cores. This is reasonable with regard to architectures where neighboring cores share a cache-module. In addition, strict pinning to cores serves the avoidance of side-effects due to the behavior of the process scheduler, which may vary dependent on the systems state and complicate an analysis of NUMA-effects.

Algorithm 2. Equal Pinning

$r = t \bmod n$
for $i = 0; i < n; i{+}{+}$ do
  if $i < r$ then
    // number of threads per node for nodes $0, \ldots, r-1$
    $tpn = \lfloor t/n \rfloor + 1$
    Assign threads $i \cdot tpn, \ldots, (i \cdot tpn) + tpn - 1$ to node i
  else
    // number of threads per node for nodes $r, \ldots, n-1$
    $tpn = \lfloor t/n \rfloor$
    Assign threads $i \cdot tpn + r, \ldots, (i \cdot tpn + r) + tpn - 1$ to node i
  end if
end for

Compact Pinning. The thread pinning policy *Compact Pinning* (CP) combines the advantages of the NUMA-aware thread pinning policies *Equal Pinning* and *Compact Ideally* (cf. [13]). The aim of CP is to avoid data transfer via the NUMA interconnect by using as few NUMA nodes as possible while avoiding the use of SMT.

Algorithm 3 shows the pseudocode for the NUMA-aware thread pinning policy CP. Let c be the total number of cores of the NUMA system and c_n be the number of cores per NUMA node excl. SMT-threads. Considering CP, threads are assigned to a single NUMA node as long as the NUMA node has got cores to which no thread is assigned to. Only if a thread is assigned to each core of a NUMA node, the next NUMA node is filled up with threads. This means especially that data transfer via the NUMA interconnect is fully avoided if $t \leq c_n$ holds. If $t \geq c$ holds, thread pinning policy EP becomes effective to reduce the usage of SMT. For this policy we apply strict pinning based on the neighboring cores principle as described in Sect. 4.2, too.

Algorithm 3. Compact Pinning

---
if $t < c$ **then**
    for $i = 0; i < n; i + +$ **do**
        Assign threads $i \cdot c_n, \ldots, (i \cdot c_n) + cn - 1$ to node i
    end for
else
    Use policy Equal Pinning
end if

CP is tailor-made for the FMM as well as for tree-structured task graphs with horizontal and vertical task dependencies in general. CP aims to keep the vertical cut through the task graph as short as possible. Hence, as few as possible task dependencies require data transfer via the NUMA interconnect.

4.3 Work Stealing Policies

Local and Remote Node. *Local and Remote Node* (LR) is the default work stealing policy that does not consider the NUMA architecture of the system at all.

Prefer Local Node. Applying the NUMA-aware work stealing policy *Prefer Local Node* (PL) means that threads preferably steal tasks from threads located on the same NUMA node. However, threads are allowed to steal tasks from threads located on remote NUMA nodes if no tasks are available on the local NUMA node.

Local Node Only. If the NUMA-aware work stealing policy *Local Node Only* (LO) is applied, threads are allowed to steal tasks from threads that are located on the same NUMA node only. According to [13], we would expect this policy to improve performance if stealing across NUMA nodes is more expensive than idling. This means that stealing a task, e.g. transferring its data, takes too long in comparison to the execution time of the task.

5 Results

5.1 Measurement Approach

The runtime measurements for the performance analysis were conducted on a 2-NUMA-node system and a 4-NUMA-node system. The 2-NUMA-node system is a single compute node of Jureca. The dual-socket system is equipped with two Intel Xeon E5-2680 v3 CPUs (Haswell) which are connected via QPI. Each CPU consists of 12 two-way SMT-cores, meaning, 24 processing units. Hence, the system provides 48 processing units overall. Each core owns an L1 data and instruction cache with 32 kB each. Furthermore, it owns an L2 cache with 256 kB. Hence, each two processing units share an L1 and an L2 cache. L3 cache and main memory are shared between all cores of a CPU.

The 4-NUMA-node system is a quad-socket system equipped with four Intel Xeon E7-4830 v4 CPUs (Haswell) which are connected via QPI. Each CPU consists of 14 two-way SMT-cores, meaning, 28 SMT-threads. Hence, the system provides 112 SMT-threads overall. Each core owns an L1 data and instruction cache with 32 kB each. Furthermore, it owns an L2 cache with 256 kB. L3 cache and main memory are shared between all cores of a CPU.

During the measurements Intel's Turbo Boost was disabled. Turbo Boost is a hardware feature that accelerates applications by varying clock frequencies dependent on the number of active cores and the workload. Even though Turbo Boost is a valuable, runtime-saving feature in production runs, it distorts scaling measurements by boosting the sequential run through a higher clock frequency.

The runtime measurements are performed with `high_resolution_clock` from `std::chrono`. FMSolvr was executed 1000× for each measuring point, with each execution covering the whole workflow of the FMM for a single time step of the simulation. Afterwards, the 75%-quantile of these measuring points was computed. This means that 75% of the measured runtimes were below the plotted value. This procedure leads to stable timing results since the influence of clock frequency variations during the starting phase of the measurements is eliminated.

As input data set a homogeneous particle ensemble with only a thousand particles is used. The values of positions and charges of the particles are defined by random numbers in range $[0, 1)$. The measurements are performed with multipole order $p = 0$ and tree depth $d = 3$. Due to the small input data set and the chosen FMM parameters, the computational effort is very low. With this setup, FMSolvr tends to be latency- and synchronization-critical. Hence, this setup is most suitable to analyze the influence of NUMA-effects on applications that aim for strong scaling and small computational effort per compute node.

5.2   Performance

NUMA-Aware Thread Pinning We consider the runtime plot of FMSolvr in Fig. 4 (top) to analyze the impact of the NUMA-aware thread pinning policies EP and CP on the 2-NUMA-node system without applying a NUMA-aware work stealing. It can be seen that both policies lead to an increase in runtime in comparison to the non-NUMA-aware base implementation for the vast majority of cases. The most considerable runtime improvement occurs for pinning policy CP at #Threads = 12 with a speed-up of 1.6. The reason for this is that data transfer via the NUMA interconnect is fully avoided by CP since all threads fit on a single NUMA node for #Threads ≤ 12. Nevertheless, the best runtime is not reached for 12, but for 47 threads with policy CP. Hence, the practically relevant speed-up in comparison to the best runtime with the base implementation is 1.19.

Figure 4 (bottom) shows the runtime plot of FMSolvr with the thread pinning policies EP and CP on the 4-NUMA-node system without applying NUMA-aware work stealing. Here, too, the most considerable runtime improvement is reached for pinning policy CP if all cores of a single NUMA node are in use. Accordingly, the highest speed-up on the 4-NUMA-node system is reached at #Threads = 14 with a value of 2.1. In this case, none of the NUMA-aware implementations of FMSolvr outperforms the base implementation. Even though the best runtime is reached by the base implementation at #Threads = 97 with 1.77 ms, we get close to this runtime with 1.83 ms using only 14 threads. Hence, we can save compute resources and energy by applying CP.

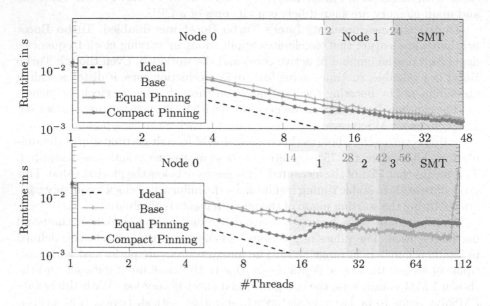

Fig. 4. Comparison of NUMA-aware thread pinning policies EP and CP with work stealing policy LR on 2-NUMA-node and 4-NUMA-node system.

NUMA-Aware Work Stealing. Figure 5 (top) shows the runtime of FMSolvr for CP in combination with the work stealing policies LR, PL and LO on the 2-NUMA-node system dependent on the number of threads. The minimal runtime of FMSolvr is reached with 1.16 ms for 48 threads if CP is applied in combination with LO. The practically relevant speed-up in comparison with the minimal runtime of the base implementation is 21%.

Figure 5 (bottom) shows the runtime plot of FMSolvr on the 4-NUMA-node for CP in combination with the work stealing policies LR, PL and LO. As already observed for the NUMA-aware thread pinning policies in Sect. 5.2, none of the implemented NUMA-awareness policies outperforms the minimal runtime of the base implementation with 1.76 ms. Hence, supplementing the NUMA-aware thread pinning policies with NUMA-aware work stealing policies is not sufficient and there is still room for the improvements described in Sect. 7.

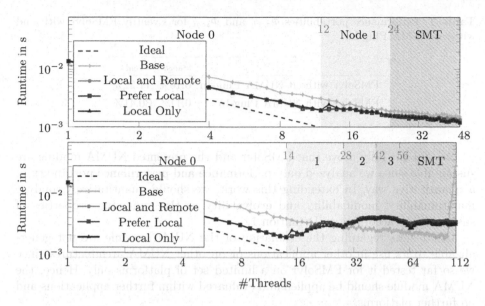

Fig. 5. Comparison of NUMA-aware work stealing policies LR, PL and LO based on NUMA-aware thread pinning policy CP on a 2-NUMA-node and 4-NUMA-node system

5.3 Performance Portability

We evaluate performance portability based on Definition 2 with two different performance efficiency metrics e: architectural efficiency for Φ_{arch} and strong scaling efficiency for Φ_{scal}.

Assumptions regarding Φ_{arch}: The theoretical double precision peak performance P_{2NN} of the 2-NUMA-node system is 960 GFLOPS (2 sockets with 480 GFLOPS [9] each), while the theoretical peak performance P_{4NN} of the 4-NUMA-node system is 1792 GFLOPS (4 sockets with 448 GFLOPS [9] each). With the considered input data set, FMSolvr executes 2.4 Million floating point operations per time step.

Assumptions regarding Φ_{scal}: The considered strong scaling efficiency is for each platform and application determined for the lowest runtime and the according amount of threads for which this runtime is reached.

As can be seen from Table 1 the determined performance portability varies greatly dependent on the applied performance efficiency metric. However, the NUMA Plug-In increases performance portability in both cases.

6 Threats to Validity

Even though we aimed at a well-defined description of our main research objective – a sustainable support of NUMA architectures for FMSolvr, the evaluation of this objective is subject to several limitations.

Table 1. Performance portabilities Φ_{arch} and Φ_{scal} for Eventify FMSolvr with and without NUMA Plug-In.

	Φ_{arch}	Φ_{scal}
FMSolvr without NUMA Plug-In	0.10%	11.22%
FMSolvr with NUMA Plug-In	0.11%	30.41%

We did not fully prove that FMSolvr and the presented NUMA module are sustainable since we analyzed only performance and performance portability in a quantitative way. In extending this work, we should quantitatively analyze maintainability, modifiability and evolvability as the remaining properties of sustainability according to Definition 1.

Our results regarding the evaluation of the NUMA module are not generalizable to its use in other applications or on other NUMA architectures since we so far tested it for FMSolvr on a limited set of platforms only. Hence, the NUMA module should be applied and evaluated within further applications and on further platforms.

The chosen input data set is very small and exhibits a very low amount of FLOPs to analyze the parallelization overhead of Eventify and the occurring NUMA effects. For lack of a performance metric for latency- and synchronization-critical applications, we applied double precision peak performance as reference value to preserve comparability with compute-bound inputs and applications. However, Eventify FMSolvr is not driven by FLOPs, e.g. it does not yet make explicit use of vectorization. In extending this work, we should reconsider the performance efficiency metrics applied to evaluate performance portability.

7   Conclusion and Future Work

Based on the properties of NUMA systems and the FMM, we described NUMA-aware data distribution and work stealing policies with respect to [13]. Furthermore, we present the NUMA-aware thread pinning policy CP based on the performance analysis provided in [13].

We found that the minimal runtime of FMSolvr is reached on the 2-NUMA-node system when thread pinning policy CP is applied in combination with work stealing policy LO. The minimal runtime is then 1.16 ms and is reached for 48 threads. However, none of the described NUMA-awareness policies outperforms the non-NUMA-aware base implementation on the 4-NUMA-node system. This is unexpected and needs further investigation since the performance analysis on another NUMA-node system provided in [13] revealed that NUMA-awareness leads to increasing speed-ups with an increasing number of NUMA nodes. Nevertheless, we can save compute resources and energy by applying CP since the policy leads to a runtime close to the minimal one with considerably less cores.

Next up on our agenda is the implementation of a NUMA-aware pool allocator, the determination of more accurate NUMA distance information and the implementation of a more balanced task graph partitioning. In this article, suitable software design paid off regarding development time and application performance. Hence, we aim at further decoupling of the parallelization layer *Eventify* and its modules to be reusable by other research software engineers.

References

1. Abduljabbar, M., Al Farhan, M., Yokota, R., Keyes, D.: Performance evaluation of computation and communication kernels of the fast multipole method on intel manycore architecture. In: Rivera, F.F., Pena, T.F., Cabaleiro, J.C. (eds.) Euro-Par 2017. LNCS, vol. 10417, pp. 553–564. Springer, Cham (2017). https://doi.org/10.1007/978-3-319-64203-1_40
2. Agullo, E., Bramas, B., Coulaud, O., Khannouz, M., Stanisic, L.: Task-based fast multipole method for clusters of multicore processors. Research Report RR-8970, Inria Bordeaux Sud-Ouest, March 2017. https://hal.inria.fr/hal-01387482
3. Agullo, E., Bramas, B., Coulaud, O., Darve, E., Messner, M., Takahashi, T.: Task-based FMM for multicore architectures. SIAM J. Sci. Comput. **36**(1), C66–C93 (2014). https://doi.org/10.1137/130915662
4. Amer, A., et al.: Scaling FMM with data-driven OpenMP tasks on multicore architectures. In: Maruyama, N., de Supinski, B.R., Wahib, M. (eds.) IWOMP 2016. LNCS, vol. 9903, pp. 156–170. Springer, Cham (2016). https://doi.org/10.1007/978-3-319-45550-1_12
5. Beatson, R., Greengard, L.: A short course on fast multipole methods. Wavelets Multilevel Methods Elliptic PDEs **1**, 1–37 (1997)
6. Greengard, L., Rokhlin, V.: A fast algorithm for particle simulations. J. Comput. Phys. **73**(2), 325–348 (1987). https://doi.org/10.1016/0021-9991(87)90140-9
7. Haensel, D.: A C++-based MPI-enabled tasking framework to efficiently parallelize fast multipole methods for molecular. Ph.D. thesis, TU Dresden (2018)
8. Haensel, D., Morgenstern, L., Beckmann, A., Kabadshow, I., Dachsel, H.: Eventify: event-based task parallelism for strong scaling. Accepted at PASC (2020)
9. Intel: APP Metrics for Intel Microprocessors (2020)
10. Kabadshow, I.: Periodic boundary conditions and the error-controlled fast multipole method. Ph.D. thesis, Bergische Universität Wuppertal (2012)
11. Koziolek, H.: Sustainability evaluation of software architectures: a systematic review. In: Proceedings of the Joint ACM SIGSOFT Conference (2011). https://doi.org/10.1145/2000259.2000263
12. Lameter, C.: NUMA (Non-Uniform Memory Access): an overview. Queue **11**(7), 40:40–40:51 (2013). https://doi.org/10.1145/2508834.2513149
13. Morgenstern, L.: A NUMA-aware task-based load-balancing scheme for the fast multipole method. Master's thesis, TU Chemnitz (2017)
14. Pennycook, S.J., Sewall, J.D., Lee, V.W.: A metric for performance portability. CoRR (2016). http://arxiv.org/abs/1611.07409
15. Ying, L., Biros, G., Zorin, D.: A kernel-independent adaptive fast multipole algorithm in two and three dimensions. J. Comput. Phys. **196**, 591–626 (2004). https://doi.org/10.1016/j.jcp.2003.11.021

Boosting Group-Level Synergies by Using a Shared Modeling Framework

Yunus Sevinchan[1]([✉])[iD], Benjamin Herdeanu[1,2][iD], Harald Mack[1][iD],
Lukas Riedel[1,3,4][iD], and Kurt Roth[1,3][iD]

[1] Institute of Environmental Physics, Heidelberg University, Heidelberg, Germany
{yunus.sevinchan,benjamin.herdeanu,harald.mack,
lukas.riedel,kurt.roth}@iup.uni-heidelberg.de
https://ts.iup.uni-heidelberg.de
[2] Heidelberg Graduate School for Physics, Heidelberg University,
Heidelberg, Germany
[3] Interdisciplinary Center for Scientific Computing, Heidelberg University,
Heidelberg, Germany
[4] Heidelberg Graduate School of Mathematical and Computational Methods
for the Sciences, Heidelberg University, Heidelberg, Germany

Abstract. Modern software engineering has established sophisticated
tools and workflows that enable distributed development of high-quality
software. Here, we present our experiences in adopting these workflows
to collectively develop, maintain, and use research software, specifically:
a modeling framework for complex and evolving systems. We exemplify
how sharing this modeling framework within our research group helped
conveying software engineering best practices, fostered cooperation, and
boosted synergies. Together, these experiences illustrate that the adoption of modern software engineering workflows is feasible in the dynamically changing academic context, and how these practices facilitate reliability, reproducibility, reusability, and sustainability of research software,
ultimately improving the quality of the resulting scientific output.

Keywords: Research software · Software quality · Reproducibility ·
Complex systems · Modeling · Scientific computing

1 Introduction

Software has become an integral part of modern research, be it for the control of
experiments, analysis of data, or computer simulations. In recent years, however,
the research community has become aware of issues relating to the reliability and
reusability of software developed and used for scientific purposes [2,6,10]. These
shortcomings are mainly attributed to the unique characteristics of scientific
software development and a resulting disconnect between the research and software engineering communities [5]. Ultimately, these issues impede the progress
of scientific research. At the same time, the Open-source Software (OSS) community has grown vigorously and is successfully developing high-quality software. This is made possible in large parts by platforms like GitHub and GitLab,

© Springer Nature Switzerland AG 2020
V. V. Krzhizhanovskaya et al. (Eds.): ICCS 2020, LNCS 12143, pp. 442–456, 2020.
https://doi.org/10.1007/978-3-030-50436-6_32

which greatly simplify collaboration on software projects, and by the adoption of software engineering workflows which have proven valuable for efficiency of development, long-term maintainability, and reliability.

Although OSS development always had a major role in scientific computing, these approaches are now gaining more traction in the wider scientific community [3]. Additionally, several institutions have been formed which aim to counteract the negative effects arising from current practices in the development and use of research software. For example, to increase the recognition and visibility of reliable and sustainable research software projects, a number of relatively young journals like the Journal of Open Research Software and the Journal of Open Source Software focus solely on publishing high-quality open-source research software. Furthermore, the extension of FAIR principles to research software is currently being discussed [8] with the goal to improve the findability, accessibility, interoperability, and reusability of software used in the scientific context.

When studying complex and evolving systems – the research field in which we operate –, computer models[1] are considered the main research method. Given the nature of the research questions in this field, models for a huge diversity of phenomena and ever-changing situations need to be conceived. This is in contrast to other research fields like weather prediction, oil recovery, and combustion optimization where the challenge is to investigate a specific scenario with the highest attainable efficiency using numerical simulations of well established systems, like flow-, transport-, and reaction-equations.

Subsequently, the implementation and analysis of these models often requires the development of custom software, thus entailing the same challenges as experienced for other research software. For instance, we observed that code was frequently written from scratch because the work involved in understanding and adapting an existing model or simulation tool was higher than writing it anew. This not only led to redundant implementations but also to the repetition of mistakes, and overall unreliable and unsustainable code. In effect, software was written by a single researcher for the purpose of their project and discarded soon afterwards, missing out on the collaborative aspect of modern research.

In this paper, we present the approaches we took on the level of our research group to address these issues. The aim was not only to improve the quality of the software we developed and used, but to also boost synergies within the group. We did so by collaboratively developing a modeling framework, using it throughout the group, and creating associated communication structures and workflows. We perceived such a framework to be the ideal focus point for promoting collaboration. At the same time, we saw this as a way to gain experience in developing reliable and sustainable software that not only adheres to best practices in software engineering, but also advocates them to young researchers.

The modeling framework resulting from this effort is called Utopia [12] and aims to be the central tool in all stages of a modeling-based research workflow,

---

[1] With the term *computer model* we denote a conceptualization of a real-world system that is investigated via computer simulations. When solely the implementation in code is concerned, we use the term *model implementation*.

providing solutions for common problems based on state-of-the-art programming practices. Utopia is publicly available[2] under an open-source license. Since the start of the project in early 2018, it formed the basis for more than thirty new modeling projects in our group and has been used as a teaching tool in graduate physics courses. In the context of this paper, the specific details of the framework are less relevant than the conceptual role it fulfilled in improving research software quality and enhancing collaboration within the group: It provided a platform to work on, a common language to communicate in, and a shared goal to motivate everyone involved.

As such, other modeling frameworks may fulfill the same role as Utopia did for us. One example is the NetLogo programming language and development environment [13], which can be used to study a wide variety of agent-based models.

In the following, we first aggregate the synergies we see arising from sharing a modeling framework (Sect. 2) and outline which software engineering workflows we regard as feasible to adopt in such a context (Sect. 3). We then describe the experiences we had in developing Utopia as a group, working with it as researchers, and using it for academic teaching (Sect. 4). By sharing these experiences, we hope to contribute to the effort of improving the responsible development and use of research software.

2 Synergies

Synergies in research environments surely are of a wide variety and depend very much on the necessities and methods of the respective fields. Here, we focus on the investigation of complex and evolving systems using computer models, where common workflows allow the use of a shared modeling framework.

In our experience, a typical workflow in this field can be represented as a four-stage process:

1. **Conceptualize** a research question into a model system
2. **Implement** the model system as a computer model
3. **Generate** simulation data using the model implementation
4. **Analyze** the simulation data, extract results, and reflect on the model

The above aims to categorize the workflow into stages that have different qualitative demands on the methods, the software tools, and the interaction with colleagues, and subsequently varying potential for synergies. In practice, such a workflow is hardly ever sequential, but an iterative and self-referential process. Furthermore, the wide diversity of phenomena under investigation requires a high flexibility in all stages of the workflow, allowing for constant redefining, redesigning, reimplementation, and analysis. While this formulation originates from observations within our group and the study of complex and evolving systems, we believe that scientific research in other areas might be represented in a similar fashion, especially when working with computer models (e.g., [4]).

---

[2] https://ts-gitlab.iup.uni-heidelberg.de/utopia/utopia.

We call a framework that covers all the stages of the typical workflow a *comprehensive modeling framework*. By modularizing the tools on the level of the whole research workflow, it abstracts away technicalities and allows researchers to think on the level of the model system they want to investigate.

Simulation Infrastructure. We identified the following software functionalities as common infrastructure during the aforementioned modeling workflow: *(i)* storing and managing simulation data, *(ii)* configuring simulations, i.e., passing parameters to the simulation, *(iii)* performing parameter sweeps, *(iv)* data processing and analysis routines, and *(v)* data visualization. In cases where the typical workflow of the involved researchers is comparable, as sketched above, sharing these infrastructure tools via a framework becomes beneficial and greatly reduces the time and effort needed to get operational in all its stages. Importantly, the framework can provide a more reliable and more sophisticated feature set, which would not be feasible to be implemented by an individual developer.

One example are parameter sweeps, as are often required for sensitivity analysis of a model's behavior. A framework-level implementation can generate the set of configurations programmatically, based on user input, and can make use of naive parallelization to speed up the generation of simulation data, while being completely independent of the model implementation itself. Having such a feature available as part of a framework is a direct benefit to all framework users.

In situations where more flexibility is required than can be offered by the framework, the use of open data standards helps avoiding a lock-in effect which would be detrimental to the use of the software and the researchers' freedom.

Modeling Techniques. The investigation of complex and evolving systems has generated a number of modeling techniques, the most notable being *(i)* cellular automata, which can be used as a representation of spatially extended systems, *(ii)* individual- or agent-based models, which represent individual entities with a varying range of autonomy, agency, and perception, and *(iii)* network-based models, which put the focus on these agents' interactions with each other.

By integrating frequently used modeling techniques into a shared framework, the knowledge about common pitfalls can be used to avoid the repetition of mistakes. Furthermore, taking into account the central importance of correctness and efficiency of these implementations, they can be thoroughly tested, reviewed by multiple developers, and optimized for performance. We see these aspects as the key arguments for an implementation on the framework level, thus achieving a higher reliability, efficiency, and usability compared to software developed in an uncoordinated manner.

A modeling framework should also allow researchers to leverage the great variety of research software already available in the scientific field, which often have been tried and tested for decades and hence attained a quality virtually unreachable by most individual researchers. These might for example provide efficient implementations of graph data structures or numerical algorithms.

A tight integration of existing solutions is often desirable, but needs to be such that particularities are abstracted away to not demand too much additional knowledge from the user of the framework.

Feature Sharing. A shared framework allows to implement new features as part of the framework rather than in an isolated manner, making them easily accessible to other users. Such a framework-level implementation often requires a higher degree of generality for the feature in order to make it useful for other users. At the expense of additional work in abstracting the functionality, this facilitates collaboration with other developers, ultimately improving reliability, usability, integration, and performance.

A Common Conceptual Language. When using the same modeling framework in a research group, not only the code base is shared, but also the conceptual language in which discussions and the exchange of ideas take place. This can be beneficial in multiple stages of the research workflow. For instance, when designing an implementation of the model system, the modeling techniques provided by the framework supply a set of abstractions that can be used when talking about the model, e.g., the concept of what constitutes an entity or the vocabulary describing a specific algorithm.

Benefits also arise from an easier understanding of code written by other researchers. The overall amount of code is reduced by the use of the simulation infrastructure and the framework-level implementations of the modeling techniques. This not only facilitates a closer and more efficient interaction between researchers on the level of the implementation, but also simplifies reusing model implementations.

Transparency. In a research context, model implementations are typically privately developed until they are made public alongside a publication. Within a research group, we see several positive effects of making the source code of model implementations openly available right from the beginning of development: By reading the code written by other members of the research group, possible implementation approaches can be exchanged and spark discussions. Furthermore, members of the research group are encouraged to learn from other developers' implementations, which pertains not only to modeling ideas but also to software engineering practices in general. This openness is an important aspect in a dynamic environment where people and problems change frequently.

While the transparency brings significant benefits, it also demands a careful consideration of collateral effects, including premature diffusion of novel ideas – even within the research group – and the potential for surveillance. We see a common understanding of intellectual property as a key requirement for this approach. Consequently, having model implementations visible probably works well within a research group, but not necessarily in a large community.

3 Adopting Software Engineering Workflows

Developing research software collaboratively allows the use of software engineering workflows that rely on divided responsibilities and interactions with other group members. In the following, we highlight the aspects we perceive as particularly beneficial for the quality of collaboratively developed research software, while not posing a prohibitively large learning or management overhead.

Effective Use of Version Control Systems. Version control systems (VCS) are not only valuable to an individual researcher but enable workflows that form the basis of efficient collaboration. Platforms like GitHub and GitLab go beyond the mere hosting of VCS repositories: they provide task management tools, communication channels, contribution workflows, automation services, and interfaces to other services that allow for custom extensions.

Research software can and should make use of these platform-level tools that have proven themselves highly valuable to a successful software engineering workflow. While such a procedure may pose additional work for a single researcher, there are considerable gains to be expected for consistent work as a group.

Code Review. Code review serves the purpose of improving the quality and maintainability of software [9]. Reviewers inspect source code changes not only with respect to potential defects that may not be found through static code checks, but also suggest improvements to the implementation, its readability, or its integration into the larger project. Furthermore, such a process can serve to transfer knowledge about the code base between the authors and the reviewers, while at the same time strengthening their team identity [1].

When working with VCS and collaborative version control platforms, code review can be conveniently carried out on so-called *Pull Requests* or *Merge Requests*, i.e., on the proposed changes to a code base, prior to merging them into the main branch of a repository. The platforms usually present the suggested changes alongside the code they will replace, identify code ownership, and allow reviewers to comment on the changes and propose improvements. This platform-assisted code review is already well-established in the OSS community. In the context of software engineering for research software, the same benefits can be expected, and would directly address issues like reliability, consistency, and general code quality. Given the often heterogeneous software engineering skills in an academic context, knowledge transfer mediated by code review may also gain an important role.

Testing and Automation. While software testing is a cornerstone of professional software development, testing of scientific software is intrinsically more difficult. Kanewala and Bieman [7] identify the challenges of testing scientific software to be either due to characteristics of the research software itself – like the oracle problem –, or to be caused by "cultural differences" between the software engineering community and scientists. They come to the conclusion that while some challenges are unique to scientific software, others can be overcome by incorporating existing testing techniques from software engineering into their development workflows.

In modeling-based research, tests are required both for the framework code and the model implementations. When collaboratively developing software, the importance of both of these can be emphasized, and the implementation of tests can be simplified to become more accessible to novice developers. At the same time, techniques that make the testing of model implementations possible can be conveyed, while also identifying in which areas further testing techniques would be required. Putting a larger emphasis on these parts of software development can also inspire the implementation workflow itself, e.g., by promoting test-driven development of models.

With growing size of a software project, automation becomes a crucial part of the development process. The term refers to a set of actions that are automatically carried out, e.g., when pushing code to the remote server or when merging a feature into the main branch. With these tools being readily available via the same platform that the software is developed on, automation becomes easily accessible. Benefits for the development of research software and the individual model developers are that associated tests are carried out automatically, taking that burden off the researcher, and making it easier to detect breaking changes in both the framework and the model implementations.

4   Utopia – Three Case Studies

The context of these case studies is our research group that focuses on the investigation of complex, chaotic, and evolving environmental systems. Development of Utopia started early 2018 with a team of four PhD candidates[3], two MSc students and two BSc students. Utopia has been used in more than thirty projects, with usual durations of one year and up to fifteen simultaneous projects. Students joining our group typically have a physics background and entry-level programming experience, mostly in Python, but are unfamiliar with version control, testing, or other software engineering workflows.

4.1   Developing the Utopia Framework

In this case study, we focus on the experiences from the collective development of the Utopia framework and the adoption of software engineering workflows throughout this process.

VCS Workflows and Code Review. Utopia was developed in a GitLab project on a self-hosted GitLab instance. We chose to use the *GitHub Flow* branching model that has little management overhead and focuses on the main branch always being in a working state: Feature branches start off the main

---

[3] YS, BH, HM, and LR.

branch and merge back into it. Each feature branch corresponds to the implementation of a single, well-isolated task, which often is planned in a GitLab Issue.

We adopted a change-based code review process for the Utopia framework. All code changes require a review by at least one other developer before being allowed to be merged into the main branch. Code review takes place in the GitLab Merge Request interface, where the author of the request gives a brief description of the changes, why they were necessary, and how they were achieved, and points out changes which require further discussion. The author then assigns a person for review, typically someone who is already involved in the task or who is familiar with the affected framework structures. The reviewer then goes through the changes and may comment directly on the code to discuss the changes or suggest improvements. Moreover, they may include other developers into the review process. If everybody involved approves of the changes, the Merge Request is merged and the corresponding Issue is closed automatically.

We perceived the review process as highly beneficial as it considerably improved the resulting code in terms of consistency and reliability. All Merge Requests that included substantial code changes or additions profited from code review, be it through the detection of potential defects or improvements relating to the robustness of the implementation. Alongside, the documentation (both of the code itself and the usage of the overall framework) profited from a thorough review process because reviewers could use it as a starting point and evaluate how well the changes could be understood by someone who did not implement the code. Furthermore, the review process helped conveying knowledge about Utopia's structure, inner workings, and agreed-upon guidelines, thus gaining an important role in increasing maintainability of the project.

These benefits greatly offset the additional time and effort that developers needed to invest into code review.

Testing and Automation. Thorough testing procedures are commonplace in software engineering and a prerequisite for projects to grow in a sustainable fashion. Accordingly, we require of all Utopia features to have a corresponding test implemented that covers the relevant use cases.

We make use of GitLab CI/CD for automated test execution and building of the framework in different build modes. Furthermore, the pipeline generates a code coverage report, and deploys a preview of the documentation and a ready-to-use Docker image of Utopia. All these automations proved highly valuable in the code review process where another requirement for merge approval is that the pipeline passes successfully and the code coverage report was inspected. That way, the pipeline provides a "ground truth" irrespective of the individual developers' systems and allows to easily detect regressions in seemingly unrelated parts of the framework.

While not feasible in all situations, this workflow also allowed test-driven development approaches which we found helpful when addressing previously undetected bugs: the bug can then first be reproduced by a new test case, which fails initially; subsequently, the code is adjusted such that the test passes.

Partaking in Framework Development. Through the focus of the framework to make the modeling workflow as convenient as possible, exposure to low-level code was inevitably reduced. Potential contributors thus developed the perception of not having enough insight into the inner workings of the framework to suggest changes or improvements to it.

As a counteraction, we organized multiple *Coding Weeks* which were preceded by a planning phase with regular meetings. During the planning phase, new contributors got to know those parts of the framework that were relevant for a particular improvement or new feature. Depending on the task, they then worked alone or in a small group to implement it, add test cases and documentation, and jointly review all of it. Overall, these events proved to be not only a group-forming experience, but also very productive, especially when it came to the coordination and implementation of larger features. One important long-term effect was that people who took part in a Coding Week were more likely to also contribute thereafter.

While it needs to be acknowledged that not everyone who is using the framework is also interested in improving it, we think that experience with these collaborative workflows has value beyond the currently undertaken project. We currently try to lower initial thresholds for new contributors by keeping everyone involved in the development of Utopia as much as possible and suggesting starting points for contributions.

4.2   Using Utopia Throughout the Research Group

In this case study, the focus is on usage of Utopia for the implementation and investigation of models of complex and evolving systems, the synergies that developed from using the shared modeling framework, and the integration of software engineering workflows into that process.

Individual model implementations were part of a single, group-internal GitLab project, separate from the framework project.

Learning Curve. To make the entry as easy as possible and thereby facilitate usage of the shared framework, we put a focus on providing detailed introductory guides. These were meant especially for students entering the research group for a BSc or MSc project, which have a typical duration of 3–4 months or about one year, respectively. The aim was to accustom these young researchers with the Utopia framework such that they can install it within a day, are capable of using essential features within a week, and can start with a model implementation soon after.

While learning to use a new framework may constitute a considerable overhead, we had positive experiences with the adoption of Utopia in the group and the speed with which models were implemented or adapted. Despite the fact that Utopia models are implemented in C++, most newcomers were quick to setup a new model using the corresponding guides, and subsequently start to implement the desired model dynamics using the abstractions and modeling techniques provided by the framework.

While we encouraged newcomers to search the documentation and the GitLab project in case of questions, some personal assistance was almost always necessary to help resolving particularities in the installation or the setup of models. Intermittently, this created a substantial additional work load for the more experienced developers. For reducing this work load, it proved crucial to reflect even seemingly minor problems in the project documentation and develop a shared corpus of knowledge among framework users.

Synergies. A central motivation for using a shared modeling framework was to boost the development of synergies within our research group. For us, this proved to be a successful and overall beneficial approach, albeit not without challenges and a certain maintenance cost. For instance, owing to the different schedules of research projects, we experienced some diffusion of knowledge and loss of experience. Thus, this approach hinges on a critical mass of users and developers carrying on the required knowledge to maintain the shared code base and offer assistance to new group members.

Simulation Infrastructure, Modeling Techniques and Feature Sharing. The aim of eliminating repeated implementation of simulation infrastructure, modeling techniques, and associated tools was largely achieved and the benefits extended over the whole modeling workflow, from conceptualization to data analysis and visualization. Sharing the simulation infrastructure was a very clear benefit, because the overlap of required software tools was very high within the research group. Also, some of the more sophisticated features like parameter sweeps with multi-node cluster support, uniform manager structures for the provided modeling techniques, or generalized data post-processing and plotting could not have been feasibly developed outside such a framework. Having these features implemented into the framework effectively addressed most of the issues we observed in solitary software development. Especially for the PhD candidates, Utopia improved work efficiency in all stages of the modeling workflow by alleviating points of friction and providing a flexible, scalable, optimized, and reliable development environment.

However, in some cases, the generalization of features required considerably more effort than if these features would have been implemented directly where they were needed; in that sense, we sometimes fell prey to "premature generalization", thinking that a framework-level feature would always provide a benefit. Also, discoverability of features proved to be more challenging than expected: Despite documentation, users sometimes did not associate the functionality they desired with the description of an already existing feature, which in some cases led to a redundant implementation by the users.

Common Language and Transparency. Having all code openly accessible for everyone in the group proved to be one important factor for improving collaboration: Group members frequently inspected existing implementations to learn about representation details and model dynamics. This not only facilitated discussions, but also provided a means of knowledge transfer, often extending to programming language features and software engineering best practices. Notably, a number of projects are under way that build on existing model implementations. The common conceptual language and the use of the framework made this process efficient by abstracting away most of the code that relates to simulation infrastructure and modeling techniques.

Regarding transparency, we observed some reluctance in newcomers to share model ideas and code with the rest of the group. We ascribe this partly to the need to learn the corresponding workflows, but also to a certain insecurity regarding code quality. This reluctance typically subsided after feeling more integrated into the group and having experienced the benefits of sharing code with other group members.

Software Engineering Workflows. Aside the synergies developing in the group, we also wanted to promote software engineering workflows during model development. Our observations varied for experienced and novice developers.

Experienced model developers profited significantly from the easy availability of the testing framework and automation, and adopted VCS workflows for their model implementations, e.g., by working on their models in a task-based fashion with frequent and granular Merge Requests.

For larger models, the implementation of tests proved indispensable. Here, the framework assisted mainly by simplifying the implementation of C++ unit tests and providing an easy-to-use interface to generate and validate simulation data; these approaches were often viable to assert the expected microscopic and macroscopic model behavior, respectively. Furthermore, by including all tests into the GitLab CI/CD alongside the framework, not only the model behavior was tested automatically but also the effect of framework changes on the models, which was crucial when working on more intricate parts of the framework.

Unlike for framework code, we did not require in-depth code review for individual model implementations, mainly due to the high work load such a policy would generate for reviewers. Nevertheless, experienced developers often requested code review from others; given the abstractions the framework provided, this was feasible for smaller Merge Requests or when the author asked for feedback on specific parts of their implementation. Despite the restrictions, this proved to be a workable compromise between a researcher's responsibility for their model and the work load for others, thus still providing mutual benefits.

For novice developers, our experiences were mixed: the adoption of SE workflows by newcomers was not as natural as we had hoped, but the benefits were typically acknowledged and, in the long run, some of the procedures found their way into their work with the framework. We observed the adoption to be impeded mostly by the learning required to understand the new procedures and their benefits. With a physics background, many of the workflows are completely new to MSc- and BSc-level students. Especially in the first weeks and months of a project, the focus typically is on the scientific literature and the model formulation and implementation itself, not on software engineering aspects. We found that adoption increases only once developers become more proficient and see these methods as solutions to new problems arising during development.

For example, learning to use VCS and GitLab and adapting personal work-flows to accommodate these was often perceived as a substantial investment without clear benefits for the students' own projects. This was despite their acknowledgment that these tools are useful in the development of the frame-work or in software development in general. In effect, for new users, a decoupling from the rest of the repository and group was quite common in the first weeks or months of a project: Once a working setup was achieved, development often continued on a local feature branch for the new model and with a fixed (i.e., outdated) version of the framework. We made similar observations regarding the implementation of tests: It was generally acknowledged that having tests would be beneficial, but their implementation was typically not a priority.

For individual model development, we deliberately decided against imposing strict workflows. Instead, we promoted their benefits and tried to reduce entry barriers by providing guides, personal assistance, and framework tools that made adoption easier. Thereby, we aimed at eliciting the motivation to learn and integrate these tools, rather than enforcing their use. This proved successful on a longer time scale and with developers becoming more experienced with Utopia: After the first few months of many MSc projects, the projects were kept up-to-date, GitLab became more regularly used, and test implementations became more frequent. We believe that without Utopia, many novice developers would not have considered adopting these software engineering techniques and would not have reached the same level in their actual research. Utopia not only promoted the utility of these tools, but also reduced the barrier of using them.

Currently, we aim to improve group-level interactions through having regular meetings for questions and discussions on the framework and model implementa-tions. Similar as with the Coding Weeks, we find these meetings to have a group-building effect, and the higher interaction having a positive effect on cooperation. As part of this process, we are also promoting pair-wise code review alongside the propositions made in [11].

4.3 Using Utopia in Teaching

Possible use cases for Utopia in academic teaching range from the generation of simulation data using existing models to the implementation and investigation of new models, as it is done in the regular research context. So far, Utopia was used for teaching in an MSc physics lecture (summer term 2019) and is currently (winter term 2019/20) used in an MSc physics seminar. With the seminar still under way at the time of writing of this paper, we present the experiences from using Utopia as part of the mandatory exercises accompanying the *Complex, Chaotic, and Evolving Environmental Systems* lecture.

There, students used Utopia to run and analyze well-established models of spatially distributed complex systems: the Forest Fire Model, the Contagious Disease Model, and the Predator-Prey Model. The focus was on the investi-gation of the behavior of the models, not on their implementation, which is why we provided these implementations alongside with a detailed description of the implemented dynamics and available configuration parameters. The exercises instructed students to generate simulation data for changing configuration parameters and subsequently use the framework to analyze the data and understand the model behavior, e.g., by extracting macroscopic fixpoints or by investigating power spectra to characterize spatial structures.

To make the setup as easy and machine-agnostic as possible, we used the automatically deployed and ready-to-use Docker image of Utopia[4], which contained all relevant model binaries. Furthermore, a Jupyter Notebook server was set up to run inside the Docker container, allowing students to work with Utopia from an interface that is more easily accessible than the command line. Setting up the Docker environment and working through a tutorial on the usage of Utopia constituted the first exercise of the lecture. We subsequently dedicated one session of the tutorials to questions arising from this exercise; this was sufficient to get students operational on their own computers.

We as teachers profited from not having to implement standalone models and bare-bones simulation infrastructure, but having a fully-fledged modeling framework available that we already had experience with. Furthermore, any effort put into model implementation or documentation enhancements was a direct benefit to other framework users.

For students, benefits included having a single and simple-to-use interface with which to investigate all provided models. While learning to use Utopia created additional challenges compared to using isolated model scripts, we see working with a modeling framework as a valuable skill and perceived this to be an appealing aspect for students interested in going into modeling-based research.

5   Conclusion

Developing flexible yet reliable software is a challenge. This is particularly true in the academic context where people remain for just a few years and typically have little experience with modern software engineering workflows. At the same time, the nature of highly dynamic research fields puts high demands on the quality of research software, e.g., in terms of reproducibility and reliability.

In the above case studies, we described our experiences with collectively developing and using Utopia, a modeling framework for complex and evolving systems. Our primary aim was to demonstrate that this approach is feasible and how it can boost synergies within the research group. Adopting these practices not only resulted in beneficial effects on the group's dynamics, but improved the quality of the developed research software and the resulting research output. We also showed that in order to sustain these workflows, the group with its ever new members has to be motivated and instructed in events like Coding Weeks and regular question-discussion meetings. We aim to further promote both this approach and the Utopia framework itself, being aware that software frameworks need to be continuously maintained and improved in order to survive.

---

[4] https://hub.docker.com/r/ccees/utopia.

Taken together, we exemplified how modern software engineering workflows allow to successfully tackle a large and rapidly changing research field in a comprehensive way which is otherwise only feasible for larger and more permanent institutions. We trust that the collaborative philosophy that is facilitated by such approaches will propagate via its users and thereby contribute to sustainable development and responsible use of research software.

Acknowledgments. We thank the reviewers of this submission for their thoughtful remarks and suggestions and Maria Blöchl for valuable feedback on an earlier version of this manuscript. We are grateful to all contributors and users of Utopia to have participated in this collective effort.

References

1. Bacchelli, A., Bird, C.: Expectations, outcomes, and challenges of modern code review. In: Proceedings of the 2013 International Conference on Software Engineering, ICSE 2013, pp. 712–721. IEEE Press (2013)
2. Hannay, J.E., MacLeod, C., Singer, J., Langtangen, H.P., Pfahl, D., Wilson, G.: How do scientists develop and use scientific software? In: 2009 ICSE Workshop on Software Engineering for Computational Science and Engineering. IEEE (2009). https://doi.org/10.1109/secse.2009.5069155
3. Heaton, D., Carver, J.C.: Claims about the use of software engineering practices in science: a systematic literature review. Inf. Softw. Technol. **67**, 207–219 (2015). https://doi.org/10.1016/j.infsof.2015.07.011
4. Helbing, D., Bialetti, S.: How to do agent-based simulations in the future: from modeling social mechanisms to emergent phenomena and interactive systems design (Oct. 2013). In: Helbing, D. (ed.) Social Self-Organization. Understanding Complex Systems. chap. Agent-Based Modeling, pp. 25–70. Springer, Heidelberg (2012). https://doi.org/10.1007/978-3-642-24004-1_2
5. Johanson, A., Hasselbring, W.: Software engineering for computational science: past, present, future. Comput. Sci. Eng. (2018). https://doi.org/10.1109/mcse.2018.108162940
6. Joppa, L.N., et al.: Troubling trends in scientific software use. Science **340**(6134), 814–815 (2013). https://doi.org/10.1126/science.1231535
7. Kanewala, U., Bieman, J.M.: Testing scientific software: a systematic literature review. Inf. Softw. Technol. **56**(10), 1219–1232 (2014). https://doi.org/10.1016/j.infsof.2014.05.006
8. Lamprecht, A.L., et al.: Towards FAIR principles for research software. Data Sci. 1–23 (2019). https://doi.org/10.3233/DS-190026
9. McIntosh, S., Kamei, Y., Adams, B., Hassan, A.E.: An empirical study of the impact of modern code review practices on software quality. Empir. Softw. Eng. **21**(5), 2146–2189 (2015). https://doi.org/10.1007/s10664-015-9381-9
10. Nguyen-Hoan, L., Flint, S., Sankaranarayana, R.: A survey of scientific software development. In: Proceedings of the 2010 ACM-IEEE International Symposium on Empirical Software Engineering and Measurement - ESEM 2010. ACM Press (2010)
11. Petre, M., Wilson, G.: Code review for and by scientists. arXiv:1407.5648v2 [cs.SE] (2014). https://arxiv.org/abs/1407.5648v2

12. Riedel, L., Herdeanu, B., Mack, H., Sevinchan, Y., Weninger, J.: Utopia: a comprehensive and collaborative modeling framework for complex and evolving systems. J. Open Sour. Softw. (2020, under review). https://joss.theoj.org/papers/8ce6d2bc26c0c6553c5ce5aff38d83c3
13. Wilensky, U.: NetLogo (1999). http://ccl.northwestern.edu/netlogo/

Testing Research Software: A Case Study

Nasir U. Eisty[1]([⊠]), Danny Perez[2], Jeffrey C. Carver[1], J. David Moulton[2],
and Hai Ah Nam[2]

[1] Department of Computer Science, University of Alabama, Tuscaloosa, AL, USA
neisty@crimson.ua.edu, carver@cs.ua.edu
[2] Los Alamos National Laboratory, Los Alamos, NM, USA
{danny_perez,moulton,hnam}@lanl.gov

Abstract. *Background*: The increasing importance of software for the
conduct of various types of research raises the necessity of proper testing
to ensure correctness. The unique characteristics of the research software
produce challenges in the testing process that require attention. *Aims*:
Therefore, the goal of this paper is to share the experience of implementing a testing framework using a statistical approach for a specific type
of research software, i.e. non-deterministic software. *Method*: Using the
ParSplice research software project as a case, we implemented a testing
framework based on a statistical testing approach called **Multinomial
Test**. *Results*: Using the new framework, we were able to test the *Par-
Splice* project and demonstrate correctness in a situation where traditional methodical testing approaches were not feasible. *Conclusions*: This
study opens up the possibilities of using statistical testing approaches
for research software that can overcome some of the inherent challenges
involved in testing non-deterministic research software.

Keywords: Research software · Testing · Software engineering

1 Introduction

Research software can enable mission-critical tasks, provide predictive capability to support decision making, and generate results for research publications.
Faults in research software can produce erroneous results, which have significant impacts including the retraction of publications [8]. There are at least two
factors leading to faults in research software: (1), the complexity of the software (often including non-determinism) presents difficulties for implementing a
standard testing process and (2) the background of people who develop research
software differ from traditional software developers.

Research software often has complex, non-deterministic computational
behavior, with many execution paths and requires many inputs. This complexity makes it difficult for developers to manually identify critical input domain
boundaries and partition the input space to identify a small but sufficient set
of test cases. In addition, some research software can produce complex outputs
whose assessment might rely on the experience of domain experts rather than on

© Springer Nature Switzerland AG 2020
V. V. Krzhizhanovskaya et al. (Eds.): ICCS 2020, LNCS 12143, pp. 457–463, 2020.
https://doi.org/10.1007/978-3-030-50436-6_33

an objective test oracle. Finally, the use of floating-point calculations can make it difficult to choose suitable tolerances for the correctness of outputs.

In addition, research software developers generally have a limited understanding of standard software engineering testing concepts [7]. Because research software projects often have difficulty obtaining adequate budget for testing activities [11], they prioritize producing results over ensuring the quality of the software that produces those results. This problem is exacerbated by the inherent exploratory nature of the software [5] and the constant focus on adding new features. Finally, researchers usually do not have training in software engineering [3], so the lack of recognition of the importance of the corresponding skills causes them to treat testing as a secondary activity [10].

To address some of the challenges with testing research software, we conducted a case study on the development of a testing infrastructure for the *ParSplice*[1] research software project. The goal of this paper is to *demonstrate the use of a statistical method for testing research software*. The key contributions of this paper are (1) an overview of available testing techniques for non-deterministic stochastic research software, (2) implementation of a testing infrastructure of a non-deterministic parallel research software, and (3) demonstration of the use of a statistical testing method to test research software that can be a role model for other research software projects.

2   Background

In a non-deterministic system, there is often no direct way for the tester (or test oracle) to exactly predetermine the expected behavior. In *ParSplice* (described in Sect. 2.1), the non-determinism stems from (1) the use of stochastic differential equations to model the physics and (2) the order in which communication between the procedures occurs (note however that even though the results from each execution depends upon message ordering, each valid order produces a statistically accurate result, which is the key requirement for the validity of *ParSplice* simulations).

In cases where development of test oracles is difficult due to the non-determinism, some potentially viable testing approaches include metamorphic testing, run-time assertions, and machine learning techniques [6]. After describing the *ParSplice* project, the remainder of this section explains these techniques along with their possible applicability to *ParSplice*.

2.1   ParSplice

ParSplice (Parallel Trajectory Splicing) [9] aims at overcoming the challenge of simulating the evolution of materials over long time scales through the time-wise parallelization of long atomistic trajectories using multiple independent producers. The key idea is that statistically accurate long-time trajectories can

---

[1] https://gitlab.com/exaalt/parsplice.

be assembled by splicing end-to-end short, independently-generated, trajectory segments. The trajectory can then grow by splicing a segment that begins in the state where the trajectory currently ends, where a state corresponds to a finite region of the configuration space of the problem. This procedure yields provably statistically accurate results, so long as the segments obey certain (relatively simple) conditions. Details can be found in the original publication [9].

The *ParSplice* code is a management layer that orchestrates a large number of calculations and does not perform the actual molecular dynamics itself. Instead, *ParSplice* uses external molecular dynamics engines. The simulations used in *ParSplice* rely on stochastic equations of motion to mimic the interaction of the system of interest with the wider environment, which introduces a first source of non-determinism.

A basic ParSplice implementation contains two types of processes: a splicer and producers. The splicer manages a database of segments, generates a trajectory by consuming segments from the database, and schedules execution of additional segments, each grouped by their respective initial state. Producers fulfill requests from the splicer and generate trajectory segments beginning in a given state; the results are then returned to the splicer. The number of segments to be scheduled for execution in any known state is determined through a predictor statistical model, built on-the-fly. Importantly, the quality of the predictor model only affects the efficiency of *ParSplice* and not the accuracy of the trajectory. This property is important because the predictor model will almost always be incomplete, as it is inferred from a finite number of simulations. The unavailability of the ground truth model (which is an extremely complex function of the underlying physical model) makes assessment of the results difficult. In addition, this type of stochastic simulation is not reproducible, adding to the difficulty of testing the code. Therefore, in this case study we create a basis for the *ParSplice* testing infrastructure using various methodical approaches and apply the test framework to the continuous integration process.

2.2 Metamorphic Testing

Metamorphic testing operates by checking whether the program under test behaves according to a set of metamorphic relations. For example, a metamorphic relation R would express a relationship among multiple inputs x1, x2,.., xN (for $N > 1$) to function f and their corresponding output values $f(x1)$, $f(x2)$,.., $f(xN)$ [2]. These relations specify how a change to an input affects the output. These metamorphic relations serve as a test oracle to determine whether a test case passes or fails. In the case of *ParSplice*, it is difficult to identify metamorphic relations because the outputs are non-deterministic. The relationship between the x's and the f's is therefore not direct but statistical in nature.

2.3 Run-Time Assertion Checking

An assertion is a boolean expression or constraint used to verify a necessary property of the program under test. Usually, testers embed assertions into the source code that evaluate when a test case is executed. Later testers use these assertions to verify whether the output is within an expected range or if there are some known relationships between program variables. In this way, a set of assertions can act as an oracle. In the context of *ParSplice*, assertions can be used to test specific functions, but not to test the overall validity of the simulations, would protect only against catastrophic failures, such as instabilities in the integration scheme.

2.4   Machine Learning Techniques

Machine learning is a useful approach for developing oracles for non-deterministic programs. Researchers have shown possibilities of both black-box features (developed using only inputs and outputs of the program) and white-box features (developed using the internal structure of the program) to train the classifier used as the oracle [1,4]. It is possible to test *ParSplice* with machine learning techniques. For example, we could fake the molecular dynamics (MD) engine with our own model to produce output data to use as a training set and consider the actual output data as a testing set. Due to the amount of effort required to use this approach in *Parsplice*, we determined that it was not feasible.

3   Case Study

To implement the testing framework, the first author spent a summer at Los Alamos National Laboratory working on the *ParSplice* project. The testing framework is based on the Multinomial testing approach (described in Sect. 3.2), implemented using a *progress tracking card* (PTC) in the Productivity and Sustainability Improvement Plan (PSIP)[2] methodology. The testing approach is integrated with the CMake/CTest tool for use in the runtime environment and continuous integration. In this section, we describe the PSIP methodology, the Multinomial test approach, and results that verify the implementation of the testing framework.

3.1   PSIP

The PSIP methodology provides a constructive approach to increase software quality. It helps decrease the cost, time, and effort required to develop and maintain software over its intended lifetime. The PSIP workflow is a lightweight, multi-step, iterative process that fits within a project's standard planning and development process. The steps of PSIP are: a) Document Project Practices, b) Set Goals, c) Construct Progress Tracking Card, d) Record Current PTC Values, e) Create Plan for Increasing PTC values, f) Execute Plan, g) Assess Progress, h) Repeat.

We created and followed a PTC containing a list of practices we were working to improve, with qualitative descriptions and values that helped set and track our progress. Our progress tracking card consists of 6 scores with a target finish date to develop the testing framework. The scores are:

---

[2] https://betterscientificsoftware.github.io/PSIP-Tools/PSIP-Overview.html.

– Score 0 - No tests or approach exists
– Score 1 - Requirement gathering and background research
– Score 2 - Develop statistical test framework
– Score 3 - Design code backend to integrate test
– Score 4 - Test framework implemented into ParSplice infrastructure
– Score 5 - Integrate into CI infrastructure

We were able to progress through these levels and obtain a score of 5 by the end of the case study.

3.2   Multinomial Test

The Multinomial test is a statistical test of the null hypothesis that the parameters of a multinomial distribution are given by specified values. In a multinomial population, the data is categorical and belongs to a collection of discrete non-overlapping classes. For instance, multinomial distributions model the probability of counts of each side for rolling a k-sided die n times. The Multinomial test uses Pearson's χ^2 test to test the null hypothesis that the observed counts are consistent with the given probabilities. The null hypothesis is rejected if the p-value of the following χ^2 test statistics is less than a given significance level. This approach enables us to test whether the observed frequency of segments starting in i and ending in j is indeed consistent with the probabilities p_{ij} given as input to the Monte Carlo backend. Our Multinomial test script uses the output file of *ParSplice* as its input and execute the test and post-processes the results by performing Pearson's χ^2 to assess whether to reject the null-hypothesis.

3.3   Results

A key insight from the theory that underpins *ParSplice* is that a random process that describes the splicing procedure should rigorously converge to a discrete time Markov chain in a discrete state space. In other words, the probability that a segment added to a trajectory currently ending in state i leaves the trajectory in state j should be a constant p_{ij} that is independent of the past history of the trajectory. One way to test *ParSplice* would be to verify that the splicing procedure is indeed Markovian (memory-less). However, taken alone, such a test would not guarantee that the splicing proceeds according to the proper Markov chain. A more powerful test would assess whether the spliced trajectory is consistent with the ground-truth Markov chain. A key obstacle to such a test is that this ground-truth model is, in practice, unknown and can only be statistically parameterized from simulation data.

To address this issue, we replaced the molecular dynamics (MD) simulation backend with a simpler Monte Carlo implementation that samples from a pre-specified, Markov chain. That is, we replaced the extremely complex model inherent to the MD backend with a known, given model of predefined probabilities. The task then becomes assessing whether the trajectory generated by *ParSplice*, as run in parallel on large numbers of cores, reproduces the statistics of the ground truth model. In this context, this technique is the ultimate test of correctness, as *ParSplice* is specifically designed to parallelize the generation of very long trajectories that are consistent with the underlying model. Statistical agreement between the trajectory and the model demonstrates that the scheduling procedure is functional (otherwise, the splicing the of trajectory would halt), the task ordering procedure is correct, the tasks executed properly, the results reduced correctly, and the splicing algorithm was correct.

The statistical assessment to test *ParSplice* can be conducted using the Multinomial test approach. Our null hypothesis was that the observed counts generated by *ParSplice* are consistent with the probabilities in the model. If the p-value from the multinomial test is less than 0.05, we reject the null hypothesis and conclude that the observed counts differ from the expected ones. Conversely, if the p-value is greater than 0.05, we do not reject the null hypothesis and can conclude that the test passes. For the sake of verifying our Multinomial test, we ran *ParSplice* in different time frames and observed the result. Figure 1 shows the p-values obtained from running *ParSplice* for 1, 2, 5, 10, 20, 40, 60, and 90 min. We can see that in all cases, the p-values are greater than 0.05, which indicates that the tests passed during these instances of the execution.

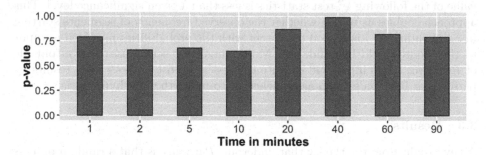

Fig. 1. p-values obtained by executing *ParSplice* for different times.

4   Conclusion

In this paper, we describe a case study of the *ParSplice* project in which we followed the PSIP methodology to develop a testing framework to address the difficulties of testing non-deterministic parallel research software. We first considered applying traditional industrial testing approaches. However, the non-determinism of *ParSplice* made these approaches unusable. Then we identified testing techniques specially designed for non-deterministic software. Once again, those techniques did not fit *ParSplice*. Finally, we identified a statistical testing approach, Multinomial Testing, that would work for *ParSplice*.

The Multinomial Testing approach is ideal for *ParSplice* given its constraints, i.e. time, non-determinism, and the existing continuous integration system. The lessons learned from this case study can be valuable to the larger research software community because, like *ParSplice*, many research software projects have stochastic behavior which produces non-deterministic results. The approach we followed to develop the test framework can be a model for other research software projects. We plan to extend the testing infrastructure in a more methodological way with as many possible testing techniques installed in the system.

Acknowledgement. This research was supported by the Exascale Computing Project (17-SC-20-SC), a collaborative effort of the U.S. Department of Energy Office of Science and the National Nuclear Security Administration (LA-UR-20-20082).

References

1. Chan, W., Cheung, S., Ho, J.C., Tse, T.: PAT: a pattern classification approach to automatic reference oracles for the testing of mesh simplification programs. J. Syst. Softw. **82**(3), 422–434 (2009)
2. Chen, T.Y., Tse, T.H., Zhou, Z.: Fault-based testing in the absence of an oracle. In: 25th Annual International Computer Software and Applications Conference. COMPSAC 2001, pp. 172–178, October 2001
3. Easterbrook, S.M., Johns, T.C.: Engineering the software for understanding climate change. Comput. Sci. Eng. **11**(6), 65–74 (2009)
4. Frounchi, K., Briand, L.C., Grady, L., Labiche, Y., Subramanyan, R.: Automating image segmentation verification and validation by learning test oracles. Inf. Softw. Technol. **53**(12), 1337–1348 (2011)
5. Heroux, M.A., Willenbring, J.M., Phenow, M.N.: Improving the development process for CSE software. In: 15th EUROMICRO International Conference on Parallel, Distributed and Network-Based Processing (PDP 2007), pp. 11–17, February 2007
6. Kanewala, U., Bieman, J.M.: Techniques for testing scientific programs without an oracle. In: Proceedings of the 5th International Workshop on Software Engineering for Computational Science and Engineering, SE-CSE 2013, pp. 48–57 (2013)
7. Kanewala, U., Bieman, J.M.: Testing scientific software: a systematic literature review. Inf. Softw. Technol. **56**(10), 1219–1232 (2014)
8. Miller, G.: A scientist's nightmare: software problem leads to five retractions. Science **314**(5807), 1856–1857 (2006). https://doi.org/10.1126/science.314.5807.1856
9. Perez, D., Cubuk, E., Waterland, A., Kaxiras, E., Voter, A.: Long-time dynamics through parallel trajectory splicing. J. Chem. Theory Comput. **12**, 18–28 (2015)
10. Segal, J.: Some problems of professional end user developers. In: IEEE Symposium on Visual Languages and Human-Centric Computing (VL/HCC), pp. 111–118 (2007)
11. Segal, J.: Software development cultures and cooperation problems: a field study of the early stages of development of software for a scientific community. Comput. Support. Coop. Work (CSCW) **18**(5), 581 (2009). https://doi.org/10.1007/s10606-009-9096-9

APE: A Command-Line Tool and API for Automated Workflow Composition

Vedran Kasalica[✉][iD] and Anna-Lena Lamprecht[✉][iD]

Department of Information and Computing Sciences, Utrecht University,
3584 CC Utrecht, The Netherlands
{v.kasalica,a.l.lamprecht}@uu.nl

Abstract. Automated workflow composition is bound to take the work with scientific workflows to the next level. On top of today's comprehensive eScience infrastructure, it enables the automated generation of possible workflows for a given specification. However, functionality for automated workflow composition tends to be integrated with one of the many available workflow management systems, and is thus difficult or impossible to apply in other environments. Therefore we have developed APE (the Automated Pipeline Explorer) as a command-line tool and API for automated composition of scientific workflows. APE is easily configured to a new application domain by providing it with a domain ontology and semantically annotated tools. It can then be used to synthesize purpose-specific workflows based on a specification of the available workflow inputs, desired outputs and possibly additional constraints. The workflows can further be transformed into executable implementations and/or exported into standard workflow formats. In this paper we describe APE v1.0 and discuss lessons learned from applications in bioinformatics and geosciences.

Keywords: Scientific workflows · Computational pipelines · Workflow management systems · Automated workflow composition · Workflow exploration

1 Introduction

Computational pipelines, or workflows, are central to contemporary computational science [5]. The international eScience community has created a comprehensive infrastructure of tools, services and platforms that support the work with scientific workflows. Numerous scientific workflow management systems exist [1,29], some of the currently most popular being Galaxy [10], KNIME [6] and Nextflow [7]. While these systems free their users from many technicalities that they would have to deal with when conventionally programming workflows, the identification of suitable computational components and their composition into executable workflows remains a manual task.

The idea of *automated workflow composition* is to let an algorithm perform this process. Based on a loose specification of the intended workflow (for example

V. V. Krzhizhanovskaya et al. (Eds.): ICCS 2020, LNCS 12143, pp. 464–476, 2020.
https://doi.org/10.1007/978-3-030-50436-6_34

in terms of available workflow inputs and desired outputs, or principal steps to take), it would automatically generate suitable, executable workflows. It has been shown that program synthesis [11] and AI planning techniques [8] can be used to implement such functionality [20,22,23]. Some workflow management systems, such as jORCA/Magallanes [15], jABC/PROPHETS [21,24] and WINGS [9], provide automated workflow composition functionality based on such techniques. However, the tight integration with the respective workflow systems makes it difficult or even impossible to use this functionality in other environments.

Therefore we have developed APE[1] (the Automated Pipeline Explorer) as a command-line tool and API for automated workflow composition. It is designed to be independent from any concrete workflow system, and thus ready to be used in other workflow management systems, tool repositories or workflow sharing platforms as needed. Internally, APE uses a SAT-based implementation of a temporal-logic process synthesis method, inspired by the approach behind the PROPHETS framework [21,27] and described in detail [17]. In a nutshell, the framework uses an extension of the well known Linear Temporal Logic (LTL) to encode the workflow specification. This specification is translated into a propositional logic formula that can be processed by an off-the-shelf SAT solver, with the resulting solutions representing possible workflows for the specification.

In this paper, we introduce APE v1.0 from an application point of view. Section 2 describes how to set it up for use by providing a semantic domain model. Section 3 focuses on the automated composition of workflows based on the domain model and custom workflow specifications. Section 4 describes how APE-composed workflows can further be transformed into executable implementations and/or exported into standard workflow formats. Section 5 discusses lessons learned from applications of APE in bioinformatics and geosciences. Section 6 concludes the paper.

2   Domain Model

The semantic domain model constitutes the knowledge base on which APE relies for the automated composition of workflows. It comprises a domain ontology and a collection of semantically annotated tools. The domain ontology provides taxonomic classifications of the data types and operations in the application domain, as a controlled vocabulary of technical terms. Tools in the domain model are semantically annotated with their inputs, outputs and operations, using terms from the ontology. Additionally, the domain model might include (temporal-logic) constraints to express further domain knowledge or rules.

For example, Fig. 1 and Table 1 show fragments of a bioinformatics domain model from a recent case study on automated workflow composition in proteomics [25]. The domain ontology (see Fig. 1) was directly derived from the popular bioinformatics data and methods ontology EDAM [12]. Table 1 shows a few tool annotations from the same case study. Each tool is semantically annotated with the operation(s) it performs and its input and output data types

---

[1] https://github.com/sanctuuary/ape.

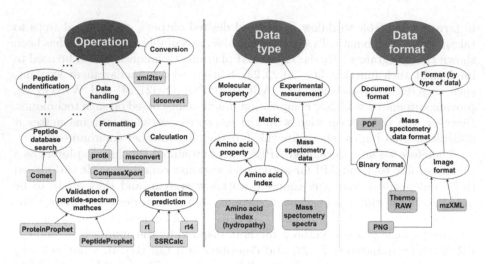

Fig. 1. Fragment of a bioinformatics domain ontology.

and formats, using terms from the respective taxonomies. These annotations were directly derived from the bio.tools registry [13,14], a large collection of EDAM-annotated bioinformatics tools. Note that in this example, two dimensions (type and format) are used for the annotation of the input and output data. Other applications need only one (e.g. format), and yet others have more than two required dimensions. Hence, APE supports the use of multiple disjoint taxonomy trees to represent the required dimensions of data characterization.

Technically, we rely on existing and (de facto) standard formalisms for the representation of the domain model. APE loads the domain ontology from a file in Web Ontology Language (OWL) format. The tool annotations are represented in JavaScript Object Notation (JSON) format, following the schema that is used in the bio.tools registry [2].

3   Automated Workflow Composition

Once the domain model has been configured, APE is ready to be used for automated workflow composition. Therefor the user specifies the workflow inputs, intended outputs and additional constraints that the workflow has to fulfill. Internally the constraints are expressed in a formal (temporal) logic, but the APE interfaces expose them in the form of intuitive natural-language templates. For example (as illustrated in Fig. 2), one workflow specification from the proteomics case study consists of "Mass spectrum" type in "Thermo RAW format" as input, "Amino acid index (hydropathy)" (in any format) as output, and constraints specifying to use tools that perform the operations "peptide identification", "validation of peptide spectrum matches" and "retention time prediction" (constraint template "Use operation X"). These operations are abstract terms

Table 1. Fragment of an annotated set of bioinformatics tools [14].

Name	Operation	Data input (type/format)	Data output (type/format)
Comet	Peptide database search	**Mass spectrum**	**Peptide identification**
		mzML or **mzXML**	**pepXML**
msconvert	Formatting Filtering	**Mass spectrum**	**Mass spectrum**
		MGF or **mzXML** or **mzML**	**MGF** or **mzXML** or **mzML**
Peptide Prophet	Peptide identification Statistical modelling	**Peptide identification**	**Peptide identification**
		pepXML or **mzIdentML**	**pepXML**
rt4	Retention time prediction	**Peptide property**	**Amino acid index (hydropathy)**
		TSV or **pepXML**	**TSV** or **XML**
xml2tsv	Conversion	**Peptide identification**	**Peptide identification**
		mzIdentML	**TSV**
SSRCalc	Retention time prediction	**Peptide property**	**Amino acid index (hydropathy)**
		Textual format or **TSV**	**Textual format**

...

from the ontology, known to scientists from the domain. This shows that formulating such constraints does not require knowledge of all available tools that fit the description. Based on the given specification APE synthesizes workflows that fulfill the specification by construction. Figure 2 shows two of many possible workflow solutions for the example specification.

Automated workflow composition with APE can be performed through its command line interface (CLI) or its application programming interface (API). While the CLI provides a simple means to interact and experiment with the system, the API provides more flexibility and control over the synthesis process. It can also be used to integrate APE's functionality into other systems.

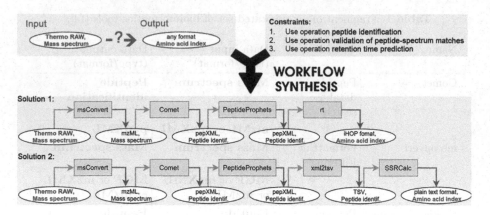

Fig. 2. Automated composition of a proteomics workflow.

3.1 Command Line Interface (CLI)

When running APE-<version>.jar from the command line, it requires a configuration file as a parameter and executes the complete automated workflow composition process accordingly. This JSON-based configuration file provides references to all therefor required information:

1. The domain model (as described in Sect. 2), provided as a pair of a well-formatted OWL and JSON files,
2. the workflow specification, provided as a list of workflow inputs/outputs and template-based workflow constraints, and
3. parameters for the synthesis execution, such as the number of desired solutions, output directory, system configurations, etc.

APE then writes the synthesized workflows into the defined output directory. Each solution consists of a text file that describes the steps of the workflow, a graphical representation, and a shell script that implements the workflow (depending on the availability of suitable shell commands in the tool annotations).

3.2 Application Programming Interface (API)

Like the CLI, the APE API relies on a configuration file that references the domain ontology, tool annotations, workflow specification and execution parameters. However, the API allows to edit this file programmatically, and thus for instance add constraints or change execution parameters dynamically. This is useful, for instance, for providing more interactive user interfaces or for systematically exploring and evaluating workflow synthesis results for varying specifications and execution parameters.

```
JSONObject apeConfig = Utils.generateGeneralConfiguration();
apeConfig.put("ontology_path", "./EDAM.owl");
apeConfig.put("tool_annotations_path", "./biotools.json");
APE apeFramework = new APE(apeConfig);
JSONObject runConfig = Utils.parseJson("./runConfig.json");
List<SolutionWorkflow> solutions = apeFramework.runSynthesis(runConfig);
apeFramework.writeSolutionToFile(solutions);
apeFramework.writeDataFlowGraphs(solutions);
```

Listing 1.1. APE API calls used to synthesize workflows and save solution.

Listing 1.1 shows a small example of using the APE API for synthesizing a set of workflows similar to the example in Fig. 2. First, the paths to the domain ontology and tool annotation files are added to the APE configuration object. Then a new instance of the APE framework is created based on the configuration, and the workflow synthesis algorithm is executed with the provided run configuration. The result of the synthesis run is a list of solutions obtained from the SAT solver, which are written into the output directory in textual and graphical (data-flow) format.

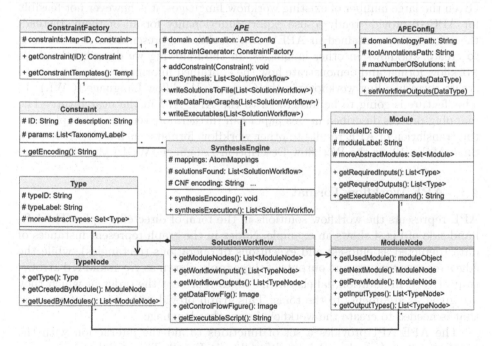

Fig. 3. Fragment of the APE API.

The APE API provides further functionality, allowing for a more fine-grained interaction with the APE framework. Figure 3 outlines the API, for brevity focusing on the most relevant fields and functions. The *ConstraintFactory* and *Constraint* classes allow for the retrieval of constraint templates and for adding new or removing existing constraints, thus further constraining or loosening the specification, respectively. As shown in the example code above, the *APE* class constitutes the main interface for interaction with the framework. It is used to define the execution parameters as well as the output formats. Once the library has generated the solutions, they are provided as a list of *SolutionWorkflows*. Each solution is represented as a directed graph that comprises type and tool nodes (internally named modules). The interface for working with the workflow solutions (further elaborated in the next section) is provided by the classes *SolutionWorkflow*, *TypeNode* (representing type instances) and *ModuleNode* (representing tool instances).

4   Workflow Implementation

As mentioned above, APE provides functionality for exporting the synthesized workflows as textual representations, in the form of (data-flow and control-flow) graphs and as executable shell scripts. In practice it is often desirable to implement workflows in one of the languages used by popular workflow management systems, in order to be able to execute them with the respective workflow engines. Given the large number of existing workflow languages, it is however not feasible for APE to provide ready-to-use export functionality for all of them. Instead, the information contained in APE's own workflow representation can be used to create workflows in other languages. In the following we describe the APE workflow format and demonstrate how the contained information can be used to create corresponding workflows in the Common Workflow Language (CWL) [4]. This feature is going to be integrated to the APE API in the near future. The mapping process described in this paper can furthermore serve as a template for the translation of APE results to other workflow formats, such as NextFlow [7], SnakeMake [19] or the Workflow Description Language (WDL) [3].

4.1   APE Workflow Format

APE represents the workflow solutions in the form of directed graphs. The left-hand side of Fig. 4 shows an example. Nodes in the graph represent instances of data (depicted as ellipses) and executions of operations (rectangles), while the edges represent inputs and outputs of these tools, shown as green and red arrows, respectively. In addition, labels on the edges represent the order in which they are given as arguments to the tools. This graph provides the trace information that is needed to create the workflow in another language.

The APE API provides a set of functions to aid the interaction with the graph structure (see class *SolutionWorkflow* in Fig. 3). The workflow inputs can simply be retrieved using the corresponding function of the *SolutionWorkflow* class, which returns it as a list of *TypeNodes*. Generally, each *TypeNode* comprises a (possibly empty) tool node that generated it as an output, a (possibly empty) list of tools that used it as an input, and a concrete data *Type* that identifies it. Further, the *SolutionWorkflow* class provides a function for retrieving the tools

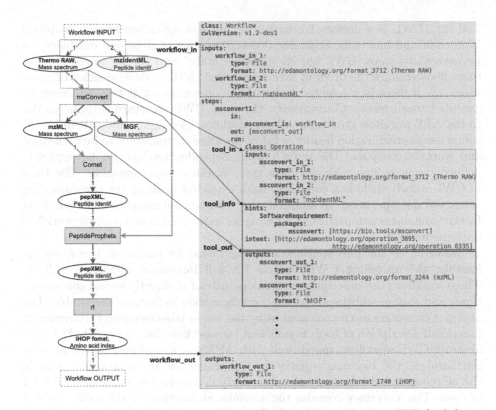

Fig. 4. Workflow in APE's native format (left) and corresponding CWL (right).

used in the workflow as list of *ModuleNodes* (sorted according to their order of execution), making it easy to iterate over all tools used in the workflow. Each *ModuleNode* provides information about the next and the previous *ModuleNode* in the sequence, the *TypeNodes* used as inputs and generated as outputs by the tool, as well as information about the actual tool (executable script, see class *Module*) that provides the information needed for its execution. Finally, the workflow outputs are provided in the same format as the initial inputs. Note that for this example the first proposed solution from Fig. 3 was artificially extended with additional inputs and outputs (depicted as gray ellipses) for illustrative purposes.

4.2 Translation to CWL

The Common Workflow Language[2] (CWL) [4] has recently emerged as an open standard for describing scientific workflows across platforms. It is increasingly adopted by the scientific community, with CWL support being added to popular scientific workflow management systems like, for example, Galaxy [10] and

---

[2] https://www.commonwl.org/.

Toil [28]. CWL is a declarative language that focuses on workflows composed from command line tools. Basically, it describes a set of steps and dependencies between those steps. CWL has its roots in "make" and similar tools, and like them it determines the order of execution based on these dependencies between tasks, i.e. if there is a required order of the operations or if they can even be executed concurrently. Conveniently, the main CWL structure is quite similar to the APE workflow structure. A basic workflow (see right-hand side of Fig. 4) comprises a configuration header, a list of workflow inputs, steps to be performed and workflow outputs. The input/output dependencies have to be explicitly defined, again in line with our data trace workflow representation. The tools in CWL usually include a command field, explicitly defining the corresponding command line operation. In addition, they can be configured to run tools from Docker containers automatically, allowing for more flexible and scalable workflow implementations.

However, as the fully automatic configuration for execution is not always feasible, the upcoming CWL version 1.2 will introduce *abstract workflows*. These workflows use descriptive containers instead of directly executable operations, and require additional (manual) configuration to become executable. The abstract containers are represented using the *intent* label (see Fig. 4). Given that functional description of tools is sufficient for workflow discovery with APE, the abstract CWL workflows match well with APE's own workflow representation. Furthermore, the bio.tools registry used as source for the tool annotations in the aforementioned bioinformatics case study is a typical example of such a set of tools. The repository contains the semantic annotations of the tools, but still might require some additional work from the user in order to execute the tool itself. Hence APE discovers workflows composed of tools that are not necessarily available on the local system, potentially requiring the installation and configuration of the tools on the execution system first.

To translate and APE workflow into CWL format, it is sufficient to 1) describe the original inputs, 2) iterate through the tools in the workflow sequence and specify the inputs used and outputs generated, and finally 3) specify the workflow output list. The right-hand side of Fig. 4 shows the CWL representation of the APE workflow on the left. To create it, first, the list of input objects is translated into a list of inputs that are annotated using their formats (see Label **workflow_in**). This means that some information about the data get lost in the translation (specifically the type description). However, as at runtime the format is sufficient to perform the execution, this is not a problem. Second, each tool in the sequence is described. The description involves a definition of the inputs, outputs and tool execution specification (mappings are annotated using labels **tool_in**, **tool_out** and **tool_info**, respectively). The most important part of the step is to keep track of the exact source of the tool inputs as well as to provide sufficient tool description that would allow for its execution. The input information is already part of the formalism, as APE keeps track of data flow traces for each data instance. The only requirement is to properly use the identifiers provided when creating the mappings to CWL. Regarding the tool descriptions, as long as the provided tool annotation file contains sufficient information, it can be translated into CWL. Third, the final workflow outputs need to be specified based on the given solution description (see Label **workflow_out**).

5   Applications and Lessons Learned

The development of APE was accompanied by three concrete application scenarios for automated workflow composition: 1) The proteomics case study mentioned earlier in this paper [25], 2) a case study on cartographic map generation [16], and 3) geospatial data transformations in the QuAnGIS project [18,26]. The experiences from these applications, in particular the feedback from the involved domain experts, influenced the design decisions that we took during the development of the APE CLI and API. While initial versions of all three application scenarios have been created with PROPHETS, they have meanwhile been migrated to APE completely and are publicly available[3].

Naturally, the quality of the workflows obtained through APE essentially depends on the quality of the semantic domain model (ontologies and functional tool annotations). Hence it is crucial to involve domain experts in the domain modeling process, or to rely on sources that have been created by expert communities, such as the EDAM ontology and bio.tools registry that we use in bioinformatics applications of APE. Essentially, the idea is that the domain model is provided and maintained by a small group of domain experts, and used by a larger and broader audience to automatically compose workflows. As a positive side effect on domain modeling, using APE for the systematic generation and evaluation of workflows from varying specifications proved to be helpful to revise and improve ontologies and annotations.

Initially we used a tabular format for the tool annotations, like the one shown in Table 1, because spreadsheets are easy to discuss with collaborators, and the corresponding CSV files easy to process programmatically. However, this approach quickly turned out to be insufficient to adequately capture non-trivial tool annotations. In the proteomics case study, we annotated tools' inputs and outputs with both data type and format terms from EDAM. As the tools have varying numbers of inputs and outputs, however, they could not be properly annotated in the tabular format with a fixed number of columns. To increase the expressiveness of APE's tool annotation template, but at the same time reuse an existing formalism, we decided to adopt the JSON-based tool annotation schema used in the bio.tools registry [2], which includes a well-defined and flexible mechanism for functional tool annotation. This has of course extremely simplified the setup of bioinformatics domain models based on bio.tools, but it has also shown to be easy to use in the other application domains.

The APE CLI and API aim to be easy-to-use, but clearly target a tech-savvy audience with a certain level coding and/or scripting confidence. To reach a broader audience, an intuitive interface that can be used without technical experience or specific training is required. As a proof of principle, we recently

---

[3] https://github.com/sanctuuary/APE_UseCases.

developed Burke (a Bio-tools and edam User interface foR automated worK-flow Exploration[4]). Preconfigured to the domain model of the proteomics case study, it provides the automated workflow composition functionality of APE through a browser-based graphical interface. Users can select input and output data types and formats, as well as constraint templates and their instantiations, from drop-down menus that are filled with the relevant EDAM terms. They can configure and run APE's synthesizer from the interface, and subsequently inspect the results, which are presented in a convenient tabular format. Feedback on Burke by APE novices has been very positive, hence we plan to develop a more sophisticated web interface for APE in the scope of future work on the framework.

A graphical interface has also the potential to overcome another limitation of the framework: Currently it is a tedious process to compare the different possible workflows generated by APE. This is however needed to make an informed decision about which of the potentially many possible workflows to select for implementation and execution. A graphical interface provides more possibilities for dynamically filtering, aggregating and displaying workflow candidates according to different criteria. Which criteria would actually provide meaningful information for workflow selection is currently an open question. This is another challenge that we are going to work on in the future.

6   Conclusion

We believe that automated workflow composition will take the work with scientific workflows to the next level. On top of today's comprehensive eScience infrastructure, it enables the automated generation of possible workflows for a given specification. In this paper we introduced APE v1.0 (the Automatic Pipeline Explorer), a command line tool and API that automates the exploration of scientific workflows. APE is under active development and continuously improving through the experiences and feedback from applications.

Future work on the APE framework will address different remaining challenges of usability and scalability. We are going to work on more end user-oriented interfaces that support better the whole life cycle of specifying, synthesizing, comparing, selecting, implementing and benchmarking computational pipelines. With growing domain models, the runtime performance of the underlying synthesis algorithm is likely to become a bottleneck. We have started to work on domain-specific search heuristics to improve synthesis performance and allow the approach to scale.

References

1. Existing Workflow systems. https://s.apache.org/existing-workflow-systems
2. bio-tools/biotoolsSchema, December 2019. https://github.com/bio-tools/biotoolsSchema, original-date: 2015-05-05T15:52:46Z

---

[4] https://github.com/sanctuuary/Burke_Docker.

3. Workflow Description Language (WDL), April 2020. https://github.com/openwdl/wdl, original-date: 2012-08-01T03:12:48Z
4. Amstutz, P., Crusoe, M.R., Tijanić, N., et al.: Common Workflow Language, v1.0, July 2016
5. Atkinson, M., Gesing, S., Montagnat, J., Taylor, I.: Scientific workflows: past, present and future. Future Gener. Comput. Syst. **75**, 216–227 (2017)
6. Berthold, M.R., et al.: Knime-the konstanz information miner: version 2.0 and beyond. AcM SIGKDD Explor. Newslett. **11**(1), 26–31 (2009)
7. Di Tommaso, P., Chatzou, M., Floden, E.W., et al.: Nextflow enables reproducible computational workflows. Nat. Biotechnol. **35**, 316–319 (2017)
8. Ghallab, M., Nau, D., Traverso, P.: Automated Planning and Acting, 1st edn. Cambridge University Press, New York (2016)
9. Gil, Y., Ratnakar, V., Kim, J., et al.: Wings: intelligent workflow-based design of computational experiments. IEEE Intell. Syst. **26**(1), 62–72 (2011)
10. Goecks, J., Nekrutenko, A., Taylor, J., et al.: Galaxy: a comprehensive approach for supporting accessible, reproducible, and transparent computational research in the life sciences. Genome Biol. **11**(8), R86 (2010)
11. Gulwani, S., Polozov, O., Singh, R.: Program Synthesis, Foundations and Trends in Programming Languages, vol. 4. now, Hanover (2017)
12. Ison, J., Kalaš, M., Jonassen, I., et al.: EDAM: an ontology of bioinformatics operations, types of data and identifiers, topics and formats. Bioinformatics **29**, 1325–1332 (2013). https://doi.org/10.1093/bioinformatics/btt113
13. Ison, J., et al.: Community curation of bioinformatics software and data resources. Brief. Bioinform. bbz075, October 2019. https://doi.org/10.1093/bib/bbz075
14. Ison, J., Rapacki, K., Ménager, H., et al.: Tools and data services registry: a community effort to document bioinformatics resources. Nucleic Acids Res. **44**(D1), D38–47 (2016)
15. Karlsson, J., Martín-Requena, V., Ríos, J., Trelles, O.: Workflow composition and enactment using jORCA. In: Margaria, T., Steffen, B. (eds.) ISoLA 2010. LNCS, vol. 6415, pp. 328–339. Springer, Heidelberg (2010). https://doi.org/10.1007/978-3-642-16558-0_28
16. Kasalica, V., Lamprecht, A.-L.: Workflow discovery through semantic constraints: a geovisualization case study. In: Misra, S., et al. (eds.) ICCSA 2019. LNCS, vol. 11621, pp. 473–488. Springer, Cham (2019). https://doi.org/10.1007/978-3-030-24302-9_34
17. Kasalica, V., Lamprecht, A.L.: Workflow Discovery with Semantic Constraints: A SAT-Based Implementation (2020). https://doi.org/10.14279/tuj.eceasst.78.1092
18. Kruiger, H., Kasalica, V., Meerlo, R., Lamprecht, A.L., Scheider, S.: Loose programming of GIS workflows with geo-analytical concepts. Transactions in GIS (2020, under review)
19. Köster, J., Rahmann, S.: Snakemake—a scalable bioinformatics workflow engine. Bioinformatics **28**(19), 2520–2522 (2012)
20. Lamprecht, A.-L. (ed.): User-Level Workflow Design - A Bioinformatics Perspective. LNCS, vol. 8311. Springer, Heidelberg (2013). https://doi.org/10.1007/978-3-642-45389-2
21. Lamprecht, A.L., Naujokat, S., Margaria, T., Steffen, B.: Synthesis-based loose programming. In: QUATIC 2010, Porto, Portugal, pp. 262–267. IEEE, September 2010
22. Lamprecht, A.L., Naujokat, S., Margaria, T., Steffen, B.: Semantics-based composition of EMBOSS services. J. Biomed. Seman. **2**(Suppl 1), S5 (2011)

23. Lamprecht, A.L., Naujokat, S., Steffen, B., Margaria, T.: Constraint-guided work-flow composition based on the EDAM ontology. In: Burger, A., Marshall, M.S., Romano, P., Paschke, A., Splendiani, A. (eds.) Proceedings of the 3rd International Workshop on Semantic Web Applications and Tools for Life Sciences (SWAT4LS 2010), vol. 698. CEUR Workshop Proceedings, December 2010
24. Naujokat, S., Lamprecht, A.-L., Steffen, B.: Loose programming with PROPHETS. In: de Lara, J., Zisman, A. (eds.) FASE 2012. LNCS, vol. 7212, pp. 94–98. Springer, Heidelberg (2012). https://doi.org/10.1007/978-3-642-28872-2_7
25. Palmblad, M., Lamprecht, A.L., Ison, J., Schwämmle, V.: Automated workflow composition in mass spectrometry-based proteomics. Bioinformatics **35**, 656–664 (2018). https://doi.org/10.1093/bioinformatics/bty646
26. Scheider, S., Meerlo, R., Kasalica, V., Lamprecht, A.L.: Ontology of core concept data types for answering geo-analytical questions. JOSIS (2020, in press). https://www.josis.org/index.php/josis/article/view/555
27. Steffen, B., Margaria, T., Freitag, B.: Module configuration by minimal model construction. Fakultät für Mathematik und Informatik, Universität Passau, Technical report (1993)
28. Vivian, J., et al.: Toil enables reproducible, open source, big biomedical data analyses. Nat. Biotechnol. **35**(4), 314–316 (2017). https://doi.org/10.1038/nbt.3772. http://www.nature.com/articles/nbt.3772
29. Wikipedia contributors: scientific workflow system – Wikipedia, the free encyclopedia (2019). https://en.wikipedia.org/w/index.php?title=Scientific_workflow_system&oldid=928001704. Accessed 3 Feb 2020

Solving Problems with Uncertainties

An Ontological Approach to Knowledge Building by Data Integration

Salvatore Flavio Pileggi[1]([✉]), Hayden Crain[1], and Sadok Ben Yahia[2]

[1] School of Information, Systems and Modelling, University of Technology Sydney, Ultimo, Australia
SalvatoreFlavio.Pileggi@uts.edu.au, Hayden.J.Crain@student.uts.edu.au
[2] Department of Software Science, Tallinn University of Technology, Tallinn, Estonia
sadok.ben@taltech.ee

Abstract. This paper discusses the uncertainty in the automation of knowledge building from heterogeneous raw datasets. Ontologies play a critical role in such a process by providing a well consolidated support to link and semantically integrate datasets via interoperability, as well as semantic enrichment and annotations. By adopting Semantic Web technology, the resulting ecosystem is fully machine consumable. However, while the manual alignment of concepts from different vocabularies is reasonable at a small scale, fully automatic mechanisms are required once the target system scales up, leading to a significant uncertainty.

Keywords: Ontology · Data integration · Semantic interoperability · Semantic Web · Uncertainty · Data engineering · Knowledge engineering

1 Introduction

Data integration, defined as *"the problem of combining data residing at different sources, and providing the user with a unified view of these data"* [21], can be considered a classic research field as could witness the myriad of contributions in literature. Its relevance is determined by the practical implications in the different applications domains.

In this respect, we rely on an ontological approach to support the data integration process. The benefits of ontology in the different application domains are well-known and have been extensively discussed from different perspectives in several contributions. The knowledge building process, as understood in this paper, is not limited to data integration but it also includes semantic enrichment and annotations. By adopting Semantic Web technology, the resulting ecosystem is fully machine consumable. However, while the manual alignment of concepts from different vocabularies is reasonable at a small scale, fully automatic mechanisms are required once the target system scales up, leading to a significant uncertainty.

V. V. Krzhizhanovskaya et al. (Eds.): ICCS 2020, LNCS 12143, pp. 479–493, 2020.
https://doi.org/10.1007/978-3-030-50436-6_35

This paper provides two key contributions:

- the manual knowledge building process is described and implemented by a tool which systematically supports data integration and semantic enrichment.
- the uncertainty introduced by the automation of the process is discussed.

2   Related Work

The scrutiny of data integration adopting an ontological approach sheds light on some key issues, to wit data semantics and uncertainty representation.

- **Ontological approach to Data Integration**: The role of semantic technology in data integration [21] has been deeply explored during the past years. The contributions currently in literature clearly show that semantic technology provides a solid support in terms of data integration and reuse via interoperability [26]. For instance, [6] proposes an ontological approach to federated databases; ontology-based integration strategies have been proposed to a range of real scientific and business issues [15], such as the integration of biomedical [33] and cancer [39] data, and the integration among systems [25]. Last but not least, ontologies are contributing significantly to an effective approach to the integration of Web Resources (normally in XML [2]) and to linked open data [16]. Ontology may be adopted to support different strategies and techniques [38] and result very effective in presence of heterogeneity [12]. For instance, central data integration assumes a global schema to provide access to information [15], while in peer-to-peer data integration there is no global point of control [15].
- **Data Semantics**: Associating formal semantics to data is a well-known problem in the fields of artificial intelligence and database management. Again, ontological structures play a key role [31] and they normally support an effective formalization of the semantics, which becomes a key asset in the context of different applications, for instance to interchange information [1,27] or to improve data quality [22]. In general, the importance of data semantics to support interoperability is gaining more and more attention within different communities, for example within the geo-spatial information [19] and within the medical community [4,20]. Moreover, the analysis of semantic data may support sophisticated data mining techniques [8,10].
- **Uncertainty Representation**: Probability theory and fuzzy logic have been used to represent uncertainty in data integration works [23]. Uncertainty management works also include possibilistic and probabilistic approaches [14]. A probabilistic approach towards ontology matching was utilized in several works, where machine learning was utilized in estimating correspondence similarity measures [14]. To refine the matcher uncertainty and improve the precision of its alignment, Gal [13] proposed a method to compute top-K alignments instead of computing a *best* single alignment, and proposed a heuristic to simultaneously compare/analyze/examine the generated top-K alignments and choose one good alignment among them. The *best* alignment is an alignment that optimizes a target function F between the two schemata.

Typical ontology matching methods commit to the *best* alignment which maximizes the sum (or average) of similarity degrees of pairwise correspondences. To model the ontology matching uncertainty, Marie and Gal [24] proposed to use *similarity matrices* as a measure of certainty. They aim at providing an answer to the question of whether there are *good* and *bad* matchers.

To represent the inherent uncertainty of the automatic schema matching, Magnani and Montesi [23] used the notion of *probabilistic uncertain semantic relationship* (pUSR), which is a pairwise correspondence defined as a tuple (E_1, E_2, R, P) where E_1 and E_2 are two elements/entities, R is a set of relationship types (equivalence, subsumption, disjointness, overlap, instantiation, *etc.*), and P is a probability distribution over R. The pUSRs form an uncertain alignment.

Dong *et al.* [9] proposed a system that models the uncertainty about the correctness of alignments by assigning each possible alignment a probability. The probabilities associated with all alignments sum up to 1. The authors define a *probabilistic schema mapping* (alignment) as a set of correspondences between a source schema and a target schema, where each uncertain mapping/alignment has an associated probability that reflects the likelihood that it is correct.

Po and Sorrentino [30] quantify uncertainties as probabilities. They define the notion of *probabilistic relationship* as a couple $(\langle t_i, t_j, R \rangle, P)$ where $\langle t_i, t_j, R \rangle$ is a relationship between t_i and t_j of the type R, and P is the probability (confidence) value (in the normalized interval [0–1]) associated to this relationship. Within the range [0–1], they can distinguish between *strong* relationships and *uncertain* relationships (*i.e.*, relationships with a low probability value). Uncertain relationships could be seen as candidate relationships that need further confirmation by a human expert.

There are several pairs of entities in different ontologies that are related to each other but not necessarily with one of the typical well-defined relationships. However, these correspondences vary in their degree of relatedness. This information is difficult to formalize. Therefore, Zhang *et al.* [40] proposed a new type of relation called *Relevance*. The latter represents relationships between entities that are not covered by a strict relation such as equivalence, subsumption or disjointness, *etc.* In this context, we think that the *relevance* relation is very similar to the *overlap* relation. The authors also presented the notion of *fuzzy ontology alignment*, that uses fuzzy set theory to model the inherent uncertainty in the alignment correspondences.

3   Knowledge Building by Data Integration

The knowledge building process is ideally composed of two sequential steps that we refer to as *physical* and *logical integration*:

- **Physical Integration: The Virtual Table Model.** As the name suggests, the physical integration aims to convert data in an interoperable format that ultimately defines the target data space. By adopting Semantic Web technology, physical integration is required only if the target dataset is not already available in a semantic format (e.g. RDF or OWL). The Virtual Table model (Fig. 1) is a simple and intuitive approach to data integration that assumes the target dataset described as one or more tables according to the classical relational model. An external dataset may be mapped into a virtual table and automatically converted in OWL. Data may be automatically retrieved from a relational table [29] or inserted manually by users through the copy&paste functions provided by the user interface as in the tool described later on in the paper.

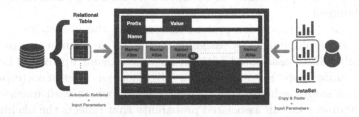

Fig. 1. Virtual Table model.

– **Logical Integration: Semantic Alignment, Internal and External Linking.** Logical integration assumes a given data set already imported within the data space and consists in the consolidation and enrichment of data semantics by specifying additional relationships, such as semantic equivalences, internal and external links. Once a data set has been imported within the semantic data space, it may need to be logically linked to other data and semantically enriched. We structure our knowledge building process by including three different kind of semantic enrichment (Fig. 2): *internal linking*, *metadata association* and *external linking*.

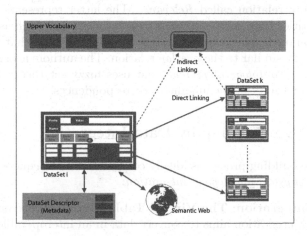

Fig. 2. Semantic linking and enrichment.

Internal linking is an ontology alignment process among the different datasets which are considered part of the data space. That is a key process to enable the effective integration at an user level of heterogeneous datasets. For instance, two attributes belonging to two different datasets may have the same meaning. Semantic technology provides simple and effective mechanisms to establish semantic equivalences among classes, instances, relationships and attributes. As discussed in the following section, these mechanisms may be used in a relatively easy way, if they properly supported by user-friendly interfaces. Semantic correspondences among ontology elements may be established directly or indirectly (Fig. 2). Direct linking, namely semantic equivalences established directly from a dataset to another, is simple from a management perspective but may result not too much effective in complex environments, i.e. within collaborative systems, or, more in general, when the scale of the system in terms of number of linked datasets becomes relevant. On the other side, indirect linking established through upper vocabularies is well-known and consolidated techniques that may result in a much more effective approach. However, it introduces an additional cost from a management perspective. The semantic infrastructure allows generic linking within the semantic space or externally. So a dataset or an element belonging to a dataset may be related with other concepts to define or extend the semantics associated. For example, a given dataset may be related to a number of keywords, to a research project or to a scientific paper by adopting the PERSWADE-CORE vocabulary [28].

A simple example of data integration involving two datasets is represented in Fig. 3. As shown, both target datasets address information related to cities. Figure 3a represents dataset in their original format, while Fig. 3b depicts the integrated space as a knowledge graph. The column city is considered like a *Web Resource* that in this case is also the primary key for both tables. Although the two datasets present some redundancies, they provide, in general, different information about cities. In this case, the integration process will enable the two original datasets within the semantic data space assuring semantic consistency among the different fields and concepts. Indeed, from a semantic perspective, even this simple use case proposes a number of potential issues that have to be addressed in order to guarantee a correct and effective integration. As shown in the figure, there are several semantic equivalences among the two datasets to be represented. They include attributes (columns in the virtual table model) that have the same name and the same meaning within their original context, as well as attributes that have different names but the same meaning. For instance, the attributes "Population" and "Residents" refer to the same concept, namely the number of people currently living in a given city. Additionally, equivalent resources have to be semantically related. In the example, "Rome" appears in both tables. This syntactic equivalence is integrated by a semantic one to properly address the reference to the city of Rome. OWL provides simple mechanisms to define equivalences among concepts.

Overall, a simplification of the scenario previously discussed can be represented by the knowledge graph in Fig. 3b that adopts OWL 2 structures. More concretely, the equivalence among classes is enforced by the OWL rule *OWL:equivalentClass*, as well as the equivalences among properties is specified by *OWL:equivalentProperty*. Similarly, an OWL statement including *OWL:sameAs* applies to instances of classes.

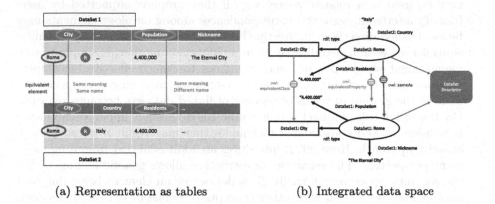

(a) Representation as tables   (b) Integrated data space

Fig. 3. An example of integration of two datasets.

3.1   A Tool for Supervised Data Integration

Our implementation supports most part of the knowledge building process as previously presented and discussed. It is based and relies on intuitive user inputs rather than on strong skills in ontology and Semantic Web technology. However, it assumes the understanding of basic concepts, i.e. the difference between an object and an attribute. The primary goal of the tool is to support the systematic conversion of a given dataset into an independent and self-contained ontology in OWL. The user interface (Fig. 4 allows to directly import a relational table. Regardless of the method used to import data (based on copy&paste in this case), the user is asked to characterize the table each column according to one of the following options:

- *ID*. It is normally equivalent to the primary key in the relational model. However, it is assumed to be an unique data field. Therefore, keys composed by multiple fields cannot be directly used and need to be encoded previously.
- *Resource*. By using this option, associated data is considered like an object, namely a Web resource in Semantic Web technology. A Web resource has an unique identifier and can be further characterized.
- *Attribute*. It's a normal data field, e.g. a text or a number.

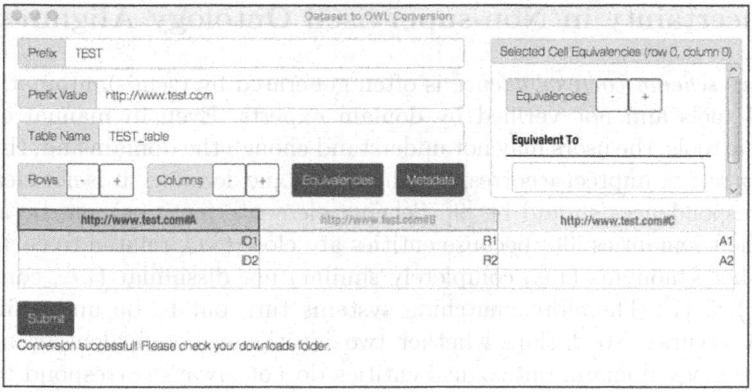

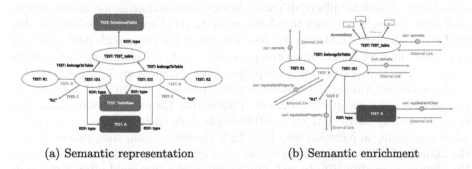

Fig. 4. User interface.

Through the provided interface, users may specifies meta data for the imported table, such as *source. license, description* and *publisher*. Last but not least, relatively friendly alienation among concepts is supported.

3.2 Semantic Representation

The output of the example proposed in Fig. 4 is represented as a knowledge graph in Fig. 5a. As shown, the IDs (as previously defined) is associated with a new class (*A* in this case) and the instances of ID (*ID1* and *ID2*) are also stated as member of the internal class *TableRaw*. This last concept identifies rows in the virtual table *TEST_table*, which is stated as a member of the class *RelationalTable*. Resources (*B* in the example) are converted in OWL Object properties, while attributes (*C* in the example) are converted in OWL data properties. The resulting schema may be semantically enriched trough concept alignment and external linking (Fig. 5b).

(a) Semantic representation   (b) Semantic enrichment

Fig. 5. Semantic representation of the integrated dataspace.

4   Uncertainty in Non-supervised Ontology Alignment

Uncertain schema correspondence is often generated by (semi-)automatic unsupervised tools and not verified by domain experts. Even in manual or semiautomatic tools, the users may not understand enough the domain and, thus, provide incorrect or imprecise correspondences. In some domains, it is not clear what the correspondences should be [9]. Schema elements (entities) can be ambiguous in their semantics [30] because entities are close (*i.e.*, related to each other) but neither synonyms (*i.e.*, completely similar) nor dissimilar (*i.e.*, completely different) [5,17]. Therefore, matching systems turn out to be uncertain, since it is not accurate to declare whether two entities are equivalent or not [40]. In the ontology domain, ontological entities do not always correspond to single physical entities, they rather share a certain amount of mutual information [40]. Indeed, real-world ontologies generally have linguistic, structural and semantic ambiguities, resulting from their heterogeneous domain conceptualizations [3]. Eventually, ambiguity and heterogeneity in ontology models/representations are carried in the process of matching and integrating ontologies [3]. Finally, *Uncertain query* is commonly associated with multiple structured queries generated by the system as candidate queries reflecting uncertainty about which is the real intent of the user.

Klir and Yuan [18] defined two basic types of uncertainty: (*i*) *Fuzziness* which is the lack of definite or sharp distinctions; and (*ii*) *Ambiguity* which is the existence of one-to-many correspondences that may introduce a disagreement in choosing among several correspondences.

There are two choices to remove (or at least reduce) the alignment uncertainty in schema matching processes: either with the support of a user (manually) or by using a threshold. According to the former approach, *aka user feedback*, users can manually select matching and non-matching correspondences from the alignment, *i.e.* in semi-automatic matching process when the system requests help [7]. The latter approach is based on a *threshold* that can be established in a semiautomatic manner (*i.e.* using user feedback cycles) or in an automatic manner (i.e. using learning approaches) in order to minimize the introduction of false correspondences. A matcher filters/discards correspondences having a confidence value that does not reach a certain threshold, assuming that correspondences with low confidence/similarity measures are less adequate than those with high similarity measures. However, separating correct from incorrect correspondences in an alignment is a hard task [14]. To find the optimal/best threshold, many trials should be made by varying/tuning the confidence value threshold [30]. In addition, different thresholds can be assigned to different applications. For example, a recommendation system may have relatively low thresholds since false positives are tolerated, while a scientific application may have high thresholds [40]. As a rule of thumb, the information loss, caused by the removal of uncertainty, leads to a worsening of the alignment quality [30]. In fact, any selection of a threshold often yields false negatives and/or false positives. Therefore, the exact alignment cannot be found by setting a threshold [13].

Generally speaking, the uncertainty generated during the matching process is lost or transformed into exact one (defuzzification) [23]. Therefore, there is a concrete need to incorporate uncertain/inaccurate correspondences and handle uncertainty in alignments [14], due to the inherent risk of losing relevant information [23].

Uncertainty Management. From a literature review, we have identified two different levels to deal with uncertainty management in schema matching: some solutions try to quantify the uncertainty of an entire alignment when there are many alignments produced for the same matching case; others try to represent and quantify the uncertainty of the correspondences of a given alignment.

Management of Ontology Alignment Uncertainty. A semantic alignment (*aka mapping*), denoted as $\mathcal{A} = \{c_1, c_2, \ldots, c_n\}$, is a set of semantic correspondences between two or more matched ontologies. It is the result/output of the ontology matching process.

The uncertainty of a matcher should be explicitly reflected in an uncertainty measurement in order to be able to choose good enough alignments [32]. The work in [9] introduced the notion of *probabilistic schema alignments*, namely a set of alignments with a probability attached to each alignment. The purpose of defining *probabilistic* alignments is to answer queries with uncertainty about (semi-)automatically created alignments [32].

Management of Correspondence Uncertainty. In general, given two matched ontologies $\mathcal{O}_1$ and $\mathcal{O}_2$, a semantic correspondence (*aka a relation* or a *relationship*) is a 4-tuple $< e_{\mathcal{O}_1}, e_{\mathcal{O}_2}, r, n >$ where $e_{\mathcal{O}_1}$ is an entity belonging to $\mathcal{O}_1$, and $e_{\mathcal{O}_2}$ is an entity belonging to $\mathcal{O}_2$, r is a semantic relation holding (or intended to hold) between $e_{\mathcal{O}_1}$ and $e_{\mathcal{O}_2}$, such as equivalence ($\equiv$), subsumption ($\sqsubseteq/\sqsupseteq$), disjointness ($\perp$), or overlap ($\between$), and n is a confidence value/measure/probability assigning a degree of trust/reliability/correctness on the identified relation and ranging typically between $[0, 1]$, where 0 represents no similarity and 1 represents full similarity. In the equivalence case, n indicates whether both entities have a high or low similarity measure/degree. The higher the confidence degree, the more likely the relation holds [11]. A matcher would be inclined to put a similarity value of 0 for each entity pair it conceives not to match, and a value higher than 0 (and probably closer to 1) for those correspondences that are conceived to be correct [13]. On the other hand, in the *crisp* correspondences (composing the *crisp* alignments), the confidence values of all correspondences are equal to 1.

Correspondences Generated by a Matcher Aggregation. Some matchers assume that similar entities are more likely to have similar names. Other matchers assume similar entities share similar domains. Other matchers assume that similar entities are more likely to have similar neighbors (parents, children, and siblings). And others assume that similar entities are more likely to have similar instances [24]. In order to combine principles by which different schema matchers judge the similarity between entities, the combined matcher aggregates the outcome (*i.e.*, the output alignment) of all matchers to produce a single alignment. It automatically computes the overall similarities of correspondences by aggregating the similarity degrees assigned by individual matchers. In the automatic process of matching, it is proven that an ensemble (a combination) of complementary matchers (*e.g.*, string-based, linguistic-based, instance-based, and structural matchers, *etc.*) outperforms the behavior of individual matchers [30] since they compensate for the weaknesses of each other [24]. In recent years, many matching tools use schema matcher ensembles to produce better results. Therefore, the similarity measure of a correspondence is generally the result of aggregating multiple similarity measures [7], and as the number of such similarity measures increase, it becomes increasingly complex to aggregate the results of the individual measures. The generated similarity degrees of correspondences are dependent to the choice of the weights of individual matchers assigned by aggregation algorithms for the similarity combination [40]. Therefore, a similarity degree of a given correspondence represents the "belief" of a matcher in the correctness of that correspondence [13]. However, the real issue in any system that manages uncertainty is whether we have reliable probabilities (degrees of similarity), because unreliable probabilities can lead us to choose erroneous or not good enough correspondences. Obtaining reliable probabilities for uncertainty management systems is one of the most interesting areas for future research [9]. Finally, disregarding semantic similarity degrees of the alignment correspondences may impede the overall integration process [5].

Correspondence Ambiguity. An ambiguous alignment [11] is a one-to-many $(1:n)$, a many-to-one $(n:1)$, or a many-to-many $(n:n)$ alignment. This means that it contains some ambiguous correspondences [11] (*i.e.*, that match the same entity from one ontology with more than one entity from the other ontology). An ambiguous correspondence is a correspondence in which at least one entity is also involved in other correspondences. Contrary to one-to-one $(1:1)$ alignments in which an entity appears in at most one correspondence.

The ambiguous correspondences are generally a source of uncertainty because they can be interpreted in two ways: (i) A first point of view considers that only a single ambiguous *equivalence* correspondence (probably the one that has the highest confidence value) truly reflects a synonym/alternative entity, while the remaining ones (having lower confidence values) rather reflect similar, related or overlapping terms, not strictly denoting equivalent entities [37]; (ii) A second point of view considers the ambiguous *equivalence* correspondences as actually *subsumption* correspondences, because an entity in one ontology can be decomposed into several entities in another ontology [13]. This happens in case where one ontology is more granular (or general) than the other one [37].

Correspondences in Coherent and Conservative Alignments: Consistency Principle. The *consistency principle* [36] states that the integrated ontology –resulting from the integration of the input ontologies– should be coherent (*i.e.*, all entities of the integrated ontology should be satisfiable), assuming that the input ontologies are also coherent (*i.e.*, the input ontologies also do not contain any unsatisfiable entities). An *unsatisfiable* entity (class or property) is an entity containing a contradiction in its description, for which it is not possible for any instance to meet all the requirements to be a member of that entity. In some applications where the logical reasoning is involved, ensuring coherence is of utmost importance since the integrated ontology must be logically/semantically correct to be really useful, otherwise it may lead to incomplete or unexpected results.

Conservativity Principle. The conservativity principle [34–36] requires that the original description (especially the *is-a* structure/class hierarchy) of an input ontology should not be altered after being integrated. Hence, the introduction of new semantic relations between entities of each matched ontology is not allowed, especially new subsumption relations causing structural changes. The conservativity principle aims that the use of the new integrated ontology –resulting from the integration of the input ontologies– does not affect the original behavior of the applications already functioning with the input ontologies (that were integrated).

Example 1 (Coherence/Conservativity Violation). Suppose that we have a class A in $\mathcal{O}_1$, two disjoint classes (B and C) in $\mathcal{O}_2$, and two correspondences c_1 and c_2 stating that A is a subclass of B and C. Formally,

$$\mathcal{O}_1 = \{A\} \quad \mathcal{O}_2 = \{B \perp C\}$$
$$\mathcal{A} = \{c_1, c_2\} \quad c_1 = <A \sqsubseteq B> \quad c_2 = <A \sqsubseteq C>$$

If a reasoning process is applied on the integrated ontology $\mathcal{O}_3$, then A will be an unsatisfiable class since it will become a subclass of two disjoint classes.

Now if we consider the following two ontologies: $\mathcal{O}_1$ has two classes A and B, and $\mathcal{O}_2$ has two classes A' and B' where B' is a subclass of A'. Formally,

$$\mathcal{O}_1 = \{A, B\} \quad \mathcal{O}_2 = \{B' \sqsubseteq A'\}$$
$$\mathcal{A} = \{c_1, c_2\} \quad c_1 = <A \equiv A'> \quad c_2 = <B \equiv B'>$$

If the ontology matching generates two correspondences c_1 and c_2 stating that A is equivalent to A', and B is equivalent to B', then the original structure of $\mathcal{O}_1$ will change in the integrated ontology $\mathcal{O}_3$ because of the addition of a new subsumption linking A and B.

Whenever an unsatisfiable entity or a conservativity violation is identified in the integrated ontology, then an alignment repair algorithm first identifies the correspondences causing these problems. The identified correspondences may actually be *erroneous* correspondences, but may also be *correct* correspondences introducing violations because of the incompatible conceptualizations of the matched ontologies. A human expert can then be notified and pointed to manually check and specify his/her opinion on these correspondences, to give his/her contribution to the matching process [23]. Otherwise, the alignment repair system can resolve these violations by automatically removing the identified correspondences and generating a repaired (coherent and conservative) output alignment. In a text annotation application, it is not necessary to ensure the coherence of the integrated ontology. However, in other applications, *e.g.*, query answering, logical errors in the integrated ontology may have a critical impact in the query answering process. Similarly, in some cases, the conservativity principle is no longer required, since the integrated ontology will be used by another specific application, *i.e.* not by the applications already using the ontologies that were integrated. Therefore, there is a need to represent/express correspondences causing (*consistency* and *conservativity*) violations in the forthcoming integrated ontology, and model them in the Alignment format [23]. The *Alignment*[1] format, (*aka* the *RDF Alignment format*), is the most consensual ontology alignment format used for representing simple pairwise alignments. In this format, we can not differentiate between a normal correspondence and a *repaired* one (involved in integration violations and identified by alignment repair systems). Therefore, there is a representation problem in the ontology alignment repair area.

5 Conclusions and Future Work

This paper presented a simple approach for knowledge building from raw datasets by adopting rich data models (ontologies). The tool developed proposes some automatic features to import data, which is mapped on virtual tables. Nevertheless, we need to automate the key mechanism to enforce semantic consistence among the different datasets is supposed. It becomes unrealistic once the scale of the system becomes significant or in presence of heterogeneity. Future work will be oriented to the automation of the whole process by particularizing existing techniques to the specific case of datasets mapped on virtual tables. Furthermore, we will include an additional virtual structure to support multi-dimensional data based on the RDF Data Cube Vocabulary[2].

Acknowledgments. The implementation of the tool described in the paper has been funded by the Faculty of Engineering and IT at the UTS through a Summer Scholarship granted to Hayden Crain. Additionally, we thank Prof. Iwona Miliszewska for encouraging and supporting the research collaboration between the UTS and TalTech.

References

1. Abdul-Ghafour, S., Ghodous, P., Shariat, B., Perna, E.: A common design-features ontology for product data semantics interoperability. In: Proceedings of the IEEE/WIC/ACM International Conference on Web Intelligence, pp. 443–446. IEEE Computer Society (2007)

---

[1] http://alignapi.gforge.inria.fr/format.html.
[2] The RDF Data Cube Vocabulary, https://www.w3.org/TR/vocab-data-cube/. Accessed: 8/01/ 2019.

2. Amann, B., Beeri, C., Fundulaki, I., Scholl, M.: Ontology-based integration of XML web resources. In: Horrocks, I., Hendler, J. (eds.) ISWC 2002. LNCS, vol. 2342, pp. 117–131. Springer, Heidelberg (2002). https://doi.org/10.1007/3-540-48005-6_11
3. Bharambe, U., Durbha, S.S., King, R.L.: Geospatial ontologies matching: an information theoretic approach. In: 2012 IEEE International Geoscience and Remote Sensing Symposium, IGARSS, pp. 2918–2921. IEEE (2012)
4. Bhatt, M., Rahayu, W., Soni, S.P., Wouters, C.: Ontology driven semantic profiling and retrieval in medical information systems. J. Web. Semant. **7**(4), 317–331 (2009)
5. Blasch, E.P., Dorion, É., Valin, P., Bossé, E.: Ontology alignment using relative entropy for semantic uncertainty analysis. In: Proceedings of the IEEE 2010 National Aerospace & Electronics Conference, pp. 140–148. IEEE (2010)
6. Buccella, A., Cechich, A., Rodríguez Brisaboa, N.: An ontology approach to data integration. J. Comput. Sci. Technol. **3**, 62–68 (2003)
7. Cross, V.: Uncertainty in the automation of ontology matching. In: Fourth International Symposium on Uncertainty Modeling and Analysis, 2003. ISUMA 2003, pp. 135–140. IEEE (2003)
8. Delgado, M., SáNchez, D., MartiN-Bautista, M.J., Vila, M.A.: Mining association rules with improved semantics in medical databases. Artif. Intell. Med. **21**(1–3), 241–245 (2001)
9. Dong, X.L., Halevy, A., Yu, C.: Data integration with uncertainty. VLDB J. **18**(2), 469–500 (2009). https://doi.org/10.1007/s00778-008-0119-9
10. Dou, D., Wang, H., Liu, H.: Semantic data mining: a survey of ontology-based approaches. In: Proceedings of the 2015 IEEE 9th International Conference on Semantic Computing, ICSC, pp. 244–251. IEEE (2015)
11. Euzenat, J., Shvaiko, P.: Ontology Matching. Springer, Heidelberg (2013). https://doi.org/10.1007/978-3-642-38721-0
12. Gagnon, M.: Ontology-based integration of data sources. In: 2007 10th International Conference on Information Fusion, pp. 1–8. IEEE (2007)
13. Gal, A.: Managing uncertainty in schema matching with Top-K schema mappings. In: Spaccapietra, S., Aberer, K., Cudré-Mauroux, P. (eds.) Journal on Data Semantics VI. LNCS, vol. 4090, pp. 90–114. Springer, Heidelberg (2006). https://doi.org/10.1007/11803034_5
14. Gal, A., Shvaiko, P.: Advances in ontology matching. In: Dillon, T.S., Chang, E., Meersman, R., Sycara, K. (eds.) Advances in Web Semantics I. LNCS, vol. 4891, pp. 176–198. Springer, Heidelberg (2008). https://doi.org/10.1007/978-3-540-89784-2_6
15. Gardner, S.P.: Ontologies and semantic data integration. Drug Discov. Today **10**(14), 1001–1007 (2005)
16. Jain, P., Hitzler, P., Sheth, A.P., Verma, K., Yeh, P.Z.: Ontology alignment for linked open data. In: Patel-Schneider, P.F., et al. (eds.) ISWC 2010. LNCS, vol. 6496, pp. 402–417. Springer, Heidelberg (2010). https://doi.org/10.1007/978-3-642-17746-0_26
17. Jan, S., Li, M., Al-Sultany, G., Al-Raweshidy, H.: Ontology alignment using rough sets. In: 2011 Eighth International Conference on Fuzzy Systems and Knowledge Discovery, FSKD, vol. 4, pp. 2683–2686. IEEE (2011)
18. Klir, G.J., Yuan, B.: Fuzzy Sets and Fuzzy Logic: Theory and Applications, p. 563. Prentice Hall, Upper Saddle River (1995)
19. Kuhn, W.: Geospatial semantics: why, of what, and how? In: Spaccapietra, S., Zimányi, E. (eds.) Journal on Data Semantics III. LNCS, vol. 3534, pp. 1–24. Springer, Heidelberg (2005). https://doi.org/10.1007/11496168_1

20. Lenz, R., Beyer, M., Kuhn, K.A.: Semantic integration in healthcare networks. Int. J. Med. Inform. **76**(2–3), 201–207 (2007)
21. Lenzerini, M.: Data integration: a theoretical perspective. In: Proceedings of the Twenty-First ACM SIGMOD-SIGACT-SIGART Symposium on Principles of Database Systems, pp. 233–246. ACM (2002)
22. Madnick, S., Zhu, H.: Improving data quality through effective use of data semantics. Data Knowl. Eng. **59**(2), 460–475 (2006)
23. Magnani, M., Montesi, D.: Uncertainty in data integration: current approaches and open problems. In: Proceedings of the First International VLDB Workshop on Management of Uncertain Data, MUD, pp. 18–32 (2007)
24. Marie, A., Gal, A.: Managing uncertainty in schema matcher ensembles. In: Prade, H., Subrahmanian, V.S. (eds.) SUM 2007. LNCS (LNAI), vol. 4772, pp. 60–73. Springer, Heidelberg (2007). https://doi.org/10.1007/978-3-540-75410-7_5
25. Mate, S., et al.: Ontology-based data integration between clinical and research systems. PLoS ONE **10**(1), e0116656 (2015)
26. Noy, N.F.: Semantic integration: a survey of ontology-based approaches. ACM Sigmod Rec. **33**(4), 65–70 (2004)
27. Patil, L., Dutta, D., Sriram, R.: Ontology-based exchange of product data semantics. IEEE Trans. Autom. Sci. Eng. **2**(3), 213–225 (2005)
28. Pileggi, S.F., Voinov, A.: Perswade-core: a core ontology for communicating socio-environmental and sustainability science. IEEE Access **7**, 127177–127188 (2019)
29. Pileggi, S., Hunter, J.: An ontology-based, linked open data framework to support the publishing, re-use and dynamic calculation of urban planning indicators. In: 15th International Conference on Computers in Urban Planning and Urban Management (2017)
30. Po, L., Sorrentino, S.: Automatic generation of probabilistic relationships for improving schema matching. Inf. Syst. **36**(2), 192–208 (2011)
31. Poggi, A., Lembo, D., Calvanese, D., De Giacomo, G., Lenzerini, M., Rosati, R.: Linking data to ontologies. In: Spaccapietra, S. (ed.) Journal on Data Semantics X. LNCS, vol. 4900, pp. 133–173. Springer, Heidelberg (2008). https://doi.org/10.1007/978-3-540-77688-8_5
32. Shvaiko, P., Euzenat, J.: Ten challenges for ontology matching. In: Meersman, R., Tari, Z. (eds.) OTM 2008. LNCS, vol. 5332, pp. 1164–1182. Springer, Heidelberg (2008). https://doi.org/10.1007/978-3-540-88873-4_18
33. Smith, B., et al.: The obo foundry: coordinated evolution of ontologies to support biomedical data integration. Nat. Biotechnol. **25**(11), 1251 (2007)
34. Solimando, A., Jiménez-Ruiz, E., Guerrini, G.: Detecting and correcting conservativity principle violations in ontology-to-ontology mappings. In: Mika, P., et al. (eds.) ISWC 2014. LNCS, vol. 8797, pp. 1–16. Springer, Cham (2014). https://doi.org/10.1007/978-3-319-11915-1_1
35. Solimando, A., Jiménez-Ruiz, E., Guerrini, G.: A multi-strategy approach for detecting and correcting conservativity principle violations in ontology alignments. In: Proceedings of the 11th International Workshop on OWL: Experiences and Directions, OWLED 2014, co-located with ISWC, pp. 13–24 (2014)
36. Solimando, A., Jimenez-Ruiz, E., Guerrini, G.: Minimizing conservativity violations in ontology alignments: algorithms and evaluation. Knowl. Inf. Syst. **51**(3), 775–819 (2017)
37. Stoilos, G., Geleta, D., Shamdasani, J., Khodadadi, M.: A novel approach and practical algorithms for ontology integration. In: Vrandečić, D., et al. (eds.) ISWC 2018. LNCS, vol. 11136, pp. 458–476. Springer, Cham (2018). https://doi.org/10.1007/978-3-030-00671-6_27

38. Wache, H., et al.: Ontology-based integration of information-a survey of existing approaches. In: OIS@ IJCAI (2001)
39. Zhang, H., et al.: An ontology-guided semantic data integration framework to support integrative data analysis of cancer survival. BMC Med. Inform. Decis. Mak. **18**(2), 41 (2018). https://doi.org/10.1186/s12911-018-0636-4
40. Zhang, Y., Panangadan, A.V., Prasanna, V.K.: UFOM: unified fuzzy ontology matching. In: Proceedings of the 2014 IEEE 15th International Conference on Information Reuse and Integration, IEEE IRI, pp. 787–794. IEEE (2014)

A Simple Stochastic Process Model for River Environmental Assessment Under Uncertainty

Hidekazu Yoshioka[1]([⊠]) [iD], Motoh Tsujimura[2] [iD],
Kunihiko Hamagami[3], and Yumi Yoshioka[1]

[1] Shimane University, Nishikawatsu-cho 1060, Matsue 690-8504, Japan
{yoshih, yyoshioka}@life.shimane-u.ac.jp
[2] Graduate School of Commerce, Doshisha University,
Karasuma-Higashi-iru, Imadegawa-dori, Kyoto 602-8580, Japan
mtsujimu@mail.doshisha.ac.jp
[3] Faculty of Agriculture, Iwate University,
3-18-8 Ueda, Morioka 020-8550, Japan
ham@iwate-u.ac.jp

Abstract. We consider a new simple stochastic single-species population dynamics model for understanding the flow-regulated benthic algae bloom in uncertain river environment: an engineering problem. The population dynamics are subject to regime-switching flow conditions such that the population is effectively removed in a high-flow regime while it is not removed at all in a low-flow regime. A focus in this paper is robust and mathematically rigorous statistical evaluation of the disutility by the algae bloom under model uncertainty. We show that the evaluation is achieved if the optimality equation derived from a dynamic programming principle is solved, which is a coupled system of non-linear and non-local degenerate elliptic equations having a possibly discontinuous coefficient. We show that the system is solvable in continuous viscosity and asymptotic senses. We also show that its solutions can be approximated numerically by a convergent finite difference scheme with a demonstrative example.

Keywords: Regime-switching stochastic process · Model uncertainty · Environmental problem · Viscosity solution

1 Introduction

This paper focuses on a population dynamics modeling of nuisance benthic algae on riverbed under uncertain environment: a common environmental problem encountered in many rivers where human regulates the flow regimes [1]. Blooms of nuisance benthic algae and macrophytes, such as *Cladophora glomerata* and *Egeria densa*, in inland water bodies are seriously affecting aquatic ecosystems [2, 3]. Such environmental problems are especially severe in dam-downstream rivers where the flow regimes are often regulated to be low, with which the nuisance algae can dominate the others [4, 5].

© Springer Nature Switzerland AG 2020
V. V. Krzhizhanovskaya et al. (Eds.): ICCS 2020, LNCS 12143, pp. 494–507, 2020.
https://doi.org/10.1007/978-3-030-50436-6_36

It has been found that the benthic algae are effectively removed when they are exposed to a sufficiently high flow discharge containing sediment particles [6]. Supplying sediment into a river environment can be achieved through transporting earth and soils from the other sites, as recently considered in Yoshioka et al. [7] focusing on a case study in Japan with a high-dimensional stochastic control model.

Assume that we could find a way to supply the sediment into a river environment where the nuisance algae are blooming. Then, a central issue is to what extent the algae bloom can be suppressed in the given environment. Hydrological studies imply that river flows are inherently stochastic and can be effectively described using a Markov-chain [8]. In the simplest case, we can classify river flow regimes into the two regimes: a high-flow regime where the nuisance algae can be effectively removed from the riverbed and a low-flow regime where they are not removed from the riverbed at all. In this view, the algae population dynamics can be considered as a stochastic dynamical system subject to a two-state regime-switching noise. To the best of our knowledge, such an attempt has been least explored despite its high engineering importance.

We approach this issue both mathematically and numerically. We formulate the algae population dynamics as a system of piecewise-deterministic system subject to a Markovian regime-switching noise [9] representing a dynamic river flow having high- and low-flow regimes. This is a system of stochastic differential equations (SDEs, Øksendal and Sulem [10]) based on a logistic model subject to the detachment during the high-flow regime [7] but with a simplification for better tractability. The model incorporates our own experimental evidence that a sudden detachment of the algae occurs when the flow regime switches from the high-flow to the low-flow. This finding introduces a non-locality into the model.

Our focus is not only on the population dynamics themselves, but also on statistical evaluation of the dynamics that can also be important in engineering applications. Namely, another focus is the evaluation of statistical indices such as a disutility caused by the population, which are given by conditional expectations of quantities related to the population. Unfortunately, it is usually difficult to accurately identify model parameters in the natural environment due to technical difficulties and poor data availability. In such cases, we must operate a model under the assumption that it is incomplete and thus uncertain (or equivalently, ambiguous). We overcome this issue by employing the concept of multiplier robust control [11], which allows us to analyze SDEs having uncertainty and further to statistically evaluate their dynamics in a worst-case robust manner. This methodology originates from economics and has been employed in finance [12] and insurance [13], but less frequently in environment and ecology [14]. With this formulation, we demonstrate that the stochastic dynamics having model uncertainty can be handled mathematically rigorously as well as efficiently.

We show that the robust evaluation of a statistical index related to the population dynamics ultimately reduces to solving a system of non-linear and non-local degenerate elliptic equations: the optimality equation having a possibly discontinuous source term. This is the governing equation of the statistical index under the worst-case. Our goal is therefore to solve the equation in some way. We show that solutions to the optimality equation are characterized in a viscosity sense [15], and that it admits a continuous viscosity solution by a comparison theorem [16]. We present an analytical asymptotic estimate of the solution as well. We finally provide a demonstrative computational example with a convergent finite difference scheme [5, 14] to show the validity of the asymptotic estimate and to deeper comprehend the behavior of the model.

2 Mathematical Model

2.1 System Dynamics

Let $(\alpha_t)_{t \geq 0}$ be a càdlàg two-state continuous-time Markov chain having a low-flow regime ($i = 0$) and a high-flow regime ($i = 1$). The switching rate from the regimes 0 to 1 (resp., 1 to 0) is a positive constant $v_{01} > 0$ (resp., $v_{10} > 0$). We assume that the regime switching occurs with the prescribed switching rates and some Poisson processes. We thus describe the SDE governing temporal evolution of $(\alpha_t)_{t \geq 0}$ as

$$d\alpha_t = \chi_{\{\alpha_{t-}=0\}} dN_t^{(01)} - \chi_{\{\alpha_{t-}=1\}} dN_t^{(10)} \text{ for } t \geq 0, \ \alpha_{0-} \in \{0,1\}, \tag{1}$$

where χ_S is the indicator function such that $\chi_S = 1$ if S is true and $\chi_S = 0$ otherwise, $\alpha_{t-} = \lim_{s \to +0} \alpha_{t-s}$ and the same representation applies to the other processes, $\left(N_t^{(01)}\right)_{t \geq 0}$ and $\left(N_t^{(10)}\right)_{t \geq 0}$ are mutually-independent standard Poisson processes with the jumping rates v_{01} and v_{10}, respectively. The switching times from the regimes 0 to 1 (resp., regimes 1 to 0) are represented by a strictly increasing sequence $\left(\tau_k^{01}\right)_{k \in \mathbb{N}}$ (resp., $\left(\tau_k^{10}\right)_{k \in \mathbb{N}}$). We assume a.s. $\tau_k^{10} \neq \tau_l^{01}$ ($k, l \in \mathbb{N}$) without loss of generality.

The algae population is represented by a continuous-time variable $(X_t)_{t \geq 0}$, which is assumed to be governed by a generalized logistic model having regime-switching coefficients. The difference between the regimes 0 and 1 in the context of the population dynamics is that the algae detachment occurs only at the regime 1 (high-flow regime). We normalize the population X_t to be valued in $[0, 1]$. We set the SDE of $(X_t)_{t \geq 0}$ as

$$dX_t = \left(\mu \left(1 - X_{t-}^{\theta} \right) - \chi_{\{\alpha_{t-}=1\}} D(X_{t-}) \right) X_{t-} dt \text{ for } t \geq 0, \ X_{0-} \in [0,1], \tag{2}$$

where $\theta \geq 1$ is the shape parameter, $\mu > 0$ is the specific growth rate and $D : [0, 1] \to [0, +\infty)$ is the detachment rate that is assumed to be non-negative and Lipschitz continuous in $[0, 1]$ and $D(1) > 0$. One of our own important experimental findings is that a sudden algae detachment occurs at the initiation of each high-flow event when there is an enough sediment supply. From the standpoint of the present model, this imposes the additional state dynamics at the switching times $X_{\tau_k^{01}} = (1 - z_k) X_{\tau_k^{01}-}$ for $k \in \mathbb{N}$, where $(z_k)_{k \in \mathbb{N}}$ is a sequence of i.i.d. stochastic variables representing the sudden detachment of the algae population. We assume that they have the common compact range $Z \subset (0, 1)$, meaning that the sudden decrease of the population does not lead to its immediate extinction, which is consistent with the empirical finding [5]. The sequence $(z_k)_{k \in \mathbb{N}}$ is assumed to be independent from $N_t^{(01)}$ and $N_t^{(10)}$. The probability density function generating $(z_k)_{k \in \mathbb{N}}$ is denoted as $g \geq 0$ with $\int_Z g(z) dz = 1$.

The stochastic nature of the detachment has been justified from our experimental results. We found that the amount of sudden detachment is different among the experimental runs even under the same experimental setting (sediment supply and water flow). This is considered due to inherently probabilistic nature of the sediment supply and microscopic difference of biological conditions (length, density, rock shape, etc.,) of the algae population. The same would be true in real river environment.

The natural filtration generated by $\left(N_t^{(01)}\right)_{t \geq 0}$, $\left(N_t^{(10)}\right)_{t \geq 0}$, and $(z_k)_{k \in \mathbb{N}}$ at time t are denoted as F_t. Set $F = (F_t)_{t \geq 0}$. Consequently, the coupled stochastic system dynamics to be observed are given as follows ($k \in \mathbb{N}$):

$$d\alpha_t = \chi_{\{\alpha_{t-}=0\}} dN_t^{(01)} - \chi_{\{\alpha_{t-}=1\}} dN_t^{(10)}, \ t \geq 0$$

$$dX_t = \left(\mu(1 - X_{t-}^\theta) - \chi_{\{\alpha_{t-}=1\}} D(X_{t-})\right) X_{t-} dt, \ t \geq 0, \ t \neq \tau_k^{01}, (\alpha_{0-}, X_{0-}) \in \{0, 1\} \times [0, 1]. \quad (3)$$

$$X_{\tau_k^{01}} = (1 - z_k) X_{\tau_k^{01}-}, \ k \in \mathbb{N}$$

2.2   Performance Index

A performance index to statistically evaluate the stochastic population dynamics (3) is formulated. To simplify the problem as much as possible, we consider the following conditional expectation evaluating the disutility caused by the algae population:

$$\Phi(i, x) = \mathbb{E}^{i,x}\left[\int_0^{+\infty} f(X_s) e^{-\delta s} ds\right], \ (i, x) \in \{0, 1\} \times [0, 1], \quad (4)$$

where $\mathbb{E}^{i,x}$ is the conditional expectation with $(\alpha_{0-}, X_{0-}) = (i, x)$, $f \geq 0$ is a bounded upper-semicontinuous function on $[0, 1]$ having at most a finite number of discontinuous points in $(0, 1)$. Such a discontinuity naturally arises if there is a threshold above which the algae bloom would severely affect the water environment. The performance index (4) is the mean value of an infinite-horizon discounted disutility. We write $\Phi(i, x) = \Phi_i(x)$ if there will be no confusion.

The dynamic programming principle [10] formally leads to the governing equation of Φ_i as the system of linear degenerate elliptic equations

$$\delta \Phi_0 - \mu(1 - x^\theta) x \frac{d\Phi_0}{dx} + v_{01} \int_Z \Delta_{01} \Phi g(z) dz - f = 0, \ x \in [0, 1], \quad (5)$$

$$\delta \Phi_1 - \left(\mu(1 - x^\theta) - D(x)\right) x \frac{d\Phi_1}{dx} + v_{10} \Delta_{10} \Phi - f = 0, \ x \in [0, 1] \quad (6)$$

with

$$\Delta_{01} \Phi(x, z) = \Phi_0(x) - \Phi_1(x(1 - z)) \text{ and } \Delta_{10} \Phi(x) = \Phi_1(x) - \Phi_0(x). \quad (7)$$

2.3   A Model with Uncertainty

We consider an uncertain counterpart, where the model uncertainty is assumed due to distortions of $\left(N_t^{(01)}\right)_{t\geq 0}$, $\left(N_t^{(10)}\right)_{t\geq 0}$, and g. Namely, we focus on hydrological uncertainties. Our formulation is based on the models with uncertain jump processes [11, 12] with our notations of the parameters and variables. Let $\mathbb{P}$ be the current probability measure and $\mathbb{Q}$ be the probability measure under the distortion. Set $\left(\phi_t^i(\cdot)\right)_{t\geq 0}$ ($i = 0, 1$) as positive and strictly bounded F-predictable random fields representing model uncertainty. There is no uncertainty when $\phi_t^0 = \phi_t^1 = 1$ ($t \geq 0$). Recall that there is no jump of X_t at each τ_k^{10}. The new and old probability measures are characterized by the Radon-Nikodym derivative $\frac{d\mathbb{Q}}{d\mathbb{P}} = \Lambda_t^0 \Lambda_t^1$, where

$$\Lambda_t^0 = \exp\left(\int_Z \left(\int_0^t \left(1 - \phi_s^0(z)\right)v_{01}\mathrm{d}s + \int_0^t \ln \phi_s^0(z)\mathrm{d}N_s^{(01)}\right)g(z)\mathrm{d}z\right), \tag{8}$$

$$\Lambda_t^1 = \exp\left(\int_0^t \left(1 - \phi_s^1\right)v_{01}\mathrm{d}s + \int_0^t \ln \phi_s^1\mathrm{d}N_s^{(10)}\right) \tag{9}$$

under the assumption that the right-hand sides are a.s. bounded for $t \geq 0$. We thus assume that Λ_t^0 and Λ_t^1 are positive and a.s. bounded for $t \geq 0$. This is true if the processes $\left(\phi_t^i(\cdot)\right)_{t\geq 0}$ ($i = 0, 1$) are strictly positive and uniformly bounded. Notice that ϕ_t^1 is actually a function only of time t, but is represented as $\phi_t^1(\cdot)$ here for convenience. Formally, at time t, the jump intensity of $N_t^{(01)}$ (resp., $N_t^{(10)}$) is modified from v_{01} to $v_{01} \int_Z \phi_t^0(z)g(z)\mathrm{d}z$ (resp., v_{10} to $v_{10}\phi_t^1(z)$) and the probability density function $g = g(z)$ to $C\phi_t^0(z)g(z)$ with a constant $C > 0$. The support of $\phi_t^0(z)g(z)$ is still Z. We impose the following normalization, so that the modified g, which is written as g^*, is truly a probability density function:

$$g^*(z) = \frac{\phi_t^0(z)g(z)}{\int_Z \phi_t^0(z)g(z)\mathrm{d}z}, \, t \geq 0. \tag{10}$$

Under the new probability measure $\mathbb{Q}$, the processes $N_t^{(01)}$ and $N_t^{(10)}$ are formally replaced by the new Poisson processes $\bar{N}_t^{(01)}$ and $\bar{N}_t^{(10)}$ having the jump intensities $v_{01} \int_Z \phi_t^0(z)g(z)\mathrm{d}z$ and $v_{10}\phi_t^1$, respectively, both of which are assumed to be strictly positive and bounded. Now, the transformed system dynamics are ($k \in \mathbb{N}$):

$$\mathrm{d}\alpha_t = \chi_{\{\alpha_{t-}=0\}}\mathrm{d}\bar{N}_t^{(01)} - \chi_{\{\alpha_{t-}=1\}}\mathrm{d}\bar{N}_t^{(10)}, \, t \geq 0$$

$$\mathrm{d}X_t = \left(\mu\left(1 - X_{t-}^\theta\right) - \chi_{\{\alpha_{t-}=1\}}D(X_{t-})\right)X_{t-}\mathrm{d}t, \, t \geq 0, \, t \neq \tau_k^{01}, \, (\alpha_{0-}, X_{0-}) \in \{0, 1\} \times [0, 1]$$

$$X_{\tau_k^{01}} = (1 - z_k)X_{\tau_k^{01}-}, \, k \in \mathbb{N}$$

$$\tag{11}$$

The performance index J is then extended to a worst-case uncertain counterpart, with which the observer can evaluate the disutility subject to an entropic penalization for the deviations between the true and distorted models:

$$\Phi(i,x) = \sup_{\left(\phi_t^{01}(\cdot), \phi_t^{10}\right)_{t \geq 0}} J\left(i,x; \phi^{01}, \phi^{10}\right), \ (i,x) \in \{0,1\} \times [0,1] \tag{12}$$

with

$$J\left(i,x; \phi^{01}, \phi^{10}\right) = \mathbb{E}^{i,x}\left[\int_0^{+\infty} f(X_s)e^{-\delta s}ds - \left(\frac{v_{01}}{\psi_0}I_0 + \frac{v_{10}}{\psi_1}I_1\right)\right], \tag{13}$$

$$I_0 = \int_0^{+\infty} e^{-\delta s}\int_Z \left(\phi_s^0(z)\ln\phi_s^0(z) - \phi_s^0(z) + 1\right)g(z)dzds, \tag{14}$$

$$I_1 = \int_0^{+\infty} e^{-\delta s}\left(\phi_s^1 \ln\phi_s^1 - \phi_s^1 + 1\right)ds, \tag{15}$$

where $\psi_0 > 0$ and $\psi_1 > 0$ are the uncertainty-aversion parameters serving as penalization parameters to constrain the allowable difference between the true and distorted models. The integrands of (14) and (15) are (discounted) relative entropy between the models with respect to ϕ_t^0 and ϕ_t^1, respectively. Each uncertainty-aversion parameter modulates the corresponding relative entropy in the way that a larger parameter allows for a larger deviation between the true and distorted models, and vice versa. This formulation reduces to the model without uncertainty as $\psi_0, \psi_1 \to +0$. In this way, the observer can flexibly presume the potential uncertainty.

The dynamic programming principle [10] leads to the governing equation of the value function Φ_i subject to model uncertainty as the coupled system of non-linear and non-local degenerate elliptic equations in $[0,1]$:

$$\delta\Phi_0 - \mu(1 - x^\theta)x\frac{d\Phi_0}{dx} - f$$
$$+ v_{01}\inf_{\phi^0(\cdot) > 0}\left\{\int_Z \left(\phi^0(z)\Delta_{01}\Phi + \frac{1}{\psi_0}\left(\phi^0(z)\ln\phi^0(z) - \phi^0(z) + 1\right)\right)g(z)dz\right\} = 0 \tag{16}$$

and

$$\delta\Phi_1 - \left(\mu(1 - x^\theta) - D(x)\right)x\frac{d\Phi_1}{dx} - f$$
$$+ v_{10}\inf_{\phi^1 > 0}\left\{\phi^1\Delta_{10}\Phi + \frac{1}{\psi_1}\left(\phi^1 \ln\phi^1 - \phi^1 + 1\right)\right\} = 0. \tag{17}$$

The minimizations in (16) and (17) are achieved by the worst-case uncertainties as

$$\phi^{0*}(x, z) = \exp(-\psi_0 \Delta_{01} \Phi(x, z)) \text{ and } \phi^{1*}(x) = \exp(-\psi_1 \Delta_{10} \Phi(x)). \tag{18}$$

Substituting (18) into (16) and (17) simplifies them to

$$\delta \Phi_0 - \mu(1 - x^\theta) x \frac{d\Phi_0}{dx} + \frac{v_{01}}{\psi_0} \int_Z (1 - \exp(-\psi_0 \Delta_{01} \Phi)) g(z) dz - f = 0, \tag{19}$$

and

$$\delta \Phi_1 - (\mu(1 - x^\theta) - D(x)) x \frac{d\Phi_1}{dx} + \frac{v_{10}}{\psi_1} (1 - \exp(-\psi_1 \Delta_{10} \Phi)) - f = 0. \tag{20}$$

We formally derive (5)–(6) under $\psi_0, \psi_1 \to +0$ as expected. The quantities in (18) represent the worst-case uncertainties conditioned on the current observations of α and X. Furthermore, by (10), the probability density function g^* under the distortion is

$$g^*(x, z) = \frac{\exp(-\psi_0 \Delta_{01} \Phi) g(z)}{\int_Z \exp(-\psi_0 \Delta_{01} \Phi) g(z) dz} \tag{21}$$

with an abuse of notations. The worst-case jump intensities from the regimes 0 to 1 and 1 to 0, which are represented as $v_{01}^* = v_{01}^*(x)$ and $v^* = v_{10}^*(x)$, are given by

$$v_{01}^*(x) = v_{01} \int_Z \exp(-\psi_0 \Delta_{01} \Phi) g(z) dz \text{ and } v_{10}^*(x) = v_{10} \exp(-\psi_1 \Delta_{10} \Phi(x)). \tag{22}$$

3 Mathematical Analysis

3.1 Viscosity Solution

Boundedness and continuity of the value function are analyzed. We firstly prove unique solvability of the system (11).

Proposition 1. *The system (11) admits a unique strong solution such that* $0 \le X_t \le 1$ *($t \ge 0$).*

Proof: A similar contradiction argument to that in the proof of Theorem 2.2 of Lungu and Øksendal [17] applies in our case. We can get unique existence of the system having coefficients extended to be Lipschitz continuous over $\mathbb{R}$, by Theorem 2.1 of Yin and Zhu [9]. With a contradiction argument [17], we obtain that the strong solution to this modified problem is bounded in $[0, 1]$.   □

By Proposition 1, we get a continuity result of the value function, with which an appropriate definition of viscosity solutions to the optimality equation (19)–(20) is found.

Proposition 2. Assume that f is Hölder continuous in $[0, 1]$. Then, we get $\Phi_i \in C[0, 1]$, $(i = 0, 1)$.

Proof: Combine the strong solution property and the boundedness result in Proposition 1 with the Hölder continuity of f. □

Now, we define viscosity solutions. Set the space of upper-semicontinuous (resp., lower-semicontinuous) functions in $[0, 1]$ as $USC[0, 1]$ (resp., $LSC[0, 1]$).

Definition 1. *A pair* $\Psi_0, \Psi_1 \in USC[0, 1]$ *is a viscosity sub-solution if for all* $x_0 \in [0, 1]$ *and* $i_0 = 0, 1$, *and for all* $\varphi_0, \varphi_1 \in C^1[0, 1]$ *s.t.* $\varphi_i - \Psi_i$ *is locally minimized at* $x = x_0$ *and* $i = i_0$ *with* $\varphi_{i_0}(x_0) = \Psi_{i_0}(x_0)$, *the following hold ((23) for* $i_0 = 0$, *(24) for* $i_0 = 1$):

$$\delta\Psi_0 - \mu(1 - x^\theta)x\frac{d\varphi_0}{dx} + \frac{v_{01}}{\psi_0}\int_Z (1 - \exp(-\psi_0\Delta_{01}\Psi))g(z)dz - f^*(x) \leq 0, \, x = x_0,$$

(23)

$$\delta\Psi_1 - (\mu(1 - x^\theta) - D(x))x\frac{d\varphi_1}{dx} + \frac{v_{10}}{\psi_1}(1 - \exp(-\psi_1\Delta_{10}\Psi)) - f^*(x) \leq 0, \, x = x_0.$$

(24)

A pair $\Psi_0, \Psi_1 \in LSC[0, 1]$ *is a viscosity super-solution if for all* $x_0 \in [0, 1]$ *and* $i_0 = 0, 1$, *and for all* $\varphi_0, \varphi_1 \in C^1[0, 1]$ *s.t.* $\varphi_i - \Psi_i$ *is locally maximized at* $x = x_0$ *and* $i = i_0$ *with* $\varphi_{i_0}(x_0) = \Psi_{i_0}(x_0)$, *the following hold ((25) for* $i_0 = 0$, *(26) for* $i_0 = 1$):

$$\delta\Psi_0 - \mu(1 - x^\theta)x\frac{d\varphi_0}{dx} + \frac{v_{01}}{\psi_0}\int_Z (1 - \exp(-\psi_0\Delta_{01}\Psi))g(z)dz - f_*(x) \geq 0, \, x = x_0$$

(25)

$$\delta\Psi_1 - (\mu(1 - x^\theta) - D(x))x\frac{d\varphi_1}{dx} + \frac{v_{10}}{\psi_1}(1 - \exp(-\psi_1\Delta_{10}\Psi)) - f_*(x) \geq 0, \, x = x_0.$$

(26)

A pair $\Psi_0, \Psi_1 \in C[0, 1]$ *is a viscosity solution if it is a viscosity sub-solution as well as a viscosity super-solution.*

Proposition 3. *Assume that* f *is Hölder continuous in* $[0, 1]$. *Then, the value function is a viscosity solution.*

Proof: Apply the Dynkin's formula and the dominated convergence theorem. □

Proposition 4. *For any viscosity sub-solution* u *and a viscosity super-solution* v, $v \geq u$ *in* Ω. *Moreover, the optimality equation (19)–(20) admits at most one viscosity solution.*

Proof: Apply a contradiction argument with the help of the monotonicity of the nonlinear terms. [Proof of Proposition 3.3 in 14] and [Proof of Theorem 11.4 in 16]. □

Notice that the uniqueness result in Proposition 4 holds true for both continuous and discontinuous f. A consequence of Propositions 3 and 4 is the next theorem.

Theorem 1. *Assume that f is Hölder continuous in* $[0, 1]$. *Then, the value function is the unique viscosity solution to the optimality equation. (19)–(20).*

3.2 Asymptotic Solution

The optimality equation is uniquely solvable, but its exact solution cannot be found analytically. Instead, we construct an asymptotic solution that close to the solution for small x. The asymptotic solution thus applies to the situation where the population is sufficiently small, which would be encountered during a high-flow regime. The next proposition can be checked by a direct calculation. The asymptotic worst-case uncertainties can also be calculated using this proposition.

Proposition 5. *Assume* $\Phi_i \in C^2[0, \varepsilon]$ $(i = 0, 1)$ *with* $0 < \varepsilon < 1$. *Assume that for small* $0 < x < \varepsilon, f(x) = x^m + O(x^{m+\upsilon})$ *and* $D(x) = d + O(x^\upsilon), d \geq 0, m, \upsilon > 0, \delta > 0$ *such that*

$$A = (\delta - m\mu + v_{01})(\delta - m(\mu - d) + v_{01}) - v_{01}v_{10}\int_Z (1 - z)^m g(z)\mathrm{d}z > 0. \qquad (27)$$

Then, we have the following asymptotic expansions for small $x > 0$:

$$\Phi_i(x) = C_i x^m + \text{higher-order terms of } x \ (i = 0, 1), \qquad (28)$$

with the positive constants

$$C_0 = \frac{\delta - m(\mu - d) + v_{01}\int_Z (1 - z)^m g(z)\mathrm{d}z + v_{10}}{A} \text{ and } C_1 = \frac{\delta - m\mu + v_{01} + v_{10}}{A}. \qquad (29)$$

4 Numerical Computation

We numerically discretize the optimality equation (19)–(20) because it is not analytically solvable and the asymptotic solution presented in the previous section can be utilized only under limited conditions. The employed numerical method here is the finite difference scheme with the Newton iteration [5], which can handle the decay and first-order differential terms of the degenerate elliptic differential equations. The non-linear and non-local terms are handled with the interpolation technique of [14]. This scheme is a version of the non-standard finite difference scheme based on local exact solutions to linearized problems [18], with which monotone, stable, and consistent discretization is established. It is thus convergent in a viscosity sense [19]. The non-local term is linearized at each iteration step to enhance computational stability.

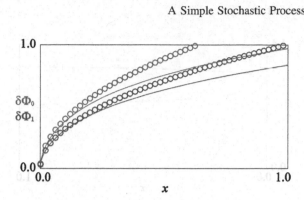

Fig. 1. The computed and asymptotic $\delta\Phi_0(x)$ (red) and $\delta\Phi_1(x)$ (blue) with $\psi = 1$. Line: computed result, Circle: asymptotic result. (Color figure online)

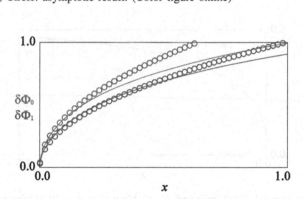

Fig. 2. The computed and asymptotic $\delta\Phi_0(x)$ (red) and $\delta\Phi_1(x)$ (blue) with $\psi = 20$. Line: computed result, Circle: asymptotic result. (Color figure online)

The computational condition here is specified as follows. For the sake of simplicity, we set $\psi_0 - \psi_1 = \psi > 0$. In addition, set $D(x) = dx$ and $f(x) = x^m$ with $d, m > 0$. The computational domain is discretized into 1,000 cells with 1,001 vertices, and the range Z of the uncertainty into 500 cells. The following parameter values are used in the computation: $m = 0.5$, $\delta = 2$, $\mu = 0.5$, $\theta = 1$, $d = 1$, $v_{01} = 0.1$, $v_{10} = 1.0$, $g(z) = 3\chi_{\{1/3 \le z \le 2/3\}}$, and $\psi = 1$ or 20. The threshold to terminate the iteration is

$$\max_{i=0,1} \frac{\left| \Phi_i^{(n+1)} - \Phi_i^{(n)} \right|}{\max\left\{ 1, \Phi_i^{(n)} \right\}} < 10^{-12} \tag{30}$$

at all the computational vertices, where the superscript represents the iteration number in the Newton iteration. The numerical solution $\Phi_i^{(n+1)}$ with the smallest n satisfying (30) is considered as the numerical solution. Each computation below requires at most 20 to 30 steps, implying satisfactory efficiency of the scheme.

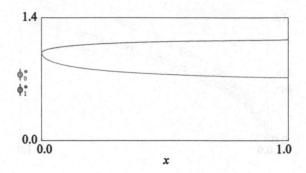

Fig. 3. The computed $\phi_0^*(x)$ (red) and $\phi_1^*(x)$ (blue) with $\psi = 1$. (Color figure online)

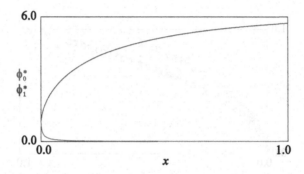

Fig. 4. The computed $\phi_0^*(x)$ (red) and $\phi_1^*(x)$ (blue) with $\psi = 20$. (Color figure online)

In Figs. 1 and 2, the numerically computed value functions (solid lines) are compared with the asymptotic results for $\psi = 1$ and $\psi = 20$, suggesting their good agreement especially for small x. We can see that the disutility assessed by the more uncertainty-averse observer (having a larger ψ) is evaluated larger. In Figs. 3 and 4, the worst-case uncertainties $\phi_0^*(x) = \int_Z \phi_0^*(z, x) g(z) \mathrm{d}z$ and $\phi_1^*(x)$ are compared for $\psi = 1$ and $\psi = 20$, suggesting the decreasing and increasing nature of the former and latter, respectively. The magnitude of decrease/increase more sharply depends on the population for larger ψ. The monotone dependence of ϕ_i^* shows that the observer considers the flood frequency smaller under larger uncertainty-aversion.

Figure 5 shows the worst-case probability density functions $g^*(z, x)$ for different values of x when $\psi = 20$. The computational results clearly show the distortion of g due to model uncertainty; larger distortion is observed for larger population. The worst-case probability density functions are positively-skewed, implying that the more uncertainty-averse observer specifies thus sudden detachment of the algae population to be smaller. In addition, their skewness significantly depends on the observed population, suggesting that the more skewed prediction is optimal when the population is large: namely, in case of the algae bloom.

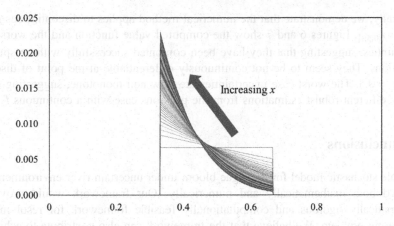

Fig. 5. The computed worst-case probability density functions $g^*(z, x)$ $(x = i/100(i = 0, 1, 2, \ldots 100))$ for $\psi = 20$: vertical (g^*) and horizontal axes (z).

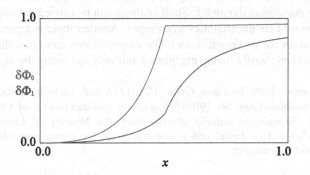

Fig. 6. The computed $\delta\Phi_0(x)$ (red) and $\delta\Phi_1(x)$ (blue) with a discontinuous f. (Color figure online)

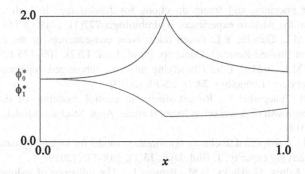

Fig. 7. The computed $\phi_0^*(x)$ (red) and $\phi_1^*(x)$ (blue) with a discontinuous f. (Color figure online)

Finally, we demonstrate that the numerical method applies to discontinuous f. We set $f = \chi_{[0.5,1]}$. Figures 6 and 7 show the computed value function and the worst-case uncertainties, suggesting that they have been computed successfully without spurious oscillations. They seem to be not continuously differentiable at the point of discontinuity $x = 0.5$. The worst-case uncertainties are now non-monotone, suggesting qualitatively different robust estimations from the previous case with a continuous f.

5 Conclusions

A simple stochastic model for the algae bloom under uncertain river environment was analyzed both mathematically and numerically. Our framework would provide a mathematically rigorous and computationally feasible framework for resolving the engineering problem. We believe that the framework can also contribute to achieving Sustainable Development Goals (SDGs) related to water environmental and ecological issues.

A future topic would be computing the probability density functions and analyzing stability of the population dynamics. Such analysis can be carried out using the derived or computed worst-case uncertainties as an input. Another topic is a partial observation modeling based on the presented model. For example, one can often directly measure water flows relatively easily (using machines), but only indirectly the algae population.

Acknowledgements. JSPS Research Grant 18K01714 and 19H03073, Kurita Water and Environment Foundation Grant No. 19B018, a grant for ecological survey of a life history of the landlocked ayu *Plecoglossus altivelis altivelis* from the Ministry of Land, Infrastructure, Transport and Tourism of Japan, and a research grant for young researchers in Shimane University support this research.

References

1. Carey, M.P., Sethi, S.A., Larsen, S.J., Rich, C.F.: A primer on potential impacts, management priorities, and future directions for *Elodea spp.* in high latitude systems: learning from the Alaskan experience. Hydrobiologia **777**(1), 1–19 (2016)
2. Gladyshev, M.I., Gubelit, Y.I.: Green tides: New consequences of the eutrophication of natural waters (Invited Review). Contemp. Probl. Ecol. **12**(2), 109–125 (2019)
3. Verhofstad, M.J., Bakker, E.S.: Classifying nuisance submerged vegetation depending on ecosystem services. Limnology **20**(1), 55–68 (2019)
4. Yoshioka, H., Yaegashi, Y.: Robust stochastic control modeling of dam discharge to suppress overgrowth of downstream harmful algae. Appl. Stochast. Models Bus. Ind. **34**(3), 338–354 (2018)
5. Yoshioka, H.: A simplified stochastic optimization model for logistic dynamics with control-dependent carrying capacity. J. Biol. Dyn. **13**(1), 148–176 (2019)
6. Hoyle, J.T., Kilroy, C., Hicks, D.M., Brown, L.: The influence of sediment mobility and channel geomorphology on periphyton abundance. Freshw. Biol. **62**(2), 258–273 (2017)

7. Yoshioka, H., Yaegashi, Y., Yoshioka, Y., Hamagami, K.: Hamilton–Jacobi–Bellman quasi-variational inequality arising in an environmental problem and its numerical discretization. Comput. Math Appl. **77**(8), 2182–2206 (2019)
8. Turner, S.W.D., Galelli, S.: Regime-shifting streamflow processes: Implications for water supply reservoir operations. Water Resour. Res. **52**(5), 3984–4002 (2016)
9. Yin, G.G., Zhu, C.: Hybrid Switching Diffusions: Properties and Applications. Springer, New York (2009). https://doi.org/10.1007/978-1-4419-1105-6
10. Øksendal, B., Sulem, A.: Applied Stochastic Control of Jump Diffusions. Springer, Cham (2019)
11. Hansen, L., Sargent, T.J.: Robust control and model uncertainty. Am. Econ. Rev. **91**(2), 60–66 (2001)
12. Cartea, A., Jaimungal, S., Qin, Z.: Model uncertainty in commodity markets. SIAM J. Financ. Math. **7**(1), 1–33 (2016)
13. Zeng, Y., Li, D., Gu, A.: Robust equilibrium reinsurance-investment strategy for a mean–variance insurer in a model with jumps. Insur. Math. Econ. **66**, 138–152 (2016)
14. Yoshioka, H., Tsujimura, M.: A model problem of stochastic optimal control subject to ambiguous jump intensity. In: The 23rd Annual International Real Options Conference London, UK (2019).http://www.realoptions.org/openconf2019/data/papers/370.pdf
15. Crandall, M.G., Ishii, H., Lions, P.L.: User's guide to viscosity solutions of second order partial differential equations. Bull. Am. Math. Soc. **27**(1), 1–67 (1992)
16. Calder, J.: Lecture Notes on Viscosity Solutions. University of Minnesota (2018). http://www-users.math.umn.edu/~jwcalder/viscosity_solutions.pdf. Accessed 22 Dec 2019
17. Lungu, E.M., Øksendal, B.: Optimal harvesting from a population in a stochastic crowded environment. Math. Biosci. **145**(1), 47–75 (1997)
18. Mbroh, N.A., Munyakazi, J.B.: A fitted operator finite difference method of lines for singularly perturbed parabolic convection–diffusion problems. Math. Comput. Simul. **165**, 156–171 (2019)
19. Barles, G., Souganidis, P.E.: Convergence of approximation schemes for fully nonlinear second order equations. Asymptotic Anal. **4**(3), 271–283 (1991)

A Posteriori Error Estimation via Differences of Numerical Solutions

Aleksey K. Alekseev[ID], Alexander E. Bondarev[(⊠)][ID],
and Artem E. Kuvshinnikov[ID]

Keldysh Institute of Applied Mathematics RAS, Moscow, Russia
aleksey.k.alekseev@gmail.com, bond@keldysh.ru,
kuvsh90@yandex.ru

Abstract. In this work we address the problem of the estimation of the approximation error that arise at a discretization of the partial differential equations. For this we take advantage of the ensemble of numerical solutions obtained by independent numerical algorithms. To obtain the approximation error, the differences between numerical solutions are treated in the frame of the Inverse Problem that is posed in the variational statement with the zero order regularization. In this work we analyse the ensemble of numerical results that is obtained by five OpenFOAM solvers for the inviscid compressible flow around a cone at zero angle of attack. We present the comparison of approximation errors that are obtained by the Inverse Problem, and the exact error that is computed as the difference of numerical solutions and a high precision solution.

Keywords: Approximation error · Ensemble of numerical solutions · Differences of solutions · Inverse problem · Euler equations · Flow around a cone · OpenFOAM

1 Introduction

An estimation of the approximation (discretization) error is a problem of the high current interest due to the need for the verification of software and numerical solutions [1–3]. Let us consider the system of the partial differential equations written in the operator form $A\tilde{u} = f$ and a discrete algorithm $A_h u_h = f_h$ that approximates the system on some grid. There are two main approaches to the estimation of the approximation error $\Delta u = u_h - \tilde{u}$ [3]. *A priori* error estimation has the appearance $\|\Delta u\| \leq Ch^n$ (h is the step of discretization over space (or time), n is the order of approximation, C is an unknown constant). It is commonly used at a design and the theoretical analysis of the finite-difference or finite element algorithms (mainly from the standpoint of the convergence order determination). Unfortunately, the approach cannot be used in applications due to an unknown constant. *A posteriori* error estimation has the form $\|\Delta u\| \leq E(u_h)$ and contains some computable error indicator $E(u_h)$ that depends on the previously computed numerical solution u_h. The approach has no unknown constants and can be easily applied to practical computations. Rather often, it has a non-intrusive form of certain postprocessor. The highly efficient technique is developed for a posteriori error estimation in the domain of the finite-element analysis [3].

© Springer Nature Switzerland AG 2020
V. V. Krzhizhanovskaya et al. (Eds.): ICCS 2020, LNCS 12143, pp. 508–519, 2020.
https://doi.org/10.1007/978-3-030-50436-6_37

Unfortunately, for the problems of Computational Fluid Dynamics, the progress in a posteriori error estimation is limited so far due to irregularity of solutions (shock waves, contact surfaces etc.). As the main tool for the verification, the standards [1, 2] recommend the Richardson extrapolation. This method provides the pointwise approximation of the error field, however, at the cost of the extremely high computational burden [6]. There exist some computationally efficient approaches for the approximation error norm estimation, for example [4, 5]. However, these approaches do not provide the pointwise information on the error.

Thus, the need for a computationally inexpensive a posteriori estimation of approximation error still exists. By this reason we consider herein the computationally cheap approach to a posteriori error estimation that is based on the ensemble of numerical solutions obtained by independent methods. By independent methods we mean the numerical algorithms with different computational properties (an inner structure or the order of the approximation). The approximation error is estimated in the point-wise approach using the difference of solutions at every grid node separately. For this purpose, the Inverse Ill-posed Problem is posed in the variational statement that includes the Tikhonov zero order regularization [7, 8]. We provide the results of the numerical tests for compressible Euler equations that demonstrate both the estimated error and the exact error (obtained by the comparison of numerical solution with the precise solution [9]). These results demonstrate the applicability of the considered approach.

2   The Estimation of Approximation Error Using the Differences of Numerical Solutions

The approximation error estimation may be performed by different methods [3–6] including the Richardson extrapolation [6] that uses the set of numerical solutions obtained *for consequently refined grids*. In contrast, we consider an ensemble of numerical solutions $u^{(i)} = \tilde{u} + \Lambda u^{(i)} (i = 1..,n)$, obtained by n independent algorithms *on the same grid*. Let us note the exact (unknown) solution as $\tilde{u}$ and the approximation error (unknown) for i-th solution as $\Delta u^{(i)}$. The differences of numerical solutions $d_{ij} = u^{(i)} - u^{(j)} = \tilde{u} + \Delta u^{(i)} - \tilde{u} - \Delta u^{(j)} = \Delta u^{(i)} - \Delta u^{(j)}$ are computable. These differences are equal to the differences of errors and, hence, contain some information regarding the unknown approximation errors $\Delta u^{(i)}$. We get $N = \frac{n(n-1)}{2}$ equations that relate unknown errors and computable differences of numerical solutions

$$D_{ij}\Delta u^{(j)} = f_i, \tag{1}$$

where D_{ij} is a rectangular $N \times n$ matrix, the summation over the repeating index is implied starting from this point of the presentation.

The approximation errors may be estimated as

$$\Delta u^{(j)} = (D_{ij})^{-1} f_i. \tag{2}$$

Formally, this system of equations may be resolved for the dimensionality n that is equal or greater three. For $n = 3$ we use the option that follows:

$$\begin{pmatrix} f_1 \\ f_2 \\ f_3 \end{pmatrix} = \begin{pmatrix} d_{12} \\ d_{13} \\ d_{23} \end{pmatrix} = \begin{pmatrix} \Delta u^{(1)} - \Delta u^{(2)} \\ \Delta u^{(1)} - \Delta u^{(3)} \\ \Delta u^{(2)} - \Delta u^{(3)} \end{pmatrix} = \begin{pmatrix} u^{(1)} - u^{(2)} \\ u^{(1)} - u^{(3)} \\ u^{(2)} - u^{(3)} \end{pmatrix} \tag{3}$$

The Eq. (1) for this case has the form

$$\begin{pmatrix} 1 & -1 & 0 \\ 1 & 0 & -1 \\ 0 & 1 & -1 \end{pmatrix} \begin{pmatrix} \Delta u^{(1)} \\ \Delta u^{(2)} \\ \Delta u^{(3)} \end{pmatrix} = \begin{pmatrix} u^{(1)} - u^{(2)} \\ u^{(1)} - u^{(3)} \\ u^{(2)} - u^{(3)} \end{pmatrix} \tag{4}$$

We numerically solve Eq. (1) for three variables in the form (4) and for five variables in the similar form, omitted for brevity, by the method considered in following section.

3 Inverse Problem for the Estimation of Approximation Error

One may easy see the solution of system (1) to be invariant relatively a shift transformation $\Delta u^{(j)} = \Delta \tilde{u}^{(j)} + b \left(\Delta \tilde{u}^{(j)} \text{ is the exact error} \right)$ for any $b \in (-\infty, \infty)$. It is caused by the usage of the difference of solutions as the input data. Thus, the problem $\Delta u^{(j)} = (D_{ij})^{-1} f_i$ is underdetermined and, hence, ill-posed. The steady and bounded solution of the ill-posed problems requires a regularization ([7, 8]) that we consider as the zero order Tikhonov one by the following reasons. It is natural to search for solutions of the minimum shift error $|b|$ (in an ideal, $|b| \to 0$). We consider the search for the minimal L_2 norm of $\Delta u^{(j)}$ that provides restrictions on the absolute value of b:

$$\min(\delta) = \min \sum_{j}^{n} (\Delta u^{(j)})^2 / 2 = \min \sum_{j}^{n} (\Delta \tilde{u}^{(j)} + b)^2 / 2 \tag{5}$$

We consider this expression as the regularizing term in further analysis.

By accounting

$$\Delta \delta(b) = \sum_{j}^{n} (\Delta \tilde{u}^{(j)} + b) \Delta b, \tag{6}$$

one may see that the minimum over b occurs at b that equals the mean error with the opposite sign:

$$b = -\frac{1}{n}\sum_{j}^{n} \Delta\tilde{u}^{(j)} = -\Delta\bar{u}. \tag{7}$$

So, the relation (5) addresses the deviation of the exact error from the mean $\Delta u^{(j)} = \Delta\tilde{u}^{(j)} - \Delta\bar{u}$ and corresponds the minimum of the error dispersion. Fortunately, the assumption of δ minimality (5) ensures the boundedness of b. Unfortunately, the maximum attainable accuracy of $\Delta u^{(j)}$ is restricted by the mean error value.

In accordance with [8] we pose the Inverse Problem for $\Delta u^{(j)}$ estimation in the variational statement that assumes both the Eq. (1) to be valid and $\|\Delta u\|_{L_2}$ to be minimal (Eq. 5). This statement implies the minimization of the functional:

$$\varepsilon(\Delta u) = 1/2(D_{ij}\Delta u^{(j)} - f_i) \cdot (D_{ik}\Delta u^{(k)} - f_i) + \alpha/2(\Delta u^{(j)} E_{jk}\Delta u^{(k)}). \tag{8}$$

Herein, the first term of (8) defines the discrepancy of the predictions and observations, the second term of (8) poses the zero order Tikhonov regularization, α is the regularization parameter, E_{jk} is the unite matrix. Equation (8) demonstrates the standard statement of the variational Inverse Problem [8] that unify the observations and the a priori information (in present case, the information regarding the boundedness of solution).

To obtain the solution that minimize the functional we use the gradient based approach:

$$\Delta u^{(j),m+1} = \Delta u^{(j),m} - \tau\nabla\varepsilon. \tag{9}$$

In this expression m is the number of the iteration. In the numerical tests that follow the gradient is obtained by the direct numerical differentiation. The iterations stop at small enough value of the functional $\varepsilon \leq \varepsilon_*$ (in these tests $\varepsilon_* = 10^{-8}$). This optimization method is very primitive one (although quite operational) and easily may be replaced by any more advanced algorithm.

The solution of Eq. (8) is dependent on the regularization parameter α. The function $\left|\Delta u^{(j)}(\alpha)\right|$ is not bounded at $\alpha = 0$, so, it is not acceptable. At $\alpha \to \infty$ the asymptotics $\left|\Delta u^{(j)}(\alpha)\right| \to 0$ occurs that is not acceptable also. However, there exists a range of α where the weak dependence of the solution on α exhibits. In this range of the regularization parameter, the solution $\Delta u^{(j)}(\alpha)$ is close to the exact one $\Delta\tilde{u}^{(j)}$ and may be accepted [8].

4 The Test Problem

We consider the approximation error estimation for the test problem governed by two dimensional compressible Euler equations. The flowfield around cone at zero angle of attack $\alpha = 0°$ (Fig. 1) in the uniform supersonic flow of ideal gas is analyzed. The test problem statement is very close to one described by [5] from the gasdynamics viewpoint, and the same solvers are used.

The precise solution by [9] is used for estimation of the exact error.

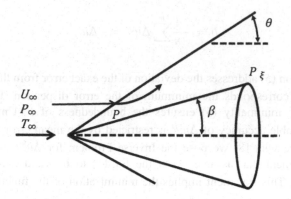

Fig. 1. Flow scheme

The cone of the half angle β = 20° is considered with the Mach number 2.

5 OpenFOAM Solvers

The following solvers from the OpenFOAM software package [10] were used similarly to [5]:

- *rhoCentralFoam* (rCF), based on a central-upwind scheme, which is a combination of central-difference and upwind schemes [11, 12]. The essence of the central-upwind schemes consists in a special choice of a control volume containing two types of domains: around the boundary points, the first type; around the center point, the second type. The boundaries of the control volumes of the first type are determined by means of local propagation velocities. The advantage of these schemes is the possibility to achieve a good resolution for discontinuous solutions in gas dynamics, using the appropriate technique for the numerical viscosity reducing.
- *sonicFoam* (sF), based on the PISO algorithm (Pressure Implicit with Splitting of Operator) [13]. The basic idea of the method is the application of two difference equations to calculate the pressure for the correction of the pressure field obtained from discrete analogs of the equations of moments and continuity. This approach takes to account that the velocities changed by the first correction may not satisfy the continuity equation, therefore, a second corrector is introduced, which enables us to calculate the velocities and pressures satisfying the linearized equations of momentum and continuity.
- *rhoPimpleFoam* (rPF), based on the PIMPLE algorithm, which is a combination of the PISO and SIMPLE (Semi-Implicit Method for Pressure-Linked Equations) algorithms. In this method, an external loop is added to the PISO algorithm, due to

which the method becomes iterative one and allows to count with the Courant number greater than 1.

- *pisoCentralFoam* (pCF), which is a combination of a Kurganov-Tadmor scheme [11] with the PISO algorithm [14].
- *QGDFoam* (QGDF), which is based on the implementation of quasi-gas dynamic equations [15].

All these solvers have the same approximation order, however, their inner structure is quite different. This circumstance engenders the hope that the exact approximation errors $\Delta \tilde{u}^{(j)}$ are independent and the mean error $\Delta \bar{u} = \frac{1}{n} \sum_j^n \Delta \tilde{u}^{(j)}$ has the acceptable magnitude and decays as the ensemble of solutions expands.

6   Numerical Results

6.1   Initial and Boundary Conditions

The computational domain and boundaries are provided in Fig. 2 similarly to [5].

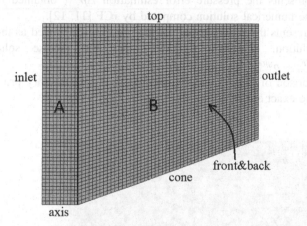

Fig. 2. Computational domain and boundaries

On the upper boundary *("top")* the zero gradient condition for the gas dynamic functions is specified. The same conditions are set on the right border (*"outlet"*). On the left border (*"inlet"*), the inflow parameters are set for Mach number $M = 2$: pressure P = 101325 Pa, temperature T = 300 K, speed $U = 694.5$ m/s. On the cone surface, the condition of zero normal gradient is posed for the pressure and the temperature, and the condition "*slip*" is posed for the speed, corresponding to the non-penetration condition.

The special "*wedge*" condition is used for the front ("*front*") and back ("*back*") borders to model the axisymmetric geometry in the OpenFoam package. The Open-Foam package also employs the special "*empty*" boundary condition. This condition is specified in cases when calculations in the given direction are not carried out. In our case, this condition is used for the "*axis*" border.

6.2 Solvers Settings

In the OpenFOAM package, there are two options for approximating differential operators: directly in the solver's code or using the fvSchemes and fvSolution configuration files. In order the comparison to be correct, we used the same parameters, where possible. The parameters are the same as in [5]. In the fvSchemes file: ddtSchemes – *Euler*, gradSchemes – *Gauss linear*, divSchemes – *Gauss linear*, laplacianSchemes – *Gauss linear corrected*, interpolationSchemes – *vanLeer*. In the fvSolution file: solver – *smoothSolver*, smoother *symGaussSeidel*, tolerance – 1e−09, nCorrectors – 2, nNonOrthogonalCorrectors – 1.

6.3 The Results of the Approximation Error Estimation in Comparison with the Exact Error

For the estimation of approximation error, we minimize the functional (8) using iterations described by the Expression (9) for all flow parameters $\{\rho, u, v, p\}$ separately at every grid point. Below, we present the results mainly for pressure, as the most important and expressive gas-dynamic function for inviscid flows with shock waves.

Figure 3 presents the pressure error estimation $\Delta p^{(IP)}$ obtained by the Inverse Problem for the numerical solution computed by rCF [11, 12].

Figure 4 presents the "exact error" for the pressure calculated as the difference of numerical solution, computed by rCF, and the precise solution by [9] $\Delta p^{(rCF)} = p^{(rCF)} - p^{(precise)}$.

The comparison of Figs. 3 and 4 demonstrate the satisfactory proximity of error estimates to the exact error.

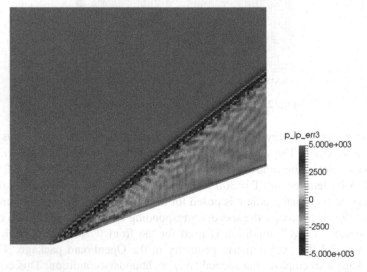

Fig. 3. The error of the pressure estimated by the Inverse Problem (for the flowfield, computed by rCF)

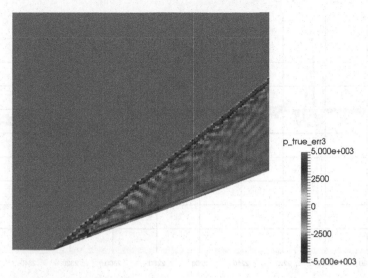

Fig. 4. The exact error of the pressure (for the flowfield, computed by rCF)

The Fig. 5 presents the piece of vectorized grid function of total pressure error (computed by rCF) obtained by the Inverse Problem (using five solutions) in comparison with the exact error. The index along abscissa axis $i = N_x(k_x - 1) + m_y$ is defined by indexes along $X(k_x)$ and $Y(m_y)$. The periodical jump of solution variables corresponds the transition through the shock wave. One may see that the error at the shock wave remains underestimated. This behavior is expected since the error at a shock tends to be singular at the mesh refinement. The scales in Figs. 3 and 4 are chosen to be smaller if compare with Fig. 5 in order to increase the visibility.

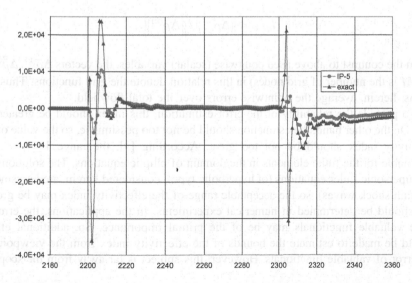

Fig. 5. The comparison of the pressure error (rCF), estimated by the Inverse Problem, with the exact error

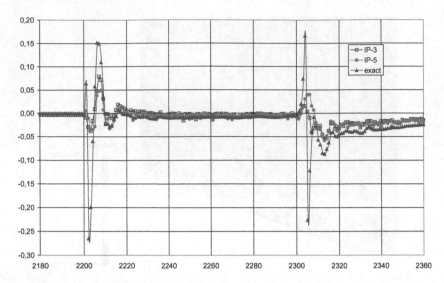

Fig. 6. The comparison of the density error (rCF), estimated by the Inverse Problem, with the exact error. IP-3 corresponds 3 solutions, IP-5 - five solutions

The Fig. 6 presents the piece of vectorized grid function of density error (for the solution computed by rCF) obtained by Inverse Problem in comparison with the exact error for the ensembles of solutions containing three and five samples. Herein, IP-3 corresponds three solutions (pCF, QGDF, rCF), IP-5 corresponds all five solutions.

In accordance with [3], the global quality of a posteriori error estimation may be described by the *effectivity index* that equals the relation of the estimated error norm to the exact error norm:

$$I_{eff,k} = \left\| \Delta \vec{\rho}^{(k)} \right\| / \left\| \Delta \vec{\tilde{\rho}}^{(k)} \right\|. \tag{10}$$

In the contrast to above used pointwise (scalar) variables, the vectors $\Delta \vec{\rho}^{(k)}, \Delta \vec{\tilde{\rho}}^{(k)} \in R^M$ (M is the number of grid nodes) in this relation denote the grid functions. Thus, the norms, herein, average the pointwise errors over the total flowfield.

To provide the reliability of the error estimation, this index should be greater the unit. On the other hand, the estimation should be not too pessimistic, so the value of the effectivity index should be not too great. According [3], the range $1 \leq I_{eff} \leq 3$ is acceptable for the finite elements in the domain of elliptic equations. The solutions for the supersonic Euler equations (of hyperbolic type), considered herein, are less smooth (contain shock waves), so the acceptable range of the effectivity index may be greater and should be determined in numerical experiments. In the applications, the error of some valuable functionals may be of the primal importance. So, additional efforts should be made to estimate the bounds of the effectivity index from the viewpoint of the error of valuable functionals. However, this subject is far away from the scope of present paper. Some information on the relation of the approximation error and the uncertainty of valuable functionals may be found in [16].

The corresponding values of the effectivity index are provided in the Table 1 for the sets of 3 and 5 solutions. L_2 norm of the global error (averaged over the total grid) is computed for the effectivity index calculation. To ensure the reliable error estimation ($I_{eff} \geq 1$) certain safety coefficient (about three) may be used.

One may expect the improvement of the error estimation quality as the number of solutions increases. This assumption is partly supported by the comparison of the effectivity index values for three and five solutions presented in the Table 1.

Table 1. Effectivity indexes of error estimation for different solutions

	I_{eff}^{pCF}	I_{eff}^{QGDF}	I_{eff}^{rCF}	I_{eff}^{rPF}	I_{eff}^{sF}
Five solutions	0.35	0.37	0.38	0.39	0.26
Three solutions	0.24	0.29	0.26	–	–

7 Discussion

Unlike the computationally cheap estimation of the error norm [4, 5], also based on the set of numerical solutions, the above discussed approach provides the pointwise error estimation.

Certainly, this approach is less accurate if compared with the Richardson extrapolation due to the presence of the irremovable error, proportional to the mean error over the ensemble of solutions. Nevertheless, in contrast to the Richardson extrapolation, the considered postprocessor is much more computationally inexpensive since it implies only single grid computations (without a mesh refinement). Additionally, it has some natural parallelization properties since the solutions are obtained by the independent codes.

It should be noted that the calculations of the set of solutions for the flow around the cone were carried out using the approach of the generalized computational experiment [17, 18]. This approach is based on the parallel solution (in the multitask mode) of the same problem for the ensemble of input data corresponding a variation of the determining parameters. This allows one to analyze a solution as the element of some ensemble engendered by the class of problems, which is set by the choice of determining parameters. The problem at hand can be considered as the generalized computational experiment, where the choice of a solver plays the role of a determining parameter.

The OpenFOAM algorithms are considered herein as independent ones from the standpoint of their logics and inner structure, however, the independence from the viewpoint of the mean error decay at the ensemble enhancement (important from the viewpoint of irremovable error diminishing) needs for a further analysis.

8 Conclusion

The Inverse Problem is stated for the estimation of the point-wise approximation error occurring at a discretization of the Partial Differential Equations. The differences between solutions, obtained by independent numerical algorithms, are used as the input data. The variational statement with the zero order regularizing term is considered for the problem.

The numerical tests demonstrate the feasibility for the estimation of the point-wise approximation error via the ensemble of numerical solutions obtained using the OpenFOAM software package.

The minimal number of numerical solutions that is necessary for the error estimation is equal to three. Five solutions provide a bit better results from the viewpoint of effectivity index.

The considered approach is less accurate in comparison with the Richardson extrapolation. However, the proposed approach is much more computationally cheap and ensures the satisfactory accuracy.

Acknowledgments. This work was supported by grant of Russian Science Foundation № 18-11-00215.

References

1. Guide for the Verification and Validation of Computational Fluid Dynamics Simulations, American Institute of Aeronautics and Astronautics, AIAA-G-077-1998, Reston, VA (1998)
2. Standard for Verification and Validation in Computational Fluid Dynamics and Heat Transfer, ASME V&V 20-2009 (2009)
3. Repin, S.I.: A posteriori estimates for partial differential equations, vol. 4. Walter de Gruyter (2008). https://doi.org/10.1515/9783110203042
4. Alekseev, A.K., Bondarev, A.E., Navon, I.M.: On Triangle Inequality Based Approximation Error Estimation. arXiv:1708.04604 [physics.comp-ph], 16 August 2017
5. Alekseev, A.K., Bondarev, A.E., Kuvshinnikov, A.E.: Verification on the ensemble of independent numerical solutions. In: Rodrigues, J.M.F., et al. (eds.) ICCS 2019. LNCS, vol. 11540, pp. 315–324. Springer, Cham (2019). https://doi.org/10.1007/978-3-030-22750-0_25
6. Alexeev, A.K., Bondarev, A.E.: On some features of richardson extrapolation for compressible inviscid flows. Mathematica Montisnigri **XL**, 42–54 (2017)
7. Tikhonov, A.N., Arsenin, V.Y.: Solutions of Ill-Posed Problems. Winston and Sons, Washington, DC (1977)
8. Alifanov, O.M., Artyukhin, E.A., Rumyantsev, S.V.: Extreme Methods for Solving Ill-Posed Problems with Applications to Inverse Heat Transfer Problems. Begell House, Danbury (1995)
9. Babenko, K.I., Voskresenskii, G.P., Lyubimov, A.N., Rusanov, V.V.: Three-Dimensional Ideal Gas Flow Past Smooth Bodies. Nauka, Moscow (1964). (in Russian)
10. OpenFOAM. http://www.openfoam.org. Accessed 30 Jan 2020
11. Kurganov, A., Tadmor, E.: New high-resolution central schemes for nonlinear conservation laws and convection-diffusion equations. J. Comput. Phys. **160**(1), 241–282 (2000). https://doi.org/10.1006/jcph.2000.6459

12. Greenshields, C., Wellerr, H., Gasparini, L., Reese, J.: Implementation of semi-discrete, non-staggered central schemes in a colocated, polyhedral, finite volume framework, for high-speed viscous flows. Int. J. Numer. Meth. Fluids **63**(1), 1–21 (2010). https://doi.org/10.1002/fld.2069

13. Issa, R.: Solution of the implicit discretized fluid flow equations by operator splitting. J. Comput. Phys. **62**(1), 40–65 (1986). https://doi.org/10.1016/0021-9991(86)90099-9

14. Kraposhin, M., Bovtrikova, A., Strijhak, S.: Adaptation of Kurganov-Tadmor numerical scheme for applying in combination with the PISO method in numerical simulation of flows in a wide range of Mach numbers. Procedia Comput. Sci. **66**, 43–52 (2015). https://doi.org/10.1016/j.procs.2015.11.007

15. Kraposhin, M.V., Smirnova, E.V., Elizarova, T.G., Istomina, M.A.: Development of a new OpenFOAM solver using regularized gas dynamic equations. Comput. Fluids **166**, 163–175 (2018). https://doi.org/10.1016/j.compfluid.2018.02.010

16. Alekseev, A.K., Bondarev, A.E., Kuvshinnikov, A.E.: On uncertainty quantification via the ensemble of independent numerical solutions. J. Comput. Sci. (2020). https://doi.org/10.1016/j.jocs.2020.101114

17. Bondarev, A.E.: On the construction of the generalized numerical experiment in fluid dynamics. Mathematica Montisnigri **XLII**, 52–64 (2018)

18. Bondarev, A.E., Galaktionov, V.A.: Generalized computational experiment and visual analysis of multidimensional data. Sci. Vis. **11**(4), 102–114 (2019). https://doi.org/10.26583/sv.11.4.09

Global Sensitivity Analysis of Various Numerical Schemes for the Heston Model

Emanouil Atanassov[1]([✉]), Sergei Kucherenko[2], and Aneta Karaivanova[1]

[1] Institute of Information and Communication Technologies, Sofia, Bulgaria
{emanouil,anet}@parallel.bas.bg
[2] Imperial College, London, UK
s.kucherenko@imperial.ac.uk

Abstract. The pricing of financial options is usually based on statistical sampling of the evolution of the underlying under a chosen model, using a suitable numerical scheme. It is widely accepted that using low-discrepancy sequences instead of pseudorandom numbers in most cases increases the accuracy. It is important to understand and quantify the reasons for this effect. In this work, we use Global Sensitivity Analysis in order to study one widely used model for pricing of options, namely the Heston model. The Heston model is an important member of the family of the stochastic volatility models, which have been found to better describe the observed behaviour of option prices in the financial markets. By using a suitable numerical scheme, like those of Euler, Milstein, Kahl-Jäckel, Andersen, one has the flexibility needed to compute European, Asian or exotic options. In any case the problem of evaluating an option price can be considered as a numerical integration problem. For the purposes of modelling and complexity reduction, one should make the distinction between the model nominal dimension and its effective dimension. Another notion of "average dimension" has been found to be more practical from the computational point of view. The definitions and methods of evaluation of effective dimensions are based on computing Sobol' sensitivity indices. A classification of functions based on their effective dimensions is also known. In the context of quantitative finance, Global Sensitivity Analysis (GSA) can be used to assess the efficiency of a particular numerical scheme. In this work we apply GSA based on Sobol sensitivity indices in order to assess the interactions of the various dimensions in using the above mentioned schemes. We observe that the GSA offers useful insight on how to maximize the advantages of using QMC in these schemes.

1 Introduction to Option Pricing Under the Heston Stochastic Volatility Model

Financial options are instruments which allow their holder to obtain certain payout, which depends on the price of the underlying security. While the payout

We acknowledge the provided access to the e-infrastructure of the NCHDC – part of the Bulgarian National Roadmap on RIs, with the financial support by the Grant No DO1-271/16.12.2019.

V. V. Krzhizhanovskaya et al. (Eds.): ICCS 2020, LNCS 12143, pp. 520–528, 2020.
https://doi.org/10.1007/978-3-030-50436-6_38

of European options only depends on the price at the time of expiration, there are options that depend on the evolution of the price throughout certain time period in a complex way. A Monte Carlo method for determining the prices of options would be obtained by sampling paths for the price of the underlying following a chosen stochastic model. In such case the price of the financial option, whose payout is defined as a function of the evolution of the price: $F(\{S_t\}_{t=0}^T)$ would be evaluated as the average of the (discounted) value of a function F' over the sampled paths, where F' uses only values of the price at certain points, determined by the discretisation used. It is obvious that such a method will have both stochastic and deterministic components of the error, i.e., the estimate would be biased.

The more sophisticated models involve also the evolution of the volatility, which is not directly observable. This evolution can be deterministic or following a stochastic process. It is widely accepted that stochastic volatility models can better explain the observed behaviours in financial markets, but pose numerical difficulties. One of the most popular stochastic volatility models is the model of Heston. The evolution of the Heston model is described by the following two equations:

$$dS_t = rS_t dt + \sqrt{v_t}S_t dW_s$$

$$dv_t = \kappa\left(\theta - v_t\right)dt + \sigma_v\sqrt{v_t}dW_v,$$

where S_t is the price and v_t is the volatility, while dW_s and dW_v are two Brownian motions that are correlated with a coefficient ρ. The parameters of the model are $\kappa, \theta, \sigma, S_0, V_0, \rho, r$. The determination of these parameters is out of scope of this paper. One can read more about the Heston model in [9]. The Monte Carlo numerical schemes are based on choosing a time step for discretization and then sampling the path of the underlying and the volatility. Brody and Kaya [7] demonstrated how it is possible to sample from the exact distributions of the price and volatility at the expense of more computational power requirements. In practice other numerical schemes require less computations and achieve sufficient accuracy when the time step is small enough. In this work we considered the schemes of Euler-Maruyama (see, e.g., [12]), Kahl-Jäckel [10], Milstein ([8]) and Andersen [5]. We denote them by the letters A, B, C, D respectively. For the scheme of Euler-Maruyama we apply the Lord's truncation method [11].

Under these schemes each time step requires the sampling of two random variables. Usually the inverse function method is used and thus we can assume that only random number uniformly distributed in $(0,1)$ are used. Thus the constructive dimensionality of the algorithm in the sense, defined by Sobol', is $2n$, where n is the number of time steps. The practitioners in Mathematical Finance also need to compute various derivatives of the option price, which are generally known as Greeks. For example, the Delta of an option is the derivative of the price with respect to the (initial) price of the underlying, while the Theta is the derivative with respect to the remaining time to expiration. Such quantities can be estimated by introducing a small number ϵ and computing the option price also for values of the corresponding parameter with added $\pm\epsilon$ and using a formula for approximate computation of the derivative.

The Global Sensitivity Analysis methodology allows to assess the importance of each variable and quantify the various interactions between variables. In the next section we shall describe how this can be applied in our problem.

2   Computation of Global Sensitivity Indices in the Context of the Numerical Schemes for the Heston Model

The Global Sensitivity Analysis is based on the computation of the Sobol' sensitivity indices, which quantify the contribution of the various terms of the ANOVA decomposition of a function f:

$$f(x) = f_0 + \sum_{i=1}^{d} f_i(x_i) + \sum_{i=1}^{d} \sum_{j=i+1}^{d} f_{ij}(x_i, x_j) \dots$$

towards it's overall variance D. Sobol' defined the coefficients

$$S_{i_1,\dots,i_k} = D_{i_1,\dots,i_k}/D,$$

so that one can evaluate the sensitivity of the function to subsets of variables, where $D_{i_1,\dots,i_k}$ denotes the variance of the corresponding term in the ANOVA decomposition.

These coefficients sum to 1. There are substantially different algorithms for computing them. In our work we followed the approach proposed in [1], where formula (15) leads to efficient Monte Carlo and consequently quasi-Monte Carlo method for computing the indices. The total sensitivity indices are also important to consider, since they can be computed efficiently with a similar formula, while providing a numerical estimation for the total contribution of a variable to the variance of the function, summing all the coefficients where it is part of the subset. Various works establish the way to use the Sobol' sensitivity indices in order to compute other useful measures, for example, the mean dimensionality in [3] or the effective dimensions in [1]. In [4] one can see how GSA can be used to improve option pricing. Since the Heston model has stochastic volatility, one has to deal with a more complex numerical schemes and consequently more heterogenious distribution of the uncertainty.

All the numerical schemes for the Heston model that we consider in this work are built upon sampling two random variables for each time step. In most cases these variables are normally distributed, while the Andersen scheme is more complex. In practice one would use a pseudorandom number generator or a generator for a low-discrepancy sequence and then use the inverse function method in order to obtain the appropriate normally distributed number. Initially we used the natural ordering of the variables, so that each time step uses one odd and one even coordinate. In this way the sampling of a path with n steps requires

$2n$ pseudorandom numbers or $2n$ coordinates of a low-discrepancy sequence. Later on we shall discuss how, based on the results obtained from the GSA, one may reorder the variables in order to improve the accuracy of the computation.

Because of the constraints on the available compute power, in our analysis of these numerical schemes we limited ourselves to compute only coefficients and total coefficients of orders one and two, which already give idea about the behaviour of the schemes. The method we use requires us to effectively double the number of coordinates used. It is a well known fact that there is substantial advantage in using low-discrepancy sequence for such kinds of problems, but using Global Sensitivity Analysis we quantify the contribution of the different dimensions and obtain suggestions about possible re-orderings of the variables in order to improve the accuracy of the computation, when using low-discrepancy sequences.

3 Numerical Results and Discussion

First of all we compared pure Monte Carlo method for computing the price of an option using the above schemes with a quasi-Monte Carlo method which utilizes either modified Halton (see, e.g., [6]) or Sobol' sequences. The results cover the calculation of the price of an Asian option. The payout of an Asian option is defined as

$$\max\left(\frac{1}{N} \sum_{i=1}^{N} X_i - K, 0 \right).$$

The parameters of the Heston model are $r = 3.19\%, \kappa = 6.21, \theta = 0.019, \sigma_v = 0.61, \rho = -0.7, S_0 = 100, V_0 = 0.010201$. One can see that with the increase in number of time steps the accuracy of the quasi-Monte Carlo method improves significantly more than that of the Monte Carlo method. In order to obtain the results in Table 1 the number of steps is fixed at 12 (corresponding to the number of months in the year). The scheme used is the Euler - Murayama scheme with the Lord full-truncation.

Table 1. Error from computation with 12 time steps, scheme A

N	MC	Sobol	Halton
256	0.60	0.23	0.13
512	0.23	0.18	0.08
1024	0.13	0.09	0.04
2048	0.08	0.01	0.02

For the computations in Table 2 the number of steps is fixed at 32, where the results for the price are on the left and the results for the delta are on the right. It is obvious that the accuracy for the delta is smaller. This time the more complex Andersen scheme (scheme D) is used. The parameters are

$$r = 0.\%, \kappa = 1.0606, \theta = 0.0733, \sigma_v = 0.3918, \rho = -0.3456,$$
$$S_0 = 100, V_0 = 0.0222, K = 100.$$

Table 2. Error from computation with 32 steps (nominal dimension 64), price (left) and delta (right), scheme D

N	MC	Sob.	Hal.	MC	Sob.	Hal.
256	0.61	0.12	0.40	1.02	.66	0.85
512	0.53	0.06	0.20	.75	0.44	0.59
1024	0.31	0.07	0.12	0.45	0.50	0.44

These results show the substantial advantage of using low-discrepancy sequences with these schemes. However, in such case we have the option to reorder of dimensions of the low-discrepancy sequence in order to position the first coordinates of the generated points to sample the most important variables. The importance of the variables is quantified by the Sobol sensitivity indices. Our first goal was to compute Sobol' one-dimensional sensitivity indices for the price under the different schemes. All the examples consider the same Heston model, with the following parameters: time steps - 52, time to expiration - 1, $\kappa = 1.0606$, interest rate $r = 0$, starting volatility $\theta_0 = 0.0222$, long-term mean volatility $\theta_v = 0.0733$, $\rho = 0.3456$, volatility of volatility $\varepsilon = 0.3918$. The starting price is $S_0 = 1$. and we consider an Asian option with strike $K = 1$. We also considered a knock out-option with a knock out level $1.2\,K$, so that the owner receives payout

$$\max{(X_N - K, 0)}$$

only if all $X_i < 1.2\,K$, otherwise the payout is zero.

On Fig. 1 one can see the one-dimensional Sobol' sensitivity indices for the Asian option under the different schemes.

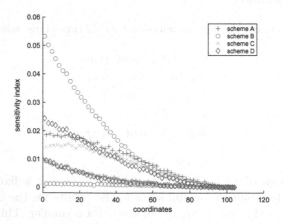

Fig. 1. One-dimensional Sobol' sensitivity indices for the price of an Asian option under the Heston model

We observe how three of the schemes lead to similar indices, while the Milstein scheme has substantially different behaviour. In all cases the indices decrease with the number of timesteps, which suggests that the leading dimensions of the low-discrepancy sequences should be used to sample the first timesteps. It is also noticeable how the even dimensions have much larger coefficients than the odd dimensions. Since this is the case for all schemes under considerations, the logical suggestion is to reorder the coordinates of the low-discrepancy sequence so that the first half of the coordinates are used to sample the odd coordinates and the second half of the coordinates are used to sample the even dimensions. We remind that two numbers are used for each timestep. These computations were carried out using the scrambled Sobol' sequence, with direction numbers provided by [2]. We used $4 \times 52 = 208$ dimensions and 2^{18} points. For the price of the knock-out option we make the same observation with regards to the odd and even dimensions (see Fig. 2). However, the last few coordinates have slightly increased importance, while also the accuracy of the computation of the indices is not as good as in the case of the Asian option.

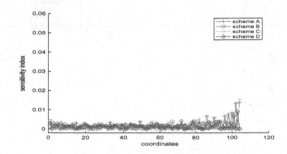

Fig. 2. One-dimensional Sobol' sensitivity indices for the price of a knock-out option under the Heston model

The accuracy improves when using the total Sobol' sensitivity indices. As we can see in Fig. 3, the Milstein scheme (scheme B) is again substantially different from the other 3 schemes in the observed behaviour of the indices. It seems that there is decreasing importance of the dimensions and the difference between odd and even dimensions is not so prounounced, except for the Milstein scheme. It would be logical to reorder the variables according to their importance as measured with the total Sobol' sensitivity indices, since they take into account the non-linear interactions.

The computation of sensitivity indices for the various "Greeks" is achieved by introducing a small parameter ϵ and using numerical differentiation formulae. Although we can expect decreased accuracy because of the differentiation, we can see in the next figure, showing the one-dimensional Sobol' sensitivity indices for the Delta of an Asian option, that sufficient accuracy is achieved and the conclusions with regards to the optimal use of the coordinates of the low-discrepancy sequence are similar to the case for the price (see Fig. 4).

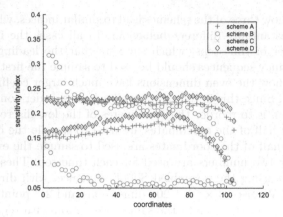

Fig. 3. One-dimensional Sobol' total sensitivity indices for the price of a knock-out option under the Heston model

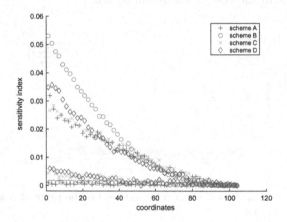

Fig. 4. One-dimensional Sobol' sensitivity indices for the Delta of an Asian option under the Heston model.

It is also possible to compute Sobol' sensitivity indices for larger subsets of variables. Unfortunately, this increases the computational complexity substantially. That is why we only show results obtained for pairs of variables. In this case, only the results for the Euler-Murayama scheme with the Lord's truncation (scheme A) are shown (see Fig. 5).

We show the interactions between indices for odd and even dimensions in 3 surfaces. As it can be expected, the highest importance is for pairs of indices that both correspond to even dimension. There is also visible decrease of the importance when increasing the index of the coordinate.

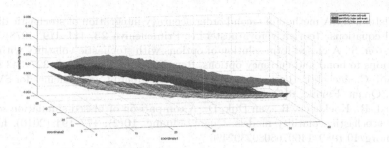

Fig. 5. Sobol' sensitivity indices for pairs of variables, when computing the price of an Asian option under the Heston model.

4 Conclusions

The various numerical schemes used for option pricing via the Heston model can benefit from use of low-discrepancy sequences. The Sobol' sequence with Owen scrambling and the modified Halton sequences proved to be effective. By studying the Sobol' sensitivity indices we can quantify the contribution of the various dimensions and their interactions to the overall variance, which gives us ideas on how to reorder the coordinates in order to increase the accuracy. Once it has been determined how much each variable contributes to the final result of the simulation, it becomes possible to optimise the parameters of the Sobol' or Halton sequences jointly with respect to a measure of quality of distribution that is weighted according to the compu. In a future work we plan to demonstrate how this approach improves the accuracy.

References

1. Kucherenko, S., Feil, B., Shah, N., Mauntz, W.: The identification of model effective dimensions using global sensitivity analysis. Reliab. Eng. Syst. Safe. **96**, 440–449 (2011)
2. Software for generating the Sobol sequences by BRODA. https://www.broda.co.uk/software.html. Accessed 7 Feb 2020
3. Liu, R., Owen, A.: Estimating mean dimensionality of analysis of variance decompositions. J. Am. Stat. Assoc. **101**(474), 712–721 (2006)
4. Bianchetti, M., Kucherenko, S., Scoleri, S.: Pricing and Risk Management with High-Dimensional Quasi Monte Carlo and Global Sensitivity Analysis. Wilmott, July, pp. 46–70 (2015)
5. Andersen, L.B.G.: Efficient simulation of the Heston stochastic volatility model, 23 January 2007. https://doi.org/10.10007/ssrn.946405
6. Atanassov, E.I., Durchova, M.K.: Generating and testing the modified halton sequences. In: Dimov, I., Lirkov, I., Margenov, S., Zlatev, Z. (eds.) NMA 2002. LNCS, vol. 2542, pp. 91–98. Springer, Heidelberg (2003). https://doi.org/10.1007/3-540-36487-0_9
7. Broadie, M., Kaya, O.: Exact simulation of stochastic volatility and other affine jump diffusion models. Oper. Res. **54**(2), 217–231 (2006)

8. Milstein, G.: A method of second order accuracy integration of stochastic differential equations. Teoriya Veroyatnostei i ee Primeneniya **23**, 414–419 (1978)
9. Heston, S.: A closed-form solution of options with stochastic volatility with applications to bond and currency options. Rev. Financial Stud. **6**(2), 327–343 (1993)
10. Kahl, C., Jackel, P.: Fast strong approximation Monte-Carlo schemes for SV models. Quant. Financ. **6**(6), 513–536 (2005)
11. Lord, R., Koekkoek, R., van Dijk, D.: A comparison of biased simulation schemes for stochastic volatility models. Quant. Financ. **10**(2), 177–194 (2010). https://doi.org/10.1080/14697680802392496
12. Kloeden, P.E., Platen, E.: Numerical Solution of Stochastic Differential Equations. Springer, Heidelberg (1992). https://doi.org/10.1007/978-3-662-12616-5. ISBN 3-540-54062-8

Robust Single Machine Scheduling with Random Blocks in an Uncertain Environment

Wojciech Bożejko[1] , Paweł Rajba[2](✉) , and Mieczysław Wodecki[3]

[1] Department of Automatics, Mechatronics and Control Systems,
Faculty of Electronics, Wrocław University of Technology,
Janiszewskiego 11-17, 50-372 Wrocław, Poland
wojciech.bozejko@pwr.wroc.pl
[2] Institute of Computer Science, University of Wrocław,
Joliot-Curie 15, 50-383 Wrocław, Poland
pawel@cs.uni.wroc.pl
[3] Telecommunications and Teleinformatics Department, Faculty of Electronics,
Wrocław University of Science and Technology,
Wybrzeże Wyspiańskiego 27, 50-370 Wrocław, Poland
mieczyslaw.wodecki@pwr.wroc.pl

Abstract. While scheduling problems in deterministic models are quite
well investigated, the same problems in an uncertain environment require
very often further exploration and examination. In the paper we consider
a single machine tabu search method with block approach in an uncertain
environment modeled by random variables with the normal distribution.
We propose a modification to the tabu search method which improves
the robustness of the obtained solutions. The conducted computational
experiments show that the proposed improvement results in a much more
robust solutions than the ones obtained in the classic block approach.

Keywords: Single machine scheduling · Uncertain parameters ·
Normal distribution · Tabu search · Block approach

1 Introduction

Uncertainty occurs in many production processes and has a direct impact on
their smooth execution. For instance it is important in construction domain to
deliver goods with no delays, but it is not easy to meet this requirement as the
transportation time depends on many external factors like weather conditions,
traffic jams, driver's condition and many others. Moreover, effective solving practical problems and taking the best approach requires also thorough knowledge
of the process or production system and values of all parameters. For example
an uncertain data of the duration of activities (operations) can be measured
and in result: approximated as deterministic ones in case the variance is small
enough, modeled by an appropriate probabilistic distribution or determined the

© Springer Nature Switzerland AG 2020
V. V. Krzhizhanovskaya et al. (Eds.): ICCS 2020, LNCS 12143, pp. 529–538, 2020.
https://doi.org/10.1007/978-3-030-50436-6_39

membership function for the fuzzy representation. So, as in practice it is difficult to clearly determine the process parameters, quite often safe ones are taken (e.g. assume longer transportation time) what is an opportunity for further improvements.

Research on scheduling problems carried out for many years is related primarily to deterministic models where the key assumption is that parameters are well defined. For those, mostly belonging to the class of strongly NP-hard problems, a number of very effective approximate algorithms have been developed. Solutions determined by these algorithms are very often only slightly worse from the optimal ones. In practice, however, as already mentioned, some parameters (e.g. operation times) may differ during the process execution from the initially assumed values. This can cause that the actual cost of execution is much bigger than expected what leads to either losing optimality or even acceptability (feasibility) of solutions.

In order to close that gap in recent years more and more research has been conducted on developing methods which find more robust solution resistant to data disturbance. Uncertain parameters are usually represented by random variables or fuzzy numbers and extensive review of methods and algorithms for solving optimization problems with random parameters is presented by Vondrák in monograph [12] and newer of Shang et al. [9], Soroush [10], Xiaoqiang et al. [14], Urgo and Vancza [11], Zhang et al. [15] and Bożejko et al. [2], [4] and [6].

In this paper we consider a single machine scheduling problem with due dates in two variants where either job execution times or due dates are represented by independent variables with normal distribution. We also present some properties of the problem (so-called block elimination properties) accelerating the review of neighborhoods in local search algorithms. The main goal is to compare the robustness of the block-based tabu search algorithm in the classic and the proposed random model and show the superiority of the latter one.

2 Deterministic Scheduling Problem

Let $\mathcal{J} = \{1, 2, \ldots, n\}$ be a set of jobs to be executed on a single machine. At any given moment a machine can execute exactly one job and all jobs must be executed without preemption. For each task $i \in \mathcal{J}$ let p_i be a *processing time*, d_i be a *due date* and w_i be a cost for tardy jobs.

Every sequence of jobs execution can be presented as a permutation $\pi = (\pi(1), \pi(2), \ldots, \pi(n))$ of items from the set $\mathcal{J}$.

Let Π be the set of all permutations of the set $\mathcal{J}$. For every permutation $\pi \in \Pi$ we define

$$C_{\pi(i)} = \sum_{j=1}^{i} p_{\pi(j)}$$

as a completion time of a job $\pi(i)$.

The cost of jobs' execution determined by the permutation π is as follows

$$\sum_{i=1}^{n} w_{\pi(i)} U_{\pi(i)}. \tag{1}$$

where

$$U_{\pi(i)} = \begin{cases} 0 & \text{for} \quad C_{\pi(i)} \leqslant d_{\pi(i)}, \\ 1 & \text{for} \quad C_{\pi(i)} > d_{\pi(i)}. \end{cases}$$

We consider the optimization problem where the goal is to find a permutation $\pi^* \in \Pi$ which minimizes the cost of jobs' execution:

$$W(\pi^*) = \min_{\pi \in \Pi} \left(\sum_{i=1}^{n} w_{\pi(i)} U_{\pi(i)} \right).$$

3   Probabilistic Jobs Times

In order to simplify the further considerations we assume w.l.o.g. that at any moment the considered solution is the natural permutation, i.e. $\pi = (1, 2, \ldots, n)$. Moreover, if X is a random variable, then F_X denotes its cumulative distribution function.

In this section we consider a TWT problem with uncertain parameters. We investigate two variants: (a) uncertain processing times and (b) uncertain due dates.

3.1   Random Processing Times

Random processing times are represented by random variables with the normal distribution $\tilde{p}_i \sim N(p_i, c \cdot p_i)$, $i \in \mathcal{J}$. Other parameters, i.e. due dates d_i and costs w_i are deterministic. Then completion times $\tilde{C}_i$ are random variables:

$$\tilde{C}_i \sim N \left(p_1 + p_2 \ldots + p_i, c \cdot \sqrt{p_1^2 + \ldots + p_i^2} \right) \tag{2}$$

and delays are random variables

$$\tilde{U}_i = \begin{cases} 0 & \text{dla} \quad \tilde{C}_i \leqslant d_i, \\ 1 & \text{dla} \quad \tilde{C}_i > d_i. \end{cases} \tag{3}$$

For each permutation $\pi \in \Pi$ the cost in the deterministic model is defined as $W(\pi) = \sum_{i=1}^{n} w_{\pi(i)} U_{\pi(i)}$ (see (1)). A corresponding cost in the random model is defined as the following random variable:

$$\widetilde{W}(\pi) = \sum_{i=1}^{n} w_i \tilde{U}_i. \tag{4}$$

In order to compare the costs of permutations from the set Π we introduce the following *comparison function* to calculate the value:

$$\mathcal{W}(\pi) = w_i E(\tilde{U}_i) \tag{5}$$

where $E(\tilde{U}_i)$ is the expected value of the random variable $\tilde{U}_i$.

3.2 Random Due Dates

Random due dates are represented by random variables with the normal distribution $\tilde{d}_i \sim N(d_i, c \cdot d_i))$, $i \in \mathcal{J}$. Other parameters, i.e. processing times p_i and costs w_i are deterministic. Delay indication is a random variable

$$\tilde{U}_i = \begin{cases} 0 & \text{dla} \quad C_i \leqslant \tilde{d}_i, \\ 1 & \text{dla} \quad C_i > \tilde{d}_i. \end{cases} \tag{6}$$

In this variant of the problem we apply the comparison function (5) defined in the previous section.

The *TWT* problem in both variants (i.e. with random processing times and random due dates) is to find a permutation for which the comparison function (5) is minimal in the set Π. We denote the probabilistic version of the problem as *TWTP*. As the deterministic version, the problem belongs to the class of *NP*-hard problems.

4 Blocks in Random Model

4.1 Random Processing Times and Due Dates

Each permutation $\pi \in \Pi$ is decomposed into m ($m \leqslant n$) subpermutations $\widetilde{B}_1, \ldots, \widetilde{B}_m$, called random blocks for π, which satisfy the following criteria:

1. $\widetilde{B}_k = (s_k, s_k + 1, \ldots, l_k - 1, l_k)$, $l_{k-1} + 1 = s_k \leqslant l_k$, $k = 1, \ldots, m$, $l_0 = 0$, $l_m = n$.
2. All jobs $j \in B_k$ satisfy either the condition

$$P(\tilde{d}_j \geqslant \widetilde{C}_{l_k}) \geqslant 1 - \epsilon \tag{7}$$

   or the condition

$$P(\tilde{d}_j \leqslant \widetilde{S}_{s_k} + \tilde{p}_j) \geqslant 1 - \epsilon. \tag{8}$$

3. $\widetilde{B}_k$ is maximal subsequence of π where all the jobs satisfy either (7) or (8).

We distinguish two types of blocks:

- E-Random Blocks, denoted as $\widetilde{B}_k^E$, the ones satisfying condition (7),
- T-Random Blocks, denoted as $\widetilde{B}_k^T$, the ones satisfying condition (8).

Theorem 1. *Let π be a permutation with a distinguished random block $\widetilde{B}$, i.e. $\pi = (1, 2, \ldots, s_k, s_k + 1, \ldots, l_k - 1, l_k, \ldots, n)$ where $A = (1, \ldots, s_k - 1)$, $B = (s_k, \ldots, l_k)$ and $C = (l_k + 1, \ldots, n)$. Estimated value of comparison function can be calculated as follows:*

$$\begin{aligned} W_{ABC} &= W_A + W_B + W_C \\ &= \sum_{i=1}^{s_k-1} w_i E(\tilde{U}_i) + \sum_{i=s_k}^{l_k} w_i E(\tilde{U}_i) + \sum_{i=l_k+1}^{n} w_i E(\tilde{U}_i) \\ &= \sum_{i \in \pi} w_i E(\tilde{U}_i). \end{aligned}$$

Now let $\mathcal{B}$ be a set of all permutations of $B = (s_k, \ldots, l_k)$ and for each $B' \in \mathcal{B}$ we define $W_{AB'C} = W_A + W'_B + W_C$ in the same way as W_{ABC}.

Then we have the following

a) if B is a random E-block, then for each $B' \in \mathcal{B}$ $W'_B \leqslant \sum_{i=s_k}^{l_k} w_i \cdot \epsilon$,

b) if B is a random T-block, then for each $B' \in \mathcal{B}$ $W'_B \geqslant \sum_{i=s_k}^{l_k} w_i \cdot (1 - \epsilon)$

what in result gives as that for each $B' \in \mathcal{B}$ there is a fixed upper bound for $W_{AB'C}$.

Proof. Let's consider the following 2 cases.

A. B is random E-block. Then we have:

$$W_B = \sum_{i=s_k}^{l_k} w_i E(\tilde{U}_i) = \sum_{i=s_k}^{l_k} w_i P(\tilde{C}_i > \tilde{d}_i) = \sum_{i=s_k}^{l_k} w_i (1 - P(\tilde{C}_i \leqslant \tilde{d}_i)).$$

Applying our assumption that B fulfills (7) (i.e. B is a random E-block) as well as by definition of $\tilde{C}_i$ and the problem formulation where every realization of $\tilde{C}_i$ will be less or equal than realization of $\tilde{C}_{l_k}$ we obtain that

$$P(\tilde{C}_i \leqslant \tilde{d}_i) \leqslant P(\tilde{C}_{l_k} \leqslant \tilde{d}_i)$$

for all $i \in \tilde{B}_k$. Having that we can proceed as follows:

$$W_B = \sum_{i=s_k}^{l_k} w_i (1 - P(\tilde{C}_{l_k} \leqslant d_i)) \leqslant \sum_{i=s_k}^{l_k} w_i (1 - 1 + \epsilon) = \sum_{i=s_k}^{l_k} w_i \cdot \epsilon$$

what leads us to the conclusion that for each permutation $B' \in \mathcal{B}$ we have

$$W'_B \leqslant \sum_{i=s_k}^{l_k} w_i \cdot \epsilon.$$

B. B is random T-block. Then we have:

$$W_B = \sum_{i=s_k}^{l_k} w_i E(\tilde{U}_i) = \sum_{i=s_k}^{l_k} w_i P(\tilde{C}_i > \tilde{d}_i).$$

By definition of S_{s_k} and $\tilde{C}_i$ ($i \in B$) we can easily observe the following:

$$\tilde{C}_i = \tilde{S}_{s_k} + \tilde{p}_{s_k} + \tilde{p}_{s_k+1} + \tilde{p}_{s_k+2} + \ldots + \tilde{p}_i \geqslant \tilde{S}_{s_k} + \tilde{p}_i.$$

Having that and applying our assumption that B fulfills (8) (i.e. B is a random T-block) we obtain that

$$P(\tilde{d}_i < \tilde{C}_i) \geqslant P(\tilde{d}_i < S_k + p_i) \geqslant 1 - \epsilon$$

what implies that

$$P(\tilde{d}_i < \tilde{C}_i) \geqslant 1 - \epsilon$$

for all $i \in \tilde{B}_k$. Having that we can proceed as follows:

$$W_B = \sum_{i=s_k}^{l_k} w_i P(\tilde{C}_i > \tilde{d}_i) \geqslant \sum_{i=s_k}^{l_k} w_i(1 - \epsilon)$$

what leads us to the conclusion that for each permutation $B' \in \mathcal{B}$ we have

$$W'_B \geqslant \sum_{i=s_k}^{l_k} w_i \cdot (1 - \epsilon).$$

That concludes the proof.

The above theorem also holds in case where we consider only random processing times or only random due dates and each case the proof is analogous.

4.2 Improving Robustness by Applying the Derived Theorem

It is easy to show the following. For the variant with random processing times $\tilde{p}_i$ we have:

$$E(\tilde{U}_{\pi(i)}) = P(\tilde{C}_{\pi(i)} > d_{\pi(i)}) = 1 - F_{\tilde{C}_{\pi(i)}}(d_{\pi(i)}).$$

and for the variant with random due dates $\tilde{d}_i$ we have:

$$E(\tilde{U}_{\pi(i)}) = P(C_{\pi(i)} > \tilde{d}_{\pi(i)}) = F_{\tilde{d}_{\pi(i)}}(C_{\pi(i)}).$$

Combining the above with assumptions expressed in (7) and (8) and adapted to respective variants, we apply the following rules to modify the base tabu search method. For the variant with random processing times $\tilde{p}_i$:

a) if B is a random E-block, then $W_B = \sum_{i=s_k}^{l_k} w_i \cdot (1 - F_{\tilde{C}_{\pi(i)}}(d_{\pi(i)}))$,

b) if B is a random T-block, then $W_B = \sum_{i=s_k}^{l_k} w_i \cdot F_{\tilde{C}_{\pi(i)}}(d_{\pi(i)})$.

For the variant with random due dates $\tilde{d}_i$:

a) if B is a random E-block, then $W_B = \sum_{i=s_k}^{l_k} w_i \cdot F_{\tilde{d}_{\pi(i)}}(C_{\pi(i)})$,

b) if B is a random T-block, then $W_B = \sum_{i=s_k}^{l_k} w_i \cdot (1 - F_{\tilde{d}_{\pi(i)}}(C_{\pi(i)}))$.

5 Computational Experiments

In this section we present the results of the robustness property comparison between the tabu search method with blocks and the tabu search method with blocks and theorem applied in a way described in Sect. 4. All tests are executed with a modified version of tabu search method described in [1]. The algorithm has been configured with the following parameters: $\pi = (1, 2, \ldots, n)$ is an initial permutation, n is the length of tabu list and n is the number of algorithm iterations where n is the tasks number.

Both methods have been tested on instances from OR-Library ([8]) where there are 125 examples for $n = 40$, 50 and 100 (in total 375 examples). For each example and each parameter $c = 0.02$, 0.04, 0.06 and 0.08 (expressing 4 levels of data disturbance) 100 randomly disturbed instances were generated according to the normal distribution defined in Sect. 3.1 (in total 400 disturbed instances per example). The full description of the method for disturbed data generation can be found in [5].

All the presented results in this section are calculated as the relative coefficient according to the following formula:

$$\delta = \frac{W - W^*}{W^*} \cdot 100\% \tag{9}$$

which expresses by what percentage the investigated solution W is worse than the reference (best known) solution W^*. Details of calculating robustness of the investigated methods can also be found in [5].

An algorithm without applied theorem we denote by $\mathcal{AD}$ and the one with applied theorem by $\mathcal{AP}$.

5.1   Results

In Tables 1 and 2 we present a complete summary of the computational experiments results. Values from columns $\mathcal{AD}$ and $\mathcal{AP}$ in both tables represent a relative distance between solutions established by a respective algorithm and the best known solution. The distance is based on (9) and it is an average of all solutions calculated for the disturbed data on a respective disturbance level expressed by the parameter c. Value from column IF (what stands for *improvement factor*) expresses the relative distance (also based on (9)) between the results obtained by $\mathcal{AD}$ and the results obtained by $\mathcal{AP}$.

Table 1. Relative distance between robustness coefficient of algorithm $\mathcal{AD}$ (or respectively $\mathcal{AP}$) and the reference value for random p_i on different disturbance levels (0.02–0.08)

N	40			50			100		
c	$\mathcal{AD}$	$\mathcal{AP}$	IF	$\mathcal{AD}$	$\mathcal{AP}$	IF(%)	$\mathcal{AD}$	$\mathcal{AP}$	IF
0.02	757.9	25.6	2863%	820.6	24.3	3275%	3625.5	11.5	31558%
0.04	1776.3	24.6	7112%	2132.2	25.8	8177%	5146.6	12.2	41952%
0.06	2442.4	26.8	9022%	3013.2	25.4	11762%	6488	13.5	47831%
0.08	2821.2	29.4	9509%	5656.3	28.7	19627%	7957.5	14.8	53661%
Avg	**1949.5**	**26.6**	**7233%**	**2905.6**	**26.0**	**11060%**	**5804.4**	**13.0**	**44525%**

Table 2. Relative distance between robustness coefficient of algorithm $\mathcal{AD}$ (or respectively $\mathcal{AP}$) and the reference value for random d_i on different disturbance levels (0.02–0.08)

N	40			50			100		
c	$\mathcal{AD}$	$\mathcal{AP}$	IF	$\mathcal{AD}$	$\mathcal{AP}$	IF(%)	$\mathcal{AD}$	$\mathcal{AP}$	IF
0.02	3000.2	70.8	4136%	4661.6	31.8	14547%	11812.1	20.1	58795%
0.04	4719.2	124.9	3678%	7948.9	181.8	4273%	17830.2	141.3	12517%
0.06	6303.4	271.6	2220%	8191.3	284.6	2777%	15180.5	264	5649%
0.08	5654.6	341.6	1555%	8109.1	511.8	1484%	7235.6	322.2	2145%
Avg	**4919.3**	**202.2**	**2332%**	**7227.7**	**252.5**	**2762%**	**13026.2**	**186.6**	**6880%**

We can easily observe that applying the theorem into the method improves results very significantly for all the investigated cases. We can also observe that robustness coefficients for random d_i are generally worse than ones for random p_i what can be explained by the fact that for random d_i there are more fluctuations in disturbed data than for random p_i what is implied by the disturbed data generation method. The other observation is related to the results on different disturbance levels (parameter c). With the bigger value of c we might expect the worse robustness coefficient. Surprisingly, we can observe a difference between results for random p_i and random d_i. For random p_i the rule generally works both for $\mathcal{AD}$ and $\mathcal{AP}$, only for $\mathcal{AP}$ there is a swap between $c = 0.02$ and $c = 0.04$ for $N = 40$ and a swap between $c = 0.04$ and $c = 0.06$ for $N = 50$, nevertheless those values are all very close to each other, so those swaps are actually meaningless. For d_i the situation is different. The rule still works for $\mathcal{AP}$, but for $\mathcal{AD}$ we are not able to observe any trend. That can be considered as an advantage of $\mathcal{AP}$ as it behaves in a more predictive and stable way than $\mathcal{AD}$ does.

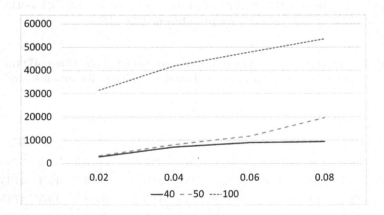

Fig. 1. Comparison of the robustness level with reference to the main methods for random p_i

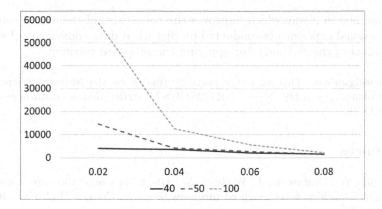

Fig. 2. Comparison of the robustness level with reference to the main methods for random d_i

It is also worth noting the difference of trends for random p_i and random d_i of the comparison between $\mathcal{AD}$ and $\mathcal{AP}$ (column IF) on different disturbance levels which are visualized on Figs. 1 and 2. While for random p_i we can see that with the increase of the parameter c the gap between $\mathcal{AD}$ and $\mathcal{AP}$ is also increasing, for random d_i we can observe exactly the opposite. On the other hand the order of magnitude of *improvement factor* (IF) is the same for both random p_i and random d_i what shows that the level of improvement introduced by the presented Theorem works similarly in both considered scenarios.

5.2 Parallelization Consideration

The tabu search method is well paralleling. According to the classification proposed by Voß [13], four models of the parallel tabu search method can be considered: SSSS (Single Starting point Single Strategy), SSMS (Single Starting point Multiple Strategies), MSSS (Multiple Starting point Single Strategy), MSMS (Multiple Starting point Multiple Strategies). They refer to the classic classification of Flynn's parallel architectures [7]. In the version proposed in this paper where block approach is applied it is natural to use MSSS or MSMS diversification strategies, because the block mechanism is quite easily parallelized (each block can be considered separately [3]). The use of 'tree' strategies using the same start solution is also possible provided the fact that the process of searching the solution space at a later stage is diversified (e.g. by using different tabu length for individual threads or through a mechanism of dynamic tabu list length change [1]).

6   Conclusions

In the paper we considered a single machine tabu search method with block approach in an uncertain environment modeled by random variables with the normal distribution. We proposed a theorem which allows to modify the base tabu search method in a way which improves the robustness of calculated solutions. Computational experiments conducted on disturbed data confirmed substantial predominant of the method after applying the proposed theorem.

Acknowledgments. This work was partially funded by the National Science Centre of Poland, grant OPUS no. 2017/25/B/ST7/02181 and a statutory subsidy 049U/0032/19.

References

1. Bożejko, W., Grabowski, J., Wodecki, M.: Block approach-tabu search algorithm for single machine total weighted tardiness problem. Comput. Ind. Eng. **50**(1–2), 1–14 (2006)
2. Bożejko, W., Rajba, P., Wodecki, M.: Stable scheduling with random processing times. In: Klempous, R., Nikodem, J., Jacak, W., Chaczko, Z. (eds.) Advanced Methods and Applications in Computational Intelligence. Topics in Intelligent Engineering and Informatics, 6th edn, pp. 61–77. Springer, Heidelberg (2014). https://doi.org/10.1007/978-3-319-01436-4_4
3. Bożejko, W., Uchroński, M., Wodecki, M.: Parallel metaheuristics for the cyclic flow shop scheduling problem. Comput. Ind. Eng. **95**, 156–163 (2016)
4. Bożejko, W., Rajba, P., Wodecki, M.: Stable scheduling of single machine with probabilistic parameters. Bull. Polish Acad. Sci. Tech. Sci. **65**(2), 219–231 (2017)
5. Bożejko, W., Rajba, P., Wodecki, M.: Robustness of the uncertain single machine total weighted tardiness problem with elimination criteria applied. In: Zamojski, W., Mazurkiewicz, J., Sugier, J., Walkowiak, T., Kacprzyk, J. (eds.) DepCoS-RELCOMEX 2018. AISC, vol. 761, pp. 94–103. Springer, Cham (2019). https://doi.org/10.1007/978-3-319-91446-6_10
6. Bożejko, W., Hejducki, Z., Wodecki, M.: Flowshop scheduling of construction processes with uncertain parameters. Arch. Civil Mech. Eng. **19**(1), 194–204 (2019)
7. Flynn, M.J.: Computer architecture: Pipelined and parallel processor design. Jones & Bartlett Learning, Burlington (1995)
8. OR-Library. http://www.brunel.ac.uk/~mastjjb/jeb/info.html. Accessed 13 April 2020
9. Shang, C., You, F.: Distributionally robust optimization for planning and scheduling under uncertainty. Comput. Chem. Eng. **110**, 53–68 (2018)
10. Soroush, H.M.: Scheduling stochastic jobs on a single machine to minimize weighted number of tardy jobs. Kuwait J. Sci. **40**(1), 123–147 (2013)
11. Urgo, M., Váncza, J.: A branch-and-bound approach for the single machine maximum lateness stochastic scheduling problem to minimize the value-at-risk. Flexible Serv. Manuf. J. **31**, 472–496 (2019)
12. Vondrák, J.: Probabilistic methods in combinatorial and stochastic optimization (Doctoral dissertation, Massachusetts Institute of Technology) (2005)
13. Voß, S.: Tabu search: Applications and prospects. In: Network Optimization Problems: Algorithms, Applications and Complexity, pp. 333–353 (1993)
14. Cai, X., Wu, X., Zhou, X.: Optimal Stochastic Scheduling. ISORMS, vol. 207. Springer, Boston (2014). https://doi.org/10.1007/978-1-4899-7405-1
15. Zhang, L., Lin, Y., Xiao, Y., Zhang, X.: Stochastic single-machine scheduling with random resource arrival times. Int. J. Mach. Learn. Cybernet. **9**(7), 1101–1107 (2018)

Empirical Analysis of Stochastic Methods of Linear Algebra

Mustafa Emre Şahin[1], Anton Lebedev[2(✉)], and Vassil Alexandrov[1]

[1] The Hartree Centre, Keckwick Ln, Warrington, UK
vassil.alexandrov@stfc.ac.uk
[2] Helmholtz Zentrum Dresden-Rossendorf, Bautzner Landstr. 400, Dresden, Germany
a.lebedev@hzdr.de

Abstract. In this paper we present the results of an empirical study of stochastic projection and stochastic gradient descent methods as means of obtaining approximate inverses and preconditioners for iterative methods. Results of numerical experiments are used to analyse scalability and overall suitability of the selected methods as practical tools for treatment of large linear systems of equations. The results are preliminary due to the code being not yet fully optimized.

Keywords: Linear algebra · Stochastic methods · High-performance computing

1 Introduction

In this paper we present an empirical evaluation of three numerical methods derived for the stochastic solution of linear algebraic systems. All methods are evaluated regarding their suitability as tools for computing an approximate inverse matrix. Furthermore, we consider their scalability and the usefulness of the approximate inverses as preconditioners for iterative solvers for linear systems of the form $Ax = b$, where $A \in \{\mathbb{R}, \mathbb{C}\}^{n \times n}$ is henceforth designated system matrix.

Solution of linear algebraic systems is of paramount importance in almost every domain of scientific computing. In case of table-top numerical experiments such systems may very well be solved using well-known methods like LU decomposition, but for a wide variety of cases the size of the matrices would prevent them from being storable even given the abundant storage space available nowadays. In such cases the matrices have, however, generally a very sparse structure, reducing their memory footprint enormously. The downside being, that an inverse of a sparse matrix is generally dense and a method like LU decomposition hence becomes unfeasible. To obtain a solution in such cases iterative methods such as generalized minimal residues (GMRES) iteration or bi-conjugate gradient stabilized (BiCGstab) iteration are employed, which solve the linear system by merit of a fixed-point iteration. Since there's no silver bullet for complexity of the problem iterative methods may suffer of slow convergence towards

© Springer Nature Switzerland AG 2020
V. V. Krzhizhanovskaya et al. (Eds.): ICCS 2020, LNCS 12143, pp. 539–549, 2020.
https://doi.org/10.1007/978-3-030-50436-6_40

the solution. This problem one attempts to ameliorate using preconditioners - matrices which modify the problem and ideally reduce the number of iterations and, ideally also the runtime to solution.

Here we consider the suitability of two methods, stochastic gradient descent with missing values (mSGD) and stochastic projection (SP) as ways to compute said preconditioners.

2   Algorithm Description

Our main focus in this work is on stochastic projection (SP), or randomized Kazmarz method, as described in [6] and on the stochastic gradient descent with missing values (mSGD) - as given in [5]. Both methods are assessed also regarding their sensitivity to the initial conditions. The latter is achieved by comparing convergence behaviour with a random initial guess at the solution to an initial guess provided by the Markov Chain Monte Carlo Matrix Inversion (MCMCMI) method described in [4]. In the following a brief description of each method is provided.

2.1   Stochastic Projection

The basic idea is fairly simple, and the implementation follows roughly eq. (2.13) of [6]. The main idea is to project the solution vector successively and orthogonally onto arbitrarily chosen subspaces until the accumulated effect leads the iteration into the subspace of the true solution. An obvious extension to block-projections has been mentioned in [6] and implemented here. The iteration step is given as follows:

$$x_{k+1} = x_k + A_i^t \left(A_i A_i^t\right)^{-1} \left(b_i - A_i x_k\right) . \tag{1}$$

Here A_i is a randomly selected block of rows of the matrix and b_i the corresponding subset of entries of the right-hand-side vector of

$$Ax = b . \tag{2}$$

The intuitive simplicity of this approach is paid for by its performance. Furthermore the computation of a matrix inverse in each step is required. Depending on the block size the computation of said inverse, or a solution of a dense system, may incur a significant cost.

2.2   Stochastic Gradient Descent

In the present modification, as proposed by Ma and Needell in [5] a gradient descent takes place not over the entirety of the column space of the matrix, but rather over a randomly selected subspace (of dimension 1). Ma and Needell derive the method under the assumption that an entry a_{ij} of the system matrix itself is missing with a probability p. This results in the following iteration rule:

$$x_{k+1} = x_k - \alpha_k \left[\frac{1}{p^2} \left(a_i \left(a_i^t x_k - pb_i \right) \right) - \frac{1-p}{p^2} diag \left(a_i a_i^t \right) x_k \right] . \tag{3}$$

Here a_i is the i-th column-vector of the matrix and the last part of the step utilises a diagonal matrix derived from the outer product of a_i with itself. Convergence of the gradient descent method depends on its step-size. The latter can be either fixed or variable and an expression for a variable step size is provided as a function of an arbitrary parameters $0 < c, r < 1$, where c is called "learning rate", the number of iterations k and the smallest eigenvalue λ_{min} of the system matrix:

$$\alpha_k = \frac{c}{|\lambda_{min}|} r^{\frac{k}{T}} . \tag{4}$$

Here T is the number of steps after which the step size will be shrunk by a factor $r < 1$. Again, the relative simplicity of the method is compensated by the existence of two parameters in (3): α_k, p. Both are, as a rule, unknown *a-priori* and have to either be guessed or, in case of the step size, computed using (4). An educated guess as to the parameters may very well result in sub-optimal choices, whereas the computation of α_k using the above equation introduces yet another pair of parameters.

2.3   Markov Chain Monte Carlo Matrix Inversion

The third solution method presented here is the approximation of an inverse using Markov Chain Monte Carlo for Matrix Inversion (MCMCMI) as presented in [4]. The fundamental idea of the method is to employ the Neumann series to compute an inverse of a diagonally-dominant matrix. To reduce the cost the Neumann series is evaluated stochastically using Markov Chains. Since the series is infinite estimates as to the number and length of chains are required for a practical application. Such an estimate was provided in [3] and the method has been implemented both for GPUs and CPUs. In the current case it stands as a stand-alone method of providing preconditioners for iterative methods, as well as a source of initial conditions for the mSGD and SP iterations.

The algorithm can be split into the following 5 phases (Notice that phases 1 and 5 are only necessary when the initial matrix is not a *diagonally dominant matrix (ddm)*): 1) Initial matrix is transformed into a ddm, 2) Transformation of ddm for suitable *Neumann series expansion*, 3) Monte Carlo method is applied to calculate sparse approximation of the inverse matrix, 4) Given 2, calculate the inverse of the ddm from 3, 5) Recovery process is applied to calculate the inverse of the original matrix due to the transformation in 1. It must be noted that the last phase requires in general $\mathcal{O}(n^3)$ operations and hence is generally neglected. Prior numerical experiments have demonstrated that it is not compulsory to obtain an effective preconditioner and is in general an impediment to an *efficient* preconditioner. The method requires two tolerance values ϵ - an error bound on the stochastic error, which determines the maximum amount of chains required - and δ, a truncation error bound which affects the length of the chains. Both have to be provided by the user and together dictate the precision of the approximation.

The major caveat of this method is that it is strictly valid only for diagonally dominant matrices if the recovery procedure is not being applied. Nevertheless, as has been demonstrated in [4], the obtained approximation of the unrecovered inverse is often sufficient as a preconditioner for iterative systems.

3   Implementation and Experiments

The effectiveness of the chosen methods has been assessed using a set of sparse matrices of varying size and structure, intended to represent multiple domains of scientific computing. It is provided in Table 1. The set contains matrices from climate simulations (nonsym_r3_a11, sym_r6_a11), semiconductor electronics (circuit5M_dc, from [2]), computational fluid dynamics (rdb2048, [2]) and simple mathematics (2DFDLaplace_20x20).

Table 1. Matrix set.

Matrix	Dimension	Non-zeros	Sparsity	Symmetry
circuit5M_dc	$3,523,317 \times 3,523,317$	19,194,193	$1.5 \cdot 10^{-6}\%$	Symmetric
nonsym_r3_a11	$20,930 \times 20,930$	638,733	0.15%	Non-symmetric
sym_r6_a11	$1,314,306 \times 1,314,306$	36,951,316	0.02%	Symmetric
rdb2048	2048×2048	12032	0.28%	Non-symmetric
2DFDLaplace_20x20	361×361	1729	1.32%	Symmetric

3.1   Implementation Details and Caveats

The methods described in the previous section have been implemented in C++. Parallelization of all of the methods for use on CPUs has been achieved by combining MPI and OpenMP (OMP) parallelisation (so-called hybrid parallelisation). A GPU implementation was available only for MCMCMI at the time of writing. This has been done using NVIDIA's CUDA programming model. To utilise multiple GPUs the implementation utilises OpenMP threads s.t. each GPU is controlled by one thread.

Parallelisation is similar for all methods and architectures. The main process reads the matrix A and the run-time parameters. It then broadcasts this data to the workers. Each worker computes the fraction of rows/columns it has to process based on its rank. At the end of the computation the resultant local block of A^{-1} is purged of entries smaller than the user-prescribed tolerance (10^{-9} resp. 10^{-14} for circuit5M_dc). Then it is collected onto the master process for storage. It is important to note that, with increasing matrix size broadcast operations introduce a non-negligible communication overhead.

In the case of the GPU implementation of MCMCMI the main process is the master thread of OpenMP and workers are the GPUs. For the hybrid implementation (MPI+OpenMP) the master process is MPI rank 0 and worker processes have rank> 0. The distribution of the matrix blocks is uniform among the processes, with the last block of rows being assigned to the master due to it being potentially smaller than the other blocks. This preferential treatment of the master is done to mask potential load imbalance by letting the master prepare the output arrays prior to collection. Within each node (MPI rank) the computation of the approximate inverse is parallelised over the rows of the block using OpenMP `dynamic` distribution to ameliorate load imbalance among the rows.

The mSGD and SP implementations follow the same structure, but acceleration on the worker processes is achieved by utilising Eigens' built-in parallelisation using OpenMP threads instead of manually distributing the columns of the inverse.

The main drawback of the CPU-oriented implementation of MCMCMI is the lack of resources which could be used to compact and sort the row of the approximate inverse, before the entries that will be retained are extracted. The current implementation of stream compaction introduces significant overhead but is still preferable to sorting an entire row of the approximate inverse if it contains only a handful of non-vanishing values.

Numerical experiments for SP and mSGD were run on the Scafell Pike system of the Hartree centre, which consists of nodes fitted with 2x XEON gold E5-6142 v5 processors resulting in 32 cores per node and due to HyperThreading in 64 threads per node.

The experiments for the assessment of MCMCMI and to generate optimized initial conditions for mSGD and SP were executed on the `hemera` system of the Helmholtz Centre Dresden-Rossendorf by A.L. The systems running the K80 experiment set consisted of 8x NVIDIA K80 GPUs with 32GB VRAM on a system with 2 Xeon E5-2630v3 CPUs with HyperThreading enabled. For hybrid MCMCMI experiments the nodes used contained 2 Xeon Gold 6148 CPUs with 20cores/40 threads each. The P100 experiments were performed on systems containing 4 P100 GPUs with 16GB VRAM each connected via the NVLink interface to a node of 2 Xeon Gold 6136 CPUs with 12 cores/24 threads each. All systems are connected via Infiniband with 56 Gb/s.

3.2   Numerical Experiments and Analysis of Results

MCMCMI. For the sake of simplicity we chose $\epsilon = \delta$ for the simulations, although this is by no means necessary. This choice ensures that neither stochastic (ϵ) nor truncation errors (δ) dominate.

From Fig. 1a one can immediately see (note that the upper horizontal axis enumerates GPUs), that the usage of the GPU many-core architecture is of little meaning if the workload is small. This is the case when the matrix to be inverted is small. In this case the GPU will spend most of the time in memory-management overhead. Hence, for the `nonsym_r3_a11` matrix an optimal number

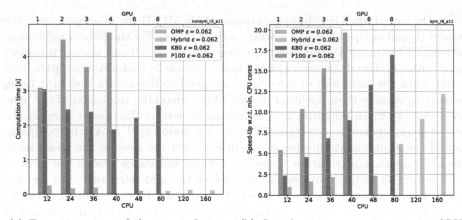

(a) Execution time of the preconditioner computation for a small matrix. Management overhead on GPUs is emphasized by the small size.

(b) Speed-up in comparison to 12 CPU cores for the sym_r6_all matrix and a precision of $\epsilon = 0.0625$.

Fig. 1. Execution time and speed-up of MCMCMI depending on the underlying architecture and matrix size. Note that the lower axis enumerates CPUs of the OpenMP or MPI+OpenMP (hybrid) implementation and the upper axis GPUs of a GPU-only implementation.

of GPUs, dependent on their architecture, appears to exist. The latter is understandable since different architectures generally feature vastly different numbers of processing units.

However, requiring a precise approximate inverse (lowering the relative error ϵ) and working with large matrices leads to a full utilization of the GPU's computing resources, resulting in a good scalability across multiple GPUs. Similarly for CPUs. Figure 1b serves to illustrate these assertions. In the case of the hybrid implementation, utilizing MPI and OpenMP parallelization, one can see that the overhead of distributing the system matrix A and the final collection of the preconditioner onto the main process inhibits scaling of the method. Increase of the tolerances beyond 0.125 is seldom useful since the execution time becomes dominated by communication and memory management overhead even for small matrices and few processes/GPUs.

mSGD Parameters. The missing element probability and learning rate parameters of mSGD are unknown and hard to estimate *a-priori*. To elucidate the sensitivity of the method to the choice of these parameters a parameter search has been performed. To this end $r = 0.5$ and $T = N_{total}/100$, with $N_{total} = 10^5$ have been chosen and c, p varied. The results of the parameter search are shown in Fig. 2. They indicate the existence of an optimal choice of parameters with a bias towards higher values for the missing probability p. The choice of an optimal parameter set will not, however, improve the convergence and errors dramatically.

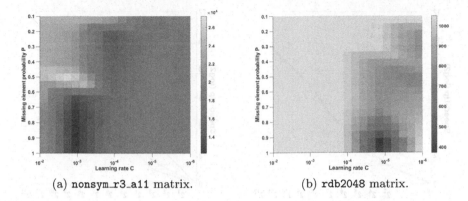

(a) `nonsym_r3_a11` matrix.   (b) `rdb2048` matrix.

Fig. 2. Dependency of the error of mSGD on the chosen parameters.

As can be seen in the figure the best reduction of the error is achieved for the `rdb2048` matrix and is $\propto 3\times$. It is also interesting to observe, that in contrast to the non-symmetric matrices of Fig. 2 the optimal parameter set for the symmetric discrete Laplacian is located towards higher learning rates. This holds, too, for the large `sym_r6_a11` matrix, too, and suggests the possibility that the learning rate may be influenced by the symmetry of the matrix (Fig. 3).

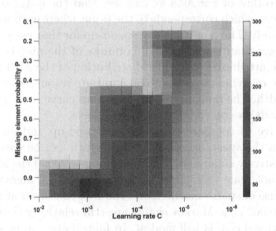

Fig. 3. Dependency of the error of mSGD on the chosen parameters for `2DFDLaplace_20x20` matrix.

Scalability. For mSGD and SP the scaling behaviour across multiple nodes using hybrid parallelization is illustrated in Fig. 4. The recline of the speed-up for the smallest matrix of the set is to be expected since each thread/worker cannot receive less than one column of the approximate inverse to process. Processes that do not receive any columns remain idle but participate in the initial matrix broadcast, hence slowing the process down.

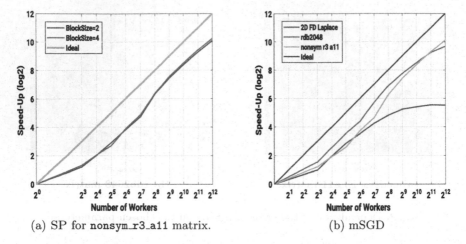

(a) SP for `nonsym_r3_a11` matrix.   (b) mSGD

Fig. 4. Speed-Up achieved by both methods for different matrices of the chosen matrix set. For the small Laplace matrix one can clearly observe a saturation. Note the logarithmic scale of the axes.

Increasing the size of the matrix leads to a better speed-up, as can be seen for the case of the `nonsym_r3_a11` matrix. Comparing the speed-up for `nonsym_r3_a11` to that of `rdb2048` we can see, that the latter reclines markedly around 2^{11} workers, which corresponds to the point where each worker has to process one column only. The non-vanishing speed-up for the case where the number of workers is larger than the number of columns of the matrix is startling, but could possibly be attributed to a wider distribution of the MPI processes on the machine, thereby relieving the pressure on compute resources such as caches and memory bandwidth. The tendency of the speed-up curve to be convex at around 2^7 workers (threads) we attribute to caching effects.

It is instructive to compare the achieved speed-up and execution times of these methods to the speed-up of MCMCMI which can be easily inferred from Fig. 1a. For a relatively small matrix of dimension ~ 20000 the MCMCMI implementation does not scale well. The reason for this is the memory management imposed by the use of sparse matrices and the fact that even at a relatively high precision of $\epsilon = 0.0625$ the Markov chains are still relatively short and hence the overall computational cost is still modest. In fact, if one compares the execution times provided in Figs. 1a, 5 one can immediately see, that although MCMCMI does not scale well it clearly has the advantage with regards to the practical requirement of short run times.

Error Convergence. As can be observed in Fig. 6 usage of the rough approximate inverse provided by MCMCMI reduces the error of both SP and mSGD significantly. Unexpectedly it appears to have no effect on the behaviour of the error for SP. This suggests a certain agnosticism of the method with regards to the starting vector/matrix for the iteration. The similarity of the behaviour of

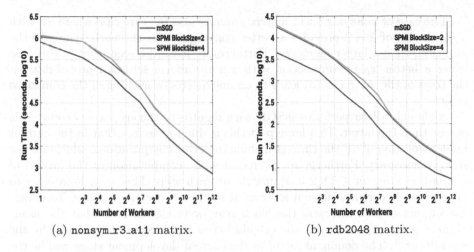

(a) `nonsym_r3_a11` matrix.   (b) `rdb2048` matrix.

Fig. 5. Total execution times of the different methods.

mSGD and SP is misleading in the case of the `nonsym_r3_a11` matrix, since for mSGD the step-size was determined using (4) with the quantities provided in Sect. 3.2. One can see that the mSGD iteration starts to converge roughly after 10^3 steps, which corresponds the point at which $\frac{k}{T} = \frac{k}{10^3}$ becomes larger than one and by merit of $0 < r < 1$ the step-size shrinks. Indeed, if one considers the error behaviour for `rdb2048` it becomes apparent that it begins to converge only after the exponential becomes > 1.

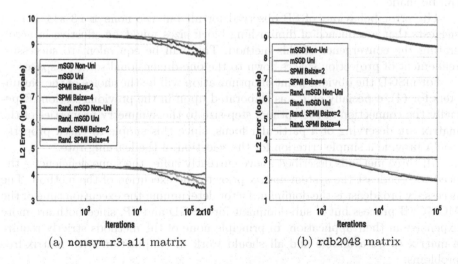

(a) `nonsym_r3_a11` matrix   (b) `rdb2048` matrix

Fig. 6. Errors achieved by SP and mSGD for different matrices using a random matrix as initial condition, or an approximate inverse provided by MCMCMI.

The case of `nonsym_r3_a11` matrix presents also the rare case where SP with a block size of 2×2 provides a better convergence in the early stages of the iteration, than a larger block size. Furthermore it can be seen, that in this case using a better x_0 in conjunction with a non-uniform selection probability for the rows of the matrix in (3) leads to an unexpected worsening of the computed error.

Although SP outperforms mSGD with regards to the error convergence it is slower than the latter. To a large part this is due to the fact that in the current implementation of SP the timings include the error computations, which are not strictly necessary. Furthermore, SP requires the computation of the inverse of a small $m \times m$, $m \in \{1, 2, 3, 4\}$ matrix at each step. This cost, however, can be neglected currently, since it is dwarfed by the error computations. To reduce the impact of this inversion the block sizes were chosen such that the usage of an explicit expression for the calculation of the inverse of the block by the used library[1]. The dominant factor at the current development stage will be the memory management, though.

4   Conclusions and Future Work

In summary we observe that in the present state of development neither of these methods is capable of competing against, for instance, MCMCMI as a method to compute an approximate inverse to be used as a preconditioner efficiently. This is chiefly due to the excessive execution times and slow error convergence. One has to note that the parameters of SP and mSGD have not been explored in full, and the analysis currently under way suggests that some sizeable improvements can be made.

The error behaviour of SP observed for all but the `nonsym_r3_a11` matrix suggests that a sequence of diminishing block sizes might be effective in accelerating the convergence of this method. This will be equivalent to successive refinement of projection spaces down to the one-dimensional solution space.

For mSGD the obvious point of optimization will be the choice of the parameters for (4), especially of T - as elaborated upon in the previous section. Especially the connection of the optimal step-size to the symmetry properties of the matrix are deserving of a particular focus, since this simple to check property would provide a simple criterion for the selection of the learning rate.

All three methods described above currently suffer the same deficiency - the need to broadcast the system matrix prior to the execution of the method. The necessary broadcast is the dominant factor determining the execution time of the MCMCMI process but is sub-dominant for mSGD and SP, since both are more expensive in their application. In principle none of the methods strictly require a matrix to be available and all should work in conjunction with matrix-free problems.

Multiple such tests are planned for the near future and will serve to assess the usefulness of the methods as preconditioners in the domain of electrodynamics

---

[1] The Eigen 3 template library.

and plasma physics. Furthermore, based on the observation that the lengths of empirical Markov Chains are much shorter than predicted by theory [1], we expect the need for further formal analysis of the MCMCMI method to provide bounds that are more precise.

References

1. Alexandrov, V.N.: Efficient parallel Monte Carlo methods for matrix computations. Math. Comput. Simul **47**, 113–122 (1998)
2. Davis, T.A., Hu, Y.: The university of Florida sparse matrix collection. ACM Trans. Math. Softw. (TOMS) **38**(1), 1 (2011). https://sparse.tamu.edu/
3. Dimov, I., Alexandrov, V.: A new highly convergent Monte Carlo method for matrix computations. Math. Comput. Simul. **47**(2–5), 165–181 (1998)
4. Lebedev, A., Alexandrov, V.: On advanced Monte Carlo methods for linear algebra on advanced accelerator architectures. In: 2018 IEEE/ACM 9th Workshop on Latest Advances in Scalable Algorithms for Large-Scale Systems (ScalA) (2018)
5. Ma, A., Needell, D.: Stochastic Gradient Descent for Linear Systems with Missing Data, February 2017. http://arxiv.org/pdf/1702.07098v4
6. Sabelfeld, K., Loshchina, N.: Stochastic iterative projection methods for large linear systems. Monte Carlo methods and applications (2010). https://doi.org/10.1515/mcma.2010.020

Wind Field Parallelization Based on Python Multiprocessing to Reduce Forest Fire Propagation Prediction Uncertainty

Gemma Sanjuan, Tomas Margalef[✉], and Ana Cortés

Computer Architecture and Operating Systems Department, Universitat Autónoma
de Barcelona, Cerdanyola del Valles, Spain
{gemma.sanjuan,tomas.margalef,ana.cortes}@uab.cat

Abstract. Forest fires provoke significant loses from the ecological,
social and economical point of view. Furthermore, the climate emergency will also increase the occurrence of such disasters. In this context,
forest fire propagation prediction is a key tool to fight against these natural hazards efficiently and mitigate the damages. However, forest fire
spread simulators require a set of input parameters that, in many cases,
cannot be measured and must be estimated indirectly introducing uncertainty in forest fire propagation predictions. One of such parameters is
the wind. It is possible to measure wind using meteorological stations
and it is also possible to predict wind using meteorological models such
as WRF. However, wind components are highly affected by the terrain
topography introducing a large degree of uncertainty in forest fire spread
predictions. Therefore, it is necessary to introduce wind field models that
estimate wind speed and direction at very high resolution to reduce such
uncertainty. Such models are time consuming models that are usually
executed under strict time constrains. So, it is critical to minimize the
execution time, taking into account the fact that in many cases it is not
possible to execute the model on a supercomputer, but must be executed
on commodity hardware available on the field or at control centers. This
work introduces a new parallelization approach for wind field calculation
based on Python multiprocessing to accelerate wind field evaluation. The
results show that the new approach reduces execution time using a single
personal computer.

Keywords: Wind field parallelization · Forest fire spread simulation ·
Python multiprocessing

1 Introduction

Forest fire propagation prediction is a key tool to fight against such disasters.
Several simulators [2,6] have been developed to provide hints on fire evolution to
guide the field means and control centres to fight against these events. Most of
these simulators are based on Rothermel's model [10], which is a semi-empirical

© Springer Nature Switzerland AG 2020
V. V. Krzhizhanovskaya et al. (Eds.): ICCS 2020, LNCS 12143, pp. 550–560, 2020.
https://doi.org/10.1007/978-3-030-50436-6_41

model that takes into account the terrain topography, the vegetation conditions and the meteorological conditions to provide forest fire expected propagation. So, this model actually requires a set of parameters that, in many cases, are not single values, but they have values over all the terrain where the fire is going on. For example, the type of vegetation changes over the terrain or the moisture contents of the same kind of vegetation can change over the terrain according to sun exposition. So, for these parameters, a field of values at high resolution is required to calculate the fire propagation. A very particular case is the wind. This parameter actually has two components: speed and direction, and it presents several features that makes it a very particular case:

- Wind speed and wind direction are, jointly with slope, the parameters that most significantly affect fire propagation [1]. Therefore, an accurate estimation of such values is critical for forest fire propagation prediction.
- The meteorological conditions change quickly and, some meteorological model, such as WRF [13], is required to estimate beforehand the values of the meteorological variables and, in particular, wind speed and direction at a surface level.
- The meteorological wind at surface level that can be measured on meteorological stations or estimated by meteorological models is affected by terrain topography, so that at each point of the terrain the wind values (speed and direction) could be different. This spatially varying wind values for a given area constitutes the so called wind field. To obtain this wind field for the underlying terrain, one should apply a diagnostic wind field model such as WindNinja [7]. WindNinja is a wind field model widely used in the forest fire simulation community, that can provide wind speed and wind direction, given a certain meteorological wind values at a surface level, in very high resolution, typically 30 m.

In this context, for each forest fire propagation simulation step, it is necessary to evaluate several values for the meteorological wind at surface level and, for each one of these meteorological winds, it is necessary to calculate the corresponding wind field. Then, once those wind fields are obtained, they must be introduced to the forest fire propagation model to obtain the forest fire propagation prediction [4]. This scheme is represented in Fig. 1. It must also be considered that the spatial resolution of the involved models, particularly wind field model, must be very high. So, the whole coupled system involves several components that must solve complex systems of equations at a very high resolution, what implies large computing requirements and long computation times. These tight needs are critical in real time emergency situations where the response time is a key factor for efficient and effective actuation. In most cases a trade off between accuracy and time must be reached to provide useful predictions in operational time. Therefore, applying high performance techniques to accelerate model execution and reduce prediction time, also contributes to increase the map resolution, reducing the uncertainty and providing more reliable predictions. So, all the efforts to accelerate the involved models have a direct impact in improving the quality and effectiveness of forest fire propagation prediction.

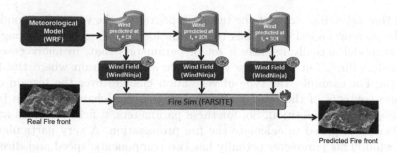

Fig. 1. Coupling meteorological, wind field and forest fire propagation models

Several efforts have been devoted to reduce the execution time of the models involved in the fore fire spread prediction process and also to improve the accuracy of the results delivered by those model [3,5,8,11,12]. In particular, this work is focused on the wind field model. As it has been previously mentioned, in this work the WindNinja wind field model is used. More precisely, a new parallelization approach based on Python multiprocessing has been done. This new parallel approach has been compared to previous parallelization schemes based on MPI (Message Passing Interface). Since one of the goals of this work is being able to developed a forest fire spread prediction system that could be brought closer to the field where the firefighter's command is taken operational decisions, as execution platform one has selected commodity hardware. The parallelization scheme proposed is based on a map partitioning strategy that has been proven to work well for this kind of problem [11] .

The rest of this paper is organised as follows. Section 2 is devoted to describe WindNinja wind field model. In Sect. 3 different WindNinja parallel implementations are introduced. Section 4 shows some preliminary results comparing the different parallel implementations described in the previous section and, finally Sect. 5 presents the main conclusions.

2   WindNinja Wind Field Simulator

As it has been mentioned above, wind speed and wind direction are critical parameters to determine forest fire propagation. In particular, meteorological wind at surface level is modified by terrain topography, so that there is a spatial distribution of wind values along the terrain map. This wind field must be determined to effectively predict forest fire propagation because there exist several high resolution wind phenomena such as, for example, wind speedup over ridges or flow channeling in valleys, that cannot be forecast otherwise. In this context, WindNinja [7] is a wind field simulator that takes the meteorological wind at a surface level and the terrain topography to determine wind speed and wind direction at each point of the underlying map grid at a given resolution, usually around 30 m. WindNinja is based on mass conservation equations that are used to generate the system of equations. In order to solve this system, the Conjugate Gradient method with Preconditioner is used (PCG). PCG is an iterative method that can only be applied when the sparse matrix representing the system is symmetric, positive define and real. It uses a matrix M as a preconditioner, which determines the convergence of the system. The native solver implementation of WindNinja includes $SSOR$ and $Jacobi$ as preconditioners. The $SSOR$ preconditioner is used by default. Furthermore, WindNinja includes an OpenMP parallelization, so that the PCG can exploit the parallelism by using the available cores in the system nodes.

WindNinja can be divided into five basic blocs, as is shown in Fig. 2, where each bloc corresponds to one particular phase of the wind field generation process. The functionality of each one of these five phases are subsequently described:

1. Discretization of the terrain map into a mesh.
2. Application of the mass conservation equations to each point of the mesh to generate the system of equations represented as $Ax = b$.
3. Generation of the CRS (Compressed Row Storage) format to store the sparse matrix A.
4. Application of the Preconditioned Conjugate Gradient (PCG) method to solve the system of equations [9].
5. Construction of the resulting wind field.

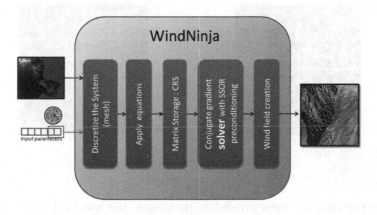

Fig. 2. WindNinja system

In the next section, the description of three parallel implementations of one single WindNinja execution are introduced. As it is subsequently explained, the proposed paralellization will not affect the way WindNinja works because they will not imply any change on WindNinja's code. In fact, the proposed parallel approaches could also benefit from the OpenMP parallelization included in WindNinja.

3   WindNinja Parallelization

In a previous work [11], a map partitioning strategy was developed to divide
the underlying terrain into equal size regions, with the aim of evaluating the
complete wind field as a composition of a set of smaller wind fields coming
from those previous regions. In particular, the map partitioning strategy divides
the map in square parts introducing an overlapping halo on each one to avoid
border effects. The optimal overlapping size has been proven to be 25 cells of the
original mesh. Figure 3 illustrates this map partitioning scheme for two different
map division, 2×2 and 4×4 map partitioning configurations.

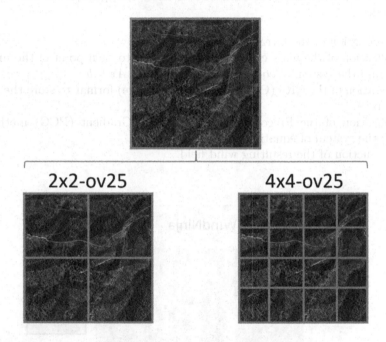

Fig. 3. Map partitioning

Once the map has been divided in partitions, the wind field corresponding
to each partition can be calculated in parallel generating as many wind field
maps as divisions of the map we have. Once all these computation has been
done, a joining process is done in order to merge those wind field maps into
one single wind field map. The main advantage of this approach is that the
individual wind field map for a given region could be evaluated independently
of the others, therefore, a straight forward parallelization scheme consists of
executing each wind field calculation in a different computing element using
the so called Master/Worker parallel programming paradigm. In the following
subsections, we present three different parallel implementation of this strategy:
MPI for C++, MPI for Python, and Python Multiprocessing.

3.1   MPI C++ Parallelization

An MPI C++ application was developed to distribute the partitions of the map among the nodes of the system to execute them independently. As it has been mentioned, for each partition one WindNinja execution has been performed using the corresponding region of the map of that partition. The size of the overlapped halo has set to 25 cells what has been proven to be enough to avoid border effects. In this implementation a set of processes is created when launching the application with MPIRUN and each process calculates the wind field for each partition of the terrain map. Moreover, the OpenMP parallelization integrated in WindNinja has also been used. Actually this implementation is more feasible on a cluster with several nodes, but nowadays desktop computers or even laptops can run this kind of hybrid application efficiently. The resulting implementation is shown in Fig. 4.

```
#include <mpi.h>
if( rank==0)
{
    splitMap #Map partitioning
    for (i = 1; i <= rank; i = i + 1)
    {
        MPI_Send #Send map partition to workers
    }
    WindNinja -worker0
}

if (rank!=0)
{
    MPI_Recv #Receive map partition
    WindNinja -worker=!0
}
```

Fig. 4. MPI C++ parallelization

3.2   MPI Python Parallelization

Currently there is a clear trend to extend the use of Python programming language on many different areas. So, many libraries and current applications have been adapted to be used on Python programs. One particular case is MPI. MPI for Python provides MPI bindings for the Python language, allowing programmers to exploit multiple processor computing systems. mpi4py is constructed on top of the MPI-1/2 specifications and provides an object oriented interface which closely follows MPI-2 C++ bindings. So, the same map partitioning approach has been implemented using this MPI binding. The implementation can be represented as shown in Fig. 5.

```
from mpi4py import MPI
…
if rank==0:
    splitMap #Map partitioning
    for i in range(size):
        if i>0:
            comm.send(data, dest=i) #Send map partition to workers
    WindNinja -worker0

if rank!=0:
    initial_time = time()
    data2 = comm.recv(source=0) #Receive map partition
    WindNinja -worker!0
```

Fig. 5. MPI Python parallelization

This approach is very similar to the previous one, but it is using Python as programming language and the MPI binding. The terrain map is split into a set of partitions and the wind field for each partition is calculated by each worker process. In this case, it is also possible to exploit the OpenMP WindNinja parallelization to calculate the wind field for each partition.

3.3 Python Multiprocessing Parallelization

Multiprocessing is a package that supports spawning processes using an API similar to the threading module. It allows the programmer to fully leverage multiple processes on a given machine. The implementation can be represented as shown in Fig. 6.

```
import multiprocessing
from multiprocessing import Process
    splitMap #Map partitioning

    for i in range(parti):
        p.append( Process(target=WindNinja, args=(nompp,))) #Spawn WindNinja subprocess
        p[i].start()
    for procesando in p:
        procesando.join()
    joinMap
```

Fig. 6. Python Multiprocessing parallelization

In this case, no MPI is used, but the Python package to manage processes is used directly. The same master-worker parallel programming paradigm that has been used before, is also used in this approach. In this case, it is also possible to exploit the OpenMP WindNinja parallelization to calculate the wind field for each partition.

4   Experimental Results

As it has been stated in the introduction of this paper, one of the objectives that we faced up when doing this work, was to be able to deploy a WindNinja parallel version that could be executed on the field during and on going event. That is, we are interested in having prediction systems that could be used in an operational fashion, instead of designing very complex forecast system that requires access to high cost computational resources either in economic terms but also in real time connectivity access possibilities. Typically, forest fires that occur in complex terrains with difficult access to the burn area, are the ones that clearly require the evaluation of the underlying wind field at high resolution because the shape of the landscape directly affects wind speed and wind direction at different points of the terrain. Since in these situations the connectivity of the computing systems in not always guaranteed, commodity hardware that could be available on the field, such a single laptop, could became an extremely useful tool. However, one cannot ignore that time constrains and accuracy of the results could not be dismissed. Thus, the experimental results outlined in this section have been obtained using a commodity hardware, in particular a single laptop based on an Intel i7-7700 processor that has 4 cores with hyperthreading, 8 MB of cache memory and works at a base frequency of 3.60 GHz. The system has 16 GB-DDR4 RAM.

As a test cases, we have used two map sizes having 775×775 and 1500×1500 cells each. The resolution in both cases has been set to 30 m and different map partitioning have been also selected. Windninja 3.5.3 version has been used for all the experiments reported.

The first experiment that has been conducted corresponds to executing WindNinja in its native version with SSOR preconditioner and using its OpenMP implementation considering 1, 2, 4 and 8 threads. Tables 1 and 2 shows, respectively, the execution time and the speed up obtained when evaluating the windfield of the 775×774 cells map. This experiment has been done as a basic parallel reference using the native OpenMP parallelization and it has been used to normalize the comparison with the proposed parallel approaches.

Table 1. WindNinja memory requirements and Execution times using native SSOR preconditioner

Map size	Memory (GB)	Execution time (seconds) # threads			
		1	2	4	8
775×775	7	205	157	136	135

Table 2. WindNinja speed up using native SSOR preconditioner

Map size	SpeedUp # threads		
	2	4	8
775 × 775	1.30	1.50	1.52

These execution times are not extremely large, but if it is considered that the wind field calculation is only one of the steps of the fire propagation prediction and, moreover, it must be executed for each value of the meteorological wind, it is mandatory to reduce the execution time as much as possible.

In order to analyze the effect of the different parallelization schemes described in Sect. 3, one have first conducted an experiment that uses the map size of 775 × 775. Two partitioning have been used: a 2 × 2 and a 4 × 4 with an overlap of 25 cells in both cases. So, in the first case, there are 4 partitions and in the second one, there are 16 partitions. For each case, Windninja has been executed using 1, 2 and 4 threads. The results are summarized in Table 3. These results can be compared with the results shown in Table 1 that reproduces the execution time for the original WindNinja implementation for a 775 × 775 cells map. It must be considered that the hardware used to run the application is just a laptop with a single processor with 4 cores and hyperthreading (8 threads). It can be observed that, for this map size, the Python multiprocessing implementation with a 2 × 2 partition and 4 OpenMP threads reaches an execution time of 50 s which represents less than 25% of the original time of 205 s and is clearly faster than the MPI C++ and the MPI for Python implementations. Actually, using just one OpenMP thread the execution time is just 60 s, which is a significant execution time improvement. However, in the case of 4 × 4 partitioning all the implementations are around 50 s (from 47 to 53) independently from the number of OpenMP threads. This is due to the fact that we start from 16 partitions and the amount of actual work involved to solve each partition is really small, and the number of cores of the system and threads support is too small. But, in any case, the execution time reached is quite good.

Table 3. WindNinja Execution time for a 775 × 775 cells map (in seconds)

775 × 775	2 × 2 # Threads				4 × 4 # Threads			
	1	2	4	8	1	2	4	8
MPI C++	115	92	91	93	48	48	50	48
MPI for Python	108	85	81	85	47	47	49	52
Python Multiprocessing	60	53	50	57	49	48	53	52

Then a 1500 × 1500 cells map with a resolution of 30 m has been considered. This map cannot be solved in the laptop used due to memory limitations

(24 GBs required), so that the map partitioning strategy also solves this problem and allows to calculate the wind field. In this case, also a 2×2 and a 4×4 partitioning have been used, and 1, 2, 4 and 8 OpenMP threads have been tested. The results are summarized in Table 4.

Table 4. WindNinja Execution time for a 1500×1500 cells map (in seconds)

1500×1500	2×2 # Threads				4×4 # Threads			
	1	2	4	8	1	2	4	8
MPI C++	262	222	225	225	187	179	169	153
MPI for Python	232	192	183	172	175	167	157	149
Python Multiprocessing	223	183	175	164	137	139	130	124

In this case, the Python Multiprocessing implementation is also the faster one and reduces the execution time up to 164 s for the 2×2 partition and 124 s for the 4×4 partition using 8 OpenMP threads in both cases.

5 Conclusions

Wind field calculation is a critical issue to provide reliable forest fire propagation predictions. However, in the case of emergency fighting there are several constraints that should be considered. These constraints include propagation prediction time and hardware availability:

- The propagation predictions must be provided well in advance to allow the control centres to manage the field means to take the appropriate actions on the adequate time.
- It is not feasible to use extremely powerful computers by the control centres, or even the field means, but it is more feasible that they have commodity hardware available.

So, the WindNinja wind field simulator has been parallelized using three approaches based on map partitioning and tested on a single laptop. All parallel implementations reduces the execution time significantly, although the Python Multiprocessing implementation is the one that reaches the best execution time, especially for large maps. Using this parallelization it is more feasible to integrate WindNinja in an operational forest fire propagation prediction system that could be used on the field, reducing the uncertainty in propagation predictions.

Acknowledgments. This research has been supported by MINECO-Spain under contract TIN2017-84553-C2-1-R, and by the Catalan government under grant 2017-SGR-313.

References

1. Abdalhaq, B., Cortés, A., Margalef, T., Luque, E.: Accelerating optimization of input parameters in wildland fire simulation. In: Wyrzykowski, R., Dongarra, J., Paprzycki, M., Waśniewski, J. (eds.) PPAM 2003. LNCS, vol. 3019, pp. 1067–1074. Springer, Heidelberg (2004). https://doi.org/10.1007/978-3-540-24669-5_138
2. Andrews, P.L.: Current status and future needs of the BehavePlus fire modeling system. Int. J. Wildland Fire **23**, 21–33 (2014). https://doi.org/10.1071/WF12167
3. Brun, C., Margalef, T., Cortés, A., Sikora, A.: Enhancing multi-model forest fire spread prediction by exploiting multi-core parallelism. J. Supercomput. **70**(2), 721–732 (2014). https://doi.org/10.1007/s11227-014-1168-z
4. Brun, C., Artés, T., Margalef, T., Cortées, A.: Coupling wind dynamics into a DDDAS forest fire propagation prediction system. Procedia Comput. Sci. **9**, 1110–1118 (2012)
5. Carrillo, C., Margalef, T., Espinosa, A., Cortés, A.: Accelerating wild fire simulator using GPU. In: Rodrigues, J., et al. (eds.) ICCS 2019. LNCS, vol. 11540, pp. 521–527. Springer, Cham (2019). https://doi.org/10.1007/978-3-030-22750-0_46
6. Finney, M.A.: Farsite: Fire area simulator–model development and evaluation. Research Paper RMRS-RP-4 Revised 236 (1998)
7. Forthofer, J.M., Shannon, K., Butler, B.W.: Simulating diurnally driven slope winds with WindNinja. In: 8th Symposium on Fire and Forest Meteorological Society (2009)
8. Ihshaish, H., Cortés, A., Senar, M.A.: Parallel multi-level genetic ensemble for numerical weather prediction enhancement. Procedia Comput. Sci. **9**, 276–285 (2012)
9. Nocedal, J., Wright, S.J.: Conjugate gradient methods. In: Sun, W., Yuan, Y.-X. (eds.) Optimization Theory and Methods. Springer Optimization and Its Applications, vol. 1. Springer, Cham (2006). https://doi.org/10.1007/0-387-24976-1_4
10. Rothermel, R.: A mathematical model for predicting fire spread in wildland fuels (US Department of Agriculture, Forest Service, Inter- mountain Forest and Range Experiment Station Ogden, UT, USA) (1972)
11. Sanjuan, G., Brun, C., Margalef, T., Cortés, A.: Wind field uncertainty in forest fire propagation prediction. Procedia Comput. Sci. **29**, 1535–1545 (2014)
12. Sanjuan, G., Margalef, T., Cortés, A.: Hybrid application to accelerate wind field calculation. J. Comput. Sci. **17**, 576–590 (2016)
13. Skamarock, W.C., et al.: A description of the advanced research WRF version 2. Technical report, DTIC Document (2005)

Risk Profiles of Financial Service Portfolio for Women Segment Using Machine Learning Algorithms

Jessica Ivonne Lozano-Medina, Laura Hervert-Escobar(✉) (iD),
and Neil Hernandez-Gress (iD)

Tecnologico de Monterrey, 64849 Monterrey, NL, Mexico
laura.hervert@tec.mx
http://tec.mx/en/

Abstract. Typically, women are scored with a lower financial risk than men. However, the understanding of variables and indicators that lead to such results, are not fully understood. Furthermore, the stochastic nature of the data makes it difficult to generate a suitable profile to offer an adequate financial portfolio to the women segment. As the amount, variety, and speed of data increases, so too does the uncertainty inherent within, leading to a lack of confidence in the results. In this research, machine learning techniques are used for data analysis. In this way, faster, more accurate results are obtained than in traditional models (such as statistical models or linear programming) in addition to their scalability.

Keywords: PCA (Principal Component Analysis) · Portfolio optimization · Machine learning

1 Introduction

In Latin America and the Caribbean (LAC), approximately one in four are poor and one in ten cannot meet their basic food needs. The latter being especially true for children. According to The World Bank [2], even the most equal country in LAC (Uruguay) is more unequal than the most unequal country of OECD (Portugal). But it has been found that women have been important in the reduction of poverty across LAC countries. According to the study [2], it was found that women made crucial contributions to extreme and moderate poverty reduction between 2000 to 2010. Growth in female income translated to 30% reduction in extreme poverty along with 39% reduction by male income. Overall, female labor market income was twice as effective in reducing the severity of poverty compared to male labor market income.

Some benefits of the inclusion of women include:

- More resiliency against poverty and better coping mechanisms during economic shocks, especially in cases of dual-income households.
- Increases in female labor income and labor market participation seem proportional to higher enrollment rates and a higher education (closer to men).

© Springer Nature Switzerland AG 2020
V. V. Krzhizhanovskaya et al. (Eds.): ICCS 2020, LNCS 12143, pp. 561–574, 2020.
https://doi.org/10.1007/978-3-030-50436-6_42

Some of those studies, have found that women tend to be more risk-adverse than men in their financial management [6]. Though there is much controversy if the difference is significant or invalid [5], the issue of understanding the financial attitudes of people is of interest for many institutions. Such institutions include governmental organizations, who wish to increase the involvement of women in the finance industry [23], and also of financial institutions that wish to classify better the profiles of their clients.

Currently, banks manage their finance portfolio clients depending on a risk profile. These profiles are unique for every client. What this means is that every client has a different risk tolerance. That is the level at which they are still comfortable risking money, and it depends on the lifestyle, age and personality of the person. Classifying these profiles correctly helps the client have decent earnings while having peace of mind [15].

In the past, many studies have been done to determine if there are differences in risk management between genders. For example, there has been numerous studies to prove or disprove such claim [6]. This included data gathering from experiments meant to measure that difference, and from experiments that were not meant to measure it. Yet these studies prove time and again that there is truth to the claim.

On the other hand, there has been previous attempts to model, the classification of woman financial habits. Studies include analysis on household income census from Italy [3], Sweden [12] and United States [21]. The majority of these studies use linear programming techniques [12,17,26] and probit models [3,18].

One of the studies has used an unsupervised machine learning technique. Simms [21] used Cluster Analysis to analyze women habits on financial advice and was able to identify two prominent groups. Cluster Analysis is useful for unlabeled data, that is instances that don't have a class assigned. And these type of techniques can benefit the study of risk profiles due to being able to create complex models, manage the amount of features and classify them with high accuracy.

Therefore as a means to understand the segment better, this research seeks to generate an adequate profile of women segment and explain the variables and indicators of the profile. If this model proves to be efficient identifying risk and classifying, it will provide a better understanding of the client risk profile of the target segment and a better-tailored product. This will help develop entrepreneurial woman in Mexico that seek to grow their financial assets and generate a reasonable income from their investments. For the financial institution it will help to provide a better product, a safer investment, and aid in its decision-making as a better way of classifying and understanding user risk profiles. The rest of the document is organized as follows: Problem definition and context are given in Sect. 2. Section 3 presents methods and approaches used. Finally Sect. 4 presents the conclusions and future work.

2 Problem Definition in Mexican Population Context

Modelling human behavior is a complicated task. Some examples are the modelling market price movements [9], manage credit card customers [10], and forecasting financial risk [24]. Most of them are done with statistical and operational techniques [22].

In finance, the modeling of the behavior of potential clients and their risk, in general, does not distinguish gender. However, analyzing the women's segment separately represents an opportunity for financial companies to expand their portfolio and contribute to gender equality.

Women are among the most vulnerable groups and provide much concern for governments in the world. According to the UN [23], at least 50% of the women in the world have a paid wage and salary employment, which is an increase of 40% compared to the 1990s. Yet women earn 24% less than men when having the same work. Therefore there is still much work to do to provide gender equality, poverty eradication and inclusive economic growth. Women make enormous contributions to economy either by working in a business, on a farm, as entrepreneurs or employees, or by doing unpaid care work at home, and still there is much to do to aid women in their development.

2.1 Mexican Financial Habits of the Population

In the context of Mexico, The National Institute of Statistics and Geography (INEGI) released the results of a survey that reveals the differences between being men or women in Mexico [11]. The survey contains information on population between 18 and 70 years old. The analysis was made for rural and urban areas and includes some topics related with habits in the expenses management. Some examples are: having a savings account, having an expense record book, source of your income (foreign or national), etc.

Saving Accounts. Mexican women overall use less financial instruments such as retirement savings accounts. Being the participation of 20% in urban areas and 15.2% in rural areas. This result is due to they have less presence in the workforce. Typically, Mexicans contribute to their retirement account through contributions from their salary. This contribution is made automatically (by law) to each salaried worker. People who do not have an employer should look for ways to save for retirement. Women overall did not have a saving account due to not having a job (43.8%) or not knowing about the benefit (18.9%).

Type of Savings. Three options were considered: formal savings (within a financial institution), informal savings and a combination of both. Overall, 6 out of 10 Mexicans have informal savings. Also, 2 out of 10 do not have savings and about the same amount have formal savings. In urban areas, men combined informal and formal savings, while women prefer informal savings. Contrary so, in rural areas, men prefer informal savings while women choose formal savings, or a combination of both. Additionally, in rural areas, women have more access to formal credits than men, while in urban areas it is the opposite.

National Business Trends. Businesses in Mexico can be divided in four categories: Commerce (46.7%), services (39.1%), manufacturing (12.2%) and other types of economic activities (2%).

While 71.8% of the workforce can be found in the areas of commerce and services in 2019, the best paid jobs can be found in the areas of manufacturing with an average of 8, 389.9 dollars per year per worker. Men have the largest presence in the areas of manufacturing, where 63.6% of the men in the workforce can be found. While women have the largest presence in the areas of commerce, with 46.9%. The states with the highest participation of women are the ones in south-center of the country, specifically in Guerrero (49.8%) and Oaxaca (48.6%). Some problems facing companies in Mexico are the insecurity (39.1%), very high costs (23.9%), and the unfair competition 21.1%.

Mexico has a dominant type of business, named micro-businesses. They are the most common type found in all the country. They give employment to more than a third of the citizens. Usually, the workforce is maximum 10 people or less. They represent 95% of all businesses, contribute 37% of the total employed workforce and 14.2% of the incomes.

From census, it has been found that most businesses are located in the State of Mexico (11.6%), Mexico City (7.3%), Puebla (6.6%), Jalisco (6.4%), and Veracruz (6%). But most of the workforce can be found in Mexico City (12.9%), State of Mexico (9.4%), Jalisco (6.9%), Nuevo León (5.7%), Guanajuato (5%), Veracruz (4.4%), and Puebla (4.4%). 36 out of 100 people work in any of these: State of Mexico, Mexico City, Jalisco or Nuevo León. This means that places with low in number of businesses but higher workforce are those with a lower proportion of micro-businesses.

So this research proposes to use this stereotype, and use it as an advantage for women in the financial industry. As their profile provide a better case study for financial portfolios.

Next section, we introduce the methods ans strategies used to generate the woman profile.

3   Methods and Approaches

High dimensional data is problematic due to high computational costs and memory usage. It can consist of irrelevant, misleading or redundant features that are the ones increasing the search space size. They make it harder to process the data and therefore do not contribute to the learning process [13]. There are methods to reduce dimensionality: Feature Selection and Feature Extraction. In the former, the features that contain the information necessary to solve the problem are the only ones selected; in the latter, it consists of methods that seeks a transformation of the input space into a lower dimensional subspace that still contains most of the relevant information. Both or only one of these methods can be used as a preprocessing step. The need for feature selection and extraction is due to the information of a dataset falling in either of three categories: relevant, irrelevant and redundant. By means of the method to remove the majority of irrelevant and redundant features, the objective is to create a subset with the least amount of features that still retains the most information and contributes to the learning accuracy [14].

Some advantages of feature selection [14] include:

- Reduce storage requirements and increase algorithm speed.
- Removes redundant, irrelevant or noisy data.
- Improves data quality.
- Increases accuracy of the resulting model.
- Reduction of feature set and therefore faster future recollections.
- Improved performance in predictive accuracy.
- Better visualization and understanding about the dataset.

There is a major difference between feature selection and feature extraction. In selection, when a dataset is complex with many relevant features, choosing only a small selection of them, means that information will be lost by the omission. Whereas in feature extraction, the size of the feature space can be decreased without discarding features, but the transformation done to linearly combine them makes them harder to interpret as individual features.

3.1 Principal Component Analysis

Among these techniques, the most common and popular one is Principal Component Analysis (PCA) developed by Karl Pearson in 1901. It consists of a linear transformation of data that minimizes the redundancy, which is measured by the covariance, and maximizes the information, which is measured by the variance. PCA uses orthogonal transformations to convert samples of correlated variables into samples of linearly uncorrelated features [13]. These new features are called principle components (PC) and describe the same or less amount of information. Principle components are new features with two properties:

- each principle component is a linear combination of the original variables.
- and the principle components are uncorrelated to each other and redundant information is discarded.

The principle components are ordered in terms of information gain. That is, the first principle component contains the most observed variability, and the last principle component contains the least. PCA reduces the number of original features by eliminating the last principle components.

The main applications of PCA are data compression, image analysis, visualization, pattern recognition, regression, and time series predictions.

PCA also has assumptions which limits its use [13]:

- relationship between variables is linear,
- variables are normalized,
- and it lacks a probabilistic model structure which is important in many statistical tests.

3.2   Unsupervised Clustering Techniques

K-Means. K-means is an unsupervised clustering algorithm whose objective is to assign data points into clusters, thus the overall distance between points and cluster centroids is minimized [8]. It tries to separate samples into groups of equal variance, minimizing a criterion knows as inertia or within-cluster sum of squares [16]. Equation 1 shows the expression K-means minimizes:

$$\sum_{i=0}^{n} min(||x_i - \mu_j||^2) \tag{1}$$

One of the advantages of this algorithm is that it scales well with data size and its flexibility allows it to be applied to many different fields.

The most common way to choose the number of clusters for K-means is using the elbow method. This is done by fitting K-means models for a range of consecutive numbers starting from 1, and plotting the total within sum of squares value for each number of clusters versus that cluster number. The elbow or turning point in the line plot, and therefore an adequate number of clusters, is where the reduction in sum of squares seems to drop off [8].

3.3   Machine Learning Algorithms

In this section, the techniques used will be discussed. Among the candidate methods for classification we used Decision Trees (DTs) which is a non-parametric supervised learning method used for classification and regression. The goal is to create a model that predicts the value of a target variable by learning simple decision rules inferred from the data features. Some advantages of decision trees are:

- Simple to understand and to interpret. Trees can be visualised.
- Requires little data preparation.
- The cost of using the tree (i.e., predicting data) is logarithmic in the number of data points used to train the tree.
- Able to handle both numerical and categorical data.
- Able to handle multi-output problems.
- Uses a white box model.
- Possible to validate a model using statistical tests.
- Performs well even if its assumptions are somewhat violated by the true model from which the data were generated.

The disadvantages of decision trees include:

- Decision-tree learners can create overfitting.
- Decision trees can be unstable.
- The problem of learning an optimal decision tree is known to be NP-complete under several aspects of optimality and even for simple concepts.
- There are concepts that are hard to learn because decision trees do not express them easily, such as XOR, parity or multiplexer problems.

– Decision tree learners create biased trees if some classes dominate. It is therefore recommended to balance the dataset prior to fitting with the decision tree.

3.4 Performance Measures

Accuracy is the most used measure to assess the performance of a credit evaluation model [7], and was seen throughout different studies [24,25] as shown in Eq. 2, 3 and 4.

$$\text{Type I accuracy} = \frac{\text{number of both observed bad and classified as bad}}{\text{number of observed bad}} \quad (2)$$

$$\text{Type II accuracy} = \frac{\text{number of both observed good and classified as good}}{\text{number of observed good}} \quad (3)$$

$$\text{Total accuracy} = \frac{\text{number of correct classification}}{\text{the number of evaluation sample}} \quad (4)$$

Other variations exist where the former measure was considered incomplete, such as in the study of Danenas and Garsvas [7]. Where the accuracy performance was represented using three different measures:

– Accuracy defined as the proportion of correct prediction shown in Eq. 5.

$$\text{accuracy} = \frac{\text{True Positives} + \text{True Negatives}}{\text{Total number of instances}} \quad (5)$$

– True Positive Rate (TPR) or Sensitivity is a ratio of True Positives over the total number of total positive instances shown in Eq. 6.

$$\text{TPR}_i = \frac{\text{True Positives}_i}{\text{True Positives}_i + \text{False Negatives}_i} \quad (6)$$

– F-Measure is defined as harmonic mean of precision and recall (True Positive) measures. It is preferred to accuracy for analysis of the classification performance case of unbalanced learning shown in Eq. 7.

$$F_i = \frac{2 \times \text{precision}_i \times \text{recall}_i}{\text{precision}_i + \text{recall}_i} \quad (7)$$

4 Results and Analysis

The population of interest in this research is focused on Mexican woman. Then, three sets of data were used in order to define the financial profile for this segment. The first dataset is from the Mexican National Institute of Statistics and Geography (INEGI). This data provide information about location and density of population in Mexico. The data was generated during the Population and Dwellings Census 2010 (ITER). A locality is any settlement where a community of people live. It can range from a small number of dwellings to a large urbanized

Fig. 1. Left: Mexican map with states. Right: Localities in a city

area. An example of how a locality looks like can be seen in Fig. 1. The dataset contains information on the localities for the thirty-two states found in Fig. 1.

The original dataset contained 196,025 localities and 200 attributes. The attributes contain demographic information of each locality. Usually in the form of counts or averages.

Among these, 23 attributes were chosen as relevant:

– 3 attributes related to coordinates
– 1 related to men-women ratio
– 1 related to child mortality
– 2 related to education
– 4 related to economic activity
– 3 related to civil status
– 2 related to household decisions
– 7 related to dwellings.

It can consist of irrelevant, misleading or redundant features that are the ones increasing the search space size. They make it harder to process the data and therefore do not contribute to the learning process.

4.1   Preprocessing

The programming language used was Python. Four steps were done to tidy and prepare the dataset for the next subsection:

– Coercing data types into floats
– Transforming attributes into proportions
– Scaling variables
– Removing nulls

By the end of the process, 93468 instances and 28 attributes remained. To reduce dimensionality, PCA was applied. 90% of the information was retained, and 11 principle components were created. The principle components percentages can be seen in Table 1. The weight of the variance for each principle component is of 0.2549, 0.2277, 0.1442, 0.0679, 0.0465, 0.0393, 0.0361, 0.0287, 0.0264, 0.0239, and 0.0200, respectively.

Table 1. Principle components percentage per attribute

PC	1	2	3	4	5	6	7	8	9	10	11
LONGITUDE	0.367	0.307	-0.371	-0.109	0.107	0.078	-0.026	0.008	-0.052	-0.103	-0.707
LATITUDE	0.389	0.492	-0.466	-0.060	-0.050	0.042	-0.021	-0.029	0.073	0.148	0.547
ALTITUDE	0.755	-0.633	0.048	0.047	-0.102	0.009	0.063	0.000	0.043	0.022	0.050
TAM_LOC	0.022	0.017	0.063	-0.071	-0.028	0.149	-0.200	0.946	-0.023	0.173	0.005
REL_H_M	0.000	0.000	-0.001	-0.001	-0.001	-0.001	-0.001	0.000	0.001	-0.001	0.000
PROM_HNV	-0.013	-0.039	-0.114	0.133	0.087	-0.015	0.029	0.017	-0.179	-0.067	-0.098
GRAPROES	0.126	0.192	0.361	-0.335	-0.120	0.018	0.132	-0.045	0.002	0.046	-0.042
GRAPROES_M	0.089	0.147	0.326	-0.311	-0.089	0.052	0.120	-0.036	-0.013	0.034	0.001
GRAPROES_F	0.154	0.220	0.358	-0.321	-0.142	-0.014	0.129	-0.050	0.011	0.051	-0.085
PEA_M	-0.011	0.085	-0.012	0.159	-0.229	-0.153	0.061	0.124	0.579	-0.376	-0.076
PEA_F	0.051	0.023	0.091	-0.093	-0.011	0.165	-0.026	0.029	0.051	-0.342	0.043
PE_INAC_M	0.017	0.021	0.035	-0.034	0.181	0.036	0.098	-0.060	-0.176	0.399	-0.012
PE_INAC_F	-0.020	0.009	0.009	0.208	0.187	-0.170	0.228	0.010	0.100	0.508	-0.147
P12YM_SOLT	-0.001	-0.020	0.031	-0.110	0.373	-0.111	0.142	0.030	0.681	0.189	-0.027
P12YM_CASA	0.034	0.128	0.069	0.289	-0.391	-0.174	0.090	0.048	-0.143	0.086	-0.194
P12YM_SEPA	0.006	0.042	0.027	0.097	0.197	0.237	0.175	0.041	-0.002	-0.151	0.039
HOGJEF_M	0.019	0.118	0.038	0.281	-0.311	-0.062	0.104	0.021	0.067	0.075	-0.059
HOGJEF_F	0.036	0.031	0.088	0.113	0.414	0.442	0.288	0.051	-0.083	-0.236	-0.005
PROM_OCUP	-0.027	-0.089	-0.067	-0.200	0.055	-0.123	-0.162	-0.032	-0.007	0.040	0.042
PRO_OCUP_C	-0.015	-0.016	-0.068	-0.032	-0.042	0.001	-0.003	-0.007	0.028	-0.024	-0.019
VPH_PISODT	0.134	0.198	0.234	0.387	-0.010	0.116	0.234	0.026	-0.045	-0.020	0.253
VPH_1CUART	-0.045	0.011	-0.028	0.000	-0.093	0.118	0.005	0.000	0.024	-0.103	0.090
VPH_2CUART	-0.109	-0.065	-0.090	0.122	-0.316	0.596	0.168	-0.095	0.190	0.275	-0.163
VPH_3YMASC	0.170	0.175	0.192	0.268	0.261	-0.388	0.143	0.125	-0.146	-0.150	-0.038
VPH_C_SERV	0.167	0.174	0.355	0.308	0.152	0.199	-0.746	-0.208	0.147	0.112	-0.072

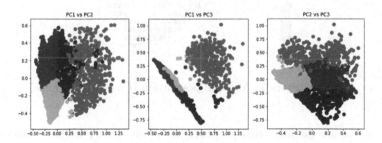

Fig. 2. PCA plot

Clustering. To identify the best number of clusters found in the data, three different scores were used: the Elbow method, the Davies-Bouldin score and the Calinski-Harabasz score. The results obtained in Fig. 3 show that four clusters was the best option overall. In the elbow method, it is were the distances start decaying; in Davies-Bouldin score is the lowest score; and in Calinski-Harabasz score is the highest value. Applying the Silhouette score gives us a value of 0.377.

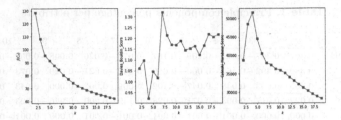

Fig. 3. Cluster scores

The reduced data then is used for clustering the population. Results show that the country is divided into four clusters as Fig. 4 shows: a North cluster (blue), a South cluster (red), a Central cluster (cyan) and Small populations clusters (yellow). By seeing them individually, such as in Fig. 5, we can appreciate the size of each cluster without the overlapping. As such, the Small Populations cluster is a very big cluster, which defining characteristic is the small size of the population in the localities. And contrary to the other clusters found, this can be found all over the country.

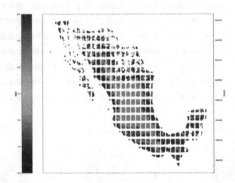

Fig. 4. Clusters for Mexican population (Color figure online)

But as we are dealing with proportions, it gives very inflated values as its population size is very small. Which also make the populations with very large populations have smaller proportions overall, and thus is excluded for further analysis to concentrate on the bigger localities.

An interesting trend found was within the proportion of household heads found in the country found in Fig. 5, where the proportion of household heads and the Economically Active Population per Locality are plotted. It is shown that the household authority of men grow proportionally with their economic activity. Whereas the household authority of women decreases with theirs. Women as mentioned in literature, are not as economically active as men, but when the economic activity of women increases the trend seems to grow positively.

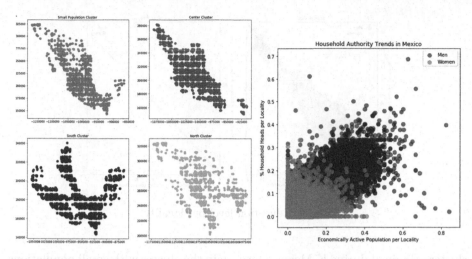

Fig. 5. Left: individual clusters. Right: household authority in Mexico (Color figure online)

Another trend can be seen in married people per locality and economic activity. The economic activity of men increases as the percentage of married men increases per locality. The case of of women is a bit trickier though as shown in Fig. 6. As the economic activity of women increases, the proportion of married women per locality stays averaged approximately in 40%.

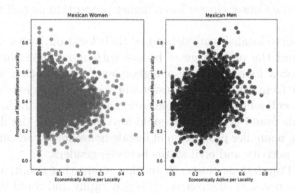

Fig. 6. Civil status in Mexico

On the other hand, the average academic grade per locality is pretty even between genders, just as mentioned in literature about LAC countries. A linear regression shows that both levels are similar just as shown in Fig. 7.

Decision Trees. The decision tree was created using an implementation found in scikit-learn. The inputs were the attributed without PCA reduction and the

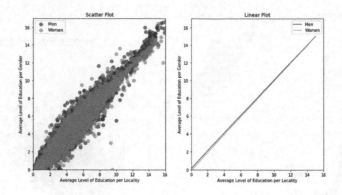

Fig. 7. Education level per locality

clusters obtained during K-Means. Once again the cluster with small populations was omitted.

A training and testing set of 70% and 30%, respectively was used. The parameters used were a criterion of entropy, a max depth of 5 and minimum samples per leaf of 30. The score obtained was of 0.955.

In most cases, the coordinates and altitude of the locality where used to separate the clusters. But in cases where overlapping existed between clusters, variables such as the level of education and percentage of houses with more than three rooms were the deciding factors. The North cluster has a higher level of education per locality and houses with more than three rooms than the other clusters, the Center cluster has a higher proportion as well than the South cluster.

We can observe localities are defined by their location, altitude, average level of education and the percentage of houses with more than three rooms. The North cluster was found to have a higher level of education and number of rooms than the Center and South cluster where overlapping existed. The Center Cluster had as well a higher level of education and rooms than the South cluster.

Many trends found in literature could also be found in the data such as an even education, a smaller proportion of female household heads and some trends with economic activity and civil status between genders.

The Small Populations cluster is another interesting result. It was the only cluster that was not defined strongly by its geography and could be found around all the country. Still, the small sample size brought along many overestimated proportions and could not be used to compare against the clusters with larger localities.

The results provide more knowledge on the partition of the Mexican population using state-of-the-art Machine Learning techniques. Prior studies made using census data on the Mexican population do not make use of them, whereas other countries [1, 4, 19, 20] have started implementing them on their own data with good results.

5   Conclusion

This research provide a methodology for analyzing several datasets using machine learning techniques in order to define a financial profile of Mexican Woman. The methodology includes first the preprocessing of the data, to then reduce the dimensionality using a PCA. Followed by a clustering analysis, and finally the use of decision trees to define the profile. The main contribution of this research is the definition of the mexican financial profile for women segment. Highlighting that women segment has been partially attended by the financial entities, missing the opportunity to offer financial products according to the risk involved. The methods applied in this research have been proof to be more flexible and accurate than traditional statistical approaches. Future work includes the risk calculator for specific financial products, as well as the extension for population of other countries of interest.

References

1. Abarca-Alvarez, F., Campos, S., Bellido, R.: Señales de gentrificación a través de la inteligencia artificial: identificación mediante el censo de vivienda. Bitácora Urbano Territorial **28**, 103–114 (2018)
2. World Bank. The effect of women's economic power in Latin America and the Caribbean. Latin America and the Caribbean poverty and labor brief, p. 48 (2012)
3. Bertocchi, G., Torricelli, C., Brunetti, M.: Portfolio choices, gender and marital status. Rivista di Politica Economica **98**, 119–154 (2008)
4. Sheng, B., Gengxin, S.: Data mining in census data with cart. In: 2010 3rd International Conference on Advanced Computer Theory and Engineering(ICACTE), vol. 3, pp. V3-260–V3-264, August 2010
5. Brindley, C.: Barriers to women achieving their entrepreneurial potential: women and risk. Int. J. Entrepreneurial Behav. Res. **11**(2), 144–161 (2005)
6. Charness, G., Gneezy, U.: Strong evidence for gender differences in risk taking. J. Econ. Behav. Organ. **83**(1), 50–58 (2012). Gender Differences in Risk Aversion and Competition
7. Danenas, P., Garsva, G.: Selection of support vector machines based classifiers for credit risk domain. Expert Syst. Appl. **42**(6), 3194–3204 (2015)
8. Flynt, A., Dean, N.: DA survey of popular R packages for cluster analysis. J. Educ. Behav. Stat. **41**, 205–225 (2016)
9. Hachicha, N., Jarboui, B., Siarry, P.: A fuzzy logic control using a differential evolution algorithm aimed at modelling the financial market dynamics. Inf. Sci. **181**(1), 79–91 (2011)
10. Hsieh, N.-C.: An integrated data mining and behavioral scoring model for analyzing bank customers. Expert Syst. Appl. **27**(4), 623–633 (2004)
11. INEGI. Mujeres y hombres en méxico 2016. Instituto Nacional de Estadística y Geografía (México), p. 250 (2016)
12. Jenny, S.-S.: Self-directed pensions: gender, risk, and portfolio choices. Scand. J. Econ. **114**(3), 705 (2012)
13. Khalid, S., Khalil, T., Nasreen, S.: A survey of feature selection and feature extraction techniques in machine learning. In 2014 Science and Information Conference, pp. 372–378, August 2014

14. Ladla, L., Deepa, T.: Feature selection methods and algorithms. Int. J. Comput. Sci. Eng. (IJCSE) **3**, 1787–1797 (2011)
15. Masthead. Factors that every financial advisor must consider when doing a client's risk profile, September 2016. https://www.masthead.co.za/newsletter/factors-that-every-financial-advisor-must-consider-when-doing-a-clients-risk-profile/
16. Pedregosa, F., et al.: Scikit-learn: machine learning in Python. J. Mach. Learn. Res. **12**, 2825–2830 (2011)
17. Petreska, B.R., Kolemisevska-Gugulovska, T.D.: A fuzzy rate-of-return based model for portfolio selection and risk estimation, pp. 1871–1877, October 2010
18. Rybczynski, K.: Gender differences in portfolio risk across birth cohort and marital status. Can. J. Econ./Revue canadienne d'économique **48**(1), 28–63 (2015)
19. Shanmuganathan, S., Li, Y.: An AI based approach to multiple census data analysis for feature selection. J. Intell. Fuzzy Syst. **31**, 859–872 (2016)
20. Sharath, R., Krishna Chaitanya, S., Nirupam, K.N., Sowmya, B.J., Srinivasa, K.G.: Data analytics to predict the income and economic hierarchy on census data. In 2016 International Conference on Computation System and Information Technology for Sustainable Solutions (CSITSS), pp. 249–254, October 2016
21. Simms, K.: Investor profiles: meaningful differences in women's use of investment advice? Financ. Serv. Rev. **23**(3), 273–286 (2014)
22. Thomas, L.C.: A survey of credit and behavioural scoring: forecasting financial risk of lending to consumers. Int. J. Forecast. **16**(2), 149–172 (2000)
23. UN. Economic Empowerment of Women. Technical report, United Nations (2013)
24. Yu, L., Wang, S., Lai, K.K.: Credit risk assessment with a multistage neural network ensemble learning approach. Expert Syst. Appl. **34**(2), 1434–1444 (2008)
25. Yu, L., Yue, W., Wang, S., Lai, K.K.: Support vector machine based multiagent ensemble learning for credit risk evaluation. Expert Syst. Appl. **37**(2), 1351–1360 (2010)
26. Zavala-Diaz, J.C., Díaz-Parra, O., Ruiz-Vanoye, J.: Analysis of risk in linear multi-objective model and its evaluation for selection of a portfolio of investment in the Mexican stock exchange. Afr. J. Bus. Manage. **5**, 7876 (2011)

Multidimensional BSDEs with Mixed Reflections and Balance Sheet Optimal Switching Problem

Rachid Belfadli[1], M'hamed Eddahbi[2], Imade Fakhouri[3]([✉]),
and Youssef Ouknine[4,5,6]

[1] Faculty of Sciences and Techniques, Cadi Ayyad University,
Bd Abdelkrim El Khattabi, B.P. 618, 40000 Guéliz, Marrakesh, Morocco
r.belfadli@uca.ma
[2] College of Sciences, Mathematics Department, King Saud University,
P.O. Box 2455, Riyadh 11451, Riyadh, Kingdom of Saudi Arabia
meddahbi@ksu.edu.sa
[3] Complex Systems Engineering and Human Systems, Mohammed VI Polytechnic
University, Lot 660, Hay Moulay Rachid, 43150 Ben Guerir, Morocco
imade.fakhouri@um6p.ma
[4] Faculty of Sciences Semlalia, Cadi Ayyad University, Bd Prince My Abdellah,
B.P. 2390, 40000 Marrakesh, Morocco
ouknine@uca.ac.ma
[5] Africa Business School, Mohammed VI Polytechnic University,
Lot 660, Hay Moulay Rachid, 43150 Ben Guerir, Morocco
youssef.ouknine@um6p.ma
[6] Hassan II Academy of Science and Technology, Rabat, Morocco

Abstract. In this paper, we study a system of multidimensional coupled reflected backward stochastic differential equations (RBSDEs) with interconnected generators and barriers and mixed reflections, i.e. oblique and normal reflections. This system of equations is arising in the context of optimal switching problem when both sides of the balance sheet are considered. This problem incorporates both the action of switching between investment modes and the action of abandoning the investment project before its maturity once it becomes unprofitable. Pricing such real options (switch option and abandon option) is equivalent to solve the system of coupled RBSDEs considered in the paper, for which we show the existence of a continuous adapted minimal solution via a Picard iteration method.

Keywords: Real options · Optimal switching · Balance sheet ·
Trade-off strategies · Merger and acquisition · Backward SDEs · Mixed
reflections

1 Introduction

Optimal switching problem (OSP) has attracted a lot of interest in the recent years (see among others [1, 2, 5–7, 11, 12]), since it can be related to many practical applications, for example the problem of valuation investment opportunities.

© Springer Nature Switzerland AG 2020
V. V. Krzhizhanovskaya et al. (Eds.): ICCS 2020, LNCS 12143, pp. 575–589, 2020.
https://doi.org/10.1007/978-3-030-50436-6_43

OSP consists in finding an optimal management strategy for a production company that can run in m, $m \geq 2$, different modes. A management strategy δ is a combination of a nondecreasing sequence of stopping times $(\tau_n)_{n \geq 0}$, and a sequence of random variables $(\epsilon_n)_{n \geq 0}$ taking values in the set of possible production modes $\Lambda = \{1, \ldots, m\}$. At time τ_n, in order to maximize the profit of the company, the manager decides to switch the production from the current mode ϵ_{n-1} to ϵ_n. When the production of the company is working under a strategy δ, it generates a gain equal to $J(\delta)$. The OSP amounts to finding an optimal management strategy δ^* such that $J(\delta^*) = \sup_{\delta} J(\delta)$. The OSP is connected with multidimensional RBSDEs with oblique reflections and interconnected barriers.

One dimensional BSDEs with normal reflections were first introduced by [10]. The multidimensional case was studied by Gegout-Petit and Pardoux [9], and then further investigated in many other works see e.g. [8,13]. Multidimensional BSDEs with oblique reflections occurring in the context of OSPs were first introduced by [12]. They consider RBSDEs with generator taking the form $f_i(\cdot, y^i, z^i)$ and barrier $\min_{j \in \Lambda^{-i}} \left(y^j + g_{i,j} \right)$ where $g_{i,j}$ are constant switching costs and $\Lambda^{-i} = \Lambda - \{i\}$. Later, Hamadène and Zhang [11] generalized the preceding work by considering general generators and barriers of the following types $f_i(\cdot, y^1, \ldots, y^m, z^i)$ and $\max_{j \in \Lambda^{-i}} h_{i,j}(., y^j)$. Xu [17] dealt with the same kind of RBS-DEs but when the generator, which is discontinuous w.r.t. y^i, and the barrier take respectively the following forms $f_i(\cdot, y^i, z^i)$ and $\max_{j \in \Lambda^{-i}} \left(y^j - g_{i,j} \right) \vee S^i$. Then, Aazizi et al. [1] extended the results of [17] to the case of generators and barriers of the form $f_i(\cdot, y^1, \ldots, y^m, z^i)$ and $\max_{j \in \Lambda^{-i}} h_{i,j}(., y^j) \vee S^i$.

In this paper, we are interested by Balance sheet OSP (BSOSP) which is a combination between the classical OSP described above and optimal stopping involving the balance sheet. BSOSP incorporates both the action of switching between modes and the action of abandoning a project once it becomes unprofitable. There are only few papers dealing with BSOSPs. Djehiche and Hamdi [4] considered the 2-modes case, i.e. $\Lambda = \{1, 2\}$. Their generators are of the form $f_i^+(\cdot, Y^{+,i}, Z^{+,i})$, $f_i^-(\cdot, Y^{-,i}, Z^{-,i})$ and their barriers of type $(Y^{+,j} - g_{i,j}(.)) \vee (Y^{-,i} - C_i(.))$ and $(Y^{-,j} + g_{i,j}(.)) \vee (Y^{+,i} + B_i(.))$, where C_i and B_i are switching costs and $j \in \Lambda^{-i}$. Recently, the BSOSP multi-modes case was solved by Eddahbi et al. [5] when the barriers are of the form $\max_{j \in \Lambda^{-i}} (Y^{+,j} - g_{i,j}(.)) \vee (Y^{-,i} + C^i(.))$ and $Y^{+,i} + B_i(.)$ (see Eddahbi et al. [6] for the mean–field case).

Now, let us describe precisely the problem studied in this paper by introducing some notations. Let $T > 0$ be a given real number, and $(\Omega, \mathcal{F}, \mathbb{P})$ is a fixed probability space endowed with a d–dimensional Brownian motion $W = (W_t)_{0 \leq t \leq T}$. $\mathbb{F} = (\mathcal{F}_t)_{0 \leq t \leq T}$ is the natural filtration of the Brownian motion augmented by the $\mathbb{P}$–null sets of $\mathcal{F}$. All the measurability notion will refer to this filtration. The euclidean norm of a vector $z \in \mathbb{R}^d$ is denoted $|z|$. Furthermore, we introduce the following spaces of processes. $\mathbb{L}^2$ is the space of $\mathbb{R}$–valued processes ξ, such that $||\xi||_{\mathbb{L}^2} := (E[|\xi|^2])^{1/2} < +\infty$. $\mathcal{S}^2$ (resp. $\mathcal{S}_c^2$) is the set of $\mathbb{R}$–valued adapted and continuous (resp. càdlàg) processes $(Y_t)_{0 \le t \le T}$ such that $||Y||_{\mathcal{S}^2}$ (resp. $||Y||_{\mathcal{S}_c^2}) := \left(E\left[\sup_{0 \le t \le T} |Y_t|^2 \right] \right)^{1/2} < +\infty$. $\mathcal{M}^{d,2}$ is the set of $\mathbb{R}^d$–valued, progressively measurable processes $(Z_t)_{0 \le t \le T}$ such that $||Z||_{\mathcal{M}^{d,2}} := \left(E\left[\int_0^T |Z_s|^2 ds \right] \right)^{1/2} < +\infty$. $\mathcal{K}^2$ (resp. $\mathcal{K}_c^2$) is the set of non-decreasing processes K, satisfying $K_0 = 0$ and that belong to $\mathcal{S}^2$ (resp. $\mathcal{S}_c^2$).

Next, to illustrate the BSOSP studied in this paper, let us deal with a concrete example. Consider a company that has m modes of production (if $m = 3$, minimal, average and maximal production modes). The manager of the company has two options. A switch option, i.e. in order to maximize its global profit, she switches the production between the modes depending on their random performances but this switching incorporates a cost called switching cost. The manager has also an abandon option i.e. stop the production once it becomes unprofitable.

More precisely, being in mode $i \in \Lambda$, one have to switch at time t to another mode $j \in \Lambda^{-i}$, once we have that the expected profit $Y^{+,i}$ in this mode falls below the following barrier

$$Y_t^{+,i} \le S_t^{+,i} := \max_{j \in \Lambda^{-i}} h_{i,j}(t, Y_t^{+,j}) \vee (Y_t^{-,i} + C^i(t)), \forall\, t \in [0,T], \qquad (1)$$

where $h_{i,j}$ is nonlinear random function (a special case is when $h_{i,j}(.,y) = y - g_{i,j}$, where $g_{i,j}$ is a switching cost from mode i to mode j), $Y^{-,i}$ is the expected cost in mode i, and C^i is the cost incurred when exiting/terminating the production while in mode i. Since we consider both sides of the balance sheet, the manager has to switch at time t to another mode $j \in \Lambda^{-i}$, as soon as the expected cost in mode i, $Y^{-,i}$ rises above the following barrier

$$Y_t^{-,i} \ge S_t^{-,i} := S_t^i \wedge \left(Y_t^{+,i} + B^i(t) \right), \forall\, t \in [0,T], \qquad (2)$$

where $S^i(t)$ is a cost of default (i.e. in this case the project is no longer profitable and thus leads to the abandon of this latter even before its maturity), and B^i is the benefit incurred when exiting/terminating the production while in mode i. It is well known that the BSOSP can be formulated using the following system of Snell envelopes

$$Y_t^{+,i} = ess \sup_{\tau \ge t} E\left[\int_t^\tau f_i^+(s)ds + S_\tau^{+,i} \mathbf{1}_{[\tau < T]} + \xi_i^+ \mathbf{1}_{[\tau = T]} | \mathcal{F}_t \right], \qquad (3)$$

$$Y_t^{-,i} = ess \inf_{\tau \ge t} E\left[\int_t^\tau f_i^-(s)ds + S_\tau^{-,i} \mathbf{1}_{[\tau < T]} + \xi_i^- \mathbf{1}_{[\tau = T]} | \mathcal{F}_t \right], \qquad (4)$$

where $\tau \in [0,T]$ are $\mathbb{F}$-stopping times which represent the exit times from the production in mode i, f_i^+ and f_i^- denote respectively the running profit and cost per unit time dt and ξ_i^+ and ξ_i^- are respectively the values at time T of the profit and the cost yields.

The BSOSP consists in showing existence and uniqueness of the processes $(Y^{+,i}, Y^{-,i})_{i \in \Lambda}$ and also proving that the following stopping times are optimal

$$\tau^{+,i} = \inf \left\{ s \geq t : Y_s^{+,i} = S_s^{+,i} \right\} \wedge T, \text{and } \tau^{-,i} = \inf \left\{ s \geq t : Y_s^{-,i} = S_s^{-,i} \right\} \wedge T.$$

Since the Snell envelope is strongly connected to RBSDEs, solving the BSOSP is equivalent to showing existence of continuous solution to the following general (since we take $f_i^+(.) = f_i^+(s, \overrightarrow{Y}_s^+, Z_s^{+,i})$ $f_i^-(s) = f_i^-(s, \overrightarrow{Y}_s^-, Z_s^{-,i})$ where $\overrightarrow{Y}^+ := (Y^{+,1}, \ldots, Y^{+,m})$, $\overrightarrow{Y}^- := (Y^{-,1}, \ldots, Y^{-,m})$) system of BSDEs with mixed reflections: for $i \in \Lambda := \{1, \ldots, m\}$

$$(S) \begin{cases} Y_t^{+,i} = \xi_i^+ + \int_t^T f_i^+(s, \overrightarrow{Y}_s^+, Z_s^{+,i}) ds - \int_t^T Z_s^{+,i} dW_s + K_T^{+,i} - K_t^{+,i}, \\[2mm] Y_t^{+,i} \geq S_t^{+,i}, \text{ and } \int_0^T \left[Y_s^{+,i} - S_s^{+,i} \right] dK_s^{+,i} = 0, \qquad (5) \\[2mm] Y_t^{-,i} = \xi_i^- + \int_t^T f_i^-(s, \overrightarrow{Y}_s^-, Z_s^{-,i}) ds - \int_t^T Z_s^{-,i} dW_s - K_T^{-,i} + K_t^{-,i}, \\[2mm] Y_t^{-,i} \leq S_t^{-,i}, \text{ and } \int_0^T \left[S_s^{-,i} - Y_s^{-,i} \right] dK_s^{-,i} = 0, \qquad (6) \end{cases}$$

where T is called the time horizon, ξ_i^+ and ξ_i^- are called the terminal conditions, the random functions $f_i^+(\omega, t, \overrightarrow{y}, z^i) : \Omega \times [0, T] \times \mathbb{R}^m \times \mathbb{R}^d \to \mathbb{R}$ and $f_i^-(\omega, t, \overrightarrow{y}, z^i) : \Omega \times [0, T] \times \mathbb{R}^m \times \mathbb{R}^d \to \mathbb{R}$ are respectively $\mathcal{F}_t$–progressively measurable for each $(\overrightarrow{y}, z^i)$, called the generators. $h_{i,j}$ is a real nonlinear random function, and $C^i := (C^i(t))_{t \in [0,T]}$, $B^i := (B^i(t))_{t \in [0,T]}$, and $S^i := (S^i(t))_{t \in [0,T]}$ are previously given $(\mathcal{F}_t)_{0 \leq t \leq T}$–adapted processes with some suitable regularity. The unknowns are the processes $(Y^{\pm,i}, Z^{\pm,i}, K^{\pm,i}) := (Y_t^{\pm,i}, Z_t^{\pm,i}, K_t^{\pm,i})_{t \in [0,T]}$ which are required to be $(\mathcal{F}_t)_{0 \leq t \leq T}$–adapted. Moreover, $K^{+,i}$ and $K^{-,i}$ are nondecreasing processes. The second condition in (5) (resp. (6)) says that the first component $Y^{+,i}$ (resp. $Y^{-,i}$) of the solution of RBSDE (5) (resp. (6)) is forced to stay above (resp. below) the barrier $S^{+,i}$ (resp. $S^{-,i}$). The role of $K^{+,i}$ (resp. $K^{-,i}$) is to push $Y^{+,i}$ (resp. $Y^{-,i}$) upwards (resp. downwards) in order to keep it above (resp. below) the respective barrier in a minimal way in the sense of the third condition of RBSDE (5) (resp. (6)) which is called the minimal boundary condition i.e. $K^{+,i}$ (resp. $K^{-,i}$) increases only when $Y^{+,i}$ (resp. $Y^{-,i}$) touches the respective barrier.

Let us make precise the notion of a solution of the system of RBSDEs (S).

Definition 1. *A 6-uplet of processes $(Y^{+,i}, Z^{+,i}, K^{+,i}, Y^{-,i}, Z^{-,i}, K^{-,i})$ is called solution of the system of RBSDEs (S) if the two triples $(Y^{+,i}, Z^{+,i}, K^{+,i})$ and $(Y^{-,i}, Z^{-,i}, K^{-,i})$ belong to $\mathcal{S}^2 \times \mathcal{M}^{d,2} \times \mathcal{K}^2$ and satisfy the system (S).*

The main contribution of our paper is to establish the existence of a continuous minimal adapted solution to system of RBSDEs (S). To this end we use a Picard iteration method (see El Karoui et al. [10] for more details). Uniqueness of the solution does not hold, since it is not verified even for the two-modes case and for a less general form of RBSDE (S) (see the counter-example in [4, subsection 3.1]).

Clearly, our results generalize the related works in the literature, since our RBSDE (S) is more general in many features. Actually, the expected profits and cost yields $\overrightarrow{Y}^{+}$ and $\overrightarrow{Y}^{-}$ are respectively interconnected in the generators f_i^{+} and f_i^{-}. This dependence can be interpreted as a nonzero-sum game problem, where the players' utilities affect each other. Furthermore, the solutions $Y^{+,i}$ and $Y^{-,i}$ are also interconnected in the barriers. Note that, the general barrier $h_{i,j}(\cdot, y)$ which is random and nonlinear, makes the dependence on the unknown process implicit. This, allows one to consider more general switching cost, for instance in the case of risk sensitive switching problem.

The remainder of the paper is organized as follows. Section 2 is devoted to the assumptions. In Sect. 3, we state and prove the main result of the paper.

2 Assumptions

Let us introduce the following assumptions:

[**H1**]: For each $i \in \Lambda$, f_i^{+} and f_i^{-} satisfies:

(i) $\mathbb{E}\left(\int_0^T \sup_{\overrightarrow{y}: y_i = 0} \left| f_i^{+}(t, \overrightarrow{y}, 0) \right|^2 dt + \int_0^T \sup_{\overrightarrow{y}: y_i = 0} \left| f_i^{-}(t, \overrightarrow{y}, 0) \right|^2 dt \right) < +\infty.$

(ii) The mappings $(t, \overrightarrow{y}, z^i) \to f_i^{+}(t, \overrightarrow{y}, z^i)$ and $(t, \overrightarrow{y}, z^i) \to f_i^{-}(t, \overrightarrow{y}, z^i)$ are Lipschitz continuous in (y^i, z^i) uniformly in t, and are continuous in y^j for $j \in \Lambda^{-i}$.

(iii) $f_i^{+}(t, \overrightarrow{y}, z^i)$ and $f_i^{-}(t, \overrightarrow{y}, z^i)$ are increasing in y^j for $j \in \Lambda^{-i}$. This assumption means that the m-players are partners.

[**H2**]: For each $i, j \in \Lambda$, $h_{i,j}$ satisfies:

(i) $h_{i,j}(t, y)$ is continuous in (t, y);
(ii) $h_{i,j}(t, y)$ is increasing in y;
(iii) $h_{i,j}(t, y) \le y$.
(iv) There is no sequence $i_2 \in \Lambda^{-i_1}, \ldots, i_k \in \Lambda^{-i_{k-1}}, i_1 \in \Lambda^{-i_k}$, and $(y^1, \ldots, y^k)$ such that $y^1 = h_{i_1, i_2}(t, y^2)$, $y^2 = h_{i_2, i_3}(t, y^3), \ldots, y_{k-1} = h_{i_{k-1}, i_k}(t, y^k)$, $y^k = h_{i_k, i_1}(t, y^1)$. This means that there is no free loop of instantaneous switchings.

[H3]: For $t \in [0, T]$ and $i \in \Lambda$, $B^i(t)$, $C^i(t)$ and $S^i(t)$ belong to $\mathcal{S}^2$.

[H4]: For any $i \in \Lambda$ the random variables ξ_i^+ and ξ_i^- are $\mathcal{F}_T$–measurable and belong to $\mathbb{L}^2$. Moreover we assume that

$$\xi_i^+ \geq \max_{j \in \Lambda^{-i}} h_{i,j}(T, \xi_j^+) \vee (\xi_i^- + C^i(T)), \quad \text{and} \quad \xi_i^- \leq S^i(T) \wedge (\xi_i^+ + B^i(T)).$$

[H5]: For every $i \in \Lambda$, the processes $(B^i(t))_{0 \leq t \leq T}$ and $(S^i(t))_{0 \leq t \leq T}$ are semi-martingales of the form $B^i(t) = B^i(0) + \int_0^t U^i(s)ds + \int_0^t V^i(s)dW_s$ and

$S^i(t) = S^i(0) + \int_0^t \bar{U}^i(s)ds + \int_0^t \bar{V}^i(s)dW_s$ where $(U^i(t), \bar{U}^i(t))$ and $(V^i(t), \bar{V}^i(t))$ are respectively $(\mathbb{R})^2$ and $(\mathbb{R}^d)^2$-valued $\mathcal{F}_t$–progressively measurable processes which are $dt \otimes d\mathbb{P}$–square integrable.

3   Main Result

Next we state and prove the main result of this paper.

Theorem 1. *Assume that* **[H1]**–**[H5]** *hold. Then for all $i \in \Lambda$, the system of RBSDEs (S) admits a continuous minimal solution $(Y^{\pm,i}, Z^{\pm,i}, K^{\pm,i})$.*

Proof. The whole proof is performed in six steps.

Step 1: Construction of Picard's sequence of solutions
Consider the following sequence of RBSDEs defined recursively, for $i \in \Lambda$ and $t \in [0, T]$, as follows: For $n = 0$ we start with the following BSDE:

$$Y_t^{+,i,0} = \xi_i^+ + \int_t^T \underline{f}_i^+(s, Y_s^{+,i,0}, Z_s^{+,i,0})ds - \int_t^T Z_s^{+,i,0}dW_s, \tag{7}$$

and RBSDE:

$$\begin{cases} Y_t^{-,i,0} = \xi_i^- + \int_t^T \underline{f}_i^-(s, Y_s^{-,i,0}, Z_s^{-,i,0})ds - \int_t^T Z_s^{-,i,0}dW_s - K_T^{-,i,0} + K_t^{-,i,0}, \\ Y_t^{-,i,0} \leq S^i(t) \wedge (Y_t^{+,i,0} + B^i(t)), \\ 0 = \int_0^T \left[S^i(t) \wedge (Y_t^{+,i,0} + B^i(t)) - Y_t^{-,i,0} \right] dK_t^{-,i,0}, \end{cases} \tag{8}$$

where $\underline{f}_i^+(s, y, z^i) = \inf_{\overrightarrow{y}:y_i=y} f_i^+(s, \overrightarrow{y}, z^i)$ and $\underline{f}_i^-(s, y, z^i) = \inf_{\overrightarrow{y}:y_i=y} f_i^-(s, \overrightarrow{y}, z^i)$.
Now, for $n = 0$ consider the following system of RBSDEs:

$$\begin{cases} Y_t^{-,i,1} = \xi_i^- - \int_t^T Z_s^{-,i,1} dW_s - K_T^{-,i,1} + K_t^{-,i,1} \\ \qquad + \int_t^T f_i^-(s, Y_s^{-,1,0}, \ldots, Y_s^{-,i-1,0}, Y_s^{-,i,1}, Y_s^{-,i+1,0}, \ldots, Y_s^{-,m,0}, Z_s^{-,i,1}) ds, \\ Y_t^{-,i,1} \leq S^i(t) \wedge \left(Y_t^{+,i,0} + B^i(t) \right), \\ 0 = \int_0^T \left[S^i(t) \wedge \left(Y_t^{+,i,0} + B^i(t) \right) - Y_t^{-,i,1} \right] dK_t^{-,i,1}, \\ Y_t^{+,i,1} = \xi_i^+ - \int_t^T Z_s^{+,i,1} dW_s + K_T^{+,i,1} - K_t^{+,i,1} \\ \qquad + \int_t^T f_i^+(s, Y_s^{+,1,0}, \ldots, Y_s^{+,i-1,0}, Y_s^{+,i,1}, Y_s^{+,i+1,0}, \ldots, Y_s^{+,m,0}, Z_s^{+,i,1}) ds, \\ Y_t^{+,i,1} \geq \max_{j \in \Lambda^{-i}} h_{i,j}(t, Y_t^{+,j,0}) \vee \left(Y_t^{-,i,1} + C^i(t) \right), \\ 0 = \int_0^T \left[Y_t^{+,i,1} - \max_{j \in \Lambda^{-i}} h_{i,j}(t, Y_t^{+,j,0}) \vee \left(Y_t^{-,i,1} + C^i(t) \right) \right] dK_t^{+,i,1}. \end{cases} \tag{9}$$

Note that by [**H1**](i) and (ii) we have that $\underline{f}_i^+$ and $\underline{f}_i^-$ are uniformly Lipschitz continuous in (y, z^i) and satisfy the following integrability condition

$$\mathbb{E} \left\{ \int_0^T \left(|\underline{f}_i^+(t, 0, 0)|^2 + |\underline{f}_i^-(t, 0, 0)|^2 \right) dt \right\} < +\infty.$$

Thus, from [14] it follows that for each $i \in \Lambda$ BSDE (7) admits a unique solution $(Y^{+,i,0}, Z^{+,i,0}) \in \mathcal{S}^2 \times \mathcal{M}^{d,2}$. Thus, there exists a constant $C > 0$ such that

$$\mathbb{E} \left[\sup_{t \in [0,T]} \left| \left(S^i(t) \wedge \left(Y_t^{+,i,0} + B^i(t) \right) \right)^+ \right|^2 \right] \leq C < +\infty, \tag{10}$$

and thus in view of [10, Proposition 2.3] we deduce that RBSDE (8) has a unique solution $(Y^{-,i,0}, Z^{-,i,0}, K^{-,i,0}) \in \mathcal{S}^2 \times \mathcal{M}^{d,2} \times \mathcal{K}^2$.

As a by product, under the assumptions [**H1**]–[**H4**], in view of [10, Proposition 2.3] the solution $(Y^{-,i,1}, Z^{-,i,1}, K^{-,i,1}) \in \mathcal{S}^2 \times \mathcal{M}^{d,2} \times \mathcal{K}^2$ exists and is unique. This in turn, in view of the following estimate, which holds due to assumption [**H2**](iii), [**H3**] and the fact that $Y^{-,i,1} \in \mathcal{S}^2$,

$$\mathbb{E} \left[\sup_{t \in [0,T]} \left| \left(\max_{j \in \Lambda^{-i}} h_{i,j}(t, Y_t^{+,j,0}) \vee \left(Y_t^{-,i,1} + C^i(t) \right) \right)^+ \right|^2 \right] \leq C < +\infty,$$

combined with [10, Proposition 2.3], leads to the existence of the unique solution $(Y^{+,i,1}, Z^{+,i,1}, K^{+,i,1}) \in \mathcal{S}^2 \times \mathcal{M}^{d,2} \times \mathcal{K}^2$.

Next, for $n \geq 1$, consider the following system

$$
\begin{cases}
Y_t^{-,i,n+1} = \xi_i^- + \int_t^T f_i^-(s, Y_s^{-,1,n}, \ldots, Y_s^{-,i-1,n}, Y_s^{-,i,n+1}, Y_s^{-,i+1,n}, \ldots \\
\qquad \ldots, Y_s^{-,m,n}, Z_s^{-,i,n+1}) ds - \int_t^T Z_s^{-,i,n+1} dW_s - K_T^{-,i,n+1} + K_t^{-,i,n+1}, \\[2mm]
Y_t^{-,i,n+1} \le S^i(t) \wedge (Y_t^{+,i,n} + B^i(t)), \\[2mm]
0 = \int_0^T \left[S^i(t) \wedge (Y_t^{+,i,n} + B^i(t)) - Y_t^{-,i,n+1} \right] dK_t^{-,i,n+1}, \\[2mm]
Y_t^{+,i,n+1} = \xi_i^+ + \int_t^T f_i^+(s, Y_s^{+,1,n}, \ldots, Y_s^{+,i-1,n}, Y_s^{+,i,n+1}, Y_s^{+,i+1,n}, \ldots \\
\qquad \ldots, Y_s^{+,m,n}, Z_s^{+,i,n+1}) ds - \int_t^T Z_s^{+,i,n+1} dW_s + K_T^{+,i,n+1} - K_t^{+,i,n+1}, \\[2mm]
Y_t^{+,i,n+1} \ge \max_{j \in \Lambda^{-i}} h_{i,j}(t, Y_t^{+,j,n}) \vee (Y_t^{-,i,n+1} + C^i(t)), \\[2mm]
0 = \int_0^T \left[Y_t^{+,i,n+1} - \max_{j \in \Lambda^{-i}} h_{i,j}(t, Y_t^{+,j,n}) \vee \left(Y_t^{-,i,n+1} + C^i(t) \right) \right] dK_t^{+,i,n+1}.
\end{cases}
\tag{11}
$$

Based on the arguments used previously, we can show by using an induction argument that for any $n \ge 2$, the system of RBSDEs (11) has a unique solution

$$(Y^{+,i,n}, Z^{+,i,n}, K^{+,i,n}, Y^{-,i,n}, Z^{-,i,n}, K^{-,i,n}) \in (\mathcal{S}^2)^2 \times (\mathcal{M}^{d,2})^2 \times (\mathcal{K}^2)^2, \ \forall i \in \Lambda.$$

Step 2: Convergence of the sequences $(Y^{\pm,i,n})_{n \ge 0}$
Let us set,

$$\widehat{f}_i^-(s, y, z^i) = \sup_{\overrightarrow{y} : y_i = y} f_i^-(s, \overrightarrow{y}, z^i) \quad \text{and} \quad \dot{f}_i^+(s, y, z^i) = \sup_{\overrightarrow{y} : y_i = y} f_i^+(s, \overrightarrow{y}, z^i).$$

Note that by **[H1]**(i) and (ii) we have that $\widehat{f}_i^-$ and $\dot{f}_i^+$ are uniformly Lipschitz continuous in (y, z^i) and satisfy the following integrability condition

$$\mathbb{E} \left\{ \int_0^T \left(|\widehat{f}_i^-(t, 0, 0)|^2 + |\dot{f}_i^+(t, 0, 0)|^2 \right) dt \right\} < +\infty. \tag{12}$$

Consider the following BSDE

$$\widehat{Y}_t = \sum_{i=1}^m |\xi_i^-| + \int_t^T \sum_{i=1}^m |\widehat{f}_i^-(s, \widehat{Y}_s, \widehat{Z}_s^i)| ds - \int_t^T \widehat{Z}_s dW_s.$$

It follows from [14] that this BSDE admits a unique solution $(\widehat{Y}_t, \widehat{Z}_t) \in \mathcal{S}^2 \times \mathcal{M}^{d,2}$. Next, let $(\dot{Y}^i, \dot{Z}^i, \dot{K}^i)$ be solutions of the following system of reflected BSDEs, for any $i \in \Lambda$ and $t \in [0, T]$, as follows

$$\begin{cases} \dot{Y}_t^i = \sum_{i=1}^m |\xi_i^+| + \sum_{i=1}^m |\xi_i^-| + |C^i(T)| + \int_t^T \sum_{i=1}^m |\dot{f}_i^+(s, \hat{Y}_s, \hat{Z}_s^i)| ds \\ \qquad - \int_t^T \dot{Z}_s dW_s + \dot{K}_T^i - \dot{K}_t^i, \\ \dot{Y}_t^i \geq \max_{j \in \Lambda^{-i}} h_{i,j}(t, \dot{Y}_t^j) \vee (\hat{Y}_t + C^i(t)), \\ \int_0^T \left[\dot{Y}_s^i - \max_{j \in \Lambda^{-i}} h_{i,j}(s, \dot{Y}_s^j) \vee (\hat{Y}_s + C^i(s)) \right] d\dot{K}_s^i = 0. \end{cases} \quad (13)$$

By using previous arguments, and thanks to the fact that $\hat{Y}_t \in \mathcal{S}^2$ and assumption [H3], applying [1, Theorem 3.1] yields that this RBSDE admits a solution $(\dot{Y}^i, \dot{Z}^i, \dot{K}^i) \in \mathcal{S}^2 \times \mathcal{M}^{d,2} \times \mathcal{K}^2$. Moreover, the following holds Next, by using an induction argument, plus a repeated use of the comparison theorem, we can easily show that for any $i \in \Lambda$, $t \in [0, T]$, $\forall n$:

$$Y_t^{-,i,0} \leq Y_t^{-,i,n} \leq Y_t^{-,i,n+1} \leq \hat{Y}_t \quad \text{and} \quad Y_t^{+,i,0} \leq Y_t^{+,i,n} \leq Y_t^{+,i,n+1} \leq \dot{Y}_t^i, \text{ a.s.} \quad (14)$$

Consequently, we deduce the following

$$\sup_{n \geq 1} \mathbb{E}[\sup_{t \in [0,T]} |Y_t^{+,i,n}|^2] \leq \mathbb{E}[\sup_{t \in [0,T]} |Y_t^{+,i,0}|^2] + \mathbb{E}[\sup_{t \in [0,T]} |\dot{Y}_t^i|^2] < \infty, \ \forall i \in \Lambda, \quad (15)$$

$$\sup_{n \geq 1} \mathbb{E}[\sup_{t \in [0,T]} |Y_t^{-,i,n}|^2] \leq \mathbb{E}[\sup_{t \in [0,T]} |Y_t^{-,i,0}|^2] + \mathbb{E}[\sup_{t \in [0,T]} |\hat{Y}_t|^2] < \infty, \ \forall i \in \Lambda. \quad (16)$$

Next, from (14) combined with (15) and (16) we deduce that the sequences $\{Y^{+,i,n}\}_{n \geq 0}$ and $\{Y^{-,i,n}\}_{n \geq 0}$ admit limits. Therefore, let $Y^{+,i}$ and $Y^{-,i}$, $i \in \Lambda$ be two optional processes which are respectively the limits of $Y^{+,i,n}$ and $Y^{-,i,n}$. Applying Fatou's Lemma and the dominated convergence theorem, we obtain

$$\mathbb{E}[\sup_{t \in [0,T]} |Y_t^{\pm,i}|^2] < \infty, \quad \lim_{n \to \infty} \mathbb{E} \int_0^T |Y_t^{\pm,i,n} - Y_t^{\pm,i}|^2 dt = 0, \ i \in \Lambda. \quad (17)$$

Step 3: Uniform estimates for the sequences. $\{(Z^{\pm,i,n}, K^{\pm,i,n})\}_{n \geq 0}, i \in \Lambda$
By [H2](iii), we obtain in view of the facts $(\hat{Y}_t, \hat{Z}_t) \in \mathcal{S}^2 \times \mathcal{M}^{d,2}$, $(\dot{Y}^i, \dot{Z}^i, \dot{K}^i) \in \mathcal{S}^2 \times \mathcal{M}^{d,2} \times \mathcal{K}^2$ and (14) combined with [H3] the following estimate of the barriers of RBSDE (11): for all $n \geq 1$ and all $i \in \Lambda$

$$\mathbb{E}\left[\sup_{0 \leq t \leq T} \left| \left(\max_{j \in \Lambda^{-i}} h_{i,j}(t, Y_t^{+,j,n}) \vee (Y_t^{-,i,n+1} + C^i(t)) \right)^+ \right|^2 \right] < +\infty, \quad (18)$$

$$\mathbb{E}\left[\sup_{0 \leq t \leq T} \left| \left(S^i(t) \wedge (Y_t^{+,i,n} + B^i(t)) \right)^+ \right|^2 \right] < +\infty. \quad (19)$$

Finally, with the estimates (15), (16), (18) and (19) at hand, applying the results in [10] we obtain that

$$\sup_{n \geq 1} \mathbb{E}[\int_0^T |Z_t^{\pm,i,n}|^2 \, dt] < \infty, \quad \sup_{n \geq 1} \mathbb{E}|K_T^{\pm,i,n}|^2 < \infty, \; i \in \Lambda. \tag{20}$$

Step 4: Continuity of the limit processes. $Y^{-,i}$ and $Y^{+,i}$, $i \in \Lambda$

To this end, let us first establish the absolute continuity of the increasing process $K^{-,i,n}$ w.r.t t for every $n \geq 0$.

We will first show that the claim holds true for $n = 0$. Let

$$\Xi_t^i := S_t^i \wedge \left(Y_t^{+,i,0} + B^i(t) \right) = Y_t^{+,i,0} + B^i(t) - \left(Y_t^{+,i,0} + B^i(t) - S_t^i \right)^+.$$

Applying Itô–Tanaka formula to Ξ_t^i, and in view of assumption [**H5**] we obtain

$$\Xi_t^i = \Xi_0^i + \int_0^t M_s^i ds + \int_0^t N_s^i dW_s - \frac{1}{2} L_t^i, \tag{21}$$

where $\{L_t^i, 0 \leq t \leq T\}$ is the local time at 0 of the continuous semimartingale $\{Y_t^{+,i,0} + B^i(t) - S^i(t)\}$,

$$M_t^i := -\mathbf{1}_{\{Y_t^{+,i,0} + B^i(t) > S^i(t)\}} \left(-\underline{f}_i^+(s, Y_s^{+,i,0}, Z_s^{+,i,0}) + U_t^i - \bar{U}_t^i \right)$$
$$- \underline{f}_i^+(s, Y_s^{+,i,0}, Z_s^{+,i,0}) + U_t^i,$$

and

$$N_t^i := Z_t^{+,i,0} + V_t^i - \mathbf{1}_{\{Y_t^{+,i,0} + B^i(t) > S^i(t)\}} \left(Z_t^{+,i,0} + V_t^i - \bar{V}_t^i \right).$$

Note that $\underline{f}_i^+$ and $\underline{f}_i^-$ are uniformly Lipschitz continuous in (y, z^i). Thus, using the fact that $(Y^{+,i,0}, Z^{+,i,0})$ and $(Y^{-,i,0}, Z^{-,i,0})$ belong respectively to $\mathcal{S}^2 \times \mathcal{M}^{d,2}$, yields that there is a constant $C > 0$ such that

$$\mathbb{E}\left\{ \int_0^T \left(|\underline{f}_i^+(t, Y_t^{+,i,0}, Z_t^{+,i,0})|^2 + |\underline{f}_i^-(t, Y_t^{-,i,0}, Z_t^{-,i,0})|^2 \right) dt \right\} \leq C. \tag{22}$$

Moreover, in view of the above and assumption [**H5**], there exists a constant $C > 0$ such that

$$E\left[\int_0^T (|M_t^i|^2 + |N_t^i|^2) dt \right] \leq C < +\infty. \tag{23}$$

Then, applying [10, Proposition 4.2] yields for all $t \leq T$

$$0 \leq dK_t^{-,i,0} \leq \mathbf{1}_{\{Y_t^{-,i,0} = \Xi_t^i\}} \left[\underline{f}_i^-(s, Y_s^{-,i,0}, Z_s^{-,i,0}) + M_t^i \right]^+ dt, \tag{24}$$

which means that, $K^{-,i,0}$ is absolutely continuous w.r.t. t. Next, in the same spirit, we can show, thanks to [10, Proposition 4.2], for all $n > 0$ that the process $K^{-,i,n+1}$ is absolutely continuous w.r.t. t. Furthermore, we can obtain that: there exists a constant $C > 0$ such that for all $n \geq 0$ and $i \in \Lambda$,

$$E\left[\int_0^T \left(\frac{dK^{-,i,n}}{dt}\right)^2 dt\right] \leq C. \tag{25}$$

Notice that, by combining [H1](ii) together with (15), (16), (20) we obtain that there is a constant $C > 0$ such that

$$\sup_{n \geq 0} E\left\{\int_0^T \left|f_i^-(t, Y_t^{-,1,n}, \ldots, Y_t^{-,i-1,n}, Y_t^{-,i,n+1}, Y_t^{-,i+1,n}, \ldots, Y_t^{-,m,n}, \right.\right.$$

$$\left.\left. Z_t^{-,i,n+1})\right|^2 dt\right\} \leq C < +\infty. \tag{26}$$

Next, in view of estimates (20), (25) and (26), we deduce that there exists a subsequence along which all $((\frac{dK_t^{-,i,n+1}}{dt})_{0 \leq t \leq T})_{n \geq 0}$, $((Z_t^{-,i,n+1})_{0 \leq t \leq T})_{n \geq 0}$ and $((f_i^-(t, Y_t^{-,1,n}, \ldots, Y_t^{-,i-1,n}, Y_t^{-,i,n+1}, Y_t^{-,i+1,n}, \ldots, Y_t^{-,m,n}, Z_t^{-,i,n+1}))_{0 \leq t \leq T})_{n \geq 0}$ converge weakly in their respective spaces $\mathcal{M}^{1,2}$, $\mathcal{M}^{d,2}$ and $\mathcal{M}^{1,2}$ to the processes $(k_t^{-,i})_{0 \leq t \leq T}$, $(Z_t^{-,i})_{0 \leq t \leq T}$ and $(\varphi^{-,i}(t))_{0 \leq t \leq T}$.

Next, for any $n \geq 0$ and any stopping time τ we have

$$Y_\tau^{-,i,n+1} = Y_0^{-,i,n+1} + K_\tau^{-,i,n+1} + \int_0^\tau Z_s^{-,i,n+1} dB_s$$

$$- \int_0^\tau f_i^-(t, Y_s^{-,1,n}, \ldots, Y_s^{-,i-1,n}, Y_s^{-,i,n+1}, Y_s^{-,i+1,n}, \ldots, Y_s^{-,m,n}, Z_s^{-,i,n+1}) ds.$$

Taking the weak limits in each side and along this subsequence yields

$$Y_\tau^{-,i} = Y_0^{-,i} - \int_0^\tau \varphi^{-,i}(s) ds + \int_0^\tau k_s^{-,i} ds + \int_0^\tau Z_s^{-,i} dB_s, \quad \mathbb{P}\text{-a.s.}$$

Since the processes appearing in each side are optional, using the Optional Section Theorem (see e.g. [3], Chapter IV pp.220), it follows that

$$Y_t^{-,i} = Y_0^{-,i} - \int_0^t \varphi^{-,i}(s) ds + \int_0^t k_s^{-,i} ds + \int_0^t Z_s^{-,i} dB_s, \; \forall \, t \leq T, \; \mathbb{P}\text{-a.s.} \tag{27}$$

Therefore, the process $Y^{-,i}$ is continuous. Relying both on Dini's Theorem and on Lebesgue's dominated convergence one (17), we also get that

$$\lim_{n \to \infty} E\left[\sup_{0 \leq t \leq T} |Y_t^{-,i,n} - Y_t^{-,i}|^2\right] = 0. \tag{28}$$

We will now focus on the continuity of the sequence of processes $(Y_t^{+,i})_{0 \leq t \leq T}$, $\forall \, i \in \Lambda$. Actually, applying Peng's Monotone Limit Theorem (see [15]) yields that for every $i \in \Lambda$, the limit process $Y^{+,i}$ is càdlàg. Based on what has been already shown in previous steps, by mimicking the arguments of [15] we can easily show that there exist two processes $K^{+,i} \in \mathcal{K}_c^2$ and $Z^{+,i} \in \mathcal{M}^{d,2}$ such that $Y^{+,i}$ satisfies the first equation of RBSDE (S). Moreover, passing to the limit in the fifth inequality of RBSDE (11), implies that $Y_t^{+,i} \geq S_t^{+,i}$, $t \in [0,T]$, $i \in \Lambda$. Thus, for $i \in \Lambda$, $(Y^{+,i}, Z^{+,i}, K^{+,i})$ satisfies

$$
\begin{cases}
Y_t^{+,i} = \xi_i^+ + \int_t^T f_i^+(s, \overrightarrow{Y}_s^{+,i}, Z_s^{+,i}) ds - \int_t^T Z_s^{+,i} dW_s + K_T^{+,i} - K_t^{+,i}, \\
Y_t^{+,i} \geq S_t^{+,i}.
\end{cases}
\tag{29}
$$

It remains to prove the minimal boundary condition. Next, consider the following RBSDE whose solution exists thanks to [16]:

$$
\begin{cases}
\widetilde{Y}_t^{+,i} = \xi_i^+ + \int_t^T f_i^+(s, Y_s^{+,1}, \ldots, Y_s^{+,i-1}, \widetilde{Y}_s^{+,i}, Y_s^{+,i+1}, \ldots, Y_s^{+,m}, \widetilde{Z}_s^{+,i}) ds \\
\qquad\qquad - \int_t^T \widetilde{Z}_s^{+,i} dW_s + \widetilde{K}_T^{+,i} - \widetilde{K}_t^{+,i}, \\
\widetilde{Y}_t^{+,i} \geq S_{t-}^{+,i}, \quad \text{and} \quad \int_0^T \left[\widetilde{Y}_{s-}^{+,i} - S_{s-}^{+,i} \right] d\widetilde{K}_s^{+,i} = 0.
\end{cases}
\tag{30}
$$

Note that RBSDEs (29) and (30) have the same lower barrier. In fact, since $\widetilde{Y}_t^{+,i}$ is the smallest f_i^+–supermartingale with lower barrier $S_t^{+,i}$, we have that for any $i \in \Lambda$, $\widetilde{Y}_t^{+,i} \leq Y_t^{+,i}$ (see [16, Theorem 2.1]). On the other hand since for any $i \in \Lambda$ and $n \geq 1$, $Y_t^{+,i,n} \leq Y_t^{+,i}$ and $Y_t^{-,i,n+1} \leq Y_t^{-,i}$, applying the comparison theorem in view of [H2](ii) yields that $Y_t^{+,i,n+1} \leq \widetilde{Y}_t^{+,i}$, and then passing to the limit implies that $Y_t^{+,i} \leq \widetilde{Y}_t^{+,i}$. Summing up we have that for any $i \in \Lambda$, $Y_t^{+,i} = \widetilde{Y}_t^{+,i}$. From the uniqueness of the Doob-Meyer decomposition, it follows that $Z_t^{+,i} = \widetilde{Z}_t^{+,i}$, $dt \times d\mathbb{P}$–a.s., and $K_t^{+,i} = \widetilde{K}_t^{+,i}$ for any $0 \leq t \leq T$, $\mathbb{P}$–a.s. Then, for $i \in \Lambda$, $(Y^{+,i}, Z^{+,i}, K^{+,i})$ satisfies RBSDE (5) but with the following minimal boundary condition

$$
\int_0^T \left[Y_{s-}^{+,i} - \max_{j \in \Lambda^{-i}} h_{i,j}(s, Y_{s-}^{+,j}) \vee (Y_s^{-,i} + C^i(s)) \right] dK_s^{+,i} = 0.
\tag{31}
$$

From the first equation of (29) and since the process $K_t^{+,i}$ is increasing, it follows that $\Delta Y_t^{+,i} = -\Delta K_t^{+,i} \leq 0$. Assume that $\Delta Y_{t^*}^{+,i_1} < 0$, for some $(i_1, t^*) \in \Lambda \times [0,T]$. Thus $\Delta K_{t^*}^{+,i_1} > 0$. From the minimality condition (31), we have

$$
Y_{t^*-}^{+,i_1} = \max_{j \in \Lambda^{-i_1}} h_{i_1,j}(t^*, Y_{t^*-}^{+,j}) \vee \left(Y_{t^*}^{-,i_1} + C^{i_1}(t^*) \right).
$$

Let $i_2 \in \Lambda^{-i_1}$ be the optimal index for which the maximum is attained. Thus,

$$
\begin{aligned}
h_{i_1,i_2}(t^*, Y_{t^*-}^{+,i_2}) \vee (Y_{t^*}^{-,i_1} + C^{i_1}(t^*)) &= Y_{t^*-}^{+,i_1} \\
&> Y_{t^*}^{+,i_1} \\
&= h_{i_1,i_2}(t^*, Y_{t^*}^{+,i_2}) \vee (Y_{t^*}^{-,i_1} + C^{i_1}(t^*)).
\end{aligned}
\tag{32}
$$

This obviously yields that $Y_{t^*-}^{+,i_1} = h_{i_1,i_2}(t^*, Y_{t^*-}^{+,i_2}) > h_{i_1,i_2}(t^*, Y_{t^*}^{+,i_2})$, and thus $\Delta Y_{t^*}^{+,i_2} < 0$. Repeating the above procedure we obtain for $i_k \in \Lambda^{-i_{k-1}}$

$$\Delta Y_{t^*}^{+,i_k} < 0, \quad \text{and} \quad Y_{t^*-}^{+,i_k} = h_{i_k,i_{k+1}}(t^*, Y_{t^*-}^{+,i_{k+1}}), \quad k = 2, \ldots, m.$$

Since each i_k can take only values in Λ which is a finite set, then there must be a loop in Λ. we may assume w.l.o.g. that $i_{k+1} = i_1$ for some $k \geq 2$ noting again that the i_k's are mutually different i.e. for each k, $i_k \in \Lambda^{-i_{k-1}}$. Therefore, we have $Y_{t-}^{+,i_1} = h_{i_1,i_2}(t^*, Y_{t^*}^{+,i_2}), \ldots, Y_{t-}^{+,i_{k-1}} = h_{i_{k-1},i_k}(t^*, Y_{t^*}^{+,i_k})$, and $Y_{t-}^{+,i_k} = h_{i_k,i_1}(t^*, Y_{t^*-}^{+,i_1})$. This contradicts assumption [**H2**](iv). Consequently, $\Delta Y_t^{+,i} = \Delta K_t^{+,i} = 0$, $t \in [0,T]$, $\forall i \in \Lambda$. Hence, the processes $Y^{+,i}$ and $K^{+,i}$, $i \in \Lambda$ are continuous.

Step 5: Identification of the limit
Next, we show that

$$\lim_{n \to \infty} \mathbb{E}\{ \sup_{0 \leq t \leq T} |Y_t^{+,i,n} - Y_t^{+,i}|^2 + |K_T^{+,i,n} - K_T^{+,i}|^2 + \int_0^T |Z_t^{+,i,n} - Z_t^{+,i}|^2 dt \} = 0.$$

Actually, since $Y^{+,i,n} \nearrow Y^{+,i}$ and $Y^{+,i}$ is continuous then relying both on Dini's Theorem and on Lebesgue's dominated convergence one (17), we get that

$$\lim_{n \to \infty} \mathbb{E}\left[\sup_{0 \leq t \leq T} |Y_t^{+,i,n} - Y_t^{+,i}|^2 \right] = 0. \tag{33}$$

Further, we can easily show by applying Itô's formula to $|Y_t^{\pm,i,n} - Y_t^{\pm,i,p}|^2$ ($n, p \geq 0$) and using standard arguments (see e.g. [10]) that

$$\lim_{n \to \infty} \mathbb{E}[\int_0^T |Z_t^{\pm,i,n} - Z_t^{\pm,i}|^2 dt + |K_T^{+,i,n} - K_T^{+,i}|^2 + |K_T^{-,i,n} - \int_0^T k_s^{-,i} ds|^2] = 0.$$

From this, and [**H1**](ii) combined with (27), (28) it holds that

$$\varphi_i^-(t) = f_i^-(t, \overrightarrow{Y}_t, Z_t^{-,i}), \quad 0 \leq t \leq T.$$

Next, passing to the limit in the second inequality of RBSDE (11), yields that $Y_t^{-,i} \leq S_t^{-,i}$, $t \in [0,T]$, $i \in \Lambda$. Furthermore, thanks to the weak convergence of $((\frac{dK_t^{-,i,n}}{dt})_{0 \leq t \leq T})_{n \geq 1}$ to the process $k^{-,i}$ and the strong convergences (28) and (33), we deduce that

$$0 = \int_0^T \left[S^i(t) \wedge \left(Y_t^{+,i,n} + B^i(t) \right) - Y_t^{-,i,n+1} \right] dK_t^{-,i,n+1}$$

$$\longrightarrow \int_0^T \left[S_t^{-,i} - Y_t^{-,i} \right] k_t^{-,i} dt = 0, \quad \text{as } n \to +\infty.$$

In fact, this implies that $(Y^{-,i}, Z^{-,i}, K^{-,i} := \int_0^\cdot k_s^{-,i} ds)$ is a solution to the second part of RBSDE (S). Summing up $(Y^{\pm,i}, Z^{\pm,i}, K^{\pm,i})$ is a solution of RBSDE (S). Finally, it remains to show that this solution is the minimal one.

Step 6: Minimality of the solution of RBSDE (S)
Let $(\bar{Y}^{+,i}, \bar{Z}^{+,i}, \bar{K}^{+,i}, \bar{Y}^{-,i}, \bar{Z}^{-,i}, \bar{K}^{-,i})$ be another solution of RBSDE (S). Since $Y^{+,i,n} \leq Y^{+,i}$ and $Y^{-,i,n} \leq Y^{-,i}$, for all $n \geq 0$, and thanks to the monotonicity of $h_{i,j}$ applying the comparison theorem yields that for each $i \in \Lambda$: $Y^{+,i,n} \leq \bar{Y}^{+,i}$ and $Y^{-,i,n} \leq \bar{Y}^{-,i}$, for all $n \geq 0$. Passing to the limit when $n \to \infty$ implies that for each $i \in \Lambda$: $Y^{+,i} \leq \bar{Y}^{+,i}$ and $Y^{-,i} \leq \bar{Y}^{-,i}$, which is the desired result. This ends the proof of Theorem 1.

4   Conclusion and Perspectives

In this paper we have proved the existence of a continuous minimal solution to RBSDE (S) which is arising from BSOSPs. Let us comment on a possible generalization of the results obtained in this paper. Actually, the full balance sheet case is still an open problem and constitutes a challenge. By the full balance sheet case we mean that we consider the two sides of the balance sheet. Indeed, in this case the expected cost in mode i, $Y^{-,i}$ should rise above the following barrier

$$\min_{j \in \Lambda^{-i}} l_{i,j}(t, Y_t^{-,j}) \wedge \left(Y_t^{+,i} + B^i(t) \right), \tag{34}$$

instead of $S_t^{-,i}$ where $l_{i,j}$ is a real nonlinear random function satisfying **[H2]**, except for **[H2]**–(iii) which is replaced by $l_{i,j}(t, y) \geq y$. We want to stress out that, the new assumption **[H2]** is satisfied when $l_{i,j}$ takes the particular form $l_{i,j}(\cdot, y) = y + g_{i,j}$ where $g_{i,j}$ is the switching cost from mode i to mode j.

A full BSOSP amounts to establishing existence of a continuous solution to the system of RBSDEs (S), but with the upper barrier (34) for $Y^{-,i}, \forall i \in \Lambda$. Note that, as in the proof of Theorem 1 (**Step 4**) the absolute continuity of the process $K_t^{-,i}$ w.r.t. t will play a primordial role to derive convergence of the corresponding approximating sequence. To do so we need to use the Itô–Tanaka formula (see **Step 4**), which makes it difficult to solve the system of RBSDEs (S) for the full balance sheet case. Note that even in the case when the functions $h_{i,j}$ and $l_{i,j}$ take the particular forms respectively $h_{i,j}(\cdot, y) = y - g_{i,j}$ and $l_{i,j}(\cdot, y) = y + g_{i,j}$ let alone the general case, the question of existence of solutions to the corresponding system of RBSDEs (S) for the full balance sheet case, is still open. This issue was discussed in [5] and in [6] in the mean field case.

Acknowledgment. The second named author extends his appreciation to the Deanship of Scientific Research at King Saud University for funding this work through research group no (RG-1441-339).

References

1. Aazizi, S., El Mellali, T., Fakhouri, I., Ouknine, Y.: Optimal switching problem and related system of BSDEs with left-Lipschitz coefficients and mixed reflections. Statist. Probab. Lett. **137**, 70–78 (2018)

2. Aazizi, S., Fakhouri, I.: Optimal switching problem and system of reflected multi-dimensional FBSDEs with random terminal time. Bull. Sci. Math. **137**(4), 523–540 (2013)
3. Dellacherie, C., Meyer, P.-A.: Probabilités et potentiels. Chapter I-IV. Hermann, Paris (1975)
4. Djehiche, B., Hamdi, A.: A full balance sheet two-mode optimal switching problem. Stochastics **87**(4), 604–622 (2015)
5. Eddahbi, M., Fakhouri, I., Ouknine, Y.: A balance sheet optimal multi-modes switching problem. Afr. Mat. **31**(2), 219–236 (2020)
6. Eddahbi, M., Fakhouri, I., Ouknine, Y.: Mean-field optimal multi-modes switching problem: a balance sheet. Stoch. Dyn. **19**(4), 1950026 (2019)
7. El Asri, B., Fakhouri, I.: Viscosity solutions for a system of PDEs and optimal switching. IMA J. Math. Control Inform. **34**(3), 937–960 (2017)
8. Fakhouri, I., Ouknine, Y., Ren, Y.: Reflected backward stochastic differential equations with jumps in time-dependent random convex domains. Stochastics **90**(2), 256–296 (2018)
9. Gegout-Petit, A., Pardoux, É.: Equations différentielles stochastiques rétrogrades réfléchies dans un convexe. Stochast. Stochast. Rep. **57**, 111–128 (1996)
10. El Karoui, N., Kapoudjian, C., Pardoux, E., Peng, S., Quenez, M.-C.: Reflected solutions of backward SDEs and related obstacle problems for PDEs. Ann. Probab. **25**(2), 702–737 (1997)
11. Hamadène, S., Zhang, J.: Switching problem and related system of reflected backward stochastic differential equations. Stochastic Process. Appl. **120**, 403–426 (2010)
12. Hu, Y., Tang, S.: Multi-dimensional BSDE with oblique reflection and optimal switching. Probab. Theory Related Fields **147**(1–2), 89–121 (2010)
13. Ouknine, Y.: Reflected backward stochastic differential equations with jumps. Stochast. Stochast. Rep. **65**, 111–125 (1998)
14. Pardoux, E., Peng, S.: Adapted solution of a backward stochastic differential equation. Syst. Control Lett. **14**, 55–61 (1990)
15. Peng, S.: Monotonic limit theorem of BSDE and nonlinear decomposition theorem of Doob- Meyer's type. Prob. Theory Relat. Fields **113**, 473–499 (1999)
16. Peng, S., Xu, M.: The smallest g-supermartingale and reflected BSDE with single and double L^2-obstacles. Ann. Inst. H. Poincaré Probab. Stat. **41**, 605–630 (2005)
17. Xu, Y.: An existence theorem for multidimensional BSDEs with mixed reflections. C. R. Acad. Sci. Paris Ser. I **354**, 1101–1108 (2016)

Teaching Computational Science

Modeling and Automatic Code Generation Tool for Teaching Concurrent and Parallel Programming by Finite State Processes

Edwin Monteiro[✉][iD], Kelvinn Pereira[iD], and Raimundo Barreto[iD]

Institute of Computing, Federal University of Amazonas, Manaus, Brazil
{edwin,kdsnp,rbarreto}@icomp.ufam.edu.br

Abstract. Understanding concurrent and parallel programming can be a very hard task on first contact by students. This paper describes the development and experimental results of the FSP2JAVA tool. The proposed method starts from concurrent systems modeling through Finite State Processes (FSP). After that, the method includes an automatic code generation from the model. This goal is achieved by a domain-specific language compiler which translates from the FSP model to Java code. The FSP2JAVA tool is available for free download in the github site. We argue that this tool helps in teaching concurrent systems, since it abstracts all complex languages concern and encourages the student to be focused at the fundamental concepts of modeling and analysis.

Keywords: FSP · Concurrent programming · Code generation · Teaching

1 Introduction

Concurrent programming is a paradigm used in building programs that make use of the simultaneous execution of multiple tasks that can be implemented as separate programs or as a single program that triggers multiple threads in parallel. The main advantage of using concurrent programming is the increased performance since it is possible to increase the number of tasks performed over a given period of time. The major challenge of concurrent programming is resource sharing, communication, and interaction between programs that run concurrently. The reason for this challenge is that parts of a program can now execute in an unpredictable order. Therefore, errors can occur depending on the order of execution of each task. However, usually such errors are difficult to find.

There are several examples of concurrent issues reported in the literature. One of them is Therac–25, which caused massive radiation overdoses. Another example was the case of Knight Capital Group, which lost $460 million in 45 min as presented in Kirilenko et al. [6]. In both cases, the systems programming had few revisions and there were no checks to certify that the software had been

© Springer Nature Switzerland AG 2020
V. V. Krzhizhanovskaya et al. (Eds.): ICCS 2020, LNCS 12143, pp. 593–607, 2020.
https://doi.org/10.1007/978-3-030-50436-6_44

developed correctly. Taking care of developing concurrent systems should be present from the initial training of professionals. However, teaching this paradigm is a major challenge in undergraduate classes since students have a hard time understanding the theory behind a problem. They usually do not take the time to think of effective solutions to problems. Instead, usually they go directly to programming attempting to get a possibly correct solution, which may lead to undetected programming errors. The problem of taking it a step further may be justified by the fact that the student deals with a complex level of programming never seen before and unrelated to the problem itself. According to [3], this may be mainly affected by the following factors: (i) a new mindset required by programming multithread; (ii) the behavior of a multithreaded program is dynamic, which makes the debugging task very difficult; and (iii) synchronization is more difficult than expected.

This paper presents FSP2JAVA, a tool for modeling concurrent systems through Finite State Processes (FSP) introduced in [8]. The goal is to prevent students to be in contact with the coding at an early time. Thus, the main aim is to facilitate the understanding of the fundamental concepts of concurrent programming in order to make the coding step easier. For methodological purposes, the high level code is shown only at the end of the modeling stage.

This paper is organized as follows. Section 2 reviews related work. In Sect. 3 we introduce the concepts of Finite State Process. Sections 4 and 5 detail the FSP2JAVA tool. Section 6 presents the planning and execution of experiments. Finally, Sect. 7 discusses the final considerations and future work.

2   Related Works

Learning the concepts of concurrent programming is essential for computer science students. The most common method of concurrent programming is one that adopts multithreaded programming. However, changing it from sequential paradigm causes significant problems for students, as concurrent programming interfaces are often more complex than necessary, causing students to spend time learning system details rather than the fundamentals as reported in [3].

Among the modeling tools found for teaching distributed systems, we can mention: (a) SPIN, (b) SCML, (c) LTSA and (d) FSP2JAVA. SPIN is an efficient verification system for distributed software system models. It was used to detect design errors in applications ranging from high-level descriptions of distributed algorithms to detailed code to control switchboards [4]. SCML is a tool that can be used to simulate a system of concurrent processes that communicate through shared variables. Mechanisms for defining non-determinism, atomic actions, and process synchronization are supported. In addition, SMCL includes a prototype for verifying basic security properties, such as mutual exclusion and deadlocks, using the model check technique [2]. The LTSA tool [7] is used at numerous universities around the world along with the book by Magee and Kramer [8]. The LTSA tool compiles the FSP specifications on a state machine and resembles the non-deterministic finite automaton. LTSA also features viewing and animating labeled transitions through graphical interfaces. Both FSP and LTSA are widely used.

The proposed tool is called FSP2JAVA. It was developed by [11] and improved by [12]. This tool receives a model in FSP and checks if the modeling is properly specified. If the model is correct, the second step is automatic code generation which consists of transforming the FSP language into Java language.

3   The Finite State Process (FSP)

The Finite State Process [9] is a formal language based on process algebra [1] to model the behavior of concurrent systems. The FSP notation is derived from the process algebra CSP and can represent a system by primitive processes (a single thread) or a composition of processes (multithreaded).

3.1   Primitive Process

The primitive process characterizes the execution of a sequential program. The term primitive is related to the basic structures of a programming language, such as the choice, guard condition, recursion, and alphabet extension. The main operations are described below:

Action Prefix (a->P): Describes a process that performs the a action, and then behaves exactly as specified in the P process. Practically, the action prefix defines a transition between states.

Choice (a->P | b->Q): Process behavior is defined by a or b. After an action is performed, subsequent behavior is described by P if the first event was a, or by Q if the first event was b.

Process STOP: Sometimes it is necessary to finish the execution of a process. When STOP is called, no further action is evaluated.

Alphabet Extension +{a,b,...,z}: A process can behave only through the actions contained in its alphabet, although the opposite is not valid. In certain situations the alphabet may be extended with actions not previously defined in the modeling. This inclusion is very common to prevent another process from performing a particular action.

Guarded Action (when B a->P): Actions of type are eligible only when Boolean condition B is satisfied, otherwise a is not a valid action.

Indexing P = (input[i:0..9]->output[i]->P). Allows the writing of processes and actions that assume multiple finite values in a simplified manner.

3.2   Composite Processes

Processes are concurrent and can perform actions in parallel. These actions may or may not be shared.

Parallel Composition (P || Q): Expresses the parallel execution of processes P and Q. Thus, actions are merged as they are executed. If two or more actions are shared, then the processes that contain them are synchronized. The || operator is the parallel composition operator.

Process Instances (a:SWITCH || b:SWITCH): Operation applied to differentiate distinct instances (in this case a and b) from the same process (in this case, SWITCH).

Set Labeling $\{a_1, ..., a_n\}$::P Add a prefix to all actions belonging to the alphabet of P. If $\{x,y,z\}$ belongs to the alphabet, then the operation results in $\{a_1.x, a_2.y, a_3.z\}$.

Re-labeling $/\{new_1/old_1, ..., new_n/old_n\}$: This operation is applied to ensure that composition of processes synchronize specific actions. Although common in composite processes, the operation can be applied to primitive processes.

4 FSP to Java Method

This paper extends the works presented in [8] and [11]. Therefore, the goal remains to evaluate an FSP model, interpret its behavior, and ultimately generate Java code from the transformation rules described later. In addition, this research increases the FSP instruction set accepted by the tool and facilitates interaction with users, as the entire process that starts at the modeling stage and ends at the code generation stage is focused exclusively on the tool.

4.1 Translation

Consider the model described below in FSP:

```
COIN=(toss->HEADS|toss->TAILS),
HEADS=(heads->COIN),
TAILS=(tails->COIN).
```

It models the tossing of a coin that assumes two states, heads or tails. Note that the choice occurs non-deterministically, since the same action implies distinct processes. In FSP models, according to the specification proposed in [8], the processes are written in uppercase and actions in lowercase. The simple distinction in writing allows us to establish the basic rule of language transformation: processes and actions are transformed into classes and methods, respectively. The following example illustrates applying this rule to generate the Java code corresponding to the COIN model:

```
public class COIN{
    public void toss_0(){
        System.out.println("toss");
    }
    public void heads_0(){
        System.out.println("heads");
    }
    public void tails_0(){
        System.out.println("tails");
    }}
```

Since they are syntactically distinct languages, some adaptations are required during the transformation. For instance, FSP algebra allows an action to be duplicated in the same process or in a new process. Thus, transforming an action into a method adds an identifier to the method name. This is done to allow the distinction between actions of the same name because in Java there are no duplicate methods that have exactly the same behavior. To represent the state transition, that is, when an action is reached in FSP, the respective methods contain the `Println` function that prints the name of the action associated with the method. This was the strategy adopted to simulate the sequence of actions FSP achieved during the execution of the respective transformed code. Another adaptation consists of local processes, for example, HEADS and TAILS are not transformed into classes. However, their methods are incorporated into the COIN class that corresponds to the main process. This identifier strategy is adopted in actions with the same name in different processes, this includes the actions of local processes, since everything will be put together to the same class.

4.2 Transformation Rules

Primitive Processes: These are transformed into classes that implement the Runnable interface. Thus, each class must contain an attribute of the class Thread. This is justified because Java does not contain multiple inheritance, so the class Thread is used without the need for inheritance. In addition to implementing the Runnable interface, the process class must contain the implementation of the **run** method that allows concurrent execution of the processes. The following example illustrates the transformation rule of primitive processes:

FSP: P = (a -> P).
Java:

```
public class P implements Runnable{
    Thread threadP;
    P(){threadP = new Thread(this);
        threadP.start();
    }
    public void a_0(){
        System.out.println("a");
    }
```

```
public void run(){
    try{
        while(true){
            Thread.sleep(1000);
            a_0();
        }
    }catch(Exception e){}
}}
```

STOP: It is analogous to the process described above with the difference that there is no infinite loop. If the STOP process is selected during model interpretation, then the Java equivalent code is represented by an interrupt method call after the last action a in this example. Then two statements are inserted before the end of the run method:

```
System.out.println("STOP");
threadP.interrupt();
```

Indexed Processes: In the case of indexed processes, the name of each method of the class, in addition to containing the action name FSP, also contains the unique action identifier and index of the respective indexed process. The following example illustrates the Indexed Processes transformation rule:

FSP:

```
const N = 1
P = P[0],
P[i:0..N] = ( a -> P[0] | b -> P[1]).
```

Java:

```
public class P implements Runnable{
    public void a_0_0(){
        System.out.println("a");
    }
    public void b_0_0(){
        System.out.println("b");
}}
```

P[i:0..N] is equivalent to writing N distinct P processes that communicate with each other from recursive calls determined by the process index that is triggered by an action. This index acts as the process selector. In the actions, the first digit represents the action identifier (this allows differentiating actions with the same name in a modeling) as explained earlier. The second digit characterizes which process the action belongs to. For example, the P process starts with the index 0, so if the first action is b, then the action will be called b_0_0. From b_0_0 process P receives the index 1. If a is selected then the action name is in the form a_1_1 because this action a is associated with P[1]. Note that for each indexed process, there is a constant amount of associated actions (in this case there are two).

Indexed Actions: The methods created from indexed actions, in addition to printing the action name, also print their index in square brackets.

FSP:

```
const N = 2
BUFF = (in[i:0..N]->out[i]->BUFF).
```

This type of indexing allows the creation of a large number of distinct actions in a few rows. Each action is named together with a value of variable i in the range 0..N. The code generation is similar to that presented for indexed processes, but what changes is the print of each method. For instance, consider the following code: System.out.println("in[0]"). In this case, there is an action that takes the form in[0] or out[0] for each indexed action such that 0 is any value in the i..N range. Unlike the previous indexing example, there is a single process and N actions so that the ith action in[i] is succeeded by the ith action out[i]. Thus, the run method will always have an in action followed by an out action of the same index.

Parallel Composition: A new class is created with the same name as the parallel composition model. This class instantiates and executes all primitive processes previously modeled in parallel composition. Consider a practical example of the roller coaster problem where the car's Mth passenger is controlled by a turnstile that allows exactly 3 passengers at a time. As long as there is room in the car, passenger boarding is cleared. When $M = 3$, the car starts rolling down and up, and a new M-capable car arrives to board new passengers. The modeling below represents the problem described in FSP:

```
const M = 3
TURNSTILE = (passenger -> TURNSTILE).
CONTROL = CONTROL[0],
CONTROL[i:0..M] = (when(i<M) passenger->CONTROL[i+1]
                 | when(i==M) depart->CONTROL[0]
                 ).
CAR = (depart->CAR).
||ROLLERCOASTER = (TURNSTILE || CONTROL || CAR).
```

In Java the ROLLERCOASTER process model is adapted to the following class:

```
public class ROLLERCOASTER{
    public static void main(String args[]){
        Monitor passenger_shared = new Monitor(4);
        Monitor depart_shared = new Monitor(2);
        TURNSTILE obj_turnstile = new
        TURNSTILE(passenger_shared);
        CONTROL obj_control = new
        CONTROL(passenger_shared, depart_shared);
        CAR obj_car = new CAR(depart_shared);
    }
}
```

The ROLLERCOASTER is the main class, because it is from it that the communication between the processes begins. Due to passenger dispute over access to the car, only three passengers are allowed to access the car at a time. To manage the seats available in the car, two monitors are created, one for passenger control and one for starting the car (releasing the car to new passengers). This release simulates the availability of a new car. The other processes, TURNSTILE, CONTROL, and CAR are instantiated in this main class. For each instance is passed a parameter of type Monitor corresponding to each process. For example, the turnstile deals with passengers. Therefore, only the passenger_shared monitor is passed as a parameter.

The TURNSTILE process is defined in terms of the following class:

```
public class TURNSTILE implements Runnable{
    Thread threadTURNSTILE;
    Monitor passenger_shared;
    TURNSTILE(Monitor passenger_shared){
        this.passenger_shared = passenger_shared;
        threadTURNSTILE = new Thread(this);
        threadTURNSTILE.start();
    }
    public synchronized void passenger_0() throws
        InterruptedException{
        passenger_shared.dec();
        if(passenger_shared.inc()){
            System.out.println("passenger");
        }
    }
    public void run(){
        try{
            while(true){
                Thread.sleep(1000);
                passenger_0();
                Thread.sleep(1000);
                passenger_0();
                Thread.sleep(1000);
                passenger_0();
                Thread.sleep(1000);
            }
        }catch(Exception e){}
}}
```

As defined in FSP, this process occurs in parallel with CONTROL and CAR, so it implements the Runnable interface to run concurrently. A thread type object is defined to concurrently execute with other model actions. In addition, a monitor type attribute is set to receive from the monitor that was passed by parameter through the class constructor (as shown in ROLLERCOASTER).

The passenger action described in the above model is defined in passenger_0 method. As shown, this method uses the keyword synchronized in order to ensure that this method is executed by only one thread at a time. The passenger_shared.inc() condition assumes two values: True or False. If True, the car contains the maximum number of passengers, so all passengers accessed the feature. If false, there is still room for more passengers in the car. For each passenger who has gained access to the car, the method prints the name "passenger" to simulate the behavior of the FSP model. Finally the run method is responsible for executing the request of M passengers. Like most FSP models, this model works continuously and never leaves the loop. The number of times a method is called on the loop run is related to the depart of a car and the arrival of a new one. For this example only one car departed, so there are 3 calls to the passenger method.

The transformation rule to the CAR and CONTROL processes is analogous to the adaptation made for TURNSTILE. It is worth noting that CONTROL will have two methods, as its constructor receives two monitors for passenger and depart actions as presented above.

5   The FSP2JAVA Tool

The FSP2JAVA (Fig. 1) is a tool designed to model real systems through the FSP language that abstracts the complexity and effort to master high level languages during the transition from sequential to concurrent paradigm. Note that the tool is available for use in the GitHub repository [12]. Based on feedback obtained from users using the previous version of the [11] tool, the current version of FSP2JAVA has a integrated user-friendly graphical interface that allows you to model, to analyse and to generate a concurrent Java code. This version also has the user manual, warning screen and error messages, interactive model interpreter and automatic code generation menu. All of these features will be described below.

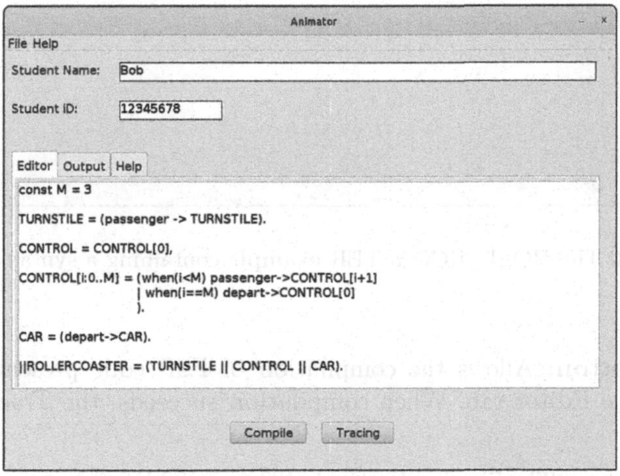

Fig. 1. FSP modeling screen.

5.1　FSP2JAVA Components

The FSP2JAVA tool contains two windows: Animator and Trace. The first of these contains the components below.

File: This menu allows to create, save or edit an FSP model. There are three options available in the File menu:

- **New:** Clears all content present in the modeling area.
- **Save:** Saves current edit area to a file with the extension ".fsp".
- **Open:** Opens a new text file.

Help: This menu provides a tool user manual in a new tab next to Output.

Editor Tab: This tab is a text editing area, in which the user writes codes in FSP that are later converted to code in Java.

Output Tab: This tab (see Fig. 2) serves to alert if the compilation of the code was successful or not. If syntax errors occur, the error location indicating the row, column and the error itself are shown.

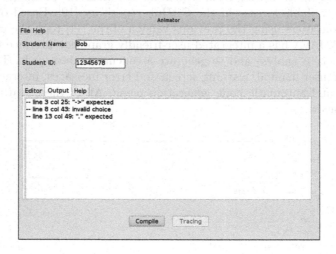

Fig. 2. The ROLLERCOASTER example containing a syntax error

Compile Button: Allows the compilation of FSP code present in the editing area of the Editor tab. When compilation succeeds, the Tracing button is enabled.

Tracing Button: The Tracing button is only available when compiling a model has been successful. In this case, the button opens the Trace window where the interactively selected actions are later added to the generated code. The tool always generates code with valid execution traces.

The components in the Trace window (see Fig. 3) are described below:

Trace History: A text area that shows all the traces produced so far.

Fig. 3. The Trace window with the successfully compiled ROLLERCOASTER code.

Checkboxes: The checkboxes correspond to the actions coming from the FSP code that was compiled. Whenever an action is eligible, the box allows its selection.

Generate Button: It allows the generation of Java code according to the actions selected so far.

View Code Button: It opens all Java code generated in the system's default text editor according to the actions selected in the checkbox. The View Code button is available only when the Generate button is clicked.

6   Experiment Planning and Application

The experiment was carried out with two undergraduate students from Software Engineering and Computer Science courses at the Federal University of Amazonas (UFAM). Although few students, they presented different profiles. One student already known about concurrent programming, while the other had no knowledge of the subject. This feature was essential, since the collection of opinions at the end of the experiment could contribute significantly to the improvement of the tool. The purpose of the experiment was to evaluate the contribution of FSP2JAVA as a mediator tool for concurrent programming teaching. In order to be able to perform the experiment, a class was taught on the topic of concurrent systems modeling by finite state processes. During the explanation of the concepts, some exercises were applied and solved in the FSP2JAVA tool with the intention of making the students more comfortable with the tool.

The FSP2JAVA experimentation took place in two steps as follows:

a) Modeling and analysis of the tool. For this purpose, the classic roller coaster problem was proposed as an exercise. In this problem, passengers arriving at the boarding platform must be registered at the roller coaster controller by a turnstile. Thus, the controller allows the car to depart only when there are enough passengers on the platform so that the car is occupied until its maximum passenger capacity M is reached. This problem was chosen because it adopts multiple threads with synchronization. Thus, all the features of the tool could be evaluated. To solve the problem, students were given one hour to use the FSP2JAVA tool without consulting the other candidate or instructor. After modeling, the students executed and evaluated the code generated by the tool.

b) Filling questionnaire - Two questionnaires were applied to gather students' perception of the tool. The first was related to the generated model, and the second to Java-generated code by FSP2JAVA. Both were applied at the end of the use of the tool. In this context, the model proposed by [10] was adapted to evaluate the motivation of the use of modeling tools based on the following aspects: intention to use, ease of use, correctness, reliability, satisfaction and usefulness. Tables 1 and 2 present the aspects evaluated by the participants, where they should report their degree of agreement by choosing the options presented in a 6-point Likert scale [5]: strongly disagree, widely disagree, partially disagree, neutral, partially agree and totally agree.

In addition to these questionnaires, some open questions were applied for the collection of perception, possible difficulties, problem detection and suggestions for improvement: (a) "What is your opinion on the tool's feedback to support your learning/performance when modeling concurrent systems?"; (b) "What made it difficult to use the supporting tool to create Finite State Process based models?"; (c) "What would you change about using the support tool to improve your learning/performance when modeling concurrent systems?";

Table 1. Items taken into consideration when evaluating modeling.

Item	Correctness	Reliability	Facility	Quality	Satisfaction	Utility
Description	When I use the support tool, it works correctly to model a distributed system	I trust the validation of the model by the support tool	The support tool is easy to use to model a distributed system	The support tool is useful for modeling distributed quality systems	I am satisfied with the validation of the model by the support tool	I would use the support tool when I wanted to model quality distributed systems

Table 2. Items taken into account to evaluate code generation.

Item	Correctness	Reliability	Facility	Quality	Satisfaction	Utility
Description	When I use the support tool, it works correctly to generate code for distributed system	I trust the code generated by the support tool	The support tool is easy to use to generate code for a distributed system	The support tool is useful for generating code for quality distributed system	I am satisfied with the code generated by the support tool	I would use the support tool when I wanted to generate code for a distributed system

(d) "What is your opinion on the feedback of the tool to support your learning/performance when generating code from concurrent systems?"; (e) "What made it difficult to use the backup tool to generate Java source code from concurrent systems?"; and (f) "What would you change about using the support tool to improve your learning/performance when generating code from concurrent systems?".

The results obtained from the experiment indicate that the tool has great potential to be adopted in the teaching of concurrent programming. The data obtained, see Fig. 4, on the use of the tool for teaching concurrent system modeling point to a broad agreement on all items in the Table 1. Students believe in the correct functioning of the tool and are satisfied with validating their models so that they would use it again to model concurrent systems. With respect to code generation, Fig. 5 allows us to understand that there is complete agreement on all items questioned in the Table 2.

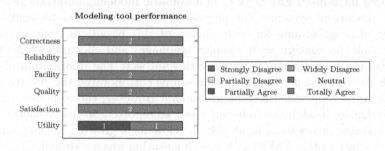

Fig. 4. Student evaluation considering the modeling aspects of FSP2JAVA.

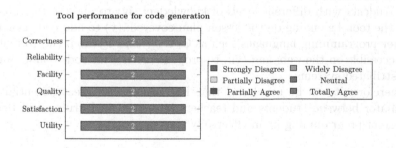

Fig. 5. Evaluation results considering the FSP2JAVA automatic code generation.

From the experiments results, it can be noted that the tool worked correctly for the designated purpose so that the students were safe and confident about its use. The analysis also allows us to conclude that the generated code behaves as expected, because the result generates satisfaction to the students regarding the use of the tool in critical situations where it is necessary to generate quality concurrent code.

In order to understand which aspects had a positive or negative impact on the students' evaluation regarding the use of FSP2JAVA, the open questions in the questionnaire were analyzed to obtain feedback and identify facts that justified the degree of agreement on the items evaluated.

The following are some positive and negative comments, where Pi corresponds to the i-th participant:

"Easy to remember, to identify things (buttons, desktop, features in general), it has a friendly interface" – Positive (P02).

"I would add some examples of FSP in the user manual" – Negative (P02).

"I would improve feedback on syntax errors generated or suggested by the system, thus helping the user to determine syntax errors faster." – Negative (P01).

7   Final Considerations

This paper introduced FSP2JAVA, an automatic modeling and code generation tool for concurrent systems. The purpose of this research was to facilitate the teaching of programming for early classes of this paradigm from a tool that abstracts all the concern with complex languages and encourages the student to master the fundamental concepts of this subject. The information obtained through the questionnaires and open questions corroborated the effectiveness of the tool in teaching concurrent programming. However, the responses collected indicate that feedback needs to be improved, as reported, so that students understand modeling errors more intuitively to allow for a growing learning curve.

Due to the fact that FSP2JAVA experimentation was performed with few students, the resuWlts obtained cannot be generalized. Therefore, in future works, it is intended: (i) to improve tool feedback; (ii) to add examples of FSP modeling and transformation rules to the manual; (iii) to apply new experiments with more students with different levels of knowledge; (iv) to observe the impact of using the tool depending on the lesson plan covered; (v) to add code generation in other programming languages such as C/C++ and Python; (vi) to make the tool accessible on the web; and (vii) to propose metrics for code quality analysis for distributed systems.

Therefore, it can be concluded that FSP2JAVA has great potential to be a mediator between students and teachers interested in learning and teaching concurrent programming in an effectively way.

Acknowledgments. This research, in accordance with Article 48 of Decree n$^{\underline{o}}$ 6.008/2006, was funded by Samsung Electronics of Amazônia Ltda, under the terms of Federal Law n$^{\underline{o}}$ 8.387/1991, through agreement n$^{\underline{o}}$ 003, signed with ICOMP/UFAM.

References

1. Baeten, J., van Beek, D.A., Rooda, J.: Process algebra. In: Handbook of Dynamic System Modeling, pp. 19–21 (2007)
2. Ben-Ari, M.: Teaching concurrency and nondeterminism with spin. In: ACM SIGCSE Bulletin, vol. 39, pp. 363–364. ACM (2007)
3. Carr, S., Mayo, J., Shene, C.K.: ThreadMentor: a pedagogical tool for multi-threaded programming. J. Educ. Resour. Comput. **3**(1), 1-es (2003)
4. Holzmann, G.J.: The SPIN Model Checker: Primer and Reference Manual, vol. 1003. Addison-Wesley Reading, Boston (2004)
5. Jamieson, S., et al.: Likert scales: how to (ab) use them. Med. Educ. **38**(12), 1217–1218 (2004)
6. Kirilenko, A.A., Lo, A.W.: Moore's law versus murphy's law: algorithmic trading and its discontents. J. Econ. Perspect. **27**(2), 51–72 (2013). https://doi.org/10.1257/jep.27.2.51
7. Lang, F., Salaün, G., Hérilier, R., Kramer, J., Magee, J.: Translating fsp into lotos and networks of automata. Formal Aspects Comput. **22**(6), 681–711 (2010)
8. Magee, J., Kramer, J.: Concurrency: State Models and Java Programs, 2nd edn. Wiley Publishing, Hoboken (2006)
9. Magee, J., Kramer, J., Giannakopoulou, D.: Analysing the behaviour of distributed software architectures: a case study. In: Proceedings 6th IEEE Computer Society Workshop on Future Trends of Distributed Computing Systems, pp. 240–245 (1997)
10. Martínez-Torres, M.R., Toral Marín, S., Garcia, F.B., Vazquez, S.G., Oliva, M.A., Torres, T.: A technological acceptance of e-learning tools used in practical and laboratory teaching, according to the european higher education area. Behav. Inf. Technol. **27**(6), 495–505 (2008)
11. Monteiro, E., Rivero, L., Barreto, R.: Uma ferramenta de suporte ao ensino de modelagem de sistemas distribuídos críticos: Uma experiência prática (in portuguese). In: Brazilian Symposium on Computers in Education, vol. 29 (2018)
12. Nunes, K., Monteiro, E., Barreto, R.: FSPTOJAVA: a modeling and automatic code generation tool (2019). https://github.com/kelvinnpereira/pibic-fsp

Automatic Feedback Provision in Teaching Computational Science

Hans Fangohr[1,2(✉)], Neil O'Brien[2,3], Ondrej Hovorka[2], Thomas Kluyver[1], Nick Hale[2], Anil Prabhakar[4], and Arti Kashyap[5]

[1] European XFEL GmbH, Schenefeld, Germany
{hans.fangohr,thomas.kluyver}@xfel.eu
[2] University of Southampton, Southampton, UK
{fangohr,nsob1c12,o.hovorka,n.w.hale}@soton.ac.uk
[3] Imaging Physics, University Hospital Southampton, Southampton, UK
[4] Department of Electrical Engineering, IIT Madras, Chennai, India
anilpr@ee.iitm.ac.in
[5] School of Basic Sciences, IIT Mandi, Mandi 175001, HP, India
arti@iitmandi.ac.in

Abstract. We describe a method of automatic feedback provision for students learning computational science and data science methods in Python. We have implemented, used and refined this system since 2009 for growing student numbers, and summarise the design and experience of using it. The core idea is to use a unit testing framework: the teacher creates a set of unit tests, and the student code is tested by running these tests. With our implementation, students typically submit work for assessment, and receive feedback by email within a few minutes after submission. The choice of tests and the reporting back to the student is chosen to optimise the educational value for the students. The system very significantly reduces the staff time required to establish whether a student's solution is correct, and shifts the emphasis of computing laboratory student contact time from assessing correctness to providing guidance. The self-paced nature of the automatic feedback provision supports a student-centred learning approach. Students can re-submit their work repeatedly and iteratively improve their solution, and enjoy using the system. We include an evaluation of the system from using it in a class of 425 students.

Keywords: Automatic assessment tools · Automatic feedback provision · Programming education · Python · Self-assessment technology · Pytest · Computational science · Data science

1 Introduction

One of the underpinning skills for computer science, software engineering and computational science is programming. A thorough treatment of the existing literature on teaching introductory programming was given by Pears et al. [1],

© Springer Nature Switzerland AG 2020
V. V. Krzhizhanovskaya et al. (Eds.): ICCS 2020, LNCS 12143, pp. 608–621, 2020.
https://doi.org/10.1007/978-3-030-50436-6_45

while a previous review focused mainly on novice programming and topics related to novice teaching and learning [2]. We provide a recent literature review in the technical report that accompanies this work (Sect. 1.3 in [3]).

Programming is a creative task: given the constraints of the programming language to be used, it is the choice of the programmer what data structure to employ, what control flow to implement, what programming paradigm to use, how to name variables and functions, how to document the code, and how to structure the code that solves the problem into smaller units (which potentially could be re-used). Experienced programmers value this freedom and gain satisfaction from developing a 'beautiful' piece of code or finding an 'elegant' solution. For beginners (and teachers) the variety of 'correct' solutions can be a challenge.

Given a particular problem (or student exercise), for example to compute the solution of an ordinary differential equation, there are a number of criteria that can be used to assess the computer program that solves the problem:

1. correctness: does the code produce the correct answer? (For numerical problems, this requires some care: for the example of the differential equation, we would expect for a well-behaved differential equation that the numerical solution converges towards the exact solution as the step-width is reduced towards zero.)
2. execution time performance: how fast is the solution computed?
3. memory usage: how much RAM is required to compute the solution?
4. robustness: how robust is the implementation with respect to missing/incorrect input values, etc?
5. elegance, readability, documentation: how long is the code? Is it easy for others to understand? Is it easy to extend? Is it well documented, or is the choice of algorithm, data structures and naming of objects sufficient to document what it does?

When teaching and providing feedback, in particular to beginners, one tends to focus on correctness of the solution. However, the other criteria 2 to 5 are also important. In particular for computational science where requirements can change rapidly and users (and readers) of code may not be fully trained computer scientists [4], the readability and documentation matter.

We demonstrate in this paper that the assessment of criteria 1 to 4 can be automated in day-to-day teaching of large groups of students. While the higher-level aspects such as elegance, readability and documentation of item 5 do require manual inspection of the code for useful feedback, we argue that the teaching of these high level aspects benefits significantly from automatic feedback as all the contact time with experienced staff can be dedicated to those points, and no time is required to manually check the criteria 1 to 4.

In this work, we describe the motivation, design, implementation and effectiveness of an automatic feedback system for Python-based exercises for computational science and data science, used in teaching undergraduate students in engineering [5] and physics, and postgraduate students from a wider range of disciplines.

We aim to address the shortcomings of the current literature as outlined in the review [6] by detailing our implementation and security model, as well as providing sample testing scripts, inputs and outputs, and usage data from the deployed system. Some of that material is made available as a technical report [3] and we make repeated reference to it in this manuscript.

In Sect. 2, we provide some historic context of how programming was taught at the University of Southampton prior the introduction of the automatic testing system described here. Section 3 introduces the new method of feedback provision, initially from a the perspective of a student, then providing more detail on design and implementation. Based on our use of the system over multiple years, we have composed results, statistics and a discussion of the system in Sect. 4, before we close with a summary in Sect. 5.

2   Traditional Delivery of Programming Education

Up to the year 2009, we taught programming and the basics of computational science in a mixture of weekly lectures that alternate with practical sessions in which the students write programs in a self-paced fashion. These programs were then reviewed and assessed individually by a demonstrator as part of the laboratory session. We estimate that 90% of the demonstrators' time went into establishing the correctness (and thus obtaining a fair assessment) of the work. Only the remaining time could be used to support students in solving the self-paced exercises and to provide feedback relating to items 2 to 5 in Sect. 1. We provide a more detailed description and discussion of the learning and teaching methods in [3, Sect. 2].

3   New Method of Automatic Feedback Provision

3.1   Overview

In 2009, we introduced an *automatic feedback provision system* that checks each student's code for correctness and provides feedback to the student within a couple of minutes of having completed the work. This takes a huge load off the demonstrators who consequently can spend most of their time helping students to do the exercises and providing additional feedback on completed and assessed solutions. Due to the introduction of the system the learning process can be supported considerably more effectively, and we could reduce the number of demonstrators from 1 per 10 students as we had pre-2009, to 1 demonstrator per 20 to 30 students, and still improve the learning experience and depth of material covered.

3.2   Student's Perspective

Once a student completes a programming exercise in the computing laboratory session, they send an email to a dedicated email account that has been created for the teaching course, and attach the file containing the code they have written. The subject line is used by the student to identify the exercise; for example "Lab 4" would identify the 4th practical session. The system receives the student's email, and tests the student's code. The student receives an email containing their assessment results and feedback. Typically, the student will receive feedback in their inbox within two to three minutes of sending their email.

Please define the following functions in the file `training1.py` and make sure they behave as expected. You also should document them suitably.

1. A function `distance(a, b)` that returns the distance between numbers a and b.
2. A function `geometric_mean(a, b)` that returns the geometric mean of two numbers, *i.e.* the edge length that a square would have so that its area equals that of a rectangle with sides a and b.
3. A function `pyramid_volume(A, h)` that computes and returns the volume of a pyramid with base area A and height h.

Fig. 1. Example exercise instructions. We focus here on question 3.

For our discussion, we use the example exercise shown in Fig. 1, which is typical of one that we might use in an introductory laboratory that introduces Python to beginners. We show a correct solution to question 3 of this example exercise in Listing 1.1. If a student submits this, along with correct responses to the other questions, by email to the system, they will receive feedback as shown in Listing 1.2.

```
def pyramid_volume(A, h):
    """Calculate and return the volume of a pyramid
    with base area A and height h."""
    return (1./3.) * A * h
```

Listing 1.1. A correct solution to question 3 of the example exercise

```
Overview
========

distance        : passed -> 100%; (weight=1)
geometric_mean  : passed -> 100%; (weight=1)
pyramid_volume  : passed -> 100%; (weight=1)

Total mark for this assignment: 3 / 3 = 100%.
(Points computed as 1 + 1 + 1 = 3)

------------------------------------------------

This message has been generated automatically. Should you feel that you
observe a malfunction of the system, or if you wish to speak to a human,
please contact the course team (course-help@uni.email.address).
```

Listing 1.2. Email response to correct submission, additional line wrapping due to column width

```
def pyramid_volume(A, h):
    """Calculate and return the volume of a pyramid
    with base area A and height h. """
    return (A * h) / 3
```

Listing 1.3. An incorrect solution to question 3 of the example exercise, using integer division

```
Overview
========

distance        : passed -> 100%; (weight=1)
geometric_mean  : passed -> 100%; (weight=1)
pyramid_volume  : failed ->   0%; (weight=1)

Total mark for this assignment: 2 / 3 = 67%.
(Points computed as 1 + 1 + 0 = 2)

Test failure report
===================

test_pyramid_volume
-------------------
def test_pyramid_volume():

    # if height h is zero, expect volume zero
    assert s.pyramid_volume(1.0, 0.0) == 0.

    # tolerance for floating point answers
    eps = 1e-14

    # if we have base area A=1, height h=1, we expect a volume of 1/3.:
    assert abs(s.pyramid_volume(1., 1.) - 1./3.) < eps

    # another example
    h = 2.
    A = 4.
    assert abs(s.pyramid_volume(A, h) - correct_pyramid_volume(A, h)) < eps

    # does this also work if arguments are integers?
>   assert abs(s.pyramid_volume(1, 1) - 1./3.) < eps
E   assert 0.3333333333333333 < 1e-14
E    + where 0.3333333333333333 = abs((0 - (1.0/3.0)))
E    +   where 0 = <function pyramid_volume at 0x7f0ce1af4e60>(1, 1)
E    +     where <function pyramid_volume at 0x7f0ce1af4e60> = s.
    pyramid_volume
```

Listing 1.4. Email response to incorrect solution

If the student submits an incorrect solution, for example with a mistake in question 3 as shown in Listing 1.3, they will instead receive the feedback shown in Listing 1.4. The submission in Listing 1.3 is incorrect because integer division is used rather than the required floating-point division. These exercises were based on Python 2, where the "/" operator represents integer division if both operands are of integer type, as is common in many programming languages. (We have since upgraded the curriculum and student exercises to Python 3.)

Within the testing feedback in Listing 1.4, the student code is visible in the name space s, i.e. the function s.pyramid_volume is the function defined in Listing 1.3. The function correct_pyramid_volume is visible to the testing system but students cannot see the implementation in the feedback they receive – this allows us to define tests that compute complicated values for comparison with those computed by the student's submission, without revealing the implementation of the reference computation to the students.

3.3   Design and Implementation of Student Code Testing

The incoming student code submissions are tested through carefully designed unit tests, as discussed below. The details of the technical design and implementation are available in [3, Sect. 3.3]. We discuss three aspects here.

Iterative Testing of Student Code. We have split each exercise on our courses into multiple questions, and arranged to test each question separately. Within a question, the testing process stops if any of the test criteria are not satisfied. This approach was picked to encourage an iterative process whereby students are guided to focus on one mistake at a time, correct it, and get further feedback, which improves the learning experience. This approach is similar to that taken by Tillmann *et al.* [7], where the iterative process of supplying code that works towards the behaviour of a model solution for a given exercise is so close to gaming that it "is viewed by users as a game, with a byproduct of learning". Our process familiarises the students with aspects of test-driven development [8] in a practical way.

Defining the Tests. Writing the tests is key to making this automatic testing system an educational success: we build on our experience before and after the introduction of the testing system, ongoing feedback from interacting with the students, and reviewing their submissions, to design the best possible unit testing for the learning experience. This includes testing for correctness but also structuring tests in a didactically meaningful order. Comments added in the testing code will be visible to the students when a test fails, and can be used to provide guidance to the learners as to what is tested for, and what the cause of any failure may be (if desired). A more detailed discussion including the test code for the example shown in Listing 1.2 and 1.4 is given in [3, Sect. 3.4.3].

Results and Feedback Provision to Students. Students receive a reply email from the testing system that provides a per-question mark, with a total mark for the submission, and then details on any errors that were encountered. In the calculation of the mark for the assessment, questions can be given different weights to reflect greater importance or challenges of particular questions. For the example shown in Listing 1.2 all questions have the same weight of 1.

We describe and illustrate a typical question, which might form part of an assignment, in Sect. 3.2. As shown in Listing 1.4 on page 5, when an error is encountered, the feedback that is sent to the student include the testing code up to the point of the failing assertion. The line that raises the exception is indicated with the > character (in this case the 6th-last line shown) as is usual for pytest output [9]. This is followed by a backtrace which illustrates that, in this case, the submitted `pyramid_volume` function returned 0 when it was expected to return an answer of $\frac{1}{3} \pm 1 \times 10^{-14}$.

All the tests above the failing line have passed, i.e. the functionality in the student code that is tested by these tests appears to be correct. The report also includes several comments, which are introduced in the testing code (shown in Listing 1.4), and assist students in working out what was being tested when the error was found. For example, the comment "does this also work if arguments are integers?" shows the learner that we are about to test their work with integer parameters; that should prompt them to check for integer division operations. If they do not succeed in doing this, they are able to show their feedback to a demonstrator or academic, who can use the feedback to locate the error in the student's code swiftly, then help the student find the problem, and discuss ways to solve the issue.

In addition, students receive a weekly summary of all their past submissions and marks. The course lecturer has access to all data in a variety of ways.

Clean Code. Our system optionally supports coding style analysis and in the majority of our tests for Python submissions, we perform an automated style check against the PEP8 coding standard [10]. We typically add $2^{-N_{\text{err}}}$ (where N_{err} is the number of stylistic errors detected) to the student's overall mark so that full compliance with the guideline is rewarded most generously.

4 Results

4.1 Testing System Deployment

The automatic testing system was first used at the University of Southampton's Highfield Campus in the academic year 2009/2010 for teaching about 85 Aerospace engineers, and has been used every year since for growing student numbers, reaching 425 students in 2014/2015. The Southampton deployment now additionally serves another cohort of students who study at the University of Southampton Malaysia Campus (USMC) and there is a further deployment at the Indian Institute of Technology (IIT) Mandi and Madras campuses, where the system has been integrated with the Moodle learning management system [3, Sect. 4.5].

The testing system has been used in a number of smaller courses at Southampton, typically of approximately 20 students, such as one-week intensive Python programming courses offered to PhD students. The feedback system has also been used in the Physics department at the University of Hamburg (Germany) to support an introduction into computational science and data science in 2019. It also serves Southampton's courses in advanced computational methods where around 100 students have submitted assignments in C (this requires an extension of the system which is beyond the scope of this contribution).

4.2 Case Study: Introduction to Computing

In this section, we present and discuss experience and pertinent statistics from the production usage of the system in teaching our first-year computing course at the University of Southampton. In 2014/15, there were about 425 students in their first semester of studying Acoustic Engineering, Aerospace Engineering, Mechanical Engineering, and Ship Science.

Course Structure. The course is delivered through weekly lectures and weekly self-paced student exercises with a completion deadline a day before the next lecture takes place (to allow the lecturer to sight submissions and provide generic feedback in the lecture the next day). Students are offered a 90 min slot (which is called "computing laboratory" in Southampton) in which they can carry out the exercises, and teaching staff are available to provide help. Students are allowed and able to start the exercise before that laboratory session, and use the submission and testing system anytime before, during and after that 90 min slot.

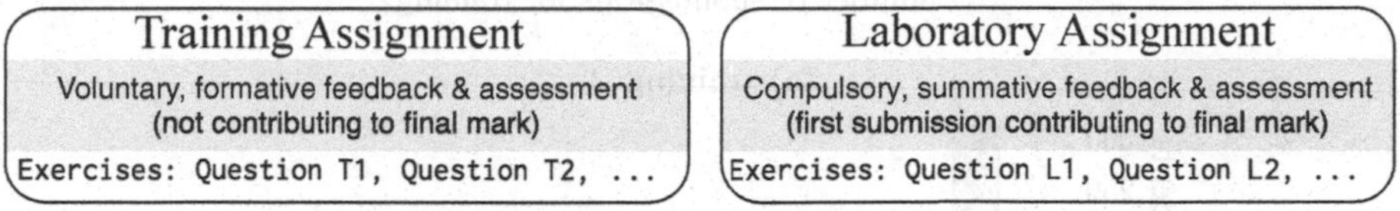

Training Assignment	Laboratory Assignment
Voluntary, formative feedback & assessment (not contributing to final mark)	Compulsory, summative feedback & assessment (first submission contributing to final mark)
Exercises: Question T1, Question T2, ...	Exercises: Question L1, Question L2, ...

Fig. 2. Overview of the structure of the weekly computer laboratory session: a voluntary set of training exercises is offered to the students as a "training" assignment, followed by a compulsory set of exercises in the same topic area as the "laboratory" assignment which contributes to each student's final mark for the course. Automatic feedback and assessment is provided for both assignments and repeat submissions are invited.

Each weekly exercise is split into two assignments: a set of "training" exercises and a set of assessed "laboratory" exercises. This is summarised in Fig. 2.

The training assignment is checked for correctness and marked using the automatic system, but whilst we record the results and feed back to the students, they do not influence the students' grades for the course. Training exercises are voluntary but the students are encouraged to complete them. Students can repeatedly re-submit their (modified) code for example until they have removed all errors from the code. Or they may wish to submit different implementations to get feedback on those.

The assessed laboratory assignment is the second part of each week's exercises. For these, the students attempt to develop a solution as perfect as possible before submitting this by email to the testing system. This "laboratory" submission is assessed, and marks and feedback are provided to the student. These marks are recorded as the student's mark for that week's exercises, and contribute to the final course mark. The student is allowed (and encouraged) to submit further solutions, which will be assessed and feedback provided, but it is the first submission that is recorded as the student's mark for that laboratory.

The main assessment of the course is done through a programming exam at the end of the semester in which students write code on a computer in a 90 min session, without Internet access but having an editor and Python interpreter to execute and debug the code they write. Each weekly assignment contributes of the order of one percent to the final mark, i.e. 10% overall for a 10 week course. Each laboratory session can be seen as a training opportunity for the exam as the format and expectations are similar.

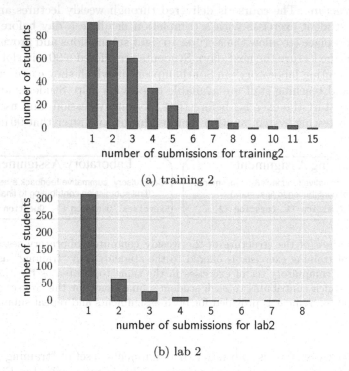

(a) training 2

(b) lab 2

Fig. 3. Histogram illustrating the distribution of submission counts per student for the (a) voluntary training and (b) assessed laboratory assignment (see text in Sect. 4.2)

Student Behaviour: Exploiting Learning Opportunities From Multiple Submissions. In Fig. 3a, we illustrate the distribution of submission counts for "training 2", which is the voluntary set of exercises from week 2 of the course. The bar labelled 1 with height 92 shows that 92 students have submitted the training assignment exactly once, the bar labelled 2 shows that 76 students submitted their training assignment exactly twice, and so on. The sum over all bars is 316 and shows the total number of students participating in this voluntary training assignment. 87 students submitted four or more times, and several students submitted 10 or more times. This illustrates that our concept of students being free to make step-wise improvements where needed and rapidly get further feedback has been successfully realised.

We can contrast this to Fig. 3b, which shows the same data for the compulsory laboratory assignment in week 2 ("lab2"). This submission attracts marks which contribute to the students' overall grades for the course. In this case the students are advised that while they are free to submit multiple times for further feedback, only the mark recorded for their first submission will count towards their score for the course. For lab 2, 423 students submitted work, of whom 314 submitted once only. However, 64 students submitted a second revised version and a significant minority of 45 students submitted three or more times to avail themselves of the benefits of further feedback after revising their submissions, even though the subsequent submissions do not affect their mark.

Significant numbers of students choose to submit their work for both voluntary and compulsory assignments repeatedly, demonstrating that the system offers the students an extended learning opportunity that the conventional cycle of submitting work once, having it marked once by a human, and moving to the next exercise, does not provide.

The proportion of students submitting multiple times for the assessed laboratory assignment (Fig. 3a) is smaller than for the training exercise (Fig. 3b) and likely to highlight the difference between the students' approaches to formative and summative assessment. It is also possible that students need more iterations to learn new concepts in the training assignment before applying the new knowledge in the laboratory assignment, contributing to the difference in resubmissions. The larger number of students submitting for the assessed assignment (423 $\approx$ 100%) over the number of students submitting for the training assignment (316 $\approx$ 74%) shows that the incentive of having a mark contribute to their overall grade is a powerful one.

Submission Behaviour Changes over Time. During the 10 practical sessions of this course, the number of assessed submissions decays slightly, and the participation in voluntary submissions decays more dramatically, before it increases slightly as the exam is approached. This is discussed in detail in [3, Sect. 4.2.3 and Figs. 4 and 5].

4.3   Feedback from Students

We invited feedback from the learners explicitly on the automatic feedback system asking for voluntary provision of (i) reasons why students liked the system and (ii) reasons why students disliked the system. The replies are not homogeneous enough to compile statistical summaries, but we provide a representative selection of comments in [3, Sect. 4.3] and discuss the main observations here.

The most frequent positive feedback from students is on the immediate feedback that the system provides. Some student comments mention explicitly the usefulness of the system's feedback which allows to identify the errors they have made more easily. In addition to these generic endorsements, some students mention explicitly advantages of the test-driven development such as re-assurance regarding correctness of code, quick feedback on refactoring, the indirect introduction of unit tests through the system, and help in writing clean code. Further student feedback welcomes the ability to re-submit code repeatedly, and the flexibility to do so at any time. One student mentions the objectiveness of the system – presumably this is based on experience with assessment systems where a set of markers manually assess submissions and naturally display some variety in the application of marking guidelines.

The most common negative feedback from the students is that the automatic testing system is hard to understand. This refers to test-failure reports such as shown in Listing 1.4. Indeed, the learning curve at the beginning of the course is quite high: the first 90 min lecture introduces Python, Hello World and functions, and demonstrates feedback from the testing system to prepare students for their self-paced exercises and the automatic feedback they will receive. However, a systematic explanation of the `assert` statements, `True` and `False` values, and exceptions, takes only place after the students have used the testing system repeatedly. The reading of error messages is of course a key skill (and the importance of this is often underestimated by our non-computer science students), and we think that the early exposure to error messages from the automatic testing is useful. In practice, most students use the hands-on computing laboratory sessions to learn and understand the error messages with the help of teaching staff before these are covered in greater detail in the lectures (see also Sect. 4.4).

We add our subjective observation from teaching the course that many students seem to regard the process of making their code pass the automatic tests as a challenge or game. The students play this game "against" the testing system, and they experience great satisfaction when they pass all the tests – be it in the first or a repeat submission. As students enjoy this game, they very much look forward to being able to start the next set of exercises which is a great motivation to actively follow and participate in all the teaching activities.

4.4 Discussion

Key Benefits of Automatic Testing. A key benefit of using the automatic testing system is to reduce the amount of repeated algorithmic work that needs to be carried out by teaching staff: establishing the correctness of student solutions, and providing basic feedback on their code solutions is virtually free as it can be done automatically.

This allowed us to very significantly increase the number of exercises that students carry out as part of the course, which helped the students to more actively engage with the content and resulted in deeper learning and greater student satisfaction.

The marking system frees teaching staff time that would otherwise have been devoted to manual marking, and which can now be used to repeat material where necessary, explain concepts, discuss elegance, cleanness, readability and effectiveness of code, and suggest alternative or advanced solution designs to those who are interested, without having to increase the number of contact hours.

Because of the more effective learning through active self-paced exercises, we have also been able to increase the breadth and depth of materials in some of our courses without increasing contact time or student time devoted to the course.

Quality of Automatic Feedback Provision. The quality of the feedback provision involves two main aspects: (i) the timeliness, and (ii) the usefulness, of the feedback.

The system typically provides feedback to students within 2 to 3 min of their submission (inclusive of an email round-trip time on the order of a couple of minutes). This speed of feedback provision allows and encourages students to iteratively improve submissions where problems are detected, addressing one issue at a time, and learning from their mistakes each time. This near-instant feedback is almost as good as one could hope for, and is a very dramatic improvement on the situation without the system in place (where the provision of feedback would be within a week of the deadline, when an academic or demonstrator is available in the next practical laboratory session).

The usefulness of the feedback depends on the student's ability to understand it, and this is a skill that takes time and practice to acquire. As we use the standard Python traceback to report errors, we suggest that it is an advantage to encourage students to develop this ability at an early stage of their learning. Students at Southampton are well-supported in acquiring these skills, including timetabled weekly laboratories and help sessions staffed by academics and demonstrators. Once the students master reading the output, the usefulness of the feedback is very good: it pinpoints exactly where the error was found, and provides – through didactic comments – the rationale for the choice of test case as well.

A third aspect of the quality of feedback and assessment is objectivity: the system also improves the objectivity of our marking compared to having several people each interpreting the mark scheme and applying their interpretations to student work.

A more detailed and thorough discussion with respect to the flexible learning opportunities that the automatic testing provides in practical use is provided within Sect. 4 of [3].

Robustness and Performance. The chosen user interface is based on sending and receiving email and is thus asynchronous. The testing server pulls emails using imap and fetchmail, and received emails are stored in a file-based queue. A "receipt email" is sent to the students, and the submitted files are tested in order of their arrival. Finally, results are communicated with a second email. (See Fig. 1 in [3] for details.) The system is robust towards interruption of the network or failures of the server as the state is captured through emails (which are inherently robust regarding network failures) and in files once the emails have arrived on the system.

The system provides scalable automatic feedback provision: we have used the system with up to 500 students in one course, and not experienced any noticeable delays in the feedback time. In the academic year 2019–2020, at Southampton the testing system was running in a Linux operating system, hosted on a virtual machine with 4 GB of RAM on a single core of an Intel Xeon E5-2697A v4 @ 2.60 GHzE5 CPU. In the earlier years of using the system, an older CPU was used and the machine had 1 GB RAM. Each test for a student submission typically takes a few seconds to complete. A test that has not completed after one minutes is interrupted: this protects from errors such as infinite loops, and ensures the testing queue cannot be blocked. An appropriate email message is fed back to the student, if such a long test is detected. Memory requirements arise from the nature of the test problem and are thus controllable, and for the courses described here low.

More detailed discussion is available [3] with respect to the dependability and resilience of the system [3, Sect. 3.4.7], flexible learning opportunities that the automatic testing provides [3, Sect. 4.8.3], support of large class teaching [3, Sect. 4.8.4], student satisfaction [3, Sect. 4.8.5], dealing with syntax errors in student submissions [3, Sect. 4.4.1], use of undeclared non-ASCII encoding [3, Sect. 4.4.2], and changing PEP8 [10] standards [3, Sect. 4.4.3]. We have also connected the system to Moodle [3, Sect. 4.5], used it to assess C code [3, Sect. 4.6], and used it to pre-mark exams the students have written [3, Sect. 4.7].

5   Summary

We have reported on the automatic marking and feedback system that we developed and deployed for teaching programming to large classes of undergraduates. We provided statistics from one year of use of our live system, illustrating that the students took good advantage of the "iterative refinement" model that the system was conceived to support, and that they also benefited from increased flexibility and choice regarding when they work on, and submit, assignments. The system has also helped reduce staff time spent on administration and manual marking duties, so that the available time can be spent more effectively supporting those students who need this. Attempting to address some of the shortcomings of other literature in the field as perceived by a recent review article, we provided copious technical details of our implementation in the supplementary report [3]. With increasing class sizes forecast for the future, we foresee this system continuing to provide us value and economy whilst giving students the benefit of prompt, efficient and impartial feedback.

Acknowledgements. This work was supported by the British Council, the Engineering and Physical Sciences Research Council (EPSRC) Doctoral Training grant EP/G03690X/1 and EP/L015382/1, and the OpenDreamKit Horizon 2020 European Research Infrastructure project (676541).

Data Availability. Data shown in this manuscript and [3] is available in Reference [11].

References

1. Pears, A., et al.: A survey of literature on the teaching of introductory programming. In: Working Group Reports on ITiCSE on Innovation and Technology in Computer Science Education, ITiCSE-WGR 2007, pp. 204–223. ACM, New York (2007)
2. Robins, A., Rountree, J., Rountree, N.: Learning and teaching programming: a review and discussion. Comput. Sci. Educ. **13**(2), 137–172 (2003)
3. Fangohr, H., O'Brien, N., Prabhakar, A., Kashyap, A.: Teaching Python programming with automatic assessment and feedback provision. Technical report, University of Southampton, IIT Madras, IIT Mandi (2015). https://arxiv.org/pdf/1509.03556.pdf
4. Johanson, A., Hasselbring, W.: Software engineering for computational science: past, present, future. Comput. Sci. Eng. **20**(2), 90–109 (2018). https://doi.org/10.1109/MCSE.2018.021651343
5. Fangohr, H.: A comparison of C, MATLAB, and Python as teaching languages in engineering. In: Bubak, M., van Albada, G.D., Sloot, P.M.A., Dongarra, J. (eds.) ICCS 2004. LNCS, vol. 3039, pp. 1210–1217. Springer, Heidelberg (2004). https://doi.org/10.1007/978-3-540-25944-2_157
6. Ihantola, P., Ahoniemi, T., Karavirta, V., Seppälä, O.: Review of recent systems for automatic assessment of programming assignments. In: Proceedings of the 10th Koli Calling International Conference on Computing Education Research, Koli Calling 2010, pp. 86–93. ACM, New York (2010)
7. Tillmann, N., de Halleux, J., Xie, T., Gulwani, S., Bishop, J.: Teaching and learning programming and software engineering via interactive gaming. In: 2013 35th International Conference on Software Engineering (ICSE), pp. 1117–1126, May 2013
8. Beck, K.: Test Driven Development: By Example, 1st edn. Addison-Wesley, Boston (2003)
9. Krekel, H., Oliveira, B., Pfannschmidt, R., Bruynooghe, F., Laugher, B., Bruhin, F.: Pytest 3.7 (2004)
10. van Rossum, G., Warsaw, B., Coghlan, N.: PEP 8 - Style Guide for Python Code (2016). https://www.python.org/dev/peps/pep-0008/. Accessed 16 Dec 2016
11. Supplementary material: data used in figures (2016). https://arxiv.org/src/1509.03556/anc

Computational Science vs. Zombies

Valerie Maxville(✉) and Brodie Sandford

Curtin University, Perth, WA, Australia
v.maxville@curtin.edu.au
http://www.curtin.edu.au/

Abstract. Computational Science attempts to solve scientific problems through the design and application of mathematical models. Researchers and research teams need domain knowledge, along with skills in computing and, increasingly, data science expertise. We have been working to draw students into STEM, Computational Science and Data Science through our Team Zombie outreach program. The program leads the students through a scenario of a disease outbreak in their local area, which turns out to a potential zombie apocalypse. They become part of Team Zombie, a multi-disciplinary response team that investigates the outbreak; models the spread and potential interventions; works towards cures or vaccines; and provides options for detection and monitoring. Throughout the activities we make reference to real-world situations where the techniques are applied. Our program has engaged students from primary school through to university level, raising awareness of the range of approaches to problem solving through models and simulations. We hope this will inspire students to choose courses and careers in computational and data science.

Keywords: Education · Computational Science · STEM

1 Introduction

Around the world, educators are working to boost student interest in Science, Technology, Engineering and Maths (STEM). According to Scitech, 'As we shift to an information-based, highly connected and technologically advanced society, STEM can empower individuals and communities with the problem-solving capabilities to meet unprecedented economic, social and environmental challenges' [1]. In addition to developing interest in the STEM subjects themselves, we can inspire students by exploring scenarios that require a combination of STEM knowledge and techniques.

Computational and data science provide many vibrant examples of STEM. Combining one or more science disciplines with computer simulations and data analytics can give interactive and realistic experiences. A key requirement for outreach activities is to be able to easily explain the context and be inclusive of a wide range of ages and backgrounds. To give better reuse and flexibility, we also look to allow for hiding or revealing details, and moving through a range of

V. V. Krzhizhanovskaya et al. (Eds.): ICCS 2020, LNCS 12143, pp. 622–633, 2020.
https://doi.org/10.1007/978-3-030-50436-6_46

models/solutions. An important point to get across is that we are working with models, which will always be wrong [2]. Inspired by The Shodor Foundation's [3] resources for mathematics and science education through the application of modeling and simulation technologies, we have previously developed outreach activities exploring supercomputing and simulation [4]. In this paper we discuss the development and results of a story/scenario-based approach where various techniques are applied through the investigation of a zombie apocalypse.

2   Approach

The initial simulation that inspired this work was an assignment set in Fundamentals of Programming, a Python programming course developed for Science and Engineering students at Curtin University [5]. The assignment gave students a simple model of disease spreading through a population, which they then had to extend to incorporate barriers, immunity/recovery and airports. Discussions around the assignment often referred to it as modelling the zombie apocalypse. The additional engagement this elicited was a sign of the potential of the scenario for outreach purposes.

To give authenticity to the resources, we considered real epidemic monitoring and policy. Table 1 provides an example of the steps taken when investigating a disease outbreak. Our scenario begins by going through the steps required to identify and characterise a disease outbreak.

Table 1. Epidemiologic steps of an outbreak investigation [7].

Step	Description
1	Prepare for field work
2	Establish the existence of an outbreak
3	Verify the diagnosis
4	Construct a working case definition
5	Find cases systematically and record information
6	Perform descriptive epidemiology
7	Develop hypotheses
8	Evaluate hypotheses epidemiologically
9	As necessary, reconsider, refine, and re-evaluate hypotheses
10	Compare and reconcile with laboratory and/or environmental studies
11	Implement control and prevention measures
12	Initiate or maintain surveillance
13	Communicate findings

Steps 1–4 would be done as preparatory work, looking at health, media and other reports. This is the settling in part of the outreach activity. Once the

zombie diagnosis is established, we need more specific information about the type of zombie outbreak we're looking at (step 5–6):

Depending on the time available, we might have an activity around researching the various types of zombies in movies and literature. From our research, these are some of the known Zombie types (transmission and behaviour):

1. Viral: An airborne virus infects the entire population. When people die, they re-animate four hours later as zombies. These zombies are gradually decomposing. They can be killed by destroying their brain. (Ref: Walking Dead)
2. Contact+Hoarding: Hoard zombies are attracted to sound, and speed up when moving towards noise. Contact from zombie to innocent bystander will turn said bystander into a zombie. Conversion takes twelve seconds. (Ref: World War Z)
3. Plant-based: A plant releases spores that turn people into zombies and hosts of the infection. The zombies transfer the spores to others. Animals can also be infected. (Ref: Last of Us)
4. Smart Zombies: Most zombies are mindless, but some can open doors and may be able to catch planes etc. [6]

With the infection information, and an understanding of the type of zombie we are dealing with, it's time to develop and validate models of the outbreak (steps 7–10). These can extend on the models used in the assignment, described earlier. Steps 11–12 look at the response and monitoring of the situation, while the final step looks at communicating the plans and progress of the response. The steps may be carried out in parallel, and may have feedback loops to update and reassess based on additional information throughout the investigation [7].

The context for the outreach tasks is given to the students by memos, briefings and/or developed websites. After consideration of the various computational tools that could be developed for battling the Zombie Apocalypse, we focused on three key demonstrations:

- Zombie Outbreak – modelling the spread of the disease
- Zombie Detector – application of computer vision and machine learning to automatically identify zombies
- Zombie Vaccine – once a potential vaccine has been identified and developed, modelling its dosage and distribution

We will now describe the approach and implementation for each of the demonstrations.

2.1   Modelling the Zombie Outbreak

As the purpose of the model is to communicate the concepts around the spread of disease, we focused on the visual and interactive elements of the implementation. The Python simulation begins with a healthy population in a 2D grid. We use the PyGame package to provide an interactive interface where the user can click on individuals to infect them. The disease spreads via a set probability of infecting

```
for row in range(NUM_ROWS):
    for col in range(NUM_COLS):
        # if person is infected, randomly infect surrounding people
        if(infected[row, col]):
            # indices of people surrounding person
            for index in [[-1,-1],[0,-1],[1,-1],[0,-1],[0,1],[-1,1],[-1,0],[-1,1]]:
                inf_row = row + index[0]
                inf_col = col + index[1]
                # ensure not out of boundary
                if(0 <= inf_row < 64 and 0 <= inf_col < 64):
                    if(bridges[inf_row, inf_col]):
                        print("trying to infect bridge")
                    # ensure not infecting water
                    if(not water[inf_row, inf_col]):
                        if(random() < infectability):
                            infect_person[inf_row, inf_col] = 1

# infect assigned grids
for row in range(NUM_ROWS):
    for col in range(NUM_COLS):
        if(infect_person[row,col] and not immune[row,col]):
            assign(row, col, "infected")
```

Fig. 1. Python code for modelling the spread of disease

a cell's neighbours in a Moore neighbourhood (eight directions). The main loop of the program is shown in Fig. 1.

Healthy and infected cells/individuals are indicated by colour (green/red in Fig. 2). A third option is a proportion of the population having immunity (yellow) which contains the spread of disease. Users can also enter a value to speed up the simulation. Possibly the most compelling part of the application is the underlying map of Perth, which localises the scenario and caused much excitement. Also included are natural and man-made barriers such as the river/ocean and bridges. Users can click on the bridges to break them down, allowing for isolation of the infected population.

2.2 Detecting Zombies

An important step in controlling a disease outbreak is the detection of infected people, preferably using a remote, automated system. When discussing zombies, a range of potential traits were considered:

- Speed of movement - slowness, gait
- Intelligence, or lack of it
- Odour - decomposing flesh
- Temperature - using thermal images
- Appearance - e.g. missing limbs, expression

Although temperature may have been the preferred approach, we chose to work with appearance as it would be engaging in an interactive display. To show the potential of computer vision, the premise was that zombies could be identified by their facial expression. We found no record of zombies smiling, so this could be a clear visual difference between healthy people and zombies.

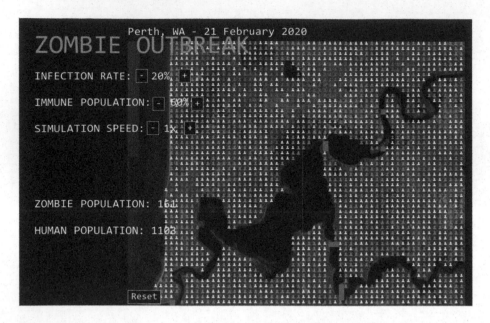

Fig. 2. Zombie Outbreak proved to be engaging and insightful, across all age groups (developed by Cleye Jensen) (Color figure online)

A Python program was developed to work with real-time camera footage to detect faces and flash up a warning if a face is determined to be a zombie (see Fig. 3). Packages used included: sklearn, open cv (cv2), dlib, numpy and csv. The implementation finds a face using HAAR Cascades [8] and puts a bounding box on the face. Then the system identifies 68 facial landmarks, and the positioning of the landmarks is used to determine if the face is human or zombie, based on a training data set.

2.3   Developing a Zombie Vaccine

The third demonstration assumes that a vaccine or medication can be developed to protect healthy humans. Once we have a vaccine, we need to calculate how much of it is required, and how often, for there to be effective protection. With this individual information, we can then consider how many people can be saved, how much medication to produce, and whether there are difficult decision to be made due to limitations in supply.

The demonstration used for communicating these concepts is based on the one-compartment model with repeat dosages described in [9]. This model has been used each semester in Fundamentals of Programming to explore simulations and parameter sweeps. In this case, we limit the interaction to adjusting key parameters to vary the half-life, absorption, dose and interval between doses. The diagram in Fig. 4 show the display of the demonstration code, which we are running in a Jupyter notebook (via Google Colab). In this case, the students see and interact with the code directly.

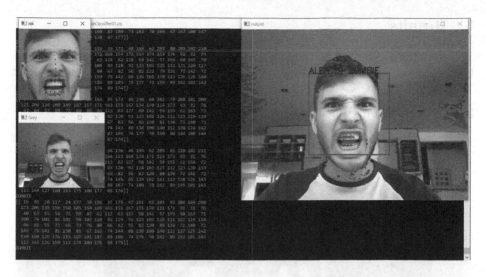

Fig. 3. Visual detection of zombies (developed by Harry Walters)

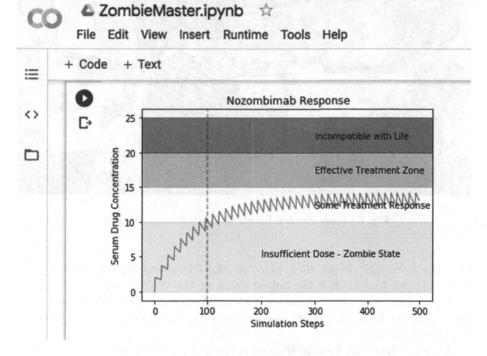

Fig. 4. Drug dosage simulation - this one needs a bit more serum to provide effective treatment (developed by Brodie Sandford, based on [9])

3 Results

The developed materials have been used in two outreach events: a local science festival and a multi-day immersive science experience.

3.1 Perth Science Festival

Fig. 5. Team Zombie booth at Perth Science Festival

This family-friendly event took place at the Claremont Showgrounds on the 24th and 25th August. It is the largest event on the WA National Science Week calendar, with over fifty exhibitors and an attendance of 8,161 people (2018 figures). In pitching for a booth, we needed to give the focus and story for our booth (see Table 2).

Ahead of the event, local media contacted us for an interview to help promote the booth and the science event. The resulting article [10] used the Team Zombie booth as the lead story to promote the Festival. The booth for Team Zombie was a standard 3 × 3 m setup with two tables. We brought multiple computers, explanatory material and zombie decorations (see Fig. 5). It was manned by Computer Science and Data Science students 10am–5pm for two days.

Table 2. Promotional text for Team Zombie booth.

Can you survive a zombie apocalypse? A zombie epidemic is breaking out and we need smart science to save the world as we know it!
Can Artificial Intelligence help us detect zombies?
Will your borders hold? Should we shut down the airports?
Can immunisation protect the population, or should we focus on finding a cure?
How much medicine do we need? How many people can we save?
Work with us as we use computer simulations and Artificial Intelligence to fight the zombies with science!

There was an excellent response from kids, parents and teachers who enjoyed the interactive displays. We were asked about school visits, our website (in progress) and possible mobile phone applications.

In terms of results and effectiveness, the attendees spoke with their feet. We had continual questions and interactions for the full day, often with people waiting to have a turn at the each of the demonstrations. Some children and parents kept coming back to chat, or to explore a different aspect of the simulations. We found they also built on the concepts in our materials, discussing herd immunity and seeing the relevance to current issues such as measles outbreaks and global epidemics.

3.2   Curtin Science Experience

This initiative has been run by Curtin Outreach for over 15 years. The four-day event has a program of science activities for Year 9 and 10 students. The premise is the give students with an interest in science a chance to participate in a range of science activities, with guidance from scientists and engineers who can share their love for their work. We were asked by the organisers to showcase Technology/Data as a link to the State Government's priority areas. We were assigned a two hour timeslot with up to fifty students attending.

The large group was split across two rooms, then into smaller teams or 4–5 to do their investigations. We scheduled the session in a collaborative learning space, giving each group a computer to share, which we prepared by pre-loading all of the software (see Fig. 6).

The lesson plan took students through setting the context; research of zombies; discussion of transmission; interaction with Zombie Outbreak model; discussion of containment strategies; Zombie Detection and Zombie Vaccine development. Google Collab notebooks were used for interacting with the Vaccine code, while the other two demonstrations were set up to run continuously. Our temporary website [11] progressed the scenario through a series of blog posts. Each team had a Google doc to record notes on their investigations.

Table 3. Examples of group investigations.

Group	Excerpt of research
Group A	HOW TO STOP A ZOMBIE APOCALYPSE Responses can be organised into four categories, the scientists say:
	– Quarantine the infected and develop a vaccine. Quarantines are difficult to maintain and vaccines take time develop
	– Hide the uninfected. A great plan if you're outnumbered, until the zombies breach your zone and find what the scientists says is a 'perfect environment' for the disease to spread
	– A selective cull. Effective if precisely orchestrated, but nearly impossible to achieve given that early cases may not show obvious symptoms
	– Eradicate the infected area. It's guaranteed to end the outbreak, but moral issues remain given the 'heavy losses of uninfected individuals'
Group B	The Generic Zombie
	– is a person who has been killed and reanimated by a pathogen, often, but not always due to a virus
	– usually aggressive and curious, but disorientated, and at a loss to fully understand their environment
	– Their most notable trait is that they kill and eat uninfected humans
Group C	– Conspiracy theories
	– Government sent zombies to solve overpopulation
	– Contaminated drinks and food
	– Team Zombie is actually the one that is infecting everybody to gain publicity and mass chaos
Group D	Possible vaccines:
	– Inject chocolate into bloodstream. (Permanent solution)
	– Make chocolate grenades
	– Make chocolate shooting guns and bullets
	– Chocolate nuke
	– Replace every water source with chocolate
	– Chocolate clothing
	– Give zombies chocolate so they become happy and not eat us (and then shoot them)

Fig. 6. Team Zombie activities at the Science Experience

A selection of excerpts are reproduced in Table 3. Note that the scenario puts forward chocolate as a potential antidote/vaccine.

Overall the session kept the students engaged and, although the scenario was a bit of fun, it happened to take place when a measles epidemic was causing the deaths of many children in Samoa. By discussing the Samoan outbreak and other epidemics (e.g. Ebola, SARS and now Coronavirus) the students clearly understood the seriousness of such situations and we hope they saw the need for a diverse team from a range of disciplines, using a variety of techniques, to best meet these challenges.

3.3   Media Coverage

Ahead of the Perth Science Festival, local media contacted us for an interview to help promote the booth and the science event. Of over fifty booths at the event, the zombie theme was considered the most exciting to attract visitors. The resulting article [10] used the Team Zombie booth as the headline and lead story to promote the Festival.

During the event, exhibitors for Scitech approached us to provide our 'expert' opinion on how zombies might spread across Australia. This evolved into a feature article for Particle [12], an online magazine producing news, stories and views. Ideas explored in the article included:

- Is it possible to mathematically work out how fast a zombie apocalypse would spread across the world?
- If they only eat brains, would they eventually run out of food?
- As an educated guess, where would it likely start? and being so isolated, does Perth/WA have an advantage?

We worked through some additional models and problems, including consideration of census data (population density) to give a prediction on the hypothetical spread of zombies across Australia. We were informed that the Particle article had strong readership figures and received many queries for more information.

4   Conclusion

We have described an outreach project intended to increase awareness of computational science and the multi-disciplinary teams that are required to address the challenges facing our planet. The project has been able to capture the imagination of students, teachers and parents through interaction with our models and simulations. Lengthy interactions, requests for additional resources and school visits indicate that we have succeeded in engaging our audience. The media interest supports our belief that we have a useful premise and plausible resources. Future work will be to develop an improved online presence and additional resources, including browser and/or app-based simulations, lesson plans and more detailed materials to expand on specific aspects of the scenario.

Acknowledgement. A squad of student volunteers joined Team Zombie to put together outreach materials and interactive displays. Our coding team were:

- Zombie Outbreak – Cleye Jensen
- Zombie Detector – Harry Walters
- Zombie Vaccine – Brodie Sandford

Cleye, Harry, Brodie, Lisa, Caitlyn and Ryan volunteered on the booth at Perth Science Festival. Team Zombie volunteers for the Science Experience were Brodie, Harry, Caitlyn, Nhan, Matthew, Cameron, Jack, Dylan, Brooklyn, Indigo, Bene and Blake.

References

1. Scitech Website: Why STEM Matters. https://www.scitech.org.au/about/why-stem-matters/. Accessed 4 Feb 2020
2. Panoff, B., Shiflet, A.B., Shiflet, G.: Introduction to Computational Thinking, Melbourne, Australia, 2011 (2013). http://messagelab.monash.edu.au/IntroductionToComputationalThinking. Accessed 7 Jan 2013
3. Shodor: SHODOR: A National Resource for Computational Science Education. http://www.shodor.org/. Accessed 4 Feb 2020
4. Maxville, V.: Introducing: computational Science. Proc. Comput. Sci. **18**, 1456–1465 (2013). https://doi.org/10.1016/j.procs.2013.05.313

5. Maxville, V.: Fundamentals of programming for science and engineering. In: 2018 IEEE/ACM 13th International Workshop on Software Engineering for Science (SE4Science), Gothenburg, Sweden, pp. 28–31 (2018)
6. Team Zombie: Zombie Identification, Activity Resources, 23 August 2019
7. CDC: Principles of Epidemiology: Home—Self-Study Course SS1978 (2012). https://www.cdc.gov/csels/dsepd/ss1978/index.html. Accessed 4 Feb 2020
8. Open CV: Face Detection using Haar Cascades. https://opencv-python-tutroals. readthedocs.io/en/latest/py_tutorials/py_objdetect/py_face_detection/py_face_ detection.html. Accessed 4 Feb 2020
9. Shiflet, A.B., Shiflet, G.W.: Introduction to Computational Science: Modeling and Simulation for the Sciences, 2nd edn. Princeton University Press, Princeton (2014). ISBN 0691160716
10. Young, E.: Nerd it up this weekend in Perth zombie apocalypse simulation, In: WA Today, 24 August 2019. https://www.watoday.com.au/national/western-australia/nerd-it-up-this-weekend-in-perth-zombie-apocalypse-simulation-20190823-p52k4e.html. Accessed 4 Feb 2020
11. Team Zombie: Team Zombie website, 11 December 2019. https://valeriemaxville. wixsite.com/zombie. Accessed 4 Feb 2020
12. McGellin, R.: Zombie a-Perth-calypse: a WA survival guide. In: Particle, 19 November 2019. https://particle.scitech.org.au/tech/zombie-a-perth-calypse-a-wa-survival-guide/. Accessed 4 Feb 2020

Supporting Education in Algorithms of Computational Mathematics by Dynamic Visualizations Using Computer Algebra System

Włodzimierz Wojas(iD) and Jan Krupa(✉)(iD)

Department of Applied Mathematics, Warsaw University of Life Sciences (SGGW),
ul. Nowoursynowska 159, 02-776 Warsaw, Poland
{wlodzimierz_wojas,jan_krupa}@sggw.pl

Abstract. Computer algebra systems (CAS) are often used programs in universities to support calculations and visualization in teaching mathematical subjects. In this paper we present some examples of dynamic visualizations which we prepared for students of Warsaw University of Life Sciences using Mathematica. Visualizations for simplex algorithm and Karush-Kuhn-Tucker algorithm are presented. We also describe a didactic experiment for students of the Production Engineering Faculty of Warsaw University of Life Sciences using dynamic visualization of the network flow problem.

Keywords: Computational mathematics · Symbolic calculations · Mathematical didactics · Linear programming · Nonlinear programming · Network flows · CAS

Subject Classifications: 97B40 · 97R20 · 90C30 · 97I60

1 Introduction

The development of computational mathematics techniques, technologies and tools in recent decades has created new educational challenges in teaching mathematics and IT subjects in higher education. On the other hand, the development of computational mathematics technology has created new educational perspectives based on the possibility of using new computational and visualization tools in education. CAS such as wxMaxima, Mathematica, Maple, Sage, and others are often used to support calculations and visualization in teaching mathematical subjects [6,7,9–11]. In teaching mathematical algorithms in the field of mathematical analysis, mathematical programming or graph theory, the possibility of symbolic calculations and visualizing the algorithm steps seems useful from an educational point of view. This allows a better and deeper understanding of the topic. In this paper we present two examples of dynamic visualizations

© Springer Nature Switzerland AG 2020
V. V. Krzhizhanovskaya et al. (Eds.): ICCS 2020, LNCS 12143, pp. 634–647, 2020.
https://doi.org/10.1007/978-3-030-50436-6_47

which we prepared for students of the Faculty of Informatics and Economet-ric of Warsaw University of Life Sciences using Mathematica [8,12] within the course of Mathematical Programming. We present dynamic visualizations for simplex algorithm for linear programming (LP) problem [2,5] and for nonlinear programming (NLP) problem which we solve using Karush-Kuhn-Tucker (KKT) algorithm [3,5]. A didactic experiment for students of the Production Engineer-ing Faculty of Warsaw University of Life Sciences using dynamic visualization of the network flow problem will also be discussed.

2 Visualization of Primal Simplex Algorithm Steps – 3D Example

In this example the authors present some new complex approach to visualization of simplex method steps.

The authors propose to use for visualization expanded Simplex Tableau which contains for each simplex step: current simplex table for this step, graph of feasible region for canonical form of LP problem with current corner point and "simplex path", level sets of goal function (hyperplanes: lines in $2D$, planes in $3D$), axis with current value of objective function for this step.

Let us solve the following LP problem:

Example 1.

$$
\begin{aligned}
\text{Maximize} \quad & z = 4x_1 + x_2 + 6x_3 \\
\text{Subject to} \quad & x_1 \qquad\qquad\qquad \le\ 9 \\
& \qquad x_2 \qquad\qquad \le\ 5 \\
& \qquad -\ x_2 + 5x_3 \le 15 \\
& -2x_1 - 3x_2 + 3x_3 \le\ 3 \\
& 5x_1 -\ x_2 + 5x_3 \le 45 \\
& x_i \ge 0 \text{ for } i = 1, 2, 3.
\end{aligned}
$$

Corresponding to it canonical form is:

$$
\begin{aligned}
\text{Maximize} \quad & z = 3x_1 + 2x_2 \\
\text{Subject to} \quad & x_1 \qquad\qquad + x_4 \qquad\qquad\qquad\qquad = 9 \\
& \qquad x_2 \qquad\qquad + x_5 \qquad\qquad\qquad = 5 \\
& \qquad -\ x_2 + 5x_3 \qquad\qquad + x_6 \qquad\qquad = 15 \\
& -2x_1 - 3x_2 + 3x_3 \qquad\qquad\qquad + x_7 \qquad = 16 \\
& 5x_1 -\ x_2 + 5x_3 \qquad\qquad\qquad\qquad + x_8 = 45 \\
& x_i \ge 0 \text{ for } i = 1, 2, \ldots, 8
\end{aligned}
$$

Let S be feasible region for this LP problem (in standard form). S is presented in each Fig. 1, 2, 3, 4 and 5. It is convex polyhedral set with vertices at: $v_1 = (0,0,0), v_2 = (9,0,0), v_3 = (9,5,0), v_4 = (0,5,0), v_5 = (0,0,1), v_6 = (3,0,3), v_7 = (6,0,3), v_8 = (9,5,1),$ $v_9 = (6,5,4),$ $v_{10} = (0,5,4), v_{11} = (0,5/2,7/2)$. In each Figs. 1, 2, 3, 4 and 5 we present expanded Simplex Tableaus for subsequent vertices of simplex path. The current vertex of simplex path and level set are in red and the previous ones are in blue. The dynamic versions of the Figs. 1, 2, 3, 4 and 5 can be found in the Electronic supplementary material: https://drive.google.com/open?id=1vgBC1ij7Z9mNL8_nmhVgrwi3qFRzzhYN.

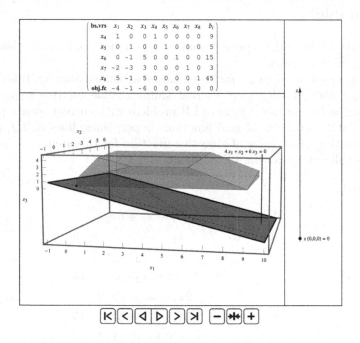

Fig. 1. First expanded Simplex Tableau.

Finally we get that $z_{\max} = z(6,5,4) = 53$.

The above example and also another $2D$ and $3D$ examples, were presented students of Informatics and Econometrics Faculty of Warsaw University of Life Science in the framework of the course of Mathematical Programming. These were introductory examples illustrating the simplex algorithm steps. Using the expanded Simplex Tableau allows students to trace the steps of simplex algorithm in quite comprehensive manner - taking into account both computational and geometric aspects of the algorithm. From Figs. 2, 3 and 4 we can see that the planes: $4x_1 + x_2 + 6x_6 = 6$, $4x_1 + x_2 + 6x_6 = 30$, $4x_1 + x_2 + 6x_6 = 42$ are not supporting planes of the feasible region for this LP problem S, but from Fig. 5 we can see that the plane $4x_1 + x_2 + 6x_6 = 53$ is supporting plane of S

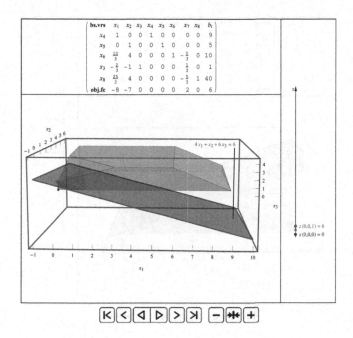

Fig. 2. Second expanded Simplex Tableau.

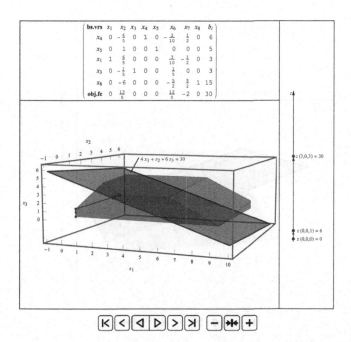

Fig. 3. Third expanded Simplex Tableau.

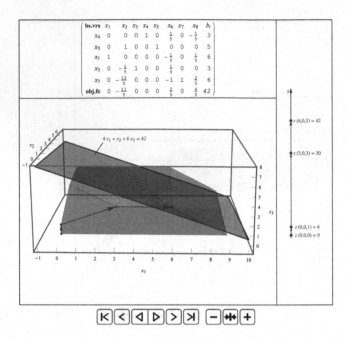

Fig. 4. Forth expanded Simplex Tableau.

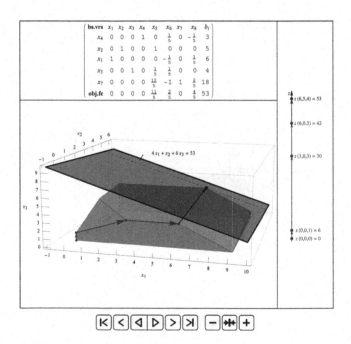

Fig. 5. Fifth expanded Simplex Tableau.

(this can be better seen in the dynamic versions of the Figs. 1, 2, 3, 4 and 5 in the Electronic supplementary material). So we may conclude independently that $z_{max} = z(6, 5, 4) = 53$. We cannot find analogical approach to present simplex algorithm steps in available literature.

3 Karush-Kuhn-Tucker Necessary Conditions

We will consider NLP problems in the following form:

$$
\begin{array}{ll}
\underset{(x_1, x_2, \dots x_n)}{\text{minimize/maximize}} & f(x_1, x_2, \dots, x_n) \\[1mm]
\text{subject to:} & g_i(x_1, x_2, \dots, x_n) \geq 0, \; i = 1, 2, \dots m, \\
& (x_1, x_2, \dots, x_n) \in X,
\end{array}
\tag{P}
$$

where n and m are positive integers, X is a subset of $\mathbb{R}^n$ and f, g_i are real-valued functions on X with at least one function of f, g_i $(i = 1, 2, \dots, m)$ being nonlinear. A feasible region of the NLP problem is defined as a set of all possible points that satisfy all the problem's constraints.

We start with the following theorem [3]:

Theorem (Karush-Kuhn-Tucker Necessary Conditions). *Let X be a nonempty open set in $\mathbb{R}^n$, and let $f : \mathbb{R}^n \mapsto \mathbb{R}$ and $g_i : \mathbb{R}^n \mapsto \mathbb{R}$ for $i = 1, \dots, m$. Consider the Problem (P) to minimize $f(x)$ subject to $g_i(x) \geq 0$ for $i = 1, \dots, m$. Let $\bar{x}$ be a feasible solution, and denote $I = \{i : g_i(\bar{x}) = 0\}$. Suppose that f and g_i, for $i = 1, 2, \dots m$ are differentiable at $\bar{x}$. Furthermore, suppose that $\nabla g_i(\bar{x})$ for $i \in I$ are linearly independent. If $\bar{x}$ solves Problem (P) locally, there exist scalars λ_i, for $i = 1, 2, \dots m$ such that:*

$$
\nabla f(\bar{x}) - \sum_{i=1}^{m} \lambda_i \nabla g_i(\bar{x}) = 0
\tag{1}
$$

$$
\lambda_i g_i(\bar{x}) = 0 \quad \text{for } i = 1, 2, \dots m
\tag{2}
$$

$$
\lambda_i \geq 0 \quad \text{for } i = 1, 2, \dots m.
\tag{3}
$$

Example 2. Using KKT necessary conditions we determine global minimum and maximum function $f(x, y) = y^2 - x$ subject to: $5 - x - y \geq 0, xy - 4 \geq 0, x - 1 \geq 0$ and next we visualise the solution graphically using dynamic plots.

Formally, KKT necessary conditions require an independent solution of two tasks - one for minimum and one for maximum. We first solve the NLP problem for minimum and next the NLP problem for maximum.

a) Consider the following problem:

$$
\begin{array}{ll}
\underset{(x,y) \in \mathbb{R}^2}{\text{minimize}} & y^2 - x \\[1mm]
\text{subject to:} & 5 - x - y \geq 0, \\
& xy - 4 \geq 0, \\
& x - 1 \geq 0.
\end{array}
\tag{4}
$$

Let $g_1(x, y) = 5 - x - y$, $g_2(x, y) = xy - 4$ and $g_3(x, y) = x - 1$.
$\nabla f(x, y) = \lambda_1 \nabla g_1(x, y) + \lambda_2 \nabla g_2(x, y) + \lambda_3 \nabla g_3(x, y)$ is equivalent to:
$[-1, 2y] = \lambda_1[-1, -1] + \lambda_2[y, x] + \lambda_3[1, 0]$. So we must solve the following system:

$$
\begin{cases}
-1 = -\lambda_1 + \lambda_2 y + \lambda_3 \\
2y = -\lambda_1 + \lambda_2 x \\
5 - x - y \geq 0 \\
xy - 4 \geq 0 \\
x - 1 \geq 0 \\
\lambda_1(5 - x - y) = 0 \\
\lambda_2(xy - 4) = 0 \\
\lambda_3(x - 1) = 0
\end{cases}
\tag{5}
$$

where $\lambda_1 \geq 0$, $\lambda_2 \geq 0$, $\lambda_3 \geq 0$.

Case 1. Let $g_1(x, y) = 0$, $g_2(x, y) = 0$ and $g_3(x, y) = 0$. Hence $5 - x - y = 0$, $xy - 4 = 0$ and $x = 1$. So $x = 1$ and $y = 4$ hence $-1 = -\lambda_1 + 4\lambda_2 + \lambda_3$ and $8 = -\lambda_1 + \lambda_2$, hence $-9 = 3\lambda_2 + \lambda_3$ contradicts nonnegativity of λ_2, λ_3.

Case 2. Let $g_1(x, y) = 0$, $g_2(x, y) = 0$ and $g_3(x, y) > 0$. Hence $5 - x - y = 0$ and $xy - 4 = 0$ and $\lambda_3 = 0$. So $x = 4$ and $y = 1$ hence $-1 = -\lambda_1 + \lambda_2$ and $2 = -\lambda_1 + 4\lambda_2$ hence $\lambda_2 = 1$ and $\lambda_1 = 2$. $\nabla g_1(4, 1) = [-1, -1]$, $\nabla g_2(4, 1) = [1, 4]$ are linearly independent.

Case 3. Let $g_1(x, y) = 0$ and $g_2(x, y) > 0$ and $g_3(x, y) = 0$. Hence $\lambda_2 = 0$, $x = 1$ and $y = 4$, but $g_2(1, 4) = 0$ contradicts assumption in current **Case 3** $g_2(x, y) > 0$.

Case 4. Let $g_1(x, y) = 0$ and $g_2(x, y) > 0$ and $g_3(x, y) > 0$. Hence $\lambda_2 = \lambda_3 = 0$ hence

$$
\begin{cases}
-1 = -\lambda_1 \\
2y = -\lambda_1 \\
5 - x - y = 0.
\end{cases}
\tag{6}
$$

It follows that $\lambda_1 = 1$. Hence $y = -1/2$ and $x = 11/2$. $g_2(11/2, -1/2) = -11/4 - 4 < 0$ contradicts assumption in current **Case 4** $g_2(x, y) > 0$.

Case 5. Let $g_1(x, y) > 0$ and $g_2(x, y) = 0$ and $g_3(x, y) = 0$. Hence $\lambda_1 = 0$ and $x = 1$ and $y = 4$, hence $-1 = 4\lambda_2 + \lambda_3$ and $8 = \lambda_2$. So $\lambda_3 = -33 < 0$ contradicts nonnegativity of λ_3.

Case 6. Let $g_1(x, y) > 0$ and $g_2(x, y) = 0$ and $g_3(x, y) > 0$. Hence $\lambda_1 = \lambda_3 = 0$ hence $-1 = \lambda_2 y$ and $2y = \lambda_2 x$ hence $\lambda_2^2 = -\frac{2}{x} < 0$ contradiction because $x > 1$.

Case 7 and 8. Let $g_1(x, y) > 0$ and $g_2(x, y) > 0$ and $g_2(x, y) \geq 0$. Hence $\lambda_1 = \lambda_2 = 0$, hence $-1 = \lambda_3$ a contradiction.

From **Cases 1, 2, 3, ..., 8** it follows that the only point which satisfies KKT conditions is $(4, 1)$, $\lambda_1 = 2$, $\lambda_2 = 1$ and $\lambda_3 = 0$.

We solve this problem also using Mathematica procedures. We present the solution below:

Listing 1.1: Mathematica code for point a):

```
In[1]:= (* min *)
{f = y² - x, g2 = xy - 4, g1 = 5 - y - x, g3 = x - 1, F = f - λ1 * g1 - λ2 * g2 - λ3 * g3}
Out[1]= {-x + y², -4 + xy, 5 - x - y, -1 + x, -λ3(-1 + x) - x - λ1(5 - x - y) + y²
        -λ2(-4 + xy)}
In[2]:= {F1 = D[F, x], F2 = D[F, y]}
Out[2]= {-1 + λ1 - λ3 - λ2y, λ1 - λ2x + 2y}
In[3]:= r1 =
    Solve[{F1 == 0, F2 == 0, λ1 * g1 == 0, λ2 * g2 == 0, λ3 * g3 == 0, λ1 >= 0, λ2 >= 0,
        λ3 >= 0, g1 >= 0, g2 >= 0, g3 >= 0}, {x, y, λ1, λ2, λ3}, Reals]
Out[3]= {{x- > 4, y- > 1, λ1- > 2, λ2- > 1, λ3- > 0}}
In[4]:= f/.r1
Out[4]= {-3}
In[5]:= Minimize[{y² - x, g1 >= 0, g2 >= 0, g3 >= 0}, {x, y}]
Out[5]= {-3, {x- > 4, y- > 1}}
```

b) Consider the following problem:

$$\begin{array}{ll}
\underset{(x,y)\in\mathbb{R}^2}{\text{maximize}} & y^2 - x \\
\text{subject to:} & 5 - x - y \geq 0, \\
& xy - 4 \geq 0, \\
& x - 1 \geq 0.
\end{array} \qquad (7)$$

Let $g_1(x, y) = -x^2 - 4y^2 + 1$, $g_2(x, y) = -1 + x + 2y$ and $g_3(x, y) = x - 1$.
$-\nabla f(x, y) = \lambda_1 \nabla g_1(x, y) + \lambda_2 \nabla g_2(x, y) + \lambda_3 g_3(x, y)$ is equivalent to:
$[1, -2y] = \lambda_1[-1, -1] + \lambda_2[y, x] + \lambda_3[1, 0]$. So we must solve the following system:

$$\begin{cases}
1 = -\lambda_1 + \lambda_2 y + \lambda_3 \\
-2y = -\lambda_1 + \lambda_2 x \\
5 - x - y \geq 0 \\
xy - 4 \geq 0 \\
x - 1 \geq 0 \\
\lambda_1(5 - x - y) = 0 \\
\lambda_2(xy - 4) = 0 \\
\lambda_3(x - 1) = 0
\end{cases} \qquad (8)$$

where $\lambda_1 \geq 0$, $\lambda_2 \geq 0$ and $\lambda_3 \geq 0$.

Case 1. Let $g_1(x, y) = 0$ and $g_2(x, y) = 0$ and $g_3(x, y) = 0$. Hence $5 - x - y = 0$ and $xy - 4 = 0$ and $x = 1$. So $x = 1$ and $y = 4$ hence $1 = -\lambda_1 + 4\lambda_2 + \lambda_3$ and $-8 = -\lambda_1 + \lambda_2$. Let $\lambda_2 = \lambda_1 - 8$ hence $\lambda_3 = 1 + \lambda_1 - 4\lambda_1 + 32 = 33 - 3\lambda_1$ hence $8 \leq \lambda_1 \leq 11$. We can choose $\lambda_1 = 8$ than $\lambda_2 = 0$ and $\lambda_3 = 9$. $\nabla g_1(1, 4) = [-1, -1]$, $\nabla g_3(1, 4) = [1, 0]$, are linearly independent.

Case 2. Let $g_1(x, y) = 0$ and $g_2(x, y) = 0$ and $g_3(x, y) > 0$. Hence $5-x-y = 0$ and $xy - 4 = 0$ and $x > 1$ and $\lambda_3 = 0$. So $x = 4$ and $y = 1$ hence $1 = -\lambda_1 + \lambda_2$ and $-2 = -\lambda_1 + 4\lambda_2$ hence $\lambda_2 = -1 < 0$ contradicts nonnegativity of λ_2.

Case 3. Let $g_1(x, y) = 0$ and $g_2(x, y) > 0$ and $g_3(x, y) = 0$. Hence $\lambda_2 = 0$, and $x = 1$ hence

$$\begin{cases} 1 = -\lambda_1 + \lambda_3 \\ -2y = -\lambda_1 \\ 5 - x - y = 0. \end{cases} \tag{9}$$

Hence $y = 4$ and $\lambda_1 = 8$ and hence $\lambda_3 = 9$ (**Case 1**).

Case 4. Let $g_1(x, y) > 0$ and $g_2(x, y) = 0$ and $g_3(x, y) = 0$. Hence $\lambda_1 = 0$ and $x = 1$ and $y = 4$, hence $1 = 4\lambda_2 + \lambda_3$ and $-8 = \lambda_2$- a contradiction.

Case 5. Let $g_1(x, y) = 0$ and $g_2(x, y) > 0$ and $g_3(x, y) > 0$. Hence $\lambda_2 = \lambda_3 = 0$ hence $\lambda_1 = -1$- a contradiction.

Case 6. Let $g_1(x, y) > 0$ and $g_2(x, y) = 0$ and $g_3(x, y) > 0$. Hence $\lambda_1 = \lambda_3 = 0$ hence $1 = \lambda_2 y$ and $-2y = \lambda_2 x$. λ_2 must be nonzero, so $y = \frac{1}{\lambda_2}$ hence $\lambda_2^2 = -\frac{2}{x} < 0$ because $x > 1$- a contradiction.

Case 7 and 8. Let $g_1(x, y) > 0$ and $g_2(x, y) > 0$ and $g_3(x, y) \geq 0$. Hence $\lambda_1 = \lambda_2 = 0$, hence $1 = \lambda_3$ and $y = 0$ hence $g_2(x, 0) = -4 < 0$ contradicts assumption in current **Case 7 and 8** $g_2(x, y) > 0$.

From **Cases 1, 2, ..., 8** it follows that the only point which satisfies KKT conditions is $(1, 4)$, $\lambda_1 = 8$, $\lambda_2 = 0$ and $\lambda_3 = 9$.

$f(1, 4) = 15$, $f(4, 1) = -3$.

The set D is compact and f is continuous on it so finally we get: the greatest value 15 and smallest value -3 of f on D are attained at points $(1, 4)$ and $(4, 1)$ respectively.

In Fig. 6 we present the solution of the above NLP problem graphically using dynamic plots. We solve the problem from Example 2 using graphical method. In the Fig. 6 we present Mathematica dynamic plot for the Example 2. The level sets is a family of parabolas $y^2 - x = c$.

The dynamic versions of the Fig. 6 can be found in the Electronic supplementary material: https://drive.google.com/open?id=1vgBC1ij7Z9mNL8_nmhVgrwi3qFRzzhYN.

Global maximum of the function $f(x, y) = y^2 - x$ at point $(1, 4)$ we can determine graphically from the level set: $y^2 - x = 15$. Similarly, global minimum of the function f at point $(4, 1)$ we can determine graphically from the level set: $y^2 - x = -3$.

This example and also another $2D$ and $3D$ examples, were dedicated students of Informatics and Econometrics Faculty of Warsaw University of Life Science within the course of Mathematical Programming. These were introductory examples illustrating the KKT method. Even relatively simple NLP problems with three or four constraints can lead to the need to consider many subcases, what is rather not suitable for hand calculations in class. The presented example shows that calculations are possible to do in class with support of computer programs such as Mathematica. Visualization presents graphical way of reaching the optimal solution of the NLP problem (for min and for max respectively)

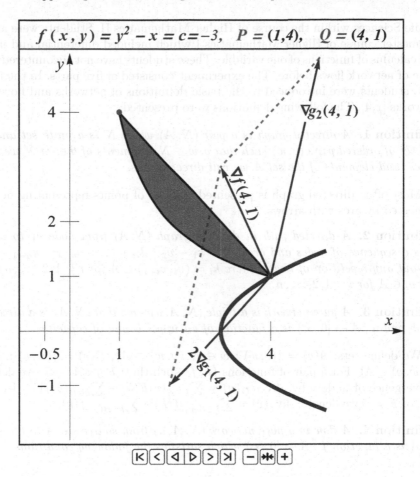

Fig. 6. Graphical method for Example 2 using dynamic plots

with graphical interpretation of KKT necessary conditions for Example 2 **a)**. In the Fig. 6 we see that: vector $\nabla f(4,1)$ is a linear combination of the vectors $\nabla g_1(4,1) = [-1,-1]$, $\nabla g_2(4,1) = [1,4]$ with coefficients $\lambda_1 = 2, \lambda_2 = 1$ ($\lambda_3 = 0$ and we see in Fig. 6 that the constraint $g_3 \geq 0$ is not active) which means that $\nabla f(4,1) = [-1,2] = 2[-1,-1] + 1[1,4]$. For graphical interpretation of KKT necessary conditions for Example 2 **b)** see Fig. 6 in Electronic supplementary material:
https://drive.google.com/open?id=1vgBC1ij7Z9mNL8_nmhVgrwi3qFRzzhYN.

4   The Didactic Experiment with Dynamic Visualization of Network Flows

This experiment was carried out in the form of a mathematics test on a group of 41 first-year students of the Production Engineering Faculty of Warsaw University

of Life Sciences within the course of Higher Mathematics II. Students were after
a semester course of Higher Mathematics I which included differential and integral calculus of functions of one variable. These students have not encountered the
topic of network flows before. The experiment consisted of five parts. In the first
part, students were introduced to the basic definitions of networks and flows in
networks [1,4]. The following definitions were presented:

Definition 1. *A directed graph is a pair (N, A) where N is a finite set and A
is a set of ordered pairs (v, w) such that $v, w \in N$. Elements of the set N we call
nodes, and elements of the set A we call directed arcs.*

Most often directed graph is presented as a set of points representing nodes
connected by arcs with arrows.

Definition 2. *A directed path in directed graph (N, A) from node v_1 to node
v_2 is a sequence of nodes and arcs: $v_1 - k_1 - v_2 - k_2 - \ldots - v_{n-1} - k_{n-1} - v_n$
without any repetition of nodes where $k_i = (v_i, v_{i+1}) \in A$ for $i = 1, 2, \ldots, n - 1$
and $v_i \in A$ for $i = 1, 2, \ldots, n$.*

Definition 3. *A pure network is a triple (N, A, u) such that (N, A) is a directed
graph and $u : A \to [0, \infty)$ is a function of an upper bound of capacity.*

We define sets: $A(v) = \{(v, w) : w \in N, (v, w) \in A\}$, $B(v) = \{(w, v) : w \in
N, (w, v) \in A\}$. For a pair of functions (F, f) such that: $F, f : A \to \mathbb{R}$, we define
a divergence of node v by: $\mathrm{div}_{(F,f)}(v) = \sum_{k \in A(v)} F(k) - \sum_{k \in B(v)} f(k)$. In the
case of $F = f$ we denote: $\mathrm{div}_f(v) = \sum_{k \in A(v)} f(k) - \sum_{k \in B(v)} f(k)$.

Definition 4. *A flow in a pure network (N, A, u) from source $s \in A$ to the sink
$t \in A$ is a function $f : A \to [0, \infty)$ which satisfies the following conditions:*

1) $0 \le f(k) \le u(k)$ for all $k \in A$,
2) $\mathrm{div}_f(s) \ge 0$,
3) $\mathrm{div}_f(v) = 0$ for all $v \in N \setminus \{s, t\}$.

A number $V = \mathrm{div}_f(s)$ we call the value of a flow f. A flow in pure network
is also called static flow.

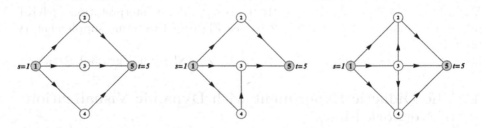

Fig. 7. Network 1   **Fig. 8.** Network 2   **Fig. 9.** Network 3

The definitions were discussed by the lecturer on the example of a simple network 1 (Fig. 7) drawn on a table with simple examples of flows in this network. The numerical values of flows on the arcs were recorded above the arcs. The first part lasted about 25 min. In the second part, students were given the first task to solve. It was based on giving an example of flow in the network 2 (Fig. 8) from source s to the sink t with the value of flow $V = 5$. Students received 10 minutes of time to solve this task. In the third part, two dynamic visualizations of flows in network were presented to the students: Example 3 and Example 4 presented in the Subsect. 4.1. These visualizations were discussed during the presentation by the lecturer. The presentation of the visualization together with the discussion lasted about 10 min. In the fourth part, students were again given a task to solve. It consisted, as before, of giving an example of flow in the network 3 (Fig. 9) from source s to outlet t with the value of flow $V = 4$. Students were given 10 minutes of time to solve this task. In the fifth part, students answered the question - to what extent dynamic flow visualizations were useful in solving the task in the fourth part. They chose one of four options: a) they were not helpful, b) they were a bit helpful, c) they were helpful, d) they were very helpful. We received the following experiment results: only 24% of students presented the correct solution to the task in the second part of the experiment, 66% of the students presented the correct solution to the task in the fourth part of the experiment (after seeing dynamic visualizations), 5% of the students chose the answer a), 19% answer b), 54% answer c) and 22% answer d).

4.1   Dynamic Visualization of Static Flow

For directed graph (N, A) presented below we define: $s = 1, t = 7, V = 5, u(k) = 3$ for all $k \in A$. We present two different static flows: static flows 1, static flows 2 in the pure network (N, A, u).

Example 3. (static flow 1)

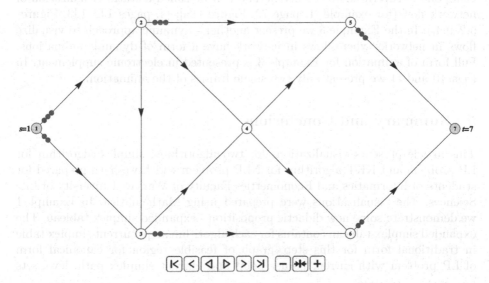

Fig. 10. Dynamic visualizations of static flow 1 in pure network (N, A, u)

Example 4. (static flow 2)

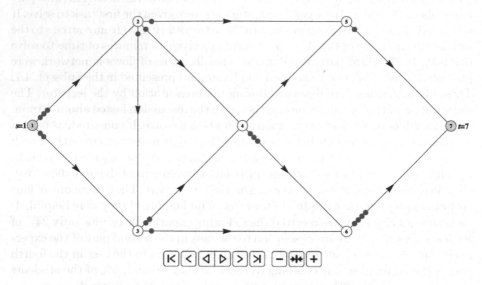

Fig. 11. Dynamic visualizations of static flow 2 in pure network (N, A, u)

The dynamic versions of the Figs. 10 and 11 can be found in the Electronic supplementary material:
https://drive.google.com/open?id=1vgBC1ij7Z9mNL8_nmhVgrwi3qFRzzhYN

A standard method of graphical presentation flows in networks presented in many academic books, is to note the flows values near the arcs of directed graph using one or several figures. This method can be called a static visualization of network flow (for example [1, page 79, Figure 3.9a]; [4, pages 142–143, Figures 9.2–9.4]). In the Example 3 we present another - dynamic approach to visualize flows in networks where flows in network have a form of dynamic animations. Full form of animation for Example 3 is presented in electronic supplement. In Figs. 10 and 11 we present only two single frames of the animation.

5   Summary and Conclusions

This article presents visualizations for two algorithms: simplex algorithm for LP problem and KKT algorithm for NLP problem which we have prepared for students of Informatics and Econometrics Faculty of Warsaw University of Life Sciences. The visualizations were prepared using Mathematica. In Example 1 we demonstrate some new didactic proposition - expanded simplex tableau. The expanded simplex tableau contains for each algorithm step: current simplex table in traditional form for this step, graph of feasible region for canonical form of LP problem with current corner point and current simplex path, level sets of goal function (hyperplanes : lines in $2D$, planes in $3D$), axis with current value of objective function for this step. In our opinion, using such visualization allows students to better understand the steps of the simplex algorithm and its geometric interpretation. In Example 2 we solve the NLP problem using KKT method. We demonstrate dynamic visualization of graphic method of obtaining the optimal solution in this example. In general, computer support (e.g. by CAS) in solving tasks by KKT method seems very useful due to the need to consider sometimes a large number of subcases difficult to consider by hand calculations. The visualizations presented in this example, seem very useful to us in a better understanding of solving NLP problems graphically. This article also presents the result of the didactic experiment carried out on a group of 41 first-year students of the Production Engineering Faculty of Warsaw University of Life Sciences. In this experiment dynamic visualizations for two examples of network flows were presented to these students. This had a significant impact on the result of the test consisting in the construction of flow in the given network. The number of correct solutions increased more than 2.5 times compared to the number of correct solutions of the analogous task before presenting this visualizations. Over 75% of students said that these visualizations were helpful or very helpful in solving the second task (after seeing dynamic visualizations).

References

1. Ahuja, R.K., Magnanti, T.L., Orlin, J.B.: Network Flows: Theory, Algorithms, and Applications, 2nd edn. Prentice Hall, Upper Saddle River (1993)
2. Bazaraa, S.M., Shetty, C.M.: Linear Programming and Network Flows, 2nd edn. Wiley, New York (1979)
3. Bazaraa, S.M., Shetty, C.M.: Nonlinear Programming. Theory and Algorithms., 2nd edn. Wiley, New York (1979)
4. Ford, L., Fulkerson, D.: Network Flows, 2nd edn. Wiley, New York (1979)
5. Griva, I., Nash, S., Sofer, A.: Linear and Nonlinear Optimization. SIAM, Philadelphia (2009)
6. Guyer, T.: Computer algebra systems as the mathematics teaching tool. World Appl. Sci. J. **3**(1), 132–139 (2008)
7. Kramarski, B., Hirsch, C.: Using computer algebra systems in mathematical classrooms. J. Comput. Assist. Learn. **19**, 35–45 (2003)
8. Ruskeepaa, H.: Mathematica Navigator: Graphics and Methods of Applied Mathematics. Academic Press, Boston (2005)
9. Wojas, W., Krupa, J.: Familiarizing students with definition of lebesgue integral: examples of calculation directly from its definition using mathematica. Math. Comput. Sci. **11**, 363–381 (2017). https://doi.org/10.1007/s11786-017-0321-5
10. Wojas, W., Krupa, J.: Some remarks on Taylor's polynomials visualization using mathematica in context of function approximation. In: Kotsireas, I.S., Martínez-Moro, E. (eds.) ACA 2015. SPMS, vol. 198, pp. 487–498. Springer, Cham (2017). https://doi.org/10.1007/978-3-319-56932-1_31
11. Wojas, W., Krupa, J.: Teaching students nonlinear programming with computer algebra system. Math. Comput. Sci. **13**(1), 297–309 (2018). https://doi.org/10.1007/s11786-018-0374-0
12. Wolfram, S.: The Mathematica Book. Wolfram Media Cambridge University Press, New York (1996)

Teaching About the Social Construction of Reality Using a Model of Information Processing

Loren Demerath[1]([⊠]), James Reid[1]([⊠]), and E. Dante Suarez[2]([⊠])

[1] Centenary College of Louisiana, Shreveport, LA 71104, USA
{ldemerath, jreid}@centenary.edu
[2] Trinity University, San Antonio, TX 78212, USA
esuarez@trinity.edu

Abstract. This paper proposes leveraging the recent fusing of Big Data, computer modeling, and complexity science for teaching students about complexity. The emergence and evolution of information orders, including social structural orders, is a product of social interactions. A simple computational model of how social interaction leads to emergent cultures is proposed for teaching students the mechanisms of complexity's development. A large-scale poker game is also described where students learn the theory and principles behind the model in the course of play. Using the model and game can help instructors achieve learning objectives that include understanding the importance of diversity, how multiple realities can exist simultaneously, and why majority group members are usually unaware of their privileges.

Keywords: Complexity · Transdisciplinarity · Information · Emergence · Teaching · Agent-based model · Information processing · Social interaction

1 Introduction

The computational capabilities of the twenty-first century have upended the way we do research. Whether it is in the modeling of the rings of Saturn, or in the use of Big Data to predict our next online purchase, computer simulations now represent an integral part of most state-of-the art research in a broad range of disciplines. The premise of this paper is that while this scientific revolution has changed how we think, conceptualize and analyze social phenomena at the highest levels of scholarship, these methodologies have not trickled down to the undergraduate classroom. We argue here that computational modeling technologies offer both new teaching techniques and new content. We attempt to marry some of the insights of traditional social science with those of complexity studies and systems thinking, proposing an agent-based model of computational sociology to serve as the cornerstone of study in an undergraduate course (or module), and a game that illustrates the model's principles.

The wave of Big Data that we reference concerns our continuous interaction with computers and the internet to generate increasingly large amounts of information on our behavior, providing researchers with data sets larger than anything seen before. Big

Data allows us to precisely measure human interactions and preferences, whether in the "real world" or online. This huge amount of heterogeneous data, integrated with an appropriate modeling methodology, forces us to reconceptualize the possibilities of the scientific endeavor. Rather than recommending the unfiltered use of traditional Big Data techniques uncoupled from solid theoretical foundations, the multidimensional nature of reality requires us to teach the ideas and modeling techniques in a way that is both rooted in the established theoretical understanding of disciplines, and also in the holistic paradigm offered by complexity science. Tolk et al. (2018) propose human simulation as a lingua franca for the computational social sciences and humanities, and we heed their call to bring these techniques into the undergraduate classroom, in an intuitive format.

Much of the development in this field has centered on agent-based models, where agents are explicitly modeled as independent computer programs that interact with each other to produce aggregate behavior that may include unexpected results. These approaches claim that the emergent upper level phenomena can be understood as a direct reflection of the interactions of the lower-level agents. Big Data, computer modeling, and complexity science are now fusing into a paradigm in which a new type of social science can be built. These ideas, however, are not yet accepted universally. In this regard, the completion of this scientific revolution may be currently stymied by the lack of imagination by scholarly leaders and those enforcing the status quo in academia. Our hopes would then lie with a new generation of researchers—no doubt, that includes current undergraduates—who may conceive of a radically new paradigm for the social sciences, rooted in part in a new kind of experimental methodology where every online human movement is measured and catalogued. Traditional social sciences were not developed with these types of capabilities for experimentation in mind. Experimenting with humans was, until relatively recently, simply not a possibility in macroeconomics or sociology. In contrast, we now face a reality in which a company such as the ride-sharing Uber can let a computer algorithm loose on the task of making its non-centralized drivers most efficient. As Gelfert (2016) points out, models are not only neutral abstractions of our world, they are the guiding light with which we conceive, mediate, contribute and enable our scientific knowledge. Just as language shapes the world we live in, scientific models both constrain and contextualize our perception of our reality.

The premise on which we base our proposal is that a common factor in some of the Big Data applications that abound today is a lack of sufficient attention to the underlying theory relating the model to the observed behavior, or to the versatility of the model and its range of possible applications. A simulation that is produced by a data-mining process may contain multiple topological aspects that are particular to the way that the model is built, and therefore not a true reflection of the real-world phenomenon at hand. With these common missteps in mind, we propose the use of a simple but interesting agent-based model of social interaction to serve as a connecting tool of instruction, in a course that teaches the benefits of model-oriented thinking, complex adaptive systems, and a holistic and interdisciplinary understanding of reality, particularly as it refers to the social sphere. As we describe in further detail, this can be a course for mathematics and computer science majors, but it can also be designed for those students rooted in the social sciences. The course may explicitly teach concepts of complexity science, or simply offer the view afforded by any agent-based model, and contrast it to linear methodologies more often discussed in an undergraduate setting. The course can be taught by a specialist in any of the functional areas considered, or co-taught by more than one professor—each possessing a limited understanding of the interdisciplinary issues addressed in the course. Finally, the course has enough material to cover a regular semester curriculum, but it can also represent a module within a larger class.

2 A Model of Agent Interaction for Introducing Computational Sociology in an Undergraduate Classroom

One of the most important concepts in complexity studies is that of emergence, often referencing how the aggregate is more than the sum of its parts, and allowing us to understand a human mind as more than a collection of neurons, or a society as more than a collection of people. If we accept the concept of emergence, we must conceive of a world where multiple realities coexist simultaneously (Suarez and Demerath 2019). Pessa (2002) and Tolk (2019) demystify the concept of emergence and separate it into two categories: epistemological and ontological. Epistemological emergence refers to situations where the system follows generalizable rules, but it is irreducible to one single level of description because of conceptual reasons. A system may therefore be epistemologically emergent if it follows rules that in principle are knowable; the unpredictability is a product of humans not being able to fully grasp these laws in their entirety. Ontological emergence, on the other hand, is where emergent properties are truly novel and cannot be explained solely by the components and their interactions. This is similar to Crutchfield's (1994) conception of intrinsic emergence as that which is independent of the external observer.

Until the second half of the twentieth century, science had no choice but to be linear. To continue advancing, most disciplines opted for the simplification of inter-acting agents, commonly describing them as homogeneous, independent and altogether exogenous. Linear science had to conceive of agents unrealistically in order to define and describe them independently. In doing so, one of the most important aspects of the system, the context in which the agents do interact, is erased. A linear model of reality that does not consider interaction between the forming parts will fail to adequately describe the structure of the system, and how that structure can change.

But how a structure forms and changes is often what needs studying. It underlies what we do when we analyze the cohesion of an army unit, the coordination of a football team, the culture of an enterprise, or the checks and balances of a government. Repeated interaction gives rise to certain structures that deserve more attention than other pre-existing ones. The cells or atoms that make up a government are far less meaningful for one's analysis of it than social factions and patterns of events. In this way (though not necessarily in all ways), a complexity perspective can be seen as anti-reductionist. Aggregates are more than the sum of their parts, and we refer to such aggregates as *emergent*. To understand them, we cannot simply reduce the system's aggregate behavior to its minimal components, study them in isolation, and then rebuild the system by multiplying the behavior of the average piece by the number of pieces in it. In the emergent, nonlinear, irreducible world of complexity, nature exists in many non-orthogonal levels, with each level potentially being governed by different laws, granularities and structure. If we believe in that proposition, then we should expect to find a world full of emergent phenomena, with distinctive levels of interaction that have unique kinds of agents, laws and granularities.

The current state of affairs in this scientific revolution rooted in complexity science and Big Data offers us the possibility to combine the strengths of both to create a truly holistic vision of reality, a vision backed by the data and models of computer scientists and supported by the bolder, more comprehensive theories offered by social scientists. The social actors that sociology has considered for decades are now within the realm of possibilities of a simple agent-based model; a model that is now freely available for curious undergraduate students. It is in this context that we offer the simplest possible model that the authors could construct that reflects this type of emergent behavior.

The goal of a simulation—particularly one presented to novice students—must be to capture the aspects of reality that are computable, whether these are linear or emergent, and to minimize aspects of the model which are not compatible with reality. Naturally, no model can ever be perfect, and thus it rests on the researcher to be upfront and straightforward about a model's limitations and applicability. In light of these precautions, our recommendation is to provide an introduction to modeling to under-graduates that stresses the aspects of reality that can be captured in the simplest pos-sible model, proceeding with more complicated models only when all aspects of a simple model have been fully explored and relatively well understood.

The model we propose for introducing computational sociology to undergraduates illustrates a theory of social emergence, as represented by simple agents interacting to create emergent social structures. Those structures are increasingly distinguishable from each other by the shared information that agents use within those structures. The model centers on two variables that affect rates at which agents self-organize from the "bottom-up" into social structures. Those variables are agent inequality in quantities of social information (which creates stability), and agent diversity in qualities of sub-jective information (which allows for connectivity). The model also includes variables that provide different representations of the information decay that varies with any environment, and which determine the rates at which structures emerge and then organize interactions "top-down." The model is able to show how agents process information by compressing that which is shared in the course of interaction through the mechanisms of quantitative social reinforcement (that could be called, "echo effects") and qualitative meaning orientation ("diversity dividends," perhaps). Overall, the model can produce characteristic behaviors of complex systems, including the emergence and evolution of both meaning and structure, reproduction, and the sub-optimization of an environment by overly dominant agents (Demerath, Reid, and Suarez, in progress).

The theoretical foundations of the model stress that novel information is found by agents interacting with some degree of mutual information and some degree of novelty, providing the probability that the interaction will be "meaningful" (Demerath 2012). The amount of mutual information determines the probability that the exchange will be comprehensible, while the relative total information of an agent determines its relia-bility. Together, the amount of shared and total information involved in the interaction help determine the sensitivity of the agent to novelty. This principle explains how those who know a lot about something are sensitive to nuances that are invisible to others, or how calibrating a system to be highly sensitive to input and precise in its output requires much information. The interaction topology and mechanisms mean that if information decays too rapidly, agents eventually have no shared information, and interaction becomes meaningless. Alternatively, if information does not decay rapidly enough, agents eventually consume all novel information in the system, leading them with no prospects for meaningful interaction. At this point, all activity stalls.

But inequality keep things going. Agents of information are oriented and motivated by the information they have relative to the overall *field* of information as embodied by other agents (Bourdieu 1984). While that accounts for the "social gravity" of larger agents, that preference is balanced by the fact that less energy is needed to interact with "smaller" agents (those with less information). This is based on a premise of the theory that processing information requires energy, explaining how homophily is balanced by the need for novelty. Such is the situation we simulate in our model, where agents chose to interact based on the probabilities they calculate for meaningful information.

As certain pieces of information become compressed over the course of interactions, "cliques" of agents emerge around their use of that information. Agents in such cliques have more energy to spend on acquiring new information, and more space for holding it. As such, agents that are part of homogeneous communities can, paradoxically, better explore for novelty, and such communities emerge and evolve accordingly.

3   Theoretical Description of the Model

The model offered here is an agent-based model that successfully produces simulations of socially distributed information processing, endogenous agency, and steadily more complex social and meaning structures. We intend this model to meet the requisites of generative emergence described by Cederman (2005) in the simplest possible topological construction. The model also illustrates specific causal mechanisms of emergence. These are not, though, mechanisms that exist at the level of the individual, as theorized by Hedstrom (2005) and modeled by Lane (2018), or to an infra-individual level as advocated by Sperber (2011). Instead, these mechanisms are interactional, as predicted by Sawyer (2011) and Demeulenaere (2011). However, our model depends on two "fields" of information: one is the social distribution of agents, and the other is distribution of meanings in their environment. We interpret the way our model functions as supporting the view that systems function by achieving a degree of operational closure (Luhmann et al. 2013) that allows them to be nested, interpenetrating, and interdependent (Bryant 2011, Cilliers 2001, Byrne and Callaghan 2013). Our model blends the advantages of Bourdieu's (1990) conception the reproduction of social structure through habitus, with Archer's (2003) account of reflexivity called for by others (e.g., Elder-Vass 2010). Details of the model are as follows:

1. Each simulation starts with at least one ambient phase space or "information space" for that simulation. This makes the knowledge of agents operating within the space comparable, and interaction possible.

2. A population of agents is distributed at the beginning of a simulation, each agent inheriting certain knowledge of the information space:
   a. bottom-level "facts," defined by their relations with each other, and
   b. higher-level "issues," defined by their relations to both facts and other issues.
3. Agents are programmed to interact with each other to reduce the entropy of their knowledge of the information space, by increasing its comprehensiveness, which they do by attempting to exchange with other agents for increasingly abstract and powerful issues in place of lower-level issues and facts.
   a. the carrying capacity of all agents is limited, and is operationalized by limiting the number of facts or issues agents can hold.
   b. the information decay rate is the same for all agents, and operationalized by having agents randomly "forget" some element of their knowledge.
4. As simulations proceed, over the course of agent interactions, we are able to observe:
   a. the development of "culture" as shared knowledge, and
   b. the development of "social structure" as agent interaction patterns.

To show how interactions become increasingly structured over a simulation, we can plot the population of agents interacting over time through the information space. Such plots illustrate long-established theoretical claims and recent empirical findings. Namely, that agents organizing according to certain shared meaning, thus illustrating a duality of culture and structure, such as argued by Archer (2003); the use of probability judgements to negotiate shared understandings that order interaction, reflecting the pragmatist philosophy of Pierce, James, and Dewey, and the sociological symbolic interactionism of Mead and Blumer (Snow 2001); and finally, the recent approach in neuroscience (e.g., Friston et al. 2017), that finds empirical support for the view that uncertainty reduction is our motivation to think and interact with the world. This model illustrates those claims by producing evolving cognitive and sociological orders.

3.1 Operational Description of the Model

The model is set up in multi-agent systems framework where agents are able to share information with each other, with the following features:

- each agent can store information
- there is a maximum amount of information that an individual agent can store
- agents can interact and share information with each other
- sharing means that the information is copied from one agent to another
- related pieces of information can be combined to form a new piece of information
- combining information decreases the amount of space the agent needs to store the information (referred to as "compressing the information")
- an agent does not have access to any information other than what it currently stores

Information is conceptualized abstractly. Each piece of information is a discrete object that takes up exactly one unit of storage, and is copied from one agent to another through interaction. We also assign each piece of information a point value (i.e. a positive integer) to reflect the degree to which it can compress information. Meanwhile, to help visualize the diversity of information, two distinct information structures can be used simultaneously with varying degrees of overlap for any agent. To observe those differences and how interactions can end up integrating the structures, information from each structure is colored blue and red respectively.

Any information structure used by the model is variable in several ways. While each piece of information is a discrete object that takes up one unit of storage, the "capacity of agents" to hold information is variable. For example, if the agent's capacity is 6, then the agent can only store 6 pieces of information at any given time. In the Net Logo implementation we offer of this model, one can use the capacity slider to adjust the capacity the agents (every agent has the same capacity). To represent the degree to which agents can compress information, we use a tree data structure. A "tree" consists of nodes (circles) and edges (lines connecting the circles), and has one root node (at the top of the tree). For our purposes, each node is one piece of information. See the illustration below as an example.

Notice that each node (except those at the bottom-level) has exactly three children. We call this the branching factor of the tree. If you start at a node at the bottom level and follow the edges (i.e. the arrows) to the root node, it takes two moves to get there. We call this the height of the tree.

The number in the node is the quality point value of that information. When the type is set to be "height-based" (vs. the options of "random," or "file"), the point values increase when we move towards the root node (higher levels give higher points). During the model's setup phase, each agent is randomly assigned pieces of information (up to their capacity) from the bottom level.

As noted above, the NetLogo implementation of the model allows two trees to be used, colored red and blue. Together, these trees form the information space of any simulation, and the basis for compressing information through information sharing. We can now discuss how agents go about sharing information with each other. Each agent performs the following three-step process:

1. choose another agent, 2. select a topic, 3. share, learn, and forget.

In every turn, an agent needs to first choose another agent, so that she can interact with it. The agents are attracted to each other through two values: similarity and novelty. This means that for every other agent, the agent must calculate a similarity and a novelty value. Both of these values are real numbers between 0.0 and 1.0.

Let us consider the interaction between two agents, which we can refer to as Alice and Bob. Assume Alice is thinking about interacting with Bob. Similarity represents how similar Bob is to Alice from Alice's point of view. Alice does not know what information Bob is storing. So, we define similarity to be:

(score of information both Alice and Bob know) / (Alice's total score).

Notice that this value is relative to Alice's score. Bob's similarity score for Alice might be different because we would divide by Bob's total score instead. In particular, if Bob knows everything Alice knows, plus some additional information, then Alice's similarity score for Bob will be 1.0 (the highest possible). However, in this case, Bob's score for Alice will be less than 1.0.

Novelty represents how much new information Alice could gain from interacting with Bob. We define novelty to be:

(score of info Bob knows that Alice doesn't know) / (Bob's total score).

For example, if Alice and Bob have no information in common, then Bob's similarity would be 0.0 and novelty would be 1.0. On the other hand, if Bob knows everything Alice knows plus some additional information, then similarity would be 1.0, but novelty is bigger than 0.0.

Finally, we need to combine these values to assign Bob a sim-nov value. In our model, Alice prefers to interact with agents with higher sim-nov values. The model is set up to do this in two different ways: "min" and "linear".

In min mode, we define the sim-nov value for Bob to be whichever value is smaller (similarity or novelty):

$$\text{sim-nov} = \min(\text{similarity}, \text{novelty}).$$

In min mode, then, Alice prefers agents with similarity and novelty values that are both as high as possible. She is less attracted to agents with a high similarity value but a low novelty value (and vice-versa).

Alternatively, in linear mode, we can set an alpha value (a number between 0.0 and 1.0) that represents how much weight Alice puts on the similarity value. 0.0 means she only cares about novelty, 1.0 means she only cares about similarity, and 0.5 means she treats each value equally. The sim-nov value for Bob is defined to be a linear combination of similarity and novelty:

$$\text{sim-nov} = \text{alpha} * \text{similarity} + (1 - \text{alpha}) * \text{novelty}.$$

Once Alice has a sim-nov value for each agent, she puts them in order from highest to lowest sim-nov. Then she uses p − agent-choice-probability to determine the probability she takes the agent at the top of the list. If she does not get an agent that is high enough on the list, she repeats this process until she chooses an agent with which to interact. In the agent-network visualization, an arrow is drawn from Alice to all agents above the one chosen in the list (including her choice). This represents her preferences in the network. Notice that if agent-choice-probability = 1.0, then she will always choose an agent with the highest sim-nov score.

The model only allows agents to share information if they share the same parent node in the information space. Once Alice has chosen Bob, she may choose any information node that is the parent of an information node. The agent knows this choice to serve as the "topic" about which to exchange information. Topics that are not yet known to an agent are referred to as topics on the frontier, since they represent the nodes she can potentially get to via compression. As agents motivated to process information, they prefer topics on the frontier. In our example, the topic selected for Alice and Bob is chosen uniformly at random from the intersection of Alice's frontier topics and Bob's frontier topics. If they have no frontier topics in common, the agents look at the following two sets:

1. the intersection of Alice's topics with Bob's information space
2. the intersection of Bob's topics with Alice's information space.

Agents then choose a topic uniformly at random from one of these sets. If these sets are both empty, then no interaction can take place. Once Alice has chosen a topic, she must choose a piece of information to share with Bob. As explained in Step 2, she can choose any information that is directly below the topic in her information space. In addition to the information sharing process, information can change by agents randomly forgetting or finding pieces of information. The forgetfulness slider represents the chance that an agent will randomly forget a piece of information. A value of 0.0 means that agents will never forget and 1.0 means that they will randomly lose a piece of information after each iteration. The fact-find-chance slider represents the chance that an agent will randomly discover a piece of information at the bottom-level of the information space. The agent will only learn this information if they are not currently at capacity. This is the only way an agent can learn new information without sharing with other agents. A value of 0.0 means the agent will never randomly learn information.

In terms of the information space parameters of the proposed model, there are three types of trees that can be selected: "random," "height-based," and "file." In addition, the blue tree can be turned off by selecting "none."

1. random: a tree is generated with random quality values for each node. The values are integers between min and max inclusive (i.e. red-min and red-max for the red tree)
2. height-based: a tree is generated with quality levels based on how far each node is from the bottom-level. The bottom-level are quality 1, next-level up are quality 2, and so on
3. file: a tree is read-in from a file.

Shown below are some of the patterns the model has produced thus far, with the first pattern showing the control settings and accompanying graphs as well:

These snapshots of simulations show social structural patterns in both classes and cliques of agents. They also show how cliques vary in color, which could be considered "cultural" in amounting to differences in shared meaning, where they reflect unique aspects of the information structure, apart from their social structure. A question that can be taken up with this model, then, is under what conditions will cultural segregation take place, and how can it hurt, or help, a society's health, or, in terms of this theoretical perspective, it's capacity to process information.

3.2   Illustrating the Model with a Game

Following Gredler (2004), and Davison and Davison (2013), we see games as potentially useful ways of illustrating computational models in the classroom. We have designed a card game that is a sort of hybrid "Go Fish" and poker for students to play as a means of becoming familiar with the basic premises of this model. It has an interactive dimension, where mutual and novel information matter, and a betting component, where the social distribution of information matters as well.

The game best played with four to five players, and the rules are as follows: each player is dealt ten cards; five of the cards arc face-up, for all to see, representing potentially mutual information as all players, since they will be able to get the cards if they have similar cards of their own. The other five cards are dealt face down, only that player seeing them, and representing potentially novel information for the other players. The cards are arranged to display potential information "trees," placing their face-up cards vertically in front of them. Behind the face-up cards they place any face-down cards if they match by number (then vertically), or by suit (horizontally). Face-down cards with no match are put aside as "unavailable for trade," until the situation changes. At each turn a player can ask for a card from another player if they share a number among their face-up cards, and for a matching card by suit if two of their face-up cards share that suit. If they have no matches, they take a card from the remaining deck. They then bet to end their turn. As players accumulate cards they order them into poker books and place them on the table to be counted. The game is won by a player being able to put all of their cards into books, or when the deck is used up, and the winner has the most points in books. The winner takes the winnings that have been bet. Players can fold at any time during the game to save their money.

As the game unfolds, there are two fields of meaning that evolve. One field is the comparative success of players over the course of hands, where successful bidding in previous hands mean they have more resources for bluffing, or calling the bluffs of others. This is akin to the social network dimension of meaning that evolves in the course of interaction. How the good fortune of being dealt good hands leads to resource advantages over time can itself be a lesson for students in how unearned advantages in resources can be reinforced, leading to increased inequality.

The other field of information that evolves in the course of play is the comparative order of players' hands. The cards are meant to represent each agents' "facts," and hands of cards are variable in their "fitness," as the degree to which they are valuable in terms of the range of possible poker hands. As in many card games, such as poker and gin-rummy, players increase the order of their hands by creating sets of cards by rank or sequence. Over multiple hands, students learn the importance of mutual information in determining how some hands are better than others in their possibilities for order. And, in processing of building such order, students learn a more intimate lesson: the pleasure they experience in ordering and reducing the entropy of their hands reveals their own nature as information processing agents.

And part of the processing is social. It is true social interactions are motivated by the "original" way information is valued in our game as the fitness of hands. However, the unpredictable outcomes of those interactions in social structural resources that provide unique contexts to the interactions is the other way information is valued. The social distribution of cards, as perceived by the players, is reflected in their interactions (strategies for asking for cards, betting patterns, etc.). As anyone familiar with poker knows, players make inferences about the hands of other players by observing how they play. They then act on those own inferences through their own betting. As such, betting patterns reflect players' estimates of the social distribution of information, and those patterns manifest a social structure of classes and cliques of actors that emerge and evolve.

The result of our go-fish poker game is that students can get a picture of how information is ordered along two different dimensions. One is the dimension of meaning in an objective reality of the card game and its rules that determines a hand's "general fitness." The other is the social network of agents of information, the distribution of which determines a hand's "social fitness," one might say. Moreover, students are able to see how the fitness of the information they develop is dependent on social contexts. Many a poker player has bemoaned the waste of a good hand when it coincides with the poor hands of others, who then fold too early to bet. Students learn that when interactions depend on shared information, the social distribution of information is a factor independent of its "objective fitness."

4   Conclusions

This essay has proposed a simple agent-based model and a card game as means of teaching the complexity mindset to undergraduates. The model can be used to describe the usefulness of modeling, and to introduce students to computational sociology. In the context of teaching complexity, referencing computational science is a necessity, as the development of its ideas have depended on computational and modeling techniques. Those have led to research avenues that were impossible a few years ago, and forced researchers to reconsider basic premises of the reductionist paradigm on which traditional linear sciences—such as neoclassical economics and individual selection theory in evolutionary biology—are built. The growing paradigm of complexity science and computer modeling has offered a new way to explore the emergent properties so central to the social sciences, and in particular to a discipline such as computational sociology.

The pedagogical methodology we are proposing can be used in team-taught courses, or for more disciplinary-based modules of instruction. For example, the authors of this article come from different disciplines, and this project has required us to find common ground and more precisely defined terminologies. It is precisely because of the rich diaspora of disciplinary and intertwined bodies of knowledge that this computational sociology model benefits interdisciplinary work and shows how modeling is a valid methodology for understanding diverse social realities. Both the model and the game show students how aggregate structures emerge and evolve through social interaction.

To what extent is the model a metaphor of reality, and to what degree is it describing an intrinsic aspect that the model and reality share? One of the subtle points this paper makes is that, in teaching students about the usefulness of modeling, it has to introduced with simple, rather than complicated models. The simplicity lets students see the inner workings of the model, illuminating how interactions can bring about emergent behavior and structures, and allowing a deeper discussion of the fabric of social reality. Using a complicated model to have such a discussion is almost invariably mired in the ad hoc aspects of such a model, thus missing the essential point that reality —however complex—can be better understood as the interplay of simpler components.

We do believe that this model shares intrinsic aspects that reflect the way our social reality is constructed, and it is because of this belief that we chose a bold title for this paper. Obviously, whether or not social reality is actually created in this emergent fashion is debatable; it represents an academic issue that may not be settled in the literature of computational sociology for years to come. The authors have their own scholarly battles to wage with peers who may agree or disagree with this view in the computational and social sciences' literatures. The benefits to the students seeing the creation and value of interdisciplinary research in action, however, are undeniable. Participating in a discussion about the way in which models can help researchers better understand reality is an experience that no student should graduate without.

References

Archer, M.: Structure, Agency and the Internal Conversation. Cambridge University Press, Cambridge (2003)

Bourdieu, P.: Distinction: A Social Critique of the Judgement of Taste. Harvard University Press, Cambridge (1984)

Bourdieu, P.: The Logic of Practice. Polity, Cambridge (1990)

Bryant, L.R.: The Democracy of Objects. Open Humanities Press, London (2011)

Byrne, D., Callaghan, G.: Complexity Theory and the Social Sciences: The State of the Art. Routledge, Abingdon (2013)

Cederman, L.E.: Computational models of social forms: advancing generative process theory. Am. J. Sociol. **110**(4), 864–893 (2005)

Cilliers, P.: Boundaries, hierarchies and networks in complex systems. Int. J. Innov. Manag. **5**(2), 135–147 (2001)

Crutchfield, J.P.: The calculi of emergence. Phys. D **75**(1–3), 11–54 (1994)

Davison, H.A., Davison, A.: Games and Simulations in Action. Routledge, Abingdon (2013)

Demerath, L.: Explaining Culture: the Social Pursuit of Subjective Order. Lexington, Lanham (2012)

Demerath, L., Reid, J., Suarez, E.D.: A model of emergence featuring social mechanisms of information compression. Working Paper (2020)

Demeulenaere, P.: Causal regularities, action and explanation. In: Analytical Sociology and Social Mechanisms, pp. 173–197 (2011)

Elder-Vass, D.: The Causal Power of Social Structures: Emergence, Structure and Agency. Cambridge University Press, Cambridge (2010)

Friston, K., FitzGerald, T., Rigoli, F., Schwartenbeck, P., Pezzulo, G.: Active inference: a process theory. Neural Comput. **29**(1), 1–49 (2017)

Gelfert, A.: How to Do Science with Models: A Philosophical Primer. Springer, Cham (2016). https://doi.org/10.1007/978-3-319-27954-1

Gredler, M.E.: Games and simulations and their relationships to learning. Handb. Res. Educ. Commun. Technol. **2**, 571–581 (2004)

Giddens, A.: Politics, Sociology and Social Theory: Encounters with Classical and Contemporary Social Thought. Polity, Cambridge (1995)

Hedstrom, P.: Dissecting the Social: On the Principles of Analytical Sociology. Cambridge University Press, Cambridge (2005)

Holland, John H.: Hidden Order: How Adaptation Builds Complexity. Addison-Wesley, Reading (1995)

Johnson, S.: Emergence: The Connected Lives of Ants, Brains, Cities, p. 19. Scribner, New York (2001). ISBN 3411040742

Lane, J.E.: The emergence of social schemas and lossy conceptual information networks: how information transmission can lead to the apparent "emergence" of culture. In: Emergent Behavior in Complex Systems Engineering: A Modeling and Simulation Approach, pp. 321–348 (2018)

Luhmann, N., Baecker, D., Gilgen, P.: Introduction to Systems Theory. Polity, Cambridge (2013)

Macy, M.W., Willer, R.: From factors to actors: computational sociology and agent-based modeling. Ann. Rev. Sociol. **28**, 143–166 (2002)

Nicolescu, B.: Methodology of transdisciplinarity. World Futures J. New Paradigm Res. **70**(3–4), 186–199 (2014)

Pessa, E.: What is emergence? In: Minati, G., Pessa, E. (eds.) Emergence in Complex, Cognitive, Social, and Biological Systems, pp. 379–382. Springer, Boston (2002). https://doi.org/10.1007/978-1-4615-0753-6_31

Sawyer, K.: Conversation as mechanism: emergence in creative groups. In: Analytical Sociology and Social Mechanisms, p. 78 (2011)

Snow, D.: Collective identity and expressive forms. CSD Center for the Study of Democracy Organized Research Unit. University of California, Irvine (2001)

Soh, L.K., Tsatsoulis, C., Sevay, H.: A satisficing, negotiated, and learning coalition formation architecture. In: Lesser, V., Ortiz, C.L., Tambe, M. (eds.) Distributed Sensor Networks. Multiagent Systems, Artificial Societies, and Simulated Organizations (International Book Series), vol. 9, pp. 109–138. Springer, Boston (2003). https://doi.org/10.1007/978-1-4615-0363-7_7

Sperber, D.: A naturalistic ontology for mechanistic explanations in the social sciences. In: Analytical Sociology and Social Mechanisms, pp. 64–77 (2011)

Suarez, E.D., Castañón-Puga, M.: Distributed agency. Int. J. Agent Technol. Syst. (IJATS) **5**(1), 32–52 (2013)

Suarez, E.D., Demerath, L.: The Cyber Creation of Social Structures. Complexity Challenges in Cyber Physical Systems: Using Modeling and Simulation (M&S) to Support Intelligence, Adaptation and Autonomy (2019)

Tolk, A., Diallo, S.Y., Padilla, J.J.: Semiotics, entropy, and interoperability of simulation systems: mathematical foundations of M&S standardization. In: Proceedings of the Winter Simulation Conference, p. 243. Winter Simulation Conference, December 2012

Tolk, A., Wildman, W.J., Shults, F.L., Diallo, S.Y.: Human simulation as the Lingua Franca for computational social sciences and humanities: potential and pitfalls. J. Cogn. Cult. **18**(5), 462–482 (2018)

Tolk, A.: Limitations and Usefulness of Computer Simulations for Complex Adaptive Systems Research. In: Sokolowski, J., Durak, U., Mustafee, N., Tolk, A. (eds.) Summer of Simulation – 50 Years of Seminal Computing Research. Simulation Foundations, Methods and Applications, pp. 77–96. Springer, Cham (2019). https://doi.org/10.1007/978-3-030-17164-3_5

Bringing Harmony to Computational Science Pedagogy

Richard Roth[⊠][iD] and William Pierce[iD]

Hood College, Frederick, MD 21701, USA
{rothr,pierce}@hood.edu

Abstract. Inherent in a musical composition are properties that are analogous to properties and laws that exist in mathematics, physics, and psychology. It follows then that computation using musical models can provide results that are analogous to results provided by mathematical, physical, and psychological models. The audible output of a carefully constructed musical model can demonstrate properties and relationships in a way that is immediately perceptible. For this reason, the study of musical computation is a worthwhile pursuit as a pedagogical tool for computational science. Proposed in this paper is a curriculum for implementing the study of musical computation within a larger computer science or computational science program at a college or university. A benefit of such a curriculum is that it provides a way to integrate artistic endeavors into a STREAM program, while maintaining the mathematical foundations of STEM. Furthermore, the study of musical computation aligns well with the arts-related components of Human-Centered Computing. The curriculum is built on the following two hypotheses: The first hypothesis is that the cognitive, creative, and structural processes involved in both musical composition and computer programming, are similar enough that skilled computer scientists, with or without musical backgrounds, can learn to use programming languages to compose interesting, expressive, and sophisticated musical works. The second hypothesis is that the links between music, mathematics, and several branches of science are strong enough that skilled computational scientists can create musical models that are able to be designed using vocabularies of mathematics and science. While the curriculum defined in this paper focuses on musical computation, the design principles behind it may be applied to other disciplines.

Keywords: Computational science · Pedagogy · Curriculum design · System modeling · Music composition · Algorithms · Sound design · Digital signal processing · Psychoacoustics · Embedded systems

1 Introduction

Outlined in this paper is a curriculum for implementing a computational music curriculum at a college or university. This curriculum is designed to be a sub-curriculum for a larger computer science or computational science program.

© Springer Nature Switzerland AG 2020
V. V. Krzhizhanovskaya et al. (Eds.): ICCS 2020, LNCS 12143, pp. 661–673, 2020.
https://doi.org/10.1007/978-3-030-50436-6_49

In this curriculum, music is studied using the vocabularies of mathematics and science. The purpose of this curriculum is to use musical models as pedagogical tools for computational science.

This paper begins with a discussion of music as it was studied in ancient Greece, as the study of numbers evolving over time. This discussion is included to provide a historical precedent for using music as a pedagogical tool for mathematics and the sciences. The paper then discusses the logic and usefulness of using shared vocabularies to study the relationships between diverse disciplines such as music, mathematics, and the sciences. Following this discussion, several examples are given to demonstrate how musical models can be effectively used in teaching multiple areas of modern mathematics and science.

The computational music curriculum is then discussed from a design perspective. This discussion is followed by a case study that provides details on how the computational music curriculum is implemented at Hood College. The paper concludes with observations and thoughts regarding the resources required to implement this curriculum, how the curriculum might be put in place at other institutions, and how computational music fits in with mathematics-based interdisciplinary trends such as Science, Technology, Engineering and Mathematics (STEM), as well as Human-Centered Computing (HCC).

2   The Laws of Musical Motion

Long before gravity and inertia were defined, the scientists, mathematicians, and philosophers of ancient Greece used the vocabulary of music to discuss the balance and beauty observed in the motion of the sun, moon, and stars. Pythagoras used the term *celestial harmony* to describe this motion. For Pythagoras, Plato, and others, an understanding of the universe was achieved through the study of the quadrivium: mathematics, geometry, music, and astronomy. The role of music in the quadrivium was the study of the evolution of numbers over time [2].

The Greek notion of celestial harmony, survived well into the seventeenth century, as evidenced by the publication of Johannes Kepler's "Harmonice Mundi" in 1619. In this text, Kepler uses his research and discoveries in astronomy, to support the theory that motions of musical voices, and the motions of the heavens, both abide by the same laws [5].

Later that century, composer, organist, and music theorist Johann Joseph Fux, defined in his text "Gradus ad Parnassum," the mathematical foundations of musical elements, and the laws governing musical counterpoint; which is the study of musical voices in motion [1]. Fux discussed musical counterpoint in a way that recalled the concepts of Harmonice Mundi and celestial motion by describing musical lines as *voices* that move through musical space, much like the way objects move through physical space.

Studying music as both an art and a science, has allowed intellectuals throughout history, with a framework, or pedagogical tool set, for studying and learning about elements and forces that may have been at their time, out of reach.

3   Shared Vocabularies

Shared vocabularies allow the vocabulary from one field of study, to describe the properties of another field of study. This is what allowed Pythagoras and Plato to study astronomy and physics using the vocabulary of music theory. It also allows us today to discuss and understand much of music theory using the vocabularies of science.

The disciplines of computer programming and music composition demonstrate how instructors can use shared vocabularies as teaching tools. Melodies for example, are described by the Harvard Dictionary of Music as "successions of pitch-plus-duration values." This concept is easily described to a student of computer science as an array of pitch-value objects.

To continue this example, instructors might describe musical counterpoint as the execution of two or more well-constructed arrays of pitch-value objects that are cycled through on separate, but concurrent threads.

The object of drawing parallels between programming and music composition in this way is not to circumnavigate the study of traditional music theory. Instead, the object of drawing these parallels is to use musical concepts such as melodies and counterpoint to help computer science students gain a deeper understanding of computer programming concepts such as arrays, loops, structures and classes, and concurrency and multi-threading.

4   Pedagogical Areas

This section provides concrete examples of how musical components and constructs are used a pedagogical tools to approach concepts of computation, algorithm design, artificial intelligence, and kinetics.

4.1   Algorithm Design

In most algorithmic compositions, the output of the program is different each time, but it is still recognizable from hearing to hearing. In the following code sample, complex musical phrases are created algorithmically from a single base pitch and an array of pitch changing values. The program derives rhythmic components by dividing a predetermined time interval (one second for example) into a variable number of parts. This code sample was written in the ChucK music programming language [6].

```
// Clarinet Solo
36 => int rootPitch;
[0,2,4,5,6,7,9,11] @=> int myPitches[];

void writeMusic(int beats, int pitches[])
// PRE: The number of measures in the composition is predefined
// FUNCTION: Determines notes and pitches for the current measure
```

```
// POST: The computer has composed and played one measure
{
    Math.random2(1,4) => int octaves; // Determine octave
    Math.random2f(-1.0,1.0) => s_pan.pan; // sound placement

    for( 0 => int j; j < beats; j++){ // generate music
        Std.mtof( (rootPitch + (12 * octaves))
        + pitches[j]) => s.freq;
        MAX / (ROUNDTRIP / beats) => float stepsize;
        // MAX 0.9, ROUNDTRIP 200

        for ( 0 => float i; i < MAX; stepsize +=> i ){
            i => s.gain;
            tempo => now;
        }
        for ( s.gain() => float i; i > 0; stepsize -=> i ){
            i => s.gain;
            tempo => now;
        }}}
```

4.2 Artificial Creativity

In this example, two arrays are used to store separate parts of a repeated drum
pattern. The main function (not shown) uses concurrent threads, thus creating
a complete pattern at run time. The program incorporates a simple algorithm
that makes decisions on how to keep the pattern interesting over time.

```
// Drum Pattern
0.0625 => float VARIETY; // Variety Factor

// SETUP PATTERN:
// Measures are divided into four groups of 16th notes.
// 1 and 2 are drum beats, 0 is silence
[1,0,0,0, 2,0,0,2, 1,0,1,0, 2,0,0,0] @=> int kickSnareBeats[];
[2,0,1,0, 2,0,1,0, 2,0,1,0, 2,0,3,0] @=> int highHatBeats[];

// ADD VARIETY: Determine the likelihood array values will change
[0,1,0,2, 0,2,0,3, 0,3,0,4, 0,4,0,5] @=> int varietyLevel

for(0 => int b_ctr; b_ctr < bNum; b_ctr++){ //Play 1 measure
beats[b_ctr] => theBeat; // Get either a drum beat or silence
vBeat(b_ctr, theBeat) => theBeat; // Maybe change the beat setting
if (theBeat == 1)
0 => k1.pos;
}
```

4.3 Geometry in Motion

Figures one through four show two excerpts from "Gradus ad Parnassum," Johann Fux's 1725 pedagogical treatise on musical counterpoint. In these figures, pitches are shown first as notes in music notation, and second as frequency in Hz on a two-dimensional graph. It may be tempting to look at this set of pitches mathematically and conclude that there are two entities; one is a sound whose fundamental vibration frequency is x, and another sound whose fundamental vibration frequency is y. With this view one can observe that there is a distance in frequency space of y − x. This is shown in Fig. 1.

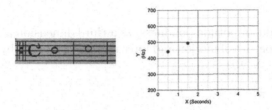

Fig. 1. The image on the left is an excerpt from Fux's Gradus ad Parnassum showing two pitches on a musical staff. The image on the right shows the fundamental frequency in Hz of the two pitches on a two-dimensional graph.

If however, an alternate view is taken, and one considers that a single sound source, or voice in musical terms, started at point x, and then moved over time, and in frequency space, to point y, then we can observe a direct parallel to kinetics; the study of geometry in motion. This is shown in Fig. 2.

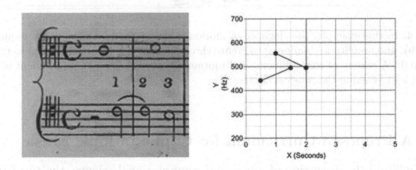

Fig. 2. The same two pitches from figure one, shown in Hz on a two-dimensional graph.

If we now consider a sound object consisting of several voices (v1, v2,...,v6) we can map the movement of an object in two dimensional space, to music as shown in Fig. 3.

Fig. 3. In this example, each leg of the tripod on the left, is represented on the right by two voices in a musical mapping. A variation on Bresenham's circle drawing algorithm is used to rotate the tripod and to generate musical information.

Lastly if we add amplitude (loudness) as a third dimension as shown in Fig. 4, we can study properties of motion and force in a way that may by seen and heard.

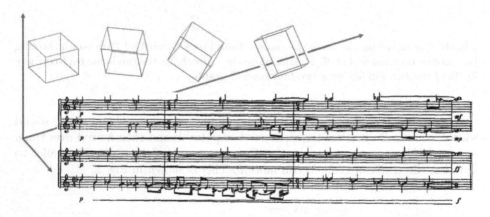

Fig. 4. In this example, the elegance of motion is translated to music using frequency (pitch), and amplitude (loudness) to create three dimensions for sound objects to travel through. A variety of three-dimensional motion algorithms drive the movement of the cube, and generate the musical content [7].

5   A Proposed Curriculum for Computational Music

To facilitate the discussion of music and computational science, the term *computational music* is introduced. This term is used in this paper to encompass the following:

- Discussions of musical properties such as harmony, melodic movement, consonance, and dissonance, in purely scientific terms.
- Creation of musical models as approachable pedagogical tools for creating models related to science and mathematics.

5.1   Modifying Existing Programs

A problem many colleges and universities face, is the challenge of implementing new elements such as computational music courses into firmly established math and science programs.

One possible way to do this might be to design a single elective course in Computational Music where students complete projects that incorporate mathematical elements such as numeric patterns or evolving functions. Other projects might explore the creation of computer models that emulate musical instruments or natural sounds. This solution has been implemented successfully by many colleges and universities. A problem with this approach, however, is the possible novelty or survey course perception that has no path forward for interested students.

Another possible solution is to add topics such as music programming and music computation to graduate and advanced undergraduate courses in topics such as artificial intelligence or robotics. This approach however, provides no preparatory support in music programming, nor the scientific and mathematical properties of music. Students might therefore be limited in what they could accomplish in a semester, and the time spent teaching the musical material, might deter from the main objectives of the course.

The solution proposed in this paper is to offer a small but broad curriculum of computational music courses that allow students from different disciplines, and different stages of their education, to find courses that align with their objectives and interests.

5.2   Design Principles

This proposed curriculum is built on three design principles. Principle one: any course in the curriculum can serve as a possible entry point. Principle two: students may complete courses in an order that suits their interests and goals provided that prerequisites are met. Principle three: allowable prerequisites for advanced courses include other courses in the curriculum, and related courses from other disciplines. This flexible approach makes the curriculum available and accessible to the largest possible number of students.

Principle one allows for students to enter the computational music curriculum in a way that aligns with their academic year, or their major. First-year or second-year undergraduate students for example, may choose a survey course with no prerequisite as their entry point. Survey courses for this curriculum might include an introductory course in computer music, or a course on the scientific properties of music.

Third and fourth-year undergraduates, as well as graduate students, may choose a more advanced course that does have prerequisites for their entry point. More advanced courses in this curriculum might include a course in digital signal processing, or a course in music programming.

Principle two is concerned with the course or courses a student may take after completing their entry-point course. Students that started with a survey course now have the option of taking another survey course, or moving to a more advanced course. Students that started with a more advanced course may take another advanced course, a survey course, or move on to a specialty course.

Principle three allows students to work their way up to advanced or specialty courses within the curriculum, or to take these courses using prerequisites earned from other disciplines. Specialty courses in a computational music curriculum might have titles such as Music and Sound for x, where x might be an area in which the college or university already specializes, such as Robotics, Cognitive Psychology, or Algorithm Design.

An example of an implementation of these three design principles in a computational music curriculum follows in the next section.

6 A Case Study: Hood College, Frederick Maryland

6.1 Hypotheses

Computational music began in the Computer Science Department at Hood College, a small traditionally liberal arts college in Frederick Maryland, with a single summer-session course entitled *Musical Computing*. The starting hypothesis behind this course was that the cognitive, creative, and structural processes involved in both musical composition, and computer programming, are similar enough that skilled computer scientists, with or without musical backgrounds, can learn to use programming languages to compose interesting, expressive, and sophisticated musical works. The overall structure of the course is shown in Fig. 5.

This single course offering expanded into a six course curriculum designed to use musical models to assist in the pedagogy of computational science, and computer science. As this curriculum evolved, the original hypothesis expanded to the following: The links between music, mathematics, and several branches of science are strong enough that skilled computational scientists can create musical models that are able to be designed using vocabularies of mathematics and science.

Table 1 lists the course titles and prerequisites for the Hood College computational music curriculum. The following sections contain descriptions of each course. The descriptions include course overviews, special topics, and explanations of how the courses support computational science pedagogy.

6.2 The Science of Music

This is a course that approaches all aspects of music using the vocabularies of science, mathematics, and computational models. Specific topics include: the physics of sound, auditory perception, derivation of intervals and scales, consonance and dissonance, and structures used in music composition. The value of this course in computational science pedagogy lies in its emphasis on physical properties. Furthermore, several lectures and class projects focus specifically on computational model creation. There are no prerequisites for this course.

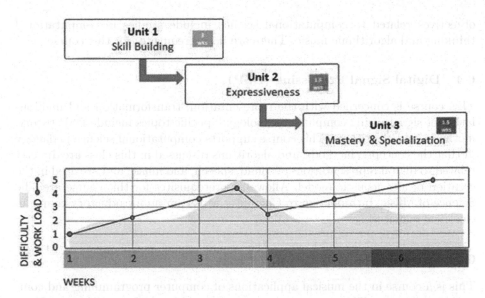

Fig. 5. A basic course structure in the music computation curriculum.

Table 1. Hood College curriculum for computational music.

Course Title	Prerequisite
The Science of Music	None
Introduction to Computer Music	None
Digital Signal Processing	The Science of Music, Intro. to Computer Music, or an introductory course in electrical engineering
Musical Computing	Intro. to Computer Music, or a previous course in a modern object-oriented programming language
Algorithms and Music Composition	Musical Computing, or a Previous course in Data Structures and Algorithms
Music and Sound in Embedded Systems and Robotics	Musical Computing, Algorithms and Music, Composition, or a previous course in embedded systems programming, or robotics

6.3   Introduction to Computer Music

This course is an introduction to the use of computers for creating, recording and editing musical information. Specific topics include: music history leading to the advent of computer music, physical properties of sound, human perception of sound, and current trends in the musical applications of computers. Pedagogical objectives related to computational science include studies in computational thinking and algorithmic music. There are no prerequisites for this course.

6.4   Digital Signal Processing (DSP)

This course is concerned with the representation, transformation and manipulation of signals using computer technology. Specific topics include DSP theory, methods, and algorithms. This course supports computational science pedagogy in that the concepts, methods, and algorithms discussed in this class are directly related to digital representations of sound, speech, and music; and are used in the creation of any acoustic model. Allowable prerequisites for this course are The Science of Music, Introduction to Computer Music, or an introductory course in electrical engineering.

6.5   Musical Computing

This is a course in the musical applications of computer programming and computational science. This course in many ways resembles the Introduction to Computer Music course described earlier. Musical Computing however, assumes a higher level of skill and experience in programming and mathematics. In this course, students use the structural elements of computer programming such as arrays, loops, classes, objects, and multi-threading, to compose music and design sounds. Students enhance their coding skills by writing programs that produce sophisticated musical compositions. This course supports computational science pedagogy through programming projects that model and support human musical composition processes. Allowable prerequisites for this course are Introduction to Computer Music, or a previous course in a modern object-oriented programming language.

6.6   Algorithms and Music Composition

This is a course in composing music using algorithms and computational models. Specific topics include: algorithmically processing data for music creation, creating algorithms that write music based on minimal input, and live algorithmic music performance. Computational science pedagogy is supported in this class through the study of creative algorithm design. Allowable prerequisites for this course are Musical Computing, or a previous course in data structures and algorithms.

6.7   Music and Sound for Embedded Systems

This is a course in the musical and sonic applications of embedded systems programming. In this course, students write programs that run on stand-alone microcomputers such as those found on Arduino and Raspberry Pi development boards. Specific topics include: efficient coding, multi-threading, use of actuators and sensors to generate and respond to sound, processing of sound information, and robotic music performance. This course supports computational science pedagogy through lectures and student projects relating to the processing of sound information. Allowable prerequisites for this course are Musical Computing, Algorithms and Music Composition, or a previous course in embedded systems programming or robotics.

7 Implementing a Computational Music Curriculum

7.1 Required Resources and Expenses

Realizing the pedagogical benefits of a computational music curriculum can start with little or no resources beyond what a typical college or university already has. These resources include computers, access to freely available software tools, and a standard classroom audio system.

As programs mature, institutions may focus on, or specialize in a specific area. At this point, additional expenses may be incurred. For example, consider a school wishing to create a psychoacoustic modeling lab. This type of modeling and experimentation uses meticulously created sound samples. Researchers accomplish this either by recording natural sounds, or by synthesizing them on a computer. The sound creation process, and subsequent playback both require that a sound's full frequency spectrum is preserved. This normally requires an acoustically treated environment that is equipped with sensitive and high quality microphones, amplifiers and speakers that can preserve a sound's full frequency spectrum. These expenses can be preventative in early stages, but may become possible as a program grows.

7.2 Textbooks

There are excellent texts on this subject, including Elaine Chew's text, "Mathematical and Computational Modeling of Tonality [3]." There is also an abundance of scholarly research in journals and conference proceedings that professors can incorporate into these courses. Nominal expenses may be incurred in subscribing to the societies and organizations that provide this information.

7.3 Faculty

Teaching this curriculum requires faculty who can discuss sound and music using the following vocabularies: traditional music theory, particularly the principles of harmony, and music composition; the sciences, particularly mathematics, engineering, and physics; as well as computer science and auditory perception.

In the Hood College curriculum, the courses are taught by professors who have fluency in music and one or more of the scientific vocabularies listed. Cross-disciplinary approaches however, are also used at Hood College and might be an effective alternative for many institutions.

8 Conclusion

The rapid growth of the computational music curriculum in the Hood College case study are promising indicators that a curriculum such as the one outlined in this paper, may resonate with prospective students who view computation, science and mathematics as creative endeavors. Due to the success of the Hood College case study, the authors conclude that computational music, if implemented as a sub-curriculum in a larger program, is an effective and motivational tool for computational science pedagogy.

There are several trends in higher education that support this conclusion, including the growing number of Ph.D programs in Human-Centered Computing (HCC) such as those at Clemson University and Georgia Tech. Evidence that computational music can play a pivotal role in HCC, is provided by Nicu Sebe in his chapter on Human-Centered Computing in the "Handbook of Ambient Intelligence and Smart Environments" [8]. In this chapter, Professor Sebe states "Perhaps one of the most exciting application areas of HCC is art." To support this statement, Sebe points to an example where computing is used to translate human gestures to music.

Other trends where computational music can play a key role are STEM and STREAM. STREAM is an extension of STEM that includes A for the arts, and R which, depending on the context, can stand for reading and writing, or religion. The authors of this paper have concluded from the case study presented in this paper, that the study of computational music not only aligns with the objectives of STREAM initiatives, but maintains the scientific and mathematical foundations on which STEM is built.

In closing, perhaps the primary reasons why computational music brings harmony to computational science pedagogy, are the effectiveness with which one can define musical properties using the vocabularies of science and mathematics; and the fact that music has been a part of scientific and mathematical analysis since ancient times.

Acknowledgement. We thank the Hood College Graduate School and our corresponding departments for their support. We are indebted to our students who embraced this curriculum and worked with us along the way while we rethought, re-planned, and refined it. Sound files that accompany this paper, are available from the Hood College Computer Science Department.

References

1. Apel, W.: Harvard Dictionary of Music. Harvard University Press, Cambridge (1950)
2. Burkholder, J., Peter, G., Jay, D., Palisca, C.V.: A History of Western Music, 9th edn. W.W. Norton & Company, New York (1999)
3. Chew, E.: Mathematical and Computational Modeling of Tonality. ISORMS, vol. 204. Springer, Boston, MA (2014). https://doi.org/10.1007/978-1-4614-9475-1
4. Chew, E.: Slicing it all ways: mathematical models for tonal induction, approximation, and segmentation using the spiral array. INFORMS J. Comput. **18**(3), 305–320 (2006)

5. Gingras, B.: Johannes Kepler's Harmonice Mundi: a "scientific" version of the harmony of the spheres. J. Roy. Astron. Soc. Can. **97**(5), 228 (2003)
6. Kapur, A., Cook, P., Salazar, S., Wang, G.: Programming for Musicians and Digital Artists: Creating Music with ChucK. Manning Publications, New York (2014)
7. Roth, R.: Music and animation toolkit: modules for multimedia composition. Comput. Math. Appl. **32**(1), 137–144 (1996)
8. Sebe, N.: Human-centered computing. In: Nakashima, H., Aghajan, H., Augusto, J.C. (eds.) Handbook of Ambient Intelligence and Smart Environments. Springer, Boston (2010). https://doi.org/10.1007/978-0-387-93808-0

UNcErtainty QUantIficatiOn for ComputationAl MdeLs

Intrusive Polynomial Chaos for CFD Using OpenFOAM

Jigar Parekh[✉] and Roel Verstappen

Bernoulli Institute for Mathematics, Computer Science and Artificial Intelligence,
University of Groningen, Groningen, The Netherlands
{j.parekh,r.w.c.p.verstappen}@rug.nl

Abstract. We present the formulation and implementation of a stochastic Computational Fluid Dynamics (CFD) solver based on the widely used finite volume library - OpenFOAM. The solver employs Generalized Polynomial Chaos (gPC) expansion to (a) quantify the uncertainties associated with the fluid flow simulations, and (b) study the non-linear propagation of these uncertainties. The aim is to accurately estimate the uncertainty in the result of a CFD simulation at a lower computational cost than the standard Monte Carlo (MC) method. The gPC approach is based on the spectral decomposition of the random variables in terms of basis polynomials containing randomness and the unknown deterministic expansion coefficients. As opposed to the mostly used non-intrusive approach, in this work, we use the intrusive variant of the gPC method in the sense that the deterministic equations are modified to directly solve for the (coupled) expansion coefficients. To this end, we have tested the intrusive gPC implementation for both the laminar and the turbulent flow problems in CFD. The results are in accordance with the analytical and the non-intrusive approaches. The stochastic solver thus developed, can serve as an alternative to perform uncertainty quantification, especially when the non-intrusive methods are significantly expensive, which is mostly true for a lot of stochastic CFD problems.

Keywords: Uncertainty quantification · Uncertainty propagation · Intrusive Polynomial Chaos · Computational Fluid Dynamics · Turbulence

1 Introduction

In simulating a physical system with a model, uncertainties may arise from various sources [15], namely, initial and boundary conditions, material properties, model parameters, etc. These uncertainties may involve significant randomness or may only be approximately known. In order to enhance the predictive reliability, it is therefore important to quantify the associated uncertainties and study its non-linear propagation especially in CFD simulations.

In order to reflect the uncertainty in the numerical solution, we need efficient Uncertainty Quantification (UQ) methods. Broadly there exists two classes of UQ methods, the intrusive method, where the original deterministic model is

© Springer Nature Switzerland AG 2020
V. V. Krzhizhanovskaya et al. (Eds.): ICCS 2020, LNCS 12143, pp. 677–691, 2020.
https://doi.org/10.1007/978-3-030-50436-6_50

replaced by its stochastic representation, and the non-intrusive method, where the original model itself is used without any modifications [11,14]. Monte Carlo (MC) sampling is one of the simplest non-intrusive approaches. However, due to its requirement of a large number of samples, MC method is computationally expensive for application in CFD. As an alternative, we can use Generalized Polynomial Chaos (gPC) representations which has been proven to be much cheaper than MC [6,15]. This approach is based on the spectral decomposition of the random variables in terms of basis polynomials containing randomness and the unknown deterministic expansion coefficients. In this paper, we focus mainly on the Intrusive Polynomial Chaos (IPC) method, where a reformulation of the original model is performed resulting in governing equations for the expansion coefficients of the model output [20].

As the model code, we use OpenFOAM [2], which is a C++ toolbox to develop numerical solvers, and pre-/post-processing utilities to solve continuum mechanics problems including CFD. OpenFOAM (a) is a highly templated code, enabling the users to customize the default libraries as needed for their applications, and, (b) gives access to most of the tensor operations (divergence, gradient, laplacian etc.) directly at the top-level code. This avails enough flexibility to implement the IPC framework for uncertainty quantification in CFD. To obtain the inner products of polynomials we use a python library called chaospy [3], as a pre-processing step to the actual stochastic simulation.

First, the idea of generalized polynomial chaos is presented with a focus on the intrusive variant. A generic differential equation is used to explain the steps involved in IPC, leading to a simple expression for the mean and variance as a function of the expansion coefficients. Next, we present the set of deterministic governing equations followed by its stochastic formulation using IPC. In particular, a Large Eddy Simulation (LES) method is used to model turbulence, which includes an uncertain model parameter. Thereafter, we discuss the algorithm and implementation steps required for the new stochastic solver in OpenFOAM. The stochastic version of the Navier-Stokes equations has a similar structure to the original system. This allows reusing the existing deterministic solver with minimal changes necessary.

The stochastic solver developed so far is tested for various standard CFD problems. Here we present two cases, the plane Poiseuille flow with uncertain kinematic viscosity, and the turbulent channel flow with uncertain LES model parameter. The results are found to be in accordance with the non-intrusive gPC method.

2   Generalized Polynomial Chaos

The Generalized Polynomial Chaos approach is based on the spectral decomposition of the random variable(s) f, in terms of basis polynomials containing randomness ψ_i (known a priori) and the unknown deterministic expansion coefficients f_i, as $f(x,q) = \sum_{i=0}^{\infty} f_i(x)\psi_i(q)$. There are two methods to determine the expansion coefficients, namely, the Intrusive Polynomial Chaos (IPC) and the

Non-intrusive Polynomial Chaos (NIPC). In IPC, a reformulation of the original model is performed resulting in governing equations for the PC mode strengths of the model output, while in NIPC, these coefficients are approximated using quadrature for numerical evaluation of the projection integrals.

The level of accuracy of these methods can be associated with the degree of gPC. To attain the same level of accuracy, particularly for a higher dimensional random space, IPC requires the solution of a much fewer number of equations that needed for NIPC. Moreover, for such a random space, the aliasing error resulting from the approximation of the exact gPC expansion in the NIPC method can become significant. This suggests that, for a multi-dimensional problem, the IPC method can deliver more accurate solutions at a much lower computational cost than the NIPC method [18].

Since the current work is based on IPC, we would introduce here the important features of the intrusive variant and we refer to the literature [5,16,19] for more details about NIPC and gPC in general.

Intrusive Polynomial Chaos. In order to demonstrate the application of IPC, we first consider a general stochastic differential equation

$$\mathcal{L}(\boldsymbol{x}, t, \omega; \boldsymbol{v}(\boldsymbol{x}, t, \omega)) = S(\boldsymbol{x}, t, \omega), \tag{1}$$

where $\mathcal{L}$ is usually a nonlinear differential operator consisting of space and/or time derivatives, $\boldsymbol{v}(\boldsymbol{x}, t, \omega)$ is the solution and $S(\boldsymbol{x}, t, \omega)$ is the source term. The random event ω represents the uncertainty in the system, introduced via uncertain parameters, the operator, source term, initial/boundary conditions, etc. The complete probability space is given by $(\Omega, \mathcal{A}, \mathcal{P})$, where Ω is the sample space such that $\omega \in \Omega$, $\mathcal{A} \subset 2^{\Omega}$ is the σ-algebra on Ω and $\mathcal{P} : \mathcal{A} \mapsto [0, 1]$ is the probability measure on $(\Omega, \mathcal{A})$.

We now employ the Galerkin polynomial chaos method, which is an IPC method for the propagation of uncertainty [16]. It provides the spectral representation of the stochastic solution and results into higher order approximations of the mean and variance. Galerkin polynomial chaos method is a non-statistical method where the uncertain parameter(s) and the solution become random variables. These random variables are approximated using the polynomial chaos (polynomial of random variables) as follow [5]

$$\boldsymbol{v}(\boldsymbol{x}, t, \omega) \approx \sum_{i=0}^{P} \boldsymbol{v}_i(\boldsymbol{x}, t) \psi_i(\boldsymbol{\xi}(\omega)). \tag{2}$$

It is worth noting that the expansion (2) is indeed the decomposition of a random variable into a deterministic component, the expansion coefficients $\boldsymbol{v}_i(\boldsymbol{x}, t)$ and a stochastic component, the random basis functions (polynomial chaoses) $\psi_i(\boldsymbol{\xi}(\omega))$. Here, $\boldsymbol{\xi}(\omega)$ is the vector of d independent random variables $\{\xi_1, ..., \xi_d\}$, corresponding to d uncertain parameters. Based on the dimension of $\boldsymbol{\xi}$ (which here is d) and the highest order n of the polynomials $\{\psi_i\}$, the infinite summation has been truncated to $P + 1 = (d + n)!/(d!\, n!)$ terms.

An important property of the basis $\{\psi_i\}$ is their orthogonality with respect to the probability density function (PDF) of the uncertain parameters, $\langle \psi_i \psi_j \rangle = \langle \psi_i^2 \rangle \delta_{ij}$. Here, δ_{ij} is the Kronecker delta and $\langle \cdot, \cdot \rangle$ denotes the inner product in the Hilbert space of the variables $\boldsymbol{\xi}$, $\langle f(\boldsymbol{\xi})g(\boldsymbol{\xi}) \rangle = \int f(\boldsymbol{\xi})g(\boldsymbol{\xi})w(\boldsymbol{\xi})d\boldsymbol{\xi}$. The weighting function $w(\boldsymbol{\xi})$ is the probability density function of the uncertain parameters. Such polynomials already exist for some standard distributions which can be found in the Askey scheme [19], for example, a Normal distribution leads to Hermite-chaos, while Legendre-chaos corresponds to a Uniform distribution. For other commonly used distributions or any arbitrary distribution, one can for example use Gram-Schmidt algorithm [17] to construct the orthogonal polynomials.

Substituting (2) in the general stochastic differential Eq. (1), we obtain

$$\mathcal{L}\left(\boldsymbol{x}, t, \omega; \sum_{i=0}^{P} \boldsymbol{v}_i \psi_i \right) \approx S. \tag{3}$$

In order to ensure that the truncation error is orthogonal to the functional space spanned by the basis polynomials $\{\psi_i\}$, a Galerkin projection of the above equation is performed onto each polynomial $\{\psi_k\}$,

$$\left\langle \mathcal{L}\left(\boldsymbol{x}, t, \omega; \sum_{i=0}^{P} \boldsymbol{v}_i \psi_i \right), \psi_k \right\rangle = \langle S, \psi_k \rangle, \qquad k = 0, 1, ..., P. \tag{4}$$

After using the orthogonality property of the polynomials, we obtain a set of $P + 1$ deterministic coupled equations for all the random modes of the solution $\{\boldsymbol{v}_0, \boldsymbol{v}_1, ..., \boldsymbol{v}_k\}$. Following the definition, the mean and the variance of the solution are given by

$$\mathbf{E}[\boldsymbol{v}] = \mu_v = \boldsymbol{v}_0(\boldsymbol{x}, t), \quad \mathbf{V}[\boldsymbol{v}] = \sigma_v^2 = \sum_{i=1}^{P} \boldsymbol{v}_i(\boldsymbol{x}, t)^2 \langle \psi_i^2 \rangle. \tag{5}$$

As the coefficients $\boldsymbol{v}_i(\boldsymbol{x}, t)$ are known, the probability distribution of the solution can be obtained.

3   Governing Equations

We first discuss the governing equations in the deterministic setting. The Navier-Stokes equations for an incompressible flow is given by

$$\frac{\partial \boldsymbol{u}}{\partial t} + (\boldsymbol{u} \cdot \boldsymbol{\nabla})\boldsymbol{u} = -\boldsymbol{\nabla} p + \boldsymbol{\nabla} \cdot (\nu \boldsymbol{\nabla} \boldsymbol{u}), \quad \boldsymbol{\nabla} \cdot \boldsymbol{u} = 0, \tag{6}$$

where $\boldsymbol{u}$ is the velocity, p is the pressure and ν is the kinematic viscosity.

In Large Eddy Simulation, the reduction in the range of scales in a simulation is achieved by applying a spatial filter to the Navier-Stokes Eqs. [12]. This results into

$$\frac{\partial \overline{\boldsymbol{u}}}{\partial t} + (\overline{\boldsymbol{u}} \cdot \boldsymbol{\nabla})\overline{\boldsymbol{u}} = -\boldsymbol{\nabla} \overline{p} + \boldsymbol{\nabla} \cdot (\nu \boldsymbol{\nabla} \overline{\boldsymbol{u}}) - \boldsymbol{\nabla} \cdot \boldsymbol{\tau}, \quad \boldsymbol{\nabla} \cdot \overline{\boldsymbol{u}} = 0 \tag{7}$$

where $\overline{u}$ is the filtered velocity, $\overline{p}$ is the filtered pressure and $\tau = \overline{uu} - \overline{u}\,\overline{u}$ is the so-called subgrid-scale (SGS) stress tensor. The subgrid-scale stress tensor represents the effect of the small (unresolved) scales on the resolved scales, and has to be modeled in order to close the filtered Navier-Stokes equations.

A popular class of SGS models is the eddy-viscosity models. In order to take account of the dissipation through the unresolved scales, the eddy-viscosity models, locally increases the viscosity by appending the molecular viscosity with the eddy viscosity. Mathematically, these models specify the anisotropic part of the subgrid-scale tensor as

$$\tau - \frac{1}{3}\operatorname{tr}(\tau) = -2\nu_t\overline{S}, \tag{8}$$

where ν_t is the eddy-viscosity and $\overline{S} = (\nabla\overline{u} + (\nabla\overline{u})^T)/2$ is the resolved strain tensor. Substituting into the filtered momentum Eq. (6), we obtain

$$\frac{\partial\overline{u}}{\partial t} + (\overline{u} \cdot \nabla)\overline{u} = -\nabla\overline{p} + \nabla \cdot ((\nu + \nu_t)\nabla\overline{u}), \tag{9}$$

where the incompressibility constraint is used to simplify the equation. The pressure here is altered to include the trace term of Eq. (8).

Smagorinsky model [13] is one of the oldest and most popular eddy-viscosity SGS model. The eddy viscosity of the Smagorinsky model is expressed as

$$\nu_t = C_s^2\Delta^2|\overline{S}|, \tag{10}$$

where C_s is the Smagorinsky coefficient, Δ is the LES filter width and $|\overline{S}| = \sqrt{2\overline{S}:\overline{S}}$. It should be noted that the coefficient C_s must be known prior to the simulation and is usually adapted to improve the results [12]. For example, $C_s = 0.2$ is used for isotropic homogeneous turbulence, while a value of $C_s = 0.1$ is used in case of channel flow. Similar values ($C_s \simeq 0.1 - 0.12$) are realized from the shear flow studies based on experiments [10].

3.1 Stochastic Formulation

Let us consider the Navier-Stokes Eqs. (6) with some uncertainty in the system. The sources of the uncertainty considered here are boundary conditions, material properties and model parameters. We employ the IPC method (see Sect. 2) by presuming the dimensionality (d) and probability density function of the uncertain random variables $\{\xi_1, \xi_2, ...\xi_d\}$ to be known, allowing us to construct the finite set of orthogonal polynomial basis $\{\psi_i\}$.

In order to obtain a rather generic formulation, unless specified otherwise, we consider uncertainty in all sources listed above. Thus, the associated polynomial chaos expansion (PCE) for kinematic viscosity is given by

$$\nu \approx \sum_{i=0}^{P} \nu_i\psi_i(\boldsymbol{\xi}). \tag{11}$$

Note that the coefficients ν_i are assumed to be known. The dependence of the flow variables, i.e. velocity and pressure, on the stochastic variables is expressed by the following PCEs

$$u(x,t) \approx \sum_{i=0}^{P} u_i(x,t)\psi_i(\xi), \quad p(x,t) \approx \sum_{i=0}^{P} p_i(x,t)\psi_i(\xi), \qquad (12)$$

where u_i and p_i are the unknown polynomial chaos mode strengths of velocity and pressure fields, respectively. For deterministic boundary conditions, u_i and p_i are all zero for $i = 1, 2, ..., P$. In case of uncertain boundary conditions with a known (or modeled) probability density function, u_i and p_i, for $i = 0, 1, ..., P$, can be estimated.

Introducing these expansions in Eqs. (6) and taking a Galerkin projection onto each polynomial $\{\psi_k\}$ while using the orthogonality of the polynomial chaos and finally dividing by $\langle \psi_k \psi_k \rangle$, results, for $k = 0, 1, ..., P$, into the following set of deterministic equations:

$$\frac{\partial u_k}{\partial t} + \sum_{i=0}^{P}\sum_{j=0}^{P}(u_i \cdot \nabla)u_j M_{ijk} = -\nabla p_k + \sum_{i=0}^{P}\sum_{j=0}^{P}\nabla \cdot (\nu_i \nabla u_j)M_{ijk}, \quad \nabla \cdot u_k = 0,$$
$$(13)$$

where $M_{ijk} = \dfrac{\langle \psi_i \psi_j \psi_k \rangle}{\langle \psi_k \psi_k \rangle}$. Note that the original system of Eqs. (6) is transformed into a system of $(P+1)$ divergence-free constraints on velocity modes and $(P+1)$ coupled equations in velocity and pressure modes. A detailed discussion on the solution procedure adopted for this large system of equations is deferred to Sect. 4.

Similarly, for the filtered momentum Eq. (7) with Smagorinsky model for turbulence, applying the above steps, results in

$$\frac{\partial \overline{u}_k}{\partial t} + \sum_{i=0}^{P}\sum_{j=0}^{P}(\overline{u}_i \cdot \nabla)\overline{u}_j M_{ijk} = -\nabla \overline{p}_k + \sum_{i=0}^{P}\sum_{j=0}^{P}\nabla \cdot (\nu_i \nabla \overline{u}_j)M_{ijk}$$
$$(14)$$
$$+ \sum_{i=0}^{P}\sum_{j=0}^{P}\sum_{l=0}^{P}\sum_{m=0}^{P}\nabla \cdot (C_{s_l}C_{s_m}\Delta^2|\overline{S}|_i \nabla \overline{u}_j)M_{ijklm},$$

where $M_{ijklm} = \dfrac{\langle \psi_i \psi_j \psi_k \psi_l \psi_m \rangle}{\langle \psi_k \psi_k \rangle}$. Note that $|\overline{S}|^2 = 2\overline{S} : \overline{S}$, and applying the IPC steps to this identity - using polynomial chaos expansion and projecting on each basis polynomial, we obtain

$$\sum_{i=0}^{P}\sum_{j=0}^{P}|\overline{S}|_i|\overline{S}|_j M_{ijk} = 2\sum_{i=0}^{P}\sum_{j=0}^{P}\overline{S}_i : \overline{S}_j M_{ijk}, \qquad (15)$$

where $\overline{S}_i$ is the resolved strain tensor based on i^{th} velocity mode. The above corresponds to system of $(P+1)$ non-linear equations in the unknown expansion coefficients of $|\overline{S}|$. This system is solved using Picard iterations with $|\overline{S}|_i = \sqrt{2\overline{S}_i : \overline{S}_i}$ as the initial guess.

4   Algorithm and Implementation

OpenFOAM uses the finite volume method (FVM) for the disretization of partial differential Eqs. [4]. Among the various fluid dynamic solvers offered by Open-FOAM, we choose the solver called *pimpleFoam* [2], which allows the use of large time-steps to solve the incompressible Navier-Stokes Eqs. (6). This solver is based on the PIMPLE algorithm for pressure-velocity coupling using Rhie and Chow type interpolation [7].

From the previous section, it can be realized that the system of governing Eqs. (14) for the evolution of the velocity and pressure modes u_k, p_k for $k = 0, 1, ..., P$, has a structure similar to the original deterministic Navier-Stokes Eqs. (6). Due to the coupling via convection and diffusion terms, the size of this new system is $P + 1$ times its deterministic version. It can be observed that the divergence-free velocity constraints are decoupled and can be solved independently. Based on this observation, a fractional step projection scheme has been previously implemented [8]. In the first fractional step, the convection and diffusion terms are integrated followed by enforcing the divergence-free constraints in the second fractional step.

Our approach of the stochastic solver is based on the development of the existing deterministic solver (*pimpleFoam*) such that it can accommodate and solve $(P + 1)$ coupled Navier-Stokes like systems in u_k, p_k for $k = 0, 1, ..., P$. We solve each of these systems sequentially, by using the initialized/updated velocity and pressure modes, and repeat until convergence. Figure 1(a), provides a graphical representation of this approach. Depending on the type of flow, the value of P and a few other parameters; it usually takes about 3–6 explicit iterations (I_e) to realize convergence at every time-step. The default value of I_e is set to $P + 1$. Following the conventions of OpenFOAM, we call this solver, *gPCPimpleFoam*. In contrast to the fractional step scheme, this approach admits better stability, stronger coupling and faster convergence; with an efficient data management and minimal changes in the exiting solver.

In Fig. 1(b), we highlight the most important steps needed to develop *gPCPimpleFoam*, over the existing solver, *pimpleFoam*. In contrast to the deterministic solver, two nested loops are introduced. The first loop is over the explicit iterations (I_e), which updates the mode strengths between the two consecutive time-steps. For every explicit iteration, the second nested loop solves the Navier-Stokes like system (u_k, p_k) for each mode strength k, while employing the existing, however modified, structure of the PIMPLE scheme. The modifications are inevitable due to the summations in the convection and the diffusion terms of the stochastic equations. It is realized that the mean mode (u_0, p_0) changes slowly as compared to the other modes. Thus, in order to increase the stability, we start solving the last system (u_P, p_P) first, and updating all the modes before solving the first system (0th mode) representing the mean.

In addition to the exiting modules, we require to either modify or create some completely new routines for turbulence, pre- and post- processing etc. Restricting the verbosity, we attempt to provide an overview of the major implementation steps: (a) Creation of new variables (vectors) for the list of mode

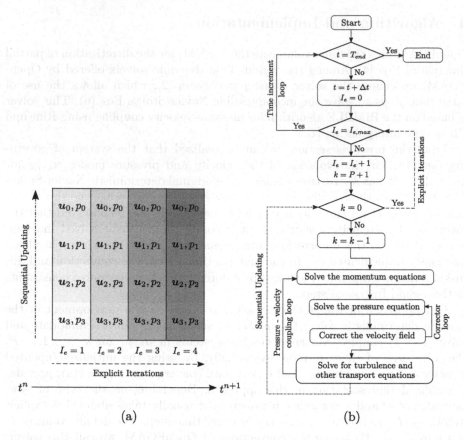

Fig. 1. (a) Evolution of modes (for $P = 3$) between two consecutive time-steps via sequential updating and explicit iterations, (b) Important steps of the algorithm implemented in the *gPCPimpleFoam* solver.

strengths of all the uncertain parameters, flow variables and derived variables for post-processing, (b) For the solver to read these inner products (obtained using *chaospy* library [3]), only a small routine is added. Another similar lines of code are added to read in the values of d, n, I_e, etc, (c) Very subtle changes are needed in the transport model in order to read in the transport properties for all the mode strengths. To accommodate for reading and initializing the turbulence models, a few minor changes are made in the LES model library. Significant changes are required specially for the Smagorinsky model which allows automatic reading of all the mode strengths of C_s from the input file, and (d) To estimate the mean and the variance of flow variables and other derived quantities, a separate routine is added to the post-processing step of the solver. We use the IPC steps to calculate the derived quantities (like Reynolds Stresses) using the resolved expansion coefficients of the flow variables and the known coefficients of other parameters.

The order of accuracy and the convergence rate are governed by the large variety of space and time disretization schemes and iterative solvers offered by the OpenFOAM library.

5   Test Cases

5.1   Plane Poiseuille Flow

We first consider a 2D steady laminar flow in long rectangular channel (with a height of 2δ) in the absence of any external forces. For a given average inlet velocity u_{avg}, the fully developed flow has an analytical solution known as the Hagen-Poiseuille solution.

$$u(x,y) = \frac{3}{2}u_{avg}\left[1 - \left(\frac{y}{\delta}\right)^2\right], \quad v(x,y) = 0, \quad \frac{\partial p}{\partial x} = -\frac{3\nu u_{avg}}{\delta^2}, \quad \frac{\partial p}{\partial y} = 0. \quad (16)$$

Therefore, the velocity field is independent of the viscosity and at $y = 0$, $u = u_{max} = \frac{3}{2}u_{avg}$. We assume the boundary condition to be deterministic and presume a known PCE for the uncertain kinematic viscosity, $\nu = \sum_{i=0}^{P}\nu_i\psi_i(\xi)$. Also, we consider a Gaussian random variable to model the viscosity, for which the associated polynomial chaoses $\psi_i(\xi)$, are the Hermite polynomials. Using PCE of pressure gradient and random viscosity, for $i = 0, 1, ..., P$, we obtain

$$\frac{\partial p_i}{\partial x} = -\frac{3u_{avg}}{\delta^2}\nu_i. \quad (17)$$

The use of polynomial chaos for incompressible laminar flow in a 2D channel has been previously investigated [8], and as a validation case, we carry out a similar study with the IPC solver developed using OpenFOAM.

A uniform velocity is used at the inlet with no-slip boundary conditions at the top and bottom walls, and the gradient of velocity is set to zero at the outlet. Note that the use of deterministic boundary condition implies, for $i = 1, ..., P$, the unknown mode strength and/or their derivatives are by default set to zero at the boundaries. The Reynolds number, $Re = 2\delta u_{avg}/\nu_0$, is set to 100. Fifth-order 1D Hermite polynomials are employed for all the PCEs, i.e. $P = 5$. The coefficient of variation ($CoV = \sigma/\mu$) for the uncertain viscosity is set to $\sim$20%, with $\nu_1/\nu_0 = 2 \times 10^{-1}$ and $\nu_2/\nu_0 = 8 \times 10^{-5}$. The remaining mode strengths of viscosity are assigned a value of zero. The simulation is performed in a domain with $L/\delta = 50$ and a 250×100 mesh with a near-wall grading. Second-order disretization schemes are used both in space and time, and time-step size $\Delta t = 10^{-2}\delta/u_{avg}$ is specified.

Figure 2 shows the profile of the mean and the standard deviation of velocity. The mean depicts the gradual transition in the flow along the channel length from a uniform inlet profile to a parabolic profile with maximum on the center-line. The uncertainty in velocity at inlet is zero, which is indeed the consequence of the deterministic boundary condition. In the developing region, a higher standard deviation is realized in the channel center as well as in the boundary layer

Fig. 2. Profiles of the mean and the standard deviation of velocity.

with two lobes close to the walls. A significant variation in the modes (and thus the standard deviation) is realized up to 10–12 channel half-widths and further downstream, these modes become less significant. Figure 3 provides the axial velocity profile with confidence region ($\pm 2\sigma$) at different locations in the downstream direction. The uncertainties tend to zero in the fully developed region, which is in accordance with the theory where the velocity is independent of the viscosity (see Eq. (16)). Figure 4 shows the estimated ratios of the modes of pressure gradient with respect to the mean pressure gradient along the channel centerline. Clearly, after the recirculating regions near the channel inlet, these ratios gradually reach their constant values further downstream. For $x/\delta > 20$, the results are identical to analytical predictions (see Eq. (17)), characterizing the uncertainty in pressure due to the uncertainty in viscosity. Velocity mode strengths are shown in Fig. 5. As evident, the results from the intrusive variant are in accordance with the non-intrusive counter-part, and due to the fast spectral convergence of the polynomial chaos representation, the magnitudes of the modes decrease as P increases.

5.2   Turbulent Channel Flow

A turbulent channel flow is a theoretical representation of a flow driven by a constant pressure gradient between two parallel planes extending infinitely. A 3D schematic is shown in Fig. 6. Since the computational domain has to be finite, in addition to channel width h, we fix the stream- and span-wise truncation lengths, l_x and l_z, respectively. The values of h, l_x and l_z are adopted from [1]. These values ensures that the computational domain is large enough to accommodate the turbulent structures in the flow.

In order to maintain an equivalent flow, instead of the pressure gradient, the bulk velocity can also be prescribed, $U_b = \frac{1}{h} \int_0^h \langle u \rangle dy$. The bulk Reynolds number is then defined as $Re_b = hU_b/\nu$. In context of turbulent channel flows, another characteristic velocity called the friction velocity is usually introduced in terms of the wall shear stress τ_w and the fluid density ρ, as $u_\tau = \sqrt{\tau_w/\rho}$. The friction Reynolds number is then defined as $Re_\tau = \delta u_\tau/\nu$, where $\delta = h/2$ is the channel half-width. Then the pressure gradient and the wall shear stress relates as $-\frac{d\tilde{p}}{dx} = \frac{\tau_w}{\delta}$, where $\tilde{p}/\rho = \overline{p}$. Thus we have a choice between prescribing the bulk Reynolds number via bulk velocity and the friction Reynolds number via pressure gradient. Since we have to study the effect of the uncertain model parameter on the flow profile, we decide to fix the pressure-gradient and compute U_b through (stochastic) simulation.

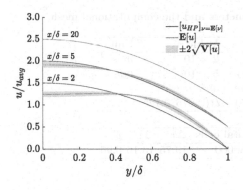

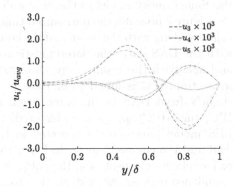

Fig. 3. Normalized axial velocity profiles at different cross-sections.

Fig. 4. Pressure gradient ratios along the centerline.

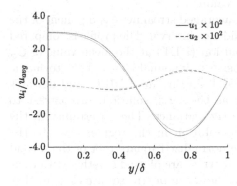

Fig. 5. Modes strengths of velocity at $x/\delta = 2$, for IPC (blue) and NIPC (red) methods. (Color figure online)

The stochastic LES Smagorinsky model, as discussed in Sect. 3, is employed to solve for the turbulence. Since the model parameter C_S may take a range of values, we assume it to be uncertain with a known PCE. We consider a Uniform random variable to model the parameter, for which the associated polynomial chaoses $\psi_i(\xi)$ are the Legendre polynomials. Table 1 summarizes the physical parameters used in the deterministic and stochastic simulations. 1D third-order Legendre polynomials are used for all the PCEs. The value of C_{S_1} is set equal to the standard deviation, while the remaining mode strengths of C_S are assigned a value of zero. Periodic boundary condition are applied in the stream- and span-wise directions, while no-slip boundary condition is used at the walls. The simulation results will be compared to the Direct Numerical Simulation (DNS) data from [9] at $Re_\tau = 395$. Based on the study of the effect of computational grid size from [1], we use a reasonably fine mesh with the details in Table 1. Note that $\Delta x^+ = \Delta x u_\tau / \nu$, $\Delta z^+ = \Delta z u_\tau / \nu$ and $y^+ = y u_\tau / \nu$, are calculated using value of u_τ, corresponding to the value of Re_τ in the DNS database. In order to capture the sharp gradients in the near-wall region, we specify a grading along

Table 1. Details of the physical parameters and the computational mesh.

Parameter	Value	Units
Kinematic viscosity (ν)	2×10^{-5}	$m^2 s^{-1}$
Pressure gradient ($-dp_0/dx$)	5×10^{-5}	ms^{-2}
Target Reynolds number (Re_τ)	395	-
Smagorinsky parameter (C_S)	$\mathcal{U}(0.075, 0.125)$	-

Mesh	Cells along x,y,z	Total cells	Δx^+	Δz^+	y^+
M1	$80 \times 100 \times 60$	480000	19.75	13.16	0.96

y direction. We use the van Driest damping function to correct the behavior of the Smagorinsky model in the near-wall region [1].

Figure 7 presents the normalized time-averaged streamwise component of the velocity along with the mean and the confidence interval. The profile is compared with the DNS data, the deterministic solution (DET) at the mean value of C_S and also with the results from non-intrusive polynomial chaos. The stochastic mean from IPC is found to be close to DET and mean of NIPC, and deviates slightly from DNS in the same manner as the DET solution. In contrast to IPC, the NIPC approach under predicts the variance. The uncertainty in the LES model parameter is reflected in the solution in the regions close to the wall and the channel center. In Fig. 8 the normalized square-root of the second order velocity moments and Reynolds shear stress are plotted together with their confidence regions. As evident, the stochastic mean of stresses are close to that of NIPC and DET solution. The deviation from DNS can mainly be attributed to the use of a relatively coarse mesh and the choice of LES model. Both the IPC and NIPC methods predicts high variance near the wall, with almost zero uncertainty in the channel center. This is expected as the Smagorinsky model parameter, when changed, usually affects significantly near the wall as compared to the channel center. While, the confidence region of intrusive method mostly

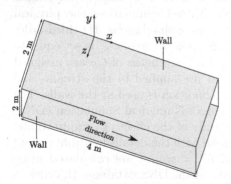

Fig. 6. Graphical representation of the turbulent channel flow.

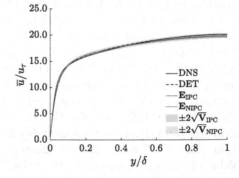

Fig. 7. Normalized time-averaged streamwise component of velocity.

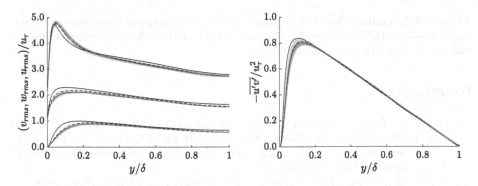

Fig. 8. Normalized square-root of the second order velocity moments (left) and normalized Reynolds shear stress (right). [Legend: Fig. 7]

overlaps with that of the non-intrusive counterpart, in some regions, NIPC still underestimates the uncertainty.

6   Conclusions

The IPC method, as it involves solving a lesser number of equations than NIPC, can be of great use when the deterministic simulation is already computationally expensive. To develop an IPC solver it is important to efficiently decouple the system of equations and ensure the overhead due to coupling is not significant.

In this work, intrusive polynomial chaos for CFD simulations using a popular finite-volume library OpenFOAM was presented. The aim was to use the existing deterministic solver for the incompressible Navier-Stokes equations, to develop a new stochastic solver for the quantification of the uncertainties involved and study its non-linear propagation.

To this end, we tested this solver for various standard CFD problems involving laminar and turbulent flows. The plane Poiseuille flow with uncertain kinematic viscosity was discussed first. Here, we realized a significant effect of the uncertainty in the re-circulation region of the flow. The results were also compared with the non-intrusive counterpart with same polynomial order. The results from IPC were found to be very close to NIPC, verifying its implementation using OpenFOAM. Next, we examined the turbulent channel flow with uncertain LES model parameter. The results were found to be in accordance with the non-intrusive gPC method. Deviations in the variance predicted by the two variants of gPC approaches can be attributed to the use of the two very different numerical methods to estimate the expansion coefficients in both variants.

Through this work, for UQ in CFD, we bring to light an alternate to the MC and the NIPC approaches. The promising results obtained from IPC method encourages to further pursue research in this direction. Pseudo-spectral methods along with early truncated expansions will be some next steps to reduce the computational cost. As a future work, Parallel-in-Time methods will be tested to overcome the saturation via space-only parallelization.

Acknowledgments. This work is supported by the Netherlands Organisation for Scientific Research (NWO) under the "Parallel-in-time methods for propagation of uncertainties in wind-farm simulations" project.

References

1. DeVilliers, E.: The potential of large eddy simulation for the modeling of wall bounded flows Eugene de Villiers. Ph.D. thesis, Imperial College of Science, Technology and Medicine (2006). http://powerlab.fsb.hr/ped/kturbo/OpenFOAM/docs/EugeneDeVilliersPhD2006.pdf
2. ESI-OpenCFD: OpenFOAM (v1806) - The Open Source CFD Toolbox (2018)
3. Feinberg, J., Langtangen, H.P.: Chaospy: an open source tool for designing methods of uncertainty quantification. J. Comput. Sci. **11**, 46–57 (2015). https://doi.org/10.1016/j.jocs.2015.08.008
4. Ferziger, J.H., Perić, M.: Computational Methods for Fluid Dynamics. Springer, Heidelberg (2002). https://doi.org/10.1007/978-3-642-56026-2
5. Ghanem, R.G., Spanos, P.D.: Stochastic Finite Elements: A Spectral Approach. Springer, New York (1991). https://doi.org/10.1007/978-1-4612-3094-6
6. Hosder, S., Walters, R.W.: Non-intrusive polynomial chaos methods for stochastic CFD-theory and applications. In: RTO Meeting Proceedings. Computational Uncertainty in Military Vehicle Design (2007). http://www.rto.nato.int
7. Issa, R.I.: Solution of the implicitly discretised fluid flow equations by operator-splitting. J. Comput. Phys. **62**(1), 40–65 (1986). https://doi.org/10.1016/0021-9991(86)90099-9
8. Le Matre, O.P., Knio, O.M., Najm, H.N., Ghanem, R.G.: A stochastic projection method for fluid flow. I. Basic formulation. J. Comput. Phys. **173**(2), 481–511 (2001). https://doi.org/10.1006/jcph.2001.6889
9. Moser, R.D., Kim, J., Mansour, N.N.: Direct numerical simulation of turbulent channel flow up to Re 590. Phys. Fluids **11**(4), 943–945 (1999). https://doi.org/10.1063/1.869966
10. O'Neil, J., Meneveau, C.: Subgrid-scale stresses and their modelling in a turbulent plane wake. J. Fluid Mech. **349**, 253–293 (1997). https://doi.org/10.1017/S0022112097006885
11. LeMaître, O.P., Knio, O.: Spectral Methods for Uncertainty Quantification with Applications to CFD. Springer, Dordrecht (2010). https://doi.org/10.1007/978-90-481-3520-2
12. Sagaut, P.: Large Eddy Simulation for Incompressible Flows. Springer, Heidelberg (1998). https://doi.org/10.1007/978-3-662-04695-1
13. Smagorinsky, J.: General circulation experiments with the primitive equations. Mon. Weather Rev. **91**(3), 99–164 (1963). https://doi.org/10.1175/1520-0493
14. Smith, R.C.: Uncertainty Quantification: Theory, Implementation, and Applications. Society for Industrial and Applied Mathematics, Philadelphia (2013)
15. Walters, R.W., Huyse, L.: Uncertainty analysis for fluid mechanics with applications. Technical report, ICASE NASA Langley Research Center Hampton (2002)
16. Wiener, N.: The homogeneous chaos. Am. J. Math. **60**(4), 897–936 (1938). https://pdfs.semanticscholar.org/21f9/b472fd25dcd75943d5da7f344cf23cfacabf.pdf

17. Witteveen, J.A., Bijl, H.: Modeling arbitrary uncertainties using Gram-Schmidt polynomial chaos. In: 44th AIAA Aerospace Sciences Meeting and Exhibit. American Institute of Aeronautics and Astronautics, Reston, Virigina, January 2006. https://doi.org/10.2514/6.2006-896. http://arc.aiaa.org/doi/10.2514/6.2006-896
18. Xiu, D.: Efficient collocational approach for parametric uncertainty analysis. Commun. Comput. Phys. 2(2), 293–309 (2007). http://www.global-sci.com/
19. Xiu, D., Karniadakis, G.E.: The Wiener-Askey polynomial chaos for stochastic differential equations. SIAM J. Sci. Comput. 24(2), 619–644 (2002). https://doi.org/10.1137/S1064827501387826
20. Xiu, D., Karniadakis, G.E.: Modeling uncertainty in flow simulations via generalized polynomial chaos. J. Computat. Phys. 187(1), 137–167 (2003). https://doi.org/10.1016/S0021-9991(03)00092-5

Distributions of a General Reduced-Order Dependence Measure and Conditional Independence Testing

Mariusz Kubkowski[1,2] (iD), Małgorzata Łazęcka[1,2] (iD), and Jan Mielniczuk[1,2](✉) (iD)

[1] Institute of Computer Science, Polish Academy of Sciences, Warsaw, Poland
{m.kubkowski,malgorzata.lazecka,miel}@ipipan.waw.pl
[2] Faculty of Mathematics and Information Sciences,
Warsaw University of Technology, Warsaw, Poland

Abstract. We study distributions of a general reduced-order dependence measure and apply the results to conditional independence testing and feature selection. Experiments with Bayesian Networks indicate that using the introduced test in the Grow and Shrink algorithm instead of Conditional Mutual Information yields promising results for Markov Blanket discovery in terms of F measure.

Keywords: Conditional Mutual Information · Asymptotic distribution · Feature selection · Markov Blanket · Reduced-order dependence measure

1 Introduction

Consider a problem of selecting a subset of all potential predictors $\{X_1, \ldots, X_p\}$ to predict an outcome Y, which consists of all predictors significantly influencing it. Selection of active predictors leads to dimension reduction and is instrumental for many machine learning and statistical procedures, in particular in structure learning of dependence networks. Commonly for this task, such methods incorporate a sequence of conditional independence tests, among which the test based on Conditional Mutual Information (CMI) is the most frequent. In the paper we consider properties of a general information-based dependence measure $J^{\beta,\gamma}(X, Y|X_S)$ introduced in [2] in a context of constructing approximations to CMI. This is a reduced-order approximation which disregards approximations of order higher than 3. It can also be considered as a measure of predictive power of X for Y when variables $X_S = (X_s, s \in S)$ have been already chosen for this task. Special cases include Mutual Information Minimization (MIM), Minimum Redundancy Maximum Relevance (MrMR) [11], Mutual Information Feature Selection (MIFS) [1], Conditional Information Feature Extraction (CIFE) [7] and Joint Mutual Information (JMI) [14] criteria. They are routinely used in nonparametric approaches to feature selection, variable importance ranking and causal discovery (see e.g. [4,13]). However, theoretical properties of such criteria

© Springer Nature Switzerland AG 2020
V. V. Krzhizhanovskaya et al. (Eds.): ICCS 2020, LNCS 12143, pp. 692–706, 2020.
https://doi.org/10.1007/978-3-030-50436-6_51

remain largely unknown hindering study of associated selection methods. Here we show that $\hat{J}^{\beta,\gamma}(X,Y|X_S)$ exhibits dichotomous behaviour meaning that its distribution can be either normal or coincides with a distribution of a certain quadratic form in normal variables. The second case is studied in detail for binary Y. In particular for two popular criteria CIFE and JMI, conditions under which their distributions converge to distributions of quadratic form are made explicit. As two cases of dichotomy differ in behaviour of the variance of $\hat{J}^{\beta,\gamma}$, its order of convergence is used to detect which case is actually valid. Then a parametric permutation test (i.e. a test based on permutations to estimate parameters of the chosen distribution) is used to check whether candidate variable X is independent of Y given X_S.

2 Preliminaries

2.1 Entropy and Mutual Information

We denote by $p(x) := P(X = x)$, $x \in \mathcal{X}$ a probability mass function corresponding to X, where $\mathcal{X}$ is a domain of X and $|\mathcal{X}|$ is its cardinality. Joint probability will be denoted by $p(x,y) = P(X = x, Y = y)$. Entropy for discrete random variable X is defined as

$$H(X) = - \sum_{x \in \mathcal{X}} p(x) \log p(x). \tag{1}$$

Entropy quantifies the uncertainty of observing random values of X. In case of discrete X, $H(X)$ is non-negative and equals 0 when the probability mass is concentrated at one point. The above definition naturally extends to the case of random vectors (i.e. X can be multivariate random variable) by using multivariate probability instead of univariate probability. In the following we will frequently consider subvectors of $X = (X_1, \ldots, X_p)$ which is a vector of all potential predictors of class index Y. The conditional entropy of X given Y is written as

$$H(X|Y) = \sum_{y \in \mathcal{Y}} p(y) H(X|Y = y) \tag{2}$$

and the mutual information (MI) between X and Y is

$$I(X,Y) = H(X) - H(X|Y) = \sum_{x,y} p(x,y) \log \frac{p(x,y)}{p(x)p(y)}. \tag{3}$$

This can be interpreted as the amount of uncertainty in X which is removed when Y is known which is consistent with an intuitive meaning of mutual information as the amount of information that one variable provides about another. MI equals zero if and only if X and Y are independent and thus it is able to discover non-linear relationships. It is easily seen that $I(X,Y) = H(X) + H(Y) - H(X,Y)$. A natural extension of MI is conditional mutual information (CMI) defined as

$$I(X,Y|Z) = H(X|Z) - H(X|Y,Z), \tag{4}$$

which measures the conditional dependence between X and Y given Z. An important property is chain rule for MI which connects $I((X_1, X_2), Y)$ to $I(X_1, Y)$:

$$I((X_1, X_2), Y) = I(X_1, Y) + I(X_2, Y|X_1). \tag{5}$$

For more properties of the basic measures described above we refer to [3]. A quantity, used in next sections, is interaction information (II) [9]. The 3-way interaction information is defined as

$$II(X_1, X_2, Y) = I(Y, X_1|X_2) - I(Y, X_1), \tag{6}$$

which is consistent with an intuitive meaning of existence of interaction as a situation in which the effect of one variable on the class variable depends on the value of another variable.

2.2 Approximations of Conditional Mutual Information

We consider a discrete class variable Y and p discrete features $X_1, \ldots, X_p$. Let X_S denote a subset of features indexed by a subset $S \subseteq \{1, \ldots, p\}$. We employ here greedy search for active features based on forward selection. Assume that S is a set of already chosen features, S^c its complement and $j \in S^c$ a candidate feature. In each step we add a feature whose inclusion gives the most significant improvement of the mutual information, i.e. we find

$$\arg\max_{j \in S^c} \left[I(X_{S \cup \{j\}}, Y) - I(X_S, Y) \right] = \arg\max_{j \in S^c} I(X_j, Y|X_S). \tag{7}$$

The equality in (7) follows from (5). Observe that (7) indicates that we select a feature that achieves the maximum association with the class given the already chosen features. For example, first-order approximation yields $I(X_j, Y)$, which is a simple univariate filter MIM, frequently used as a pre-processing step in high-dimensional data analysis. However, this method suffers from many drawbacks as it does not take into account possible interactions between features and redundancy of some features. When the second order approximation is used, the dependence score for candidate feature is

$$\begin{aligned} J(X_j) &= I(X_j, Y) + \sum_{i \in S} II(X_i, X_j, Y) \\ &= I(X_j, Y) + \sum_{i \in S} [I(X_i, X_j|Y) - I(X_i, X_j)]. \end{aligned} \tag{8}$$

The second equality uses (6). In literature (8) is known as CIFE (Conditional Infomax Feature Extraction) [7] criterion. Observe that in (8) we take into account not only relevance of the candidate feature, but also its possible interactions with the already selected features. However, frequently it is useful to scale down the corresponding term [2]. Among such modifications the most popular is JMI

$$J(X_j) = I(X_j, Y) + \frac{1}{|S|} \sum_{i \in S} [I(X_i, X_j|Y) - I(X_i, X_j)] = \frac{1}{|S|} \sum_{i \in S} I(X_j, Y|X_i),$$

where the second equality follows from (5). JMI was also proved to be an approximation of CMI under certain dependence assumptions [13]. Data-adaptive version of JMI will be considered in Sect. 4. In [2] it is proposed to consider a general information-theoretic dependence measure

$$J^{\beta,\gamma}(X_j, Y|X_S) = I(X_j, Y) - \beta \sum_{i \in S} I(X_i, X_j) + \gamma \sum_{i \in S} I(X_i, X_j|Y), \quad (9)$$

where β, γ are some positive constants usually depending in decreasing manner on the size $|S| = k$ of set S. Several frequently used selection criteria are special cases of (9). MrMR criterion [11] corresponds to $(\beta, \gamma) = (|S|^{-1}, 0)$ whereas more general MIFS (Mutual Information Feature Selection) criterion [1] corresponds to pair $(\beta, 0)$. Obviously, the simplest criterion MIM corresponds to $(0, 0)$ pair. CIFE defined above in (8) is obtained for $(1, 1)$ pair, whereas $(\beta, \gamma) = (1/|S|, 1/|S|)$ leads to JMI. In the following we consider asymptotic distributions of the sample version of $J^{\beta,\gamma}(X_j)$, namely

$$\hat{J}^{\beta,\gamma}(X_j) = \hat{I}(X_j, Y) - \beta \sum_{i \in S} \hat{I}(X_i, X_j) + \gamma \sum_{i \in S} \hat{I}(X_i, X_j|Y), \quad (10)$$

and show how the distribution depends on underlying parameters. In this way we gain a more clear idea what is an influence of β and γ on the behaviour of $\hat{J}^{\beta,\gamma}$. Sample version in (10) is obtained by plugging in fractions of observations instead of probabilities in (3) and (4).

3   Distributions of a General Dependence Measure

In the following we will state our theoretical results which study asymptotic distributions of $\hat{J}^{\beta,\gamma}(X, Y|Z)$ where $Z = (Z_1, \ldots, Z_{|S|})$ is possible multivariate discrete vector and then we apply it to previously introduced framework by putting $X := X_j$ and $Z := (X_1, \ldots, X_{|S|})$. We will show that its distribution is either approximately normal or, if the asymptotic variance vanishes, is approximately equal to distribution of quadratic form of normal variables. Let $p = (p(x, y, z))_{x,y,z}$ be a vector of probabilities for (X, Y, Z) and we assume whence forth that $p(x, y, z) > 0$ for any triple of (x, y, z) values in the range of (X, Y, Z). Moreover, $f(p)$ equals $J^{\beta,\gamma}(X, Y|Z)$ treated as a function of p, Df denotes a derivative of function f and $\xrightarrow{d}$ convergence in distribution. The special case of the result below for CIFE criterion has been proved in [6].

Theorem 1. *(i) We have*

$$n^{1/2}(\hat{J}^{\beta,\gamma}(X, Y|Z) - J^{\beta,\gamma}(X, Y|Z)) \xrightarrow{d} N(0, \sigma_j^2), \quad (11)$$

where $\sigma_j^2 = Df(p)^T \Sigma Df(p) = \mathrm{Var}(Df(p)^T \hat{p})$ and $\Sigma = n\mathrm{Var}(\hat{p} - p)$.
(ii) If $\sigma_j^2 = 0$ then

$$2n(\hat{J}^{\beta,\gamma}(X, Y|Z) - J^{\beta,\gamma}(X, Y|Z)) \xrightarrow{d} V^T H V, \quad (12)$$

where V follows $N(0, \Sigma)$ distribution, $\Sigma^{x'y'z'}_{xyz} = p(x', y', z')(I(x = x', y = y', z = z') - p(x, y, z))/n$ and $H = D^2 f(p)$ is a Hessian of f.

Proof. Note that $f(p) = J^{\beta, \gamma}(X, Y | Z)$ equals

$$I(X, Y) - \sum_{s \in S} (\beta I(X, Z_s) - \gamma I(X, Z_s | Y)) = \sum_{x,y,z} p(x, y, z) \left(\ln \left(\frac{p(x, y)}{p(x)p(y)} \right) \right.$$

$$\left. - \sum_{s \in S} \left(\beta \ln \left(\frac{p(x, z_s)}{p(x)p(z_s)} \right) - \gamma \ln \left(\frac{p(x, y, z_s)p(y)}{p(x, y)p(y, z_s)} \right) \right) \right).$$

After some calculations one obtains that $\frac{\partial f(p)}{\partial p(x,y,z)}$ equals for $z = (z_1, \ldots, z_{|s|})$

$$\ln \left(\frac{p(x, y)}{p(x)p(y)} \right) - \beta \sum_{s \in S} \left(\ln \left(\frac{p(x, z_s)}{p(x)p(z_s)} \right) - 1 \right)$$

$$+ \gamma \sum_{s \in S} \ln \left(\frac{p(x, y, z_s)p(y)}{p(x, y)p(y, z_s)} \right) - 1. \tag{13}$$

Let $\hat{p}(x, y, z) = n(x, y, z)/n$, $\hat{p} = (\hat{p}(x, y, z))_{x,y,z}$. Then $\hat{J}^{\beta, \gamma}(X, Y | Z) = f(\hat{p})$. The remaining part of the proof relies on Taylor's formula for $f(\hat{p}) - f(p)$. Details are given in supplemental material [5].

We characterize the case when $\sigma_{\hat{j}}^2 = 0$ in more detail for binary Y and $\beta = \gamma \neq 0$ which encompasses CIFE and JMI criteria. Note that binary Y case covers an important situation of distinguishing between cases $(Y = 1)$ and control $(Y = 0)$. We define two scenarios:

- Scenario 1 (S1): $X \perp Y | Z_s$ for any $s \in S$ and $X \perp Y$ ($X \perp Y | Z$ denotes conditional independence of X and Y given Z).
- Scenario 2 (S2): $\exists W \subset S$ such that $W \neq \emptyset$ and for $s \in W$ $Z_s \perp Y | X$, $X \not\perp Y | Z_s$ and for $s \in W^c$ we have $X \perp Y | Z_s$.

Define W as

$$W = \left\{ s \in S : \exists_{x,y,z_s} \frac{p(x, y, z_s)p(z_s)}{p(x, z_s)p(y, z_s)} \neq 1 \right\}. \tag{14}$$

We will study in detail the case when $\sigma_{\hat{j}}^2 = 0$ and either $\beta = \gamma \neq 0$ or at least one of the parameters β, γ equal 0. We note that all cases of used information-based criteria fall in one of these categories [2]. We have

Theorem 2. *Assume that $\sigma_{\hat{j}}^2 = 0$ and $\beta = \gamma \neq 0$. Then we have:*

(i) *If $|S| > 1$ and $\beta^{-1} \in \{1, 2, \ldots, |S| - 1\}$ then one of the above scenarios holds with W defined in (14).*

(ii) *If $\beta^{-1} = |S|$ or $\beta^{-1} \notin \{1, 2, \ldots, |S| - 1\}$ then Scenario 1 is valid.*

The analogous result can be stated for the case when at least one of the parameters β or γ equals 0 (details are given in supplement [5]).

3.1   Special Case: JMI

We state below corollary for criterion JMI. Note that in view of Theorem 2 Scenario 2 holds for JMI. Let

$$\sigma^2_{\widehat{JMI}} = \sum_{x,y,z} p(x,y,z) \left(\frac{1}{|S|} \sum_{s \in S} \ln \frac{p(x,y,z_s)p(z_s)}{p(x,z_s)p(y,z_s)} \right)^2 - (JMI)^2. \qquad (15)$$

Corollary 1. *Let Y be binary. (i) If $\sigma^2_{\widehat{JMI}} \neq 0$ then*

$$n^{1/2}(\widehat{JMI} - JMI) \xrightarrow{d} N(0, \sigma^2_{\widehat{JMI}}).$$

(ii) If $\sigma^2_{\widehat{JMI}} = 0$ then $JMI = 0$ and

$$2n\widehat{JMI} \xrightarrow{d} V^T HV,$$

where V and H are defined in Theorem 1. Moreover in this case Scenario 1 holds.

Note that $\sigma^2_{\widehat{JMI}} = 0$ implies $JMI = 0$ as in this case Scenario 1 holds. The result for CIFE is analogous (see supplemental material [5]).

In both cases we can infer the type of limiting distribution if the corresponding theoretical value of the statistic is nonzero. Namely, if $JMI \neq 0$ $(CIFE \neq 0)$ then $\sigma^2_{\widehat{JMI}} \neq 0$ (respectively, $\sigma^2_{\widehat{CIFE}} \neq 0$) and the limiting distribution is normal. Checking that $JMI \neq 0$ is simpler than $CIFE \neq 0$ as it is implied by $X \not\perp Y | Z_s$ for at least one $s \in S$. Actually, $JMI = 0$ is equivalent to conditional independence of X and Y given Z_s for any $s \in S$ which in its turn is equivalent to $\sigma^2_{\widehat{JMI}} = 0$. In the next section we will use a behaviour of the variance to decide which distribution to use as a benchmark for testing conditional independence. In a nutshell, the corresponding switch which is constructed in data-adaptive way and is based on different order of convergence of the variance to 0 in both cases. This is exemplified in the Fig. 1 which shows boxplots of the empirical variance of JMI multiplied by sample size in two cases, when the theoretical variance is 0 (model M2 discussed below) or not (model M1). The Figure clearly indicates that the switch can be based on the behaviour of the variance.

4   JMI-Based Conditional Independence Test and Its Behaviour

4.1   JMI-Based Conditional Independence Test

In the following we use $\hat{J} = \widehat{JMI}$ as a test statistic for testing conditional independence hypothesis

$$H_0 : X \perp Y | X_S. \qquad (16)$$

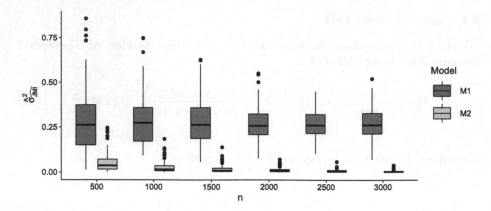

Fig. 1. Behaviour of the empirical variance multiplied by n in the case when corresponding value of $\sigma^2_{\widehat{JMI}}$ is zero (yellow) or not (blue). Models: M1, M2 (see text), JMI = $JMI(X_1^{(1)}, Y | X_1, \ldots, X_5)$, $n = 1000$, $\rho = 0$, $\gamma = 1$. (Color figure online)

where X_S denotes set of X_i with $i \in S$. A standard way of testing it is to use Conditional Mutual Information (CMI) as a test statistic and its asymptotic distribution to construct the rejection region. However, it is widely known that such test loses power when the size of conditioning set grows due to inadequate estimation of $p(x, y | X_S = x_S)$ for all strata $\{X_S = x_S\}$. Here we use as a test statistic $\hat{J}$ which does not suffer from this drawback as it involves conditional probabilities given univariate strata $\{X_s = x_s\}$ for $s \in S$. As behaviour of $\hat{J}$ is dichotomous on (16) we consider a data-dependent way of determining which of the two distributions: normal or distribution of quadratic form (abbreviated to d.q.f. further on) is closer to distribution of $\hat{J}$. Here we propose a switch based on the connection between distribution of the statistics and its variance (see Theorem 1). We consider the test based on JMI as in this case $\sigma^2_{\widehat{JMI}} = 0$ is equivalent to $JMI = 0$. Namely, it is seen from Theorem 1 that normality of asymptotic distribution corresponds to the case when the asymptotic variance calculated for samples of size n and $n/2$ should be approximately the same and should be strictly smaller for a larger sample otherwise. For each strata $X_S = x_S$ we permute corresponding n_{X_S} values of X B times and for each permutation we obtain value of $\widehat{JMI}$ as well as an estimator of its asymptotic variance v_n. The permutation scheme is repeated for randomly chosen subsamples of original sample of size $n/2$ and B values of $v_{n/2}$ are calculated. We than compare the mean of v_n with the mean of $v_{n/2}$ using t-test. If the equality of the means is not rejected we bet on normality of asymptotic distribution, in the opposite case d.q.f. is chosen. Note that permuting samples for a given value $X_S = x_S$ we generate data (X_{perm}, Y, X_S) which follows null hypothesis (16) while keeping the distribution $P_{X|X_S} = P_{X_{perm}|X_S}$ unchanged. In Fig. 2 we show that when conditional independence hypothesis is satisfied then distribution of estimated variance $\hat{\sigma}^2_{\widehat{JMI}}$ based on permuted samples follows closely distribution of $\hat{\sigma}^2_{\widehat{JMI}}$

based on independent samples. Thus indeed using permutation scheme described above we can approximate the distribution of the variance of JMI under H_0 for a fixed conditional distribution $\hat{\sigma}^2_{\widetilde{JMI}}$.

Now we approximate sample distribution of $\widehat{JMI}$ by $N(\hat{\mu}, \hat{\sigma}^2)$ when normal distribution has been picked or when d.q.f. has been picked approximation is $\chi^2_{\hat{\mu}}$ (with $\hat{\mu}$ being the empirical mean of $\widehat{JMI}$) or scaled chi square $\hat{\alpha}\chi^2_{\hat{d}} + \hat{\beta}$ where parameters are based on three first empirical moments of the permuted samples [15]. Then the observed value $\widehat{JMI}$ is compared to quantile of the above benchmark distribution and conditional independence is rejected when this quantile is exceeded. Note that as parametric permutation test is employed we need much smaller B than in the case of non-parametric permutation test and we use $B = 50$. Algorithm will be denoted by JMI(norm/chi) or JMI(norm/chi_scale) depending on whether chi square or scaled chi square is considered in the switch. The pseudocode of the algorithm is given below in Algorithm 1 and the code itself is available in [5]. For comparison we consider two tests: asymptotic test for CMI (called CMI) and semi-parametric permutation test (called CMI(sp)) proposed in [12]. In CMI(sp) the permutation test is used to estimate the number of degrees of freedom of reference chi square distribution.

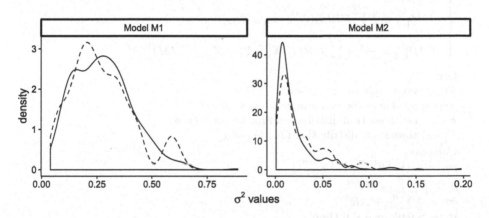

Fig. 2. Comparison of variances' distributions under conditional independence hypothesis. SIM corresponds to distribution of $\hat{\sigma}^2_{\widetilde{JMI}}$ based on $N = 500$ simulated samples. PERM is based on $N = 50$ simulated samples. For each of them X was permuted on strata ($B = 1$) and $\hat{\sigma}^2_{\widetilde{JMI}}$ was calculated. Models: M1, M2 (see text), JMI $= JMI(X_1^{(1)}, Y|X_1, \ldots, X_5)$, $n = 1000$, $\rho = 0$, $\gamma = 1$

Algorithm 1: $JMI(chi/norm)$

Input   : Training data $D_0 = (X, Y, Z)$ of size n (Z with p columns),
                number of permutations B.

Let:
$$CRIT_i(X, Y, Z) := (JMI(X, Y|Z))_{i=1}^n = \sum_{j=1}^{p} \log \frac{\hat{p}(x_i, y_i, z_{i,j})\hat{p}(z_i)}{(\hat{p}(y_i, z_{i,j})\hat{p}(x_i, z_{i,j})}$$

Compute:
$$JMI^{(0)} = \frac{1}{n} \sum_{i=1}^{n} CRIT_i(X, Y, Z)$$

for $b = 1, \ldots, B$ **do**

> Randomly permute X (on each strata on Z) to obtain permuted sample
> $D^{(b)} = (X^{(b)}, Y, Z)$
>
> Compute:
> $$JMI^{(b)} = \frac{1}{n} \sum_{i=1}^{n} CRIT_i(X^{(b)}, Y, Z),$$
> $$VAR^{(b)} = \frac{1}{n-1} \sum_{i=1}^{n} (CRIT_i(X^{(b)}, Y, Z) - JMI^{(b)})^2$$

for $b = 1, \ldots, B$ **do**

> Randomly permute X (on each strata on Z) and randomly choose $[n/2]$
> observations to obtain permuted sample $D_{1/2}^{(b)} = (X_{1/2}^{(b)}, Y_{1/2}, Z_{1/2})$
>
> Compute:
> $$JMI_{1/2}^{(b)} = \frac{2}{n} \sum_{i=1}^{n/2} CRIT_i(X_{1/2}^{(b)}, Y_{1/2}, Z_{1/2}),$$
> $$VAR_{1/2}^{(b)} = \frac{1}{n/2-1} \sum_{i=1}^{n/2} (CRIT_i(X_{1/2}^{(b)}, Y_{1/2}, Z_{1/2}) - JMI_{1/2}^{(b)})^2$$

Let:
$T(\cdot, \cdot)$ two sample t-test statistic
$p_T(\cdot, \cdot)$ p-value of the two sample t-test statistic
$F_{N(\hat{\mu}, \hat{\sigma})}(s)$ theoretical distribution function of $N(\hat{\mu}, \hat{\sigma})$
$F_{\chi_{\hat{\mu}}^2}(s)$ theoretical distribution function of $\chi_{\hat{\mu}}^2$
Compute:
$p_T := p_T(VAR^{(1:B)}, VAR_{1/2}^{(1:B)})$
$\hat{\mu} := \frac{1}{B} \sum_{b=1}^{B} JMI^{(b)}$
$\hat{\sigma}^2 := \frac{1}{B} \sum_{b=1}^{B} VAR^{(b)}$
if $p_T > 0.05$ *or* $\hat{\mu} \leq 0$ **then**

> $p = 1 - F_{N(\sqrt{n}\hat{\mu}, \hat{\sigma})}(\sqrt{n}JMI^{(0)})$

else

> $p = 1 - F_{\chi_{2n\hat{\mu}}^2}(2nJMI^{(0)})$

Output   : p-value p

4.2   Numerical Experiments

We investigate the behaviour of the proposed test in two generative tree models
shown in the left and the right panel of Fig. 3 which will be called M1 and
M2. Note that in model M1 $X_1^{(1)} \perp Y|X_1, \ldots, X_k$ whereas for model M2 the

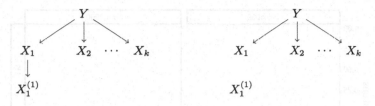

Fig. 3. Models under consideration in an experiment I. The models in the left and right panel will be called M1 and M2.

stronger condition $X_1^{(1)} \perp (Y, X_1, \ldots, X_k)$ holds. We consider the performance of JMI based test for testing hypothesis $H_{01} : X_1^{(1)} \perp Y | X_1, \ldots, X_k$ when the sample size and parameters of the model vary. As H_{01} is satisfied in both models this contributes to the analysis of the size of the test.

Observations in M1 are generated as follows: first, Y is chosen from Bernoulli distribution with success probability $P(Y = 1) = 0.5$. Then $(Z_1, \ldots, Z_k)$ are generated from $N(0, \Sigma)$ given $Y = 0$ and $N(\gamma, \Sigma)$ given $Y = 1$, where elements of Σ are equal $\sigma_{ij} = \rho^{|i-j|}$ and $\gamma = (1, \gamma, \ldots, \gamma^{k-1})^T$ with $0 \leq \rho < 1$ and $0 < \gamma \leq 1$ some chosen values. Then Z values are discretised to two values (0 and 1) to obtain $X_1, \ldots X_k$. In the next step $Z_1^{(1)}$ is generated from conditional distribution $N(X_1, 1)$ given X_1 and then $Z_1^{(1)}$ is discretised to $X_1^{(1)}$. We note that such method of generation yields that $Z_1^{(1)}$ and Y are conditionally independent given X_1 and the same is true for $X_1^{(1)}$. Observations in M2 are generated similarly, the only difference being that $Z_1^{(1)}$ is now generated independently of $(Y, X_1, \ldots, X_k)$.

We will also check the power of the tests in M1 for testing hypotheses $H_{02} : X_1^{(1)} \perp Y | X_2, \ldots, X_k$ and $H_{03} : X_1 \perp Y | X_2, \ldots, X_k$ as neither of them is satisfied in M1. Note however, that since information $I(X_1^{(1)}, Y | X_2, \ldots, X_k)$ and $I(X_1, Y | X_2, \ldots, X_k)$ decreases when k (or γ, ρ) increases the task becomes more challenging for larger k (or γ, ρ, respectively) which will result in a loss of power for large k when sample size is fixed.

Estimated tests sizes and powers are based on $N = 200$ times repeated simulations.

We first check how the switch behaves for JMI test while testing H_{01} (see Fig. 4). In M1 for $k = 1$ as $X_1^{(1)} \perp Y$ given X_1 and $JMI = I(X_1^{(1)}, Y | X_1) = 0$ asymptotic distribution is d.q.f. and we expect switching to d.q.f. which indeed happens in almost 100%. For $k \geq 2$, $JMI \neq 0$ asymptotic distribution is normal which is reflected by the fact that the normal distribution is chosen with large probability. Note that this probability increases with n as summands $\hat{I}(X_1^{(1)}, Y | X_i)$ of $\widehat{JMI}$ for $i \geq 2$ converge to normal distributions due to Central Limit Theorem. The situation is even more clear-cut for M2 where $JMI = 0$ for all k and the switch should choose d.q.f.

Figure 5 shows the empirical sizes of the test when theoretical size has been fixed at $\alpha = 0.05$ and $\rho = 0$ and $\gamma = 1$. We see that empirical size is controlled fairly well for CMI(sp) and for the proposed methods, with the switch

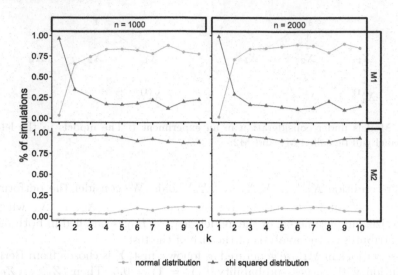

Fig. 4. The behaviour of the switch for testing H_{01} in M1 and M2 models ($\rho = 0$, $\gamma = 1$, $n = 1000$).

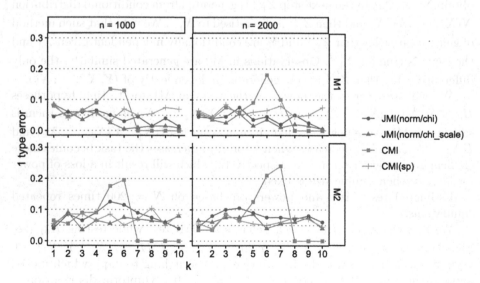

Fig. 5. Test sizes for testing H_{01} in M1 and M2 models ($\rho = 0$, $\gamma = 1$) for fixed $\alpha = 0.05$.

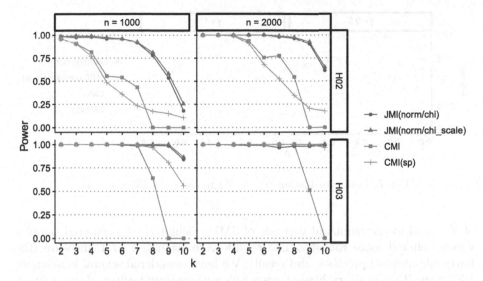

Fig. 6. Power for testing H_{02} and H_{03} in M1 model ($\rho = 0$, $\gamma = 1$).

(norm/chi_scale) working better than the switch (norm/chi). A superiority of the former is even more pronounced for $0 < \gamma < 1$ and when $X_1, \ldots, X_k$ are dependent (not shown). Note erratic behaviour of size for CMI, which significantly exceeds 0.1 for certain k and then drops to 0. Figures 6 and 7 show the power of the considered methods for hypotheses H_{02} and H_{03}. It is seen that for $\gamma = 1$, $\rho = 0$ the expected decrease of power with respect to k is much more moderate for the proposed methods than for CMI and CMI(sp). JMI (norm/chi_scale) works in most cases slightly better than JMI (norm/chi). For H_{03} power of CMI(sp) is similar to that of CMI but it exceeds it for large k, however, it is significantly smaller than the power of both proposed methods. For H_{03} superiority of JMI-based tests is visible only for large k when n is moderate ($n = 500, 1000$), whereas for H_{02} it is also evident for small k. With changing ρ and γ superiority of the proposed methods is still evident (see Fig. 7). Note that for fixed γ the power of all methods decreases when ρ increases.

5   Application to Feature Selection

Finally, we illustrate how the proposed test can be applied for Markov Blanket (MB, see e.g. [10]) discovery of Bayesian Networks (BN). MB for a target Y is defined as the minimal set of predictors given which Y and remaining predictors are conditionally independent [2]. We have used the JMI test (with normal/scaled chi square switch) in the Grow and Shrink (GS, see e.g. [8]) algorithm for MB discovery and compared it with GS using CMI and CMi(sp). GS algorithm finds a large set of potentially active features in the Grow phase and then whittles it down in the Shrink phase. In the real data experiments we used another estimator

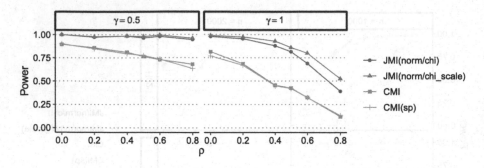

Fig. 7. Power for testing H_{02} in M1 model ($n = 1000$, $k = 4$).

of σ^2 equal to the empirical variance of JMIs calculated for permuted samples which behaved more robustly. The results were evaluated by F measure (the harmonic mean of precision and recall). We have considered several benchmark BNs from BN repository https://www.bnlearn.com/bnrepository (asia, cancer, child, earthquake, sachs, survey). For each of them Y has been chosen as the variable having the largest MB. The results are given in Table 1. It is seen that with respect to F in the majority of cases GS-JMI method is the winner and ties with one of the other methods and the more detailed analysis indicates that this is due to its largest recall in comparison with GS-CMI and GS-CMI(sp) (see supplement [5]). This agrees with our initial motivation of considering such method which was the lack of power (i.e. missing important variables) by CMI-based tests.

Table 1. Values of F measure for GS algorithm using JMI, CMI and CMIsp tests. The winner is in bold.

Dataset	Y	MB size	JMI	CMI(sp)	CMI
Asia	Either	5	**0.58**	0.57	**0.58**
Cancer	Cancer	4	**0.78**	0.65	0.56
Child	Disease	8	0.55	**0.74**	0.55
Earthquake	Alarm	4	**0.87**	**0.87**	0.76
Sachs	PKA	7	0.83	**0.88**	0.59
Survey	E	4	**0.81**	0.52	0.54

6   Conclusions

We have proposed a new test of conditional independence based on approximation JMI of the conditional mutual information CMI and its asymptotic distributions. We have shown using synthetic data that the introduced test is more powerful than tests based on asymptotic or permutation distributions of

CMI when a conditioning set is large. In our analysis of real data sets we have indicated that the proposed test used in GS algorithm yields promising results in MB discovery problem. Drawback of such a test is that it disregards interactions between predictors and target variables of order higher than 3. Further research topics include systematic study of $\hat{J}^{\beta,\gamma}$ and especially how its parameters influence the power of the associated tests and feature selection procedures. Moreover, studying tests based on extended JMI including higher order terms is worthwhile.

References

1. Battiti, R.: Using mutual information for selecting features in supervised neural-net learning. IEEE Trans. Neural Netw. **5**(4), 537–550 (1994)
2. Brown, G., Pocock, A., Zhao, M., Luján, M.: Conditional likelihood maximisation: a unifying framework for information theoretic feature selection. J. Mach. Learn. Res. **13**(1), 27–66 (2012)
3. Cover, T.M., Thomas, J.A.: Elements of Information Theory (Wiley Series in Telecommunications and Signal Processing). Wiley-Interscience, New York (2006)
4. Guyon, I., Elyseeff, A.: An introduction to feature extraction. In: Guyon, I., Nikravesh, M., Gunn, S., Zadeh, L.A. (eds.) Feature Extraction. Studies in Fuzziness and Soft Computing, vol. 207, pp. 1–25. Springer, Heidelberg (2006). https://doi.org/10.1007/978-3-540-35488-8_1
5. Kubkowski, M., Łazęcka, M., Mielniczuk, J.: Distributions of a general reduced-order dependence measure and conditional independence testing: supplemental material (2020). http://github.com/lazeckam/JMI_CondIndTest
6. Kubkowski, M., Mielniczuk, J., Teisseyre, P.: How to gain on power: novel conditional independence tests based on short expansion of conditional mutual information (2019, submitted)
7. Lin, D., Tang, X.: Conditional infomax learning: an integrated framework for feature extraction and fusion. In: Leonardis, A., Bischof, H., Pinz, A. (eds.) ECCV 2006. LNCS, vol. 3951, pp. 68–82. Springer, Heidelberg (2006). https://doi.org/10.1007/11744023_6
8. Margaritis, D., Thrun, S.: Bayesian network induction via local neighborhoods. In: Proceedings of the 12th International Conference on Neural Information Processing Systems, NIPS 1999, pp. 505–511 (1999)
9. McGill, W.J.: Multivariate information transmission. Psychometrika **19**(2), 97–116 (1954). https://doi.org/10.1007/BF02289159
10. Pena, J.M., Nilsson, R., Bjoerkegren, J., Tegner, J.: Towards scalable and data efficient learning of Markov boundaries. Int. J. Approximate Reasoning **45**(2), 211–232 (2007)
11. Peng, H., Long, F., Ding, C.: Feature selection based on mutual information criteria of max-dependency, max-relevance and min-redundancy. IEEE Trans. Pattern Anal. Mach. Intell. **27**(1), 1226–1238 (2005)
12. Tsamardinos, I., Borboudakis, G.: Permutation testing improves bayesian network learning. In: Balcázar, J.L., Bonchi, F., Gionis, A., Sebag, M. (eds.) ECML PKDD 2010. LNCS (LNAI), vol. 6323, pp. 322–337. Springer, Heidelberg (2010). https://doi.org/10.1007/978-3-642-15939-8_21

13. Vergara, J.R., Estévez, P.A.: A review of feature selection methods based on mutual information. Neural Comput. Appl. **24**(1), 175–186 (2013). https://doi.org/10. 1007/s00521-013-1368-0
14. Yang, H.H., Moody, J.: Data visualization and feature selection: new algorithms for nongaussian data. In: Advances in Neural Information Processing Systems, vol. 12, pp. 687–693 (1999)
15. Zhang, J.T.: Approximate and asymptotic distributions of chi-squared type mixtures with applications. J. Am. Stat. Assoc. **100**, 273–285 (2005)

MCMC for Bayesian Uncertainty Quantification from Time-Series Data

Philip Maybank[1]() (iD), Patrick Peltzer[2], Uwe Naumann[2], and Ingo Bojak[3] (iD)

[1] Numerical Algorithms Group Ltd (NAG), Oxford, UK
philip.maybank@nag.co.uk
[2] Software and Tools for Computational Engineering (STCE), RWTH Aachen
University, Aachen, Germany
info@stce.rwth-aachen.de
[3] School of Psychology and Clinical Language Sciences, University of Reading,
Reading, UK
i.bojak@reading.ac.uk

Abstract. In computational neuroscience, Neural Population Models
(NPMs) are mechanistic models that describe brain physiology in a range
of different states. Within computational neuroscience there is growing
interest in the inverse problem of inferring NPM parameters from record-
ings such as the EEG (Electroencephalogram). Uncertainty quantification is essential in this application area in order to infer the mechanistic
effect of interventions such as anaesthesia.

This paper presents C++ software for Bayesian uncertainty quantification in the parameters of NPMs from approximately stationary data
using Markov Chain Monte Carlo (MCMC). Modern MCMC methods
require first order (and in some cases higher order) derivatives of the
posterior density. The software presented offers two distinct methods of
evaluating derivatives: finite differences and exact derivatives obtained
through Algorithmic Differentiation (AD). For AD, two different implementations are used: the open source Stan Math Library and the commercially licenced dco/c++ tool distributed by NAG (Numerical Algorithms Group). The use of derivative information in MCMC sampling is
demonstrated through a simple example, the noise-driven harmonic oscillator. And different methods for computing derivatives are compared.
The software is written in a modular object-oriented way such that it
can be extended to derivative based MCMC for other scientific domains.

Keywords: Uncertainty quantification · Algorithmic Differentiation ·
Computational neuroscience

1 Introduction

Bayesian uncertainty quantification is useful for calibrating physical models to
observed data. As well as inferring the parameters that produce the best fit
between a model and observed data, Bayesian methods can also identify the

© Springer Nature Switzerland AG 2020
V. V. Krzhizhanovskaya et al. (Eds.): ICCS 2020, LNCS 12143, pp. 707–718, 2020.
https://doi.org/10.1007/978-3-030-50436-6_52

range of parameters that are consistent with observations and allow for prior beliefs to be incorporated into inferences. This means that not only can predictions be made but also their uncertainty can be quantified. In demography, Bayesian analysis is used to forecast the global population [7]. In defence systems, Bayesian analysis is used to track objects from radar signals [1]. And in computational neuroscience Bayesian analysis is used to compare different models of brain connectivity and to estimate physiological parameters in mechanistic models [13]. Many more examples can be found in the references of [2,6,15].

We focus here on problems which require the use of Markov Chain Monte Carlo (MCMC), a widely applicable methodology for generating samples approximately drawn from the posterior distribution of model parameters given observed data. MCMC is useful for problems where a parametric closed form solution for the posterior distribution cannot be found. MCMC became popular in the statistical community with the re-discovery of Gibbs sampling [26], and the development of the BUGS software [15]. More recently it has been found that methods which use derivatives of the posterior distribution with respect to model parameters, such as the Metropolis Adjusted Langevin Algorithm (MALA) and Hamiltonian Monte Carlo (HMC) tend to generate samples more efficiently than methods which do not require derivatives [12]. HMC is used in the popular Stan software [4]. From the perspective of a C++ programmer, the limitations of Stan are as follows: it may take a significant investment of effort to get started. Either the C++ code has to be translated into the Stan modelling language. Or, alternatively, C++ code can be called from Stan, but it may be challenging to (efficiently) obtain the derivatives that Stan needs in order to sample efficiently.

The software that we present includes (i) our own implementation of a derivative-based MCMC sampler called simplified manifold MALA (smMALA). This sampler can be easily be used in conjunction with C++ codes for Bayesian data analysis, (ii) Stan's MCMC sampler with derivatives computed using dco/c++, an industrial standard tool for efficient derivative computation.

An alternative approach to the one presented in this paper would be simply to use Stan as a stand-alone tool without the smMALA sampler and without dco/c++. Determining the most efficient MCMC sampler for a given problem is still an active area of research, but at least within computational neuroscience, it has been found the smMALA performs better than HMC methods on certain problems [25]. Determining the most appropriate method for computing derivatives will depend on both the user and the problem at hand. In many applications Algorithmic Differentiation (AD) is needed for the reasons given in Sect. 2.3. The Stan Math Library includes an open-source AD tool. Commercial AD tools such as dco/c++ offer a richer set of features than open-source tools, and these features may be needed in order to optimize derivative computations. For example, the computations done using the Eigen linear algebra library [10] can be differentiated either using the Stan Math Library or using dco/c++ but there are cases where dco/c++ computes derivatives more efficiently than the Stan Math Library [22]. The aim of the software we present is to offer a range of options that both make it easy to get started and to tune performance.

2 Methods for Spectral Time-Series Analysis

2.1 Whittle Likelihood

The software presented in this paper is targeted at differential equation models with a stable equilibrium point and stochastic input. We refer to such models as stable SDEs. The methods that are implemented assume that the system is operating in a regime where we can approximate the dynamics through linearization around the stable fixed point. If the time-series data is stationary this is a reasonable assumption. Note that the underlying model may be capable of operating in nonlinear regimes such as limit cycles or chaos in addition to approximately linear dynamics. However, parameter estimation using data in nonlinear regimes quickly becomes intractable - see Chapter 2 of [16]. The stability and linearity assumptions are commonly made in the computational neuroscience literature, see for example [18].

In order to simplify the presentation we illustrate the software using a linear state-space model, which is of the form,

$$d\mathbf{X}(t) = A\,\mathbf{X}(t)dt + \mathbf{P}(t), \tag{1}$$

where the term $A\,\mathbf{X}(t)$ represents the deterministic evolution of the system and $\mathbf{P}(t)$ represents the noisy input. The example we analyze in this paper is the noise-driven harmonic oscillator, which is a linear state-space model with

$$A = \begin{pmatrix} 0 & 1 \\ -\omega_0^2 & -2\zeta\omega_0 \end{pmatrix}, \quad P(t) = \begin{pmatrix} 0 \\ dW(t) \end{pmatrix}, \tag{2}$$

and where $dW(t)$ represents a white noise process with variance σ_{in}^2. The observations are modelled as $Y_k = X_0(k \cdot \Delta t) + \epsilon_k$ with $\epsilon_k \sim N(0, \sigma_{obs}^2)$.

Our aim is to infer model parameters $(\omega_0, \zeta, \sigma_{in})$ from time-series data. This could be done in the time-domain using a Kalman filter, but it is often more computationally efficient to do inference in the frequency domain [3,17]. In the case where we only have a single output (indexed by i) and a single input (indexed by j), we can compute a likelihood of the parameters θ in the frequency domain through the following steps.

1. Compute the (left and right) eigendecomposition of A, such that,

$$A\mathcal{R} = \Lambda\mathcal{R}, \quad \mathcal{L}A = \mathcal{L}\Lambda, \quad \mathcal{L}\mathcal{R} = \text{diag}(c) \tag{3}$$

   where $\text{diag}(c)$ is a diagonal matrix, such that c_i is the dot product of the ith left eigenvector with the ith right eigenvector.
2. Compute ijth element of the transfer matrix for frequencies $\omega_1, \ldots, \omega_K$,

$$\mathcal{T}(\omega) = \mathcal{R}\,\text{diag}\left[\frac{1}{c_k(i\omega - \lambda_k)}\right]\mathcal{L}. \tag{4}$$

3. Evaluate the spectral density for component i of $\mathbf{X}(t)$, $f_{X_i}(\omega)$, and the spectral density for the observed time-series, $f_Y(\omega)$,

$$f_{X_i}(\omega) = |T_{ij}(\omega)|^2 f_{P_j}(\omega), \tag{5}$$

$$f_Y(\omega) = f_{X_i}(\omega) + \sigma_{obs}^2 \Delta t, \tag{6}$$

where $f_{P_j}(\omega)$ is the spectral density for component j of $\mathbf{P}(t)$.

4. Evaluate the Whittle likelihood,

$$p(y_0, \ldots, y_{n-1}|\theta) = p(S_0, \ldots, S_{n-1}|\theta) \approx \prod_{k=1}^{n/2-1} \frac{1}{f_Y(\omega_k)} \exp\left[-\frac{S_k}{f_Y(\omega_k)}\right], \tag{7}$$

where $\{S_k\}$ is the Discrete Fourier Transform of $\{y_k\}$. Note that θ represents a parameter set (e.g. specific values of $\omega_0, \zeta, \sigma_{in}$) that determines the spectral density.

The matrix A that parameterizes a linear state-space model is typically non-symmetric, which means that eigenvectors and eigenvalues will be complex-valued. We use Eigen-AD [22], a fork of the C++ linear algebra library Eigen [10]. Eigen is templated which facilitates the application of AD by overloading tools and Eigen-AD provides further optimizations for such tools. The operations above require an AD tool that supports differentiation of complex variables. AD of complex variables is considered in [24]. It is currently available in the feature/0123-complex-var branch of the Stan Math Library and in dco/c++ from release 3.4.3.

2.2   Markov Chain Monte Carlo

In the context of Bayesian uncertainty quantification, we are interested in generating samples from the following probability distribution,

$$p(\theta|y_0, \ldots, y_{n-1}) \propto p(y|\theta)p(\theta), \tag{8}$$

where $p(y|\theta)$ is the likelihood of the parameter set θ given observed data $y_0, \ldots, y_{n-1}$, and $p(\theta)$ is the prior distribution of the parameters. In many application where the likelihood $p(y|\theta)$ is based on some physical model we cannot derive a closed form expression for the posterior density $p(\theta|y)$. Markov Chain Monte Carlo (MCMC) has emerged over the last 30 years as one of the most generally applicable and widely used framework for generating samples from the posterior distribution [6,9]. The software used in this paper makes use of two MCMC algorithms: the No U-Turn Sampler (NUTS) [12] and the simplified manifold Metropolis Adjusted Langevin Algorithm (smMALA) [8]. NUTS is called via the Stan environment [4]. It is a variant of Hamiltonian Monte Carlo (HMC), which uses the gradient (first derivative) of the posterior density, whereas smMALA uses the gradient and Hessian (first and second derivatives) of the posterior density, smMALA is described in Algorithm 1.

The error in estimates obtained from MCMC is approximately $C/\sqrt{N}$, where N is the number of MCMC iterations and C is some problem-dependent constant. In general it is not possible to demonstrate that MCMC has converged, but there are several diagnostics that can indicate non-convergence, see Section 11.4 of [6] for more detail. Briefly, there are two phases of MCMC sampling: burn-in and the stationary phase. Burn-in is finished when we are in the region of the parameter space containing the true parameters. In this paper we restrict ourselves to synthetic data examples. In this case it is straight-forward to assess whether the sampler is burnt in by checking whether the true parameters used to simulate the data are contained in the credible intervals obtained from the generated samples. In real data applications, it is good practice to test MCMC sampling on a synthetic data problem that is analogous to the real data problem. During the stationary phase we assess convergence rate using the effective sample size, N Eff. If we were able to generate independent samples from the posterior then the constant C is $\mathcal{O}(1)$. MCMC methods generate correlated samples, in which case C may be $\gg 1$. A small N Eff (relative to N) indicates that this is the case.

If we are sampling from a multivariate target distribution, N Eff for the ith component is equal to,

$$\frac{S}{1 + 2\sum_k \hat{\rho}_i(k)}, \tag{9}$$

where S is the number of samples obtained during the stationary period, and $\hat{\rho}_i(k)$, is an estimate of the autocorrelation at lag k for the ith component of the samples. This expression can be derived from writing down the variance of the average of a correlated sequence (Chapter 11 of [6]). The key point to note is that if the autocorrelation decays slowly N Eff will be relatively small.

Algorithm 1: smMALA

Input: Data, y; Initial value for $\theta = (\theta_1, \ldots, \theta_N)$;
       Likelihood $l(y|\theta)$; Prior distribution $p(\theta)$.
Parameters : Step size, h; Number of iterations, I
for $i = 2, \ldots, I$ **do**

> Evaluate gradient and Hessian of the unnormalized log posterior,
> $$g_\theta = \nabla\left[\log[\ l(y|\theta)\ p(\theta)\ ]\right] \text{ and } \mathbf{H}_\theta;$$
> Set $C = h^2\mathbf{H}_\theta^{-1}$, and $m = \theta + \frac{1}{2}C g_\theta$;
> Propose $\theta^* \sim q(\theta^*|\theta) = N(m, C)$;
> Evaluate acceptance ratio, $\alpha(\theta, \theta^*) = \min\left[1, \dfrac{l(y|\theta^*)p(\theta^*)q(\theta|\theta^*)}{l(y|\theta)p(\theta)q(\theta^*|\theta)}\right]$;
> Draw $u \sim \text{Uniform}(0, 1)$;
> **if** $u < \alpha(\theta, \theta^*)$ **then** Set $\theta = \theta^*$

end

2.3 Derivative Computation

A finite difference approximation to the first and second derivatives of the function $F(\theta) : \mathbb{R}^N \to \mathbb{R}^M$ can be computed as,

$$\frac{\partial F}{\partial \theta_i} = \frac{F(\theta + he_i) - F(\theta)}{h}, \qquad \frac{\partial^2 F}{\partial \theta_i \partial \theta_j} = \frac{\frac{\partial F}{\partial \theta_i}(\theta + he_j) - \frac{\partial F}{\partial \theta_i}(\theta)}{h},$$

where e_i is the ith Cartesian basis vector, and h is a user-defined step-size. For first-order derivatives the default value we use is $\sqrt{\epsilon}|\theta_i|$, where ϵ is machine epsilon for double types, i.e., if θ_i is $\mathcal{O}(1)$, $h \approx 10^{-8}$. For second derivatives the default value is $\epsilon^{1/3}|\theta_i|$, so that $h \approx 5 \cdot 10^{-6}$. More details on the optimal step size in finite difference approximations can be found in [20]. First derivatives computed using finite differences require $\mathcal{O}(N)$ function evaluations, where N is the number of input variables. Second derivatives require $\mathcal{O}(N^2)$ function evaluations.

Derivatives that are accurate to machine precision can be computed using AD tools. AD tools generally use one of two modes: tangent (forward) or adjoint (reverse). For a function with N inputs and M outputs, tangent mode requires $\mathcal{O}(N)$ function evaluations and adjoint mode requires $\mathcal{O}(M)$ function evaluations. In statistical applications our output is often a scalar probability density, so adjoint AD will scale better with N than either tangent mode or finite differences in terms of total computation time. Adjoint AD is implemented as follows using the Stan Math Library, [5].

```
// Define and init matrix with Stan's adjoint type
Matrix<stan::math::var,Dynamic,1> theta = input_values;

// Compute primal and gradient
stan::math::var lp = computation(theta);
lp.grad();

// Extract gradient
Matrix<double,Dynamic,1> grad_vec(theta.size());
for(int j=0; j<theta.size(); j++) grad_vec(j) = theta(j).adj();
```

In the code above the function is evaluated using the `stan::math:var` scalar type rather than `double`. During function evaluation, derivative information is recorded. When the `grad()` function is called this derivative information is interpreted and derivative information is then accessed by calling `adj()` on the input variables. Similarly to the Stan Math Library, exact derivatives can be computed using the NAG `dco/c++` tool developed in collaborations with RWTH Aachen University's STCE group, [14].

```
// Define and init matrix with dco's adjoint type
typedef dco::ga1s<double> DCO_MODE;
```

```
typedef DCO_MODE::type Scalar;
Matrix<Scalar,Dynamic,1> theta = input_values;

// Enable dynamic activity analysis
for(int j=0; j<theta.size(); j++)
  DCO_MODE::global_tape->register_variable(theta(j));

// Compute primal and gradient
Scalar lp = computation(theta);
dco::derivative(lp) = 1.0;
DCO_MODE::global_tape->interpret_adjoint();

// Extract gradient
Matrix<double,Dynamic,1> grad_vec(theta.size());
for(int j=0; j<theta.size(); j++)
  grad_vec(j) = dco::derivative(theta(j));
```

dco/c++ provides its adjoint type using the `dco::ga1s<T>::type` typedef, where T s the corresponding primal type (e.g. `double`). Higher order adjoints can be achieved by recursively nesting the adjoint type. Using this type, the program is first executed in the augmented forward run, where derivative information is stored in the `global_tape` data structure. The `register_variable` function initialises recording of the tape and facilitates dynamic varied analysis [11]. Derivative information recorded during the augmented primal run is then interpreted using the `interpret_adjoint()` function and `dco::derivative` is used to access the propagated adjoints.

3   Results for Noise Driven Harmonic Oscillator

We now present the results of MCMC sampling using smMALA to estimate parameters of the noise driven harmonic oscillator. Then we demonstrate that, for this particular model, MCMC sampling can be accelerated by using NUTS. We also compare run times and sampling efficiency when AD is used to evaluate derivatives. Here we estimate parameters using synthetic data (i.e. data generated from the model). This is a useful check that the MCMC sampler is working correctly: we should be able to recover the parameters that were used to simulate the data. We generate a pair of datasets representing different conditions (which we label c_1 and c_2). In Table 1 we show that the 95% credible intervals for each parameter include the actual parameter values for all 5 parameters. These results came from one MCMC run of 10, 000 iterations. Figure 1 shows 95% confidence intervals for the spectral density obtained using the Welch method alongside 95% credible intervals for the spectral density estimated from 10, 000 MCMC iterations. This is another useful check: the spectral density predictions generated by sampled parameter sets are consistent with non-parametric estimates of the spectral density.

Table 1. Estimated quantiles of posterior distribution for noise driven harmonic oscillator. Synthetic data was generated by simulating from the model: duration = 20.0, time-step = 0.01. Two datasets (c_1 and c_2) were generated with different parameter values for ω_0 and σ_{in} in each dataset (values are given in 'actual' column). The quantiles are estimated from a sequence of 10, 000 MCMC samples from the joint (posterior) distribution of all the parameters given all of the data. For example, 2.5% of the MCMC samples had $\omega_0(c_1)$ values less than 77.3.

	Actual	Quantile		
		0.025	0.50	0.975
$\omega_0(c_1)$	80	77.3	80.3	81.9
$\omega_0(c_2)$	40	36.8	38.8	41.3
$\sigma_{in}(c_1)$	100	92	101	111
$\sigma_{in}(c_2)$	10	9.81	10.8	12.4
ζ	0.2	0.164	0.193	0.223

Table 2 uses the same model and the same datasets as above to compare smMALA with NUTS, and finite differences with AD. The AD implementation used in smMALA was dco's tangent over adjoint mode (i.e. `dco::gt1s` combined with `dco::ga1s`). In NUTS we used the dco's tangent mode for computing the derivatives of the spectral density and Stan's adjoint mode (`stan::math::var`) for the rest of the computation. Given the most recent developments in the `feature/0123-complex-var` branch of the Stan Math Library it would likely be possible to use Stan's adjoint mode for the whole computation. However this was not the case when we started writing the code. In general users may find that they need the more advanced functionality of `dco/c++` for part of their computation in order to obtain good performance.

The MCMC samplers were each run for 1,000 iterations. The results were analyzed using Stan's `stansummary` command-line tool. To account for correlation between MCMC samples we use the *N Eff* diagnostic defined in Sect. 2.2 to measure sampling efficiency, min *N Eff* is the minimum over the 5 model parameters. Table 2 shows that, for the noise driven harmonic oscillator, we can accelerate MCMC sampling by a factor of around 3–5 by using NUTS rather than smMALA. We also see that the *N Eff/s* is higher for finite differences than for AD, because of the small number of input variables. However, even for this simple model the NUTS min *N Eff* is higher for AD than for finite differences suggesting that the extra accuracy in the derivatives results in more efficient sampling per MCMC iteration.

In the context of Bayesian uncertainty quantification what we are interested in is whether the MCMC samples are an accurate representation of the true posterior distribution. A necessary condition for posterior accuracy is that the true parameters are (on average) contained in the credible intervals derived from the MCMC samples. This is the case for all the different variants of MCMC that we tested. The main differences we found between the different variants was

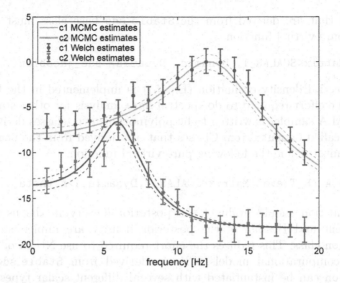

Fig. 1. Spectral density estimates for noise driven harmonic oscillator for the datasets described in Table 1.

the sampling efficiency. As discussed in Sect. 2.2, a sampling efficiency that is 3–5 time greater means a reduction in the value of C in the Monte Carlo error $(C/\sqrt{N})$ by that factor. Another way of interpreting this is that we could reduce the number of MCMC iterations by a factor of 3–5 and expect to obtain the same level of accuracy in the estimated posterior distribution.

Table 2. Noise-driven harmonic oscillator benchmarking results.

MCMC sampler	Derivative implementation	CPU time (s)	min N Eff	min N Eff/s
smMALA	Finite differences	3.3	152	46
smMALA	dco/c++	5.2	152	29
NUTS	Finite differences	3.2	485	153
NUTS	Stan and dco/c++	3.9	506	130

4   Discussion

The Whittle likelihood function is written to be polymorphic over classes derived from the a base class called `Stable_sde`. The function signature includes a reference to the base class.

```
template<typename SCALAR_T>
SCALAR_T log_whittle_like(Stable_sde<SCALAR_T> & m, ...);
```

Classes that are derived from the `Stable_sde` base class must define the following pure virtual function.

```
virtual Matrix<SCALAR_T, Dynamic, Dynamic> sde_jacobian(...) = 0;
```

The spectral density evaluation (Eq. (5)) is implemented in the base class, reducing the effort required to do spectral data analysis for other stable SDEs. The smMALA sampler is written to be polymorphic over classes derived from a base class called `Computation`. Classes that are derived from the `Computation` base class must define the following pure virtual function.

```
virtual SCALAR_T eval(Matrix<SCALAR_T,Dynamic,1>& theta,...) = 0;
```

This function should evaluate the posterior density in the user's model. The gradient and Hessian of the posterior density are implemented in the `Computation` class. This reduces the effort required to use NUTS or smMALA for other computational models. Classes derived from `Stable_sde` or from `Computation` can be instantiated with several different scalar types including `double`, `stan::math::var`, and a number of `dco/c++` types. This enables automatic evaluation of first and second derivatives. The gradient and Hessian functions need a template specialization to be defined in the base class for each different scalar type that is instantiated.

We conclude with some comments regarding the choice of MCMC sampler and the method for evaluating derivatives. Our software only implements derivative-based MCMC as this tends to be more computationally efficient than other MCMC methods [12, 23]. Derivative-based MCMC samplers can be further subdivided into methods that use higher-order derivatives, such as smMALA (sometimes referred to as Riemannian) and methods that only require first-order derivatives (such as NUTS). Riemannian methods tend to be more robust to complexity in the posterior distribution, but first-order methods tend to be more computationally efficient for problems where the posterior geometry is relatively simple. We recommend using smMALA in the first instance, then NUTS as a method that may provide acceleration in MCMC sampling, in terms of effective samples per second, *N Eff/s*. Regarding the derivative method, finite differences often results in adequate performance for problems with moderate input dimension (e.g. 10–20), at least with smMALA. But for higher-dimensional problems (e.g. Partial Differential Equations or Monte Carlo simulation) we recommend accelerating derivative computation using adjoint AD [19, 21]. The software presented enables users to implement and benchmark all these alternatives so that the most appropriate methods for a given problem can be chosen.

References

1. Arulampalam, M.S., Maskell, S., Gordon, N., Clapp, T.: A tutorial on particle filters for online nonlinear/non-Gaussian Bayesian tracking. IEEE Trans. Signal Process. **50**(2), 174–188 (2002)
2. Bishop, C.M.: Pattern Recognition and Machine Learning. Springer, New York (2006)
3. Bojak, I., Liley, D.: Modeling the effects of anesthesia on the electroencephalogram. Phys. Rev. E **71**(4), 041902 (2005)
4. Carpenter, B., et al.: Stan: A probabilistic programming language. J. Stat. Softw. **76**(1), 1–32 (2017)
5. Carpenter, B., Hoffman, M.D., Brubaker, M., Lee, D., Li, P., Betancourt, M.: The Stan math library: reverse-mode automatic differentiation in C++. arXiv preprint arXiv:1509.07164 (2015)
6. Gelman, A., Carlin, J.B., Stern, H.S., Dunson, D.B., Vehtari, A., Rubin, D.B.: Bayesian Data Analysis. CRC Press, Boca Raton (2013)
7. Gerland, P., et al.: World population stabilization unlikely this century. Science **346**(6206), 234–237 (2014)
8. Girolami, M., Calderhead, B.: Riemann manifold Langevin and Hamiltonian Monte Carlo methods. J. Roy. Stat. Soc.: Ser. B (Stat. Methodol.) **73**(2), 123–214 (2011)
9. Green, P.J., Łatuszyński, K., Pereyra, M., Robert, C.P.: Bayesian computation: a summary of the current state, and samples backwards and forwards. Stat. Comput. **25**(4), 835–862 (2015)
10. Guennebaud, G., Jacob, B., et al.: Eigen v3 (2010). http://eigen.tuxfamily.org
11. Hascoët, L., Naumann, U., Pascual, V.: "To be recorded" analysis in reverse-mode automatic differentiation. Future Gen. Comput. Syst. **21**(8), 1401–1417 (2005)
12. Hoffman, M.D., Gelman, A.: The No-U-Turn sampler: adaptively setting path lengths in Hamiltonian Monte Carlo. J. Mach. Learn. Res. **15**(1), 1593–1623 (2014)
13. Kiebel, S.J., Garrido, M.I., Moran, R.J., Friston, K.J.: Dynamic causal modelling for EEG and MEG. Cogn. Neurodyn. **2**(2), 121 (2008)
14. Leppkes, K., Lotz, J., Naumann, U.: Derivative code by overloading in C++ (dco/c++): introduction and summary of features. Technical report AIB-2016-08, RWTH Aachen University, September 2016. http://aib.informatik.rwth-aachen.de/2016/2016-08.pdf
15. Lunn, D., Jackson, C., Best, N., Thomas, A., Spiegelhalter, D.: The BUGS Book: A Practical Introduction to Bayesian Analysis. CRC Press, Boca Raton (2012)
16. Maybank, P.: Bayesian inference for stable differential equation models with applications in computational neuroscience. Ph.D. thesis, University of Reading (2019)
17. Maybank, P., Bojak, I., Everitt, R.G.: Fast approximate Bayesian inference for stable differential equation models. arXiv preprint arXiv:1706.00689 (2017)
18. Moran, R.J., Stephan, K.E., Seidenbecher, T., Pape, H.C., Dolan, R.J., Friston, K.J.: Dynamic causal models of steady-state responses. NeuroImage **44**(3), 796–811 (2009)
19. NAG: NAG algorithmic differentiation software. https://www.nag.com/content/algorithmic-differentiation-software. Accessed 27 Jan 2020
20. NAG: OptCorner: the price of derivatives - using finite differences. https://www.nag.co.uk/content/optcorner-price-derivatives-using-finite-differences. Accessed 27 Jan 2020
21. Naumann, U., du Toit, J.: Adjoint algorithmic differentiation tool support for typical numerical patterns in computational finance. J. Comput. Finan. **21**(4), 23–57 (2018)

22. Peltzer, P., Lotz, J., Naumann, U.: Eigen-AD: Algorithmic differentiation of the Eigen library. arXiv preprint arXiv:1911.12604 (2019)
23. Penny, W., Sengupta, B.: Annealed importance sampling for neural mass models. PLoS Comput. Biol. **12**(3), e1004797 (2016)
24. Pusch, G., Bischof, C., Carle, A.: On Automatic Differentiation of Codes with Complex Arithmetic with Respect to Real Variables, September 1995
25. Sengupta, B., Friston, K.J., Penny, W.D.: Gradient-based MCMC samplers for dynamic causal modelling. NeuroImage **125**, 1107–1118 (2016)
26. Smith, A.F., Roberts, G.O.: Bayesian computation via the Gibbs sampler and related Markov chain Monte Carlo methods. J. Roy. Stat. Soc.: Ser. B (Methodol.) **55**(1), 3–23 (1993)

Uncertainty Quantification for Multiscale Fusion Plasma Simulations with VECMA Toolkit

Jalal Lakhlili[⊠], Olivier Hoenen, Onnie O. Luk, and David P. Coster

Max-Planck Institute for Plasma Physics,
Boltzmannstrasse 2, 85748 Garching, Germany
jalal.lakhlili@ipp.mpg.de

Abstract. Within VECMAtk platform we perform Uncertainty Quantification (UQ) for multiscale fusion plasmas simulations. The goal of VECMAtk is to enable modular and automated tools for a wide range of applications to archive robust and actionable results. Our aim in the current paper is to incorporate suitable features to build UQ workflow over the existing fusion codes and to tackle simulations on high performance parallel computers.

Keywords: Fusion plasmas · Uncertainty quantification · Sensitivity analysis · Polynomial Chaos Expansion · Quasi-Monte Carlo method · Multiscale modeling · High performance computing

1 Introduction

Thermonuclear fusion can potentially provide a carbon free and safe solution to the provision of base load electricity. Heat and particle transport will play the key roles in determining the size and efficiency of future fusion power reactor, so understanding their mechanisms is an important milestone towards the realization of fusion-based electricity. Our present understanding is that turbulence at small space and time scales is responsible for much of this transport, but the profiles of temperature and density evolve over much larger space and time scales. To study these phenomena, multiscale and multiphysics applications have been developed throughout the years. Notable efforts in Europe lean toward implementing such applications by following a workflow approach, in which several *single-scale* (or single-physics) codes are coupled together [1,2].

When one element in such a workflow is a turbulence code, whose outputs are inherently noisy due to the chaotic nature of the turbulence (intrinsic uncertainties), or if input data such as external sources contain uncertainties (extrinsic), these need to be propagated through different models of the workflow. The goal, therefore, is to produce temperature and density profiles, along with their confidence intervals. This information will then allow for an improved validation of the simulation results against the experimental measurements that come with error bars.

© Springer Nature Switzerland AG 2020
V. V. Krzhizhanovskaya et al. (Eds.): ICCS 2020, LNCS 12143, pp. 719–730, 2020.
https://doi.org/10.1007/978-3-030-50436-6_53

Uncertainty quantification in the existing fusion workflow is implemented with the VECMA open source toolkit[1], which provides tools to facilitate the verification, validation and uncertainty quantification (VVUQ) process in multiscale, multiphysics applications [3]. So far we have integrated two elements from this toolkit: the EasyVVUQ library [4] to incorporate uncertainty quantification (UQ) and sensitivity analysis (SA) in a non-intrusive manner, and the QCG Pilot-Job middleware [5] to execute efficiently the large number of tasks or samples prepared by EasyVVUQ.

In this work we focus on the different sampling methods used to quantify the uncertainties, their results and performance for two different black box use cases: on the entire workflow for extrinsic uncertainties coming from the heating source terms and boundary conditions for temperature profiles (on electron and ions), and on the turbulence model for uncertain ions and electron temperatures values and gradients. This paper is organized as follows. After a brief overview on the background theory for uncertainty quantification, we present the various sampling methods used in our application that are also available in the EasyVVUQ. We then describe our two use cases together with UQ and SA results for the method that demonstrates the best performance. Finally, we conclude with future works related to more intrusive approaches.

2 Theory Overview

We consider a numerical model $\mathcal{M}$ that has a number d of uncertain input parameters $Q_1, Q_2, ..., Q_d$, and gives the output Y:

$$Y = \mathcal{M}(x, Q). \tag{1}$$

Here, Q is the vector $[Q_1, Q_2, ..., Q_d]$ and x is a space variable. The model can also depends on other variables (e.g. time) and on additional but invariant input parameters. In this study we focus on the uncertain parameters $(Q_i)_{i=1}^d$ and how the uncertainty is propagated through the model $\mathcal{M}$. That way we can quantify the uncertainty in the output Y and identify the relative contribution of each Q_i into this uncertain result.

First, we assume that the uncertain parameters are statistically independent components, and each one has an univariate probability density function f_{Q_i}, hence the vector Q can be described by the joint probability density function:

$$f_Q = \prod_{i=1}^d f_{Q_i}. \tag{2}$$

Thereafter, we seek relevant properties of the output Y that characterize the uncertainty propagation and where possible, we build an approximation of the unknown output distribution f_Y.

---

[1] www.vecma-toolkit.eu.

2.1   Statistical Moments

The principal aim of UQ is to describe the output distribution f_Y through useful statistics [6,7]. Several metrics are based on statistical moments, and among these the most commonly used are the mean $\mathbb{E}$ and the variance $\mathbb{V}$:

$$\mathbb{E}[Y] = \int_{\Omega_Y} y f_Y(y) dy, \tag{3}$$

$$\mathbb{V}[Y] = \int_{\Omega_Y} (y - \mathbb{E}[Y])^2 f_Y(y) dy. \tag{4}$$

Here, Ω_Y is the output space.

The mean indicates the expected value of the model output Y and the variance measures how much the output varies around the mean. In general, instead of using the variance, we take its square root to obtain the standard deviation σ:

$$\sigma[Y] = \sqrt{\mathbb{V}[Y]}. \tag{5}$$

The advantage of using σ instead of the variance is that, like the mean, σ has the same units as Y and that makes it easier to interpret.

Another useful measure for UQ is percentiles which can be used to define confidence intervals [6].

2.2   Sensitivity Analysis

The principal aim of SA is to evaluate the contribution of different inputs parameters $(Q_i)_{i=1}^{d}$ to the uncertainty in output Y. Among the various sensitivity metrics that exist, we choose in this study the variance-based sensitivity analysis by computing the Sobol indices [8,9]. It is the most commonly used metric as it allows to measure sensitivity across the whole input space and can cope with nonlinear outputs.

There are several orders of Sobol indices. Here, we select two of the most-used indices in UQ: the first-order Sobol index S_i and the total order index S_{Ti}. They measure the direct effect of the parameter Q_i on the variance of Y and the interactions between Q_i and other parameters, respectively:

$$S_i = \frac{\mathbb{V}[\mathbb{E}[Y|Q_i]]}{\mathbb{V}[Y]}, \tag{6}$$

$$S_{Ti} = 1 - \frac{\mathbb{V}[\mathbb{E}[Y|Q_{-i}]]}{\mathbb{V}[Y]}. \tag{7}$$

We denote by $\mathbb{E}[Y|Q_i]$ the conditional mean, which represents the expected value of Y for a fixed value of Q_i, and by Q_{-i} all uncertain parameters except Q_i.

The first-order Sobol index, also known as the main effect index, cannot exceed one, and if it is close to zero we can claim that its input parameter have a small or no effect on the output. The higher the index value is, the more influence the associated parameter has on the resulting uncertainty. For the total Sobol indices, their sum is equal to or greater than one, and it is only equal to one if there is no interaction between the parameters [10].

3   Sampling Approaches

Sampling is the usual computational method used to estimate uncertainty measures, statistical moments and sensitivity as they have been described in the previous section. We can choose between several widely used sampling methods for a non-intrusive (or black box) approach: Quasi-Monte Carlo (QMC) [11] or Polynomial Chaos Expansion (PCE) [12]. This choice can be motivated by the numerical cost of such estimation, which is proportional to the number of samples and mainly depends on the number of uncertain input parameters.

3.1   Quasi-Monte Carlo

Monte Carlo is one of the most widespread method due to its simplicity and ease of implementation. The basic idea of the method is a purely random selection of sample evaluations without assumptions about the model. The number of samples is independent of the number of uncertain parameters, but a very high number is required in order to obtain reliable statistics, thus UQ estimations can rapidly become computationally expensive.

QMC methods are different approaches that aim at improving Monte Carlo method. They are based on variance-reduction techniques to reduce the number of model evaluations needed [11,13]. In this work, instead of random selection of samples from f_Q, we use Sobol sequences which have a low-discrepancy sequence [8]. Using N_s samples to evaluate the model $\mathcal{M}$, which gives the output results $Y = [Y_1, Y_2, ..., Y_{N_s}]$, the mean and the variance can approximated by:

$$\mathbb{E}[Y] \approx \frac{1}{N_s} \sum_{i=1}^{N_s} Y_i, \tag{8}$$

$$\mathbb{V}[Y] \approx \frac{1}{N_s - 1} \sum_{i=1}^{N_s} (Y_i - \mathbb{E}[Y])^2. \tag{9}$$

For the SA, we use Saltelli's procedure [9,14]. It is also based on Sobol sequences and to obtain a full set of main and total sensitivity indices, it is sufficient to take:

$$N_s = \frac{(d+2)N}{2}, \tag{10}$$

where N is the number of samples required to get a given accuracy.

3.2   Polynomial Chaos Expansion

The PCE method is one of most competitive alternatives to (Quasi-)Random methods. The output Y is expanded into a series of orthogonal polynomials $(P_j)_j$ of degree p, which are functions of the input parameters Q and are chosen such that they are orthogonal to the input distributions [7,15–17]:

$$Y \approx \widehat{Y}(Q) = \sum_{j=1}^{N_p} c_j P_j(Q). \tag{11}$$

Here, N_p is the number of expansion factors, which depends on both the degree p and the number of the uncertain parameters d, and is given by the binomial coefficient:

$$N_p = \binom{d+p}{d}; \tag{12}$$

And $(c_j)_j$ are the expansion coefficients, which can be estimated using tensor product quadrature or linear regression. These two methods that will be detailed later in this section.

By exploiting the properties of the orthogonal polynomials, the expectation value and the variance for the output model are:

$$\mathbb{E}[Y] \approx c_1, \tag{13}$$

$$\mathbb{V}[Y] \approx \sum_{j=2}^{N_p} \gamma_j c_j \tag{14}$$

Here, γ_j is a normalization factor and it is defined as: $\gamma_j = \mathbb{E}[P_j^2(Q)]$.

For the sensitivity analysis, the first and total-order Sobol indices can also be computed using Sobol's variance decomposition and by exploiting the polynomial chaos expansion properties [16,18].

The estimation of the expansion coefficients is the most important step in the PCE construction. As mentioned before, we use two of the most common methods: spectral projection with tensor product quadrature and linear regression.

Spectral Projection. In this PCE variant, each expansion coefficient c_j is estimated by projecting the solution onto the output space Ω_Y. This is analogous to the calculation of Fourier coefficients in the approximation of periodic function. By exploiting the orthogonality of the polynomial, Eq. 11 yields:

$$c_j = \frac{1}{H_j} \int_{\Omega_Y} Y(q) P_j(q) f_Q(q) dq, \tag{15}$$

where H_j is a normalization factor associated with the polynomial P_j.

To compute the integral from Eq. 15, we use quadrature generators with weights $(\omega_k)_k$ and nodes $(q_k)_k$, and then we evaluate the outputs at those nodes. As the integrands are smooth polynomials, c_j can be approximated with a good accuracy by the expression:

$$c_j \approx \frac{1}{H_j} \sum_{k=1}^{N_s} Y(q_k) P_j(q_k) \omega_k. \tag{16}$$

The number of samples N_s in this case is determined by the quadrature rule. For our applications, we use tensored quadrature with Gaussian schemes and in this case N_s is:

$$N_s = (p+2)^d. \tag{17}$$

As the required number of samples increases exponentially with the number of uncertain parameters, this method is beneficial only for a relatively small number of parameters.

Regression. This variant, also known as point collocation, uses a single linear least square approximation of the form:

$$\begin{bmatrix} P_1(q_1) & P_2(q_1) & \dots & P_{N_p}(q_1) \\ P_1(q_2) & P_2(q_2) & \dots & P_{N_p}(q_2) \\ \vdots & \vdots & \ddots & \vdots \\ P_1(q_{N_s}) & P_2(q_{N_s}) & \dots & P_{N_p}(q_{N_s}) \end{bmatrix} \begin{bmatrix} c_1 \\ c_2 \\ \vdots \\ c_{N_p} \end{bmatrix} = \begin{bmatrix} Y(q_1) \\ Y(q_2) \\ \vdots \\ Y(q_{N_s}) \end{bmatrix} \tag{18}$$

The set of linear equations above results from the expansion of Eq. 11 at a set of collocation nodes $(q_k)_{k=1}^{N_s}$. To get stable nodes, we use pseudo-random samples generated from the input distribution f_Q with specific rules as indicated in Chaospy [19]. To avoid an under-determined system, we should choose a number of samples such that $N_s \geq N_p$. According to [20], it is recommended that we use at least $2N_p$ samples. Therefore, we set:

$$N_s = 2\frac{(p+d)!}{p!d!}. \tag{19}$$

In this case, we obtain a good least square approximation which becomes significantly more affordable than the full tensor product quadrature used in the spectral projection.

3.3   Implementation and Choice

All these sampling methods and variants are available, among others, in the EasyVVUQ[2] library; it provides a simple decomposition of UQ and SA as generic steps (*sampling, encoder, decoder, collation, analysis*) in a modular workflow approach. Sampling and analysis methods are mainly based on existing Python packages such as Chaospy [19], which provides probability distributions, statistics, quadrature generators and SA for PCE, and SALib [21] for SA with the QMC method. This decomposition allows users to exchange sampling methods easily without impacting the overall structure of the application.

    In terms of sample sizes, we can use PCE with spectral projection as long as the number of uncertain parameters is low, typically smaller than 6. For a moderate number of parameters, PCE with regression becomes advantageous, and for a high number of parameters, quasi-Monte Carlo is the only method that is applicable on current computing platforms. In terms of accuracy, for smooth problems, PCE with spectral projection is expected to be more efficient compared with the regression variant or QMC. A more in-depth analysis and comparison between these methods and others can be found in [16,19,20].

---

[2] github.com/UCL-CCS/EasyVVUQ.

4 Numerical Applications and Results

To perform UQ and SA on any multiscale application that couples single-scale models as a workflow, various level of intrusiveness can be considered. In a non-intrusive approach, either the entire workflow is treated as a black box, or only one single-scale model is being studied and treated as a black box. In a semi-intrusive approach, each single scale code is placed inside a black box, and UQ and SA are performed on the entire workflow.

In this section, we will present the two use cases for non-intrusive approach with the fusion multiscale application, and discuss their results as well as some performance considerations.

4.1 Multiscale Fusion Workflow as a Black Box

The multiscale fusion workflow[3] was created to study the effects of turbulence happening at the micro time and space scales on the evolution of temperature profiles at the macro level of the fusion device. Currently, the workflow combines three main models as shown in Fig. 1: an equilibrium code that describes the plasma geometry, a turbulence code that approximates transport coefficient, and a transport solver that evolves temperature profiles at the macro time scale.

When considering the entire workflow as a black box, we use a simpler analytical model for the turbulence to reduce drastically the computational cost of the simulation. The extrinsic uncertainties we consider are:

- Boundary conditions defining the electron temperature at the plasma edge.
- Simplified heating sources, for which electron heating power Gaussian distributions are characterized by their amplitude, width, and position.

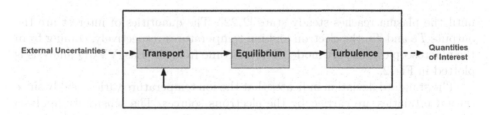

Fig. 1. Sketch of the targeted fusion workflow diagram

We select the PCE method, described in Sect. 3, and assume that each of the uncertain parameters has a normal distribution in the range of ±20% around its original value. The workflow runs within a black-box for multiple time iterations

---

[3] github.com/vecma-ipp/MFW.

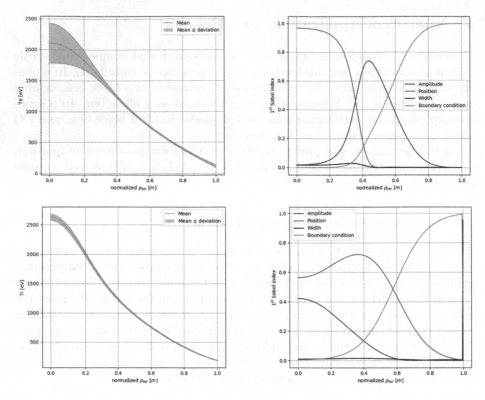

Fig. 2. Descriptive statistics and sensitivity analysis for electron and ion temperatures: on the left hand-side, the expected value and standard deviation are shown with respect to the normalized toroidal flux coordinate ρ_{tor}, and on the right hand-side the first-order Sobol indices are plotted for each of the uncertain parameters.

until the plasma reaches steady state [2, 22]. The quantities of interest are the outputs Te and Ti, the electron and ion temperatures respectively, coming from the macroscopic transport model. The outcome from the EasyVVUQ analysis is plotted in Fig. 2.

The standard deviation indicates that the ion temperature varies weakly since the uncertainties are carried by the electrons sources. The sensitivity analysis reveals that the variance in the electron and ion temperatures is mainly due to the uncertainty from three parameters: the position and amplitude parameters of the sources at core region of the plasma ($\rho_{tor} = 0$) and, as expected, boundary condition parameter at the edge region ($\rho_{tor} = 1$). The parameter width has no direct effect on the variance of the two quantities, so according to [23], this parameter can be neglected and then the number of samples can be reduced while keeping the same variance behavior.

We did the same simulation using uncertain parameters from heating sources and temperature boundary conditions for ions, and both electrons and ions. In all cases the UQ and SA results are qualitatively the same.

This use case was performed using both PCE variants: spectral projection and regression. They required 1296 and 140 samples, respectively, because we have 4 uncertain parameters and use quadratic polynomials. The results are qualitatively the same, and the differences between output means and variances of each method are at the order of $\mathcal{O}(10^{-6})$.

4.2   3D Gyrofluid Turbulence as a Black Box

In the workflow, given a plasma state and the geometry, the turbulence code provides the transport coefficients by computing the turbulent fluxes of particles and energy. Due to the chaotic nature of the turbulence model, these outputs are inherently noisy. And running such stochastic model is computationally expensive. Therefore, we begin the UQ study by isolating a 3D gyrofluid turbulence model that computes heat and particle fluxes [24] and consider one flux tube. We vary the electron and ion temperatures, Te and Ti, and their respective gradients, ∇Te and ∇Ti, at the position of the flux tube.

We introduce four uncertain input parameters into the black box. Similar to the previous study, we assume that each parameter has a normal distribution in the range of $\pm 20\%$ around its original value. Finally, we perform the sensitivity analysis to explore how these parameters affect the resulting heat fluxes that determine the transport coefficients. It will also allow us to reduce the number of uncertain inputs that have negligible effect on the output variance [16,25].

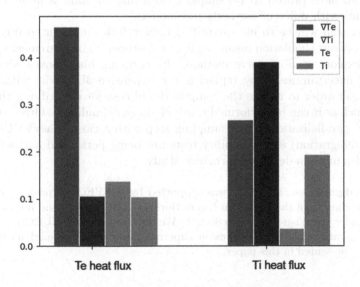

Fig. 3. First-order Sobol indices electron and ion heat fluxes.

As indicated in Fig. 3, the variance in the electrons and ion heat fluxes are mainly due to the uncertainty in temperature gradients. However, we should be cautious about the impact coming from temperature uncertainties, especially Ti on ion heat flux.

In this case, we also use PCE method with spectral projection and with regression. As in the previous case, we also set the polynomial order to 4 and have 4 uncertain parameters, hence, the number of samples required are 1296 and 140, respectively. The parallel execution of the samples, whether the black box contains serial or parallel codes, is done via QCG PilotJob [5]. This middleware considers each sample as an independent job, but schedule them dynamically through a single batch allocation. For example, using 32 cores for each sample, we notice that the time to run UQ simulation with regression is 10 times less than the spectral projection.

5   Conclusion

In this paper, we presented results from the introduction of uncertainty quantification and sensitivity analysis methods in our multiscale fusion workflow application. We focused so far on two black box use cases: the entire coupled workflow with simple analytical model for turbulence, and singled-out 3D gyrofluid turbulence model. Integration of the non-intrusive methods was done with the EasyVVUQ library and the parallel execution of samples was performed in a single batch allocation with QCG-PilotJob. Both tools, available as part of the VECMA toolkit and in conjunction with the coupled workflow nature of our application, have proven to be simple to use and to allow a large variety of experiments with different methods and models.

A next step will be to use quantified uncertainties in order to improve the validation of our simulation results against experimental measurements. We also want to explore semi-intrusive methods, by coupling black boxes with propagation of uncertainties or by replacing the expensive 3D model with a cheap surrogate, in order to reduce the computational cost for quantifying the uncertainties such as it can be performed routinely in our simulation. Other optimizations (e.g. parallelization of the sampling step with a closer EasyVVUQ–QCG-PilotJob integration) and scalability tests are being performed and will be the subject of a more in-depth performance study.

Acknowledgements. This work was supported by the VECMA project, which has received funding from the European Union Horizon 2020 research and innovation programme under grant agreement No 800925. We also acknowledge EUROfusion for the computing grant on MARCONI-Fusion supercomputer which allowed us to run all simulations presented in this paper.

References

1. Falchetto, G.L., et al.: The European Integrated Tokamak Modelling (ITM) effort: achievements and first physics results. Nucl. Fusion **54**(4), 043018 (2014)

2. Luk, O., Hoenen, O., Bottino, A., Scott, B., Coster, D.: ComPat framework for multiscale simulations applied to fusion plasmas. Comput. Phys. Commun. **239**, 126–133 (2019)
3. Groen, D., et al.: Introducing VECMAtk - verification, validation and uncertainty quantification for multiscale and HPC simulations. In: Rodrigues, J., et al. (eds.) ICCS 2019. LNCS, vol. 11539, pp. 479–492. Springer, Cham (2019). https://doi.org/10.1007/978-3-030-22747-0_36
4. Richardson, R.A., Wright, D.W., Jancauskas, V., Lakhlili, J., Edeling, W.: EasyVVUQ: a library for verification, validation and uncertainty quantification in high performance computing. J. Open Res. Softw. **8**, 11 (2019)
5. Piontek, T., et al.: Development of science gateways using QCG – lessons learned from the deployment on large scale distributed and HPC infrastructures. J. Grid Comput. **14**(4), 559–573 (2016). https://doi.org/10.1007/s10723-016-9384-9
6. Schmidt, P.C.: C. W. Gardiner: handbook of stochastic methods for physics, chemistry and the natural sciences, Springer-verlag, Berlin, Heidelberg, New York, Tokyo 1983. 442 seiten, preis: Dm 115,-. Berichte der Bunsengesellschaft für physikalische Chemie **89**(6), 721–721 (1985). https://doi.org/10.1002/bbpc.19850890629
7. Xiu, D.: Numerical Methods for Stochastic Computations: A Spectral Method Approach. Princeton University Press, Princeton (2010)
8. Sobol, I.: On quasi-Monte Carlo integrations. Math. Comput. Simul. **47**(2), 103–112 (1998)
9. Saltelli, A., Annoni, P.: Sensitivity analysis. In: Lovric, M. (ed.) International Encyclopedia of Statistical Science. Springer, Heidelberg (2011). https://doi.org/10.1007/978-3-642-04898-2
10. Saltelli, A.: Global Sensitivity Analysis: The Primer. Wiley, Chichester (2008)
11. Lemieux, C.: Monte Carlo and Quasi-Monte Carlo Sampling. SSS. Springer, New York (2009). https://doi.org/10.1007/978-0-387-78165-5
12. Sullivan, T.J.: Introduction to Uncertainty Quantification. TAM, vol. 63. Springer, Cham (2015). https://doi.org/10.1007/978-3-319-23395-6
13. Hammersley, J.M.: Monte Carlo methods for solving multivariable problems. Ann. N. Y. Acad. Sci. **86**(3), 844–874 (1960)
14. Saltelli, A.: Making best use of model evaluations to compute sensitivity indices. Comput. Phys. Commun. **145**(2), 280–297 (2002)
15. Xiu, D., Karniadakis, G.E.: The Wiener-Askey polynomial chaos for stochastic differential equations. SIAM J. Sci. Comput. **24**(2), 619–644 (2002)
16. Eck, V.G., et al.: A guide to uncertainty quantification and sensitivity analysis for cardiovascular applications. Int. J. Num. Methods Biomed. Eng. **32**(8), e02755 (2016). cnm.2755
17. Preuss, R., von Toussaint, U.: Uncertainty quantification in ion-solid interaction simulations. Nucl. Instrum. Methods Phys. Res. Sect. B Beam Interact. Mater. Atoms **393**, 26–28 (2017). Computer Simulation of Radiation effects in Solids Proceedings of the 13 COSIRES Loughborough, UK, June 19–24 2016
18. Sudret, B.: Global sensitivity analysis using polynomial chaos expansions. Reliab. Eng. Syst. Safe. **93**(7), 964–979 (2008). Bayesian Networks in Dependability
19. Feinberg, J., Langtangen, H.P.: Chaospy: an open source tool for designing methods of uncertainty quantification. J. Comput. Sci. **11**, 46–57 (2015)
20. Hosder, S., Walters, R., Balch, M.: Efficient sampling for non-intrusive polynomial chaos applications with multiple uncertain input variables. In: 48th AIAA/ASME/ASCE/AHS/ASC Structures, Structural Dynamics, and Materials Conference, April 2007

21. Herman, J., Usher, W.: SALib: an open-source python library for sensitivity analysis. J. Open Source Softw. **2**(9), 97 (2017)
22. Coster, D.P.: Members of the task force on integrated tokamak modelling: the European transport solver. IEEE Trans. Plasma Sci. **38**(9), 2085–2092 (2010)
23. Nikishova, A., Veen, L., Zun, P., Hoekstra, A.G.: Semi-intrusive multiscale metamodeling uncertainty quantification with application to a model of in-stent restenosis. Philos. Trans. R. Soc. A **377**, 20180154 (2018)
24. Scott, B.D.: Free-energy conservation in local gyrofluid models. Phys. Plasmas **12**(10), 102307 (2005)
25. Nikishova, A., Veen, L., Zun, P., Hoekstra, A.G.: Uncertainty quantification of a multiscale model for in-stent restenosis. Cardiovasc. Eng. Technol. **9**(4), 761–774 (2018)

Sensitivity Analysis of Soil Parameters in Crop Model Supported with High-Throughput Computing

Mikhail Gasanov[1]([✉])(iD), Anna Petrovskaia[1](iD), Artyom Nikitin[1](iD),
Sergey Matveev[1,2](iD), Polina Tregubova[1](iD), Maria Pukalchik[1](iD),
and Ivan Oseledets[1,2](iD)

[1] Skolkovo Institute of Science and Technology,
Bolshoy Boulevard 30, bld. 1, Moscow 121205, Russia
Mikhail.Gasanov@skoltech.ru
[2] Marchuk Institute of Numerical Mathematics, RAS,
Gubkin st. 8, Moscow 119333, Russia
http://www.skoltech.ru, http://www.inm.ras.ru

Abstract. Uncertainty of input parameters in crop models and high costs of their experimental evaluation provide an exciting opportunity for sensitivity analysis, which allows identifying the most significant parameters for different crops. In this research, we perform a sensitivity analysis of soil parameters which play an essential role in plant growth for the MONICA agro-ecosystem model. We utilize Sobol' sensitivity indices to estimate the importance of main soil parameters for several crop cultures (soybeans, sugar beet and spring barley). High-throughput computing allows us to speed up the computations by more than thirty times and increase the number of sampling points significantly. We identify soil indicators that play an essential role in crop yield productivity and show that their influence is the highest in the topsoil layer.

Keywords: Crop model · Sobol' indices · Soil parameters UQ

1 Introduction

Numerical digital crop models are used for crop yield prediction worldwide nowadays [24] and have applications in decision-support systems for farmers [10,15]. These models require soil, environmental and agro-management input data to establish plant growth simulation. The most widespread crop models, such as CENTURY [13], APSIM [5], DNDC [2] and MONICA [11] include modules of soil processes, climate and crop properties and allow to improve model's forecast with the calibration of internal parameters. Unfortunately, measurements of soil parameters for spatial modeling might be expensive and time-consuming, especially in countries where agrochemical data is not freely available.

Various approaches to reduce the number of parameters in environmental models have been proposed [14]. One of the most promising tools is evaluation

© Springer Nature Switzerland AG 2020
V. V. Krzhizhanovskaya et al. (Eds.): ICCS 2020, LNCS 12143, pp. 731–741, 2020.
https://doi.org/10.1007/978-3-030-50436-6_54

of the performance of complex environmental models through sensitivity analysis (SA) [22]. Sensitivity analysis simplifies the process of modeling by identifying and removing unnecessary elements from the structure of the model. There is a series of recent publications regarding the assessment of soil and plant sensitivity indicators conducted by Krishnan [8], Zhang [26], Karki [7], Gunarathna [3].

In practice, sensitivity analysis involves a) sampling of the multidimensional input parameter space; and b) subsequent simulations of the model. To obtain reliable confidence intervals, it may require millions of simulation runs, which may be infeasible for general-purpose computers. In previous works, the number of input samples was limited to 2000–30 000 points [12,23], which may be insufficient in other settings, where the amount of varied parameters is much larger. A natural way to speed up the simulations is to distribute them into independent blocks and perform parallel computations using a supercomputer [6].

In this work, we develop an effective and fast method for computing more than 500 000 agro-ecosystem MONICA model simulations per hour. It allows us to consider a much broader class of problems of practical interest. In particular, we increase the number of sampling points for sensitivity analysis of parameters up to 100 000, efficiently distribute calculations using a supercomputer and perform 2 000 000 model runs.

2   Materials and Methods

In this section, we describe the materials and methods that we use in our work.

2.1   MONICA Agro-Ecosystem Model

There is a variety of commercial and open-source models for crop growth simulations and yield prediction. In our research, we choose an open-source process-based agro-ecosystem model MONICA [11] that has been developed by ZALF institute during the last decades (Müncheberg, Germany). As input, MONICA requires soil parameters, crop rotation, fertilization schedule, and climate data.

Even though MONICA was developed for Western Europe soil conditions and climate, it can be optimized to other crop types by using model parameters for physiology and plant development. The MONICA model includes more than 120 parameters in soil hydrology processes, soil nutrients and organic matter turnover, plant physiology, and other blocks responsible for different processes that influence crop yield. MONICA receives soil data as different depth horizons (layers of soil with relatively uniform properties), which can be set up by a user in the format of a JSON-file.

2.2   Soil Parameters

The selection of parameters and their bounds play an essential role in sensitivity analysis [17]. In our research, we use agricultural data from a field experiment in the Russian Chernozem region. Among the great variety of climate conditions and soil types in Russia, the Chernozem region has special significance because of its potential productivity due to the highest nutrition and carbon content. We examine the commercial field in Kursk, Russia (51°52′20″N, 37°50′52″E) with six years crop-rotation from 2011 to 2017. The crop rotation consists of three different crops, namely sugar beet (*Beta vulgaris*) for years 2011, 2014, 2017, spring-barley (*Hordeum vulgare*) in 2012, and soybean (*Glycine max*) in 2015. The soil profile consists of several layers (or horizons). The upper arable horizon is especially crucial for the growth and development of crops. Subsoil layers may take part in hydrology regime and affect plant growth as well.

MONICA model requires more than ten different parameters for simulation within each soil layer. We select six main soil parameters (see Table 1) for sensitivity analysis (Soil organic carbon, Soil pH, Clay content, Sand content, Carbon:Nitrogen ratio, Bulk density). These parameters represent significant soil properties and have a considerable impact on crop yield. The value boundaries for the parameters were taken from the Russian Soil database [1] and represent the actual values for chernozem soils. In our research, we concentrate on crop yield ($kgDryMatter * ha^{-1}$) as an output of the MONICA model for sensitivity analysis. Prediction of yield is a complicated task because the yield depends on almost all processes in an agricultural system.

Table 1. Soil parameters of MONICA model and their min/max values used in SA.

Parameter	Description	Unit	Min.	Max.
SOC	Soil organic carbon	%	2.58	6.20
Sand	Soil sand fraction	$kg * kg^{-1}$	0.01	0.30
Clay	Soil clay fraction	$kg * kg^{-1}$	0.01	0.30
pH	Soil pH value	–	4.6	6.9
CN	Soil carbon:nitrogen ratio	–	10.9	12.4
BD	Soil bulk density	$kg * m^{-3}$	900.0	1350.0

To identify the most critical horizons, we evaluate the sensitivity indices of soil parameters at various depths. MONICA model allows us to set up soil layers with various thickness and parameters. We set nine layers with different thickness typical for agricultural soils of the Chernozem region as follows: topsoil 30 cm, seven horizons with 10 cm depth and the subsoil layer with 100 cm depth. We iteratively select each parameter from the Table 1 and evaluate how changes of this parameter in each soil layer affect the model's predictions. After identifying the most influential (in terms of crop yield) horizon, we perform sensitivity analysis of all six soil parameters for this layer specifically.

2.3 The Sobol' Sensitivity Analysis

Sensitivity analysis is a methodology of qualitative investigation of a model and its parameters which helps to identify parameters affecting the output of the model. It is possible to distinguish local and global sensitivity analyses. In our research, we choose the method developed by Sobol' [19] for global SA.

Consider the model output as

$$Y = f(X) = f(X_1, \ldots, X_p),$$

where f in our case depicts MONICA simulator, X are p varied input parameters and Y is the predicted crop yield. Following the techniques by Sobol' [21] we represent the multi-variate random function f using Hoeffding decomposition:

$$f(X_1, \ldots, X_p) = f_0 + \sum_i^p f_i + \sum_i^p \sum_{j>i}^p f_{ij} + \cdots + f_{1\ldots p}, \tag{1}$$

where f_0 is a constant term, $f_i = f_i(X_i)$ denotes main effects, $f_{ij} = f_{ij}(X_i, X_j)$ and others describe higher-order interactions. These terms can be written as

$$f_0 = E(Y),$$
$$f_i = E_{X_{\sim i}}(Y|X_i) - E(Y),$$
$$f_{ij} = E_{X_{\sim ij}}(Y|X_i, X_j) - f_i - f_j - f_0,$$
$$\ldots$$

where E is mathematical expectation and $X_{\sim i}$ denotes all parameters except i^{th}. Under the assumption that the input parameters are independent, total variance $V(Y)$ of the crop yield can be decomposed as follows:

$$V(Y) = \sum_i^p V_i + \sum_i^p \sum_{j>i}^p V_{ij} + \cdots + V_{12\ldots p},$$

where partial variances are

$$V_i = V[f_i(X_i)] = V_{X_i}\left[E_{X_{\sim i}}(Y|X_i)\right],$$
$$V_{ij} = V[f_{ij}(X_i, X_j)] = V_{X_i X_j}\left[E_{X_{\sim ij}}(Y|X_i, X_j)\right] - V_i - V_j,$$
$$\ldots$$

This way, sensitivity indices (SI) can be introduced as

$$S_i = \frac{V_i}{V(Y)}, \quad S_{ij} = \frac{V_{ij}}{V(Y)}, \quad \ldots \tag{2}$$

One can note the total sum of the indices is normalized to 1. The value of the Sobol' index corresponds to the "order" of sensitivity of f to the change of the corresponding input parameter or their group (see the details in [19] or [21]). Analogously to Eq. 1, first-order indices denote variance induced by changes of a single parameter without any interactions; second-order indices consider second-order interactions between the parameters; etc. In order to incorporate all of the interactions for a particular parameter, one can compute the total effect index:

$$S_{T_i} = \frac{E_{X_{\sim i}}\left[V_{X_i}(Y|X_{\sim i})\right]}{V(Y)} = 1 - \frac{V_{X_{\sim i}}\left[E_{X_i}(Y|X_{\sim i})\right]}{V(Y)} \tag{3}$$

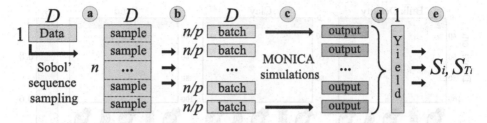

Fig. 1. Distributed computations comprising 5 steps: a) Sobol' sequence sampling from the initial input data, where D is the number of perturbed input parameters, $n = N \times (2 \times D + 2)$ is the total number of simulations and N is sampling size; b) mapping of acquired samples in batches to p HTC nodes; c) running n/p parallel MONICA simulations on each node; d) aggregating yield values from simulation results; and e) calculation of Sobol' sensitivity indices.

Evaluation of Sobol' indices requires us to perform a random sampling of the parameter hyperspace. In our work, we utilize quasi-random sampling approach. In general, such methods add new points into the sequence taking into account previously added points and may create a uniformly filled parameter space in the unit hypercube. In our work, we use the classical approach also proposed by Sobol' [18,20], which helps to achieve a convergence rate of confidence intervals almost $O(N^{-1})$, where N is the number of samples [9].

2.4 Crop Simulation and High-Throughput Computing

Sensitivity analysis requires the results of a significant number of simulations with various parameters. The number of simulations necessary for the convergence of sensitivity indices can be computationally expensive. In our work, we use "Zhores" supercomputer to tackle this problem [25]. Figure 1 outlines our approach. First, the initial values of D parameters are used to generate the $n \times D$ matrix of samples using Sobol' quasi-random sequence, more particularly, its extension proposed by Saltelli [16], where $n = N \times (2 \times D + 2)$ and N is the sample size. Second, these samples are grouped into batches of size $n/p \times D$ and each batch is then mapped to one of p HTC nodes. Third, each node performs n/p MONICA simulations in parallel. Then, yield predictions are extracted from output results and aggregated back to a vector of size n. Finally, these yield values are used to calculate Sobol' sensitivity indices using SALib Python library [4]. The most computationally expensive step is running the simulations, whereas the generation of samples and sensitivity analysis are negligible (several minutes in our experiments compared to hours of simulations).

To obtain the convergence of confidence intervals for sensitivity indices we use a different number of input sample points (from 10 to 100 000) to find the optimal amount needed for SA. To evaluate the acceleration, we compare the time spent for 2 000 000 simulations on a single core and $p = 96$ *Intel Xeon C6140* cores. This set of simulations is the main time-consuming part

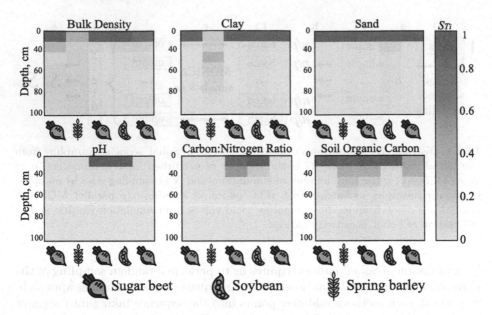

Fig. 2. Heatmap of S_{T_i} index for six soil parameters across different horizons in the five-year crop rotation. For the majority of crops, the most considerable role in yield variation is played by the change of parameters in the upper horizon. (Color figure online)

of computational work. It takes almost ∼112 h to calculate this on one core, and 3.5 h on 96 cores. The acceleration is 32.5 times and limited mostly by the performance of the file system. The speedup is defined by $S = t_s/t_p$, where t_s and t_p are the time for serial and parallel model simulations. We could have achieved additional acceleration of computations via the storage of simulation input files and technical outputs in RAM instead of direct creation and removal of files on hard-drives. We plan to do it in our future work.

3   Results and Discussion

In this section, we investigate the effect of input soil data on crop yields and provide our experimental results. For this purpose, we select six principal soil parameters important for plant growth, develop and evaluate the sensitivity of the model for each indicator at different soil depths. To demonstrate which soil horizons have the most significant impact on crop productivity, first, we divide the soil profile into nine horizons of different thickness and, second, perform separate sensitivity analyses for each of the parameters from the Table 1.

We present the obtained results in Fig. 2 in the form of heatmaps displaying the soil profile. Crop rotation and the depth of soil horizons are represented with X and Y axes, respectively. We use a sample size N equal to 100 000 and conduct

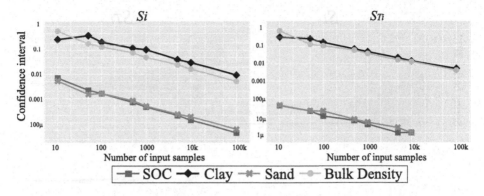

Fig. 3. The convergence of S_i (left) and S_{T_i} (right) confidence intervals for the topsoil layer of Sugar beet crop (2017 year) with different input sample sizes N. It can be seen that the convergence rate achieved is equivalent to $O(N^{-1})$.

2 000 000 simulations for each soil parameter, which allows us to obtain suitable confidence intervals. Color depicts the values of the total-order Sobol' SI, where light yellow color indicates no impact of parameter variation on the crop yield, and purple color indicates significant influence. We conclude that for most of the considered crops the main influence of parameters on final yield concentrates in the top horizon. However, the clay fraction and soil organic carbon affect the yield of barley in the entire soil profile. The effect of soil organic matter content on sugar beet yields changes during crop rotation. The content of organic matter in the upper horizon affects the yield only at the beginning of crop rotation. The transformation of organic matter in the soil leads to the distribution of organic compounds along the profile, and the subsurface horizons begin to affect the yield of sugar beet. For further analysis of soil parameters, we concentrate only on the upper horizon of 0–30 cm.

To identify the parameters in the topsoil layer that have the most significant impact on crop yield, we analyze first-order S_i and total effect S_{T_i} indices. One of the necessary conditions for successful SA is the convergence of the obtained SI values. As noted above, we use quasi-random sampling method proposed by Sobol' to increase the convergence rate of sensitivity indices values with a sample size N varying from 10 to 100 000. Figure 3 demonstrates convergence of the confidence intervals for S_i and S_{T_i} indices of the main six parameters, which achieves the rate of $O(N^{-1})$ for Sugar beet crop (2017). For other crops, we obtain the same qualitative results.

Figure 4 shows S_i and S_{T_i} values and additionally their confidence intervals for different cultures: spring barley (2012), soybean (2015) and sugar beet (2017). We exclude the plots for two other years of Sugar beat because they are qualitatively the same as in 2017. Soil parameters with sensitivity indices close to zero achieved stable values faster than the parameters with higher indices. Significant model parameters (which have strongly nonzero sensitivity indices)

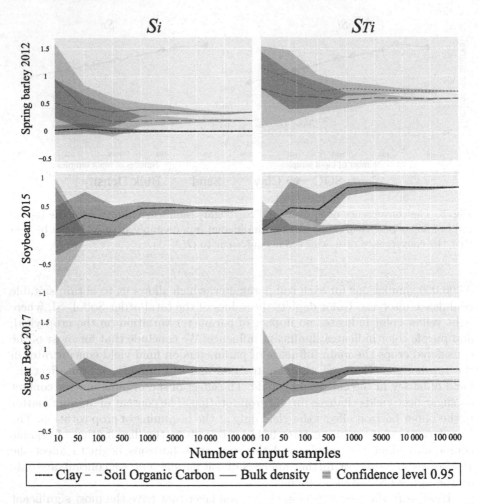

Fig. 4. Values of S_i (left) and S_{T_i} (right) indices for different crops, sample sizes N and soil parameters. Filled regions depict confidence intervals of the respective indices. Some of the parameters have rapid convergence of their confidence intervals because their Sobol' indices are very close to zero.

required input samples size from 1000 to 5000. It can be seen that different soil parameters have different importance for crop yield. Soil organic carbon content plays an essential role in all crops. The change in bulk density and soil organic carbon has a more significant impact on spring barley yields than on other indicators. Almost all the difference in soybean yield is due to the change in the soil clay fraction. The yield of sugar beet depends on the content of soil clay and bulk density, which coincides with real data, since beets are demanding to water nutrition, and clay content and bulk density can affect water regime. The values S_i and S_{T_i} for soil pH and carbon-nitrogen ratio in soil organic matter were almost equal to zero. It seems strange that they did not affect the crop yield.

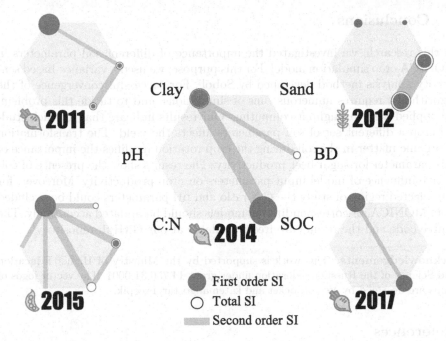

Fig. 5. First-order, second-order and total Sobol' sensitivity indices' for the six soil parameters and three crops: sugar beet (2011, 2014, 2017), spring barley (2012) and soybean (2015).

Considering that pH condition of soil determines the availability of nutrients for plants, and carbon to nitrogen ratio shows the quality of organic matter, it could influence the activity of soil microbial community. We plan to provide a more detailed analysis of corresponding MONICA submodels in our future research.

To represent second-order effects we employ diagrams in Fig. 5. It shows S_i, S_{ij} and S_{T_i} for different crop rotations, where gray lines describe interactions between soil parameters, black and white circles denote first-order and total sensitivity indices, respectively. There are total SI values for sand content and organic matter for soybean, which indicates the importance of the interactions of these parameters for yield. The figure demonstrates that clay, soil organic carbon, and bulk density are the parameters that have the highest value of first-order SI for the majority of years. As the first-order SI measures the fractional contribution of a single parameter to the output variance, we conclude that these parameters have the most significant impact on yield in our case. Second-order sensitivity indices in the figure show which parameter interactions play the biggest role in yield prediction. The soybean yield is affected by second-order interactions between almost all indicators. In contrast, spring barley yield is affected only by the coupled effect of soil density and organic matter content. From Fig. 5 we also see that the importance of SOC parameter and values of the second-order Sobol' indices change in time due to the transformation of soil organic matter during the crop rotation.

4 Conclusions

In this research, we investigated the importance of different soil parameters in MONICA crop simulation model. For this purpose, we used a variance-based sensitivity analysis method developed by Sobol'. For successful convergence of the algorithm it requires numerous runs of simulations, and to tackle this problem, we applied high throughput computing. Our results indicate that for each studied crop a different set of soil parameters affects the yield. The transformation of organic matter in the soil during the crop rotation modifies the importance of this parameter for sugar beet productivity. The results show the presence of collective influence of model input parameters on crop productivity. Moreover, for the selected region of study the C:N ratio and pH parameters could be excluded from MONICA, or corresponding submodels should be updated accordingly. The source code and the results are freely available at our GitHub repository[1].

Acknowledgements. This work is supported by the Ministry of Higher Education and Science of the Russian Federation under grant 14.756.31.0001. The vector logos of crops are designed by Vectorpocket and katemangostar/Freepik.

References

1. Russian Soil Database, Soil Science Institute named by V.V. Dokuchaev (2010). http://egrpr.esoil.ru/content/norm.html
2. Giltrap, D.L., Li, C., Saggar, S.: DNDC: a process-based model of greenhouse gas fluxes from agricultural soils. Agr. Ecosyst. Environ. **136**(3–4), 292–300 (2010)
3. Gunarathna, M., Sakai, K., Nakandakari, T., Momii, K., Kumari, M.: Sensitivity analysis of plant-and cultivar-specific parameters of APSIM-sugar model: variation between climates and management conditions. Agronomy **9**(5), 242 (2019)
4. Herman, J.D., Usher, W.: SALib: an open-source python library for sensitivity analysis. J. Open Source Softw. **2**(9), 97 (2017)
5. Holzworth, D.P., et al.: APSIM-evolution towards a new generation of agricultural systems simulation. Environ. Model Softw. **62**, 327–350 (2014)
6. Huang, X., et al.: A dynamic agricultural prediction system for large-scale drought assessment on the Sunway Taihulight supercomputer. Comput. Electron. Agric. **154**, 400–410 (2018)
7. Karki, R., Srivastava, P., Bosch, D.D., Kalin, L., Lamba, J., Strickland, T.C.: Multi-variable sensitivity analysis, calibration, and validation of afield-scale SWAT model: building stakeholder trust in hydrologic/water quality modeling. Trans. ASABE **63**, 523–539 (2020)
8. Krishnan, P., Aggarwal, P.: Global sensitivity and uncertainty analyses of a web based crop simulation model (web InfoCrop wheat) for soil parameters. Plant Soil **423**(1–2), 443–463 (2018). https://doi.org/10.1007/s11104-017-3498-0
9. Kucherenko, S., Rodriguez-Fernandez, M., Pantelides, C., Shah, N.: Monte Carlo evaluation of derivative-based global sensitivity measures. Reliab. Eng. Syst. Safe. **94**(7), 1135–1148 (2009)

---

[1] https://github.com/mishagrol/SA_agro_model.

10. Lavik, M.S., Hardaker, J.B., Lien, G., Berge, T.W.: A multi-attribute decision analysis of pest management strategies for Norwegian crop farmers. Agric. Syst. **178**, 102741 (2020)
11. Nendel, C., et al.: The MONICA model: testing predictability for crop growth, soil moisture and nitrogen dynamics. Ecol. Model. **222**(9), 1614–1625 (2011)
12. Nossent, J., Elsen, P., Bauwens, W.: Sobol' sensitivity analysis of a complex environmental model. Environ. Model Softw. **26**(12), 1515–1525 (2011)
13. Parton, W.J., Stewart, J.W., Cole, C.V.: Dynamics of C, N, P and S in grassland soils: a model. Biogeochemistry **5**(1), 109–131 (1988). https://doi.org/10.1007/BF02180320
14. Razavi, S., Gupta, H.V.: What do we mean by sensitivity analysis? The need for comprehensive characterization of "global" sensitivity in Earth and environmental systems models. Water Resour. Res. **51**(5), 3070–3092 (2015)
15. Rurinda, J., et al.: Science-based decision support for formulating crop fertilizer recommendations in sub-Saharan Africa. Agric. Syst. **180**, 102790 (2020)
16. Saltelli, A., Annoni, P., Azzini, I., Campolongo, F., Ratto, M., Tarantola, S.: Variance based sensitivity analysis of model output. Design and estimator for the total sensitivity index. Comput. Phys. Commun. **181**(2), 259–270 (2010)
17. Saltelli, A., Tarantola, S., Campolongo, F., Ratto, M.: Sensitivity Analysis in Practice: A Guide to Assessing Scientific Models, vol. 1. Wiley, New York (2004)
18. Sobol, I.M.: Uniformly distributed sequences with an additional uniform property. USSR Comput. Math. Math. Phys. **16**(5), 236–242 (1976)
19. Sobol, I.M.: Global sensitivity indices for nonlinear mathematical models and their Monte Carlo estimates. Math. Comput. Simul. **55**(1–3), 271–280 (2001)
20. Sobol', I.M.: On the distribution of points in a cube and the approximate evaluation of integrals. Zhurnal Vychislitel'noi Matematiki i Matematicheskoi Fiziki **7**(4), 784–802 (1967)
21. Sobol', I.M.: On sensitivity estimation for nonlinear mathematical models. Matematicheskoe modelirovanie **2**(1), 112–118 (1990)
22. Varella, H., Guérif, M., Buis, S.: Global sensitivity analysis measures the quality of parameter estimation: the case of soil parameters and a crop model. Environ. Model Softw. **25**(3), 310–319 (2010)
23. Vazquez-Cruz, M., Guzman-Cruz, R., Lopez-Cruz, I., Cornejo-Perez, O., Torres-Pacheco, I., Guevara-Gonzalez, R.: Global sensitivity analysis by means of EFAST and Sobol' methods and calibration of reduced state-variable TOMGRO model using genetic algorithms. Comput. Electron. Agric. **100**, 1–12 (2014)
24. Webber, H., Hoffmann, M., Rezaei, E.E.: Crop models as tools for agroclimatology. Agroclimatol. Link. Agric. Clim. **60**, 519–546 (2020)
25. Zacharov, I., et al.: "Zhores"–petaflops supercomputer for data-driven modeling, machine learning and artificial intelligence installed in Skolkovo Institute of Science and Technology. Open Eng. **9**(1), 512–520 (2019)
26. Zhang, Y., Arabi, M., Paustian, K.: Analysis of parameter uncertainty in model simulations of irrigated and rainfed agroecosystems. Environ. Model Softw. **126**, 104642 (2020)

A Bluff-and-Fix Algorithm for Polynomial Chaos Methods

Laura Lyman[1(✉)] and Gianluca Iaccarino[2]

[1] Institute for Computational and Mathematical Engineering, Stanford University,
475 Via Ortega, Stanford, CA 94305, USA
lymanla@stanford.edu
[2] Department of Mechanical Engineering and Institute for Computational
Mathematical Engineering, Stanford University,
Building 500, Stanford, CA 94305, USA
jops@stanford.edu

Abstract. Stochastic Galerkin methods can be used to approximate the solution to a differential equation in the presence of uncertainties represented as stochastic inputs or parameters. The strategy is to express the resulting stochastic solution using $M + 1$ terms of a polynomial chaos expansion and then derive and solve a deterministic, coupled system of PDEs with standard numerical techniques. One of the critical advantages of this approach is its provable convergence as M increases. The challenge is that the solution to the M system cannot easily reuse an already-existing computer solution to the $M - 1$ system. We present a promising iterative strategy to address this issue. Numerical estimates of the accuracy and efficiency of the proposed algorithm (bluff-and-fix) demonstrate that it can be more effective than using monolithic methods to solve the whole $M + 1$ system directly.

Keywords: Polynomial chaos · Galerkin projections · Stochastic differential equations · Numerical PDE solvers · Spectral methods

1 Introduction

Uncertainty quantification (UQ) in physical models governed by systems of partial differential equations is important to build confidence in the resulting predictions. A common approach is to represent the sources of uncertainty as stochastic variables; in this context the solution to the original differential equations becomes random. Stochastic Galerkin schemes (SGS) are used to approximate the solution to parametrized differential equations. In particular, they utilize a functional basis on the parameter to express the solution and then derive and solve a *deterministic* system of PDEs with standard numerical techniques [2]. A Galerkin method projects the randomness in a solution onto a finite-dimensional basis, making deterministic computations possible. SGS are part of a broader class known as *spectral methods*.

Supported by the US Department of Energy under the PSAAP II program.

V. V. Krzhizhanovskaya et al. (Eds.): ICCS 2020, LNCS 12143, pp. 742–756, 2020.
https://doi.org/10.1007/978-3-030-50436-6_55

The most common UQ strategies involve Monte Carlo (MC) algorithms, which suffer from a slow convergence rate proportional to the inverse square root of the number of samples [5]. If each sample evaluation is expensive — as is often the case for the solution of a PDE—this slow convergence can make obtaining tens of thousands of samples computationally infeasible. Initial spectral method applications to UQ problems showed orders-of-magnitude reductions in the cost needed to estimate statistics with comparable accuracy [3].

In the present approach for SGS, the unknown quantities are expressed as an infinite series of orthogonal polynomials in the space of the random input variable. This representation has its roots in the work of Wiener [7], who expressed a Gaussian process as an infinite series of Hermite polynomials. Ghanem and Spanos [4] truncated Wiener's representation and used the resulting finite series as a key ingredient in a stochastic finite element method. SGS based on polynomial expansions are often referred to as *polynomial chaos* approaches.

Let $\mathcal{D} = [0, 1] \times [0, T]$ be a subset of the spatial and time domain $\mathbb{R}_x \times \mathbb{R}_{t \geq 0}$.[1] Then let $u : \mathcal{D} \to \mathbb{R}$ be continuous and differential in its space and time components; further, let $u \in \mathcal{L}^2(\mathcal{D})$, and $u(0, t) = 0$. This u represents the solution to a differential equation,

$$\mathcal{F}(u, x, t) = 0. \tag{1}$$

Here $\mathcal{F}$ is a general differential operator that contains both spatial and temporal derivatives. Often $\mathcal{F}$ is assumed to be nonlinear.

Let ξ be a zero mean, square-integrable, real random variable. We assume uncertainty is present in the initial condition $u(\,\cdot\,, 0)$ and represent it by setting

$$u(x, 0; \xi) : \mathbb{R}_x \to \mathbb{R} \qquad u(x, 0; \xi) = f(x, \xi)$$

where f is a known function of x and ξ. Accordingly, the solution $u(x, t; \xi)$ to $\mathcal{F}(u, x, t)$ is now a random variable indexed by $(x, t) \in \mathcal{D}$, meaning $u(x, t; \xi)$ is a stochastic process.

As statistics of ξ, we require that both $u(x, t; \xi)$ and $f(x, \xi)$ have existing second moments — and in accordance with $u(0, t; \xi) = 0$, we assume $f(0, \xi) = 0$ as well.[2] Note that these are the only restrictions; namely, even though f is chosen as sinusoidal in the example of Sect. 2, we do not require f to be periodic, bounded over the real line, zero on the whole boundary $\partial \mathcal{D}$, etc.

We consider a polynomial chaos expansion (PCE) and *separate* the deterministic and random components of u by writing

$$u(x, t; \xi) = \sum_{k=0}^{\infty} u_k(x, t) \Psi_k(\xi).$$

The $u_k : \mathcal{D} \to \mathbb{R}$ output deterministic coefficients, while the Ψ_k are orthogonal polynomials with respect to the measure $d\xi$ induced by ξ. Let $\langle \cdot, \cdot \rangle$ denote

---

[1] For convenience we set $\mathcal{D} = [0, 1] \times [0, T]$, though all of the presented results follow immediately when $\mathcal{D} = [a, b] \times \mathcal{T}$ for some arbitrary interval $[a, b] \subset \mathbb{R}_x$ and bounded $\mathcal{T} \subset \mathbb{R}_{t \geq 0}$ such that $0 \in \mathcal{T}$.

[2] These assumptions ensure that the needed conditions for applying the Cameron-Martin theorem [1] are met.

the inner product mapping $(\phi, \psi) \mapsto \int \phi(\xi)\psi(\xi)\,d\xi$. The triple-product notation $\langle \phi\psi\varphi \rangle$ is understood as $\langle \phi, \psi\varphi \rangle = \langle \phi\psi, \varphi \rangle = \int \phi(\xi)\psi(\xi)\varphi(\xi)\,d\xi$ and the singleton $\langle \phi \rangle$ as $\int \phi(\xi)\,d\xi$. Then we require the Ψ_k to satisfy the properties [2]

$$\langle \Psi_k \rangle = \mathbb{E}_\xi(\Psi_k) = \delta_{k,0} \qquad \langle \Psi_i \Psi_j \rangle = c_i\delta_{i,j}$$

where the $c_i \in \mathbb{R}$ are nonzero and $\delta_{i,j}$ is the Kronecker delta.

By the Cameron-Martin theorem [1], the PCE of this random quantity converges in mean square,

$$\sum_{k=0}^{M} u_k(x,t)\Psi_k(\xi) \xrightarrow[\mathcal{L}^2]{M\to\infty} u(x,t;\xi).$$

This justifies the PCE and its truncation to a finite number of terms for the sake of computation. Substituting the truncation into Eq. (1), we have

$$\mathcal{F}\left(\sum_{k=0}^{M} u_k(x,t)\Psi_k(\xi), x, t \right) = 0. \tag{2}$$

Furthermore, we can determine the initial conditions for the deterministic component functions. Multiplying $u(x,0;\xi)$ by any Ψ_k and integrating with respect to the ξ-measure $d\xi$ yields

$$\sum_{i=0}^{\infty} u_i(x,0) \int \Psi_i(\xi)\Psi_k(\xi)\,d\xi = \int f(x,\xi)\Psi_k(\xi)\,d\xi = \mathbb{E}_\xi[f(x,\xi)\Psi_k(\xi)]$$

$$u_k(x,0) = \frac{1}{c_k}\mathbb{E}_\xi[f(x,\xi)\Psi_k(\xi)].$$

The scalars c_k of course are dependent on the choice of polynomial Ψ_k. Similarly, we can "integrate away" the randomness in Eq. (2) by projecting onto each basis polynomial. This is discussed in detail in the next section.

2   Inviscid Burgers' Equation

Our choice of orthogonal polynomials Ψ_k will rely on the distribution of the ξ random variable. Throughout this paper, we will choose $\xi \sim \mathcal{N}(0,1)$ and the Ψ_k to be Hermite polynomials; however, many of the results apply almost identically to other distributions and their corresponding polynomials.

Note that Hermite polynomials satisfy $\langle \Psi_k, \Psi_j \rangle = (k!)\delta_{kj}$ and by [6],

$$\langle \Psi_i \Psi_j \Psi_k \rangle = \begin{cases} 0 & \text{if } i+j+k \text{ is odd or } \max(i,j,k) > s \\ \frac{i!j!k!}{(s-i)!(s-j)!(s-k)!} & \text{else} \end{cases}$$

where $s = (i+j+k)/2$. Now let $u : \mathcal{D} \to \mathbb{R}$ be the solution to the inviscid Burgers' equation

$$\frac{\partial u}{\partial t} + u\frac{\partial u}{\partial x} = 0 \qquad u(x,0;\xi) = \xi\sin(x) \qquad u(0,t;\xi) = 0. \tag{3}$$

The goal is to determine the solution u given the randomness in the initial conditions. Substituting in the finite PCE to (3), projecting on the polynomial basis, and integrating with respect to the ξ-measure yields

$$\sum_{i=0}^{M} \frac{\partial u_i}{\partial t} \langle \Psi_i \Psi_k \rangle + \sum_{i=0}^{M} \sum_{j=0}^{M} u_i \frac{\partial u_j}{\partial x} \langle \Psi_i \Psi_j \Psi_k \rangle = 0 \quad \text{for every } k \in \{0, \dots, M\}. \quad (4)$$

This is a system of $(M + 1)$ non-linear and coupled PDEs, which we will refer to as an M *system*. Solving the original problem with randomness has now been transformed into solving a system that is completely *deterministic*.

For instance, the system for $M = 2$ is

$$\frac{\partial u_0}{\partial t} = -u_0 \frac{\partial u_0}{\partial x} - u_1 \frac{\partial u_1}{\partial x} - 2u_2 \frac{\partial u_2}{\partial x}$$

$$\frac{\partial u_1}{\partial t} = -u_1 \frac{\partial u_0}{\partial x} - (u_0 + 2u_2) \frac{\partial u_1}{\partial x} - 2u_1 \frac{\partial u_2}{\partial x}$$

$$\frac{\partial u_2}{\partial t} = -u_2 \frac{\partial u_0}{\partial x} - u_1 \frac{\partial u_1}{\partial x} - (u_0 + 4u_2) \frac{\partial u_2}{\partial x}.$$

As shown in Sect. 1, the initial conditions are

$$u_k(x,0) = \frac{1}{k!} \sin(x) \int \xi \Psi_k(\xi) \, d\xi = \frac{1}{k!} \sin(x) \mathbb{E}_\xi[\xi \Psi_k(\xi)] \quad \text{for every } k \in \{0, \dots, M\}.$$

These initial conditions are easily computed. For example, when $0 \le k \le 10$ the only nonzero $u_k(x,0)$ is $u_1(x,0) = \sin(x)$.

For any $k \in \{0, \dots, M\}$, define the symmetric matrix $\boldsymbol{\Psi}_k^{(M)} \in \mathbb{R}^{M+1 \times M+1}$ such that

$$(\boldsymbol{\Psi}_k^{(M)})_{ij} = \langle \Psi_i \Psi_j \Psi_k \rangle \quad \text{for } i, j \in \{0, \dots, M\}. \quad (5)$$

For the case $k = M$, we also name the particular matrix $L_M := \boldsymbol{\Psi}_M^{(M)}$ and let $(L_M)_{k\bullet}$ denote its kth row. When the Ψ_k are *Hermite* polynomials, we can easily prove that

$$L_M = \begin{bmatrix} 0 & \cdots & 0 & \langle \Psi_0 \Psi_M \Psi_M \rangle \\ \vdots & & \ddots & \vdots \\ 0 & \ddots & & \vdots \\ \langle \Psi_M \Psi_0 \Psi_M \rangle & \cdots & \cdots & \langle \Psi_M \Psi_M \Psi_M \rangle \end{bmatrix}.$$

Lemma 1. *Let* $\mathbf{u} = (u_0, \dots, u_M)^T$ *be the vector of* $M + 1$ *functions,* $\mathbf{v} = (u_0, \dots, u_{M-1})^T$, *and* $w = u_M$. *Adopting the shorthand* $(u_j)_x := \frac{\partial u_j}{\partial x}$ *and* $(\mathbf{u})_x := \frac{\partial \mathbf{u}}{\partial x}$, *we can rewrite Eq. (4) as*

$$\frac{\partial u_k}{\partial t} + \frac{1}{k!} \mathbf{u}^T \boldsymbol{\Psi}_k^{(M)} (\mathbf{u})_x = 0 \quad \text{for every } k \in \{0, \dots, M\} \quad (6)$$

and when $k < M$ *as*

$$\frac{\partial u_k}{\partial t} + \frac{1}{k!} \mathbf{v}^T \boldsymbol{\Psi}_k^{(M-1)} (\mathbf{v})_x + \frac{1}{k!} (L_M)_{k\bullet} \frac{\partial}{\partial x} \left[w \left(\frac{\mathbf{v}}{\frac{1}{2}w} \right) \right] = 0 \quad (7)$$

recalling that $(L_M)_{k\bullet}$ *is the kth row of* L_M.

Let D_M be the diagonal $M \times M$ matrix such that $D_{kk} = \frac{1}{k!}$ and $\tilde{L}_M$ be the $M \times (M+1)$ matrix comprised of the first M rows of L_M. Then let $\mathbf{\Psi}^{(M-1)}$ denote the $M^2 \times M$ matrix

$$\mathbf{\Psi}^{(M-1)} := \begin{bmatrix} \mathbf{\Psi}_0^{(M-1)} \\ \vdots \\ \mathbf{\Psi}_{M-1}^{(M-1)} \end{bmatrix}.$$

Then (7) can be expressed in aggregate (i.e. for all $k < M$) as

$$\frac{\partial \mathbf{v}}{\partial t} + (D_M \otimes \mathbf{v}^T)\mathbf{\Psi}^{(M-1)}(\mathbf{v})_x + D_M \tilde{L}_M \frac{\partial}{\partial x}\left[w\left(\frac{\mathbf{v}}{\frac{1}{2}(w)} \right) \right] = 0 \qquad (8)$$

where $\otimes$ is the Kronecker product.

Proof. See Appendix.

Equations (6) and (7), along with the last term in the LHS of Eq. (8), will be used to inform an algorithm to solve the M system using the solution to the $M - 1$ system. This is described in the following section.

3   Bluff-and-Fix (BNF) Algorithm

We will use the superscript $\mathbf{u}^{(M)}$ to denote the $(M+1) \times 1$ vector of functions that is the solution to the M system. Similarly, $u_k^{(M)}$ is the kth component function of $\mathbf{u}^{(M)}$. When a component function $u_k^{(M)}(x, t)$ is written without the superscript i.e. as $u_k(x, t)$, the value of M is considered to be a fixed but arbitrary positive integer.

Suppose we have computed $\mathbf{u}^{(M-1)}$. How can we incorporate this information into solving for $\mathbf{u}^{(M)}$? Firstly, the M system has the added equation for $\frac{\partial u_M}{\partial t}$, namely

$$\frac{\partial u_M^{(M)}}{\partial t} = -\frac{1}{M!} \sum_{i=0}^{M} \sum_{j=0}^{M} \langle \Psi_i \Psi_j \Psi_M \rangle u_i^{(M)} \frac{\partial u_j^{(M)}}{\partial x}. \qquad (9)$$

If we had the first M coefficients $u_0^{(M)}, \ldots, u_{M-1}^{(M)}$ of $\mathbf{u}^{(M)}$ corresponding to the M system rather than to the $M-1$ system, we could substitute directly into the RHS of (9) and simply integrate. However, the numerical solutions $u_0^{(M-1)}, \ldots, u_{M-1}^{(M-1)}$ will differ from $u_0^{(M)}, \ldots, u_{M-1}^{(M)}$. To see why, recall Lemma 1 and its notation to observe that $u_k^{(M-1)}$ is the solution to

$$\frac{\partial v_k}{\partial t} = -\frac{1}{k!} \mathbf{v}^T \mathbf{\Psi}_k^{(M-1)}(\mathbf{v})_x := F_{M-1}(\mathbf{v}, w) \qquad (10)$$

and by (7), $u_k^{(M)}$ for $k < M$ is the solution to

$$\frac{\partial v_k}{\partial t} = -\frac{1}{k!}\left(\mathbf{v}^T\mathbf{\Psi}_k^{(M-1)}(\mathbf{v})_x + (L_M)_{k\bullet}\frac{\partial}{\partial x}\left[w\left(\frac{\mathbf{v}}{\frac{1}{2}w}\right)\right]\right) := F_M(\mathbf{v}, w). \quad (11)$$

Thus, the numerical solutions $u_k^{(M-1)}$ and $u_k^{(M)}$ are different, because the right hand sides of (10) and (11) differ by the *function*

$$G_k(\mathbf{v}, w) := -\frac{1}{k!}(L_M)_{k\bullet}\frac{\partial}{\partial x}\left[w\left(\frac{\mathbf{v}}{\frac{1}{2}w}\right)\right]. \quad (12)$$

We can write all of the G_k functions together via the third term on the LHS of (8); that is,

$$\mathbf{G} := (G_0, \ldots, G_{M-1})^T \text{ so that } \mathbf{G}(\mathbf{v}, w) = -D_M\widetilde{L}_M\frac{\partial}{\partial x}\left[w\left(\frac{\mathbf{v}}{\frac{1}{2}(w)}\right)\right]. \quad (13)$$

3.1 One Step Bluff-and-Fix

From (12), we know there is a discrepancy between $u_k^{(M-1)}$ and $u_k^{(M)}$ for $k < M$. Regardless, we can *bluff* and take the solutions we have $u_0^{(M-1)}, \ldots, u_{M-1}^{(M-1)}$ to solve for some approximation of $u_M^{(M)}$, which we can call $\widehat{u}_M^{(M)}$, and then back-substitute $\widehat{u}_M^{(M)}$ into the previous M equations in the M system to solve for $\widehat{u}_0^{(M)}, \ldots, \widehat{u}_{M-1}^{(M)}$. This approach is potentially more efficient than calculating the solution of the M system directly via classic monolithic methods. However, we will opt for an algorithm with even less computation time.

A workable strategy is based on a similar idea. Instead of re-computing all of the $\widehat{u}_0^{(M)}, \ldots, \widehat{u}_{M-1}^{(M)}$ after obtaining $\widehat{u}_M^{(M)}$, we re-solve for the *least accurate* $\widehat{u}_k^{(M)}$ at the same time as solving for $\widehat{u}_M^{(M)}$. That is, we only correct the $\widehat{u}_k^{(M)}$ that we believe will be the worst approximations of their corresponding $u_k^{(M)}$. The $\mathcal{I}$ in Algorithm 1 is the collection of *correction indices* i.e. the indices k denoting which $u_k^{(M)}$ approximations are corrected.

An algorithm to use the solution to an $M - 1$ system to solve an M system.

Algorithm 1: one step bluff-and-fix(c, $\mathbf{u}^{(M-1)}$)

input:

- correction size $c \in \{1, \ldots, M\}$
- $\mathbf{u}^{(M-1)}$ obtained by standard monolithic methods

---

select correction indices $\mathcal{I} \subseteq \{0, \ldots, M-1\}$ such that $|\mathcal{I}| = c - 1$
$\mathcal{I} \leftarrow \mathcal{I} \cup \{M\}$
(bluff) set approximate solutions to the M system equal to those of the $(M-1)$ system, if those approximations are not getting corrected:

$$\widehat{u}_k^{(M)} \leftarrow u_k^{(M-1)} \text{ for all } k \in \{0, \ldots, M\} \setminus \mathcal{I}$$

(fix) solve the coupled system $\{u_k^{(M)}\}_{k \in \mathcal{I}}$ of size c to obtain $\widehat{\mathbf{u}}^{(M)}$

Algorithm 1 poses two immediate questions.

1. How large should we make the correction size $c \in \{1, \ldots, M\}$?
2. How should we choose the correction indices $\mathcal{I}$? That is, how do we pick which $\widehat{u}_k^{(M)}$ approximations should be fixed?

To address question 1, note that *if $c = M$, then the one step bluff-and-fix (BNF) is equivalent to solving the entire M system directly*. In this case, we use $\mathbf{u}^{(M-1)}$ to set *none* of the approximations $\widehat{u}_k^{(M)}$ (no bluffing), and the entire coupled system $\{\widehat{u}_k^{(M)}\}_{k=0}^{M}$ of $M+1$ equations is solved by standard numeric techniques.

For the other extreme, choosing $c = 1$ corresponds to ignoring the difference between the $M - 1$ and M system solutions as much as possible; we set $\{\widehat{u}_k^{(M)}\}_{k=0}^{M-1} \leftarrow \mathbf{u}^{(M-1)}$ and solve only a single PDE for $\widehat{u}_M^{(M)}$. Heuristically, c can represent a trade-off between accuracy and efficiency. The hypothesis is that by choosing which $\widehat{u}_k^{(M)}$ to correct judiciously, we can still well-approximate $\mathbf{u}^{(M)}$ when $c < M$.

This brings us to the second posed question. To determine the correction indices $\mathcal{I}$, we target the $u_k^{(M)}$ such that $u_k^{(M-1)}$ and $u_k^{(M)}$ are very different so that the approximation $\widehat{u}_k^{(M)} = u_k^{(M-1)}$ is a poor one. While we cannot examine $\|\widehat{u}_k^{(M)} - u_k^{(M)}\|$ directly, we can see from (12) that the difference between the numeric solutions $u_k^{(M)}$ and $u_k^{(M-1)}$ arises from the function G_k. So we would like $\mathcal{I}$ to ideally include $k^* = \arg\max_{k \in \{0, \ldots M-1\}} \|G_k\|$, where $\|\cdot\|$ denotes some function norm over $(x, t) \in \mathcal{D}$. From the definition of G_k, we do not know what values its input functions $(\mathbf{v}, w)$ or their derivatives will take over $(x, t) \in \mathcal{D}$; however, we *do* know the entries of the matrix L_M.

Now there is some choice. The function G_k is the difference between F_M and F_{M-1} from Eqs. (10) and (11), and all three of these equations are scaled by the $\frac{1}{k!}$ factor. We can keep this $\frac{1}{k!}$ factor and select the G_k that is large in an "absolute error" sense. Alternatively, we can ignore this scaling and choose the $\widehat{u}_k^{(M)}$ to fix such that the difference function G_k is significant *relative* to its corresponding F_M and F_{M-1}.

The former approach (call it the *absolute version*) targets $k_1^* = \arg\max_{0 \le k \le M-1} \frac{1}{k!}\|(\widetilde{L}_M)_{k\bullet}\|_\infty = \arg\max_{0 \le k \le M-1} \frac{1}{k!}\sum_{j=0}^{M}|(\widetilde{L}_M)_{k,j}|$. Equivalently, by Eq. (13), this is selecting $k_1^* = \arg\max_{0 \le k \le M-1}\sum_{j=1}^{M}|(D_M\widetilde{L}_M)_{kj}|$ i.e. the k_1^* indexing the row of $D_M\widetilde{L}_M$ with largest absolute row sum — where we recall that $\widetilde{L}_M$ is the matrix of the first M rows of L_M. The latter approach (call it the *relative version*) simply picks the row indexing the largest row sum in $\widetilde{L}_M$ itself (i.e. not in $D_M\widetilde{L}_M$).

Selecting the row of $\widetilde{L}_M$ with maximal absolute row sum is simple; it is straight-forward to verify that row $M - 1$ obtains the maximum (though may not do so uniquely) and that the row sums of $\widetilde{L}_M$ are non-decreasing as the row index k increases. The structure of the $D_L\widetilde{L}_M$ matrix does not lend itself to as obvious of a pattern.

We opt for the *relative* version when constructing our algorithm, since its numeric results are overwhelmingly promising (as discussed in Sect. 4) and its implementation avoids an additional row sorting step; however, this is a potential area for future investigation. Since we require $|\mathcal{I}| = c$ for a given correction size parameter c, we simply pick the indices corresponding to the last c rows in the M system i.e. $\mathcal{I} = \{M - c + 1, \ldots, M\}$.

3.2   Iterative Bluff-and-Fix

An assumption of the one step bluff-and-fix algorithm is that you already have solved the fully coupled $M - 1$ system via some explicit time-stepping scheme (e.g. fourth-order Runge-Kutta). Realistically, we likely only want to solve a fully coupled M_0 system for when M_0 is small (say 2 or 3). How can we then use this information for approximating $\mathbf{u}^{(M)}$ for a larger $M > M_0$?

An algorithm to use the solution to an M_0 system to solve an M system for general $M > M_0$.

Algorithm 2: iterative bluff-and-fix$(c, \mathbf{u}^{(M_0)})$

input:

- correction size $c \in \{1, \ldots, M\}$
- $\mathbf{u}^{(M_0)}$ obtained by standard monolithic methods

---

$\widehat{\mathbf{u}}^{(M_0)} \leftarrow \mathbf{u}^{(M_0)}$

for $m = M_0 + 1, \ldots, M$ **do**

    select correction indices $\mathcal{I} \subseteq \{0, \ldots, m - 1\}$ such that $|\mathcal{I}| = c - 1$

    $\mathcal{I} \leftarrow \mathcal{I} \cup \{m\}$

    (bluff) **set** approximate solutions to the m system equal to those of the $(m - 1)$ system, if those solutions are not getting corrected:

$$\widehat{u}_k^{(m)} \leftarrow u_k^{(m-1)} \text{ for all } k \in \{0, \ldots, m\} \setminus \mathcal{I}$$

    (fix) **solve** the coupled system $\{u_k^{(m)}\}_{k \in \mathcal{I}}$ to obtain $\widehat{\mathbf{u}}^{(m)}$

end

The iterative bluff-and-fix algorithm uses some baseline $\mathbf{u}^{(M_0)}$ solution to get an approximation $\widehat{\mathbf{u}}^{(M_0+1)}$ of $\mathbf{u}^{(M_0+1)}$. The approximation $\widehat{\mathbf{u}}^{(M_0+1)}$ is then re-fed into the one step bluff-and-fix algorithm, instead of the "true" $\mathbf{u}^{(M_0+1)}$, to determine $\widehat{\mathbf{u}}^{(M_0+2)}$, and the process continues.

4   Numerical Results

We report solutions for Burgers' equation with uncertain initial conditions, which are $u(x, 0; \xi) = \xi \sin(x)$ for $\xi \sim \mathcal{N}(0, 1)$. The equation is solved for $x \in [0, 3]$ on a uniform grid with $\Delta_x = 0.05$. Time integration is based on the Runge-Kutta 4-step (RK4) scheme with $\Delta_t = 0.001$.

Throughout this discussion, we will define *error* as the deviation of the solution approximation $\widehat{\mathbf{u}}^{(M)}$ produced by bluff-and-fix from the solution yielded from solving the full M system via RK4. *That is, our computations are not being compared with a true analytic solution but instead with the numerical solution from a standard monolithic method.* In particular,

$$\text{absolute error in } \widehat{u}_k(\mathbf{x}, \mathbf{t}) := \|\widehat{u}_k^{\text{BNF}}(\mathbf{x}, \mathbf{t}) - u_k^{\text{RK4}}(\mathbf{x}, \mathbf{t})\|_2 \qquad (14)$$

$$\text{relative error in } \widehat{u}_k(\mathbf{x}, \mathbf{t}) := \frac{\|\widehat{u}_k^{\text{BNF}}(\mathbf{x}, \mathbf{t}) - u_k^{\text{RK4}}(\mathbf{x}, \mathbf{t})\|_2}{\|u_k^{\text{RK4}}(\mathbf{x}, \mathbf{t})\|_2}. \qquad (15)$$

In (14) and (15), $u_k(\mathbf{x}, \mathbf{t})$ is the matrix $(u_k(x_i, t_j))_{i,j}$ of u_k evaluated on the uniform grid of position $\mathbf{x}$ and time $\mathbf{t}$ points. The $\|\cdot\|_2$ denotes the matrix 2-norm. When reporting the *average* absolute (resp. relative) error for an approximation $\widehat{\mathbf{u}}^{\text{BNF}}$ for an M system, we mean taking the average of the absolute (resp. relative) errors of each $\widehat{u}_k^{\text{BNF}}$ over all k indices.

4.1   Results for One Step Version

We compare the solution obtained by solving the full M system via RK4 against the computations obtained by inputting the $M - 1$ solution $\mathbf{u}^{(M-1)}$ into the one step BNF algorithm. We test

1. correction size $c = 1$ (only $\widehat{u}_M^{(M)}$ is computed per step),
2. correction size $c = 2$ ($\widehat{u}_M^{(M)}$ is computed and $\widehat{u}_{M-1}^{(M)}$ is fixed per step), and
3. correction size $c = 3$ ($\widehat{u}_M^{(M)}$ is computed and both $\widehat{u}_{M-2}^{(M)}$ and $\widehat{u}_{M-1}^{(M)}$ are fixed per step).

The results are shown in Table 1.

Picking a small correction size (say one or two) can be sufficient for producing accurate solution approximations. For instance, using the solution $\mathbf{u}^{(4)}$ to produce $\widehat{\mathbf{u}}^{(5)}$ has average absolute error of under 1% for all c, around 4% average relative error for $c = 1$, and under 1% average relative error for $c = 2, 3$, as shown in the $M = 5$ row of Table 1. Figure 1 displays how the approximate solution $\widehat{u}_5^{(5)}$ in this system converges to the RK4 solution as the correction size c increases from 1 to 2. Similarly, using the solution $\mathbf{u}^{(6)}$ to produce $\widehat{\mathbf{u}}^{(7)}$ has average absolute error of under 1% for all c values, around 6% average relative error for $c = 1$, and about 2.5% average relative error for $c = 2$, as shown in the $M = 7$ row of Table 1.

What if average or relative error can be further reduced by choosing a different subset of $\widehat{u}_k^{(M)}$ to fix? To evaluate how well the correction indices $\mathcal{I}$ were selected, we can examine the absolute and relative error in the approximation $\widehat{u}_k^M = u_k^{(M-1)}$ for all k over various M values. Then we can see whether the least accurate $\widehat{u}_k^{(M)}$ for every M were appropriately chosen for correction. These results are displayed in Table 2.

Table 1. Average absolute and relative errors for correction sizes $c \in \{1, 2, 3\}$ in **one step** bluff-and-fix for $M = 3, \ldots, 8$.

M	Avg. Abs. Error			Avg. Rel. Error		
	$c = 1$	$c = 2$	$c = 3$	$c = 1$	$c = 2$	$c = 3$
3	0.1515	0.05593	0.01114	0.03196	0.004364	0.001683
4	0.03481	0.01823	0.007423	0.03711	0.005768	0.0008361
5	0.009536	0.006371	0.003599	0.04300	0.009232	0.001656
6	0.003279	0.002619	0.001859	0.05112	0.01530	0.003765
7	0.001330	0.001195	0.0009746	0.05986	0.02541	0.008166
8	0.0006404	0.0006118	0.0005623	0.06080	0.02630	0.01116

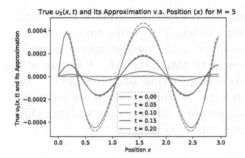 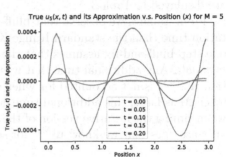

Fig. 1. The standard ("true") RK4 solution $u_5^{(5)}$ (solid lines) against the $\widehat{u}_5^{(5)}$ produced by one step bluff-and-fix (dashed lines) at time points $t = 0, 0.05, 0.10, 0.15, 0.20$ sec. for $0 \leq x \leq 3$. LHS figure shows $\widehat{u}_5^{(5)}$ for $c = 1$, and RHS figure shows the $\widehat{u}_5^{(5)}$ improvement for $c = 2$.

Table 2. Selection of correction indices $\mathcal{I}$ ranked by priority in one step bluff-and-fix against the ideal $\widehat{u}_k^{(M)}$ to correct (in this numeric example) for $M \in \{3, \ldots, 8\}$. Matching indices are shown in blue.

M	Ranking of $k < M$ by Numeric Error (Descending) in $\widehat{u}_k^{(M)} = u_k^{(M-1)}$		Order of Selection in $\mathcal{I}$
	Absolute Error	Relative Error	
3	2, 1, 0	2, 0, 1	2, 1, 0
4	3, 2, 1, 0	3, 2, 0, 1	3, 2, 1, 0
5	4, 3, 2, 1, 0	4, 3, 2, 0, 1	4, 3, 2, 1, 0
6	3, 4, 5, 2, 1, 0	5, 4, 3, 2, 0, 1	5, 4, 3, 2, 1, 0
7	3, 4, 5, 2, 6, 1, 0	6, 5, 4, 3, 2, 0, 1	6, 5, 4, 3, 2, 1, 0
8	3, 4, 2, 5, 1, 6, 7, 0	7, 6, 5, 4, 3, 2, 0, 1	7, 6, 5, 4, 3, 2, 1, 0

We see from Table 2 that one step bluff-and-fix is often spot-on for guessing which approximations $\widehat{u}_k^{(M)}$ are least accurate. When defining the "worst" approximation by relative error, BNF always selects correctly for $c < M - 1$. While the ideal indices when $c = M - 1$ or $c = M$ are not all chosen, this is likely no issue, as in practice a correction size that large would not be used. (Recall that $c = M$ is equivalent to using regular RK4.) Also, it should be noted that bluff-and-fix can *still* produce an accurate solution approximation when an "incorrect" index is chosen — it just might not be the best approximation *possible* given that value of c.

In addition, we test the computational cost of one step and iterative bluff-and-fix for different correction sizes and M values. The runtimes are measured using the %timeit command in iPython with parameters -r 10 -n 10 to obtain an average with standard deviation over 100 realizations. All computations are performed on a machine with a 1.8 GHz Intel Dual-Core processor. The results are displayed in Table 3.

From Table 3, we observe that bluff-and-fix *consistently* requires less computation time than the standard Runge-Kutta 4-step. That being said, recall that one step bluff-and-fix assumes that the *solution to the smaller system is readily available*. Any additional time that was spent to obtain this $M - 1$ system solution $\mathbf{u}^{(M-1)}$ is not accounted for when obtaining runtime measurements — and when M is large, this additional time is not trivial. Such concerns are addressed when timing the iterative version of the algorithm in the following section, which only assumes that we know $\mathbf{u}^{(M_0)}$ for some small M_0 value.

Table 3. Runtimes of **one step** bluff-and-fix to approximate $\mathbf{u}^{(M)}$ when given $\mathbf{u}^{(M-1)}$ compared with runtimes of solving full M systems via RK4. Tested over $M = 3, \ldots, 8$ with correction sizes $c = 2, 3$. Each time measurement is averaged over 100 loops to provide a confidence interval.

M	Avg. One Step BNF Runtime		Avg. RK4 Runtime
	$c = 2$	$c = 3$	
3	146 ms $\pm$ 7.81 ms	175 ms $\pm$ 14.1 ms	181 ms $\pm$ 7.81 ms
4	162 ms $\pm$ 13.2 ms	183 ms $\pm$ 22.5 ms	222 ms $\pm$ 22.4 ms
5	167 ms $\pm$ 21.3 ms	184 ms $\pm$ 4.42 ms	260 ms $\pm$ 15.2 ms
6	167 ms $\pm$ 5.42 ms	189 ms $\pm$ 3.96 ms	315 ms $\pm$ 37.6 ms
7	191 ms $\pm$ 21.1 ms	202 ms $\pm$ 6.01 ms	336 ms $\pm$ 13.1 ms
8	216 ms $\pm$ 31.5 ms	217 ms $\pm$ 5.38 ms	377 ms $\pm$ 3.77 ms

4.2   Results for Iterative Version

Now we present results for Algorithm 2 when using the baseline solution $\mathbf{u}^{(M_0)}$ for $M_0 = 2$ to approximate solutions to the M systems for $M = 3, \ldots, 8$. Correction sizes $c = 1, 2, 3$ are all tested. The results are displayed in Table 4.

We observe how quickly the solution approximation from iterative bluff-and-fix converges to the "true" RK4 solution as the correction size is increased (Fig. 2).

For example, the average relative error in $\widehat{\mathbf{u}}^{(7)}$ drops from $\sim27\%$ ($c = 1$) to $\sim5.8\%$ ($c = 2$) to $\sim 1.4\%$ ($c = 3$), and the average absolute error falls from $\sim8.7\%$ to $\sim3.5\%$ to $\sim1\%$ (Table 4).

Furthermore, choosing a small correction size, such as $c = 2$, is highly accurate even when M is as large as 8, with average relative error in this case always below 8.5% and average absolute error always below 6%. Future work will be focused on testing the algorithm for larger values of M to assess whether such a small c value can maintain these promising results.

Table 4. Average absolute and relative errors in **iterative** bluff-and-fix when using the baseline solution $M_0 = 2$ to approximate solutions to the M systems for $M = 3, \ldots, 8$. Results for correction sizes $c = 1, 2, 3$ are shown.

M	Avg. Abs. Error			Avg. Rel. Error		
	$c = 1$	$c = 2$	$c = 3$	$c = 1$	$c = 2$	$c = 3$
3	0.1515	0.05593	0.01115	0.03196	0.004363	0.001682
4	0.1337	0.05301	0.01420	0.07467	0.01019	0.001924
5	0.1145	0.04634	0.01357	0.1313	0.02034	0.003412
6	0.09894	0.04033	0.01227	0.2013	0.03642	0.006904
7	0.08676	0.03547	0.01096	0.2740	0.05793	0.01400
8	0.07711	0.03147	0.009704	0.3586	0.08464	0.02020

As before, we test the computational cost via `%timeit` in `iPython` for different correction sizes and M values, averaging each measurement over 100 realizations to provide a confidence interval. For solving the $M = 8$ system with iterative bluff-and-fix with $M_0 = 2$, the computation time is 853 ms ± 39.5 ms, 1.12 s ± 121 ms, and 1.28 s ±63.5 ms for $c = 1, 2, 3$, respectively. We can see that the added cost from correcting $\ddot{u}_{M-1}^{(M)}$, as opposed to just solving for $\widehat{u}_M^{(M)}$, is on average only 267 ms. This suggests that the reduction in error between $c = 1$ and $c = 2$ potentially comes at a cheap cost, especially given that average absolute error and relative error is under 6% and 8.5% (respectively) after this transition.

Using RK4 to *only* solve the $M = 8$ system is faster than using iterative bluff-and-fix from the baseline $M_0 = 2$ solution: the former spends 426 ms ± 48.6 ms per loop. However, it is possible that for larger M values, the iterative bluff-and-fix algorithm will eventually out-perform RK4 in terms of runtime. This is an area for future investigation.

Furthermore, iterative bluff-and-fix has the advantage of producing approximate solutions to all of the M systems for $M = 3, \ldots, 8$ along the way. When repeatedly solving the full M system via RK4 for $M = 3, \ldots, 8$, the average runtime is 2.05 s ± 145 ms per loop — meaning bluff-and-fix with correction sizes $c = 1$ (averaged at 853 ms), $c = 2$ (averaged at 1.21 s), and $c = 3$ (averaged at 1.28 s) is *far* more efficient for this type of goal.

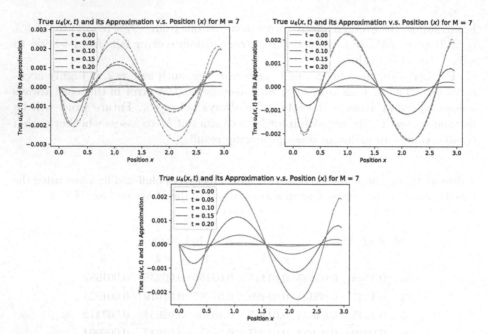

Fig. 2. The standard ("true") RK4 solution $u_4^{(7)}$ (solid lines) against the $\widehat{u}_4^{(7)}$ produced by iterative bluff-and-fix (dashed lines) from a baseline $M_0 = 2$ solution. Solved over a uniform grid with $0 \leq x \leq 3$ and $0 \leq t \leq 0.25$, with solutions at times $t = 0, 0.05, 0.10, 0.15, 0.20$ sec. displayed. We can see how the approximation $\widehat{u}_4^{(7)}$ converges to the RK4 solution as the correction size is increases from one (top left figure) to two (top right figure) to three (bottom figure).

5   Conclusion

Polynomial chaos (PC) methods are effective for incorporating and quantifying uncertainties in problems governed by partial differential equations. In this paper, we present a promising algorithm (one step bluff-and-fix) for utilizing the solution to a polynomial chaos $M-1$ system arising from inviscid Burgers' equation to approximate the solution to the corresponding M system. We expand the algorithm to an iterative version, which utilizes the solution to an M_0 system to approximate the solution to an M system for a general $M > M_0$. Bluff-and-fix is shown to be effective in producing accurate approximations, even when its correction size parameter is small, for both its one step and iterative versions. In the one step version, these approximations are produced more efficiently than doing so with a standard monolithic numeric scheme. While iterative bluff-and-fix initialized from some baseline M_0 can be less efficient than solving the full M system directly, it has the advantage of producing approximations to all of the m systems along the way for $M_0 \leq m \leq M$ — and does so faster than the monolithic method solves all of the full m systems one by one. In general, it could be beneficial to know the solution to an M system for a consecutive range of M values, because then one could observe when the difference between consecutive system solutions is small, which provides a rough sense of the M value sufficient for solution convergence.

Future work will be focused on generalizing and testing the algorithm on other nonlinear PDEs with uncertain initial conditions. We also plan to investigate different choices of the uncertainty representation ξ. It is expected that the priority ordering for which solutions to "fix" first — that is, the choice of the correction indices — will change, since that ordering depends on ranking the absolute row sums of L_M, and the precise structure of L_M depends on the choice of orthogonal polynomial.

A   Appendix

Proof of Lemma 1. The computation for (6) is straight-forward. To prove (7) write $\sum_{i=0}^{M}\sum_{j=0}^{M} u_i \frac{\partial u_j}{\partial x}\langle \Psi_i \Psi_j \Psi_k\rangle$ in (4) as

$$\sum_{i,j=0}^{M-1}\langle \Psi_i \Psi_j \Psi_k\rangle u_i (u_j)_x + \sum_{j=0}^{M-1}\langle \Psi_M \Psi_j \Psi_k\rangle [u_M u_j]_x + \langle \Psi_M^2 \Psi_k\rangle u_M (u_M)_x$$

$$= \sum_{i,j=0}^{M-1}(\Psi_k^{(M-1)})_{ij} u_i (u_j)_x + \sum_{j=0}^{M-1}(L_M)_{k,j}\left[u_M u_j\right]_x + (L_M)_{k,M}\left[\tfrac{1}{2} u_M (u_M)\right]_x$$

$$= \mathbf{v}^T \Psi_k^{(M-1)}(\mathbf{v})_x + (L_M)_{k\bullet}\frac{\partial}{\partial x}\left[w\left(\frac{\mathbf{v}}{\frac{1}{2}w}\right)\right]$$

where the chain rule and symmetry of the triple-product were applied in the first equality. Finally, note that the first and third terms on the LHS of (7) in aggregate are clearly $\frac{\partial \mathbf{v}}{\partial t}$ and $D_M \widetilde{L}_M \frac{\partial}{\partial x}\left[w\left(\frac{\mathbf{v}}{\frac{1}{2}w}\right)\right]$, respectively. By definition of the Kronecker product, $D_M \otimes \mathbf{v}^T$ is the $M \times M^2$ matrix

$$D_M \otimes \mathbf{v}^T = \begin{bmatrix} \frac{1}{(0)!}\mathbf{v}^T & & \\ & \ddots & \\ & & \frac{1}{(M-1)!}\mathbf{v}^T \end{bmatrix}$$

and so $(D_M \otimes \mathbf{v}^T)\Psi^{(M-1)}$ is the $M \times M$ matrix whose kth row is $\frac{1}{k!}\mathbf{v}^T \Psi_k^{(M-1)}$. Thus, $(D_M \otimes \mathbf{v}^T)\Psi^{(M-1)}(\mathbf{v})_x$ is the desired $M \times 1$ vector whose kth entry is $\frac{1}{k!}\mathbf{v}^T \Psi_k^{(M-1)}(\mathbf{v})_x$, completing the proof of (8).

References

1. Cameron, R., Martin, W.: The orthogonal development of non-linear functionals in series of Fourier-Hermite functionals. Ann. Math. **48**, 385–392 (1947)
2. Constantine, P.: A primer on stochastic Galerkin methods. Academic homepage. https://www.cs.colorado.edu/~paco3637/docs/constantine2007primer.pdf. Accessed 6 Feb 2020

3. Constantine, P.: Spectral methods for parametrized matrix equations. Ph.D. thesis, Stanford University (2009)
4. Ghanem, R., Spanos, P.: Stochastic Finite Elements: A Spectral Approach. Springer, New York (1991). https://doi.org/10.1007/978-1-4612-3094-6
5. Owen, A.: Monte Carlo theory, methods and examples, Ch. 1. Academic homepage. https://statweb.stanford.edu/~owen/mc/. Accessed 15 April 2020
6. Szegö, G.: Orthogonal Polynomials, 2nd edn. American Mathematical Society, New York (1959)
7. Wiener, N.: The homogenous chaos. Amer. J. Math. **60**(4), 897–936 (1938)
8. Xiu, D., Karniadakis, G.: The Wiener-Askey polynomial chaos for stochastic differential equations. SIAM J. Sci. Comput. **24**, 619–644 (2002)

Markov Chain Monte Carlo Methods for Fluid Flow Forecasting in the Subsurface

Alsadig Ali[1], Abdullah Al-Mamun[1,2], Felipe Pereira[1],
and Arunasalam Rahunanthan[3(✉)]

[1] Department of Mathematical Sciences, The University of Texas at Dallas,
Richardson, TX 75080, USA
[2] Institute of Natural Sciences, United International University, Dhaka, Bangladesh
[3] Department of Mathematics and Computer Science, Central State University,
Wilberforce, OH 45384, USA
aRahunanthan@centralstate.edu

Abstract. Accurate predictions in subsurface flows require the forecasting of quantities of interest by applying models of subsurface fluid flow with very little available data. In general a Bayesian statistical approach along with a Markov Chain Monte Carlo (MCMC) algorithm can be used for quantifying the uncertainties associated with subsurface parameters. However, the complex nature of flow simulators presents considerable challenges to accessing inherent uncertainty in all flow simulator parameters of interest. In this work we focus on the transport of contaminants in a heterogeneous permeability field of a aquifer. In our problem the limited data comes in the form of contaminant fractional flow curves at monitoring wells of the aquifer. We then employ a Karhunen-Loève expansion to truncate the stochastic dimension of the permeability field and thus the expansion helps reducing the computational burden. Aiming to reduce the computational burden further, we code our numerical simulator using parallel programming procedures on Graphics Processing Units (GPUs). In this paper we mainly present a comparative study of two well-known MCMC methods, namely, two-stage and Differential Evolution Adaptive Metropolis (DREAM), for the characterization of the two-dimensional aquifer. A thorough statistical analysis of ensembles of the contaminant fractional flow curves from both MCMC methods is presented. The analysis indicates that although the average fractional flow curves are quite similar, both time-dependent ensemble variances and posterior analysis are considerably distinct for both methods.

Keywords: Porous media · Contaminant transport · Two-stage ·
MCMC · DREAM · MPSRF

1 Introduction

In subsurface characterization and flow forecasting, one could characterize the subsurface using a Bayesian framework. The Bayesian framework consists essentially of a Markov Chain Monte Carlo (MCMC) algorithm, in which we repeatedly solve a flow numerical simulator that models the porous media problem of

© Springer Nature Switzerland AG 2020
V. V. Krzhizhanovskaya et al. (Eds.): ICCS 2020, LNCS 12143, pp. 757–771, 2020.
https://doi.org/10.1007/978-3-030-50436-6_56

interest [11]. In this paper we use a single-phase flow numerical simulator that simulates the contaminant transport in an aquifer. In the MCMC algorithm we characterize the heterogeneous permeability field of the aquifer using the simulator. Our C/C++ flow simulator takes advantage of Graphics Processing Units (GPUs), which have the computational capacity to speedup the single-phase flow simulation [12]. The GPU flow simulator runs on a computational grid of the permeability field that is divided into several thousand elements. However, populating each computational element by a random permeability value and changing those values in each MCMC iteration is impracticable. In this case the dimension of the stochastic space is that of the computational domain and the dimension reduction is achieved by a Karhunen-Loève expansion (KLE) [15]. Due to the serial nature of the MCMC algorithm, the computational burden in solving the problem recurrently using the GPU numerical simulator is still huge. This can make the Bayesian framework less attractive for our problem. Parallelizations of the MCMC algorithm to speedup the characterization were considered in [10,12]. The simulation of several parallel MCMCs reduces the computational cost drastically. However, the convergence of such parallel MCMCs should be carefully analyzed.

In this paper we consider two-stage and DiffeRential Evolution Adaptive Metropolis (DREAM) MCMC methods for the subsurface characterization. The two-stage MCMC was introduced in [6,8] and it has been investigated more recently (see [7,14] and references therein). The two-stage procedure is of particular interest to subsurface flow problems because samples can be rejected with inexpensive simulations on coarse grids. The DREAM is an extension of Differential Evolution (DE) MCMC, which integrates the essential ideas of DE genetic algorithm and MCMC algorithm [3]. Inspired by DE MCMC, the DREAM was proposed in [22] and applied to several interesting problems (see [19–21] and references therein). The DREAM is well known to converge relatively faster when compared to earlier procedures. It employs subspace sampling and outlier chain correction to accelerate the convergence towards the stationary distribution.

In the current work we run several parallel simulations for each MCMC method. In this approach, we need to determine when it is safe to stop the MCMC simulations for a reliable characterization of the permeability field. There are several convergence diagnostics available for this purpose and those diagnostics fall into two categories: the first category of diagnostics entirely depends on the output values of the MCMC simulation and those in the second category use not only the output values but also the information on the target distribution. In the first category, Brooks and Gelman [4] proposed a convergence diagnostic that uses the Multivariate Potential Scale Reduction Factor (MPSRF) to decide when to terminate MCMC simulations. Very recently in [2] we proposed a stopping criterion using a statistical analysis for single-phase flow prediction. In the analysis we considered ensembles of fractional flow curves to decide when it is safe to stop MCMC simulations for a reliable characterization and prediction. In this work we use the criterion in [2] for both MCMC methods and compare them for the characterization of the permeability field of the aquifer. Our results show that the two-stage MCMC provides a good estimate for the average of the quantity of interest (fractional flow curves), but the DREAM MCMC method reveals that the posterior distribution is not well characterized by the two-stage method.

We organize the present study as follows. We discuss the computational modeling of the single-phase flow of the aquifer in Sect. 2. The reduction of the parameter space dimension by the KLE is discussed in Sect. 3. Section 4 contains the statistical framework using two-stage and DREAM MCMC methods for the characterization of the permeability field in the aquifer. The MPSRF, a frequently used criterion to estimate convergence of the MCMC methods, is presented in Sect. 5. Results obtained from our numerical experiments are discussed in Sect. 6. Concluding remarks appear in Sect. 7.

2   Computational Physical Model

We consider a unit square-shaped subsurface aquifer Ω with a heterogeneous permeability field. The aquifer contains two monitoring wells: one of which is located at the top right corner of the domain (corner well) and the other well (center well) is placed at the middle of the right boundary. An accidental contamination occurs at the spill site and the contaminated water flows naturally through the aquifer. The spill well is positioned at the bottom left corner through which the contaminated (or tracer) water is discharged (Fig. 1). In this single-phase flow model we assume a relatively low concentrations of the contaminant, thus it does not affect the velocity field.

Let us denote the Darcy velocity and the pressure of the fluid by $v(x,t)$, and $p(x,t)$, respectively, where $x \in \Omega$ is the location at time t. We also denote the absolute permeability of the rock, porosity of the rock, fluid viscosity, and contaminant concentration in the fluid by $k(x)$, $\phi(x)$, μ, and $\rho(x,t)$, respectively. We consider that the pore space of the aquifer is filled by water. Applying Darcy's law and mass conservation, the governing equations describing the single-phase flow can be written as the following [5]:

$$\nabla \cdot v = 0, \quad \text{where} \quad v = -\frac{k(x)}{\mu}\nabla p, \quad x \in \Omega,$$

$$\phi(x)\frac{\partial \rho(x,t)}{\partial t} + \nabla \cdot (\rho(x,t)v) = 0. \tag{1}$$

In this study the porosity of the rock is considered a constant throughout the domain with $\phi(x) = 0.2$. We assume no-flow boundary conditions and we take $\rho(x, t = 0) = 0$. The system of partial differential equations in (1) does not contain any source or sink because all three wells are modeled through appropriate boundary conditions. The coupled system consists of an elliptic problem and a hyperbolic problem. After applying an operator splitting technique, we solve each problem separately by an appropriate numerical scheme [13,18].

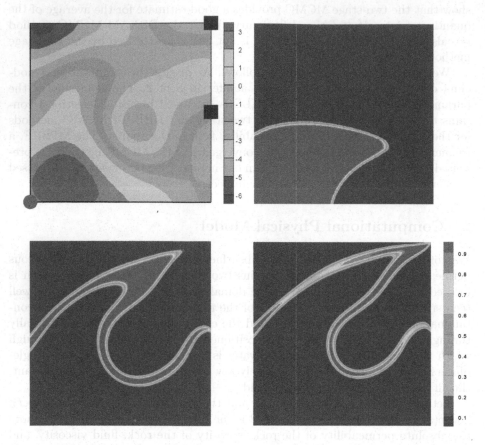

Fig. 1. Top: On the left, the reference permeability field for a unit square-shaped region, where circle and squares on the boundary denote the spill and monitoring wells, respectively. On the right, contaminant flow at $t = 0.25$ PVI. Bottom: Contaminant flow at $t = 0.75$ PVI on the left and $t = 1.0$ PVI on the right.

The permeability field is characterized by using measured data in the form of fractional flow curves, $F(t)$ which are defined as

$$F(t) = 1 - \frac{\int_{\partial \Omega_{\text{out}}} v_n(\boldsymbol{x},t)\rho(\boldsymbol{x},t) \, ds}{\int_{\partial \Omega_{\text{out}}} v_n(\boldsymbol{x},t) \, ds}, \tag{2}$$

where $\partial \Omega_{\text{out}}$ represents the well outflow boundary, and $v_n(\boldsymbol{x}, t)$ is the component of the velocity field normal to the well boundary. The dimensionless time is denoted by t, which is measured in Pore Volume Injected (PVI) and is computed using the following integral:

$$\text{PVI} = \int_0^T V_p^{-1} \int_{\partial \Omega_{\text{out}}} v_n(\boldsymbol{x},t) \, ds \, d\tau, \tag{3}$$

where V_p denotes the total pore volume of the reservoir and T represents the total time the contaminated water entered through the spill well.

3   Reduction in Parameter Space

If we consider the aquifer in a 128×128 computational domain, we have 16, 384 elements. In our Bayesian framework the numerical simulator uses the permeability value in each element. Therefore, we need to start with a random permeability value in each element and change one or more of those values in each MCMC iteration. Thus, the dimension of the parameter space is 16, 384 and it presents a far-fetched framework for the characterization of the permeability field. In order to reduce the number of uncertainty parameters, we therefore use the KLE. Below we reduce the parameter space from 16, 384 to d, which is 20 for our study, using the KLE. The KLE has been explained in [10–12]. A very short discussion on the KLE is being presented below for the sake of completeness.

Suppose $\log[k(\boldsymbol{x})] = Y^k(\boldsymbol{x})$, where $\boldsymbol{x} \in \Omega \subset \mathbf{R}^2$, $Y^k(\boldsymbol{x})$ is a Gaussian field and the covariance function $\mathrm{Cov}(Y^k(\boldsymbol{x}_1), Y^k(\boldsymbol{x}_2))$ is given by the following formula:

$$R(\boldsymbol{x}_1, \boldsymbol{x}_2) = \sigma_Y^2 \exp\left(-\frac{|x_1 - x_2|^2}{2L_x^2} - \frac{|y_1 - y_2|^2}{2L_y^2}\right), \qquad (4)$$

where L_x and L_y are the correlation lengths in $x-$ and $y-$direction, respectively, and $\sigma_Y^2 = \mathrm{Var}[(Y^k)^2]$. It is assumed that $Y^k(\boldsymbol{x})$ is a second-order stochastic process and $E[(Y^k)^2] = 0$. Thus, for a given orthonormal basis $\{\varphi_i\}$ in $L^2(\Omega)$, $Y^k(\boldsymbol{x})$ can be expressed as the following:

$$Y^k(\boldsymbol{x}) = \sum_{i=1}^{\infty} Y_i^k \varphi_i(\boldsymbol{x}), \quad \text{with} \quad Y_i^k = \int_{\Omega} Y^k(\boldsymbol{x})\varphi_i(\boldsymbol{x})d\boldsymbol{x}, \qquad (5)$$

where Y_i^k are random coefficients. On the other hand, since L^2 is a complete space, thus $\varphi_i(\boldsymbol{x})$ is an eigenfunction satisfying

$$\int_{\Omega} R(\boldsymbol{x}_1, \boldsymbol{x}_2)\varphi_i(\boldsymbol{x}_2)d\boldsymbol{x}_2 = \lambda_i\varphi_i(\boldsymbol{x}_1), \quad i = 1, 2, ..., \qquad (6)$$

and the corresponding eigenvalue $\lambda_i = E[(Y_i^k)^2] > 0$. By using the assumption $\theta_i = Y_i^k/\sqrt{\lambda_i}$, the KLE in Eq. (5) can be expressed as the following:

$$Y^k(\boldsymbol{x}) = \sum_{i=1}^{\infty} \sqrt{\lambda_i}\theta_i\varphi_i(\boldsymbol{x}). \qquad (7)$$

If the eigenvalues decrease, a truncated KLE can be written as

$$Y_d^k(\boldsymbol{x}) = \sum_{i=1}^{d} \sqrt{\lambda_i}\theta_i\varphi_i(\boldsymbol{x}). \qquad (8)$$

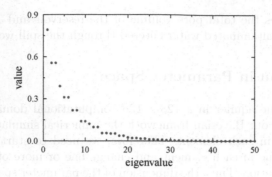

Fig. 2. Decay of eigenvalues in KLE

If the eigenvalues in (8) decay quickly, the Y_d^k will be a good approximation of Y^k.

We set $L_x = L_y = 0.2$ and $\sigma_Y^2 = 4$ in (4). Figure 2 shows that the eigenvalues decay fast for these values. In this paper we thus consider the first twenty eigenvalues in the KLE to model the permeability field.

4   Bayesian Framework

4.1   Exploration of the Posterior

In this subsection we discuss how to characterize the permeability field using the fractional flow curves at the monitoring wells of the aquifer. Let us denote the fractional flow data by F_m and the corresponding permeability field by ψ. Using the Bayes' theorem we can write the posterior probability

$$P(\psi|F_m) \propto P(F_m|\psi)P(\psi), \tag{9}$$

where $P(\psi)$ denotes the prior distribution and the normalizing constant is ignored due to the iterative search in the MCMC algorithm. The ψ is generated through the KLE, for which the vector θ is used as input in the expansion. In the remainder of the discussion we use $\psi = \text{KLE}[\theta]$. Moreover, we consider a Gaussian distribution as in [9] for the likelihood function, i.e.,

$$P(F_m|\psi) \propto \exp\left(-(F_m - F_\psi)^\top \Sigma (F_m - F_\psi)\right), \tag{10}$$

where the simulated fractional flow data F_ψ is obtained by the numerical solution from the GPU simulator for each permeability distribution ψ in the MCMC algorithm. We denote the covariance matrix by Σ, which is defined as $\Sigma = I/2\sigma_F^2$ where I and σ_F^2 are the identity matrix and the precision parameter, respectively.

We sample data from the posterior by using the Metropolis-Hasting (MH) MCMC and create a Markov chain, which has the posterior distribution as target distribution. We consider an instrumental distribution $q(\psi_p|\psi)$, where ψ

denotes the previously accepted proposal, in order to propose $\boldsymbol{\psi}_p = \mathrm{KLE}[\boldsymbol{\theta}_p]$ at every iteration. We use the following acceptance probability in MH MCMC and the probability value is computed by solving the forward problem for a given permeability distribution on the numerical simulator:

$$\alpha(\boldsymbol{\psi}, \boldsymbol{\psi}_p) = \min\left(1, \frac{q(\boldsymbol{\psi}|\boldsymbol{\psi}_p)P(\boldsymbol{\psi}_p|F_m)}{q(\boldsymbol{\psi}_p|\boldsymbol{\psi})P(\boldsymbol{\psi}|F_m)}\right). \tag{11}$$

We now describe the two-stage and DREAM MCMCs that use the MH algorithm.

4.2   Two-Stage MCMC

Here we present the two-stage MCMC method. The method has been widely used for porous media applications [8,16]. The two-stage MCMC consists of a screening procedure, which relies on a coarse-scale model approximating the governing equations (1). The coarse-scale discretization is done in a similar way as in the fine-scale discretization. The main idea lies on a rigorous projection of k on the coarse-scale that is obtained from the fine-scale resolution. For this reason, an upscaling method is used so that the effective permeability values on the coarse-scale yield a similar average response as that of the underlying fine-scale problem locally [11]. We then run the numerical simulator on the coarse-scale model and get the numerical solution F_c. Now we can compute the coarse-scale and fine-scale acceptance probabilities

$$\alpha_c(\boldsymbol{\psi}, \boldsymbol{\psi}_p) = \min\left(1, \frac{q(\boldsymbol{\psi}|\boldsymbol{\psi}_p)P_c(\boldsymbol{\psi}_p|F_m)}{q(\boldsymbol{\psi}_p|\boldsymbol{\psi})P_c(\boldsymbol{\psi}|F_m)}\right) \text{ and}$$
$$\alpha_f(\boldsymbol{\psi}, \boldsymbol{\psi}_p) = \min\left(1, \frac{P_f(\boldsymbol{\psi}_p|F_m)P_c(\boldsymbol{\psi}|F_m)}{P_f(\boldsymbol{\psi}|F_m)P_c(\boldsymbol{\psi}_p|F_m)}\right), \text{ respectively.} \tag{12}$$

In the two-stage MCMC the following Random Walk Sampler (RWS) is used:

$$\boldsymbol{\theta}_p = \beta\,\boldsymbol{\theta} + \sqrt{1-\beta^2}\,\boldsymbol{\epsilon}, \tag{13}$$

where $\boldsymbol{\theta}$ and $\boldsymbol{\theta}_p$ represent the previously accepted proposal and the current proposal, respectively. The symbol $\boldsymbol{\epsilon}$ stands for a $\mathcal{N}(0,1)$-random variable and β $(= 0.75)$ is a tuning parameter [1]. The two-stage MCMC algorithm is presented in Algorithm 1. The convergence diagnostic to break the "for loop" in the algorithm is described in Sect. 5.

4.3   DREAM MCMC

The DREAM, which is an extension of DE MCMC [3], runs multiple MCMCs simultaneously for a thorough exploration of the posterior, and has an in-built mechanism to adapt the scale and orientation of the proposal distribution during the evolution to the posterior distribution [22]. We describe the DREAM for our

Algorithm 1. Two-stage MCMC

1: Given covariance function R generate KLE.
2: **for** $p = 1$ to M_{mcmc} **do**
3:     At $\psi = \mathrm{KLE}[\theta]$ using (13) generate $\psi_p = \mathrm{KLE}[\theta_p]$.
4:     Compute the upscaled permeability on the coarse-scale using ψ_p.
5:     Solve the forward problem on the coarse-scale to get F_c.
6:     Compute the coarse-scale acceptance probability $\alpha_c(\psi, \psi_p)$.
7:     **if** ψ_p is accepted **then**
8:         Use ψ_p in the fine-scale simulation to get F_f.
9:         Compute the fine-scale acceptance probability $\alpha_f(\psi, \psi_p)$.
10:         **if** ψ_p is accepted **then** $\psi = \psi_p$.
11:         **end if**
12:     **end if**
13: **end for**

application below. In Algorithm 2, m denotes the number of parallel chains that we run simultaneously.

Algorithm 2. DREAM MCMC

1: Given covariance function R generate KLE.
2: **for** $c = 1$ to m **do**
3:     **for** $p = 1$ to M_{mcmc} **do**
4:         At $\psi = \mathrm{KLE}[\theta]$ using equation (14) generate $\psi_p = \mathrm{KLE}[\theta_p]$.

$$\theta_p = \theta + (I_d + f)\,\gamma(\delta, d') \left(\sum_{j=1}^{\delta} \theta^{r_1(j)} - \sum_{k=1}^{\delta} \theta^{r_2(k)} \right) + \epsilon, \tag{14}$$

where δ and d' denote the number of pairs that are used to generate the proposal and the number of parameters that are updated jointly in each iteration, respectively. Two randomly chosen chains are denoted by r_1 and r_2. Moreover, f and ϵ are drawn independently from $\mathrm{CU}[-b, b]$ and $\mathcal{N}(0, b^*)$, respectively, where $\mathrm{CU}[a, b]$ represents the continuous uniform distribution on the interval $[a, b]$.

Algorithm 3. DREAM MCMC (continued)

5:         Use ψ_p to get F_f (on the fine-scale).
6:         Compute the acceptance probability $\alpha(\psi, \psi_p)$ using (11).
7:         **if** ψ_p is accepted **then** $\psi = \psi_p$.
8:         **end if**
9:     **end for**
10: **end for**

We simultaneously run the DREAM MCMC with $m = 11$ parallel chains. In (14) we set $\delta = 5$, $b = 0.1$ and $b^* = 10^{-6}$. The jump rate is given by $\gamma = \frac{2.38\beta_0}{\sqrt{2\delta d'}}$, where the constant $\beta_0 = \frac{1}{16}$. The user should select the value of β_0 in such a way the MCMC method has an acceptance rate between 15–35%. In the present study we set $d' = 5$, i.e., we update five parameters at every iteration. In addition to setting those parameters, we also set $\gamma = 1.0$ at every tenth iteration to encourage a jump between two disconnected posterior modes.

5   Convergence Diagnostics of MCMC Methods

The DREAM MCMC requires a parallel simulation of multiple chains simultaneously. However, the two-stage MCMC does not require the same. Due to the computational burden in repeatedly solving the numerical simulator, we still run the parallel simulation of several two-stage MCMCs. Now we need to investigate the convergence of multiple chains in each method for a reliable characterization of the permeability field of the aquifer. In this section we discuss convergence diagnostics for that purpose. The Potential Scale Reduction Factor (PSRF) and its multivariate extension MPSRF are used to measure the convergence of multiple MCMCs [4].

We set the number of parameters as $d = 20$ in $\boldsymbol{\theta}$ as discussed in Sect. 3. Let us consider m chains and n posterior draws of $\boldsymbol{\theta}$ in each chain. $\boldsymbol{\theta}_i^t$ refers to the vector $\boldsymbol{\theta}$ at iteration t in the ith chain of multiple MCMCs. Then the posterior variance-covariance matrix in higher dimensions is computed by

$$\widehat{\mathbf{V}} = \frac{n-1}{n}\mathbf{W} + \left(1 + \frac{1}{m}\right)\frac{\mathbf{B}}{n}. \tag{15}$$

The within-covariance-matrix $\mathbf{W}$ is given by

$$\mathbf{W} = \frac{1}{m(n-1)} \sum_{i=1}^{m} \sum_{t=1}^{n} \left(\boldsymbol{\theta}_i^t - \bar{\boldsymbol{\theta}}_{i\cdot}\right)\left(\boldsymbol{\theta}_i^t - \bar{\boldsymbol{\theta}}_{i\cdot}\right)', \tag{16}$$

and the between-chain-covariance-matrix $\mathbf{B}$ is computed by

$$\mathbf{B} = \frac{n}{m-1} \sum_{i=1}^{m} \left(\bar{\boldsymbol{\theta}}_{i\cdot} - \bar{\boldsymbol{\theta}}_{\cdot\cdot}\right)\left(\bar{\boldsymbol{\theta}}_{i\cdot} - \bar{\boldsymbol{\theta}}_{\cdot\cdot}\right)', \tag{17}$$

where $\bar{\boldsymbol{\theta}}_{i\cdot}$ denotes the mean of θs within the chain and $\bar{\boldsymbol{\theta}}_{\cdot\cdot}$ represents the mean of θs in all the chains. The PSRF is defined as follows:

$$\text{PSRF}_p = \sqrt{\frac{\text{diag}(\widehat{\mathbf{V}})_p}{\text{diag}(\mathbf{W})_p}}, \quad \text{where } p = 1, 2, ..., d. \tag{18}$$

The PSRF values close to one indicate that the samples in multiple chains are being generated from the same limiting distribution and thus confirm the convergence. The MPSRF is computed as follows [17]:

$$\text{MPSRF} = \sqrt{\left(\frac{n-1}{n} + \left(\frac{m+1}{m}\right)\lambda_1\right)}, \tag{19}$$

where λ_1 is the greatest eigenvalue of $\mathbf{W}^{-1}\mathbf{B/n}$. If the MPSRF approaches one for a reasonably large n, the convergence of the multiple chains is ensured. Moreover, a relationship between the PSRFs and MPSRF is given by [4]

$$\text{max of PSRFs} \leq \text{MPSRF}. \tag{20}$$

6   Numerical Results

6.1   Simulation Study

In this subsection we present a simulation study of the characterization of the heterogeneous permeability field of the aquifer. See Fig. 1. The aquifer is not contaminated initially. The contaminated water flows through a spill well into the aquifer at a rate of one pore-volume every five years. The synthetic reference permeability field is constructed on a fine-grid of size 128×128. The fractional flow curves for this reference field at the monitoring wells are shown in Fig. 5. These curves are generated by running the numerical simulator on the reference field until $t = 1.0$ PVI. The time evolution of the contaminant flow on the reference permeability field is shown in Fig. 1. We use the two-stage and DREAM MCMC methods to generate samples from the posterior. A coarse-grid of size 32×32 is chosen for the two-stage algorithm, which runs four times faster than the fine-grid simulation, but still manages to capture the general trend of the flow. See Fig. 3.

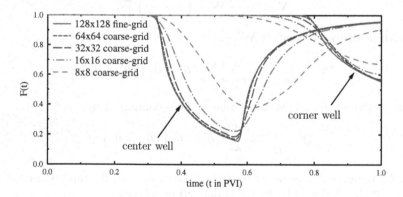

Fig. 3. A comparison of simulated fractional flows obtained from coarse-scale models and the fine-scale model.

6.2   Convergence Analysis

This subsection contains the MPSRF and PSRF analysis for both two-stage and DREAM MCMCs. We run twelve and eleven chains for two-stage and DREAM MCMCs, respectively. Using an equal number of total proposals in all chains, we then compute the maximum of PSRFs and the MPSRF against the number of iterations. Figure 4 shows that the MPSRF for each method is the upper bound of the maximum of the PSRFs, which is consistent with the inequality shown in (20). Moreover, it is also observed that both MPSRF and the maximum of PSRFs for the DREAM MCMC have a faster downward trend than the two-stage MCMC method. Thus, we can conclude that the DREAM MCMC samples from the posterior faster than the two-stage MCMC and converges faster towards the stationary distribution. However, Fig. 4 demonstrates that both MCMC methods need a large number of iterations to achieve a complete convergence. To achieve a complete convergence, the curves should approach the numerical value of one.

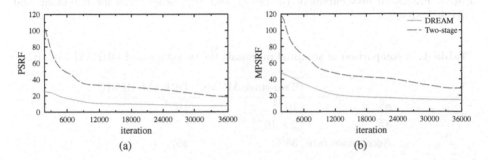

(a)   (b)

Fig. 4. The maximum of PSRFs and the MPSRF for DREAM and two-stage MCMCs

Next we focus on a statistical analysis of two ensembles of accepted fractional flow curves. In the analysis we consider the variances of the ensembles of fractional flow curves as well as the posterior distributions.

6.3 Variance Analysis

We compute the average production curves using 24000 accepted proposals and compare those curves with the reference fractional flow curves. See Fig. 5. From the comparison we can say that both two-stage and DREAM MCMCs produce very similar fractional flow curves. Table 1 shows the acceptance rates for both MCMCs, where σ_F^2 and σ_C^2 are precision parameters in (10) for coarse- and fine-scale simulations, respectively. Note that the acceptance rates for both MCMC methods are the same, however, the convergence rate in the DREAM MCMC is considerably higher than that in the two-stage MCMC (see Fig. 4).

We now construct two ensembles by taking the same number of fractional flow curves in each MCMC method: The first ensemble contains 24000 samples and the second one has 36000 samples. Figure 6 displays the variances of the fractional flow curves for the center and corner wells of the aquifer. The variance curves differ between not only both ensembles but also both MCMC methods.

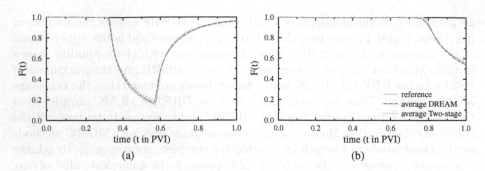

Fig. 5. Fractional flow curves of (a) center and (b) corner wells for two-stage and DREAM MCMCs.

Table 1. A comparison of accepted proposals for two-stage and DREAM MCMCs

	Two-stage MCMC	DREAM MCMC
σ_F^2	10^{-4}	10^{-4}
σ_C^2	2×10^{-4}	–
Acceptance rate	35%	35%

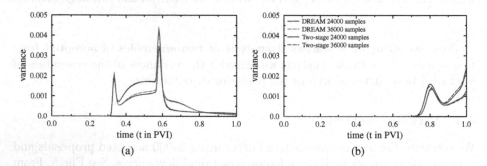

Fig. 6. Variance plots of fractional flow curves at (a) the center well and (b) corner well for two-stage and DREAM MCMCs.

Furthermore, we sketch the posterior densities in Fig. 7 for the same ensembles that we considered for the variance analysis. The normalized frequencies reveal that the posterior densities are also different for both MCMC methods.

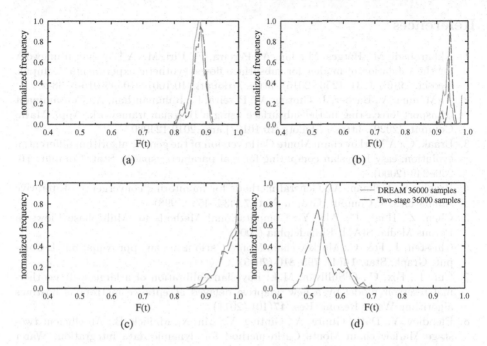

Fig. 7. Normalized frequency plots of fractional flow curves for two-stage and DREAM MCMCs. Top: Center well. Bottom: Corner well. Left: At $t = 0.78$ PVI. Right: At $t = 0.98$ PVI.

7 Conclusions

Using a GPU-based single-phase flow simulator we have compared two frequently used MCMCs, the two-stage algorithm based on a random walk sampler and the DREAM. We have confirmed that the DREAM converges much faster than the two-stage MCMC by comparing the corresponding PRSF and MPRSF curves. Moreover, a careful statistical analysis of ensembles of accepted fractional flow curves, produced by the two MCMCs, reveals that such ensembles share essentially the same average behavior. However, significant differences have been observed in the time-dependent variance curves as well as in the posterior distributions for those ensembles. This provides an indication that, for the purpose of making Monte Carlo predictive simulations, one might observe considerable differences in the results. One could combine both MCMC methods in a two-stage version of DREAM to achieve good convergence along with reduced computational cost in line with the work of [14]. As a future work we intend to combine both methods for our problem and study the convergence.

Acknowledgment. A. Rahunanthan's research was supported by National Science Foundation under Grant No. HRD-1600818. The computer simulations were performed on the NSF-funded GPU computing cluster housed in the Department of Mathematics and Computer Science at Central State University.

References

1. Akbarabadi, M., Borges, M., Jan, A., Pereira, F., Piri, M.: A Bayesian framework for the validation of models for subsurface flows: synthetic experiments. Comput. Geosci. **19**(6), 1231–1250 (2015). https://doi.org/10.1007/s10596-015-9538-z
2. Al-Mamun, A., Barber, J., Ginting, V., Pereira, F., Rahunanthan, A.: Contaminant transport forecasting in the subsurface using a Bayesian framework. Appl. Math. Comput. (2020). https://doi.org/10.1016/j.amc.2019.124980
3. Braak, C.: A Markov chain Monte Carlo version of the genetic algorithm differential evolution: easy Bayesian computing for real parameter spaces. Stat. Comput. **16**, 239–249 (2006)
4. Brooks, S., Gelman, A.: General methods for monitoring convergence of iterative simulations. J. Comput. Graph. Stat. **7**, 434–455 (1998)
5. Chen, Z., Huan, G., Ma, Y.: Computational Methods for Multiphase Flows in Porous Media. SIAM, Philadelphia (2006)
6. Christen, J., Fox, C.: Markov chain Monte Carlo using an approximation. J. Comput. Graph. Stat. **14**(4), 795–810 (2005)
7. Cui, T., Fox, C., O'Sullivan, M.J.: Bayesian calibration of a large-scale geothermal reservoir model by a new adaptive delayed acceptance metropolis hastings algorithm. Water Resour. Res. **47**(10) (2011)
8. Efendiev, Y., Datta-Gupta, A., Ginting, V., Ma, X., Mallick, B.: An efficient two-stage Markov chain Monte Carlo method for dynamic data integration. Water Resour. Res. **41**(12) (2005)
9. Efendiev, Y., Hou, T., Luo, W.: Preconditioning Markov chain Monte Carlo simulations using coarse-scale models. SIAM J. Sci. Comput. **28**(2), 776–803 (2006)
10. Ginting, V., Pereira, F., Rahunanthan, A.: Multiple Markov chains Monte Carlo approach for flow forecasting in porous media. Procedia Comput. Sci. **9**, 707–716 (2012)
11. Ginting, V., Pereira, F., Rahunanthan, A.: A multi-stage Bayesian prediction framework for subsurface flows. Int. J. Uncertain. Quantif. **3**(6), 499–522 (2013)
12. Ginting, V., Pereira, F., Rahunanthan, A.: A prefetching technique for prediction of porous media flows. Comput. Geosci. **18**(5), 661–675 (2014). https://doi.org/10.1007/s10596-014-9413-3
13. Ginting, V., Pereira, F., Rahunanthan, A.: Multi-physics Markov chain Monte Carlo methods for subsurface flows. Math. Comput. Simul. **118**, 224–238 (2015)
14. Laloy, E., Rogiers, B., Vrugt, J.A., Mallants, D., Jacques, D.: Efficient posterior exploration of a high-dimensional groundwater model from two-stage Markov chain Monte Carlo simulation and polynomial chaos expansion. Water Resour. Res. **49**(5), 2664–2682 (2013)
15. Loève, M.: Probability Theory. Springer, Heidelberg (1997)
16. Ma, X., Al-Harbi, M., Datta-Gupta, A., Efendiev, Y.: An efficient two-stage sampling method for uncertainty quantification in history matching geological models. SPE J. **13**(1), 77–87 (2008)
17. Malyshkina, N.: Markov switching models: an application to roadway safety. Ph.D. thesis, Purdue University (2008)
18. Pereira, F., Rahunanthan, A.: A semi-discrete central scheme for the approximation of two-phase flows in three space dimensions. Math. Comput. Simul. **81**(10), 2296–2306 (2011)
19. Shockley, E.M., Vrugt, J.A., Lopez, C.F.: PyDREAM: high-dimensional parameter inference for biological models in python. Bioinformatics **34**(4), 695–697 (2017)

20. Vrugt, J.A., ter Braak, C.J.F.: Dream$_{(D)}$: an adaptive Markov Chain Monte Carlo simulation algorithm to solve discrete, noncontinuous, and combinatorial posterior parameter estimation problems. Hydrol. Earth Syst. Sci. **15**(12), 3701–3713 (2011)
21. Vrugt, J.: Markov chain Monte Carlo simulation using the DREAM software package: theory, concepts, and MATLAB implementation. Environ. Model Softw. **75**, 273–316 (2016)
22. Vrugt, J.A., ter Braak, C.J.F., Clark, M.P., Hyman, J.M., Robinson, B.A.: Treatment of input uncertainty in hydrologic modeling: doing hydrology backward with Markov chain Monte Carlo simulation. Water Resour. Res. **44**(12) (2008)

Author Index

Printed in the United States
By Bookmasters